Fundamentals of
Power Electronics

SECOND EDITION

D1331549

Fundamentals of Power Electronics

SECOND EDITION

Robert W. Erickson
Dragan Maksimović
University of Colorado
Boulder, Colorado

Fundamentals of Power Electronics
Second Edition

Authors: Robert W. Erickson and Dragan Maksimović

ISBN 978-1-4757-0559-1   ISBN 978-0-306-48048-5 (eBook)
DOI 10.1007/978-0-306-48048-5

Library of Congress Control Number: 00-052569

Printed on acid free paper

15 14 13

springer.com

Dedicated to
Linda, William, and Richard
Lidija, Filip, Nikola, and Stevan

Contents

Preface

The objective of the First Edition was to serve as a textbook for introductory power electronics courses where the fundamentals of power electronics are defined, rigorously presented, and treated in sufficient depth so that students acquire the knowledge and skills needed to design practical power electronic systems. The First Edition has indeed been adopted for use in power electronics courses at a number of schools. An additional goal was to contribute as a reference book for engineers who practice power electronics design, and for students who want to develop their knowledge of the area beyond the level of introductory courses. In the Second Edition, the basic objectives and philosophy of the First Edition have not been changed. The modifications include addition of a number of new topics aimed at better serving the expanded audience that includes students of introductory and more advanced courses, as well as practicing engineers looking for a reference book and a source for further professional development. Most of the chapters have been significantly revised and updated. Major additions include a new Chapter 10 on input filter design, a new Appendix B covering simulation of converters, and a new Appendix C on Middlebrook's Extra Element Theorem. In addition to the introduction of new topics, we have made major revisions of the material to improve the flow and clarity of explanations and to provide additional specific results, in chapters covering averaged switch modeling, dynamics of converters operating in discontinuous conduction mode, current mode control, magnetics design, pulse-width modulated rectifiers, and resonant and soft-switching converters.

A completely new Chapter 10 covering input filter design has been added to the second addition. The problem of how the input filter affects the dynamics of the converter, often in a manner that degrades stability and performance of the converter system, is explained using Middlebrook's Extra Element Theorem. This design-oriented approach is explained in detail in the new Appendix C. Simple conditions are derived to allow filter damping so that converter transfer functions are not changed. Complete results for optimum filter damping are presented. The chapter concludes with a discussion about the design of multiple-section filters, illustrated by a design example.

Computer simulation based on the averaged switch modeling approach is presented in Appendix B, including PSpice models for continuous and discontinuous conduction mode, and current-mode control. Extensive simulation examples include: finding the dc conversion ratio and efficiency of a SEPIC, plotting the transient response of a buck-boost converter, comparing the control-to-output transfer functions of a SEPIC operating in CCM and DCM, determining the loop gain, line-to-output transfer function, and load transient response of a closed-loop buck voltage regulator, finding the input current waveform and THD of a DCM boost rectifier, and comparing the transfer functions and output imped-ances of buck converters operating with current programmed control and with duty cycle control. The major purpose of Appendix B is to supplement the text discussions, and to enable the reader to effec-tively use averaged models and simulation tools in the design process. The role of simulation as a design verification tool is emphasized. In our experience of teaching introductory and more advanced power electronics courses, we have found that the use of simulation tools works best with students who have mastered basic concepts and design-oriented analytical techniques, so that they are able to make correct interpretations of simulation results and model limitations. This is why we do not emphasize simulation in introductory chapters. Nevertheless, Appendix B is organized so that simulation examples can be introduced together with coverage of the theoretical concepts of Chapters 3, 7, 9, 10, 11, 12, and 18.

Middlebrook's Extra Element Theorem is presented in Appendix C, together with four tutorial examples. This valuable design-oriented analytical tool allows one to examine effects of adding an extra element to a linear system, without solving the modified system all over again. The theorem has many practical applications in the design of electronic circuits, from solving circuits by inspection, to quickly finding effects of unmodeled parasitic elements. In particular, in the Second Edition, Middlebrook's Extra Element Theorem is applied to the input filter design of Chapter 10, and to resonant inverter design in Chapter 19.

In Chapter 7, we have revised the section on circuit averaging and averaged switch modeling. The process of circuit averaging and deriving averaged switch models has been explained to allow read-ers not only to use the basic models, but also to construct averaged models for other applications of inter-est. Examples of extensions of the averaged switch modeling approach include modeling of switch conduction and switching losses. Related to the revision of Chapter 7, in Appendix B we have included new material on simulation of converters based on the averaged switch modeling approach.

Chapter 8 contains a new substantial introduction that explains the engineering design process and the need for design-oriented analysis. The discussions of design-oriented methods for construction of frequency response have been revised and expanded. A new example has been added, involving approximate analysis of a damped input filter.

Chapter 11 on dynamics of DCM (discontinuous conduction mode) converters, and Chapter 12 on current-mode control, have been thoroughly revised and updated. Chapter 11 includes a simplified derivation of DCM averaged switch models, as well as an updated discussion of high-frequency DCM dynamics. Chapter 12 includes a new, more straightforward explanation and discussion of current-mode dynamics, as well as new complete results for transfer functions and model parameters of all basic con-verters.

The chapters on magnetics design have been significantly revised and reorganized. Basic mag-netics theory necessary for informed design of magnetic components in switching power converters is presented in Chapter 13. The description of the proximity effect has been completely revised, to explain this important but complex subject in a more intuitive manner. The design of magnetic components based on the copper loss constraint is described in Chapter 14. A new step-by-step design procedure is given for multiple-winding inductors, and practical design examples are included for the design of filter induc-tors, coupled inductors and flyback transformers. The design of magnetic components (transformers and ac inductors) based on copper and core loss considerations is described in Chapter 15.

To improve their logical flow, the chapters covering pulse-width modulated rectifiers have been combined into a single Chapter 18, and have been completely reorganized. New sections on current con-trol based on the critical conduction mode, as well as on operation of the CCM boost and DCM flyback as PWM rectifiers, have been added.

Part V consists of Chapter 19 on resonant converters and Chapter 20 on soft-switching convert-ers. The discussion of resonant inverter design, a topic of importance in the field of high-frequency elec-tronic ballasts, has been expanded and explained in a more intuitive manner. A new resonant inverter design example has also been added to Chapter 19. Chapter 20 contains an expanded tutorial explanation of switching loss mechanisms, new charts illustrating the characteristics of quasi-square-wave and multi-resonant converters, and new up-to-date sections about soft-switching converters, including the zero-voltage transition full-bridge converter, the auxiliary switch approach, and the auxiliary resonant commutated pole approach for dc–dc converters and dc–ac inverters.

The material of the Second Edition is organized so that chapters or sections of the book can be selected to offer an introductory one-semester course, but yet enough material is provided for a sequence of more advanced courses, or for individual professional development. At the University of Colorado, we cover the material from the Second Edition in a sequence of three semester-long power electronics courses. The first course, intended for seniors and first-year graduate students, covers Chapters 1 to 6, Sections 7.1, 7.2, 7.5, and 7.6 from Chapter 7, Chapters 8 and 9, and Chapters 13 to 15. A project-oriented power electronics design laboratory is offered in parallel with this course. This course serves as a prerequisite for two follow-up courses. The second course starts with Section 7.4, proceeds to Appendices B and C, Chapters 10, 11 and 12, and concludes with the material of Chapters 16 to 18. In the third course we cover resonant and soft-switching techniques of Chapters 19 and 20.

The website for the Second Edition contains comprehensive supporting materials for the text, including solved problems and slides for instructors. Computer simulation files can be downloaded from this site, including a PSpice library of averaged switch models, and simulation examples.

This text has evolved from courses developed over seventeen years of teaching power electronics at the University of Colorado. These courses, in turn, were heavily influenced by our previous experiences as graduate students at the California Institute of Technology, under the direction of Profs. Slobodan Ćuk and R. D. Middlebrook, to whom we are grateful. We appreciate the helpful suggestions of Prof. Arthur Witulski of the University of Arizona. We would also like to thank the many readers of the First Edition, students, and instructors who offered their comments and suggestions, or who pointed out errata. We have attempted to incorporate these suggestions wherever possible.

ROBERT W. ERICKSON
DRAGAN MAKSIMOVIĆ
Boulder, Colorado

1

Introduction

1.1 INTRODUCTION TO POWER PROCESSING

The field of power electronics is concerned with the processing of electrical power using electronic devices [1–7]. The key element is the *switching converter*, illustrated in Fig. 1.1. In general, a switching converter contains power input and control input ports, and a power output port. The raw input power is processed as specified by the control input, yielding the conditioned output power. One of several basic functions can be performed [2]. In a *dc–dc converter*, the dc input voltage is converted to a dc output voltage having a larger or smaller magnitude, possibly with opposite polarity or with isolation of the input and output ground references. In an ac–dc *rectifier*, an ac input voltage is rectified, producing a dc output voltage. The dc output voltage and/or ac input current waveform may be controlled. The inverse process, dc–ac *inversion*, involves transforming a dc input voltage into an ac output voltage of controllable magnitude and frequency. Ac–ac *cycloconversion* involves converting an ac input voltage to a given ac output voltage of controllable magnitude and frequency.

Control is invariably required. It is nearly always desired to produce a well-regulated output

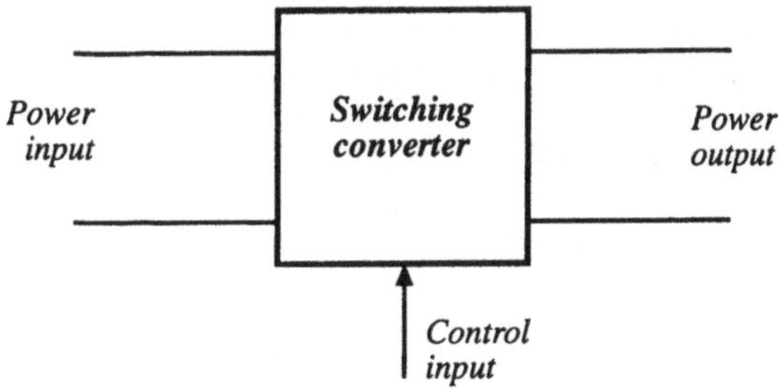

Fig. 1.1 The switching converter, a basic power processing block.

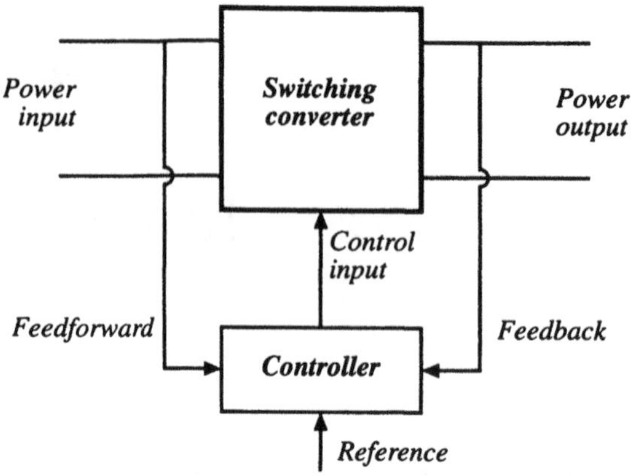

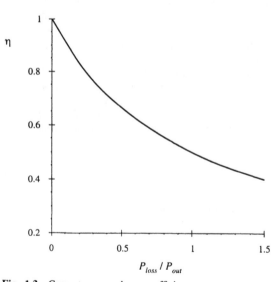

Fig. 1.2   A controller is generally required.

voltage, in the presence of variations in the input voltage and load current. As illustrated in Fig. 1.2, a controller block is an integral part of any power processing system.

High efficiency is essential in any power processing application. The primary reason for this is usually not the desire to save money on one's electric bills, nor to conserve energy, in spite of the nobility of such pursuits. Rather, high efficiency converters are necessary because construction of low-efficiency converters, producing substantial output power, is impractical. The efficiency of a converter having output power P_{out} and input power P_{in} is

$$\eta = \frac{P_{out}}{P_{in}} \tag{1.1}$$

The power lost in the converter is

$$P_{loss} = P_{in} - P_{out} = P_{out}\left(\frac{1}{\eta} - 1\right) \tag{1.2}$$

Equation (1.2) is plotted in Fig. 1.3. In a converter that has an efficiency of 50%, power P_{loss} is dissipated by the converter elements and this is equal to the output power, P_{out}. This power is converted into heat, which must be removed from the converter. If the output power is substantial, then so is the loss power. This leads to a large and expensive cooling system, it causes the electronic elements within the converter to operate at high temperature, and it reduces the system reliability. Indeed, at high output powers, it may be impossible to adequately cool the converter elements using current technology.

Increasing the efficiency is the key to obtaining higher output powers. For example, if the converter efficiency is 90%, then the converter loss power is equal to only 11%

Fig. 1.3   Converter power loss vs. efficiency.

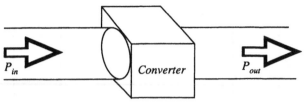

Fig. 1.4 A goal of current converter technology is to construct converters of small size and weight, which process substantial power at high efficiency.

of the output power. Efficiency is a good measure of the success of a given converter technology. Figure 1.4 illustrates a converter that processes a large amount of power, with very high efficiency. Since very little power is lost, the converter elements can be packaged with high density, leading to a converter of small size and weight, and of low temperature rise.

How can we build a circuit that changes the voltage, yet dissipates negligible power? The various conventional circuit elements are illustrated in Fig. 1.5. The available circuit elements fall broadly into the classes of resistive elements, capacitive elements, magnetic devices including inductors and transformers, semiconductor devices operated in the linear mode (for example, as class *A* or class *B* amplifiers), and semiconductor devices operated in the switched mode (such as in logic devices where transistors operate in either saturation or cutoff). In conventional signal processing applications, where efficiency is not the primary concern, magnetic devices are usually avoided wherever possible, because of their large size and the difficulty of incorporating them into integrated circuits. In contrast, capacitors and magnetic devices are important elements of switching converters, because ideally they do not consume power. It is the resistive element, as well as the linear-mode semiconductor device, that is avoided [2]. Switched-mode semiconductor devices are also employed. When a semiconductor device operates in the off state, its current is zero and hence its power dissipation is zero. When the semiconductor device operates in the on (saturated) state, its voltage drop is small and hence its power dissipation is also small. In either event, the power dissipated by the semiconductor device is low. So capacitive and inductive elements, as well as switched-mode semiconductor devices, are available for synthesis of high-efficiency converters.

Let us now consider how to construct the simple dc-dc converter example illustrated in Fig. 1.6. The input voltage V_g is 100 V. It is desired to supply 50 V to an effective 5 Ω load, such that the dc load current is 10 A.

Introductory circuits textbooks describe a low-efficiency method to perform the required function: the voltage divider circuit illustrated in Fig. 1.7(a). The dc–dc converter then consists simply of a

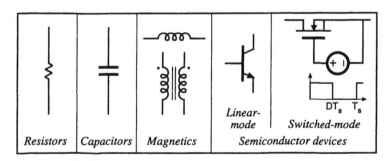

Resistors | *Capacitors* | *Magnetics* | *Linear-mode* / *Switched-mode* / *Semiconductor devices*

Fig. 1.5 Devices available to the circuit designer [2].

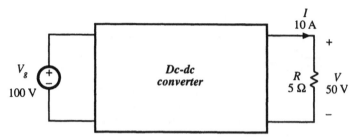

Fig. 1.6   A simple power processing example: construction of a 500 W dc–dc converter.

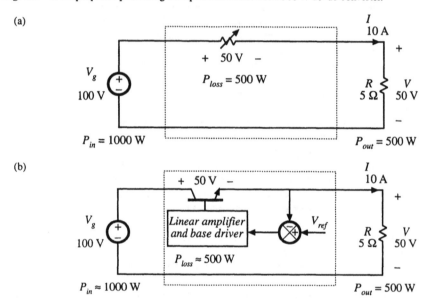

Fig. 1.7   Changing the dc voltage via dissipative means: (a) voltage divider, (b) series pass regulator.

variable resistor, whose value is adjusted such that the required output voltage is obtained. The load current flows through the variable resistor. For the specified voltage and current levels, the power P_{loss} dissipated in the variable resistor equals the load power $P_{out} = 500$ W. The source V_g supplies power $P_{in} = 1000$ W. Figure 1.7(b) illustrates a more practical implementation known as the linear series-pass regulator. The variable resistor of Fig. 1.7(a) is replaced by a linear-mode power transistor, whose base current is controlled by a feedback system such that the desired output voltage is obtained. The power dissipated by the linear-mode transistor of Fig. 1.7(b) is approximately the same as the 500 W lost by the variable resistor in Fig. 1.7(a). Series-pass linear regulators generally find modern application only at low power levels of a few watts.

Figure 1.8 illustrates another approach. A single-pole double-throw (SPDT) switch is connected as shown. The switch output voltage $v_s(t)$ is equal to the converter input voltage V_g when the switch is in position 1, and is equal to zero when the switch is in position 2. The switch position is varied periodically, as illustrated in Fig. 1.9, such that $v_s(t)$ is a rectangular waveform having frequency f_s and period $T_s = 1/f_s$. The duty cycle D is defined as the fraction of time in which the switch occupies position 1. Hence, $0 \leq D \leq 1$. In practice, the SPDT switch is realized using switched-mode semiconductor devices,

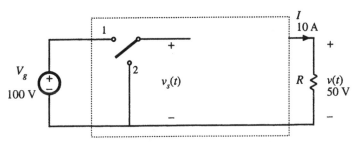

Fig. 1.8 Insertion of SPDT switch which changes the dc component of the voltage.

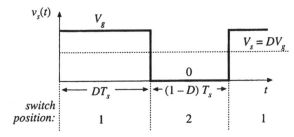

Fig. 1.9 Switch output voltage waveform $v_s(t)$.

which are controlled such that the SPDT switching function is attained.

The switch changes the dc component of the voltage. Recall from Fourier analysis that the dc component of a periodic waveform is equal to its average value. Hence, the dc component of $v_s(t)$ is

$$V_s = \frac{1}{T_s} \int_0^{T_s} v_s(t)\,dt = DV_g \tag{1.3}$$

Thus, the switch changes the dc voltage, by a factor equal to the duty cycle D. To convert the input voltage $V_g = 100$ V into the desired output voltage of $V = 50$ V, a duty cycle of $D = 0.5$ is required.

Again, the power dissipated by the switch is ideally zero. When the switch contacts are closed, then their voltage is zero and hence the power dissipation is zero. When the switch contacts are open, then the current is zero and again the power dissipation is zero. So we have succeeded in changing the dc voltage component, using a device that is ideally lossless.

In addition to the desired dc component V_s, the switch output voltage waveform $v_s(t)$ also contains undesirable harmonics of the switching frequency. In most applications, these harmonics must be removed, such that the output voltage $v(t)$ is essentially equal to the dc component $V = V_s$. A low-pass filter can be employed for this purpose. Figure 1.10 illustrates the introduction of a single-section L–C low-pass filter. If the filter corner frequency f_0 is sufficiently less than the switching frequency f_s, then the filter essentially passes only the dc component of $v_s(t)$. To the extent that the switch, inductor, and capacitor elements are ideal, the efficiency of this dc–dc converter can approach 100%.

In Fig. 1.11, a control system is introduced for regulation of the output voltage. Since the output voltage is a function of the switch duty cycle, a control system can be constructed that varies the duty cycle to cause the output voltage to follow a given reference. Figure 1.11 also illustrates a typical way in which the SPDT switch is realized using switched-mode semiconductor devices. The converter power stage developed in Figs. 1.8 to 1.11 is called the *buck converter*, because it reduces the dc voltage.

Converters can be constructed that perform other power processing functions. For example, Fig.

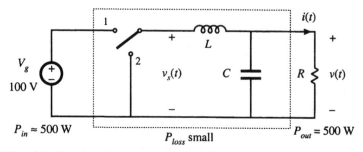

Fig. 1.10   Addition of *L–C* low-pass filter, for removal of switching harmonics.

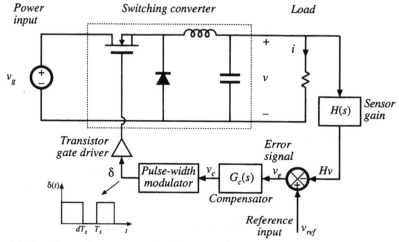

Fig. 1.11   Addition of control system to regulate the output voltage.

Fig. **1.12** The boost converter: (a) ideal converter circuit, (b) output voltage *V* vs. transistor duty cycle *D*.

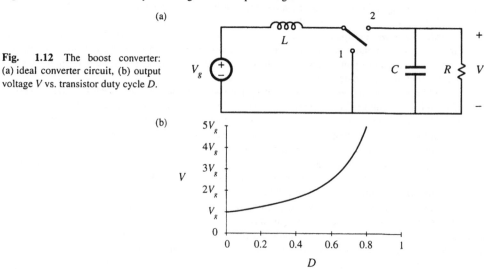

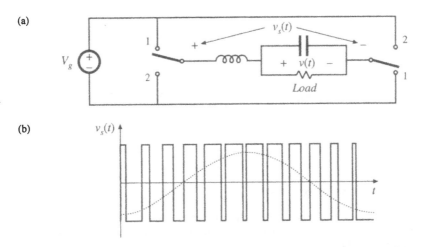

Fig. 1.13   A bridge-type dc-1øac inverter: (a) ideal inverter circuit, (b) typical pulse-width-modulated switch voltage waveform $v_s(t)$, and its low-frequency component.

1.12 illustrates a circuit known as the *boost converter*, in which the positions of the inductor and SPDT switch are interchanged. This converter is capable of producing output voltages that are greater in magnitude than the input voltage. In general, any given input voltage can be converted into any desired output voltage, using a converter containing switching devices embedded within a network of reactive elements.

Figure 1.13(a) illustrates a simple dc–1øac inverter circuit. As illustrated in Fig. 1.13(b), the switch duty cycle is modulated sinusoidally. This causes the switch output voltage $v_s(t)$ to contain a low-frequency sinusoidal component. The L–C filter cutoff frequency f_0 is selected to pass the desired low-frequency components of $v_s(t)$, but to attenuate the high-frequency switching harmonics. The controller modulates the duty cycle such that the desired output frequency and voltage magnitude are obtained.

1.2   SEVERAL APPLICATIONS OF POWER ELECTRONICS

The power levels encountered in high-efficiency switching converters range from (1) less than one watt, in dc–dc converters within battery-operated portable equipment, to (2) tens, hundreds, or thousands of watts in power supplies for computers and office equipment, to (3) kilowatts to Megawatts, in variable-speed motor drives, to (4) roughly 1000 Megawatts in the rectifiers and inverters that interface dc transmission lines to the ac utility power system. The converter systems of several applications are illustrated in this section.

A power supply system for a laptop computer is illustrated in Fig. 1.14. A lithium battery powers the system, and several dc–dc converters change the battery voltage into the voltages required by the loads. A buck converter produces the low-voltage dc required by the microprocessor. A boost converter increases the battery voltage to the level needed by the disk drive. An inverter produces high-voltage high-frequency ac to drive lamps that light the display. A charger with transformer isolation converts the ac line voltage into dc to charge the battery. The converter switching frequencies are typically in the vicinity of several hundred kilohertz; this leads to substantial reductions in the size and weight of the reactive elements. *Power management* is used, to control sleep modes in which power consumption is reduced and battery life is extended. In a *distributed power system*, an intermediate dc voltage appears at the computer backplane. Each printed circuit card contains high-density dc–dc converters that produce

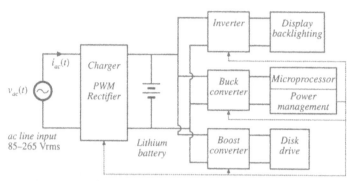

Fig. 1.14 A laptop computer power supply system.

locally-regulated low voltages. Commercial applications of power electronics include off-line power systems for computers, office and laboratory equipment, uninterruptable ac power supplies, and electronic ballasts for gas discharge lighting.

Figure 1.15 illustrates a power system of an earth-orbiting spacecraft. A solar array produces the main power bus voltage V_{bus}. DC–DC converters convert V_{bus} to the regulated voltages required by the spacecraft payloads. Battery charge/discharge controllers interface the main power bus to batteries; these controllers may also contain dc–dc converters. Aerospace applications of power electronics include the power systems of aircraft, spacecraft, and other aerospace vehicles.

Figure 1.16 illustrates an electric vehicle power and drive system. Batteries are charged by a converter that draws high power-factor sinusoidal current from a single-phase or three-phase ac line. The batteries supply power to variable-speed ac motors to propel the vehicle. The speeds of the ac motors are controlled by variation of the electrical input frequency. Inverters produce three-phase ac output voltages of variable frequency and variable magnitude, to control the speed of the ac motors and the vehicle. A dc–dc converter steps down the battery voltage to the lower dc levels required by the electronics of the system. Applications of motor drives include speed control of industrial processes, such as control of compressors, fans, and pumps; transportation applications such as electric vehicles, subways, and locomotives; and motion control applications in areas such as computer peripherals and industrial robots.

Power electronics also finds application in other diverse industries, including dc power supplies,

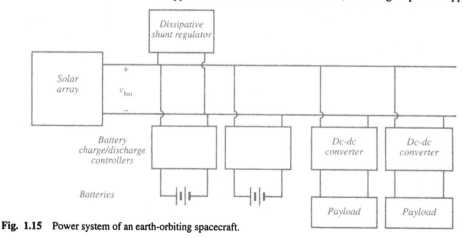

Fig. 1.15 Power system of an earth-orbiting spacecraft.

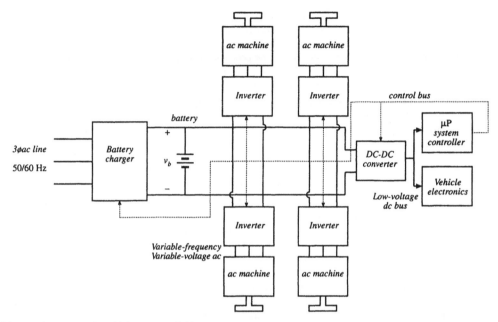

Fig. 1.16   An electric vehicle power and drive system.

uninterruptable power supplies, and battery chargers for the telecommunications industry; inverter systems for renewable energy generation applications such as wind and photovoltaic power; and utility power systems applications including high-voltage dc transmission and static *VAR* (reactive volt-ampere) compensators.

1.3   ELEMENTS OF POWER ELECTRONICS

One of the things that makes the power electronics field interesting is its incorporation of concepts from a diverse set of fields, including:

- analog circuits
- electronic devices
- control systems
- power systems
- magnetics
- electric machines
- numerical simulation

Thus, the practice of power electronics requires a broad electrical engineering background. In addition, there are fundamental concepts that are unique to the power electronics field, and that require specialized study.

   The presence of high-frequency switching makes the understanding of switched-mode converters not straightforward. Hence, converter modeling is central to the study of power electronics. As introduced in Eq. (1.3), the dc component of a periodic waveform is equal to its average value. This ideal can be generalized, to predict the dc components of all converter waveforms via averaging. In Part I of this book, averaged equivalent circuit models of converters operating in steady state are derived. These models not only predict the basic ideal behavior of switched-mode converters, but also model efficiency and losses. Realization of the switching elements, using power semiconductor devices, is also discussed.

Design of the converter control system requires models of the converter dynamics. In Part II of this book, the averaging technique is extended, to describe low-frequency variations in the converter waveforms. Small-signal equivalent circuit models are developed, which predict the control-to-output and line-to-transfer functions, as well as other ac quantities of interest. These models are then employed to design converter control systems and to lend an understanding of the well-known current-programmed control technique.

The magnetic elements are key components of any switching converter. The design of high-power high-frequency magnetic devices having high efficiency and small size and weight is central to most converter technologies. High-frequency power magnetics design is discussed in Part III.

Pollution of the ac power system by rectifier harmonics is a growing problem. As a result, many converter systems now incorporate low-harmonic rectifiers, which draw sinusoidal currents from the utility system. These modern rectifiers are considerably more sophisticated than the conventional diode bridge: they may contain high-frequency switched-mode converters, with control systems that regulate the ac line current waveform. Modern rectifier technology is treated in Part IV.

Resonant converters employ quasi-sinusoidal waveforms, as opposed to the rectangular waveforms of the buck converter illustrated in Fig. 1.9. These resonant converters find application where high-frequency inverters and converters are needed. Resonant converters are modeled in Part V. Their loss mechanisms, including the processes of zero-voltage switching and zero-current switching, are discussed.

REFERENCES

[1]   W. E. NEWELL, "Power Electronics—Emerging from Limbo," *IEEE Power Electronics Specialists Conference*, 1973 Record, pp. 6-12.

[2]   R. D. MIDDLEBROOK, "Power Electronics: An Emerging Discipline," *IEEE International Symposium on Circuits and Systems*, 1981 Proceedings, April 1981.

[3]   R. D. MIDDLEBROOK, "Power Electronics: Topologies, Modeling, and Measurement," *IEEE International Symposium on Circuits and Systems*, 1981 Proceedings, April 1981.

[4]   S. CUK, "Basics of Switched-Mode Power Conversion: Topologies, Magnetics, and Control," in *Advances in Switched-Mode Power Conversion,* vol. 2, pp. 279--310, Irvine: Teslaco, 1981.

[5]   N. MOHAN, "Power Electronics Circuits: An Overview," *IEEE IECON*, 1988 Proceedings, pp. 522-527.

[6]   B. K. BOSE, "Power Electronics—A Technology Review," *Proceedings of the IEEE*, vol. 80, no. 8, August 1992, pp. 1303-1334.

[7]   M. NISHIHARA, "Power Electronics Diversity," *International Power Electronics Conference* (Tokyo), 1990 Proceedings, pp. 21-28.

Part I

Converters in Equilibrium

2

Principles of Steady-State Converter Analysis

2.1   INTRODUCTION

In the previous chapter, the buck converter was introduced as a means of reducing the dc voltage, using only nondissipative switches, inductors, and capacitors. The switch produces a rectangular waveform $v_s(t)$ as illustrated in Fig. 2.1. The voltage $v_s(t)$ is equal to the dc input voltage V_g when the switch is in position 1, and is equal to zero when the switch is in position 2. In practice, the switch is realized using

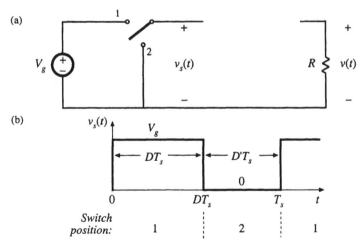

Fig. 2.1   Ideal switch, (a), used to reduce the voltage dc component, and (b) its output voltage waveform $v_s(t)$.

Fig. 2.2 Determination of the switch output voltage dc component, by integrating and dividing by the switching period.

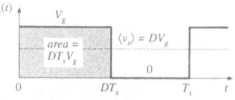

power semiconductor devices, such as transistors and diodes, which are controlled to turn on and off as required to perform the function of the ideal switch. The switching frequency f_s, equal to the inverse of the switching period T_s, generally lies in the range of 1 kHz to 1 MHz, depending on the switching speed of the semiconductor devices. The duty ratio D is the fraction of time that the switch spends in position 1, and is a number between zero and one. The complement of the duty ratio, D', is defined as $(1 - D)$.

The switch reduces the dc component of the voltage: the switch output voltage $v_s(t)$ has a dc component that is less than the converter dc input voltage V_g. From Fourier analysis, we know that the dc component of $v_s(t)$ is given by its average value $\langle v_s \rangle$, or

$$\langle v_s \rangle = \frac{1}{T_s} \int_0^{T_s} v_s(t) dt \tag{2.1}$$

As illustrated in Fig. 2.2, the integral is given by the area under the curve, or $DT_s V_g$. The average value is therefore

$$\langle v_s \rangle = \frac{1}{T_s} \left(DT_s V_g \right) = DV_g \tag{2.2}$$

So the average value, or dc component, of $v_s(t)$ is equal to the duty cycle times the dc input voltage V_g. The switch reduces the dc voltage by a factor of D.

What remains is to insert a low-pass filter as shown in Fig. 2.3. The filter is designed to pass the dc component of $v_s(t)$, but to reject the components of $v_s(t)$ at the switching frequency and its harmonics. The output voltage $v(t)$ is then essentially equal to the dc component of $v_s(t)$:

$$v \approx \langle v_s \rangle = DV_g \tag{2.3}$$

The converter of Fig. 2.3 has been realized using lossless elements. To the extent that they are ideal, the inductor, capacitor, and switch do not dissipate power. For example, when the switch is closed, its voltage drop is zero, and the current is zero when the switch is open. In either case, the power dissipated by the switch is zero. Hence, efficiencies approaching 100% can be obtained. So to the extent that the components are ideal, we can realize our objective of changing dc voltage levels using a lossless network.

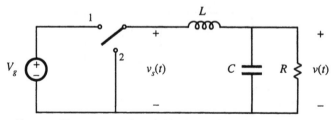

Fig. 2.3 Insertion of low-pass filter, to remove the switching harmonics and pass only the dc component of $v_s(t)$ to the output.

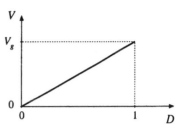

Fig. 2.4   Buck converter dc output voltage V vs. duty cycle D.

The network of Fig. 2.3 also allows control of the output. Figure 2.4 is the control characteristic of the converter. The output voltage, given by Eq. (2.3), is plotted vs. duty cycle. The buck converter has a linear control characteristic. Also, the output voltage is less than or equal to the input voltage, since $0 \leq D \leq 1$. Feedback systems are often constructed that adjust the duty cycle D to regulate the converter output voltage. Inverters or power amplifiers can also be built, in which the duty cycle varies slowly with time and the output voltage follows.

The buck converter is just one of many possible switching converters. Two other commonly used converters, which perform different voltage conversion functions, are illustrated in Fig. 2.5. In the boost converter, the positions of the inductor and switch are reversed. It is shown later in this chapter that the boost converter steps the voltage up: $V \geq V_g$. Another converter, the buck-boost converter, can either increase or decrease the magnitude of the voltage, but the polarity is inverted. So with a positive input voltage, the ideal buck-boost converter can produce a negative output voltage of any magnitude. It may at first be surprising that dc output voltages can be produced that are greater in magnitude than the input, or that have opposite polarity. But it is indeed possible to produce any desired dc output voltage using a passive network of only inductors, capacitors, and embedded switches.

In the above discussion, it was possible to derive an expression for the output voltage of the buck converter, Eq. (2.3), using some simple arguments based on Fourier analysis. However, it may not be immediately obvious how to directly apply these arguments to find the dc output voltage of the boost, buck–boost, or other converters. The objective of this chapter is the development of a more general method for analyzing any switching converter comprised of a network of inductors, capacitors, and switches [1-8].

The principles of *inductor volt-second balance* and *capacitor charge balance* are derived; these can be used to solve for the inductor currents and capacitor voltages of switching converters. A useful approximation, the *small–ripple* or *linear–ripple approximation*, greatly facilitates the analysis. Some simple methods for selecting the filter element values are also discussed.

2.2   INDUCTOR VOLT-SECOND BALANCE, CAPACITOR CHARGE BALANCE, AND THE SMALL-RIPPLE APPROXIMATION

Let us more closely examine the inductor and capacitor waveforms in the buck converter of Fig. 2.6. It is impossible to build a perfect low-pass filter that allows the dc component to pass but completely removes the components at the switching frequency and its harmonics. So the low-pass filter must allow at least some small amount of the high-frequency harmonics generated by the switch to reach the output. Hence, in practice the output voltage waveform $v(t)$ appears as illustrated in Fig. 2.7, and can be expressed as

$$v(t) = V + v_{ripple}(t) \tag{2.4}$$

So the actual output voltage $v(t)$ consists of the desired dc component V, plus a small undesired ac com-

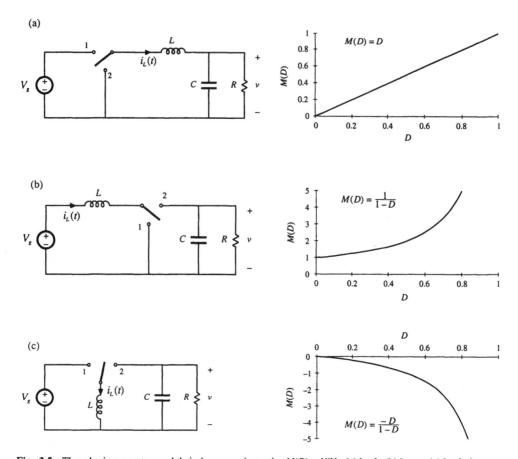

Fig. 2.5   Three basic converters and their dc conversion ratios $M(D) = V/V_g$: (a) buck, (b) boost, (c) buck–boost.

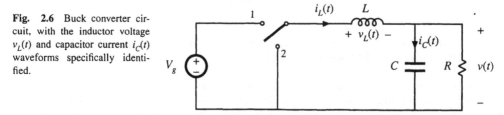

Fig. 2.6 Buck converter circuit, with the inductor voltage $v_L(t)$ and capacitor current $i_C(t)$ waveforms specifically identified.

ponent $v_{ripple}(t)$ arising from the incomplete attenuation of the switching harmonics by the low-pass filter. The magnitude of $v_{ripple}(t)$ has been exaggerated in Fig. 2.7.

The output voltage switching ripple should be small in any well-designed converter, since the object is to produce a dc output. For example, in a computer power supply having a 3.3 V output, the switching ripple is normally required to be less than a few tens of millivolts, or less than 1% of the dc component V. So it is nearly always a good approximation to assume that the magnitude of the switching

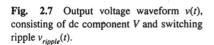

Fig. 2.7 Output voltage waveform $v(t)$, consisting of dc component V and switching ripple $v_{ripple}(t)$.

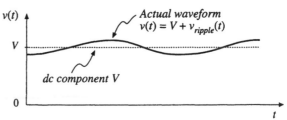

ripple is much smaller than the dc component:

$$\left| v_{ripple} \right| \ll V \tag{2.5}$$

Therefore, the output voltage $v(t)$ is well approximated by its dc component V, with the small ripple term $v_{ripple}(t)$ neglected:

$$v(t) \approx V \tag{2.6}$$

This approximation, known as the small-ripple approximation, or the linear-ripple approximation, greatly simplifies the analysis of the converter waveforms and is used throughout this book.

Next let us analyze the inductor current waveform. We can find the inductor current by integrating the inductor voltage waveform. With the switch in position 1, the left side of the inductor is connected to the input voltage V_g, and the circuit reduces to Fig. 2.8(a). The inductor voltage $v_L(t)$ is then given by

$$v_L = V_g - v(t) \tag{2.7}$$

As described above, the output voltage $v(t)$ consists of the dc component V, plus a small ac ripple term $v_{ripple}(t)$. We can make the small ripple approximation here, Eq. (2.6), to replace $v(t)$ with its dc component V:

$$v_L \approx V_g - V \tag{2.8}$$

So with the switch in position 1, the inductor voltage is essentially constant and equal to $V_g - V$, as shown in Fig. 2.9. By knowledge of the inductor voltage waveform, the inductor current can be found by use of the definition

$$v_L(t) = L \frac{di_L(t)}{dt} \tag{2.9}$$

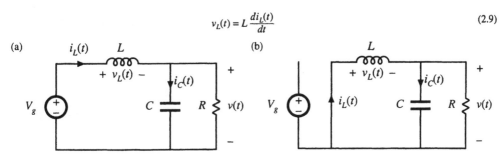

Fig. 2.8 Buck converter circuit: (a) while the switch is in position 1, (b) while the switch is in position 2.

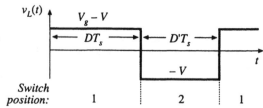

Fig. 2.9 Steady-state inductor voltage waveform, buck converter.

Thus, during the first interval, when $v_L(t)$ is approximately $(V_g - V)$, the slope of the inductor current waveform is

$$\frac{di_L(t)}{dt} = \frac{v_L(t)}{L} \approx \frac{V_g - V}{L} \tag{2.10}$$

which follows by dividing Eq. (2.9) by L, and substituting Eq. (2.8). Since the inductor voltage $v_L(t)$ is essentially constant while the switch is in position 1, the inductor current slope is also essentially constant and the inductor current increases linearly.

Similar arguments apply during the second subinterval, when the switch is in position 2. The left side of the inductor is then connected to ground, leading to the circuit of Fig. 2.8(b). It is important to consistently define the polarities of the inductor current and voltage; in particular, the polarity of $v_L(t)$ is defined consistently in Figs. 2.7, 2.8(a), and 2.8(b). So the inductor voltage during the second subinterval is given by

$$v_L(t) = -v(t) \tag{2.11}$$

Use of the small ripple approximation, Eq. (2.6), leads to

$$v_L(t) \approx -V \tag{2.12}$$

So the inductor voltage is also essentially constant while the switch is in position 2, as illustrated in Fig. 2.9. Substitution of Eq. (2.12) into Eq. (2.9) and solution for the slope of the inductor current yields

$$\frac{di_L(t)}{dt} \approx -\frac{V}{L} \tag{2.13}$$

Hence, during the second subinterval the inductor current changes with a negative and essentially constant slope.

We can now sketch the inductor current waveform (Fig. 2.10). The inductor current begins at some initial value $i_L(0)$. During the first subinterval, with the switch in position 1, the inductor current increases with the slope given in Eq. (2.10). At time $t = DT_s$, the switch changes to position 2. The current then decreases with the constant slope given by Eq. (2.13). At time $t = T_s$, the switch changes back to

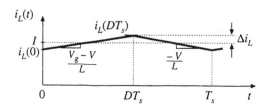

Fig. 2.10 Steady-state inductor current waveform, buck converter.

position 1, and the process repeats.

It is of interest to calculate the inductor current ripple Δi_L. As illustrated in Fig. 2.10, the peak inductor current is equal to the dc component I plus the peak-to-average ripple Δi_L. This peak current flows through not only the inductor, but also through the semiconductor devices that comprise the switch. Knowledge of the peak current is necessary when specifying the ratings of these devices.

Since we know the slope of the inductor current during the first subinterval, and we also know the length of the first subinterval, we can calculate the ripple magnitude. The $i_L(t)$ waveform is symmetrical about I, and hence during the first subinterval the current increases by $2\Delta i_L$ (since Δi_L is the peak ripple, the peak-to-peak ripple is $2\Delta i_L$). So the change in current, $2\Delta i_L$, is equal to the slope (the applied inductor voltage divided by L) times the length of the first subinterval (DT_s):

$$\text{(change in } i_L) = \text{(slope)(length of subinterval)}$$

$$\left(2\Delta i_L\right) = \left(\frac{V_g - V}{L}\right)\left(DT_s\right) \tag{2.14}$$

Solution for Δi_L yields

$$\Delta i_L = \frac{V_g - V}{2L} DT_s \tag{2.15}$$

Typical values of Δi_L lie in the range of 10% to 20% of the full-load value of the dc component I. It is undesirable to allow Δi_L to become too large; doing so would increase the peak currents of the inductor and of the semiconductor switching devices, and would increase their size and cost. So by design the inductor current ripple is also usually small compared to the dc component I. The small-ripple approximation $i_L(t) \approx I$ is usually justified for the inductor current.

The inductor value can be chosen such that a desired current ripple Δi_L is attained. Solution of Eq. (2.15) for the inductance L yields

$$L = \frac{V_g - V}{2\Delta i_L} DT_s \tag{2.16}$$

This equation is commonly used to select the value of inductance in the buck converter.

It is entirely possible to solve converters exactly, without use of the small-ripple approximation. For example, one could use the Laplace transform to write expressions for the waveforms of the circuits of Figs. 2.8(a) and 2.8(b). One could then invert the transforms, match boundary conditions, and find the periodic steady-state solution of the circuit. Having done so, one could then find the dc components of the waveforms and the peak values. But this is a great deal of work, and the results are nearly always intractable. Besides, the extra work involved in writing equations that exactly describe the ripple is a waste of time, since the ripple is small and is undesired. The small-ripple approximation is easy to apply, and quickly yields simple expressions for the dc components of the converter waveforms.

The inductor current waveform of Fig. 2.10 is drawn under steady-state conditions, with the converter operating in equilibrium. Let's consider next what happens to the inductor current when the converter is first turned on. Suppose that the inductor current and output voltage are initially zero, and an input voltage V_g is then applied. As shown in Fig. 2.11, $i_L(0)$ is zero. During the first subinterval, with the switch in position 1, we know that the inductor current will increase, with a slope of $(V_g - v)/L$ and with v initially zero. Next, with the switch in position 2, the inductor current will change with a slope of $- v/L$; since v is initially zero, this slope is essentially zero. It can be seen that there is a net increase in inductor current over the first switching period, because $i_L(T_s)$ is greater than $i_L(0)$. Since the inductor current

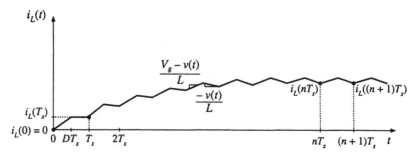

Fig. 2.11   Inductor current waveform during converter turn-on transient.

flows to the output, the output capacitor will charge slightly, and v will increase slightly. The process repeats during the second and succeeding switching periods, with the inductor current increasing during each subinterval 1 and decreasing during each subinterval 2.

As the output capacitor continues to charge and v increases, the slope during subinterval 1 decreases while the slope during subinterval 2 becomes more negative. Eventually, the point is reached where the increase in inductor current during subinterval 1 is equal to the decrease in inductor current during subinterval 2. There is then no net change in inductor current over a complete switching period, and the converter operates in steady state. The converter waveforms are periodic: $i_L(nT_s) = i_L((n+1)T_s)$. From this point on, the inductor current waveform appears as in Fig. 2.10.

The requirement that, in equilibrium, the net change in inductor current over one switching period be zero leads us to a way to find steady-state conditions in any switching converter: the principle of *inductor volt-second balance*. Given the defining relation of an inductor:

$$v_L(t) = L \frac{di_L(t)}{dt} \tag{2.17}$$

Integration over one complete switching period, say from $t = 0$ to T_s, yields

$$i_L(T_s) - i_L(0) = \frac{1}{L} \int_0^{T_s} v_L(t)\,dt \tag{2.18}$$

This equation states that the net change in inductor current over one switching period, given by the left-hand side of Eq. (2.18), is proportional to the integral of the applied inductor voltage over the interval. In steady state, the initial and final values of the inductor current are equal, and hence the left-hand side of Eq. (2.18) is zero. Therefore, in steady state the integral of the applied inductor voltage must be zero:

$$0 = \int_0^{T_s} v_L(t)\,dt \tag{2.19}$$

The right-hand side of Eq. (2.19) has the units of volt-seconds or flux-linkages. Equation (2.19) states that the total area, or net volt-seconds, under the $v_L(t)$ waveform must be zero.

An equivalent form is obtained by dividing both sides of Eq. (2.19) by the switching period T_s:

$$0 = \frac{1}{T_s} \int_0^{T_s} v_L(t)\,dt = \langle v_L \rangle \tag{2.20}$$

The right-hand side of Eq. (2.20) is recognized as the average value, or dc component, of $v_L(t)$. Equation

Fig. 2.12 The principle of inductor volt-second balance: in steady state, the net volt-seconds applied to an inductor (i.e., the total area λ) must be zero.

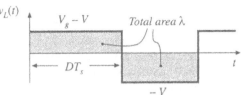

(2.20) states that, in equilibrium, the applied inductor voltage must have zero dc component.

The inductor voltage waveform of Fig. 2.9 is reproduced in Fig. 2.12, with the area under the $v_L(t)$ curve specifically identified. The total area λ is given by the areas of the two rectangles, or

$$\lambda = \int_0^{T_s} v_L(t)dt = \left(V_g - V\right)\left(DT_s\right) + \left(-V\right)\left(D'T_s\right) \tag{2.21}$$

The average value is therefore

$$\langle v_L \rangle = \frac{\lambda}{T_s} = D\left(V_g - V\right) + D'\left(-V\right) \tag{2.22}$$

By equating $\langle v_L \rangle$ to zero, and noting that $D + D' = 1$, one obtains

$$0 = DV_g - \left(D + D'\right)V = DV_g - V \tag{2.23}$$

Solution for V yields

$$V = DV_g \tag{2.24}$$

which coincides with the result obtained previously, Eq. (2.3). So the principle of inductor volt-second balance allows us to derive an expression for the dc component of the converter output voltage. An advantage of this approach is its generality—it can be applied to any converter. One simply sketches the applied inductor voltage waveform, and equates the average value to zero. This method is used later in this chapter, to solve several more complicated converters.

Similar arguments can be applied to capacitors. The defining equation of a capacitor is

$$i_C(t) = C\frac{dv_C(t)}{dt} \tag{2.25}$$

Integration of this equation over one switching period yields

$$v_C(T_s) - v_C(0) = \frac{1}{C}\int_0^{T_s} i_C(t)dt \tag{2.26}$$

In steady state, the net change over one switching period of the capacitor voltage must be zero, so that the left-hand side of Eq. (2.26) is equal to zero. Therefore, in equilibrium the integral of the capacitor current over one switching period (having the dimensions of amp-seconds, or charge) should be zero. There is no net change in capacitor charge in steady state. An equivalent statement is

$$0 = \frac{1}{T_s}\int_0^{T_s} i_C(t)\,dt = \langle i_C\rangle \tag{2.27}$$

The average value, or dc component, of the capacitor current must be zero in equilibrium.

This should be an intuitive result. If a dc current is applied to a capacitor, then the capacitor will charge continually and its voltage will increase without bound. Likewise, if a dc voltage is applied to an inductor, then the flux will increase continually and the inductor current will increase without bound. Equation (2.27), called the principle of *capacitor amp-second balance* or *capacitor charge balance*, can be used to find the steady-state currents in a switching converter.

2.3   BOOST CONVERTER EXAMPLE

The boost converter, Fig. 2.13(a), is another well-known switched-mode converter that is capable of producing a dc output voltage greater in magnitude than the dc input voltage. A practical realization of the switch, using a MOSFET and diode, is shown in Fig. 2.13(b). Let us apply the small-ripple approximation and the principles of inductor volt-second balance and capacitor charge balance to find the steady-state output voltage and inductor current for this converter.

With the switch in position 1, the right-hand side of the inductor is connected to ground, resulting in the network of Fig. 2.14(a). The inductor voltage and capacitor current for this subinterval are given by

$$\begin{aligned} v_L &= V_g \\ i_C &= -\frac{v}{R} \end{aligned} \tag{2.28}$$

Use of the linear ripple approximation, $v \approx V$, leads to

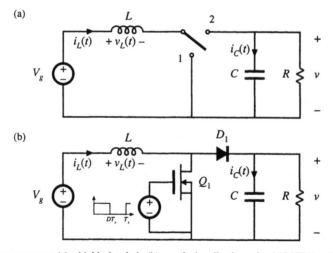

Fig. 2.13 Boost converter: (a) with ideal switch, (b) practical realization using MOSFET and diode.

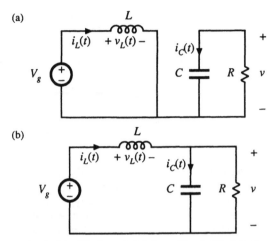

Fig. 2.14 Boost converter circuit, (a) while the switch is in position 1, (b) while the switch is in position 2.

$$v_L = V_g$$
$$i_C = -\frac{V}{R} \qquad (2.29)$$

With the switch in position 2, the inductor is connected to the output, leading to the circuit of Fig. 2.14(b). The inductor voltage and capacitor current are then

$$v_L = V_g - v$$
$$i_C = i_L - \frac{v}{R} \qquad (2.30)$$

Use of the small-ripple approximation, $v \approx V$ and $i_L \approx I$, leads to

$$v_L = V_g - V$$
$$i_C = I - \frac{V}{R} \qquad (2.31)$$

Equations (2.29) and (2.31) are used to sketch the inductor voltage and capacitor current waveforms of Fig. 2.15.

Fig. 2.15 Boost converter voltage and current waveforms.

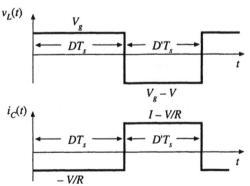

Fig. 2.16 Dc conversion ratio $M(D)$ of the boost converter.

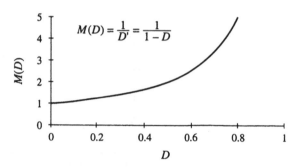

It can be inferred from the inductor voltage waveform of Fig. 2.15(a) that the dc output voltage V is greater than the input voltage V_g. During the first subinterval, $v_L(t)$ is equal to the dc input voltage V_g, and positive volt-seconds are applied to the inductor. Since, in steady-state, the total volt-seconds applied over one switching period must be zero, negative volt-seconds must be applied during the second subinterval. Therefore, the inductor voltage during the second subinterval, $(V_g - V)$, must be negative. Hence, V is greater than V_g.

The total volt-seconds applied to the inductor over one switching period are:

$$\int_0^{T_s} v_L(t)dt = \left(V_g\right)DT_s + \left(V_g - V\right)D'T_s \tag{2.32}$$

By equating this expression to zero and collecting terms, one obtains

$$V_g(D + D') - VD' = 0 \tag{2.33}$$

Solution for V, and by noting that $(D + D') = 1$, yields the expression for the output voltage,

$$V = \frac{V_g}{D'} \tag{2.34}$$

The voltage conversion ratio $M(D)$ is the ratio of the output to the input voltage of a dc-dc converter. Equation (2.34) predicts that the voltage conversion ratio is given by

$$M(D) = \frac{V}{V_g} = \frac{1}{D'} = \frac{1}{1-D} \tag{2.35}$$

This equation is plotted in Fig. 2.16. At $D = 0$, $V = V_g$. The output voltage increases as D increases, and in the ideal case tends to infinity as D tends to 1. So the ideal boost converter is capable of producing any output voltage greater than the input voltage. There are, of course, limits to the output voltage that can be produced by a practical boost converter. In the next chapter, component nonidealities are modeled, and it is found that the maximum output voltage of a practical boost converter is indeed limited. Nonetheless, very large output voltages can be produced if the nonidealities are sufficiently small.

The dc component of the inductor current is derived by use of the principle of capacitor charge balance. During the first subinterval, the capacitor supplies the load current, and the capacitor is partially discharged. During the second subinterval, the inductor current supplies the load and, additionally, recharges the capacitor. The net change in capacitor charge over one switching period is found by integrating the $i_C(t)$ waveform of Fig. 2.15(b),

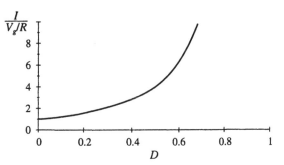

Fig. 2.17 Variation of inductor current dc component I with duty cycle, boost converter.

$$\int_0^{T_s} i_C(t)\,dt = \left(-\frac{V}{R}\right)DT_s + \left(I - \frac{V}{R}\right)D'T_s \tag{2.36}$$

Collecting terms, and equating the result to zero, leads the steady-state result

$$-\frac{V}{R}\left(D + D'\right) + ID' = 0 \tag{2.37}$$

By noting that $(D + D') = 1$, and by solving for the inductor current dc component I, one obtains

$$I = \frac{V}{D'R} \tag{2.38}$$

So the inductor current dc component I is equal to the load current, V/R, divided by D'. Substitution of Eq. (2.34) to eliminate V yields

$$I = \frac{V_g}{D'^2 R} \tag{2.39}$$

This equation is plotted in Fig. 2.17. It can be seen that the inductor current becomes large as D approaches 1.

This inductor current, which coincides with the dc input current in the boost converter, is greater than the load current. Physically, this must be the case: to the extent that the converter elements are ideal, the converter input and output powers are equal. Since the converter output voltage is greater than the input voltage, the input current must likewise be greater than the output current. In practice, the inductor current flows through the semiconductor forward voltage drops, the inductor winding resistance, and other sources of power loss. As the duty cycle approaches one, the inductor current becomes very large and these component nonidealities lead to large power losses. In consequence, the efficiency of the boost converter decreases rapidly at high duty cycle.

Next, let us sketch the inductor current $i_L(t)$ waveform and derive an expression for the inductor current ripple Δi_L. The inductor voltage waveform $v_L(t)$ has been already found (Fig. 2.15), so we can sketch the inductor current waveform directly. During the first subinterval, with the switch in position 1, the slope of the inductor current is given by

$$\frac{di_L(t)}{dt} = \frac{v_L(t)}{L} = \frac{V_g}{L} \tag{2.40}$$

Likewise, when the switch is in position 2, the slope of the inductor current waveform is

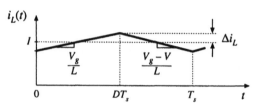

Fig.   2.18   Boost converter inductor current waveform $i_L(t)$.

$$\frac{di_L(t)}{dt} = \frac{v_L(t)}{L} = \frac{V_g - V}{L} \tag{2.41}$$

The inductor current waveform is sketched in Fig. 2.18. During the first subinterval, the change in inductor current, $2\Delta i_L$, is equal to the slope multiplied by the length of the subinterval, or

$$2\Delta i_L = \frac{V_g}{L} DT_s \tag{2.42}$$

Solution for Δi_L leads to

$$\Delta i_L = \frac{V_g}{2L} DT_s \tag{2.43}$$

This expression can be used to select the inductor value L such that a given value of Δi_L is obtained.

Likewise, the capacitor voltage $v(t)$ waveform can be sketched, and an expression derived for the output voltage ripple peak magnitude Δv. The capacitor current waveform $i_C(t)$ is given in Fig. 2.15. During the first subinterval, the slope of the capacitor voltage waveform $v(t)$ is

$$\frac{dv_C(t)}{dt} = \frac{i_C(t)}{C} = \frac{-V}{RC} \tag{2.44}$$

During the second subinterval, the slope is

$$\frac{dv_C(t)}{dt} = \frac{i_C(t)}{C} = \frac{I}{C} - \frac{V}{RC} \tag{2.45}$$

The capacitor voltage waveform is sketched in Fig. 2.19. During the first subinterval, the change in capacitor voltage, $-2\Delta v$, is equal to the slope multiplied by the length of the subinterval:

$$-2\Delta v = \frac{-V}{RC} DT_s \tag{2.46}$$

Solution for Δv yields

Fig.   2.19   Boost converter output voltage waveform $v(t)$.

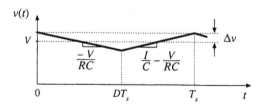

$$\Delta v = \frac{V}{2RC} DT_s \tag{2.47}$$

This expression can be used to select the capacitor value C to obtain a given output voltage ripple peak magnitude Δv.

2.4   ĆUK CONVERTER EXAMPLE

As a second example, consider the Ćuk converter of Fig. 2.20(a). This converter performs a dc conversion function similar to the buck-boost converter: it can either increase or decrease the magnitude of the dc voltage, and it inverts the polarity. A practical realization using a transistor and diode is illustrated in Fig. 2.20(b).

This converter operates via capacitive energy transfer. As illustrated in Fig. 2.21, capacitor C_1 is connected through L_1 to the input source while the switch is in position 2, and source energy is stored in C_1. When the switch is in position 1, this energy is released through L_2 to the load.

The inductor currents and capacitor voltages are defined, with polarities assigned somewhat arbitrarily, in Fig. 2.20(a). In this section, the principles of inductor volt-second balance and capacitor charge balance are applied to find the dc components of the inductor currents and capacitor voltages. The voltage and current ripple magnitudes are also found.

During the first subinterval, while the switch is in position 1, the converter circuit reduces to Fig. 2.21(a). The inductor voltages and capacitor currents are:

$$
\begin{aligned}
v_{L1} &= V_g \\
v_{L2} &= -v_1 - v_2 \\
i_{C1} &= i_2 \\
i_{C2} &= i_2 - \frac{v_2}{R}
\end{aligned}
\tag{2.48}
$$

(a)

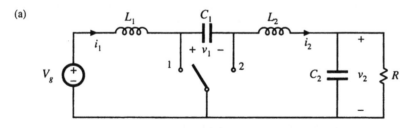

(b)

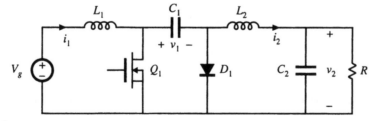

Fig. 2.20 Ćuk converter: (a) with ideal switch, (b) practical realization using MOSFET and diode.

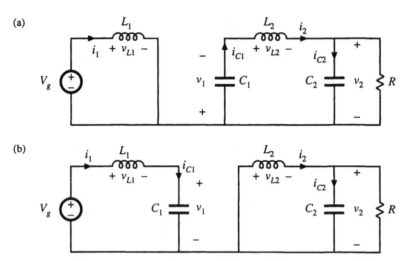

Fig. 2.21 Ćuk converter circuit: (a) while switch is in position 1, (b) while switch is in position 2.

We next assume that the switching ripple magnitudes in $i_1(t)$, $i_2(t)$, $v_1(t)$, and $v_2(t)$ are small compared to their respective dc components I_1, I_2, V_1, and V_2. We can therefore make the small-ripple approximation, and Eq. (2.48) becomes

$$\begin{aligned} v_{L1} &= V_g \\ v_{L2} &= -V_1 - V_2 \\ i_{C1} &= I_2 \\ i_{C2} &= I_2 - \frac{V_2}{R} \end{aligned} \qquad (2.49)$$

During the second subinterval, with the switch in position 2, the converter circuit elements are connected as in Fig. 2.21(b). The inductor voltages and capacitor currents are:

$$\begin{aligned} v_{L1} &= V_g - v_1 \\ v_{L2} &= -v_2 \\ i_{C1} &= i_1 \\ i_{C2} &= i_2 - \frac{v_2}{R} \end{aligned} \qquad (2.50)$$

We again make the small-ripple approximation, and hence Eq. (2.50) becomes

$$\begin{aligned} v_{L1} &= V_g - V_1 \\ v_{L2} &= -V_2 \\ i_{C1} &= I_1 \\ i_{C2} &= I_2 - \frac{V_2}{R} \end{aligned} \qquad (2.51)$$

Equations (2.49) and (2.51) are used to sketch the inductor voltage and capacitor current waveforms in Fig. 2.22.

The next step is to equate the dc components, or average values, of the waveforms of Fig. 2.22

(a) $v_{L1}(t)$

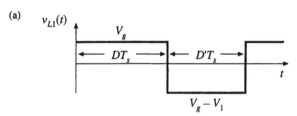

(b) $v_{L2}(t)$

Fig. 2.22 Ćuk converter waveforms:
(a) inductor voltage $v_{L1}(t)$, (b) inductor
voltage $v_{L2}(t)$, (c) capacitor current $i_{C1}(t)$,
(d) capacitor current $i_{C2}(t)$.

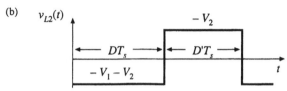

(c) $i_{C1}(t)$

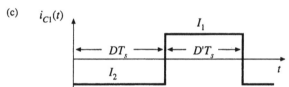

(d) $i_{C2}(t)$

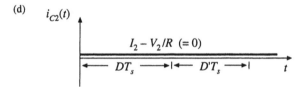

to zero, to find the steady-state conditions in the converter. The results are:

$$\begin{aligned}
\langle v_{L1}\rangle &= DV_g + D'\left(V_g - V_1\right) = 0\\
\langle v_{L2}\rangle &= D\left(-V_1 - V_2\right) + D'\left(-V_2\right) = 0\\
\langle i_{C1}\rangle &= DI_2 + D'I_1 = 0\\
\langle i_{C2}\rangle &= I_2 - \frac{V_2}{R} = 0
\end{aligned} \tag{2.52}$$

Solution of this system of equations for the dc components of the capacitor voltages and inductor currents leads to

$$\begin{aligned}
V_1 &= \frac{V_g}{D'}\\
V_2 &= -\frac{D}{D'}V_g\\
I_1 &= -\frac{D}{D'}I_2 = \left(\frac{D}{D'}\right)^2\frac{V_g}{R}\\
I_2 &= \frac{V_2}{R} = -\frac{D}{D'}\frac{V_g}{R}
\end{aligned} \tag{2.53}$$

The dependence of the dc output voltage V_2 on the duty cycle D is sketched in Fig. 2.23.
The inductor current waveforms are sketched in Fig. 2.24(a) and 2.24(b), and the capacitor C_1

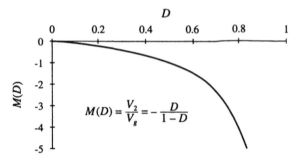

Fig.   2.23 Dc conversion ratio $M(D) = -V/V_g$ of the Ćuk converter.

$$M(D) = \frac{V_2}{V_g} = -\frac{D}{1-D}$$

voltage waveform $v_1(t)$ is sketched in Fig. 2.24(c). During the first subinterval, the slopes of the waveforms are given by

$$\frac{di_1(t)}{dt} = \frac{v_{L1}(t)}{L_1} = \frac{V_g}{L_1}$$
$$\frac{di_2(t)}{dt} = \frac{v_{L2}(t)}{L_2} = \frac{-V_1 - V_2}{L_2} \qquad (2.54)$$
$$\frac{dv_1(t)}{dt} = \frac{i_{C1}(t)}{C_1} = \frac{I_2}{C_1}$$

Equation (2.49) has been used here to substitute for the values of v_{L1}, v_{L2}, and i_{C1} during the first subinterval. During the second interval, the slopes of the waveforms are given by

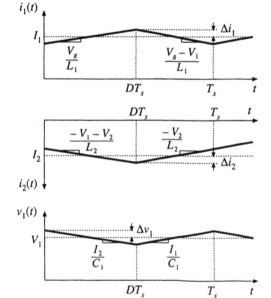

Fig. 2.24 Ćuk converter waveforms: (a) inductor current $i_1(t)$, (b) inductor current $i_2(t)$, (c) capacitor voltage $v_1(t)$.

$$\frac{di_1(t)}{dt} = \frac{v_{L1}(t)}{L_1} = \frac{V_g - V_1}{L_1}$$

$$\frac{di_2(t)}{dt} = \frac{v_{L2}(t)}{L_2} = \frac{-V_2}{L_2} \qquad (2.55)$$

$$\frac{dv_1(t)}{dt} = \frac{i_{C1}(t)}{C_1} = \frac{I_1}{C_1}$$

Equation (2.51) was used to substitute for the values of v_{L1}, v_{L2}, and i_{C1} during the second subinterval.

During the first subinterval, the quantities $i_1(t)$, $i_2(t)$, and $v_1(t)$ change by $2\Delta i_1$, $-2\Delta i_2$, and $-2\Delta v_1$, respectively. These changes are equal to the slopes given in Eq. (2.54), multiplied by the subinterval length DT_s, yielding

$$\Delta i_1 = \frac{V_g DT_s}{2L_1}$$

$$\Delta i_2 = \frac{V_1 + V_2}{2L_2} DT_s \qquad (2.56)$$

$$\Delta v_1 = \frac{-I_2 DT_s}{2C_1}$$

The dc relationships, Eq. (2.53), can now be used to simplify these expressions and eliminate V_1, V_2, and I_1, leading to

$$\Delta i_1 = \frac{V_g DT_s}{2L_1}$$

$$\Delta i_2 = \frac{V_g DT_s}{2L_2} \qquad (2.57)$$

$$\Delta v_1 = \frac{V_g D^2 T_s}{2D'RC_1}$$

These expressions can be used to select values of L_1, L_2, and C_1, such that desired values of switching ripple magnitudes are obtained.

Similar arguments cannot be used to estimate the switching ripple magnitude in the output capacitor voltage $v_2(t)$. According to Fig. 2.22(d), the current $i_{C2}(t)$ is continuous: unlike v_{L1}, v_{L2}, and i_{C1}, the capacitor current $i_{C2}(t)$ is nonpulsating. If the switching ripple of $i_2(t)$ is neglected, then the capacitor current $i_{C2}(t)$ does not contain an ac component. The small-ripple approximation then leads to the conclusion that the output switching ripple Δv_2 is zero.

Of course, the output voltage switching ripple is not zero. To estimate the magnitude of the output voltage ripple in this converter, we must not neglect the switching ripple present in the inductor current $i_2(t)$, since this current ripple is the only source of ac current driving the output capacitor C_2. A simple way of doing this in the Ćuk converter and in other similar converters is discussed in the next section.

2.5 ESTIMATING THE OUTPUT VOLTAGE RIPPLE IN CONVERTERS CONTAINING TWO-POLE LOW-PASS FILTERS

A case where the small ripple approximation is not useful is in converters containing two-pole low-pass filters, such as in the output of the Ćuk converter (Fig. 2.20) or the buck converter (Fig. 2.25). For these

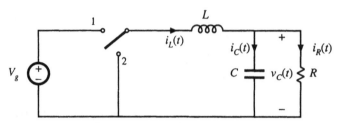

Fig. 2.25 The buck converter contains a two-pole output filter.

converters, the small-ripple approximation predicts zero output voltage ripple, regardless of the value of the output filter capacitance. The problem is that the only component of output capacitor current in these cases is that arising from the inductor current ripple. Hence, inductor current ripple cannot be neglected when calculating the output capacitor voltage ripple, and a more accurate approximation is needed.

An improved approach that is useful for this case is to estimate the capacitor current waveform $i_C(t)$ more accurately, accounting for the inductor current ripple. The capacitor voltage ripple can then be related to the total charge contained in the positive portion of the $i_C(t)$ waveform.

Consider the buck converter of Fig. 2.25. The inductor current waveform $i_L(t)$ contains a dc component I and linear ripple of peak magnitude Δi_L, as shown in Fig. 2.10. The dc component I must flow entirely through the load resistance R (why?), while the ac switching ripple divides between the load resistance R and the filter capacitor C. In a well-designed converter, in which the capacitor provides significant filtering of the switching ripple, the capacitance C is chosen large enough that its impedance at the switching frequency is much smaller than the load impedance R. Hence nearly all of the inductor current ripple flows through the capacitor, and very little flows through the load. As shown in Fig. 2.26, the capacitor current waveform $i_C(t)$ is then equal to the inductor current waveform with the dc component removed. The current ripple is linear, with peak value Δi_L.

When the capacitor current $i_C(t)$ is positive, charge is deposited on the capacitor plates and the capacitor voltage $v_C(t)$ increases. Therefore, between the two zero-crossings of the capacitor current waveform, the capacitor voltage changes between its minimum and maximum extrema. The waveform is symmetrical, and the total change in v_C is the peak-to-peak output voltage ripple, or $2\Delta v$.

This change in capacitor voltage can be related to the total charge q contained in the positive

Fig. 2.26 Output capacitor voltage and current waveforms, for the buck converter in Fig. 2.25.

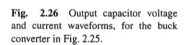

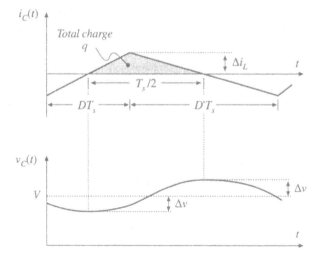

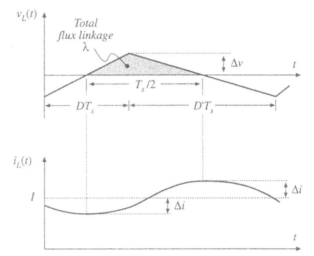

Fig. 2.27 Estimating inductor current ripple when the inductor voltage waveform is continuous.

portion of the capacitor current waveform. By the capacitor relation $Q = CV$,

$$q = C(2\Delta v) \tag{2.58}$$

As illustrated in Fig. 2.26, the charge q is the integral of the current waveform between its zero crossings. For this example, the integral can be expressed as the area of the shaded triangle, having a height Δi_L. Owing to the symmetry of the current waveform, the zero crossings occur at the centerpoints of the DT_s and $D'T_s$ subintervals. Hence, the base dimension of the triangle is $T_s/2$. So the total charge q is given by

$$q = \tfrac{1}{2}\Delta i_L \frac{T_s}{2} \tag{2.59}$$

Substitution of Eq. (2.58) into Eq. (2.59), and solution for the voltage ripple peak magnitude Δv yields

$$\Delta v = \frac{\Delta i_L T_s}{8C} \tag{2.60}$$

This expression can be used to select a value for the capacitance C such that a given voltage ripple Δv is obtained. In practice, the additional voltage ripple caused by the capacitor equivalent series resistance (esr) must also be included.

Similar arguments can be applied to inductors. An example is considered in Problem 2.9, in which a two-pole input filter is added to a buck converter as in Fig. 2.32. The capacitor voltage ripple cannot be neglected; doing so would lead to the conclusion that no ac voltage is applied across the input filter inductor, resulting in zero input current ripple. The actual inductor voltage waveform is identical to the ac portion of the input filter capacitor voltage, with linear ripple and with peak value Δv as illustrated in Fig. 2.27. By use of the inductor relation $\lambda = Li$, a result similar to Eq. (2.60) can be derived. The derivation is left as a problem for the student.

2.6   SUMMARY OF KEY POINTS

1.   The dc component of a converter waveform is given by its average value, or the integral over one switching period, divided by the switching period. Solution of a dc-dc converter to find its dc, or steady-state, voltages and currents therefore involves averaging the waveforms.

2.   The linear- (or small-) ripple approximation greatly simplifies the analysis. In a well-designed converter, the switching ripples in the inductor currents and capacitor voltages are small compared to the respective dc components, and can be neglected.

3.   The principle of inductor volt-second balance allows determination of the dc voltage components in any switching converter. In steady state, the average voltage applied to an inductor must be zero.

4.   The principle of capacitor charge balance allows determination of the dc components of the inductor currents in a switching converter. In steady state, the average current applied to a capacitor must be zero.

5.   By knowledge of the slopes of the inductor current and capacitor voltage waveforms, the ac switching ripple magnitudes may be computed. Inductance and capacitance values can then be chosen to obtain desired ripple magnitudes.

6.   In converters containing multiple-pole filters, continuous (nonpulsating) voltages and currents are applied to one or more of the inductors or capacitors. Computation of the ac switching ripple in these elements can be done using capacitor charge and/or inductor flux-linkage arguments, without use of the small-ripple approximation.

7.   Converters capable of increasing (boost), decreasing (buck), and inverting the voltage polarity (buck-boost and Ćuk) have been described. Converter circuits are explored more fully in the problems and in a later chapter.

REFERENCES

[1]   S. ĆUK, "Basics of Switched-Mode Power Conversion: Topologies, Magnetics, and Control," in *Advances in Switched-Mode Power Conversion*, Vol. 2, pp. 279-310, Irvine, CA: Teslaco, 1981.

[2]   N. MOHAN, T. UNDELAND, and W. ROBBINS, *Power Electronics: Converters, Applications, and Design*, 2nd edit., New York: John Wiley & Sons, 1995.

[3]   J. KASSAKIAN, M. SCHLECHT, and G. VERGESE, *Principles of Power Electronics*, Reading, MA: Addison-Wesley, 1991.

[4]   R. SEVERNS and G. E. BLOOM, *Modern Dc-to-dc Switch Mode Power Converter Circuits*, New York: Van Nostrand Reinhold, 1985.

[5]   D. HART, *Introduction to Power Electronics*, New York: Prentice Hall, 1997.

[6]   M. RASHID, *Power Electronics: Circuits, Devices, and Applications*, 2nd edit., New York: Prentice Hall, 1993.

[7]   P. KREIN, *Elements of Power Electronics*, New York: Oxford University Press, 1998.

[8]   K. KIT SUM, *Switch Mode Power Conversion—Basic Theory and Design*, New York: Marcel Dekker, 1984.

PROBLEMS

2.1   Analysis and design of a buck-boost converter: A buck-boost converter is illustrated in Fig. 2.28(a), and a practical implementation using a transistor and diode is shown in Fig. 2.28(b).

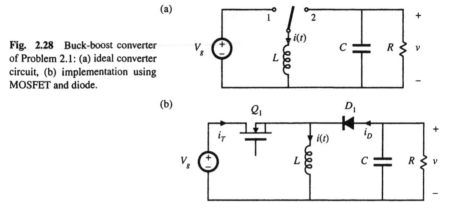

Fig. 2.28 Buck-boost converter of Problem 2.1: (a) ideal converter circuit, (b) implementation using MOSFET and diode.

(a)   Find the dependence of the equilibrium output voltage V and inductor current I on the duty ratio D, input voltage V_g, and load resistance R. You may assume that the inductor current ripple and capacitor voltage ripple are small.

(b)   Plot your results of part (a) over the range $0 \le D \le 1$.

(c)   Dc design: for the specifications

$$V_g = 30 \text{ V} \qquad\qquad V = -20 \text{ V}$$
$$R = 4 \text{ } \Omega \qquad\qquad f_s = 40 \text{ kHz}$$

(*i*)   Find D and I

(*ii*)   Calculate the value of L that will make the peak inductor current ripple Δi equal to ten percent of the average inductor current I.

(*iii*)   Choose C such that the peak output voltage ripple Δv is 0.1 V.

(d)   Sketch the transistor drain current waveform $i_T(t)$ for your design of part (c). Include the effects of inductor current ripple. What is the peak value of i_T? Also sketch $i_T(t)$ for the case when L is decreased such that Δi is 50% of I. What happens to the peak value of i_T in this case?

(e)   Sketch the diode current waveform $i_D(t)$ for the two cases of part (d).

2.2   In a certain application, an unregulated dc input voltage can vary between 18 and 36 V. It is desired to produce a regulated output of 28 V to supply a 2 A load. Hence, a converter is needed that is capable of both increasing and decreasing the voltage. Since the input and output voltages are both positive, converters that invert the voltage polarity (such as the basic buck-boost converter) are not suited for this application.

One converter that is capable of performing the required function is the nonisolated SEPIC (single-ended primary inductance converter) shown in Fig. 2.29. This converter has a conversion ratio $M(D)$ that can both buck and boost the voltage, but the voltage polarity is not inverted. In the normal converter operating mode, the transistor conducts during the first subinterval $(0 < t < DT_s)$, and the diode conducts during the second subinterval $(DT_s < t < T_s)$. You may assume that all elements are ideal.

(a)   Derive expressions for the dc components of each capacitor voltage and inductor current, as functions of the duty cycle D, the input voltage V_g, and the load resistance R.

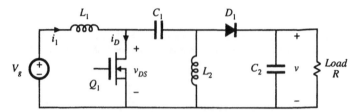

Fig. 2.29 SEPIC of Problems 2.2 and 2.3.

(b)   A control circuit automatically adjusts the converter duty cycle D, to maintain a constant output voltage of $V = 28$ V. The input voltage slowly varies over the range $18 \text{ V} \le V_g \le 36$ V. The load current is constant and equal to 2 A. Over what range will the duty cycle D vary? Over what range will the input inductor current dc component I_1 vary?

2.3   For the SEPIC of Problem 2.2,

(a)   Derive expressions for each inductor current ripple and capacitor voltage ripple. Express these quantities as functions of the switching period T_s; the component values L_1, L_2, C_1, C_2; the duty cycle D; the input voltage V_g; and the load resistance R.

(b)   Sketch the waveforms of the transistor voltage $v_{DS}(t)$ and transistor current $i_D(t)$, and give expressions for their peak values.

2.4   The switches in the converter of Fig. 2.30 operate synchronously: each is in position 1 for $0 < t < DT_s$, and in position 2 for $DT_s < t < T_s$. Derive an expression for the voltage conversion ratio $M(D) = V/V_g$. Sketch $M(D)$ vs. D.

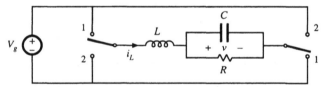

Fig. 2.30 H-bridge converter of Problems 2.4 and 2.6.

2.5   The switches in the converter of Fig. 2.31 operate synchronously: each is in position 1 for $0 < t < DT_s$, and in position 2 for $DT_s < t < T_s$. Derive an expression for the voltage conversion ratio $M(D) = V/V_g$. Sketch $M(D)$ vs. D.

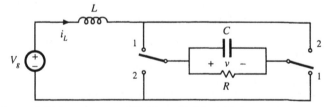

Fig. 2.31 Current-fed bridge converter of Problems 2.5, 2.7, and 2.8.

2.6   For the converter of Fig. 2.30, derive expressions for the inductor current ripple Δi_L and the capacitor voltage ripple Δv_C.

2.7   For the converter of Fig. 2.31, derive an analytical expression for the dc component of the inductor current, I, as a function of D, V_g, and R. Sketch your result vs. D.

2.8 For the converter of Fig. 2.31, derive expressions for the inductor current ripple Δi_L and the capacitor voltage ripple Δv_C.

2.9 To reduce the switching harmonics present in the input current of a certain buck converter, an input filter consisting of inductor L_1 and capacitor C_1 is added as shown in Fig. 2.32. Such filters are commonly used to meet regulations limiting conducted electromagnetic interference (EMI). For this problem, you may assume that all inductance and capacitance values are sufficiently large, such that all ripple magnitudes are small.

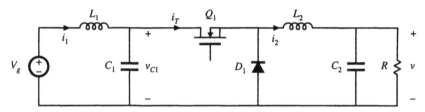

Fig. 2.32 Addition of *L–C* input filter to buck converter, Problem 2.9.

(a) Sketch the transistor current waveform $i_T(t)$.

(b) Derive analytical expressions for the dc components of the capacitor voltages and inductor currents.

(c) Derive analytical expressions for the peak ripple magnitudes of the input filter inductor current and capacitor voltage.

(d) Given the following values:

Input voltage	$V_g = 48$ V
Output voltage	$V = 36$ V
Switching frequency	$f_s = 100$ kHz
Load resistance	$R = 6\ \Omega$

Select values for L_1 and C_1 such that (*i*) the peak voltage ripple on C_1, Δv_{C1}, is two percent of the dc component V_{C1}, and (*ii*) the input peak current ripple Δi_1 is 20 mA.

Extra credit problem: Derive exact analytical expressions for (*i*) the dc component of the output voltage, and (*ii*) the peak-to-peak inductor current ripple, of the ideal buck-boost converter operating in steady state. Do not make the small-ripple approximation.

3

Steady-State Equivalent Circuit Modeling, Losses, and Efficiency

Let us now consider the basic functions performed by a switching converter, and attempt to represent these functions by a simple equivalent circuit. The designer of a converter power stage must calculate the network voltages and currents, and specify the power components accordingly. Losses and efficiency are of prime importance. The use of equivalent circuits is a physical and intuitive approach which allows the well-known techniques of circuit analysis to be employed. As noted in the previous chapter, it is desirable to ignore the small but complicated switching ripple, and model only the important dc components of the waveforms.

The dc transformer is used to model the ideal functions performed by a dc-dc converter [1–4]. This simple model correctly represents the relationships between the dc voltages and currents of the converter. The model can be refined by including losses, such as semiconductor forward voltage drops and on-resistances, inductor core and copper losses, etc. The resulting model can be directly solved, to find the voltages, currents, losses, and efficiency in the actual nonideal converter.

3.1   THE DC TRANSFORMER MODEL

As illustrated in Fig. 3.1, any switching converter contains three ports: a power input, a power output, and a control input. The input power is processed as specified by the control input, and then is output to the load. Ideally, these functions are performed with 100% efficiency, and hence

$$P_{in} = P_{out} \qquad (3.1)$$

or,

$$V_g I_g = VI \qquad (3.2)$$

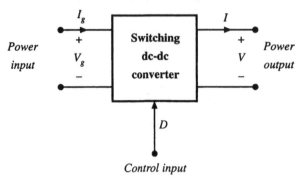

Fig. 3.1 Switching converter terminal quantities.

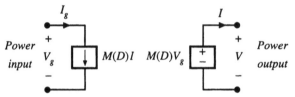

Fig. 3.2 A switching converter equivalent circuit using dependent sources, corresponding to Eqs. (3.3) and (3.4).

These relationships are valid only under equilibrium (dc) conditions: during transients, the net stored energy in the converter inductors and capacitors may change, causing Eqs. (3.1) and (3.2) to be violated.

In the previous chapter, we found that we could express the converter output voltage in an equation of the form

$$V = M(D)V_g \tag{3.3}$$

where $M(D)$ is the equilibrium conversion ratio of the converter. For example, $M(D) = D$ for the buck converter, and $M(D) = 1/(1 - D)$ for the boost converter. In general, for ideal PWM converters operating in the continuous conduction mode and containing an equal number of independent inductors and capacitors, it can be shown that the equilibrium conversion ratio M is a function of the duty cycle D and is independent of load.

Substitution of Eq. (3.3) into Eq. (3.2) yields

$$I_g = M(D)I \tag{3.4}$$

Hence, the converter terminal currents are related by the same conversion ratio.

Equations (3.3) and (3.4) suggest that the converter could be modeled using dependent sources, as in Fig. 3.2. An equivalent but more physically meaningful model (Fig. 3.3) can be obtained through the realization that Eqs. (3.1) to (3.4) coincide with the equations of an ideal transformer. In an ideal transformer, the input and output powers are equal, as stated in Eqs. (3.1) and (3.2). Also, the output voltage is equal to the turns ratio times the input voltage. This is consistent with Eq. (3.3), with the turns ratio taken to be the equilibrium conversion ratio $M(D)$. Finally, the input and output currents should be related by the same turns ratio, as in Eq. (3.4).

Thus, we can model the ideal dc-dc converter using the ideal dc transformer model of Fig. 3.3.

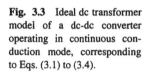

Fig. 3.3 Ideal dc transformer model of a dc-dc converter operating in continuous conduction mode, corresponding to Eqs. (3.1) to (3.4).

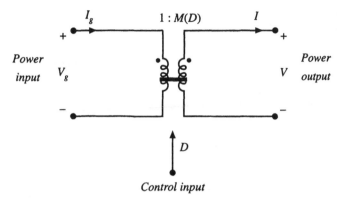

Control input

This symbol represents the first-order dc properties of any switching dc-dc converter: transformation of dc voltage and current levels, ideally with 100% efficiency, controllable by the duty cycle D. The solid horizontal line indicates that the element is ideal and capable of passing dc voltages and currents. It should be noted that, although standard magnetic-core transformers cannot transform dc signals (they saturate when a dc voltage is applied), we are nonetheless free to define the idealized model of Fig. 3.3 for the purpose of modeling dc-dc converters. Indeed, the absence of a physical dc transformer is one of the reasons for building a dc-dc switching converter. So the properties of the dc-dc converter of Fig. 3.1 can be modeled using the equivalent circuit of Fig. 3.3. An advantage of this equivalent circuit is that, for constant duty cycle, it is time invariant: there is no switching or switching ripple to deal with, and only the important dc components of the waveforms are modeled.

The rules for manipulating and simplifying circuits containing transformers apply equally well to circuits containing dc-dc converters. For example, consider the network of Fig. 3.4(a), in which a resistive load is connected to the converter output, and the power source is modeled by a Thevenin-equivalent voltage source V_1 and resistance R_1. The converter is replaced by the dc transformer model in Fig. 3.4(b). The elements V_1 and R_1 can now be pushed through the dc transformer as in Fig. 3.4(c); the volt-

(a)

Fig. 3.4 Example of use of the dc transformer model: (a) original circuit; (b) substitution of switching converter dc transformer model; (c) simplification by referring all elements to secondary side.

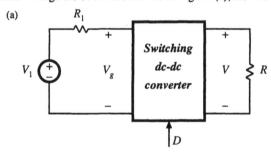

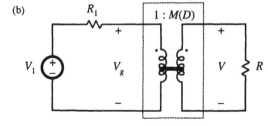

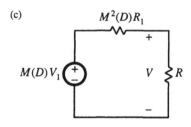

age source V_1 is multiplied by the conversion ratio $M(D)$, and the resistor R_1 is multiplied by $M^2(D)$. This circuit can now be solved using the voltage divider formula to find the output voltage:

$$V = M(D)V_1 \frac{R}{R + M^2(D)R_1} \tag{3.5}$$

It should be apparent that the dc transformer/equivalent circuit approach is a powerful tool for understanding networks containing converters.

3.2   INCLUSION OF INDUCTOR COPPER LOSS

The dc transformer model of Fig. 3.3 can be extended, to model other properties of the converter. Non-idealities, such as sources of power loss, can be modeled by adding resistors as appropriate. In later chapters, we will see that converter dynamics can be modeled as well, by adding inductors and capacitors to the equivalent circuit.

Let us consider the inductor copper loss in a boost converter. Practical inductors exhibit power loss of two types: (1) *copper loss*, originating in the resistance of the wire, and (2) *core loss*, due to hysteresis and eddy current losses in the magnetic core. A suitable model that describes the inductor copper loss is given in Fig. 3.5, in which a resistor R_L is placed in series with the inductor. The actual inductor then consists of an ideal inductor, L, in series with the copper loss resistor R_L.

Fig. 3.5 Modeling inductor copper loss via series resistor R_L.

The inductor model of Fig. 3.5 is inserted into the boost converter circuit in Fig. 3.6. The circuit can now be analyzed in the same manner as used for the ideal lossless converter, using the principles of inductor volt-second balance, capacitor charge balance, and the small-ripple approximation. First, we draw the converter circuits during the two subintervals, as in Fig. 3.7.

For $0 < t < DT_s$, the switch is in position 1 and the circuit reduces to Fig. 3.7(a). The inductor voltage $v_L(t)$, across the ideal inductor L, is given by

$$v_L(t) = V_g - i(t)R_L \tag{3.6}$$

and the capacitor current $i_C(t)$ is

$$i_C(t) = -\frac{v(t)}{R} \tag{3.7}$$

Next, we simplify these equations by assuming that the switching ripples in $i(t)$ and $v(t)$ are small compared to their respective dc components I and V. Hence, $i(t) \approx I$ and $v(t) \approx V$, and Eqs. (3.6) and (3.7)

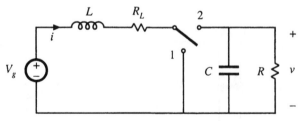

Fig. 3.6   Boost converter circuit, including inductor copper resistance R_L.

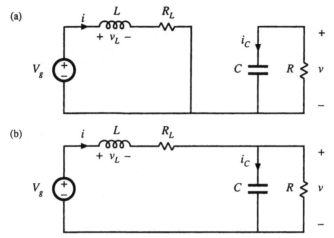

Fig. 3.7 Boost converter circuits during the two subintervals, including inductor copper loss resistance R_L: (a) with the switch in position 1, (b) with the switch in position 2.

become

$$v_L(t) = V_g - IR_L$$
$$i_C(t) = -\frac{V}{R}$$

(3.8)

For $DT_s < t < T_s$, the switch is in position 2 and the circuit reduces to Fig. 3.7(b). The inductor current and capacitor voltage are then given by

$$v_L(t) = V_g - i(t)R_L - v(t) \approx V_g - IR_L - V$$
$$i_C(t) = i(t) - \frac{v(t)}{R} \approx I - \frac{V}{R}$$

(3.9)

We again make the small-ripple approximation.

The principle of inductor volt-second balance can now be invoked. Equations (3.8) and (3.9) are used to construct the inductor voltage waveform $v_L(t)$ in Fig. 3.8. The dc component, or average value, of the inductor voltage $v_L(t)$ is

$$\langle v_L(t) \rangle = \frac{1}{T_s} \int_0^{T_s} v_L(t)dt = D\left(V_g - IR_L\right) + D'\left(V_g - IR_L - V\right)$$

(3.10)

By setting $\langle v_L \rangle$ to zero and collecting terms, one obtains

$$0 = V_g - IR_L - D'V$$

(3.11)

(recall that $D + D' = 1$). It can be seen that the inductor winding resistance R_L adds another term to the inductor volt-second balance equation. In the ideal boost converter ($R_L = 0$) example of Chapter 2, we were able to solve this equation directly for the voltage conversion ratio V/V_g. Equation (3.11) cannot be immediately solved in this manner, because the inductor current I is unknown. A second equation is needed, to eliminate I.

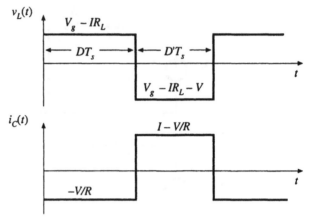

Fig. 3.8 Inductor voltage and capacitor current waveforms, for the nonideal boost converter of Fig. 3.6.

The second equation is obtained using capacitor charge balance. The capacitor current $i_C(t)$ waveform is given in Fig. 3.8. The dc component, or average value, of the capacitor current waveform is

$$\langle i_C(t) \rangle = D\left(-\frac{V}{R}\right) + D'\left(I - \frac{V}{R}\right) \tag{3.12}$$

By setting $\langle i_C \rangle$ to zero and collecting terms, one obtains

$$0 = D'I - \frac{V}{R} \tag{3.13}$$

We now have two equations, Eqs. (3.11) and (3.13), and two unknowns, V and I. Elimination of I and solution for V yields

$$\frac{V}{V_g} = \frac{1}{D'} \frac{1}{\left(1 + \dfrac{R_L}{D'^2 R}\right)} \tag{3.14}$$

This is the desired solution for the converter output voltage V. It is plotted in Fig. 3.9 for several values of R_L/R. It can be seen that Eq. (3.14) contains two terms. The first, $1/D'$, is the ideal conversion ratio, with $R_L = 0$. The second term, $1/(1 + R_L/D'^2 R)$, describes the effect of the inductor winding resistance. If R_L is much less than $D'^2 R$, then the second term is approximately equal to unity and the conversion ratio is approximately equal to the ideal value $1/D'$. However, as R_L is increased in relation to $D'^2 R$, then the second term is reduced in value, and V/V_g is reduced as well.

As the duty cycle D approaches one, the inductor winding resistance R_L causes a major qualitative change in the V/V_g curve. Rather than approaching infinity at $D = 1$, the curve tends to zero. Of course, it is unreasonable to expect that the converter can produce infinite voltage, and it should be comforting to the engineer that the prediction of the model is now more realistic. What happens at $D = 1$ is that the switch is always in position 1. The inductor is never connected to the output, so no energy is transferred to the output and the output voltage tends to zero. The inductor current tends to a large value, limited only by the inductor resistance R_L. A large amount of power is lost in the inductor winding resistance, equal to V_g^2/R_L, while no power is delivered to the load; hence, we can expect that the converter efficiency tends to zero at $D = 1$.

Another implication of Fig. 3.9 is that the inductor winding resistance R_L limits the maximum

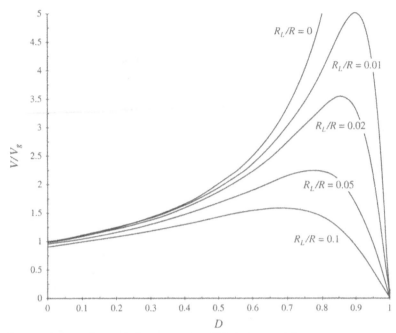

Fig. 3.9 Output voltage vs. duty cycle, boost converter with inductor copper loss.

voltage that the converter can produce. For example, with $R_L/R = 0.02$, it can be seen that the maximum V/V_g is approximately 3.5. If it is desired to obtain $V/V_g = 5$, then according to Fig. 3.9 the inductor winding resistance R_L must be reduced to less than 1% of the load resistance R. The only problem is that decreasing the inductor winding resistance requires building a larger, heavier, more expensive inductor. So it is usually important to optimize the design, by correctly modeling the effects of loss elements such as R_L, and choosing the smallest inductor that will do the job. We now have the analytical tools needed to do this.

3.3 CONSTRUCTION OF EQUIVALENT CIRCUIT MODEL

Next, let us refine the dc transformer model, to account for converter losses. This will allow us to determine the converter voltages, currents, and efficiency using well-known techniques of circuit analysis.

In the previous section, we used the principles of inductor volt-second balance and capacitor charge balance to write Eqs. (3.11) and (3.13), repeated here:

$$\langle v_L \rangle = 0 = V_g - IR_L - D'V$$
$$\langle i_C \rangle = 0 = D'I - \frac{V}{R} \tag{3.15}$$

These equations state that the dc components of the inductor voltage and capacitor current are equal to zero. Rather than algebraically solving the equations as in the previous section, we can reconstruct a circuit model based on these equations, which describes the dc behavior of the boost converter with inductor copper loss. This is done by constructing a circuit whose Kirchoff loop and node equations are

Fig. 3.10 Circuit whose loop equation is identical to Eq. (3.16), obtained by equating the average inductor voltage $\langle v_L \rangle$ to zero.

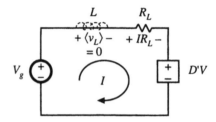

identical to Eqs. (3.15).

3.3.1 Inductor Voltage Equation

$$\langle v_L \rangle = 0 = V_g - IR_L - D'V$$

(3.16)

This equation was derived by use of Kirchoff's voltage law to find the inductor voltage during each subinterval. The results were averaged and set to zero. Equation (3.16) states that the sum of three terms having the dimensions of voltage are equal to $\langle v_L \rangle$, or zero. Hence, Eq. (3.16) is of the same form as a loop equation; in particular, it describes the dc components of the voltages around a loop containing the inductor, with loop current equal to the dc inductor current I.

So let us construct a circuit containing a loop with current I, corresponding to Eq. (3.16). The first term in Eq. (3.16) is the dc input voltage V_g, so we should include a voltage source of value V_g as shown in Fig. 3.10. The second term is a voltage drop of value IR_L, which is proportional to the current I in the loop. This term corresponds to a resistance of value R_L. The third term is a voltage $D'V$, dependent on the converter output voltage. For now, we can model this term using a dependent voltage source, with polarity chosen to satisfy Eq. (3.16).

3.3.2 Capacitor Current Equation

$$\langle i_C \rangle = 0 = D'I - \frac{V}{R}$$

(3.17)

This equation was derived using Kirchoff's current law to find the capacitor current during each subinterval. The results were averaged, and the average capacitor current was set to zero.

Equation (3.17) states that the sum of two dc currents are equal to $\langle i_C \rangle$, or zero. Hence, Eq. (3.17) is of the same form as a node equation; in particular, it describes the dc components of currents

Fig. 3.11 Circuit whose node equation is identical to Eq. (3.17), obtained by equating the average capacitor current $\langle i_C \rangle$ to zero.

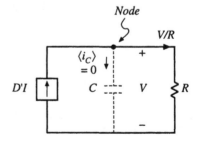

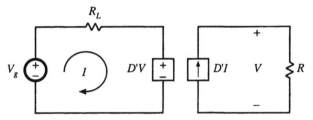

Fig. 3.12 The circuits of Figs. 3.10 and 3.11, drawn together.

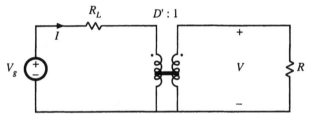

Fig. 3.13 Equivalent circuit model of the boost converter, including a $D':1$ dc transformer and the inductor winding resistance R_L.

flowing into a node connected to the capacitor. The dc capacitor voltage is V.

So now let us construct a circuit containing a node connected to the capacitor, as in Fig. 3.11, whose node equation satisfies Eq. (3.17). The second term in Eq. (3.17) is a current of magnitude V/R, proportional to the dc capacitor voltage V. This term corresponds to a resistor of value R, connected in parallel with the capacitor so that its voltage is V and hence its current is V/R. The first term is a current $D'I$, dependent on the dc inductor current I. For now, we can model this term using a dependent current source as shown. The polarity of the source is chosen to satisfy Eq. (3.17).

3.3.3 Complete Circuit Model

The next step is to combine the circuits of Figs. 3.10 and 3.11 into a single circuit, as in Fig. 3.12. This circuit can be further simplified by recognizing that the dependent voltage and current sources constitute an ideal dc transformer, as discussed in Section 3.1. The $D'V$ dependent voltage source depends on V, the voltage across the dependent current source. Likewise, the $D'I$ dependent current source depends on I, the current flowing through the dependent voltage source. In each case, the coefficient is D'. Hence, the dependent sources form a circuit similar to Fig. 3.2; the fact that the voltage source appears on the primary rather than the secondary side is irrelevant, owing to the symmetry of the transformer. They are therefore equivalent to the dc transformer model of Fig. 3.3, with turns ratio $D':1$. Substitution of the ideal dc transformer model for the dependent sources yields the equivalent circuit of Fig. 3.13.

The equivalent circuit model can now be manipulated and solved to find the converter voltages and currents. For example, we can eliminate the transformer by referring the V_g voltage source and R_L resistance to the secondary side. As shown in Fig. 3.14, the voltage source value is divided by the effective turns ratio D', and the resistance R_L is divided by the square of the turns ratio, D'^2. This circuit can be solved directly for the output voltage V, using the voltage divider formula:

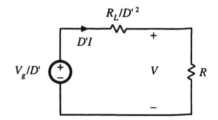

Fig. 3.14 Simplification of the equivalent circuit of Fig. 3.13, by referring all elements to the secondary side of the transformer.

$$V = \frac{V_g}{D'} \frac{R}{R + \dfrac{R_L}{D'^2}} = \frac{V_g}{D'} \frac{1}{1 + \dfrac{R_L}{D'^2 R}} \tag{3.18}$$

This result is identical to Eq. (3.14). The circuit can also be solved directly for the inductor current I, by referring all elements to the transformer primary side. The result is:

$$I = \frac{V_g}{D'^2 R + R_L} = \frac{V_g}{D'^2 R} \frac{1}{1 + \dfrac{R_L}{D'^2 R}} \tag{3.19}$$

3.3.4   Efficiency

The equivalent circuit model also allows us to compute the converter efficiency η. Figure 3.13 predicts that the converter input power is

$$P_{in} = (V_g)\,(I) \tag{3.20}$$

The load current is equal to the current in the secondary of the ideal dc transformer, or $D'I$. Hence, the model predicts that the converter output power is

$$P_{out} = (V)\,(D'I) \tag{3.21}$$

Therefore, the converter efficiency is

$$\eta = \frac{P_{out}}{P_{in}} = \frac{(V)\,(D'I)}{(V_g)\,(I)} = \frac{V}{V_g}D' \tag{3.22}$$

Substitution of Eq. (3.18) into Eq. (3.22) to eliminate V yields

$$\eta = \frac{1}{1 + \dfrac{R_L}{D'^2 R}} \tag{3.23}$$

This equation is plotted in Fig. 3.15, for several values of R_L/R. It can be seen from Eq. (3.23) that, to obtain high efficiency, the inductor winding resistance R_L should be much smaller that $D'^2 R$, the load resistance referred to the primary side of the ideal dc transformer. This is easier to do at low duty cycle, where D' is close to unity, than at high duty cycle where D' approaches zero. It can be seen from Fig.

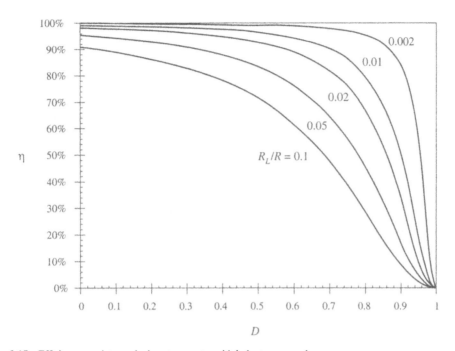

Fig. 3.15 Efficiency vs. duty cycle, boost converter with inductor copper loss.

3.15 that the efficiency is typically high at low duty cycles, but decreases rapidly to zero near $D = 1$.

Thus, the basic dc transformer model can be refined to include other effects, such as the inductor copper loss. The model describes the basic properties of the converter, including (a) transformation of dc voltage and current levels, (b) second-order effects such as power losses, and (c) the conversion ratio M. The model can be solved to find not only the output voltage V, but also the inductor current I and the efficiency η. All of the well-known techniques of circuit analysis can be employed to solve the model, making this a powerful and versatile approach.

The example considered so far is a relatively simple one, in which there is only a single loss element, R_L. Of course, real converters are considerably more complicated, and contain a large number of loss elements. When solving a complicated circuit to find the output voltage and efficiency, it behooves the engineer to use the simplest and most physically meaningful method possible. Writing a large number of simultaneous loop or node equations is not the best approach, because its solution typically requires several pages of algebra, and the engineer usually makes algebra mistakes along the way. The practicing engineer often gives up before finding the correct solution. The equivalent circuit approach avoids this situation, because one can simplify the circuit via well-known circuit manipulations such as pushing the circuit elements to the secondary side of the transformer. Often the answer can then be written by inspection, using the voltage divider rule or other formulas. The engineer develops confidence that the result is correct, and does not contain algebra mistakes.

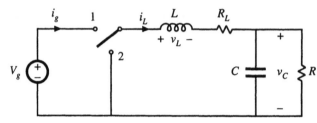

Fig. 3.16 Buck converter example.

3.4   HOW TO OBTAIN THE INPUT PORT OF THE MODEL

Let's try to derive the model of the buck converter of Fig. 3.16, using the procedure of Section 3.3. The inductor winding resistance is again modeled by a series resistor R_L.

The average inductor voltage can be shown to be

$$\langle v_L \rangle = 0 = DV_g - I_L R_L - V_C \tag{3.24}$$

This equation describes a loop with the dc inductor current I_L. The dc components of the voltages around this loop are: (*i*) the DV_g term, modeled as a dependent voltage source, (*ii*) a voltage drop $I_L R_L$, modeled as resistor R_L, and (*iii*) the dc output voltage V_C.

The average capacitor current is

$$\langle i_C \rangle = 0 = I_L - \frac{V_C}{R} \tag{3.25}$$

This equation describes the dc currents flowing into the node connected to the capacitor. The dc component of inductor current, I_L, flows into this node. The dc load current V_C/R (i.e., the current flowing through the load resistor R) flows out of this node. An equivalent circuit that models Eqs. (3.24) and (3.25) is given in Fig. 3.17. This circuit can be solved to determine the dc output voltage V_C.

What happened to the dc transformer in Fig. 3.17? We expect the buck converter model to contain a dc transformer, with turns ratio equal to the dc conversion ratio, or 1:*D*. According to Fig. 3.2, the secondary of this transformer is equivalent to a dependent voltage source, of value DV_g. Such a source does indeed appear in Fig. 3.17. But where is the primary? From Fig. 3.2, we expect the primary of the dc transformer to be equivalent to a dependent current source. In general, to derive this source, it is necessary to find the dc component of the converter input current $i_g(t)$.

The converter input current waveform $i_g(t)$ is sketched in Fig. 3.18. When the switch is in position 1, $i_g(t)$ is equal to the inductor current. Neglecting the inductor current ripple, we have $i_g(t) \approx I_L$. When the switch is in position 2, $i_g(t)$ is zero. The dc component, or average value, of $i_g(t)$ is

Fig. 3.17 Equivalent circuit derived from Eqs. (3.24) and (3.25).

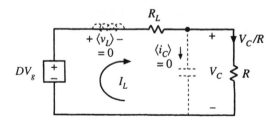

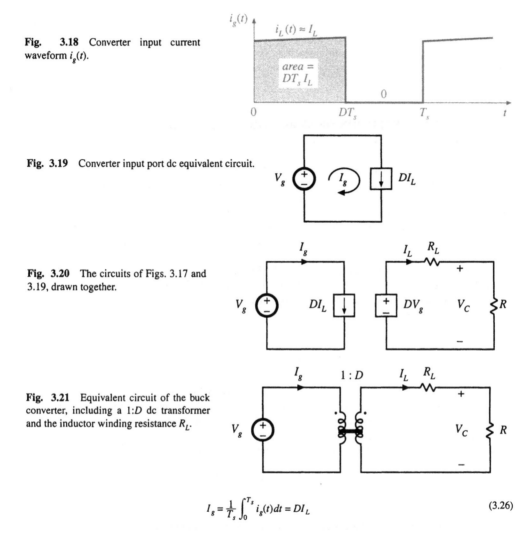

Fig. 3.18 Converter input current waveform $i_g(t)$.

Fig. 3.19 Converter input port dc equivalent circuit.

Fig. 3.20 The circuits of Figs. 3.17 and 3.19, drawn together.

Fig. 3.21 Equivalent circuit of the buck converter, including a 1:D dc transformer and the inductor winding resistance R_L.

$$I_g = \frac{1}{T_s}\int_0^{T_s} i_g(t)\,dt = DI_L \tag{3.26}$$

The integral of $i_g(t)$ is equal to the area under the $i_g(t)$ curve, or $DT_s I_L$ according to Fig. 3.18. The dc component I_g is therefore $(DT_s I_L)/T_s = DI_L$. Equation (3.26) states that I_g, the dc component of current drawn by the converter out of the V_g source, is equal to DI_L. An equivalent circuit is given in Fig. 3.19.

A complete model for the buck converter can now be obtained by combining Figs. 3.17 and 3.19 to obtain Fig. 3.20. The dependent current and voltage sources can be combined into a dc transformer, since the DV_g dependent voltage source has value D times the voltage V_g across the dependent current source, and the current source is the same constant D times the current I_L through the dependent voltage source. So, according to Fig. 3.2, the sources are equivalent to a dc transformer with turns ratio 1:D, as shown in Fig. 3.21.

In general, to obtain a complete dc equivalent circuit that models the converter input port, it is necessary to write an equation for the dc component of the converter input current. An equivalent circuit corresponding to this equation is then constructed. In the case of the buck converter, as well as in other converters having pulsating input currents, this equivalent circuit contains a dependent current source which becomes the primary of a dc transformer model. In the boost converter example of Section 3.3, it was unnecessary to explicitly write this equation, because the input current $i_g(t)$ coincided with the inductor current $i(t)$, and hence a complete equivalent circuit could be derived using only the inductor voltage and capacitor current equations.

3.5   EXAMPLE: INCLUSION OF SEMICONDUCTOR CONDUCTION LOSSES IN THE BOOST CONVERTER MODEL

As a final example, let us consider modeling semiconductor conduction losses in the boost converter of Fig. 3.22. Another major source of power loss is the conduction loss due to semiconductor device forward voltage drops. The forward voltage of a metal oxide semiconductor field-effect transistor (MOSFET) or bipolar junction transistor (BJT) can be modeled with reasonable accuracy as an on-resistance R_{on}. In the case of a diode, insulated-gate bipolar transistor (IGBT), or thyristor, a voltage source plus an on-resistance yields a model of good accuracy; the on-resistance may be omitted if the converter is being modeled at a single operating point.

When the gate drive signal is high, the MOSFET turns on and the diode is reverse-biased. The circuit then reduces to Fig. 3.23(a). In the conducting state, the MOSFET is modeled by the on-resistance R_{on}. The inductor winding resistance is again represented as in Fig. 3.5. The inductor voltage and capacitor current are given by

$$v_L(t) = V_g - iR_L - iR_{on} \approx V_g - IR_L - IR_{on}$$
$$i_C(t) = -\frac{v}{R} \approx -\frac{V}{R} \tag{3.27}$$

The inductor current and capacitor voltage have again been approximated by their dc components.

When the gate drive signal is low, the MOSFET turns off. The diode becomes forward-biased by the inductor current, and the circuit reduces to Fig. 3.23(b). In the conducting state, the diode is modeled in this example by voltage source V_D and resistance R_D. The inductor winding resistance is again modeled by resistance R_L. The inductor voltage and capacitor current for this subinterval are

$$v_L(t) = V_g - iR_L - V_D - iR_D - v \approx V_g - IR_L - V_D - IR_D - V$$
$$i_C(t) = i - \frac{v}{R} \approx I - \frac{V}{R} \tag{3.28}$$

The inductor voltage and capacitor current waveforms are sketched in Fig. 3.24.

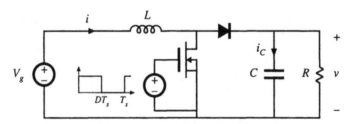

Fig. 3.22   Boost converter example.

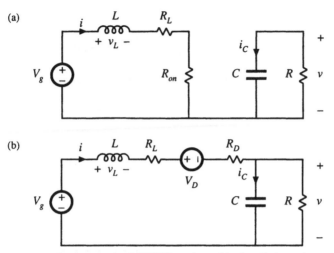

Fig. 3.23 Boost converter circuits: (a) when MOSFET conducts, (b) when diode conducts.

The dc component of the inductor voltage is given by

$$\langle v_L \rangle = D\left(V_g - IR_L - IR_{on}\right) + D'\left(V_g - IR_L - V_D - IR_D - V\right) = 0 \tag{3.29}$$

By collecting terms and noting that $D + D' = 1$, one obtains

$$V_g - IR_L - IDR_{on} - D'V_D - ID'R_D - D'V = 0 \tag{3.30}$$

This equation describes the dc components of the voltages around a loop containing the inductor, with loop current equal to the dc inductor current I. An equivalent circuit is given in Fig. 3.25.

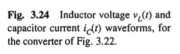

Fig. 3.24 Inductor voltage $v_L(t)$ and capacitor current $i_C(t)$ waveforms, for the converter of Fig. 3.22.

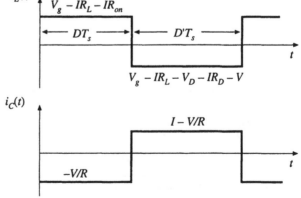

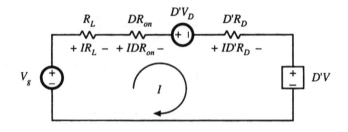

Fig. 3.25 Equivalent circuit corresponding to Eq. (3.30).

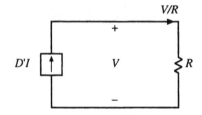

Fig. 3.26 Equivalent circuit corresponding to Eq. (3.32).

The dc component of the capacitor current is

$$\langle i_C \rangle = D\left(-\frac{V}{R}\right) + D'\left(I - \frac{V}{R}\right) = 0 \tag{3.31}$$

Upon collecting terms, one obtains

$$D'I - \frac{V}{R} = 0 \tag{3.32}$$

This equation describes the dc components of the currents flowing into a node connected to the capacitor, with dc capacitor voltage equal to V. An equivalent circuit is given in Fig. 3.26.

The two circuits are drawn together in 3.27. The dependent sources are combined into an ideal $D':1$ transformer in Fig. 3.28, yielding the complete dc equivalent circuit model.

Solution of Fig. 3.28 for the output voltage V yields

$$V = \left(\frac{1}{D'}\right)\left(V_g - D'V_D\right)\left(\frac{D'^2R}{D'^2R + R_L + DR_{on} + D'R_D}\right) \tag{3.33}$$

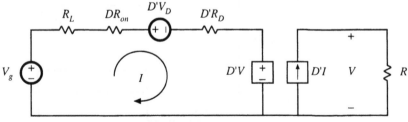

Fig. 3.27 The circuits of Figs. 3.25 and 3.26, drawn together.

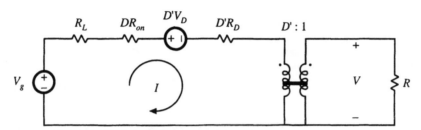

Fig. 3.28 Equivalent circuit model of the boost converter of Fig. 3.22, including ideal dc transformer, inductor winding resistance, and MOSFET and diode conduction losses.

Dividing by V_g gives the voltage conversion ratio:

$$\frac{V}{V_g} = \left(\frac{1}{D'}\right)\left(1 - \frac{D'V_D}{V_g}\right)\left(\frac{1}{1 + \frac{R_L + DR_{on} + D'R_D}{D'^2 R}}\right) \tag{3.34}$$

It can be seen that the effect of the loss elements V_D, R_L, R_{on}, and R_D is to decrease the voltage conversion ratio below the ideal value $(1/D')$.

The efficiency is given by $\eta = P_{out}/P_{in}$. From Fig. 3.28, $P_{in} = V_g I$ and $P_{out} = VD'I$. Hence,

$$\eta = D'\frac{V}{V_g} = \frac{\left(1 - \frac{D'V_D}{V_g}\right)}{\left(1 + \frac{R_L + DR_{on} + D'R_D}{D'^2 R}\right)} \tag{3.35}$$

For high efficiency, we require

$$\begin{aligned} V_g/D' &\gg V_D \\ D'^2 R &\gg R_L + DR_{on} + D'R_D \end{aligned} \tag{3.36}$$

It may seem strange that the equivalent circuit model of Fig. 3.28 contains effective resistances DR_{on} and $D'R_D$, whose values vary with duty cycle. The reason for this dependence is that the semiconductor on-resistances are connected in the circuit only when their respective semiconductor devices conduct. For example, at $D = 0$, the MOSFET never conducts, and the effective resistance DR_{on} disappears from the model. These effective resistances correctly model the average power losses in the elements. For instance, the equivalent circuit predicts that the power loss in the MOSFET on-resistance is $I^2 DR_{on}$. In the actual circuit, the MOSFET conduction loss is $I^2 R_{on}$ while the MOSFET conducts, and zero while the MOSFET is off. Since the MOSFET conducts with duty cycle D, the average conduction loss is $DI^2 R_{on}$, which coincides with the prediction of the model.

In general, to predict the power loss in a resistor R, we must calculate the root-mean-square current I_{rms} through the resistor, rather than the average current. The average power loss is then given by $I_{rms}^2 R$. Nonetheless, the average model of Fig. 3.28 correctly predicts average power loss, provided that the inductor current ripple is small. For example, consider the MOSFET conduction loss in the buck converter. The actual transistor current waveform is sketched in Fig. 3.29, for several values of inductor current ripple Δi. Case (a) corresponds to use of an infinite inductance L, leading to zero inductor current ripple. As shown in Table 3.1, the MOSFET conduction loss is then given by $I_{rms}^2 R_{on} = DI^2 R_{on}$, which

Fig. 3.29 Transistor current waveform, for various filter inductor values: (a) with a very large inductor, such that $\Delta i \approx 0$; (b) with a typical inductor value, such that $\Delta i = 0.1I$; (c) with a small inductor value, chosen such that $\Delta i = I$.

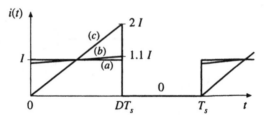

agrees exactly with the prediction of the average model. Case (b) is a typical choice of inductance L, leading to an inductor current ripple of $\Delta i = 0.1I$. The exact MOSFET conduction loss, calculated using the rms value of MOSFET current, is then only 0.33% greater than the prediction of the average model. In the extreme case (c) where $\Delta i = I$, the actual conduction loss is 33% greater than that predicted by the average model. Thus, the dc (average) model correctly predicts losses in the component nonidealities, even though rms currents are not calculated. The model is accurate provided that the inductor current ripple is small.

Table 3.1 Effect of inductor current ripple on MOSFET conduction loss

Inductor current ripple	MOSFET rms current	Average power loss in R_{on}
(a) $\Delta i = 0$	$I\sqrt{D}$	$DI^2 R_{on}$
(b) $\Delta i = 0.1i$	$(1.00167)I\sqrt{D}$	$(1.0033)DI^2 R_{on}$
(c) $\Delta i = I$	$(1.155)I\sqrt{D}$	$(1.3333)DI^2 R_{on}$

3.6   SUMMARY OF KEY POINTS

1. The dc transformer model represents the primary functions of any dc-dc converter: transformation of dc voltage and current levels, ideally with 100% efficiency, and control of the conversion ratio M via the duty cycle D. This model can be easily manipulated and solved using familiar techniques of conventional circuit analysis.

2. The model can be refined to account for loss elements such as inductor winding resistance and semiconductor on-resistances and forward voltage drops. The refined model predicts the voltages, currents, and efficiency of practical nonideal converters.

3. In general, the dc equivalent circuit for a converter can be derived from the inductor volt-second balance and capacitor charge balance equations. Equivalent circuits are constructed whose loop and node equations coincide with the volt-second and charge balance equations. In converters having a pulsating input current, an additional equation is needed to model the converter input port; this equation may be obtained by averaging the converter input current.

REFERENCES

[1]   R. D. MIDDLEBROOK, "A Continuous Model for the Tapped-Inductor Boost Converter," *IEEE Power Electronics Specialists Conference*, 1975 Record, pp. 63-79, June 1975.

[2]   S. M. ĆUK, "Modeling, Analysis, and Design of Switching Converters," Ph.D. thesis, California Institute of Technology, November 1976.

[3]   G. WESTER and R. D. MIDDLEBROOK, "Low-Frequency Characterization of Switched Dc–Dc Converters," *IEEE Transactions on Aerospace and Electronic Systems*, Vol. AES-9, pp. 376–385, May 1973.

[4]   R. D. MIDDLEBROOK and S. M. ĆUK, "Modeling and Analysis Methods for Dc-to-Dc Switching Converters," *IEEE International Semiconductor Power Converter Conference*, 1977 Record, pp. 90-111.

PROBLEMS

3.1   In the buck-boost converter of Fig. 3.30, the inductor has winding resistance R_L. All other losses can be ignored.

    (a)   Derive an expression for the nonideal voltage conversion ratio V/V_g.

    (b)   Plot your result of part (a) over the range $0 \le D \le 1$, for $R_L/R = 0$, 0.01, and 0.05.

    (c)   Derive an expression for the efficiency. Manipulate your expression into a form similar to Eq. (3.35)

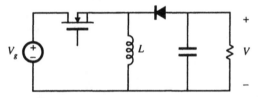

Fig. 3.30   Nonideal buck–boost converter, Problems 3.1 and 3.2.

3.2   The inductor in the buck-boost converter of Fig. 3.30 has winding resistance R_L. All other losses can be ignored. Derive an equivalent circuit model for this converter. Your model should explicitly show the input port of the converter, and should contain two dc transformers.

3.3   In the converter of Fig. 3.31, the inductor has winding resistance R_L. All other losses can be ignored. The switches operate synchronously: each is in position 1 for $0 < t < DT_s$, and in position 2 for $DT_s < t < T_s$.

    (a)   Derive an expression for the nonideal voltage conversion ratio V/V_g.

    (b)   Plot your result of part (a) over the range $0 \le D \le 1$, for $R_L/R = 0$, 0.01, and 0.05.

    (c)   Derive an expression for the efficiency. Manipulate your expression into a form similar to Eq. (3.35)

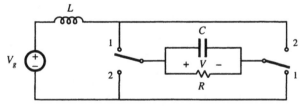

Fig. 3.31   Nonideal current-fed bridge converter, Problems 3.3 and 3.4.

3.4   The inductor in the converter of Fig. 3.31 has winding resistance R_L. All other losses can be ignored. Derive an equivalent circuit model for this converter.

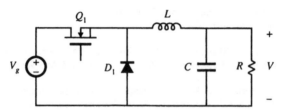

Fig. 3.32   Nonideal buck converter, Problem 3.5.

3.5   In the buck converter of Fig. 3.32, the MOSFET has on-resistance R_{on} and the diode forward voltage drop can be modeled by a constant voltage source V_D. All other losses can be neglected.

   (a)   Derive a complete equivalent circuit model for this converter.

   (b)   Solve your model to find the output voltage V.

   (c)   Derive an expression for the efficiency. Manipulate your expression into a form similar to Eq. (3.35).

3.6   To reduce the switching harmonics present in the input current of a certain buck converter, an input filter is added as shown in Fig. 3.33. Inductors L_1 and L_2 contain winding resistances R_{L1} and R_{L2}, respectively. The MOSFET has on-resistance R_{on}, and the diode forward voltage drop can be modeled by a constant voltage V_D plus a resistor R_D. All other losses can be ignored.

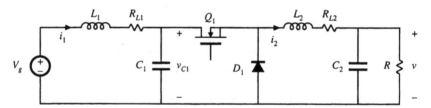

Fig. 3.33   Buck converter with input filter, Problem 3.6.

   (a)   Derive a complete equivalent circuit model for this circuit.

   (b)   Solve your model to find the dc output voltage V.

   (c)   Derive an expression for the efficiency. Manipulate your expression into a form similar to Eq. (3.35).

3.7   A 1.5 V battery is to be used to power a 5 V, 1 A load. It has been decided to use a buck-boost converter in this application. A suitable transistor is found with an on-resistance of 35 mΩ, and a Schottky diode is found with a forward drop of 0.5 V. The on-resistance of the Schottky diode may be ignored. The power stage schematic is shown in Fig. 3.34.

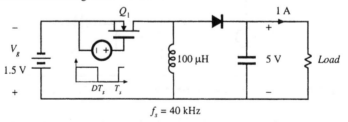

$$f_s = 40 \text{ kHz}$$

Fig. 3.34   Nonideal buck-boost converter powering a 5 V load from a 1.5 V battery, Problem 3.7

(a)   Derive an equivalent circuit that models the dc properties of this converter. Include the transistor and diode conduction losses, as well as the inductor copper loss, but ignore all other sources of loss. Your model should correctly describe the converter dc input port.

(b)   It is desired that the converter operate with at least 70% efficiency under nominal conditions (i.e., when the input voltage is 1.5 V and the output is 5 V at 1 A). How large can the inductor winding resistance be? At what duty cycle will the converter then operate? *Note:* there is an easy way and a not-so-easy way to analytically solve this part.

(c)   For your design of part (b), compute the power loss in each element.

(d)   Plot the converter output voltage and efficiency over the range $0 \leq D \leq 1$, using the value of inductor winding resistance which you selected in part (b).

(e)   Discuss your plot of part (d). Does it behave as you expect? Explain.

For Problems 3.8 and 3.9, a transistor having an on-resistance of 0.5 Ω is used. To simplify the problems, you may neglect all losses other than the transistor conduction loss. You may also neglect the dependence of MOS-FET on-resistance on rated blocking voltage. These simplifying assumptions reduce the differences between converters, but do not change the conclusions regarding which converter performs best in the given situations.

3.8   It is desired to interface a 500 V dc source to a 400 V, 10 A load using a dc-dc converter. Two possible approaches, using buck and buck-boost converters, are illustrated in Fig. 3.35. Use the assumptions described above to:

(a)   Derive equivalent circuit models for both converters, which model the converter input and output ports as well as the transistor conduction loss.

(b)   Determine the duty cycles that cause the converters to operate with the specified conditions.

(c)   Compare the transistor conduction losses and efficiencies of the two approaches, and conclude which converter is better suited to the specified application.

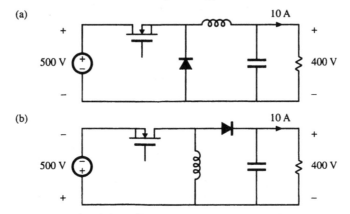

Fig. 3.35   Problem 3.8: interfacing a 500 V source to a 400 V load, using (a) a buck converter, (b) a buck-boost converter.

3.9   It is desired to interface a 300 V battery to a 400 V, 10 A load using a dc-dc converter. Two possible approaches, using boost and buck-boost converters, are illustrated in Fig. 3.36. Using the assumptions described above (before Problem 3.8), determine the efficiency and power loss of each approach. Which converter is better for this application?

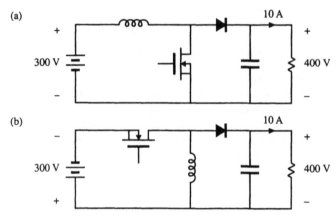

Fig. 3.36 Problem 3.9: interfacing a 300 V battery to a 400 V load, using: (a) a boost converter, (b) a buck–boost converter.

3.10   A buck converter is operated from the rectified 230 V ac mains, such that the converter dc input voltage is

$$V_g = 325 \text{ V} \pm 20\%$$

A control circuit automatically adjusts the converter duty cycle D, to maintain a constant dc output voltage of $V = 240$ V dc. The dc load current I can vary over a 10:1 range:

$$10 \text{ A} \leq I \leq 1 \text{ A}$$

The MOSFET has an on-resistance of 0.8 Ω. The diode conduction loss can be modeled by a 0.7 V source in series with a 0.2 Ω resistor. All other losses can be neglected.

(a)   Derive an equivalent circuit that models the converter input and output ports, as well as the loss elements described above.

(b)   Given the range of variation of V_g and I described above, over what range will the duty cycle vary?

(c)   At what operating point (i.e., at what value of V_g and I) is the converter power loss the largest? What is the value of the efficiency at this operating point?

3.11   In the Ćuk converter of Fig. 3.37, the MOSFET has on-resistance R_{on} and the diode has a constant forward voltage drop V_D. All other losses can be neglected.

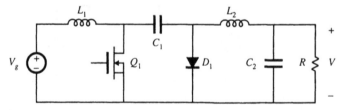

Fig. 3.37 Ćuk converter, Problem 3.11.

(a)   Derive an equivalent circuit model for this converter. *Suggestion*: if you don't know how to handle some of the terms in your dc equations, then temporarily leave them as dependent sources. A more physical representation of these terms may become apparent once dc transformers are incorporated into the model.

(b)   Derive analytical expressions for the converter output voltage and for the efficiency.

(c)   For $V_D = 0$, plot V/V_g vs. D over the range $0 \leq D \leq 1$, for (*i*) $R_{on}/R = 0.01$, and (*ii*) $R_{on}/R = 0.05$.

(d)   For $V_D = 0$, plot the converter efficiency over the range $0 \leq D \leq 1$, for (*i*) $R_{on}/R = 0.01$, and (*ii*) $R_{on}/R = 0.05$.

4

Switch Realization

We have seen in previous chapters that the switching elements of the buck, boost, and several other dc-dc converters can be implemented using a transistor and diode. One might wonder why this is so, and how to realize semiconductor switches in general. These are worthwhile questions to ask, and switch implementation can depend on the power processing function being performed. The switches of inverters and cycloconverters require more complicated implementations than those of dc-dc converters. Also, the way in which a semiconductor switch is implemented can alter the behavior of a converter in ways not predicted by the ideal-switch analysis of the previous chapters—an example is the discontinuous conduction mode treated in the next chapter. The realization of switches using transistors and diodes is the subject of this chapter.

Semiconductor power devices behave as single-pole single-throw (SPST) switches, represented ideally in Fig. 4.1. So, although we often draw converter schematics using ideal single-pole double-throw (SPDT) switches as in Fig. 4.2(a), the schematic of Fig. 4.2(b) containing SPST switches is more realistic. The realization of a SPDT switch using two SPST switches is not as trivial as it might at first seem, because Fig. 4.2(a) and 4.2(b) are not exactly equivalent. It is possible for both SPST switches to be simultaneously in the on state or in the off state, leading to behavior not predicted by the SPDT switch of Fig. 4.2(a). In addition, it is possible for the switch state to depend on the applied voltage or current waveforms—a familiar example is the diode. Indeed, it is common for these phenomena to occur in converters operating at light load, or occasionally at heavy load, leading to the discontinuous conduction mode previously mentioned. The converter properties are then significantly modified.

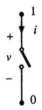

Fig. 4.1 SPST switch, with defined voltage and current polarities.

How an ideal switch can be realized using semiconductor devices depends on the polarity of the voltage that the devices must block in the off state, and on the polarity of the current that the devices

(a)

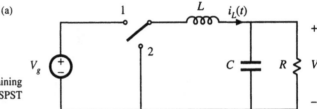

Fig. 4.2 Buck converter: (a) containing SPDT switch, (b) containing two SPST switches.

(b)

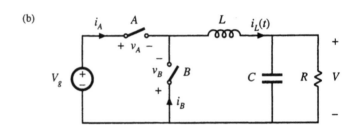

must conduct in the on state. For example, in the dc–dc buck converter of Fig. 4.2(b), switch A must block positive voltage V_g when in the off state, and must conduct positive current i_L when in the on state. If, for all intended converter operating points, the current and blocking voltage lie in a single quadrant of the plane as illustrated in Fig. 4.3, then the switch can be implemented in a simple manner using a transistor or a diode. Use of single-quadrant switches is common in dc–dc converters. Their operation is discussed briefly here.

In inverter circuits, two-quadrant switches are required. The output current is ac, and hence is sometimes positive and sometimes negative. If this current flows through the switch, then its current is ac, and the semiconductor switch realization is more complicated. A two-quadrant SPST switch can be realized using a transistor and diode. The dual case also sometimes occurs, in which the switch current is always positive, but the blocking voltage is ac. This type of two-quadrant switch can be constructed using a different arrangement of a transistor and diode. Cycloconverters generally require four-quadrant switches, which are capable of blocking ac voltages and conducting ac currents. Realizations of these elements are also discussed in this chapter.

Next, the synchronous rectifier is examined. The reverse-conducting capability of the metal oxide semiconductor field-effect transistor (MOSFET) allows it to be used where a diode would normally be required. If the MOSFET on-resistance is sufficiently small, then its conduction loss is less than that obtained using a diode. Synchronous rectifiers are sometimes used in low-voltage high-current applications to obtain improved efficiency. Several basic references treating single-, two-, and four-quadrant

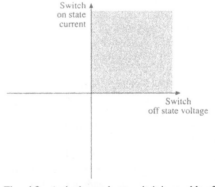

Fig. 4.3 A single-quadrant switch is capable of conducting currents of a single polarity, and of blocking voltages of a single polarity.

switches are listed at the end of this chapter [1–8].

Several power semiconductor devices are briefly discussed in Section 4.2. Majority-carrier devices, including the MOSFET and Schottky diode, exhibit very fast switching times, and hence are preferred when the off state voltage levels are not too high. Minority-carrier devices, including the bipolar junction transistor (BJT), insulated-gate bipolar transistor (IGBT), and thyristors [gate turn-off (GTO) and MOS-controlled thyristor (MCT)] exhibit high breakdown voltages with low forward voltage drops, at the expense of reduced switching speed.

Having realized the switches using semiconductor devices, switching loss can next be discussed. There are a number of mechanisms that cause energy to be lost during the switching transitions [11]. When a transistor drives a clamped inductive load, it experiences high instantaneous power loss during the switching transitions. Diode stored charge further increases this loss, during the transistor turn-on transition. Energy stored in certain parasitic capacitances and inductances is lost during switching. Parasitic ringing, which decays before the end of the switching period, also indicates the presence of switching loss. Switching loss increases directly with switching frequency, and imposes a maximum limit on the operating frequencies of practical converters.

4.1   SWITCH APPLICATIONS

4.1.1   Single-Quadrant Switches

The ideal SPST switch is illustrated in Fig. 4.1. The switch contains power terminals 1 and 0, with current and voltage polarities defined as shown. In the on state, the voltage v is zero, while the current i is zero in the off state. There is sometimes a third terminal C, where a control signal is applied. Distinguishing features of the SPST switch include the control method (active vs. passive) and the region of the i–v plane in which they can operate.

A passive switch does not contain a control terminal C. The state of the switch is determined by the waveforms $i(t)$ and $v(t)$ applied to terminals 0 and 1. The most common example is the diode, illustrated in Fig. 4.4. The ideal diode requires that $v(t) \leq 0$ and $i(t) \geq 0$. The diode is off ($i = 0$) when $v < 0$, and is on ($v = 0$) when $i > 0$. It can block negative voltage but not positive voltage. A passive SPST switch can be realized using a diode provided that the intended operating points [i.e., the values of $v(t)$ and $i(t)$ when the switch is in the on and off states] lie on the diode characteristic of Fig. 4.4(b).

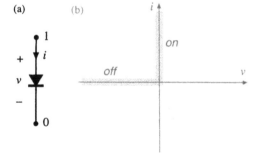

Fig. 4.4   Diode symbol (a), and its ideal characteristic (b).

The conducting state of an active switch is determined by the signal applied to the control terminal C. The state does not directly depend on the waveforms $v(t)$ and $i(t)$ applied to terminals 0 and 1. The BJT, MOSFET, IGBT, GTO, and MCT are examples of active switches. Idealized characteristics $i(t)$ vs. $v(t)$ for the BJT and IGBT are sketched in Fig. 4.5. When the control terminal causes the transistor to be in the off state, $i = 0$ and the device is capable of blocking positive voltage: $v \geq 0$. When the control terminal causes the transistor to be in the on state, $v = 0$ and the device is capable of conducting positive current: $i \geq 0$. The reverse-conducting and

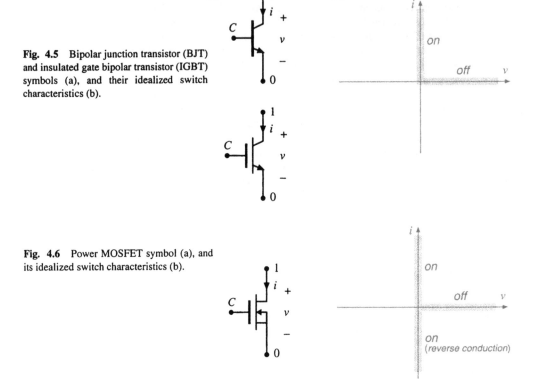

Fig. 4.5 Bipolar junction transistor (BJT) and insulated gate bipolar transistor (IGBT) symbols (a), and their idealized switch characteristics (b).

Fig. 4.6 Power MOSFET symbol (a), and its idealized switch characteristics (b).

reverse-blocking characteristics of the BJT and IGBT are poor or nonexistent, and have essentially no application in the power converter area. The power MOSFET (Fig. 4.6) has similar characteristics, except that it is able to conduct current in the reverse direction. With one notable exception (the synchronous rectifier discussed later), the MOSFET is normally operated with $i \geq 0$, in the same manner as the BJT and IGBT. So an active SPST switch can be realized using a BJT, IGBT, or MOSFET, provided that the intended operating points lie on the transistor characteristic of Fig. 4.5(b).

To determine how to implement an SPST switch using a transistor or diode, one compares the switch operating points with the i–v characteristics of Figs. 4.4(b), 4.5(b), and 4.6(b). For example, when it is intended that the SPDT switch of Fig. 4.2(a) be in position 1, SPST switch A of Fig. 4.2(b) is closed, and SPST switch B is opened. Switch A then conducts the positive inductor current, $i_A = i_L$, and switch B must block negative voltage, $v_B = -V_g$. These switch operating points are illustrated in Fig. 4.7. Likewise, when it is intended that the SPDT switch of Fig. 4.2(a) be in position 2, then SPST switch A is opened and switch B is closed. Switch B then conducts the positive inductor current, $i_B = i_L$, while switch A blocks positive voltage, $v_A = V_g$.

By comparison of the switch A operating points of Fig. 4.7(a) with Figs. 4.5(b) and 4.6(b), it can be seen that a transistor (BJT, IGBT, or MOSFET) could be used, since switch A must block positive voltage and conduct positive current. Likewise, comparison of Fig. 4.7(b) with Fig. 4.4(b) reveals that switch B can be implemented using a diode, since switch B must block negative voltage and conduct positive current. Hence a valid switch realization is given in Fig. 4.8.

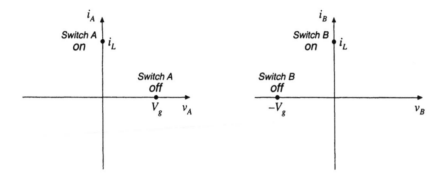

Fig. 4.7 Operating points of switch A, (a), and switch B, (b), in the buck converter of Fig. 4.2(b).

Figure 4.8 is an example of a single-quadrant switch realization: the devices are capable of conducting current of only one polarity, and blocking voltage of only one polarity. When the controller turns the transistor on, the diode becomes reverse-biased since $v_B = -V_g$. It is required that V_g be positive; otherwise, the diode will be forward-biased. The transistor conducts current i_L. This current should also be positive, so that the transistor conducts in the forward direction.

When the controller turns the transistor off, the diode must turn on so that the inductor current can continue to flow. Turning the transistor off causes the inductor current $i_L(t)$ to decrease. Since $v_L(t) = L \, di_L(t)/dt$, the inductor voltage becomes sufficiently negative to forward-bias the diode, and the diode turns on. Diodes that operate in this manner are sometimes called *freewheeling diodes*. It is required that i_L be positive; otherwise, the diode cannot be forward-biased since $i_B = i_L$. The transistor blocks voltage V_g; this voltage should be positive to avoid operating the transistor in the reverse blocking mode.

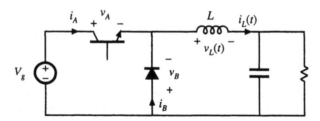

Fig. 4.8 Implementation of the SPST switches of Fig. 4.2(b) using a transistor and diode.

4.1.2 Current-Bidirectional Two-Quadrant Switches

In any number of applications such as dc-ac inverters and servo amplifiers, it is required that the switching elements conduct currents of both polarities, but block only positive voltages. A current-bidirectional two-quadrant SPST switch of this type can be realized using a transistor and diode, connected in an antiparallel manner as in Fig. 4.9.

The MOSFET of Fig. 4.6 is also a two-quadrant switch. However, it should be noted here that

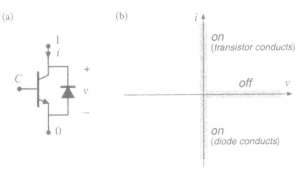

Fig. 4.9 A current-bidirectional two-quadrant SPST switch: (a) implementation using a transistor and antiparallel diode, (b) idealized switch characteristics.

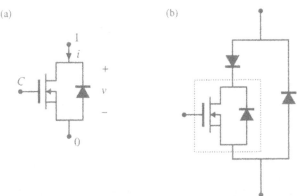

Fig. 4.10 The power MOSFET inherently contains a built-in body diode: (a) equivalent circuit, (b) addition of external diodes to prevent conduction of body diode.

practical power MOSFETs inherently contain a built-in diode, often called the *body diode*, as illustrated in Fig. 4.10. The switching speed of the body diode is much slower than that of the MOSFET. If the body diode is allowed to conduct, then high peak currents can occur during the diode turn-off transition. Most MOSFETs are not rated to handle these currents, and device failure can occur. To avoid this situation, external series and antiparallel diodes can be added as in Fig. 4.10(b). Power MOSFETs can be specifically designed to have a fast-recovery body diode, and to operate reliably when the body diode is allowed to conduct the rated MOSFET current. However, the switching speed of such body diodes is still somewhat slow, and significant switching loss due to diode stored charge (discussed later in this chapter) can occur.

A SPDT current-bidirectional two-quadrant switch can again be derived using two SPST switches as in Fig. 4.2(b). An example is given in Fig. 4.11. This converter operates from positive and negative dc supplies, and can produce an ac output voltage $v(t)$ having either polarity. Transistor Q_2 is driven with the complement of the Q_1 drive signal, so that Q_1 conducts during the first subinterval $0 < t < DT_s$, and Q_2 conducts during the second subinterval $DT_s < t < T_s$.

It can be seen from Fig. 4.11 that the switches must block voltage $2V_g$. It is required that V_g be positive; otherwise, diodes D_1 and D_2 will conduct simultaneously, shorting out the source.

It can be shown via inductor volt-second balance that

$$v_0 = (2D - 1)V_g \tag{4.1}$$

This equation is plotted in Fig. 4.12. The converter output voltage v_0 is positive for $D > 0.5$, and negative for $D < 0.5$. By sinusoidal variation of the duty cycle,

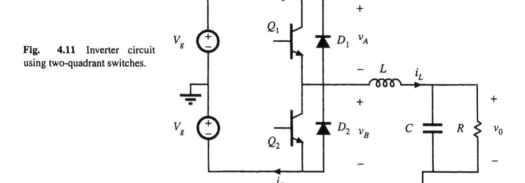

Fig.   4.11 Inverter circuit
using two-quadrant switches.

$$D(t) = 0.5 + D_m\sin(\omega t) \tag{4.2}$$

with D_m being a constant less that 0.5, the output voltage becomes sinusoidal. Hence this converter could
be used as a dc–ac inverter.

The load current is given by v_0/R; in equilibrium, this current coincides with the inductor current i_L,

$$i_L = \frac{v_0}{R} = (2D - 1)\frac{V_g}{R} \tag{4.3}$$

The switches must conduct this current. So the switch current is also positive when $D > 0.5$, and negative
when $D < 0.5$. With high-frequency duty cycle variations, the L–C filter may introduce a phase lag into
the inductor current waveform, but it is nonetheless true that switch currents of both polarities occur. So
the switch must operate in two quadrants of the plane, as illustrated in Fig. 4.13. When i_L is positive, Q_1
and D_2 alternately conduct. When i_L is negative, Q_2 and D_1 alternately conduct.

A well-known dc–3øac inverter circuit, the *voltage-source inverter* (VSI), operates in a similar
manner. As illustrated in Fig. 4.14, the VSI contains three two-quadrant SPDT switches, one per phase.
These switches block the dc input voltage V_g, and must conduct the output ac phase currents i_a, i_b, and i_c,

Fig. 4.12 Output voltage vs. duty cycle, for the
inverter of Fig. 4.11. This converter can produce
both positive and negative output voltages.

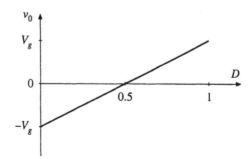

Fig. 4.13 The switches in the inverter of Fig. 4.11 must be capable of conducting both positive and negative current, but need block only positive voltage.

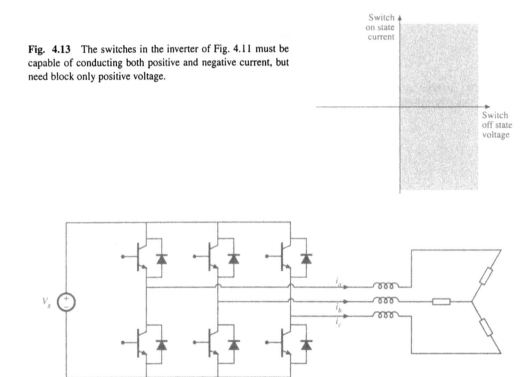

Fig. 4.14 The dc-3øac voltage-source inverter requires two-quadrant switches.

respectively.

Another current-bidirectional two-quadrant switch example is the bidirectional battery charger/discharger illustrated in Fig. 4.15. This converter can be used, for example, to interface a battery to the main power bus of a spacecraft. Both the dc bus voltage v_{bus} and the battery voltage v_{batt} are always positive. The semiconductor switch elements block positive voltage v_{bus}. When the battery is being charged, i_L is positive, and Q_1 and D_2 alternately conduct current. When the battery is being discharged, i_L is negative, and Q_2 and D_1 alternately conduct. Although this is a dc–dc converter, it requires two-quadrant switches because the power can flow in either direction.

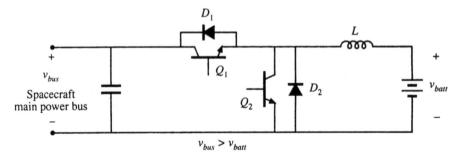

Fig. 4.15 Bidirectional battery charger/discharger, based on the dc-dc buck converter.

Fig. 4.16 Voltage-bidirectional two-quadrant switch properties.

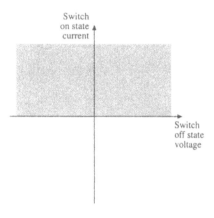

Fig. 4.17 A voltage-bidirectional two-quadrant SPST switch: (a) implementation using a transistor and series diode, (b) idealized switch characteristics.

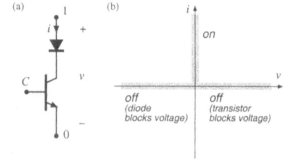

4.1.3 Voltage-Bidirectional Two-Quadrant Switches

Another type of two-quadrant switch, having the voltage-bidirectional properties illustrated in Fig. 4.16, is sometimes required. In applications where the switches must block both positive and negative voltages, but conduct only positive current, an SPST switch can be constructed using a series-connected transistor and diode as in Fig. 4.17. When it is intended that the switch be in the off state, the controller turns the transistor off. The diode then blocks negative voltage, and the transistor blocks positive voltage. The series connection can block negative voltages up to the diode voltage rating, and positive voltages up to the transistor voltage rating. The silicon-controlled rectifier is another example of a voltage-bidirectional two-quadrant switch.

A converter that requires this type of two-quadrant switch is the dc-3øac buck-boost inverter shown in Fig. 4.18 [4]. If the converter functions in inverter mode, so that the inductor current $i_L(t)$ is always positive, then all switches conduct only positive current. But the switches must block the output ac line-to-line voltages, which are sometimes positive and sometimes negative. Hence voltage-bidirectional two-quadrant switches are required.

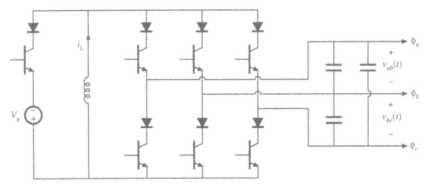

Fig. 4.18   Dc–3øac buck–boost inverter.

4.1.4   Four-Quadrant Switches

The most general type of switch is the four-quadrant switch, capable of conducting currents of either polarity and blocking voltages of either polarity, as in Fig. 4.19. There are several ways of constructing a four-quadrant switch. As illustrated in Fig. 4.20(b), two current-bidirectional two-quadrant switches described in Section 4.1.2 can be connected back-to-back. The transistors are driven on and off simultaneously. Another approach is the antiparallel connection of two voltage-bidirectional two-quadrant switches described in Section 4.1.3, as in Fig. 4.20(a). A third approach, using only one transistor but additional diodes, is given in Fig. 4.20(c).

Cycloconverters are a class of converters requiring four-quadrant switches. For example, a 3øac-to-3øac matrix converter is illustrated in Fig. 4.21. Each of the nine SPST switches is realized using one of the semiconductor networks of Fig. 4.20. With proper control of the switches, this converter can produce a three-phase output of variable fre-

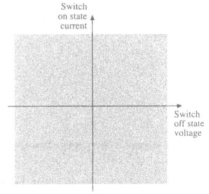

Fig. 4.19   A four-quadrant switch can conduct either polarity of current, and can block either polarity of voltage.

Fig. 4.20   Three ways of implementing a four-quadrant SPST switch.

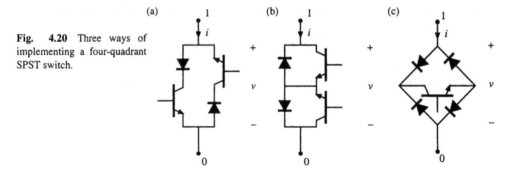

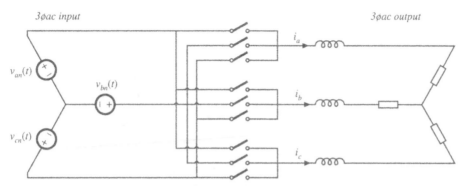

3øac input

$v_{an}(t)$

$v_{bn}(t)$

$v_{cn}(t)$

i_a

i_b

i_c

3øac output

Fig. 4.21 A 3øac–3øac matrix converter, which requires nine SPST four-quadrant switches.

quency and voltage, from a given three-phase ac input. Note that there are no dc signals in this converter: all of the input and output voltages and currents are ac, and hence four-quadrant switches are necessary.

4.1.5 Synchronous Rectifiers

The ability of the MOSFET channel to conduct current in the reverse direction makes it possible to employ a MOSFET where a diode would otherwise be required. When the MOSFET is connected as in Fig. 4.22(a) [note that the source and drain connections are reversed from the connections of Fig. 4.6(a)], the characteristics of Fig. 4.22(b) are obtained. The device can now block negative voltage and conduct positive current, with properties similar to those of the diode in Fig. 4.4. The MOSFET must be controlled such that it operates in the on state when the diode would normally conduct, and in the off state when the diode would be reverse-biased.

Thus, we could replace the diode in the buck converter of Fig. 4.8 with a MOSFET, as in Fig. 4.23. The BJT has also been replaced with a MOSFET in the figure. MOSFET Q_2 is driven with the complement of the Q_1 control signal.

The trend in computer power supplies is reduction of output voltage levels, from 5 V to 3.3 V and lower. As the output voltage is reduced, the diode conduction loss increases; in consequence, the diode conduction loss is easily the largest source of power loss in a 3.3 V power supply. Unfortunately, the diode junction contact potential limits what can be done to reduce the forward voltage drop of diodes. Schottky diodes having reduced junction potential can be employed; nonetheless, low-voltage power

(a)   (b)

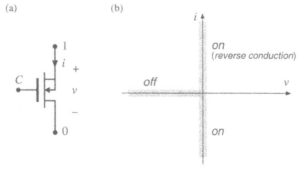

Fig. 4.22 Power MOSFET connected as a synchronous rectifier, (a), and its idealized switch characteristics, (b).

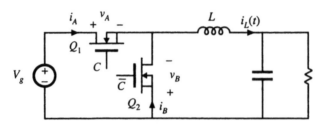

Fig. 4.23   Buck converter, implemented using a synchronous rectifier.

supplies containing diodes that conduct the output current must have low efficiency.

A solution is to replace the diodes with MOSFETs operated as synchronous rectifiers. The conduction loss of a MOSFET having on-resistance R_{on} and operated with rms current is I_{rms}, is $I_{rms}^2 R_{on}$. The on-resistance can be decreased by use of a larger MOSFET. So the conduction loss can be reduced as low as desired, if one is willing to pay for a sufficiently large device. Synchronous rectifiers find widespread use in low-voltage power supplies.

4.2   A BRIEF SURVEY OF POWER SEMICONDUCTOR DEVICES

The most fundamental challenge in power semiconductor design is obtaining a high breakdown voltage, while maintaining low forward voltage drop and on-resistance. A closely related issue is the longer switching times of high-voltage low–on-resistance devices. The tradeoff between breakdown voltage, on-resistance, and switching times is a key distinguishing feature of the various power devices.

The breakdown voltage of a reverse-biased *p–n* junction and its associated depletion region is a function of doping level: obtaining a high breakdown voltage requires low doping concentration, and hence high resistivity, in the material on at least one side of the junction. This high-resistivity region is usually the dominant contributor to the on-resistance of the device, and hence high-voltage devices must have higher on-resistance than low-voltage devices. In *majority carrier* devices, including the MOSFET and Schottky diode, this accounts for the first-order dependence of on-resistance on rated voltage. However, *minority carrier* devices, including the diffused-junction *p–n* diode, the bipolar junction transistor (BJT), the insulated-gate bipolar transistor (IGBT), and the thyristor family (SCR, GTO, MCT), exhibit another phenomenon known as *conductivity modulation*. When a minority-carrier device operates in the on state, minority carriers are injected into the lightly doped high-resistivity region by the forward-biased *p–n* junction. The resulting high concentration of minority carriers effectively reduces the apparent resistivity of the region, reducing the on-resistance of the device. Hence, minority-carrier devices exhibit lower on-resistances than comparable majority-carrier devices.

However, the advantage of decreased on-resistance in minority-carrier devices comes with the disadvantage of decreased switching speed. The conducting state of any semiconductor device is controlled by the presence or absence of key charge quantities within the device, and the turn-on and turn-off switching times are equal to the times required to insert or remove this controlling charge. Devices operating with conductivity modulation are controlled by their injected minority carriers. The total amount of controlling minority charge in minority-carrier devices is much greater than the charge required to control an equivalent majority-carrier device. Although the mechanisms for inserting and removing the controlling charge of the various devices can differ, it is nonetheless true that, because of their large amounts of minority charge, minority-carrier devices exhibit switching times that are significantly longer than those of majority-carrier devices. In consequence, majority-carrier devices find application at lower voltage levels and higher switching frequencies, while the reverse is true of minority-carrier devices.

Modern power devices are fabricated using up-to-date processing techniques. The resulting small feature size allows construction of highly interdigitated devices, whose unwanted parasitic elements are less significant. The resulting devices are more rugged and well-behaved than their predecessors.

A detailed description of power semiconductor device physics and switching mechanisms is beyond the scope of this book. Selected references on power semiconductor devices are listed in the reference section [9-19].

4.2.1   Power Diodes

As discussed above, the diffused-junction p–n diode contains a lightly doped or intrinsic high-resistivity region, which allows a high breakdown voltage to be obtained. As illustrated in Fig. 4.24(a), this region comprises one side of the p–n^- junction (denoted n^-); under reverse-biased conditions, essentially all of the applied voltage appears across the depletion region inside the n^- region. On-state conditions are illustrated in Fig. 4.24(b). Holes are injected across the forward-biased junction, and become minority carriers in the n^- region. These minority carriers effectively reduce the apparent resistivity of the n^- region via conductivity modulation. Essentially all of the forward current $i(t)$ is comprised of holes that diffuse across the p–n region, and then recombine with electrons from the n region.

Typical switching waveforms are illustrated in Fig. 4.25. The familiar exponential i–v character-

(a)

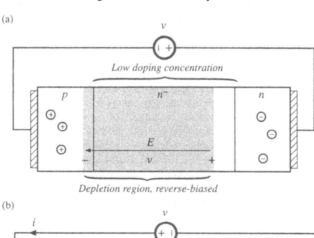

Fig. 4.24 Power diode: (a) under reverse-bias conditions, (b) under forward-bias conditions.

(b)

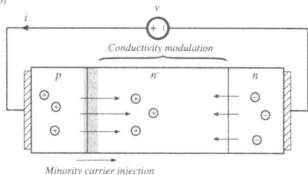

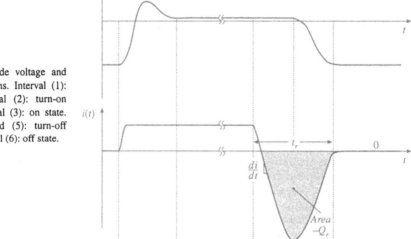

Fig. 4.25 Diode voltage and current waveforms. Interval (1): off state. Interval (2): turn-on transition. Interval (3): on state. Intervals (4) and (5): turn-off transition. Interval (6): off state.

istic of the p–n diode is an equilibrium relation. During transients, significant deviations from the exponential characteristic are observed; these deviations are associated with changes in the stored minority charge. As illustrated in Fig. 4.25, the diode operates in the off state during interval (1), with zero current and negative voltage. At the beginning of interval (2), the current increases to some positive value. This current charges the effective capacitance of the reverse-biased diode, supplying charge to the depletion region and increasing the voltage $v(t)$. Eventually, the voltage becomes positive, and the diode junction becomes forward-biased. The voltage may rise to a peak value of several volts, or even several tens of volts, reflecting the somewhat large resistance of the lightly doped n^- region. The forward-biased p–n^- junction continues to inject minority charge into the n^- region. As the total minority charge in the n^- region increases, conductivity modulation of the n^- region causes its effective resistance to decrease, and hence the forward voltage drop $v(t)$ also decreases. Eventually, the diode reaches equilibrium, in which the minority carrier injection rate and recombination rate are equal. During interval (3), the diode operates in the on state, with forward voltage drop given by the diode static i–v characteristic.

The turn-off transient is initiated at the beginning of interval (4). The diode remains forward-biased while minority charge is present in the vicinity of the diode p–n^- junction. Reduction of the stored minority charge can be accomplished either by active means, via negative terminal current, or by passive means, via recombination. Normally, both mechanisms occur simultaneously. The charge Q_r contained in the negative portion of the diode turn-off current waveform is called the *recovered charge*. The portion of Q_r occurring during interval (4) is actively-removed minority charge. At the end of interval (4), the stored minority charge in the vicinity of the p–n^- junction has been removed, such that the diode junction becomes reverse-biased and is able to block negative voltage. The depletion region effective capacitance is then charged during interval (5) to the negative off-state voltage. The portion of Q_r occurring during interval (5) is charge supplied to the depletion region, as well as minority charge that is actively removed from remote areas of the diode. At the end of interval (5), the diode is able to block the entire applied reverse voltage. The length of intervals (4) and (5) is called the *reverse recovery time* t_r. During interval (6), the diode operates in the off state. The diode turn-off transition, and its influence on switching loss in a PWM converter, is discussed further in Section 4.3.2.

Table 4.1 Characteristics of several commercial power rectifier diodes

Part number	Rated maximum voltage	Rated average current	V_F (typical)	t_r (max)
Fast recovery rectifiers				
1N3913	400 V	30 A	1.1 V	400 ns
SD453N25S20PC	2500 V	400 A	2.2 V	3 μs
Ultra-fast recovery rectifiers				
MUR815	150 V	8 A	0.975 V	35 ns
MUR1560	600 V	15 A	1.2 V	60 ns
RHRU100120	1200 V	100 A	2.6 V	60 ns
Schottky rectifiers				
MBR6030L	30 V	60 A	0.48 V	
444CNQ045	45 V	440 A	0.69 V	
30CPQ150	150 V	30 A	1.19 V	

Diodes are rated according to the length of their reverse recovery time t_r. *Standard recovery* rectifiers are intended for 50 Hz or 60 Hz operation; reverse recovery times of these devices are usually not specified. *Fast recovery* rectifiers and *ultrafast recovery* rectifiers are intended for use in converter applications. The reverse recovery time t_r, and sometimes also the recovered charge Q_r, are specified by manufacturers of these devices. Ratings of several commercial devices are listed in Table 4.1.

Schottky diodes are essentially majority-carrier devices whose operation is based on the rectifying characteristic of a metal-semiconductor junction. These devices exhibit negligible minority stored charge, and their switching behavior can be adequately modeled simply by their depletion-region capacitance and equilibrium exponential *i–v* characteristic. Hence, an advantage of the Schottky diode is its fast switching speed. An even more important advantage of Schottky diodes is their low forward voltage drops, especially in devices rated 45 V or less. Schottky diodes are restricted to low breakdown voltages; very few commercial devices are rated to block 100 V or more. Their off-state reverse currents are considerably higher than those of *p–n* junction diodes. Characteristics of several commercial Schottky rectifiers are also listed in Table 4.1.

Another important characteristic of a power semiconductor device is whether its on-resistance and forward voltage drop exhibit a positive temperature coefficient. Such devices, including the MOSFET and IGBT, are advantageous because multiple chips can be easily paralleled, to obtain high-current modules. These devices also tend to be more rugged and less susceptible to hot-spot formation and second-breakdown problems. Diodes cannot be easily connected in parallel, because of their negative temperature coefficients: an imbalance in device characteristics may cause one diode to conduct more current than the others. This diode becomes hotter, which causes it to conduct even more of the total current. In consequence, the current does not divide evenly between the paralleled devices, and the current rating of one of the devices may be exceeded. Since BJTs and thyristors are controlled by a diode junction, these devices also exhibit negative temperature coefficients and have similar problems when operated in parallel. Of course, it is possible to parallel any type of semiconductor device; however, use of matched devices, a common thermal substrate, and/or external circuitry may be required to cause the on-state currents of the devices to be equal.

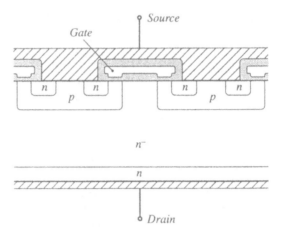

Fig. 4.26 Cross-section of DMOS n-channel power MOSFET structure. Crosshatched regions are metallized contacts. Shaded regions are insulating silicon dioxide layers.

4.2.2   Metal Oxide Semiconductor Field-Effect Transistor (MOSFET)

The power MOSFET is a modern power semiconductor device having gate lengths close to one micron. The power device is comprised of many small parallel-connected enhancement-mode MOSFET cells, which cover the surface of the silicon die. A cross-section of one cell is illustrated in Fig. 4.26. Current flows vertically through the silicon wafer: the metallized drain connection is made on the bottom of the chip, while the metallized source connection and polysilicon gate are on the top surface. Under normal operating conditions, in which $v_{ds} \geq 0$, both the $p-n$ and $p-n^-$ junctions are reverse-biased. Figure 4.27(a) illustrates operation of the device in the off state. The applied drain-to-source voltage then appears across the depletion region of the $p-n^-$ junction. The n^- region is lightly doped, such that the desired breakdown voltage rating is attained. Figure 4.27(b) illustrates operation in the on state, with a sufficiently large positive gate-to-source voltage. A channel then forms at the surface of the p region, underneath the gate. The drain current flows through the n^- region, channel, n region, and out through the source contact. The on-resistance of the device is the sum of the resistances of the n^- region, the channel, the source and drain contacts, etc. As the breakdown voltage is increased, the on-resistance becomes dominated by the resistance of the n^- region. Since there are no minority carriers to cause conductivity modulation, the on-resistance increases rapidly as the breakdown voltage is increased to several hundred volts and beyond.

The $p-n^-$ junction is called the *body diode*; as illustrated in Fig. 4.27(c), this junction forms an effective diode in parallel with the MOSFET channel. The body diode can become forward-biased when the drain-to-source voltage $v_{ds}(t)$ is negative. This diode is capable of conducting the full rated current of the MOSFET. However, most MOSFETs are not optimized with respect to the speed of their body diodes, and the large peak currents that flow during the reverse recovery transition of the body diode can cause device failure. Several manufacturers produce MOSFETs that contain fast recovery body diodes; these devices are rated to withstand the peak currents during the body diode reverse recovery transition.

Typical n-channel MOSFET static switch characteristics are illustrated in Fig. 4.28. The drain current is plotted as a function of the gate-to-source voltage, for various values of drain-to-source voltage. When the gate-to-source voltage is less than the threshold voltage V_{th}, the device operates in the off state. A typical value of V_{th} is 3 V. When the gate-to-source voltage is greater than 6 or 7 V, the device operates in the on state; typically, the gate is driven to 12 or 15 V to ensure minimization of the forward voltage drop. In the on state, the drain-to-source voltage V_{DS} is roughly proportional to the drain current

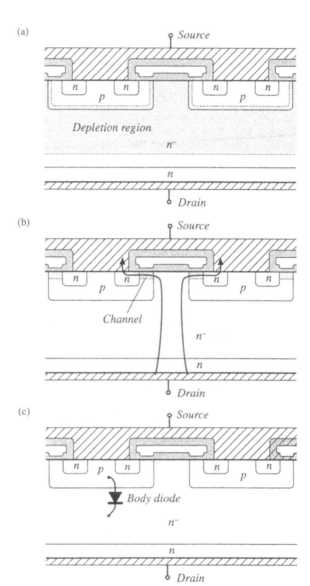

Fig. 4.27 Operation of the power MOSFET: (a) in the off state, v_{ds} appears across the depletion region in the n^- region; (b) current flow through the conducting channel in the on state; (c) body diode due to the $p–n^-$ junction.

I_D. The MOSFET is able to conduct peak currents well in excess of its average current rating, and the nature of the static characteristics is unchanged at high current levels. Logic-level power MOSFETs are also available, which operate in the on state with a gate-to-source voltage of 5 V. A few p-channel devices can be obtained, but their properties are inferior to those of equivalent n-channel devices.

The on-resistance and forward voltage drop of the MOSFET have a positive temperature coefficient. This property makes it relatively easy to parallel devices. High current MOSFET modules are available, containing several parallel-connect chips.

The major capacitances of the MOSFET are illustrated in Fig. 4.29. This model is sufficient for

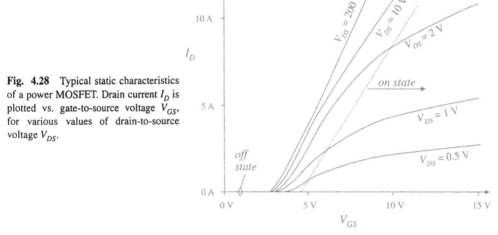

Fig. 4.28 Typical static characteristics of a power MOSFET. Drain current I_D is plotted vs. gate-to-source voltage V_{GS}, for various values of drain-to-source voltage V_{DS}.

qualitative understanding of the MOSFET switching behavior; more accurate models account for the parasitic junction field-effect transistor inherent in the DMOS geometry. Switching times of the MOSFET are determined essentially by the times required for the gate driver to charge these capacitances. Since the drain current is a function of the gate-to-source voltage, the rate at which the drain current changes is dependent on the rate at which the gate-to-source capacitance is charged by the gate drive circuit. Likewise, the rate at which the drain voltage changes is a function of the rate at which the gate-to-drain capacitance is charged. The drain-to-source capacitance leads directly to switching loss in PWM converters, since the energy stored in this capacitance is lost during the transistor turn-on transition. Switching loss is discussed in Section 4.3.

The gate-to-source capacitance is essentially linear. However, the drain-to-source and gate-to-drain capacitances are strongly nonlinear: these incremental capacitances vary as the inverse square root of the applied capacitor voltage. For example, the dependence of the incremental drain-to-source capacitance can be written in the form

$$C_{ds}(v_{ds}) = \frac{C_0}{\sqrt{1 + \frac{v_{ds}}{V_0}}} \tag{4.4}$$

where C_0 and V_0 are constants that depend on the construction of the device. These capacitances can easily vary by several orders of magnitude as v_{ds} varies over its normal operating range. For $v_{ds} \gg V_0$, Eq.

Fig. 4.29 MOSFET equivalent circuit which accounts for the body diode and effective terminal capacitances.

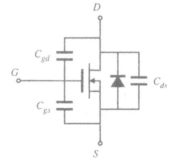

Table 4.2 Characteristics of several commercial *n*-channel power MOSFETs

Part number	Rated maximum voltage	Rated average current	R_{on}	Q_g (typical)
IRFZ48	60 V	50 A	0.018 Ω	110 nC
IRF510	100 V	5.6 A	0.54 Ω	8.3 nC
IRF540	100 V	28 A	0.077 Ω	72 nC
APT10M25BNR	100 V	75 A	0.025 Ω	171 nC
IRF740	400 V	10 A	0.55 Ω	63 nC
MTM15N40E	400 V	15 A	0.3 Ω	110 nC
APT5025BN	500 V	23 A	0.25 Ω	83 nC
APT1001RBNR	1000 V	11 A	1.0 Ω	150 nC

(4.4) can be approximated as

$$C_{ds}(v_{ds}) \approx C_0 \sqrt{\frac{V_0}{v_{ds}}} = \frac{C_0'}{\sqrt{v_{ds}}} \tag{4.5}$$

These expressions are used in Section 4.3.3 to determine the switching loss due to energy stored in C_{ds}.

Characteristics of several commercially available power MOSFETs are listed in Table 4.2. The gate charge Q_g is the charge that the gate drive circuit must supply to the MOSFET to raise the gate voltage from zero to some specified value (typically 10 V), with a specified value of off state drain-to-source voltage (typically 80% of the rated V_{DS}). The total gate charge is the sum of the charges on the gate-to-drain and the gate-to-source capacitance. The total gate charge is to some extent a measure of the size and switching speed of the MOSFET.

Unlike other power devices, MOSFETs are usually not selected on the basis of their rated average current. Rather, on-resistance and its influence on conduction loss are the limiting factors, and MOSFETs typically operate at average currents somewhat less than the rated value.

MOSFETs are usually the device of choice at voltages less than or equal to approximately 400 to 500 V. At these voltages, the forward voltage drop is competitive or superior to the forward voltage drops of minority-carrier devices, and the switching speed is significantly faster. Typical switching times are in the range 50 ns to 200 ns. At voltages greater than 400 to 500 V, minority-carrier devices having lower forward voltage drops, such as the IGBT, are usually preferred. The only exception is in applications where the high switching speed overrides the increased cost of silicon required to obtain acceptably low conduction loss.

4.2.3 Bipolar Junction Transistor (BJT)

A cross-section of an NPN power BJT is illustrated in Fig. 4.30. As with other power devices, current flows vertically through the silicon wafer. A lightly doped n^- region is inserted in the collector, to obtain the desired voltage breakdown rating. The transistor operates in the off state (cutoff) when the *p–n* base-emitter junction and the *p–n*$^-$ base-collector junction are reverse-biased; the applied collector-to-emitter voltage then appears essentially across the depletion region of the *p–n*$^-$ junction. The transistor operates in the on state (saturation) when both junctions are forward-biased; substantial minority charge is then present in the *p* and n^- regions. This minority charge causes the n^- region to exhibit a low on-

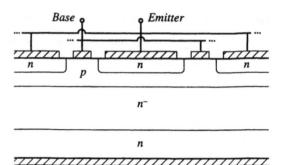

Fig. 4.30 Power BJT structure. Crosshatched regions are metallized contacts.

resistance via the conductivity modulation effect. Between the off state and the on state is the familiar active region, in which the *p–n* base-emitter junction is forward-biased and the *p–n⁻* base-collector junction is reverse-biased. When the BJT operates in the active region, the collector current is proportional to the base region minority charge, which in turn is proportional (in equilibrium) to the base current. There is in addition a fourth region of operation known as *quasi-saturation*, occurring between the active and saturation regions. Quasi-saturation occurs when the base current is insufficient to fully saturate the device; hence, the minority charge present in the *n⁻* region is insufficient to fully reduce the *n⁻* region resistance, and high transistor on-resistance is observed.

Consider the simple switching circuit of Fig. 4.31. Figure 4.32 contains waveforms illustrating the BJT turn-on and turn-off transitions. The transistor operates in the off state during interval (1), with the base-emitter junction reverse-biased by the source voltage $v_s(t) = - V_{s1}$. The turn-on transition is initiated at the beginning of interval (2), when the source voltage changes to $v_s(t) = + V_{s2}$. Positive current is then supplied by source v_s to the base of the BJT. This current first charges the capacitances of the depletion regions of the reverse-biased base-emitter and base-collector junctions. At the end of interval (2), the base-emitter voltage exceeds zero sufficiently for the base-emitter junction to become forward-biased. The length of interval (2) is called the *turn-on delay time*. During interval (3), minority charge is injected across the base-emitter junction from the emitter into the base region; the collector current is proportional to this minority base charge. Hence during interval (3), the collector current increases. Since the transistor drives a resistive

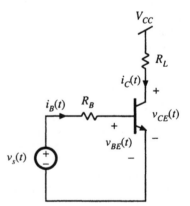

Fig. 4.31 Circuit for BJT switching time example.

load R_L, the collector voltage also decreases during interval (3). This causes the voltage to reduce across the reverse-biased base-collector depletion region (Miller) capacitance. Increasing the base current I_{B1} (by reducing R_B or increasing V_{s2}) increases the rate of change of both the base region minority charge and the charge in the Miller capacitance. Hence, increased I_{B1} leads to a decreased turn-on switching time.

Near or at the end of interval (3), the base-collector *p–n⁻* junction becomes forward-biased. Minority carriers are then injected into the *n⁻* region, reducing its effective resistivity. Depending on the device geometry and the magnitude of the base current, a *voltage tail* [interval (4)] may be observed as

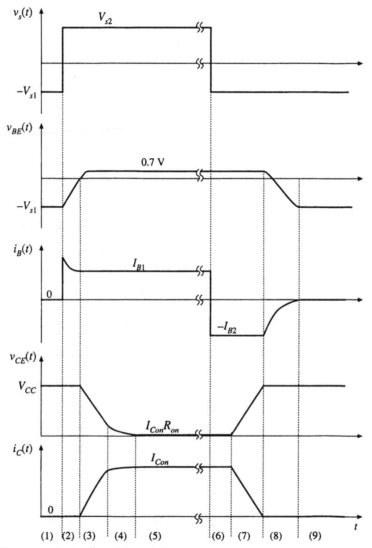

Fig. 4.32 BJT turn-on and turn-off transition waveforms.

the apparent resistance of the n^- region decreases via conductivity modulation. The BJT reaches on-state equilibrium at the beginning of interval (5), with low on-resistance and with substantial minority charge present in both the n^- and p regions. This minority charge significantly exceeds the amount necessary to support the active region conduction of the collector current I_{Con}; its magnitude is a function of $I_{B1} - I_{Con}/\beta$, where β is the active-region current gain.

The turn-off process is initiated at the beginning of interval (6), when the source voltage changes to $v_s(t) = -V_{s1}$. The base-emitter junction remains forward-biased as long as minority carriers are present in its vicinity. Also, the collector current continues to be $i_C(t) = I_{Con}$ as long as the minority

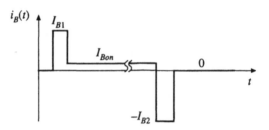

Fig. 4.33   Ideal base current waveform for minimization of switching times.

charge exceeds the amount necessary to support the active region conduction of I_{Con}, that is, as long as *excess charge* is present. So during interval (6), a negative base current flows equal to $-I_{B2} = (-V_{s1} - v_{BE}(t))/R_B$. This negative base current actively removes the total stored minority charge. Recombination further reduces the stored minority charge. Interval (6) ends when all of the excess minority charge has been removed. The length of interval (6) is called the *storage time*. During interval (7), the transistor operates in the active region. The collector current $i_C(t)$ is now proportional to the stored minority charge. Recombination and the negative base current continue to reduce the minority base charge, and hence the collector current decreases. In addition, the collector voltage increases, and hence the base current must charge the Miller capacitance. At the end of interval (7), the minority stored charge is equal to zero, and the base-emitter junction can become reverse-biased. The length of interval (7) is called the turn-off time or *fall time*. During interval (8), the reverse-biased base-emitter junction capacitance is discharged to voltage $-V_{s1}$. During interval (9), the transistor operates in equilibrium, in the off state.

It is possible to turn the transistor off using $I_{B2} = 0$; for example, we could let V_{s1} be approximately zero. However, this leads to very long storage and turn-off switching times. If $I_{B2} = 0$, then all of the stored minority charge must be removed passively, via recombination. From the standpoint of minimizing switching times, the base current waveform of Fig. 4.33 is ideal. The initial base current I_{B1} is large in magnitude, such that charge is inserted quickly into the base, and the turn-on switching times are short. A compromise value of equilibrium on state current I_{Bon} is chosen, to yield a reasonably low collector-to-emitter forward voltage drop, while maintaining moderate amounts of excess stored minority charge and hence keeping the storage time reasonably short. The current $-I_{B2}$ is large in magnitude, such that charge is removed quickly from the base and hence the storage and turn-off switching times are minimized.

Unfortunately, in most BJTs, the magnitudes of I_{B1} and I_{B2} must be limited because excessive values lead to device failure. As illustrated in Fig. 4.34, the base current flows laterally through the *p*

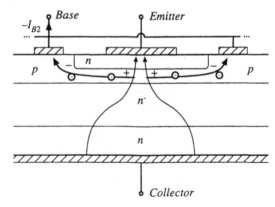

Fig. 4.34   A large I_{B2} leads to focusing of the emitter current away from the base contacts, due to the voltage induced by the lateral base region current.

region. This current leads to a voltage drop in the resistance of the p material, which influences the voltage across the base-emitter junction. During the turn-off transition, the base current $-I_{B2}$ causes the base-emitter junction voltage to be greater in the center of the base region, and smaller at the edges near the base contacts. This causes the collector current to focus near the center of the base region. In a similar fashion, a large I_{B1} causes the collector current to crowd near the edges of the base region during the turn-on transition. Since the collector-to-emitter voltage and collector current are simultaneously large during the switching transitions, substantial power loss can be associated with current focusing. Hence hot spots are induced at the center or edge of the base region. The positive temperature coefficient of the base-emitter junction current (corresponding to a negative temperature coefficient of the junction voltage) can then lead to thermal runaway and device failure. Thus, to obtain reliable operation, it may be necessary to limit the magnitudes of I_{B1} and I_{B2}. It may also be necessary to add external *snubber* networks which the reduce the instantaneous transistor power dissipation during the switching transitions.

Steady-state characteristics of the BJT are illustrated in Fig. 4.35. In Fig. 4.35(a), the collector current I_C is plotted as a function of the base current I_B, for various values of collector-to-emitter voltage V_{CE}. The cutoff, active, quasi-saturation, and saturation regions are identified. At a given collector cur-

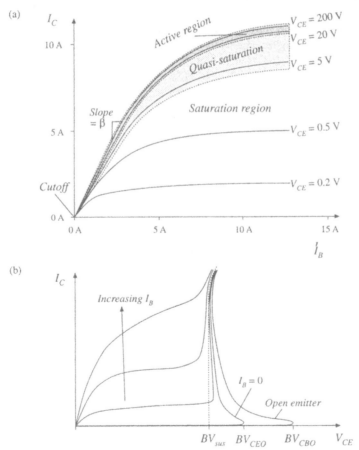

Fig. 4.35 BJT static characteristics: (a) I_C vs. I_B, illustrating the regions of operation; (b) I_C vs. V_{CE}, illustrating voltage breakdown characteristics.

rent I_C, to operate in the saturation region with minimum forward voltage drop, the base current I_B must be sufficiently large. The slope dI_C/dI_B in the active region is the current gain β. It can be seen that β decreases at high current—near the rated current of the BJT, the current gain decreases rapidly and hence it is difficult to fully saturate the device. Collector current I_C is plotted as a function of collector-to-emitter voltage V_{CE} in Fig. 4.35(b), for various values of I_B. The breakdown voltages BV_{sus}, BV_{CEO}, and BV_{CBO} are illustrated. BV_{CBO} is the avalanche breakdown voltage of the base-collector junction, with the emitter open-circuited or with sufficiently negative base current. BV_{CEO} is the somewhat smaller collector-emitter breakdown voltage observed when the base current is zero; as avalanche breakdown is approached, free carriers are created that have the same effect as a positive base current and that cause the breakdown voltage to be reduced. BV_{sus} is the breakdown voltage observed with positive base current. Because of the high instantaneous power dissipation, breakdown usually results in destruction of the BJT. In most applications, the off state transistor voltage must not exceed BV_{CEO}.

High-voltage BJTs typically have low current gain, and hence Darlington-connected devices (Fig. 4.36) are common. If transistors Q_1 and Q_2 have current gains β_1 and β_2, respectively, then the Darlington-connected device has the substantially increased current gain $\beta_1 + \beta_2 + \beta_1\beta_2$. In a monolithic Darlington device, transistors Q_1 and Q_2 are integrated on the same silicon wafer. Diode D_1 speeds up the turn-off process, by allowing the base driver to actively remove the stored charge of both Q_1 and Q_2 during the turn-off transition.

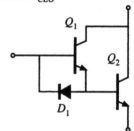

At voltage levels below 500 V, the BJT has been almost entirely replaced by the MOSFET in power applications. It is also being displaced in higher voltage applications, where new designs utilize faster IGBTs or other devices.

Fig.   4.36 Darlington-connected BJTs, including diode for improvement of turn-off times.

4.2.4   Insulated Gate Bipolar Transistor (IGBT)

A cross-section of the IGBT is illustrated in Fig. 4.37. Comparison with Fig. 4.26 reveals that the IGBT and power MOSFET are very similar in construction. The key difference is the p region connected to the collector of the IGBT. So the IGBT is a modern four-layer power semiconductor device having a MOS gate.

Fig.   4.37 IGBT structure. Crosshatched regions are metallized contacts. Shaded regions are insulating silicon dioxide layers.

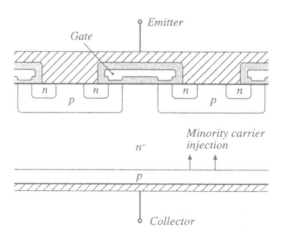

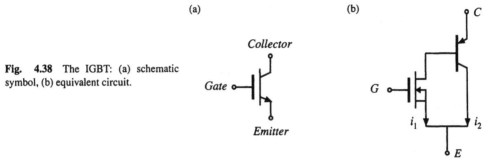

Fig. 4.38 The IGBT: (a) schematic symbol, (b) equivalent circuit.

The function of the added p region is to inject minority charges into the n^- region while the device operates in the on state, as illustrated in Fig. 4.37. When the IGBT conducts, the p–n^- junction is forward-biased, and the minority charges injected into the n^- region cause conductivity modulation. This reduces the on-resistance of the n^- region, and allows high-voltage IGBTs to be constructed which have low forward voltage drops. As of 1999, IGBTs rated as low as 600 V and as high as 3300 V are readily available. The forward voltage drops of these devices are typically 2 to 4 V, much lower than would be obtained in equivalent MOSFETs of the same silicon area.

Several schematic symbols for the IGBT are in current use; the symbol illustrated in Fig. 4.38(a) is the most popular. A two-transistor equivalent circuit for the IGBT is illustrated in Fig. 4.38(b). The IGBT functions effectively as an n-channel power MOSFET, cascaded by a PNP emitter-follower BJT. The physical locations of the two effective devices are illustrated in Fig. 4.39. It can be seen that there are two effective currents: the effective MOSFET channel current i_1, and the effective PNP collector current i_2.

The price paid for the reduced voltage drop of the IGBT is its increased switching times, especially during the turn-off transition. In particular, the IGBT turn-off transition exhibits a phenomenon known as *current tailing*. The effective MOSFET can be turned off quickly, by removing the gate charge such that the gate-to-emitter voltage is negative. This causes the channel current i_1 to quickly become zero. However, the PNP collector current i_2 continues to flow as long as minority charge is present in the n^- region. Since there is no way to actively remove the stored minority charge, it slowly decays via recombination. So i_2 slowly decays in proportion to the minority charge, and a current tail is observed. The length of the current tail can be reduced by introduction of recombination centers in the n^- region, at

Fig. 4.39 Physical locations of the effective MOSFET and PNP components of the IGBT.

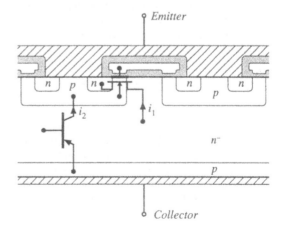

Table 4.3   Characteristics of several commercial IGBTs

Part number	Rated maximum voltage	Rated average current	V_F (typical)	t_f (typical)
Single-chip devices				
HGTP12N60A4	600 V	23 A	2.0 V	70 ns
HGTG32N60E2	600 V	32 A	2.4 V	0.62 μs
HGTG30N120D2	1200 V	30 A	3.2 V	0.58 μs
Multiple-chip modules				
CM400HA-12E	600 V	400 A	2.7 V	0.3 μs
CM300HA-24E	1200 V	300 A	2.7 V	0.3 μs
CM800HA-34H	1700 V	800 A	3.3 V	0.6 μs
High voltage modules				
CM 800HB-50H	2500 V	800 A	3.15 V	1.0 μs
CM 600HB-90H	4500 V	900 A	3.3 V	1.2 μs

the expense of a somewhat increased on-resistance. The current gain of the effective PNP transistor can also be minimized, causing i_1 to be greater than i_2. Nonetheless, the turn-off switching time of the IGBT is significantly longer than that of the MOSFET, with typical turn-off times in the range 0.5 μs to 5 μs. Switching loss induced by IGBT current tailing is discussed in Section 4.3.1. The switching frequencies of PWM converters containing IGBTs are typically in the range 1 to 30 kHz.

The added p–n^- diode junction of the IGBT is not normally designed to block significant voltage. Hence, the IGBT has negligible reverse voltage-blocking capability.

Since the IGBT is a four-layer device, there is the possibility of SCR-type latchup, in which the IGBT cannot be turned off by gate voltage control. Recent devices are not susceptible to this problem. These devices are quite robust, hot-spot and current crowding problems are nonexistent, and the need for external snubber circuits is minimal.

The on-state forward voltage drop of the IGBT can be modeled by a forward-biased diode junction, in series with an effective on-resistance. The temperature coefficient of the IGBT forward voltage drop is complicated by the fact that the diode junction voltage has a negative temperature coefficient, while the on-resistance has a positive temperature coefficient. Fortunately, near rated current the on-resistance dominates, leading to an overall positive temperature coefficient. In consequence, IGBTs can be easily connected in parallel, with a modest current derating. Large modules are commercially available, containing multiple parallel-connected chips.

Characteristics of several commercially available single-chip IGBTs and multiple-chip IGBT modules are listed in Table 4.3.

4.2.5   Thyristors (SCR, GTO, MCT)

Of all conventional semiconductor power devices, the silicon-controlled rectifier (SCR) is the oldest, has the lowest cost per rated kVA, and is capable of controlling the greatest amount of power. Devices having voltage ratings of 5000 to 7000 V and current ratings of several thousand amperes are available. In utility dc transmission line applications, series-connected light-triggered SCRs are employed in inverters and rectifiers that interface the ac utility system to dc transmission lines which carry roughly 1 kA and 1 MV. A single large SCR fills a silicon wafer that is several inches in diameter, and is mounted in a hockey-puck-style case.

(a)   (b)

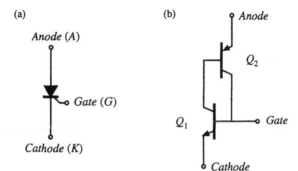

Fig. 4.40 The SCR: (a) schematic symbol, (b) equivalent circuit.

The schematic symbol of the SCR is illustrated in Fig. 4.40(a), and an equivalent circuit containing NPN and PNP BJT devices is illustrated in Fig. 4.40(b). A cross-section of the silicon chip is illustrated in Fig. 4.41. Effective transistor Q_1 is composed of the n, p, and n^- regions, while effective transistor Q_2 is composed of the p, n^-, and p regions as illustrated.

The device is capable of blocking both positive and negative anode-to-cathode voltages. Depending on the polarity of the applied voltage, one of the p–n^- junctions is reverse-biased. In either case, the depletion region extends into the lightly doped n^- region. As with other devices, the desired voltage breakdown rating is obtained by proper design of the n^- region thickness and doping concentration.

The SCR can enter the on state when the applied anode-to-cathode voltage v_{AK} is positive. Positive gate current i_G then causes effective transistor Q_1 to turn on; this in turn supplies base current to effective transistor Q_2, and causes it to turn on as well. The effective connections of the base and collector regions of transistors Q_1 and Q_2 constitute a positive feedback loop. Provided that the product of the current gains of the two transistors is greater than one, then the currents of the transistors will increase regeneratively. In the on state, the anode current is limited by the external circuit, and both effective transistors operate fully saturated. Minority carriers are injected into all four regions, and the resulting conductivity modulation leads to very low forward voltage drop. In the on state, the SCR can be modeled as a forward-biased diode junction in series with a low-value on-resistance. Regardless of the gate current, the SCR is latched in the on state: it cannot be turned off except by application of negative anode current or negative anode-to-cathode voltage. In phase controlled converters, the SCR turns off at the zero crossing of the converter ac input or output waveform. In forced commutation converters, external commuta-

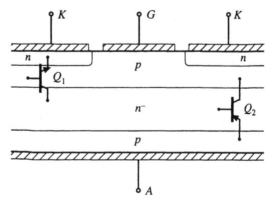

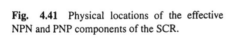

Fig. 4.41 Physical locations of the effective NPN and PNP components of the SCR.

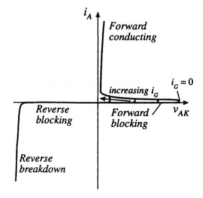

Fig. 4.42   Static i_A–v_{AK} characteristics of the SCR.

tion circuits force the controlled turn-off of the SCR, by reversing either the anode current or the anode-to-cathode voltage.

Static i_A–v_{AK} characteristics of the conventional SCR are illustrated in Fig. 4.42. It can be seen that the SCR is a voltage-bidirectional two-quadrant switch. The turn-on transition is controlled actively via the gate current. The turn-off transition is passive.

During the turn-off transition, the rate at which forward anode-to-cathode voltage is reapplied must be limited, to avoid retriggering the SCR. The turn-off time t_q is the time required for minority stored charge to be actively removed via negative anode current, and for recombination of any remaining minority charge. During the turn-off transition, negative anode current actively removes stored minority charge, with waveforms similar to diode turn-off transition waveforms of Fig. 4.25. Thus, after the first zero crossing of the anode current, it is necessary to wait for time t_q before reapplying positive anode-to-cathode voltage. It is then necessary to limit the rate at which the anode-to-cathode voltage increases, to avoid retriggering the device. Inverter-grade SCRs are optimized for faster switching times, and exhibit smaller values of t_q.

Conventional SCR wafers have large feature size, with coarse or nonexistent interdigitation of the gate and cathode contacts. The parasitic elements arising from this large feature size lead to several limitations. During the turn-on transition, the rate of increase of the anode current must be limited to a safe value. Otherwise, cathode current focusing can occur, which leads to formation of hot spots and device failure.

The coarse feature size of the gate and cathode structure is also what prevents the conventional SCR from being turned off by active gate control. One might apply a negative gate current, in an attempt to actively remove all of the minority stored charge and to reverse-bias the p–n gate-cathode junction. The reason that this attempt fails is illustrated in Fig. 4.43. The large negative gate current flows laterally through the adjoining the p region, inducing a voltage drop as shown. This causes the gate-cathode junction voltage to be smaller near the gate contact, and relatively larger away from the gate contact. The negative gate current is able to reverse-bias only the portion of the gate-cathode junction in the vicinity of the gate contact; the remainder of the gate-cathode junction continues to be forward-biased, and cathode current continues to flow. In effect, the gate contact is able to influence only the nearby portions of the cathode.

The gate turn off thyristor, or GTO, is a modern power device having small feature size. The gate and cathode contacts highly interdigitated, such that the entire gate-cathode p–n junction can be reverse-biased via negative gate current during the turn-off transition. Like the SCR, a single large GTO can fill an entire silicon wafer. Maximum voltage and current ratings of commercial GTOs are lower than those of SCRs.

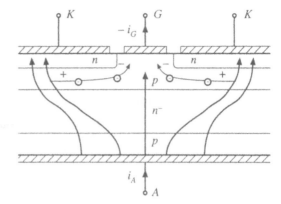

Fig. 4.43 Negative gate current is unable to completely reverse-bias the gate-cathode junction. The anode current focuses away from the gate contact.

The turn-off gain of a GTO is the ratio of on-state current to the negative gate current magnitude required to switch the device off. Typical values of this gain are 2 to 5, meaning that several hundred amperes of negative gate current may be required to turn off a GTO conducting 1000 A. Also of interest is the maximum controllable on-state current. The GTO is able to conduct peak currents significantly greater than the rated average current; however, it may not be possible to switch the device off under gate control while these high peak currents are present.

The MOS-controlled thyristor, or MCT, is a recent power device in which MOSFETs are integrated onto a highly interdigitated SCR, to control the turn-on and turn-off processes. Like the MOSFET and IGBT, the MCT is a single-quadrant device whose turn-on and turn-off transitions are controlled by a MOS gate terminal. Commercial MCTs are p-type devices. Voltage-bidirectional two-quadrant MCTs, and n-type MCTs, are also possible.

A cross-section of an MCT containing MOSFETs for control of the turn-on and turn-off transitions is illustrated in Fig. 4.44. An equivalent circuit which explains the operation of this structure is given in Fig. 4.45. To turn the device on, the gate-to-anode voltage is driven negative. This forward-biases p-channel MOSFET Q_3, forward-biasing the base-emitter junction of BJT Q_1. Transistors Q_1 and Q_2 then latch in the on-state. To turn the device off, the gate-to-anode voltage is driven positive. This forward-biases n-channel MOSFET Q_4, which in turn reverse-biases the base-emitter junction of BJT Q_2. The BJTs then turn off. It is important that the on-resistance of the n-channel MOSFET be small enough

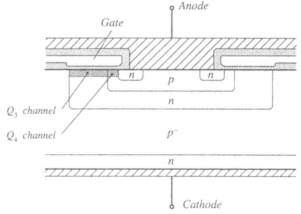

Fig. 4.44 MCT structure. Crosshatched regions are metallized contacts. Lightly shaded regions are insulating silicon dioxide layers.

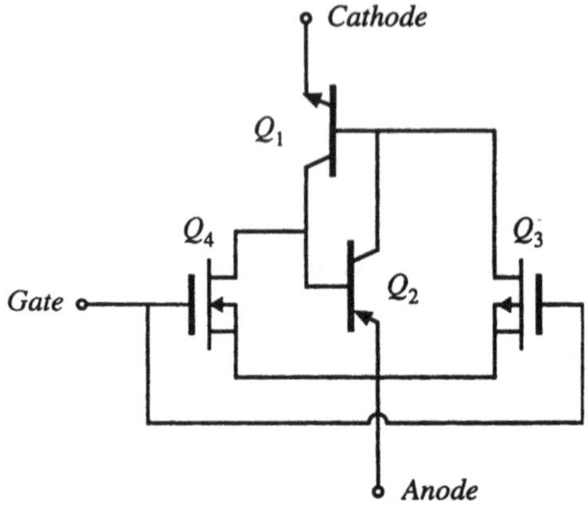

Fig. 4.45   Equivalent circuit for the MCT.

that sufficient influence on the cathode current is exerted—this limits the maximum controllable on state current (i.e., the maximum current that can be interrupted via gate control).

High-voltage MCTs exhibit lower forward voltage drops and higher current densities than IGBTs of similar voltage ratings and silicon area. However, the switching times are longer. Like the GTO, the MCT can conduct considerable surge currents; but again, the maximum current that can be interrupted via gate control is limited. To obtain a reliable turn-off transition, external snubbers are required to limit the peak anode-to-cathode voltage. A sufficiently fast gate-voltage rise time is also required. To some extent, the MCT is still an emerging device—future generations of MCTs may exhibit considerable improvements in performance and ratings.

4.3   SWITCHING LOSS

Having implemented the switches using semiconductor devices, we can now discuss another major source of loss and inefficiency in converters: switching loss. As discussed in the previous section, the turn-on and turn-off transitions of semiconductor devices require times of tens of nanoseconds to microseconds. During these switching transitions, very large instantaneous power loss can occur in the semiconductor devices. Even though the semiconductor switching times are short, the resulting average power loss can be significant.

Semiconductor devices are charge controlled. For example, the conducting state of a MOSFET is determined by the charge on its gate and in its channel, and the conducting state of a silicon diode or a BJT is determined by the presence or absence of stored minority charge in the vicinity of the semiconductor junctions inside the device. To switch a semiconductor device between the on and off states, the controlling charge must be inserted or removed; hence, the amount of controlling charge influences both the switching times and the switching loss. Charge, and energy, are also stored in the output capacitances of semiconductor devices, and energy is stored in the leakage and stray inductances in the circuit. In most converter circuits, these stored energies are also lost during the switching transitions.

In this section the major sources of switching loss are described, and a simple method for estimation of their magnitudes is given. For clarity, conduction losses and semiconductor forward voltage drops are neglected throughout this discussion.

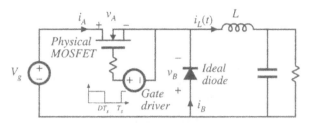

Fig. 4.46 MOSFET driving a clamped inductive load, buck converter example.

4.3.1 Transistor Switching with Clamped Inductive Load

Let's consider first the switching waveforms in the buck converter of Fig. 4.46. Let us treat the diode as ideal, and investigate only the switching loss due to the MOSFET switching times. The MOSFET drain-to-source capacitance is also neglected.

The diode and inductor present a clamped inductive load to the transistor. With such a load, the transistor voltage $v_A(t)$ and current $i_A(t)$ do not change simultaneously. For example, a magnified view of the transistor turn-off-transition waveforms is given in Fig. 4.47. For simplicity, the waveforms are

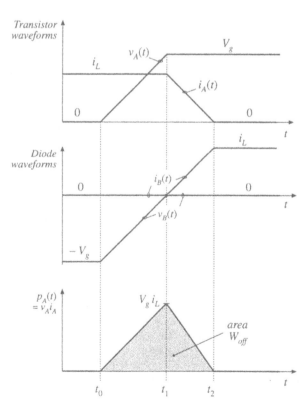

Fig. 4.47 Magnified view of transistor turn-off transition waveforms for the circuit of Fig. 4.46.

approximated as piecewise-linear. The switching times are short, such that the inductor current $i_L(t)$ is essentially constant during the entire switching transition $t_0 < t < t_2$. No current flows through the diode while the diode is reverse-biased, and the diode cannot become forward-biased while its voltage $v_B(t)$ is negative. So first, the voltage $v_A(t)$ across the transistor must rise from zero to V_g. The interval length $(t_1 - t_0)$ is essentially the time required for the gate driver to charge the MOSFET gate-to-drain capacitance. The transistor current $i_A(t)$ is constant and equal to i_L during this interval.

The diode voltage $v_B(t)$ and current $i_B(t)$ are given by

$$v_B(t) = v_A(t) - V_g$$
$$i_A(t) + i_B(t) = i_L$$

(4.6)

At time $t = t_1$, when $v_A = V_g$, the diode becomes forward-biased. The current i_L now begins to commute from the transistor to the diode. The interval length $(t_2 - t_1)$ is the time required for the gate driver to discharge the MOSFET gate-to-source capacitance down to the threshold voltage which causes the MOSFET to be in the off state.

The instantaneous power $p_A(t)$ dissipated by the transistor is equal to $v_A(t)i_A(t)$. This quantity is also sketched in Fig. 4.47. The energy W_{off} lost during the transistor turn-off transition is the area under this waveform. With the simplifying assumption that the waveforms are piecewise-linear, then the energy lost is the area of the shaded triangle:

$$W_{off} = \tfrac{1}{2} V_g i_L (t_2 - t_0)$$

(4.7)

This is the energy lost during each transistor turn-off transition in the simplified circuit of Fig. 4.46.

The transistor turn-on waveforms of the simplified circuit of Fig. 4.46 are qualitatively similar to those of Fig. 4.47, with the time axis reversed. The transistor current must first rise from 0 to i_L. The diode then becomes reverse-biased, and the transistor voltage can fall from V_g to zero. The instantaneous transistor power dissipation again has peak value $V_g i_L$, and if the waveforms are piecewise linear, then the energy lost during the turn-on transition W_{on} is given by $0.5 V_g i_L$ multiplied by the transistor turn-on time.

Thus, during one complete switching period, the total energy lost during the turn-on and turn-off transitions is $(W_{on} + W_{off})$. If the switching frequency is f_s, then the average power loss incurred due to switching is

$$P_{sw} = \frac{1}{T_s} \int_{\substack{\text{switching} \\ \text{transitions}}} p_A(t)dt = \left(W_{on} + W_{off}\right) f_s$$

(4.8)

So the switching loss P_{sw} is directly proportional to the switching frequency.

An example where the loss due to transistor switching times is particularly significant is the current tailing phenomenon observed during the turn-off transition of the IGBT. As discussed in Section 4.2.4, current tailing occurs due to the slow recombination of stored minority charge in the n^- region of the IGBT. This causes the collector current to slowly decay after the gate voltage has been removed.

A buck converter circuit containing an ideal diode and nonideal (physical) IGBT is illustrated in Fig. 4.48. Turn-off transition waveforms are illustrated in Fig. 4.49; these waveforms are similar to the MOSFET waveforms of Fig. 4.47. The diode is initially reverse-biased, and the voltage $v_A(t)$ rises from approximately zero to V_g. The interval length $(t_1 - t_0)$ is the time required for the gate drive circuit to charge the IGBT gate-to-collector capacitance. At time $t = t_1$, the diode becomes forward-biased, and current begins to commute from the IGBT to the diode. The interval $(t_2 - t_1)$ is the time required for the gate drive circuit to discharge the IGBT gate-to-emitter capacitance to the threshold value which causes

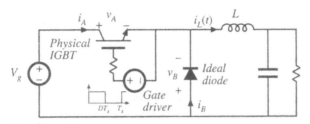

Fig. 4.48 IGBT switching loss example.

the effective MOSFET in Fig. 4.38(b) to be in the off state. This time can be minimized by use of a high-current gate drive circuit which discharges the gate capacitance quickly. However, switching off the effective MOSFET does not completely interrupt the IGBT current $i_A(t)$: current $i_2(t)$ continues to flow through the effective PNP bipolar junction transistor of Fig. 4.38(b) as long as minority carriers continue to exist within its base region. During the interval $t_2 < t < t_3$, the current is proportional to this stored minority charge, and the current tail interval length $(t_3 - t_2)$ is equal to the time required for this remaining stored minority charge to recombine.

The energy W_{off} lost during the turn-off transition of the IGBT is again the area under the instantaneous power waveform, as illustrated in Fig. 4.49. The switching loss can again be evaluated using Eq. (4.8).

The switching times of the IGBT are typically in the vicinity of 0.2 to 2 μs, or several times

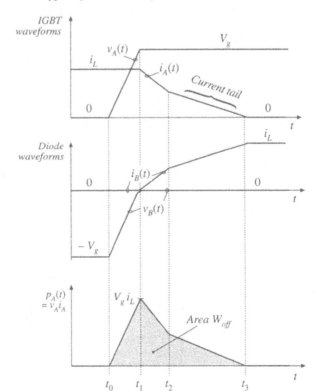

Fig. 4.49 IGBT turn-off transition waveforms for the circuit of Fig. 4.48.

longer than those of the power MOSFET. The resulting switching loss limits the maximum switching frequencies of conventional PWM converters employing IGBTs to roughly 1 to 30 kHz.

4.3.2   Diode Recovered Charge

As discussed previously, the familiar exponential $i\text{–}v$ characteristic of the diffused-junction $p\text{–}n$ diode is an equilibrium relationship. During switching transients, significant deviations from this characteristic are observed, which can induce transistor switching loss. In particular, during the diode turn-off transient, its stored minority charge must be removed, either actively via negative current $i_B(t)$, or passively via recombination inside the device. The diode remains forward-biased while minority charge is present in the vicinity of the diode semiconductor junction. The initial amount of minority charge is a function of the forward current, and its rate of change, under forward-biased conditions. The turn-off switching time is the time required to remove all of this charge, and to establish a new reverse-biased operating point. This process of switching the diode from the forward-biased to reverse-biased states is called *reverse recovery*.

Again, most diffused-junction power diodes are actually $p\text{–}n^-\text{–}n^+$ or $p\text{–}i\text{–}n$ devices. The lightly doped or intrinsic region (of the diode and other power semiconductor devices as well) allows large breakdown voltages to be obtained. Under steady-state forward-biased conditions, a substantial amount of stored charge is present in this region, increasing its conductivity and leading to a low diode on-resistance. It takes time to insert and remove this charge, however, so there is a tradeoff between high breakdown voltage, low on-resistance, and fast switching times.

To understand how the diode stored charge induces transistor switching loss, let us consider the buck converter of Fig. 4.50. Assume for this discussion that the transistor switching times are much faster than the switching times of the diode, such that the diode reverse recovery mechanism is the only significant source of switching loss. A magnified view of the transistor-turn-on transition waveforms under these conditions is given in Fig. 4.51.

Initially, the diode conducts the inductor current, and hence some amount of stored minority charge is present in the diode. The transistor is initially in the off state. When the transistor turns on, a negative current flows through the diode; this current actively removes some or most of the diode stored minority charge, while the remainder of the minority charge recombines within the diode. The rate of change of the current is typically limited by the package inductance and other stray inductances present in the external circuit; hence, the peak magnitude of the reverse current depends on the external circuit, and can be many times larger than the forward current i_L. The area within the negative portion of the diode current waveform is the recovered stored charge Q_r, while the interval length $(t_2 - t_0)$ is the reverse recovery time t_r. The magnitude of Q_r is a function of the on state forward current i_L at the initiation of the turn-off process, as well as the circuit-limited rate-of-change of the diode current, $di_B(t)/dt$. During

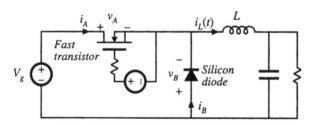

Fig. 4.50   Example, switching loss induced by diode stored charge.

the interval $t_0 < t < t_1$, the diode remains forward-biased, and hence the transistor voltage is V_g. At time $t = t_1$, the stored charge in the vicinity of the p–n^- or p–i junction is exhausted. This junction becomes reverse-biased, and begins to block voltage. During the interval $t_1 < t < t_2$, the diode voltage decreases to $-V_g$. Some negative diode current continues to flow, removing any remaining stored minority charge as well as charging the depletion layer capacitance. At time $t = t_2$, this current is essentially zero, and the diode operates in steady state under reverse-biased conditions.

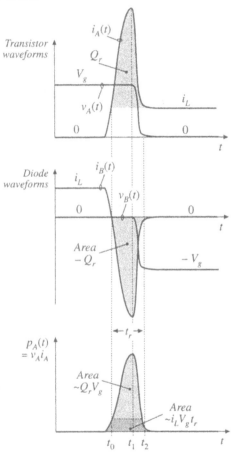

Fig. 4.51 Transistor-turn-on transition waveforms for the circuit of Fig. 4.50.

Diodes in which the interval length $(t_2 - t_1)$ is short compared to $(t_1 - t_0)$ are called abrupt-recovery or "snappy" diodes. Soft recovery diodes exhibit larger values of $(t_2 - t_1)/(t_1 - t_0)$. When significant package and/or stray inductance is present in series with the diode, ringing of the depletion region capacitance with the package and stray inductances may be observed. If severe, this ringing can cause excess reverse voltage that leads to device failure. External R–C snubber circuits are sometimes necessary for reliable operation. The reverse-recovery characteristics of soft recovery diodes are intended to exhibit less ringing and voltage overshoot. Snubbing of these diodes can be reduced or eliminated.

The instantaneous power $p_A(t)$ dissipated in the transistor is also sketched in Fig. 4.51. The energy lost during the turn-on transition is

$$W_D = \int_{\substack{\text{switching} \\ \text{transition}}} v_A(t) i_A(t) dt \qquad (4.9)$$

For an abrupt-recovery diode in which $(t_2 - t_1) \ll (t_1 - t_0)$, this integral can be evaluated in a simple manner. The transistor voltage $v_A(t)$ is then equal to V_g for essentially the entire diode recovery interval. In addition, $i_A = i_L - i_B$. Equation (4.9) then becomes

$$W_D \approx \int_{\substack{\text{switching} \\ \text{transition}}} V_g \big(i_L - i_B(t) \big) dt \qquad (4.10)$$

$$= V_g i_L t_r + V_g Q_r$$

where the recovered charge Q_r is defined as the integral of the diode current $-i_B(t)$ over the interval $t_0 < t < t_2$. Hence, the diode reverse recovery process leads directly to switching loss $W_D f_s$. This is often the largest single component of switching loss in a conventional switching converter. It can be reduced by use of faster diodes, designed for minimization of stored minority charge.

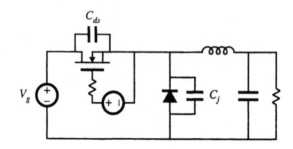

Fig. 4.52 The energy stored in the semiconductor output capacitances is lost during the transistor turn-on transition.

4.3.3   Device Capacitances, and Leakage, Package, and Stray Inductances

Reactive elements can also lead to switching loss. Capacitances that are effectively in parallel with switching elements are shorted out when the switch turns on, and any energy stored in the capacitance is lost. The capacitances are charged without energy loss when the switching elements turn off, and the transistor turn-off loss W_{off} computed in Eq. (4.7) may be reduced. Likewise, inductances that are effectively in series with a switching element lose their stored energy when the switch turns off. Hence, series inductances lead to additional switching loss at turn-off, but can reduce the transistor turn-on loss.

The stored energies of the reactive elements can be summed to find the total energy loss per switching period due to these mechanisms. For linear capacitors and inductors, the stored energy is

$$W_C = \sum_{\substack{\text{capacitive} \\ \text{elements}}} \tfrac{1}{2} C_i V_i^2$$

$$W_L = \sum_{\substack{\text{inductive} \\ \text{elements}}} \tfrac{1}{2} L_j I_j^2 \tag{4.11}$$

A common source of this type of switching loss is the output capacitances of the semiconductor switching devices. The depletion layers of reverse-biased semiconductor devices exhibit capacitance which stores energy. When the transistor turns on, this stored energy is dissipated by the transistor. For example, in the buck converter of Fig. 4.52, the MOSFET exhibits drain-to-source capacitance C_{ds}, and the reverse-biased diode exhibits junction capacitance C_j. During the switching transitions these two capacitances are effectively in parallel, since the dc source V_g is effectively a short-circuit at high frequency. To the extent that the capacitances are linear, the energy lost when the MOSFET turns on is

$$W_C = \tfrac{1}{2} \left(C_{ds} + C_j \right) V_g^2 \tag{4.12}$$

Typically, this type of switching loss is significant at voltage levels above 100 V. The MOSFET gate drive circuit, which must charge and discharge the MOSFET gate capacitances, also exhibits this type of loss.

As noted in Section 4.2.2, the incremental drain-to-source capacitance C_{ds} of the power MOSFET is a strong function of the drain-to-source voltage v_{ds}. $C_{ds}(v_{ds})$ follows an approximate inverse-square-root dependence of v_{ds}, as given by Eq. (4.5). The energy stored in C_{ds} at $v_{ds} = V_{DS}$ is

$$W_{Cds} = \int v_{ds} i_C \, dt = \int_0^{V_{DS}} v_{ds} C_{ds}(v_{ds}) \, dv_{ds} \tag{4.13}$$

where $i_C = C_{ds}(v_{ds})\, dv_{ds}/dt$ is the current in C_{ds}. Substitution of Eq. (4.5) into (4.13) yields

$$W_{Cds} = \int_0^{V_{DS}} C_0'(v_{ds})\sqrt{v_{ds}}\, dv_{ds} = \tfrac{2}{3} C_{ds}(V_{DS}) V_{DS}^2 \qquad (4.14)$$

This energy is lost each time the MOSFET switches on. From the standpoint of switching loss, the drain-to-source capacitance is equivalent to a linear capacitance having the value $\tfrac{4}{3} C_{ds}(V_{DS})$.

The Schottky diode is essentially a majority-carrier device, which does not exhibit a reverse-recovery transient such as in Fig. 4.51. Reverse-biased Schottky diodes do exhibit significant junction capacitance, however, which can be modeled with a parallel capacitor C_j as in Fig. 4.52, and which leads to energy loss at the transistor turn-on transition.

Common sources of series inductance are transformer leakage inductances in isolated converters (discussed in Chapter 6), as well as the inductances of interconnections and of semiconductor device packages. In addition to generating switching loss, these elements can lead to excessive peak voltage stress during the transistor turn-off transition. Interconnection and package inductances can lead to significant switching loss in high-current applications, and leakage inductance is an important source of switching loss in many transformer-isolated converters.

Diode stored minority charge can induce switching loss in the (nonideal) converter reactive elements. As an example, consider the circuit of Fig. 4.53, containing an ideal voltage source $v_i(t)$, an inductor L, a capacitor C (which may represent the diode junction capacitance, or the junction capacitance in parallel with an external capacitor), and a silicon diode. The diode switching processes of many converters can be modeled by a circuit of this form. Many rectifier circuits containing SCRs exhibit similar waveforms. The voltage source produces the rectangular waveform $v_i(t)$ illustrated in Fig. 4.54. This voltage is initially positive, causing the diode to become

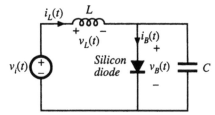

Fig. 4.53 A circuit in which the diode stored charge induces ringing, and ultimately switching loss, in (nonideal) reactive elements.

forward-biased and the inductor current $i_L(t)$ to increase linearly with slope V_1/L. Since the current is increasing, the stored minority charge inside the diode also increases. At time $t = t_1$, the source voltage $v_i(t)$ becomes negative, and the inductor current decreases with slope $di_L/dt = -V_2/L$. The diode stored charge also decreases, but at a slower rate that depends not only on i_L but also on the minority carrier recombination lifetime of the silicon material in the diode. Hence, at time $t = t_2$, when $i_L(t)$ reaches zero, some stored minority charge remains in the diode. So the diode continues to be forward-biased, and the inductor current continues to decrease with the same slope. The negative current for $t > t_2$ constitutes a reverse diode current, which actively removes diode stored charge. At some time later, $t = t_3$, the diode stored charge in the vicinity of the diode junction becomes zero, and the diode junction becomes reverse-biased. The inductor current is now negative, and must flow through the capacitor. The inductor and capacitor then form a series resonant circuit, which rings with decaying sinusoidal waveforms as shown. This ringing is eventually damped out by the parasitic loss elements of the circuit, such as the inductor winding resistance, inductor core loss, and capacitor equivalent series resistance.

The diode recovered charge induces loss in this circuit. During the interval $t_2 < t < t_3$, the minority stored charge Q_r recovered from the diode is

$$Q_r = -\int_{t_2}^{t_3} i_L(t)\, dt \qquad (4.15)$$

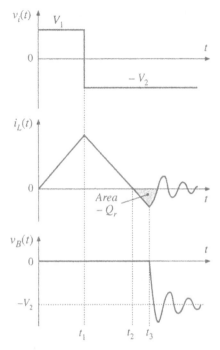

Fig. 4.54 Waveforms of the circuit of Fig. 4.53.

This charge is directly related to the energy stored in the inductor during this interval. The energy W_L stored in the inductor is the integral of the power flowing into the inductor:

$$W_L = \int_{t_2}^{t_3} v_L(t) i_L(t)\, dt \qquad (4.16)$$

During this interval, the applied inductor voltage is

$$v_L(t) = L \frac{di_L(t)}{dt} = -V_2 \qquad (4.17)$$

Substitution of Eq. (4.17) into Eq. (4.16) leads to

$$W_L = \int_{t_2}^{t_3} L \frac{di_L(t)}{dt} i_L(t)\, dt = \int_{t_2}^{t_3} \left(-V_2\right) i_L(t)\, dt \qquad (4.18)$$

Evaluation of the integral on the left side yields the stored inductor energy at $t = t_3$, or $L i_L{}^2(t_3)/2$. The right-side integral is evaluated by noting that V_2 is constant and by substitution of Eq. (4.15), yielding $V_2 Q_r$. Hence, the energy stored in the inductor at $t = t_3$ is

$$W_L = \tfrac{1}{2} L i_L^2(t_3) = V_2 Q_r \qquad (4.19)$$

or, the recovered charge multiplied by the source voltage. For $t > t_3$, the ringing of the resonant circuit formed by the inductor and capacitor causes this energy to be circulated back and forth between the inductor and capacitor. If parasitic loss elements in the circuit cause the ringing amplitude to eventually decay to zero, then the energy becomes lost as heat in the parasitic elements.

So diode stored minority charge can lead to loss in circuits that do not contain an active switching element. Also, ringing waveforms that decay before the end of the switching period indicate the presence of switching loss.

4.3.4   Efficiency vs. Switching Frequency

Suppose next that we add up all of the energies lost due to switching, as discussed above:

$$W_{tot} = W_{on} + W_{off} + W_D + W_C + W_L + \ldots \qquad (4.20)$$

This is the energy lost in the switching transitions of one switching period. To obtain the average switching power loss, we must multiply by the switching frequency:

$$P_{sw} = W_{tot} f_{sw} \qquad (4.21)$$

Other losses in the converter include the conduction losses P_{cond}, modeled and solved as in Chapter 3,

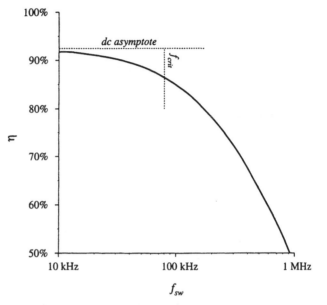

Fig. 4.55 Efficiency vs. switching frequency, based on Eq. (4.22), using arbitrary choices for the values of loss and load power. Switching loss causes the efficiency to decrease rapidly at high frequency.

and other frequency-independent fixed losses P_{fixed}, such as the power required to operate the control circuit. The total loss is therefore

$$P_{loss} = P_{cond} + P_{fixed} + W_{tot} f_{sw} \tag{4.22}$$

which increases linearly with frequency. At the critical frequency

$$f_{crit} = \frac{P_{cond} + P_{fixed}}{W_{tot}} \tag{4.23}$$

the switching losses are equal to the other converter losses. Below this critical frequency, the total loss is dominated by the conduction and fixed loss, and hence the total loss and converter efficiency are not strong functions of switching frequency. Above the critical frequency, the switching loss dominates the total loss, and the converter efficiency decreases rapidly with increasing switching frequency. Typical dependence of the full-load converter efficiency on switching frequency is plotted in Fig. 4.55, for an arbitrary choice of parameter values. The critical frequency f_{crit} can be taken as a rough upper limit on the switching frequency of a practical converter.

4.4   SUMMARY OF KEY POINTS

1.   How an SPST ideal switch can be realized using semiconductor devices depends on the polarity of the voltage that the devices must block in the off state, and on the polarity of the current which the devices must conduct in the on state.

2.   Single-quadrant SPST switches can be realized using a single transistor or a single diode, depending on the relative polarities of the off state voltage and on state current.

3. Two-quadrant SPST switches can be realized using a transistor and diode, connected in series (bidirectional-voltage) or in antiparallel (bidirectional-current). Several four-quadrant schemes are also listed here.

4. A "synchronous rectifier" is a MOSFET connected to conduct reverse current, with gate drive control as necessary. This device can be used where a diode would otherwise be required. If a MOSFET with sufficiently low R_{on} is used, reduced conduction loss is obtained.

5. Majority carrier devices, including the MOSFET and Schottky diode, exhibit very fast switching times, controlled essentially by the charging of the device capacitances. However, the forward voltage drops of these devices increases quickly with increasing breakdown voltage.

6. Minority carrier devices, including the BJT, IGBT, and thyristor family, can exhibit high breakdown voltages with relatively low forward voltage drop. However, the switching times of these devices are longer, and are controlled by the times needed to insert or remove stored minority charge.

7. Energy is lost during switching transitions, owing to a variety of mechanisms. The resulting average power loss, or switching loss, is equal to this energy loss multiplied by the switching frequency. Switching loss imposes an upper limit on the switching frequencies of practical converters.

8. The diode and inductor present a "clamped inductive load" to the transistor. When a transistor drives such a load, it experiences high instantaneous power loss during the switching transitions. An example where this leads to significant switching loss is the IGBT and the "current tail" observed during its turn-off transition.

9. Other significant sources of switching loss include diode stored charge and energy stored in certain parasitic capacitances and inductances. Parasitic ringing also indicates the presence of switching loss.

REFERENCES

[1]   R. D. MIDDLEBROOK, S. ĆUK, and W. BEHEN, "A New Battery Charger/Discharger Converter," *IEEE Power Electronics Specialists Conference*, 1978 Record, pp. 251-255, June 1978.

[2]   H. MATSUO and F. KUROKAWA, "New Solar Cell Power Supply System Using a Boost Type Bidirectional Dc-Dc Converter," *IEEE Power Electronics Specialists Conference*, 1982 Record, pp. 14-19, June 1982.

[3]   M. VENTURINI, "A New Sine-Wave-In Sine-Wave-Out Conversion Technique Eliminates Reactive Elements," *Proceedings Seventh International Solid-State Power Conversion Conference* (Powercon 7), pp. E3.1-E3.13, 1980.

[4]   K. D. T. NGO, S. ĆUK, and R. D. MIDDLEBROOK, "A New Flyback Dc-to-Three-Phase Converter with Sinusoidal Outputs," *IEEE Power Electronics Specialists Conference*, 1983 Record, pp. 377–388.

[5]   S. ĆUK, "Basics of Switched-Mode Power Conversion: Topologies, Magnetics, and Control," *Advances in Switched-Mode Power Conversion*, Vol. II, Irvine CA: Teslaco, pp. 279-310, 1983.

[6]   L. GYUGI and B. PELLY, *Static Power Frequency Changers: Theory, Performance, and Applications*, New York: Wiley-Interscience, 1976.

[7]   R. S. KAGAN and M. CHI, "Improving Power Supply Efficiency with MOSFET Synchronous Rectifiers," *Proceedings Ninth International Solid-State Power Conversion Conference* (Powercon 9), pp. D4.1-D4.9, July 1982.

[8]   R. BLANCHARD and P. E. THIBODEAU, "The Design of a High Efficiency, Low Voltage Power Supply Using MOSFET Synchronous Rectification and Current Mode Control," *IEEE Power Electronics Special-*

ists Conference, 1985 Record, pp. 355-361, June 1985.

[9]   N. MOHAN, T. UNDELAND, and W. ROBBINS, *Power Electronics: Converters, Applications, and Design*, 2nd edit., New York: John Wiley & Sons, 1995, Chapters 19-26.

[10]   C. L. MA and P. O. LAURITZEN, "A Simple Power Diode Model with Forward and Reverse Recovery," *IEEE Power Electronics Specialists Conference*, 1991 Record, pp. 411-415, June 1991.

[11]   M. SCHLECHT and L. CASEY, "A Comparison of the Square Wave and Quasi-Resonant Topologies," *IEEE Applied Power Electronics Conference*, 1987 Record, pp. 124-134, March 1987.

[12]   B. J. BALIGA, *Modern Power Devices*, New York: John Wiley & Sons, 1987.

[13]   P. GRAY, D. DEWITT, A. BOOTHROYD, and J. GIBBONS, *Physical Electronics and Circuit Models of Transistors*, Semiconductor Electronics Education Committee, Vol. 2, New York: John Wiley & Sons, 1964.

[14]   E. OXNER, *Power FETs and Their Applications*, Englewood, New Jersey: Prentice-Hall, 1982.

[15]   M. RASHID, *Power Electronics: Circuits, Devices, and Applications*, 2nd edit., Englewood, New Jersey: Prentice Hall, 1993, Chapters 3, 4, and 8.

[16]   B. J. BALIGA, M. S. ADLER, R. P. LOVE, P. V. GRAY, and N. D. ZAMMER, "The Insulated Gate Transistor —A New Three Terminal MOS-Controlled Bipolar Power Device," *IEEE Transactions on Electron Devices*, Vol. 31, No. 6, pp. 821-828, June 1984.

[17]   V. TEMPLE, "MOS-Controlled Thyristors—A New Class of Power Devices," *IEEE Transactions on Electron Devices*, Vol. 33, No. 10, pp. 1609-1618, October 1986.

[18]   S. SUL, F. PROFUMO, G. CHO, and T. LIPO, "MCTs and IGBTs: A Comparison of Performance in Power Electronics Circuits," *IEEE Power Electronics Specialists Conference*, 1989 Record, pp. 163-169, June 1989.

[19]   V. TEMPLE, S. ARTHUR, D. WATROUS, R. DE DONCKER, and H. METHA, "Megawatt MOS Controlled Thyristor for High Voltage Power Circuits," *IEEE Power Electronics Specialists Conference*, 1992 Record, pp. 1018-1025, June 1992.

PROBLEMS

In Problems 4.1 to 4.6, the input voltage V_g is dc and positive with the polarity shown. Specify how to implement the switches using a minimal number of diodes and transistors, such that the converter operates over the entire range of duty cycles $0 \leq D \leq 1$. The switch states should vary as shown in Fig. 4.56. You may assume that the inductor current ripples and capacitor voltage ripples are small.

For each problem, do the following:

(a)   Realize the switches using SPST ideal switches, and explicitly define the voltage and current of each switch.

(b)   Express the on-state current and off-state voltage of each SPST switch in terms of the converter inductor currents, capacitor voltages, and/or

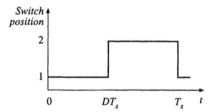

Fig. 4.56 Switch control method for Problems 4.1 to 4.6.

input source voltage.

(c)   Solve the converter to determine the inductor currents and capacitor voltages, as in Chapter 2.

(d)   Determine the polarities of the switch on-state currents and off-state voltages. Do the polarities vary with duty cycle?

(e)   State how each switch can be realized using transistors and/or diodes, and whether the realization requires single-quadrant, current-bidirectional two-quadrant, voltage-bidirectional two-quadrant, or four-quadrant switches.

4.1

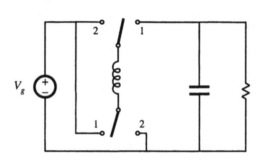

4.2

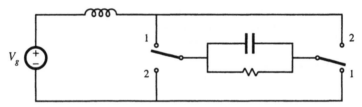

4.3

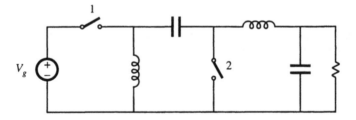

4.4

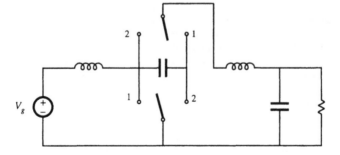

4.5

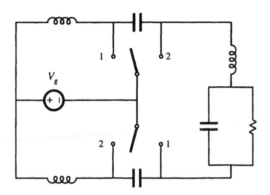

4.6

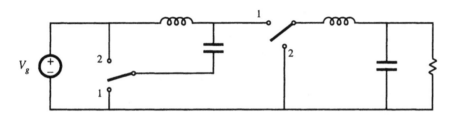

4.7 An IGBT and a silicon diode operate in a buck converter, with the IGBT waveforms illustrated in Fig. 4.57. The converter operates with input voltage $V_g = 400$ V, output voltage $V = 200$ V, and load current $I = 10$ A.

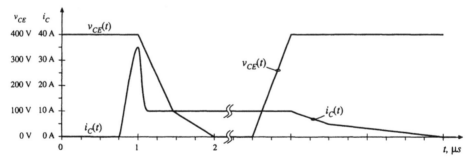

Fig. 4.57 IGBT voltage and current waveforms, Problem 4.7.

(a) Estimate the total energy lost during the switching transitions.

(b) The forward voltage drop of the IGBT is 2.5 V, and the diode has forward voltage drop 1.5 V. All other sources of conduction loss and fixed loss can be neglected. Estimate the semiconductor conduction loss.

(c) Sketch the converter efficiency over the range of switching frequencies 1 kHz $\leq f_s \leq$ 100 kHz, and label numerical values.

4.8 Two MOSFETs are employed as current-bidirectional two-quadrant switches in a bidirectional battery charger/discharger based on the dc-dc buck converter. This converter interfaces a 16 V battery to a 28 V main power bus. The maximum battery current is 40 A. The MOSFETs have on-resistances of 35 mΩ.

Their body diodes have forward voltage drops of 1.0 V, and exhibit recovered charge Q_r of 25 μC and reverse recovery times t_r of 200 ns in the given circuit. You may assume that all diodes in this problem have "snappy" reverse recovery characteristics, and also assume that diode stored charge is the dominant cause of switching loss in this circuit. You may neglect all losses other than the semiconductor conduction losses and the switching loss induced by diode stored charge.

The current-bidirectional two-quadrant switches are realized as in Fig. 4.10(a), utilizing the MOSFET body diodes.

(a)   Estimate the switching energy loss, conduction loss, and converter efficiency, when the battery is being charged at the maximum rate. The switching frequency is 100 kHz.

External diodes are now added as illustrated in Fig. 4.10(b). These diodes have forward voltage drops of 1.0 V, and exhibit recovered charge Q_r of 5 μC and reverse recovery times t_r of 40 ns in the given circuit.

(b)   Repeat the analysis of Part (a), for this case.

(c)   Over what range of switching frequencies does the addition of the external diodes improve the converter efficiency?

4.9   A switching converter operates with a switching frequency of 100 kHz. The converter waveforms exhibit damped sinusoidal ringing, initiated by the transistor turn-off transition, which decays slowly but eventually reaches zero before the end of the switching period. This ringing occurs in a series resonant circuit formed by parasitic inductances and capacitances in the circuit. The frequency of the ringing is 5 MHz. During the first period of sinusoidal ringing, the ac inductor current reaches a peak magnitude of 0.5 A, and the ac capacitor voltage reaches a peak magnitude of 200 V. Determine the following quantities:

(a)   the value of the total parasitic inductance,

(b)   the value of the total parasitic capacitance,

(c)   the energy lost per switching period, associated with this ringing, and

(d)   the switching loss associated with this ringing.

(e)   Derive a general expression for the switching loss, as a function of the switching frequency, ringing frequency, and the ringing voltage and current peak magnitudes during the first period of ringing.

5

The Discontinuous Conduction Mode

When the ideal switches of a dc-dc converter are implemented using current-unidirectional and/or voltage-unidirectional semiconductor switches, one or more new modes of operation known as *discontinuous conduction modes* (DCM) can occur. The discontinuous conduction mode arises when the switching ripple in an inductor current or capacitor voltage is large enough to cause the polarity of the applied switch current or voltage to reverse, such that the current- or voltage-unidirectional assumptions made in realizing the switch with semiconductor devices are violated. The DCM is commonly observed in dc–dc converters and rectifiers, and can also sometimes occur in inverters or in other converters containing two-quadrant switches.

The discontinuous conduction mode typically occurs with large inductor current ripple in a converter operating at light load and containing current-unidirectional switches. Since it is usually required that converters operate with their loads removed, DCM is frequently encountered. Indeed, some converters are purposely designed to operate in DCM for all loads.

The properties of converters change radically in the discontinuous conduction mode. The conversion ratio M becomes load-dependent, and the output impedance is increased. Control of the output may be lost when the load is removed. We will see in a later chapter that the converter dynamics are also significantly altered.

In this chapter, the origins of the discontinuous conduction mode are explained, and the mode boundary is derived. Techniques for solution of the converter waveforms and output voltage are also described. The principles of inductor volt-second balance and capacitor charge balance must always be true in steady state, regardless of the operating mode. However, application of the small ripple approximation requires some care, since the inductor current ripple (or one of the inductor current or capacitor voltage ripples) is not small.

Buck and boost converters are solved as examples. Characteristics of the basic buck, boost, and buck-boost converters are summarized in tabular form.

5.1   ORIGIN OF THE DISCONTINUOUS CONDUCTION MODE, AND MODE BOUNDARY

Let us consider how the inductor and switch current waveforms change as the load power is reduced. Let's use the buck converter (Fig. 5.1) as a simple example. The inductor current $i_L(t)$ and diode current $i_D(t)$ waveforms are sketched in Fig. 5.2 for the continuous conduction mode. As described in Chapter 2, the inductor current waveform contains a dc component I, plus switching ripple of peak amplitude Δi_L. During the second subinterval, the diode current is identical to the inductor current. The minimum diode current during the second subinterval is equal to $(I - \Delta i_L)$; since the diode is a single-quadrant switch, operation in the continuous conduction mode requires that this current remain positive. As shown in Chapter 2, the inductor current dc component I is equal to the load current:

$$I = \frac{V}{R} \tag{5.1}$$

since no dc current flows through capacitor C. It can be seen that I depends on the load resistance R. The

Fig. 5.1 Buck converter example.

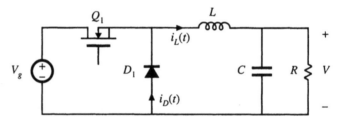

(a)

Fig. 5.2 Buck converter waveforms in the continuous conduction mode: (a) inductor current $i_L(t)$, (b) diode current $i_D(t)$.

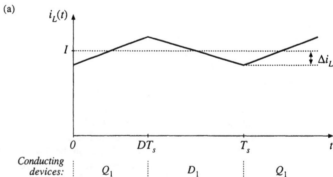

(b)

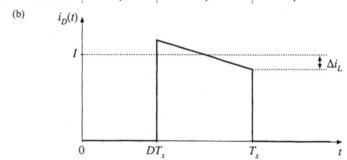

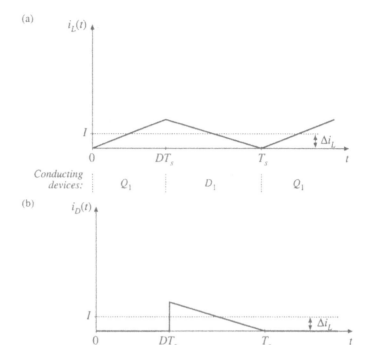

Fig. 5.3 Buck converter waveforms at the boundary between the continuous and discontinuous conduction modes: (a) inductor current $i_L(t)$, (b) diode current $i_D(t)$.

switching ripple peak amplitude is:

$$\Delta i_L = \frac{(V_g - V)}{2L} DT_s = \frac{V_g DD'T_s}{2L} \tag{5.2}$$

The ripple magnitude depends on the applied voltage $(V_g - V)$, on the inductance L, and on the transistor conduction time DT_s. But it does not depend on the load resistance R. The inductor current ripple magnitude varies with the applied voltages rather than the applied currents.

Suppose now that the load resistance R is increased, so that the dc load current is decreased. The dc component of inductor current I will then decrease, but the ripple magnitude Δi_L will remain unchanged. If we continue to increase R, eventually the point is reached where $I = \Delta i_L$, illustrated in Fig. 5.3. It can be seen that the inductor current $i_L(t)$ and the diode current $i_D(t)$ are both zero at the end of the switching period. Yet the load current is positive and nonzero.

What happens if we continue to increase the load resistance R? The diode current cannot be negative; therefore, the diode must become reverse-biased before the end of the switching period. As illustrated in Fig. 5.4, there are now three subintervals during each switching period T_s. During the first subinterval of length $D_1 T_s$ the transistor conducts, and the diode conducts during the second subinterval of length $D_2 T_s$. At the end of the second subinterval the diode current reaches zero, and for the remainder of the switching period neither the transistor nor the diode conduct. The converter operates in the discontinuous conduction mode.

Figure 5.3 suggests a way to find the boundary between the continuous and discontinuous conduction modes. It can be seen that, for this buck converter example, the diode current is positive over the entire interval $DT_s < t < T_s$ provided that $I > \Delta i_L$. Hence, the conditions for operation in the continuous and discontinuous conduction modes are:

(a)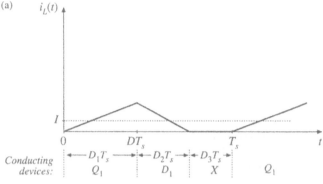

Fig.   5.4 Buck converter waveforms in the discontinuous conduction mode: (a) inductor current $i_L(t)$, (b) diode current $i_D(t)$.

(b)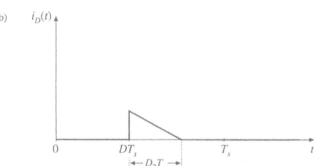

$$I > \Delta i_L \quad \text{for CCM}$$
$$I < \Delta i_L \quad \text{for DCM} \tag{5.3}$$

where I and Δi_L are found assuming that the converter operates in the continuous conduction mode. Insertion of Eqs. (5.1) and (5.2) into Eq. (5.3) yields the following condition for operation in the discontinuous conduction mode:

$$\frac{DV_g}{R} < \frac{DD'T_s V_g}{2L} \tag{5.4}$$

Simplification leads to

$$\frac{2L}{RT_s} < D' \tag{5.5}$$

This can also be expressed

$$K < K_{crit}(D) \quad \text{for DCM} \tag{5.6}$$

where

$$K = \frac{2L}{RT_s} \qquad \text{and} \qquad K_{crit}(D) = D'$$

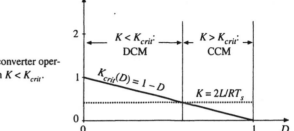

Fig. 5.5 Buck converter $K_{crit}(D)$ vs. D. The converter operates in CCM when $K > K_{crit}$, and in DCM when $K < K_{crit}$.

The dimensionless parameter K is a measure of the tendency of a converter to operate in the discontinuous conduction mode. Large values of K lead to continuous mode operation, while small values lead to the discontinuous mode for some values of duty cycle. The critical value of K at the boundary between modes, $K_{crit}(D)$, is a function of duty cycle, and is equal to D' for the buck converter.

The critical value $K_{crit}(D)$ is plotted vs. duty cycle D in Fig. 5.5. An arbitrary choice of K is also illustrated. For the values shown, it can be seen that the converter operates in DCM at low duty cycle, and in CCM at high duty cycle. Figure 5.6 illustrates what happens with heavier loading. The load resistance R is reduced in value, such that K is larger. If K is greater than one, then the converter operates in the continuous conduction mode for all duty cycles.

It is natural to express the mode boundary in terms of the load resistance R, rather than the dimensionless parameter K. Equation (5.6) can be rearranged to directly expose the dependence of the mode boundary on the load resistance:

$$R < R_{crit}(D) \quad \text{for CCM}$$
$$R > R_{crit}(D) \quad \text{for DCM}$$

(5.7)

where

$$R_{crit}(D) = \frac{2L}{D'T_s}$$

So the converter enters the discontinuous conduction mode when the load resistance R exceeds the critical value R_{crit}. This critical value depends on the inductance, the switching period, and the duty cycle. Note that, since $D' \leq 1$, the minimum value of R_{crit} is $2L/T_s$. Therefore, if $R < 2L/T_s$, then the converter will operate in the continuous conduction mode for all duty cycles.

These results can be applied to loads that are not pure linear resistors. An effective load resis-

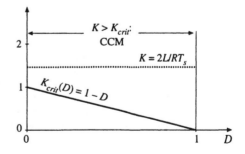

Fig. 5.6 Comparison of K with $K_{crit}(D)$, for a larger value of K. Since $K > 1$, the converter operates in CCM for all D.

Table 5.1 CCM-DCM mode boundaries for the buck, boost, and buck-boost converters

Converter	$K_{crit}(D)$	$\max\limits_{0 \leq D \leq 1} (K_{crit})$	$R_{crit}(D)$	$\min\limits_{0 \leq D \leq 1} (R_{crit})$
Buck	$(1 - D)$	1	$\dfrac{2L}{(1 - D)T_s}$	$2\dfrac{L}{T_s}$
Boost	$D(1 - D)^2$	$\dfrac{4}{27}$	$\dfrac{2L}{D(1 - D)^2 T_s}$	$\dfrac{27}{2}\dfrac{L}{T_s}$
Buck–boost	$(1 - D)^2$	1	$\dfrac{2L}{(1 - D)^2 T_s}$	$2\dfrac{L}{T_s}$

tance R is defined as the ratio of the dc output voltage to the dc load current: $R = V/I$. This effective load resistance is then used in the above equations.

A similar mode boundary analysis can be performed for other converters. The boost converter is analyzed in Section 5.3, while analysis of the buck-boost converter is left as a homework problem. The results are listed in Table 5.1, for the three basic dc–dc converters. In each case, the dimensionless parameter K is defined as $K = 2L/RT_s$, and the mode boundary is given by

$$K > K_{crit}(D) \quad \text{or} \quad R < R_{crit}(D) \quad \text{for CCM}$$

$$K < K_{crit}(D) \quad \text{or} \quad R > R_{crit}(D) \quad \text{for DCM}$$

(5.8)

5.2   ANALYSIS OF THE CONVERSION RATIO $M(D, K)$

With a few modifications, the same techniques and approximations developed in Chapter 2 for the steady-state analysis of the continuous conduction mode may be applied to the discontinuous conduction mode.

(a) *Inductor volt-second balance.* The dc component of the voltage applied to an inductor must be zero:

$$\langle v_L \rangle = \frac{1}{T_s} \int_0^{T_s} v_L(t)dt = 0$$

(5.9)

(b) *Capacitor charge balance.* The dc component of current applied to a capacitor must be zero:

$$\langle i_C \rangle = \frac{1}{T_s} \int_0^{T_s} i_C(t)dt = 0$$

(5.10)

These principles must be true for any circuit that operates in steady state, regardless of the operating mode.

(c) *The linear ripple approximation.* Care must be used when employing the linear ripple approximation in the discontinuous conduction mode.

(i)   *Output capacitor voltage ripple.* Regardless of the operating mode, it is required that the output voltage ripple be small. Hence, for a well-designed converter operating in the discontinuous conduction mode, the peak output voltage ripple Δv should be much smaller in magnitude than the output voltage dc component V. So the linear ripple approximation applies to the output voltage waveform:

$$v(t) \approx V \qquad\qquad (5.11)$$

(*ii*) *Inductor current ripple.* By definition, the inductor current ripple is not small in the discontinuous conduction mode. Indeed, Eq. (5.3) states that the inductor current ripple Δi_L is greater in magnitude than the dc component I. So neglecting the inductor current ripple leads to inaccurate results. In other converters, several inductor currents, or a capacitor voltage, may contain large switching ripple which should not be neglected.

The equations necessary for solution of the voltage conversion ratio can be obtained by invoking volt-second balance for each inductor voltage, and charge balance for each capacitor current, in the network. The switching ripple is ignored in the output capacitor voltage, but the inductor current switching ripple must be accounted for in this buck converter example.

Let us analyze the conversion ratio $M = V/V_g$ of the buck converter of Eq. (5.1). When the transistor conducts, for $0 < t < D_1 T_s$, the converter circuit reduces to the network of Fig. 5.7(a). The inductor voltage and capacitor current are given by

$$v_L(t) = V_g - v(t)$$
$$i_C(t) = i_L(t) - \frac{v(t)}{R} \qquad\qquad (5.12)$$

By making the linear ripple approximation, to ignore the output capacitor voltage ripple, one obtains

(a)

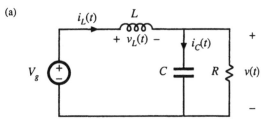

(b)

Fig. 5.7 Buck converter circuits for operation in the discontinuous conduction mode: (a) during subinterval 1, (b) during subinterval 2, (c) during subinterval 3.

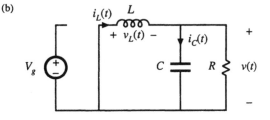

(c)

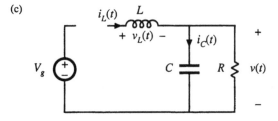

$$v_L(t) \approx V_g - V$$
$$i_C(t) \approx i_L(t) - \frac{V}{R} \tag{5.13}$$

Note that the inductor current ripple has not been ignored.

The diode conducts during subinterval 2, $D_1 T_s < t < (D_1 + D_2)T_s$. The circuit then reduces to Fig. 5.7(b). The inductor voltage and capacitor current are given by

$$v_L(t) = - v(t)$$
$$i_C(t) = i_L(t) - \frac{v(t)}{R} \tag{5.14}$$

By neglecting the ripple in the output capacitor voltage, one obtains

$$v_L(t) \approx - V$$
$$i_C(t) \approx i_L(t) - \frac{V}{R} \tag{5.15}$$

The diode becomes reverse-biased at time $t = (D_1 + D_2)T_s$. The circuit is then as shown in Fig. 5.7(c), with both transistor and diode in the off state. The inductor voltage and inductor current are both zero for the remainder of the switching period $(D_1 + D_2)T_s < t < T_s$. The network equations for the third subinterval are given by

$$v_L = 0, \quad i_L = 0$$
$$i_C(t) = i_L(t) - \frac{v(t)}{R} \tag{5.16}$$

Note that the inductor current is constant and equal to zero during the third subinterval, and therefore the inductor voltage must also be zero in accordance with the relationship $v_L(t) = L\, di_L(t)/dt$. In practice, parasitic ringing is observed during this subinterval. This ringing occurs owing to the resonant circuit formed by the inductor and the semiconductor device capacitances, and typically has little influence on the converter steady-state properties. Again ignoring the output capacitor voltage ripple, one obtains

$$v_L(t) = 0$$
$$i_C(t) = - \frac{V}{R} \tag{5.17}$$

Equations (5.13), (5.15), and (5.17) can now be used to plot the inductor voltage waveform as in Fig. 5.8. According to the principle of inductor volt-second balance, the dc component of this waveform must be zero. Since the waveform is rectangular, its dc component (or average value) is easily evaluated:

$$\langle v_L(t) \rangle = D_1 \big(V_g - V \big) + D_2 \big(- V \big) + D_3 \big(0 \big) = 0 \tag{5.18}$$

Solution for the output voltage yields

$$V = V_g \frac{D_1}{D_1 + D_2} \tag{5.19}$$

The transistor duty cycle D (which coincides with the subinterval 1 duty cycle D_1) is the control input to the converter, and can be considered known. But the subinterval 2 duty cycle D_2 is unknown, and hence

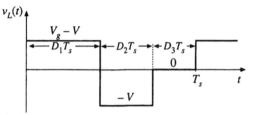

Fig. 5.8 Inductor voltage waveform $v_L(t)$, buck converter operating in discontinuous conduction mode.

another equation is needed to eliminate D_2 and solve for the output voltage V.

The second equation is obtained by use of capacitor charge balance. The connection of the capacitor to its adjacent components is detailed in Fig. 5.9. The node equation of this network is

$$i_L(t) = i_C(t) + \frac{v(t)}{R} \tag{5.20}$$

By capacitor charge balance, the dc component of capacitor current must be zero:

$$\langle i_C \rangle = 0 \tag{5.21}$$

Therefore, the dc load current must be supplied entirely by the other elements connected to the node. In particular, for the case of the buck converter, the dc component of inductor current must be equal to the dc load current:

$$\langle i_L \rangle = \frac{V}{R} \tag{5.22}$$

So we need to compute the dc component of the inductor current.

Since the inductor current ripple is not small, determination of the inductor current dc component requires that we examine the current waveform in detail. The inductor current waveform is sketched in Fig. 5.10. The current begins the switching period at zero, and increases during the first subinterval with a constant slope, given by the applied voltage divided by the inductance. The peak inductor current i_{pk} is equal to the constant slope, multiplied by the length of the first subinterval:

$$i_L(D_1 T_s) = i_{pk} = \frac{V_g - V}{L} D_1 T_s \tag{5.23}$$

The dc component of the inductor current is again the average value:

$$\langle i_L \rangle = \frac{1}{T_s} \int_0^{T_s} i_L(t)\,dt \tag{5.24}$$

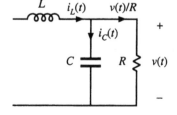

Fig. 5.9 Connection of the output capacitor to adjacent components.

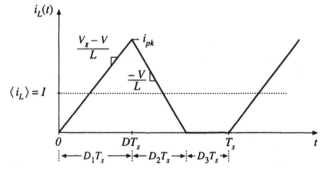

Fig. 5.10 Inductor current waveform $i_L(t)$, buck converter operating in discontinuous conduction mode.

The integral, or area under the iL(t) curve, is the area of the triangle having height i_{pk} and base dimension $(D_1 + D_2)T_s$. Use of the triangle area formula yields

$$\int_0^{T_s} i_L(t)dt = \tfrac{1}{2}i_{pk}\big(D_1 + D_2\big)T_s \tag{5.25}$$

Substitution of Eqs. (5.23) and (5.25) into Eq. (5.24) leads to

$$\langle i_L \rangle = \big(V_g - V\big)\left(\frac{D_1 T_s}{2L}\right)\big(D_1 + D_2\big) \tag{5.26}$$

Finally, by equating this result to the dc load current, according to Eq. (5.22), we obtain

$$\frac{V}{R} = \frac{D_1 T_s}{2L}\big(D_1 + D_2\big)\big(V_g - V\big) \tag{5.27}$$

Thus, we have two unknowns, V and D_2, and we have two equations. The first equation, Eq. (5.19), was obtained by inductor volt-second balance, while the second equation, Eq. (5.27), was obtained using capacitor charge balance. Elimination of D_2 from the two equations, and solution for the voltage conversion ratio $M(D_1, K) = V/V_g$, yields

$$\frac{V}{V_g} = \frac{2}{1 + \sqrt{1 + \dfrac{4K}{D_1^2}}} \tag{5.28}$$

where $K = 2L/RT_s$

valid for $K < K_{crit}$

This is the solution of the buck converter operating in discontinuous conduction mode.

The complete buck converter characteristics, including both continuous and discontinuous conduction modes, are therefore

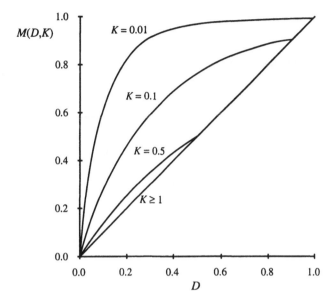

Fig. 5.11   Voltage conversion ratio $M(D, K)$, buck converter.

$$M = \begin{cases} D & \text{for } K > K_{crit} \\[2ex] \dfrac{2}{1 + \sqrt{1 + \dfrac{4K}{D^2}}} & \text{for } K < K_{crit} \end{cases} \qquad (5.29)$$

where the transistor duty cycle D is identical to the subinterval 1 duty cycle D_1 of the above derivation. These characteristics are plotted in Fig. 5.11, for several values of K. It can be seen that the effect of the discontinuous conduction mode is to cause the output voltage to increase. As K tends to zero (the unloaded case), M tends to unity for all nonzero D. The characteristics are continuous, and Eq. (5.28) intersects the CCM characteristic $M = D$ at the mode boundary.

5.3   BOOST CONVERTER EXAMPLE

As a second example, consider the boost converter of Fig. 5.12. Let's determine the boundary between modes, and solve for the conversion ratio in the discontinuous conduction mode. Behavior of the boost converter operating in the continuous conduction mode was analyzed previously, in Section 2.3, and expressions for the inductor current dc component I and ripple peak magnitude Δi_L were found.

When the diode conducts, its current is identical to the inductor current $i_L(t)$. As can be seen from Fig. 2.18, the minimum value of the inductor current during the diode conduction subinterval $DT_s < t < T_s$ is $(I - \Delta i_L)$. If this minimum current is positive, then the diode is forward-biased for the entire subinterval $DT_s < t < T_s$, and the converter operates in the continuous conduction mode. So the conditions for operation of the boost converter in the continuous and discontinuous conduction modes are:

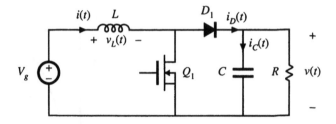

Fig. 5.12   Boost converter example.

$$I > \Delta i_L \quad \text{for CCM}$$
$$I < \Delta i_L \quad \text{for DCM} \tag{5.30}$$

which is identical to the results for the buck converter. Substitution of the CCM solutions for I and Δi_L, Eqs. (2-39) and (2-43), yields

$$\frac{V_g}{D'^2 R} > \frac{DT_s V_g}{2L} \quad \text{for CCM} \tag{5.31}$$

This equation can be rearranged to obtain

$$\frac{2L}{RT_s} > DD'^2 \quad \text{for CCM} \tag{5.32}$$

which is in the standard form

$$K > K_{crit}(D) \quad \text{for CCM}$$
$$K < K_{crit}(D) \quad \text{for DCM} \tag{5.33}$$

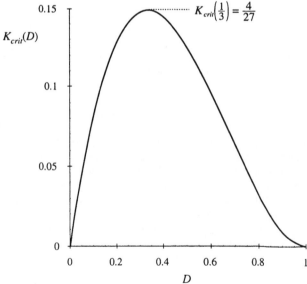

Fig. 5.13   Boost converter $K_{crit}(D)$ vs. D.

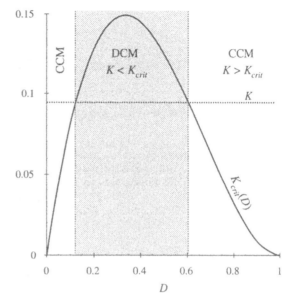

Fig. 5.14   Comparison of K with $K_{crit}(D)$.

where

$$K = \frac{2L}{RT_s} \quad \text{and} \quad K_{crit}(D) = DD'^2$$

The conditions for operation in the continuous or discontinuous conduction modes are of similar form to those for the buck converter; however, the critical value $K_{crit}(D)$ is a different function of the duty cycle D. The dependence of $K_{crit}(D)$ on the duty cycle D is plotted in Fig. 5.13. $K_{crit}(D)$ is zero at $D = 0$ and at $D = 1$, and has a maximum value of 4/27 at $D = 1/3$. Hence, if K is greater than 4/27, then the converter operates in the continuous conduction mode for all D. Figure 5.14 illustrates what happens when K is less than 4/27. The converter then operates in the discontinuous conduction mode for some intermediate range of values of D near $D = 1/3$. But the converter operates in the continuous conduction mode near $D = 0$ and $D = 1$. Unlike the buck converter, the boost converter must operate in the continuous conduction mode near $D = 0$ because the ripple magnitude approaches zero while the dc component I does not.

Next, let us analyze the conversion ratio $M = V/V_g$ of the boost converter. When the transistor conducts, for the subinterval $0 < t < D_1 T_s$, the converter circuit reduces to the circuit of 5.15(a). The inductor voltage and capacitor current are given by

$$v_L(t) = V_g$$
$$i_C(t) = -\frac{v(t)}{R} \tag{5.34}$$

Use of the linear ripple approximation, to ignore the output capacitor voltage ripple, leads to

$$v_L(t) \approx V_g$$
$$i_C(t) \approx -\frac{V}{R} \tag{5.35}$$

During the second subinterval $D_1 T_s < t < (D_1 + D_2)T_s$, the diode conducts. The circuit then reduces to

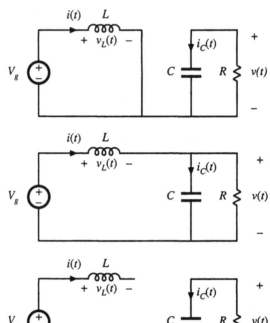

Fig. 5.15 Boost converter circuits: (a) during subinterval 1, $0 < t < D_1 T_s$, (b) during subinterval 2, $D_1 T_s < t < (D_1 + D_2)T_s$, (c) during subinterval 3, $(D_1 + D_2)T_s < t < T_s$.

Fig. 5.15(b). The inductor voltage and capacitor current are given by

$$v_L(t) = V_g - v(t)$$
$$i_C(t) = i(t) - \frac{v(t)}{R} \tag{5.36}$$

Neglect of the output capacitor voltage ripple yields

$$v_L(t) \approx V_g - V$$
$$i_C(t) \approx i(t) - \frac{V}{R} \tag{5.37}$$

The inductor current ripple has not been neglected.

During the third subinterval, $(D_1 + D_2)T_s < t < T_s$, both transistor and diode are in the off state, and Fig. 5.15(c) is obtained. The network equations are:

$$v_L = 0, \quad i = 0$$
$$i_C(t) = -\frac{v(t)}{R} \tag{5.38}$$

Use of the small-ripple approximation yields

Fig. 5.16 Inductor voltage waveform $v_L(t)$, boost converter operating in discontinuous conduction mode.

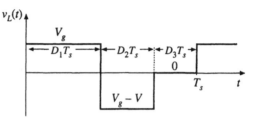

$$v_L(t) = 0$$
$$i_C(t) = -\frac{V}{R}$$
(5.39)

Equations (5.35), (5.37), and (5.39) are now used to sketch the inductor voltage waveform as in Fig. 5.16. By volt-second balance, this waveform must have zero dc component when the converter operates in steady state. By equating the average value of this $v_L(t)$ waveform to zero, one obtains

$$D_1 V_g + D_2\left(V_g - V\right) + D_3\left(0\right) = 0$$
(5.40)

Solution for the output voltage V yields

$$V = \frac{D_1 + D_2}{D_2} V_g$$
(5.41)

The diode duty cycle D_2 is again an unknown, and so a second equation is needed for elimination of D_2 before the output voltage V can be found.

We can again use capacitor charge balance to obtain the second equation. The connection of the output capacitor to its adjacent components is detailed in Fig. 5.17. Unlike the buck converter, the diode in the boost converter is connected to the output node. The node equation of Fig. 5.17 is

$$i_D(t) = i_C(t) + \frac{v(t)}{R}$$
(5.42)

where $i_D(t)$ is the diode current. By capacitor charge balance, the capacitor current $i_C(t)$ must have zero dc component in steady state. Therefore, the diode current dc component $\langle i_D \rangle$ must be equal to the dc component of the load current:

$$\langle i_D \rangle = \frac{V}{R}$$
(5.43)

So we need to sketch the diode current waveform, and find its dc component.

Fig. 5.17 Connection of the output capacitor to adjacent components in the boost converter.

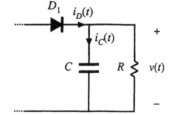

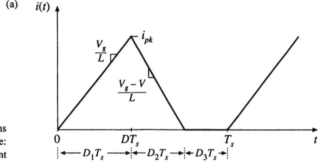

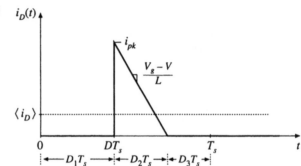

Fig. 5.18 Boost converter waveforms in the discontinuous conduction mode: (a) inductor current $i(t)$, (b) diode current $i_D(t)$.

The waveforms of the inductor current $i(t)$ and diode current $i_D(t)$ are illustrated in Fig. 5.18. The inductor current begins at zero, and rises to a peak value i_{pk} during the first subinterval. This peak value i_{pk} is equal to the slope V_g/L, multiplied by the length of the first subinterval, $D_1 T_s$:

$$i_{pk} = \frac{V_g}{L} D_1 T_s \tag{5.44}$$

The diode conducts during the second subinterval, and the inductor current then decreases to zero, where it remains during the third subinterval. The diode current $i_D(t)$ is identical to the inductor current $i(t)$ during the second subinterval. During the first and third subintervals, the diode is reverse-biased and hence $i_D(t)$ is zero.

The dc component of the diode current, $\langle i_D \rangle$, is:

$$\langle i_D \rangle = \frac{1}{T_s} \int_0^{T_s} i_D(t)\, dt \tag{5.45}$$

The integral is the area under the $i_D(t)$ waveform. As illustrated in Fig. 5.18(b), this area is the area of the triangle having peak value i_{pk} and base dimension $D_2 T_s$:

$$\int_0^{T_s} i_D(t)\, dt = \tfrac{1}{2} i_{pk} D_2 T_s \tag{5.46}$$

Substitution of Eqs. (5.44) and (5.46) into Eq. (5.45) leads to the following expression for the dc component of the diode current:

$$\langle i_D \rangle = \frac{1}{T_s} \left(\frac{1}{2} i_{pk} D_2 T_s \right) = \frac{V_g D_1 D_2 T_s}{2L} \tag{5.47}$$

By equating this expression to the dc load current as in Eq. (5.43), one obtains the final result

$$\frac{V_g D_1 D_2 T_s}{2L} = \frac{V}{R} \tag{5.48}$$

So now we have two unknowns, V and D_2. We have two equations: Eq. (5.41) obtained via inductor volt-second balance, and Eq. (5.48) obtained using capacitor charge balance. Let us now eliminate D_2 from this system of equations, and solve for the output voltage V. Solution of Eq. (5.41) for D_2 yields

$$D_2 = D_1 \frac{V_g}{V - V_g} \tag{5.49}$$

By inserting this result into Eq. (5.48), and rearranging terms, one obtains the following quadratic equation:

$$V^2 - VV_g - \frac{V_g^2 D_1^2}{K} = 0 \tag{5.50}$$

Use of the quadratic formula yields

$$\frac{V}{V_g} = \frac{1 \pm \sqrt{1 + \frac{4D_1^2}{K}}}{2} \tag{5.51}$$

The quadratic equation has two roots: one of the roots of Eq. (5.51) is positive, while the other is negative. We already know that the output voltage of the boost converter should be positive, and indeed, from Eq. (5.41), it can be seen that V/V_g must be positive since the duty cycles D_1 and D_2 are positive. So we should select the positive root:

$$\frac{V}{V_g} = M(D_1, K) = \frac{1 + \sqrt{1 + \frac{4D_1^2}{K}}}{2} \tag{5.52}$$

$$\text{where} \qquad K = 2L/RT_s$$

$$\text{valid for} \qquad K < K_{crit}(D)$$

This is the solution of the boost converter operating in the discontinuous conduction mode.

The complete boost converter characteristics, including both continuous and discontinuous conduction modes, are

$$M = \begin{cases} \dfrac{1}{1 - D} & \text{for } K > K_{crit} \\[2ex] \dfrac{1 + \sqrt{1 + \dfrac{4D^2}{K}}}{2} & \text{for } K < K_{crit} \end{cases} \tag{5.53}$$

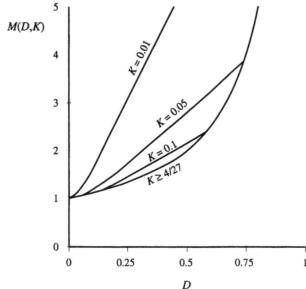

Fig. 5.19 Voltage conversion ratio $M(D, K)$ of the boost converter, including both continuous and discontinuous conduction modes

These characteristics are plotted in Fig. 5.19, for several values of K. As in the buck converter, the effect of the discontinuous conduction mode is to cause the output voltage to increase. The DCM portions of the characteristics are nearly linear, and can be approximated as

$$M \approx \frac{1}{2} + \frac{D}{\sqrt{K}}$$

$$(5.54)$$

5.4   SUMMARY OF RESULTS AND KEY POINTS

The characteristics of the basic buck, boost, and buck-boost are summarized in Table 5.2. Expressions for $K_{crit}(D)$, as well as for the solutions of the dc conversion ratios in CCM and DCM, and for the DCM diode conduction duty cycle D_2, are given.

The dc conversion ratios of the DCM buck, boost, and buck-boost converters are compared in

Table 5.2 Summary of CCM-DCM characteristics for the buck, boost, and buck-boost converters

Converter	$K_{crit}(D)$	DCM $M(D, K)$	DCM $D_2(D, K)$	CCM $M(D)$
Buck	$(1 - D)$	$\dfrac{2}{1 + \sqrt{1 + 4K/D^2}}$	$\dfrac{K}{D} M(D, K)$	D
Boost	$D(1 - D)^2$	$\dfrac{1 + \sqrt{1 + 4D^2/K}}{2}$	$\dfrac{K}{D} M(D, K)$	$\dfrac{1}{1 - D}$
Buck-boost	$(1 - D)^2$	$-\dfrac{D}{\sqrt{K}}$	$\sqrt{K}$	$-\dfrac{D}{1 - D}$

with $K = 2L/RT_s$,   DCM occurs for $K < K_{crit}$.

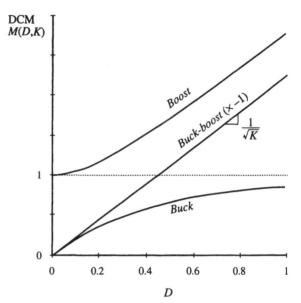

Fig. 5.20 Comparison of dc conversion ratios of the buck–boost, buck, and boost converters operated in the discontinuous conduction mode.

Fig. 5.20. The buck-boost characteristic is a line with slope $1/\sqrt{K}$. The characteristics of the buck and the boost converters are both asymptotic to this line, as well as to the line $M = 1$. Hence, when operated deeply into the discontinuous conduction mode, the boost converter characteristic becomes nearly linear with slope $1/\sqrt{K}$, especially at high duty cycle. Likewise, the buck converter characteristic becomes nearly linear with the same slope, when operated deeply into discontinuous conduction mode at low duty cycle.

The following are the key points of this chapter:

1. The discontinuous conduction mode occurs in converters containing current- or voltage-unidirectional switches, when the inductor current or capacitor voltage ripple is large enough to cause the switch current or voltage to reverse polarity.

2. Conditions for operation in the discontinuous conduction mode can be found by determining when the inductor current or capacitor voltage ripples and dc components cause the switch on state current or off state voltage to reverse polarity.

3. The dc conversion ratio M of converters operating in the discontinuous conduction mode can be found by application of the principles of inductor volt-second and capacitor charge balance.

4. Extra care is required when applying the small-ripple approximation. Some waveforms, such as the output voltage, should have small ripple which can be neglected. Other waveforms, such as one or more inductor currents, may have large ripple that cannot be ignored.

5. The characteristics of a converter changes significantly when the converter enters DCM. The output voltage becomes load-dependent, resulting in an increase in the converter output impedance.

PROBLEMS

5.1   The elements of the buck-boost converter of Fig. 5.21 are ideal: all losses may be ignored. Your results for parts (a) and (b) should agree with Table 5.2.

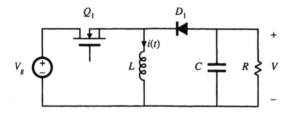

Fig. 5.21   Buck–boost converter of Problems 5.1 and 5.13.

    (a)   Show that the converter operates in discontinuous conduction mode when $K < K_{crit}$, and derive expressions for K and K_{crit}.

    (b)   Derive an expression for the dc conversion ratio V/V_g of the buck–boost converter operating in discontinuous conduction mode.

    (c)   For $K = 0.1$, plot V/V_g over the entire range $0 \le D \le 1$.

    (d)   Sketch the inductor voltage and current waveforms for $K = 0.1$ and $D = 0.3$. Label salient features.

    (e)   What happens to V at no load ($R \rightarrow \infty$)? Explain why, physically.

5.2   A certain buck converter contains a synchronous rectifier, as described in Section 4.1.5.

    (a)   Does this converter operate in the discontinuous conduction mode at light load? Explain.

    (b)   The load resistance is disconnected ($R \rightarrow \infty$), and the converter is operated with duty cycle 0.5. Sketch the inductor current waveform.

5.3   An unregulated dc input voltage V_g varies over the range 35 V $\le V_g \le$ 70 V. A buck converter reduces this voltage to 28 V; a feedback loop varies the duty cycle as necessary such that the converter output voltage is always equal to 28 V. The load power varies over the range 10 W $\le P_{load} \le$ 1000 W. The element values are:

    $L = 22~\mu H$ $C = 470~\mu F$ $f_s = 75$ kHz
    Losses may be ignored.

    (a)   Over what range of V_g and load current does the converter operate in CCM?

    (b)   Determine the maximum and minimum values of the steady-state transistor duty cycle.

5.4   The transistors in the converter of Fig. 5.22 are driven by the same gate drive signal, so that they turn on and off in synchronism with duty cycle D.

    (a)   Determine the conditions under which this converter operates in the discontinuous conduction mode, as a function of the steady-state duty ratio D

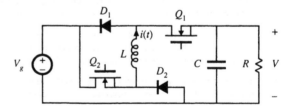

Fig. 5.22   Watkins-Johnson converter of Problem 5.4.

and the dimensionless parameter $K = 2L/RT_s$.

(b)   What happens to your answer to Part (a) for $D < 0.5$?

(c)   Derive an expression for the dc conversion ratio $M(D, K)$. Sketch M vs. D for $K = 10$ and for $K = 0.1$, over the range $0 \le D \le 1$.

5.5   DCM mode boundary analysis of the Ćuk converter of Fig. 5.23. The capacitor voltage ripples are small.

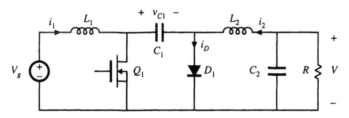

Fig. 5.23   Ćuk converter, Problems 5.5, 5.6, 5.11, and 5.12

(a)   Sketch the diode current waveform for CCM operation. Find its peak value, in terms of the ripple magnitudes Δi_{L1}, Δi_{L2}, and the dc components I_1 and I_2, of the two inductor currents $i_{L1}(t)$ and $i_{L2}(t)$, respectively.

(b)   Derive an expression for the conditions under which the Ćuk converter operates in the discontinuous conduction mode. Express your result in the form $K < K_{crit}(D)$, and give formulas for K and $K_{crit}(D)$.

5.6   DCM conversion ratio analysis of the Ćuk converter of Fig. 5.23.

(a)   Suppose that the converter operates at the boundary between CCM and DCM, with the following element and parameter values:

$$D = 0.4 \qquad\qquad f_s = 100 \text{ kHz}$$
$$V_g = 120 \text{ V} \qquad\qquad R = 10 \text{ }\Omega$$
$$L_1 = 54 \text{ }\mu\text{H} \qquad\qquad L_2 = 27 \text{ }\mu\text{H}$$
$$C_1 = 47 \text{ }\mu\text{F} \qquad\qquad C_2 = 100 \text{ }\mu\text{F}$$

Sketch the diode current waveform $i_D(t)$, and the inductor current waveforms $i_1(t)$ and $i_2(t)$. Label the magnitudes of the ripples and dc components of these waveforms.

(b)   Suppose next that the converter operates in the discontinuous conduction mode, with a different choice of parameter and element values. Derive an analytical expression for the dc conversion ratio $M(D, K)$.

(c)   Sketch the diode current waveform $i_D(t)$, and the inductor current waveforms $i_1(t)$ and $i_2(t)$, for operation in the discontinuous conduction mode.

5.7   DCM mode boundary analysis of the SEPIC of Fig. 5.24

(a)   Sketch the diode current waveform for CCM operation. Find its peak value, in terms of the ripple magnitudes Δi_{L1}, Δi_{L2}, and the dc components I_1 and I_2, of the two inductor currents $i_{L1}(t)$ and $i_{L2}(t)$, respectively.

(b)   Derive an expression for the conditions under which the SEPIC operates in the discontinuous conduction mode. Express your result in the form $K < K_{crit}(D)$, and give formulas for K and $K_{crit}(D)$.

5.8   DCM conversion ratio analysis of the SEPIC of Fig. 5.24.

(a)   Suppose that the converter operates at the boundary between CCM and DCM, with the follow-

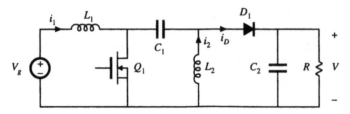

Fig. 5.24   SEPIC, Problems 5.7 and 5.8.

ing element and parameter values:

$D = 0.225$	$f_s = 100\,\text{kHz}$
$V_g = 120\,\text{V}$	$R = 10\,\Omega$
$L_1 = 50\,\mu\text{H}$	$L_2 = 75\,\mu\text{H}$
$C_1 = 47\,\mu\text{F}$	$C2 = 200\,\mu\text{F}$

Sketch the diode current waveform $i_D(t)$, and the inductor current waveforms $i_1(t)$ and $i_2(t)$. Label the magnitudes of the ripples and dc components of these waveforms.

(b)   Suppose next that the converter operates in the discontinuous conduction mode, with a different choice of parameter and element values. Derive an analytical expression for the dc conversion ratio $M(D, K)$.

(c)   Sketch the diode current waveform $i_D(t)$, and the inductor current waveforms $i_1(t)$ and $i_2(t)$, for operation in the discontinuous conduction mode.

5.9   An L–C input filter is added to a buck converter as illustrated in Fig. 5.25. Inductors L_1 and L_2 and capacitor C_2 are large in value, such that their switching ripples are small. All losses can be neglected.

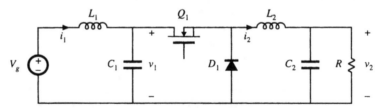

Fig. 5.25   Buck converter with input filter, Problems 5.9 and 5.10.

(a)   Sketch the capacitor C_1 voltage waveform $v_1(t)$, and derive expressions for its dc component V_1 and peak ripple magnitude Δv_1.

(b)   The load current is increased (R is decreased in value) such that Δv_1 is greater than V_1.

   (i)   Sketch the capacitor voltage waveform $v_1(t)$.

   (ii)   For each subinterval, determine which semiconductor devices conduct.

   (iii)   Determine the conditions under which the discontinuous conduction mode occurs. Express your result in the form $K < K_{crit}(D)$, and give formulas for K and $K_{crit}(D)$.

5.10   Derive an expression for the conversion ratio $M(D, K)$ of the DCM converter described in the previous problem. Note: D is the transistor duty cycle.

5.11   In the Cuk converter of Fig. 5.23, inductors L_1 and L_2 and capacitor C_2 are large in value, such that their switching ripples are small. All losses can be neglected.

(a)   Assuming that the converter operates in CCM, sketch the capacitor C_1 voltage waveform $v_{C1}(t)$, and derive expressions for its dc component V_1 and peak ripple magnitude Δv_{C1}.

(b)   The load current is increased (R is decreased in value) such that Δv_{C1} is greater than V_1.

(i)   Sketch the capacitor voltage waveform $v_{C1}(t)$.

(ii)   For each subinterval, determine which semiconductor devices conduct.

(iii) Determine the conditions under which the discontinuous conduction mode occurs. Express your result in the form $K < K_{crit}(D)$, and give formulas for K and $K_{crit}(D)$.

5.12   Derive an expression for the conversion ratio $M(D, K)$ of the DCM Ćuk converter described in the previous problem. Note: D is the transistor duty cycle.

5.13   A DCM buck-boost converter as in Fig. 5.21 is to be designed to operate under the following conditions:

$$136 \text{ V} \leq V_g \leq 204 \text{ V}$$
$$5 \text{ W} \leq P_{load} \leq 100 \text{ W}$$
$$V = -150 \text{ V}$$
$$f_s = 100 \text{ kHz}$$

You may assume that a feedback loop will vary to transistor duty cycle as necessary to maintain a constant output voltage of −150 V.

Design the converter, subject to the following considerations:

- The converter should operate in the discontinuous conduction mode at all times
- Given the above requirements, choose the element values to minimize the peak inductor current
- The output voltage peak ripple should be less than 1V.

Specify:

(a)   The inductor value L

(b)   The output capacitor value C

(c)   The worst-case peak inductor current i_{pk}

(d)   The maximum and minimum values of the transistor duty cycle D

5.14   A DCM boost converter as in Fig. 5.12 is to be designed to operate under the following conditions:

$$18 \text{ V} \leq V_g \leq 36 \text{ V}$$
$$5 \text{ W} \leq P_{load} \leq 100 \text{ W}$$
$$V = 48 \text{ V}$$
$$f_s = 150 \text{ kHz}$$

You may assume that a feedback loop will vary to transistor duty cycle as necessary to maintain a constant output voltage of 48 V.

Design the converter, subject to the following considerations:

- The converter should operate in the discontinuous conduction mode at all times. To ensure an adequate design margin, the inductance L should be chosen such that K is no greater than 75% of K_{crit} at all operating points.
- Given the above requirements, choose the element values to minimize the peak inductor current.
- The output voltage peak ripple should be less than 1V.

Specify:

(a)   The inductor value L

(b)   The output capacitor value C

(c)   The worst-case peak inductor current i_{pk}

(d)   The maximum and minimum values of the transistor duty cycle D.

(e)   The values of D, K and K_{crit} at the following operating points: (*i*) $V_g = 18$ V and $P_{load} = 5$ W; (*ii*) $V_g = 36$ V and $P_{load} = 5$ W; (*iii*) $V_g = 18$ V and $P_{load} = 100$ W; (*iv*) $V_g = 36$ V and $P_{load} = 100$ W.

5.15   In dc–dc converters used in battery-powered portable equipment, it is sometimes required that the converter continue to regulate its load voltage with high efficiency while the load is in a low-power "sleep" mode. The power required by the transistor gate drive circuitry, as well as much of the switching loss, is dependent on the switching frequency but not on the load current. So to obtain high efficiency at very low load powers, a variable-frequency control scheme is used, in which the switching frequency is reduced in proportion to the load current.

Consider the boost converter system of Fig. 5.26(a). The battery pack consists of two nickel-cadmium cells, which produce a voltage of $V_g = 2.4$ V $\pm$ 0.4 V. The converter boosts this voltage to a regulated 5 V. As illustrated in Fig. 5.26(b), the converter operates in the discontinuous conduction mode, with constant transistor on-time t_{on}. The transistor off-time t_{off} is varied by the controller to regulate the output voltage.

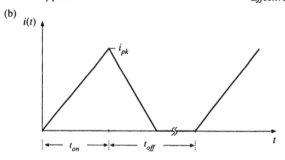

(a)

Fig. 5.26 Boost converter employed in portable battery-powered equipment with sleep mode, Problem 5.15: (a) converter circuit, (b) inductor current waveform.

(a)   Write the equations for the CCM-DCM boundary and conversion ratio $M = V/V_g$, in terms of t_{on}, t_{off}, L, and the effective load resistance R.

For parts (b) and (c), the load current can vary between 100 μA and 1 A. The transistor on time is fixed: $t_{on} = 10$ μs.

(b)   Select values for L and C such that:
  •   The output voltage peak ripple is no greater than 50 mV,
  •   The converter always operates in DCM, and
  •   The peak inductor current is as small as possible.

(c)   For your design of part (b), what are the maximum and minimum values of the switching frequency?

6

Converter Circuits

We have already analyzed the operation of a number of different types of converters: buck, boost, buck–boost, Ćuk, voltage-source inverter, etc. With these converters, a number of different functions can be performed: step-down of voltage, step-up, inversion of polarity, and conversion of dc to ac or vice-versa.

It is natural to ask, Where do these converters come from? What other converters occur, and what other functions can be obtained? What are the basic relations between converters? In this chapter, several different circuit manipulations are explored, which explain the origins of the basic converters. Inversion of source and load transforms the buck converter into the boost converter. Cascade connection of converters, and simplification of the resulting circuit, shows how the buck–boost and Ćuk converters are based on the buck and the boost converters. Differential connection of the load between the outputs of two or more converters leads to a single-phase or polyphase inverter. A short list of some of the better known converter circuits follows this discussion.

Transformer-isolated dc–dc converters are also covered in this chapter. Use of a transformer allows isolation and multiple outputs to be obtained in a dc-dc converter, and can lead to better converter optimization when a very large or very small conversion ratio is required. The transformer is modeled as a magnetizing inductance in parallel with an ideal transformer; this allows the analysis techniques of the previous chapters to be extended to cover converters containing transformers. A number of well-known isolated converters, based on the buck, boost, buck–boost, single-ended primary inductance converter (SEPIC), and Ćuk, are listed and discussed.

Finally, the evaluation, selection, and design of converters to meet given requirements are considered. Important performance-related attributes of transformer-isolated converters include: whether the transformer reset process imposes excessive voltage stress on the transistors, whether the converter can supply a high-current output without imposing excessive current stresses on the secondary-side components, and whether the converter can be well-optimized to operate with a wide range of operating points,

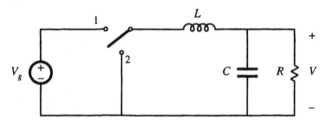

Fig. 6.1 The basic buck converter.

that is, with large tolerances in V_g and P_{load}. Switch utilization is a simplified figure-of-merit that measures the ratio of the converter output power to the total transistor voltage and current stress. As the switch utilization increases, the converter efficiency increases while its cost decreases. Isolated converters with large variations in operating point tend to utilize their power devices more poorly than nonisolated converters which function at a single operating point. Computer spreadsheets are a good tool for optimization of power stage designs and for trade studies to select a converter topology for a given application.

6.1   CIRCUIT MANIPULATIONS

The buck converter (Fig. 6.1) was developed in Chapter 1 using basic principles. The switch reduces the voltage dc component, and the low-pass filter removes the switching harmonics. In the continuous conduction mode, the buck converter has a conversion ratio of $M = D$. The buck converter is the simplest and most basic circuit, from which we will derive other converters.

6.1.1   Inversion of Source and Load

Let us consider first what happens when we interchange the power input and power output ports of a converter. In the buck converter of Fig. 6.2(a), voltage V_1 is applied at port 1, and voltage V_2 appears at port 2. We know that

$$V_2 = DV_1 \tag{6.1}$$

This equation can be derived using the principle of inductor volt-second balance, with the assumption that the converter operates in the continuous conduction mode. Provided that the switch is realized such that this assumption holds, then Eq. (6.1) is true regardless of the direction of power flow.

So let us interchange the power source and load, as in Fig. 6.2(b). The load, bypassed by the capacitor, is connected to converter port 1, while the power source is connected to converter port 2. Power now flows in the opposite direction through the converter. Equation (6.1) must still hold; by solving for the load voltage V_1, one obtains

$$V_1 = \frac{1}{D} V_2 \tag{6.2}$$

So the load voltage is greater than the source voltage. Figure 6.2(b) is a boost converter, drawn backwards. Equation 6.2 nearly coincides with the familiar boost converter result, $M(D) = 1/D'$, except that D' is replaced by D.

Since power flows in the opposite direction, the standard buck converter unidirectional switch

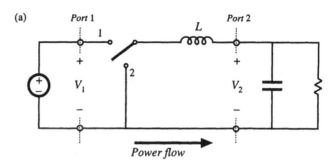

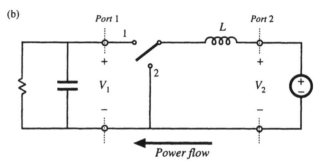

Fig. 6.2 Inversion of source and load transforms a buck converter into a boost converter: (a) buck converter, (b) inversion of source and load, (c) realization of switch.

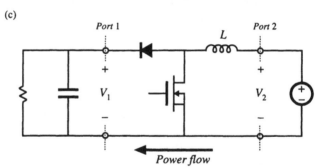

realization cannot be used with the circuit of Fig. 6.2(b). By following the discussion of Chapter 4, one finds that the switch can be realized by connecting a transistor between the inductor and ground, and a diode from the inductor to the load, as shown in Fig. 6.2(c). In consequence, the transistor duty cycle D becomes the fraction of time which the single-pole double-throw (SPDT) switch of Fig. 6.2(b) spends in position 2, rather than in position 1. So we should interchange D with its complement D' in Eq. (6.2), and the conversion ratio of the converter of Fig. 6.2(c) is

$$V_1 = \frac{1}{D'} V_2 \tag{6.3}$$

Thus, the boost converter can be viewed as a buck converter having the source and load connections exchanged, and in which the switch is realized in a manner that allows reversal of the direction of power flow.

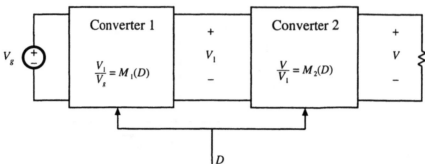

Fig. 6.3   Cascade connection of converters.

6.1.2   Cascade Connection of Converters

Converters can also be connected in cascade, as illustrated in Fig. 6.3 [1,2]. Converter 1 has conversion ratio $M_1(D)$, such that its output voltage V_1 is

$$V_1 = M_1(D)V_g \tag{6.4}$$

This voltage is applied to the input of the second converter. Let us assume that converter 2 is driven with the same duty cycle D applied to converter 1. If converter 2 has conversion ratio $M_2(D)$, then the output voltage V is

$$V = M_2(D)V_1 \tag{6.5}$$

Substitution of Eq. (6.4) into Eq. (6.5) yields

$$\frac{V}{V_g} = M(D) = M_1(D)M_2(D) \tag{6.6}$$

Hence, the conversion ratio $M(D)$ of the composite converter is the product of the individual conversion ratios $M_1(D)$ and $M_2(D)$.

Let us consider the case where converter 1 is a buck converter, and converter 2 is a boost converter. The resulting circuit is illustrated in Fig. 6.4. The buck converter has conversion ratio

$$\frac{V_1}{V_g} = D \tag{6.7}$$

The boost converter has conversion ratio

$$\frac{V}{V_1} = \frac{1}{1-D} \tag{6.8}$$

So the composite conversion ratio is

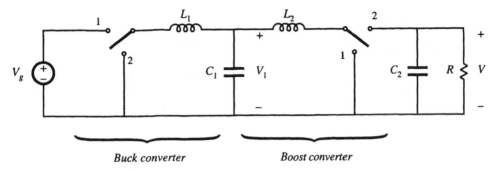

Fig. 6.4 Cascade connection of buck converter and boost converter.

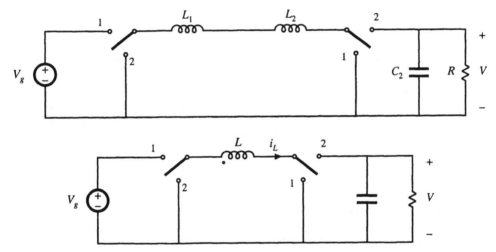

Fig. 6.5 Simplification of the cascaded buck and boost converter circuit of Fig. 6.4: (a) removal of capacitor C_1, (b) combining of inductors L_1 and L_2.

$$\frac{V}{V_g} = \frac{D}{1-D} \tag{6.9}$$

The composite converter has a noninverting buck–boost conversion ratio. The voltage is reduced when $D < 0.5$, and increased when $D > 0.5$.

The circuit of Fig. 6.4 can be simplified considerably. Note that inductors L_1 and L_2, along with capacitor C_1, form a three-pole low-pass filter. The conversion ratio does not depend on the number of poles present in the low-pass filter, and so the same steady-state output voltage should be obtained when a simpler low-pass filter is used. In Fig. 6.5(a), capacitor C_1 is removed. Inductors L_1 and L_2 are now in series, and can be combined into a single inductor as shown in Fig. 6.5(b). This converter, the noninverting buck–boost converter, continues to exhibit the conversion ratio given in Eq. (6.9).

The switches of the converter of Fig. 6.5(b) can also be simplified, leading to a negative output voltage. When the switches are in position 1, the converter reduces to Fig. 6.6(a). The inductor is connected to the input source V_g, and energy is transferred from the source to the inductor. When the

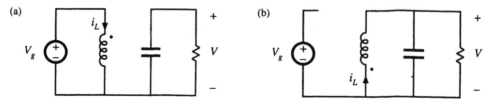

Fig. 6.6 Connections of the circuit of Fig. 6.5(b): (a) while the switches are in position 1, (b) while the switches are in position 2.

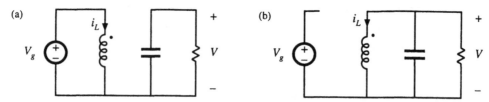

Fig. 6.7 Reversal of the output voltage polarity, by reversing the inductor connections while the switches are in position 2: (a) connections with the switches in position 1, (b) connections with the switches in position 2.

switches are in position 2, the converter reduces to Fig. 6.6(b). The inductor is then connected to the load, and energy is transferred from the inductor to the load. To obtain a negative output, we can simply reverse the polarity of the inductor during one of the subintervals (say, while the switches are in position 2). The individual circuits of Fig. 6.7 are then obtained, and the conversion ratio becomes

$$\frac{V}{V_g} = -\frac{D}{1-D} \tag{6.10}$$

Note that one side of the inductor is now always connected to ground, while the other side is switched between the input source and the load. Hence only one SPDT switch is needed, and the converter circuit of Fig. 6.8 is obtained. Figure 6.8 is recognized as the conventional buck–boost converter.

Thus, the buck–boost converter can be viewed as a cascade connection of buck and boost converters. The properties of the buck–boost converter are consistent with this viewpoint. Indeed, the equivalent circuit model of the buck-boost converter contains a $1{:}D$ (buck) dc transformer, followed by a $D'{:}1$ (boost) dc transformer. The buck–boost converter inherits the pulsating input current of the buck converter, and the pulsating output current of the boost converter.

Other converters can be derived by cascade connections. The Ćuk converter (Fig. 2.20) was originally derived [1,2] by cascading a boost converter (converter 1), followed by a buck (converter 2). A negative output voltage is obtained by reversing the polarity of the internal capacitor connection during one of the subintervals; as in the buck–boost converter, this operation has the additional benefit of reducing the number of switches. The equivalent circuit model of the Ćuk converter contains a $D'{:}1$ (boost)

Fig. 6.8 Converter circuit obtained from the subcircuits of Fig. 6.7.

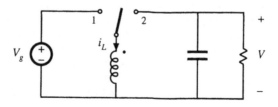

ideal dc transformer, followed by a 1:*D* (buck) ideal dc transformer. The Ćuk converter inherits the non-pulsating input current property of the boost converter, and the nonpulsating output current property of the buck converter.

6.1.3 Rotation of Three-Terminal Cell

The buck, boost, and buck–boost converters each contain an inductor that is connected to a SPDT switch. As illustrated in Fig. 6.9(a), the inductor-switch network can be viewed as a basic cell having the three terminals labeled *a*, *b*, and *c*. It was first pointed out in [1,2], and later in [3], that there are three distinct ways to connect this cell between the source and load. The connections *a–A b–B c–C* lead to the buck converter. The connections *a–C b–A c–B* amount to inversion of the source and load, and lead to the boost converter. The connections *a–A b–C c–B* lead to the buck-boost converter. So the buck, boost, and buck–boost converters could be viewed as being based on the same inductor-switch cell, with different source and load connections.

A dual three-terminal network, consisting of a capacitor-switch cell, is illustrated in Fig. 6.9(b). Filter inductors are connected in series with the source and load, such that the converter input and output currents are nonpulsating. There are again three possible ways to connect this cell between the source and load. The connections *a–A b–B c–C* lead to a buck converter with *L–C* input low-pass filter. The connections *a–B b–A c–C* coincide with inversion of source and load, and lead to a boost converter with an added output *L–C* filter section. The connections *a–A b–C c–B* lead to the Ćuk converter.

Rotation of more complicated three-terminal cells is explored in [4].

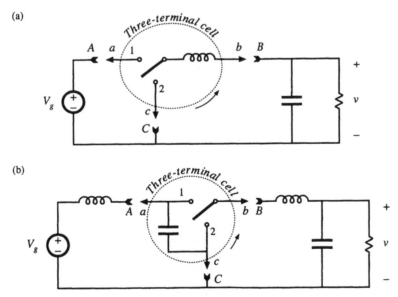

Fig. 6.9 Rotation of three-terminal switch cells: (a) switch/inductor cell, (b) switch/capacitor cell.

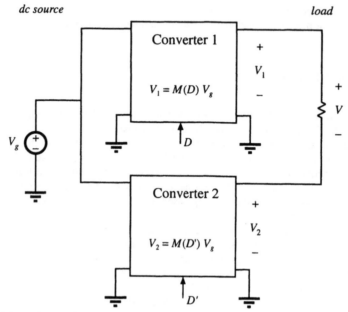

dc source   load

Fig. 6.10   Obtaining a bipolar output by differential connection of load.

6.1.4   Differential Connection of the Load

In inverter applications, where an ac output is required, a converter is needed that is capable of producing an output voltage of either polarity. By variation of the duty cycle in the correct manner, a sinusoidal output voltage having no dc bias can then be obtained. Of the converters studied so far in this chapter, the buck and the boost can produce only a positive unipolar output voltage, while the buck–boost and Ćuk converter produce only a negative unipolar output voltage. How can we derive converters that can produce bipolar output voltages?

A well-known technique for obtaining a bipolar output is the differential connection of the load across the outputs of two known converters, as illustrated in Fig. 6.10. If converter 1 produces voltage V_1, and converter 2 produces voltage V_2, then the load voltage V is given by

$$V = V_1 - V_2 \tag{6.11}$$

Although V_1 and V_2 may both individually be positive, the load voltage V can be either positive or negative. Typically, if converter 1 is driven with duty cycle D, then converter 2 is driven with its complement, D', so that when V_1 increases, V_2 decreases, and vice versa.

Several well-known inverter circuits can be derived using the differential connection. Let's realize converters 1 and 2 of Fig. 6.10 using buck converters. Figure 6.11(a) is obtained. Converter 1 is driven with duty cycle D, while converter 2 is driven with duty cycle D'. So when the SPDT switch of converter 1 is in the upper position, then the SPDT switch of converter 2 is in the lower position, and vice-versa. Converter 1 then produces output voltage $V_1 = DV_g$, while converter 2 produces output volt-

(a)

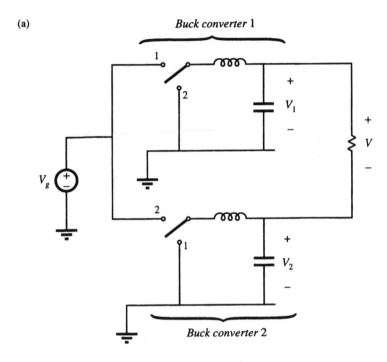

(b)

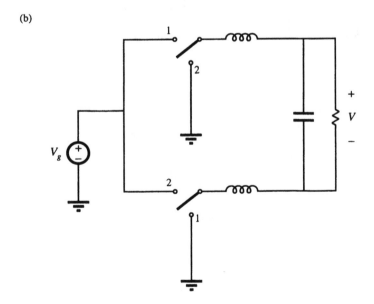

Fig. 6.11 Derivation of bridge inverter (H-bridge): (a) differential connection of load across outputs of buck converters, (b) bypassing load by capacitor, (c) combining series inductors, (d) circuit (c) redrawn in its usual form.

(c)

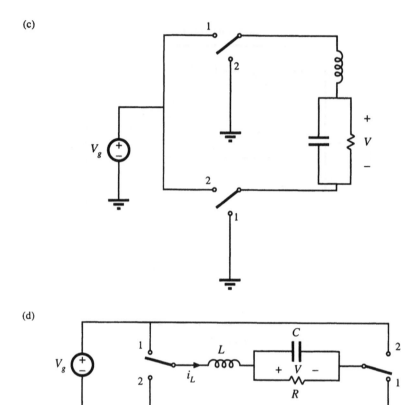

(d)

Fig. 6.11   Continued

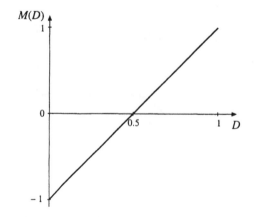

Fig. 6.12   Conversion ratio of the H-bridge inverter circuit.

age $V_2 = D'V_g$. The differential load voltage is

$$V = DV_g - D'V_g \tag{6.12}$$

Simplification leads to

$$V = (2D - 1)V_g \tag{6.13}$$

This equation is plotted in Fig. 6.12. It can be seen the output voltage is positive for $D > 0.5$, and negative for $D < 0.5$. If the duty cycle is varied sinusoidally about a quiescent operating point of 0.5, then the output voltage will be sinusoidal, with no dc bias.

The circuit of Fig. 6.11(a) can be simplified. It is usually desired to bypass the load directly with a capacitor, as in Fig. 6.11(b). The two inductors are now effectively in series, and can be combined into a single inductor as in Fig. 6.11(c). Figure 6.11(d) is identical to Fig. 6.11(c), but is redrawn for clarity. This circuit is commonly called the H-bridge, or bridge inverter circuit. Its use is widespread in servo amplifiers and single-phase inverters. Its properties are similar to those of the buck converter, from which it is derived.

Polyphase inverter circuits can be derived in a similar manner. A three-phase load can be connected differentially across the outputs of three dc–dc converters, as illustrated in Fig. 6.12. If the three-phase load is balanced, then the neutral voltage V_n will be equal to the average of the three converter output voltages:

$$V_n = \tfrac{1}{3}\left(V_1 + V_2 + V_3\right) \tag{6.14}$$

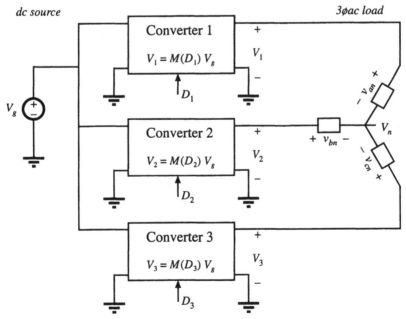

Fig. 6.12 Generation of dc–3ϕac inverter by differential connection of 3ϕ load.

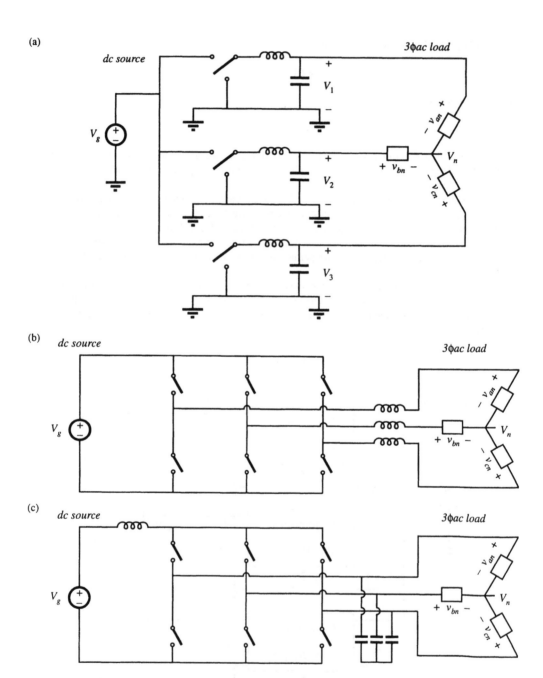

Fig. 6.13 Dc–3φac inverter topologies: (a) differential connection of 3φ load across outputs of buck converters; (b) simplification of low-pass filters to obtain the dc–3φac voltage-source inverter; (c) the dc–3φac current-source inverter.

If the converter output voltages V_1, V_2, and V_3 contain the same dc bias, then this dc bias will also appear at the neutral point V_n. The phase voltages V_{an}, V_{bn}, and V_{cn} are given by

$$
\begin{aligned}
V_{an} &= V_1 - V_n \\
V_{bn} &= V_2 - V_n \\
V_{cn} &= V_3 - V_n
\end{aligned}
\tag{6.15}
$$

It can be seen that the dc biases cancel out, and do not appear in V_{an}, V_{bn}, and V_{cn}.

Let us realize converters 1, 2, and 3 of Fig. 6.12 using buck converters. Figure 6.13(a) is then obtained. The circuit is re-drawn in Fig. 6.13(b) for clarity. This converter is known by several names, including the *voltage-source inverter* and the buck-derived three-phase bridge.

Inverter circuits based on dc–dc converters other than the buck converter can be derived in a similar manner. Figure 6.13(c) contains a three-phase current-fed bridge converter having a boost-type voltage conversion ratio, also known as the *current-source inverter*. Since most inverter applications require the capability to reduce the voltage magnitude, a dc–dc buck converter is usually cascaded at the dc input port of this inverter. Several other examples of three-phase inverters are given in [5–7], in which the converters are capable of both increasing and decreasing the voltage magnitude.

6.2   A SHORT LIST OF CONVERTERS

An infinite number of converters are possible, and hence it is not feasible to list them all. A short list is given here.

Let's consider first the class of single-input single-output converters, containing a single inductor. There are a limited number of ways in which the inductor can be connected between the source and load. If we assume that the switching period is divided into two subintervals, then the inductor should be connected to the source and load in one manner during the first subinterval, and in a different manner during the second subinterval. One can examine all of the possible combinations, to derive the complete set of converters in this class [8–10]. By elimination of redundant and degenerate circuits, one finds that there are eight converters, listed in Fig. 6.14. How the converters are counted can actually be a matter of semantics and personal preference; for example, many people in the field would not consider the noninverting buck–boost converter as distinct from the inverting buck–boost. Nonetheless, it can be said that a converter is defined by the connections between its reactive elements, switches, source, and load; by how the switches are realized; and by the numerical range of reactive element values.

The first four converters of Fig. 6.14, the buck, boost, buck-boost, and the noninverting buck-boost, have been previously discussed. These converters produce a unipolar dc output voltage. With these converters, it is possible to increase, decrease, and/or invert a dc voltage.

Converters 5 and 6 are capable of producing a bipolar output voltage. Converter 5, the H–bridge, has previously been discussed. Converter 6 is a nonisolated version of a push–pull current-fed converter [11–15]. This converter can also produce a bipolar output voltage; however, its conversion ratio $M(D)$ is a nonlinear function of duty cycle. The number of switch elements can be reduced by using a two-winding inductor as shown. The function of the inductor is similar to that of the flyback converter, discussed in the next section. When switch 1 is closed the upper winding is used, while when switch 2 is closed, current flows through the lower winding. The current flows through only one winding at any given instant, and the total ampere-turns of the two windings are a continuous function of time. Advantages of this converter are its ground-referenced load and its ability to produce a bipolar output voltage using only two SPST current-bidirectional switches. The isolated version and its variants have found

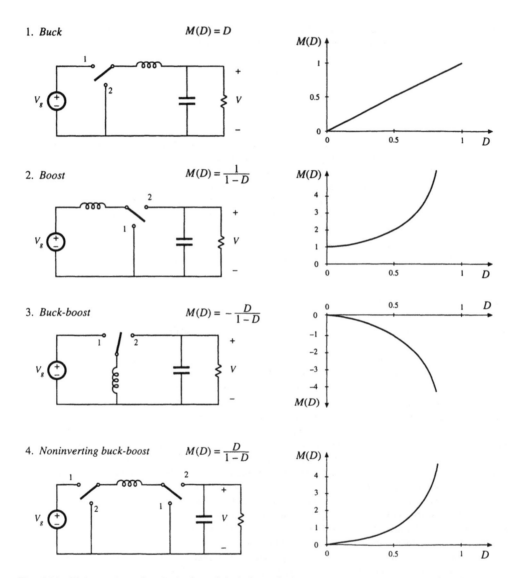

Fig. 6.14 Eight members of the basic class of single-input single-output converters containing a single inductor.

application in high-voltage dc power supplies.

Converters 7 and 8 can be derived as the inverses of converters 5 and 6. These converters are capable of interfacing an ac input to a dc output. The ac input current waveform can have arbitrary waveshape and power factor.

The class of single-input single-output converters containing two inductors is much larger. Several of its members are listed in Fig. 6.15. The Ćuk converter has been previously discussed and ana-

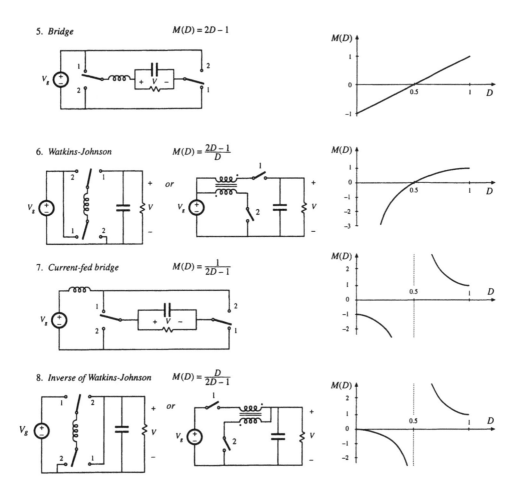

Fig. 6.14 Continued

lyzed. It has an inverting buck-boost characteristic, and exhibits nonpulsating input and output terminal currents. The SEPIC (single-ended primary inductance converter) [16], and its inverse, have noninverting buck–boost characteristics. The Ćuk and SEPIC also exhibit the desirable feature that the MOSFET source terminal is connected to ground; this simplifies the construction of the gate drive circuitry. Two-inductor converters having conversion ratios $M(D)$ that are biquadratic functions of the duty cycle D are also numerous. An example is converter 4 of Fig. 6.15 [17]. This converter can be realized using a single transistor and three diodes. Its conversion ratio is $M(D) = D^2$. This converter may find use in nonisolated applications that require a large step-down of the dc voltage, or in applications having wide variations in operating point.

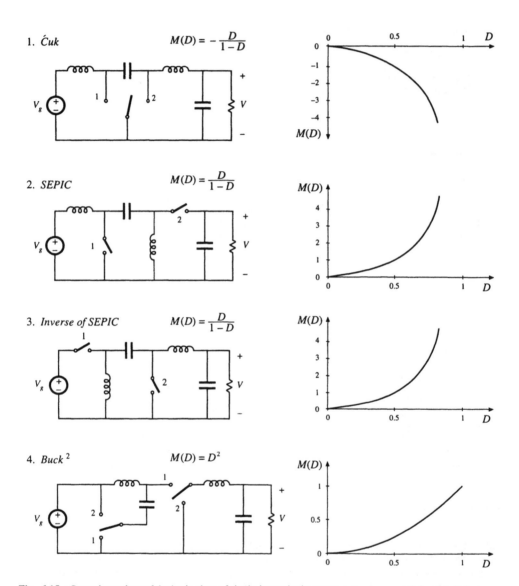

Fig. 6.15 Several members of the basic class of single-input single-output converters containing two inductors.

6.3   TRANSFORMER ISOLATION

In a large number of applications, it is desired to incorporate a transformer into a switching converter, to obtain dc isolation between the converter input and output. For example, in off-line applications (where the converter input is connected to the ac utility system), isolation is usually required by regulatory agen-

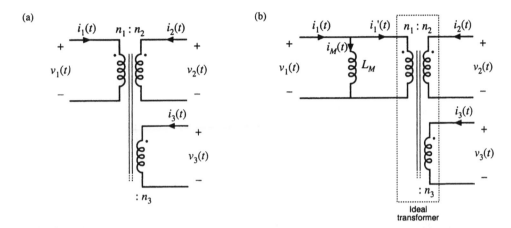

Fig. 6.16 Simplified model of a multiple-winding transformer: (a) schematic symbol, (b) equivalent circuit containing a magnetizing inductance and ideal transformer.

cies. Isolation could be obtained in these cases by simply connecting a 50 Hz or 60 Hz transformer at the converter ac input. However, since transformer size and weight vary inversely with frequency, significant improvements can be made by incorporating the transformer into the converter, so that the transformer operates at the converter switching frequency of tens or hundreds of kilohertz.

When a large step-up or step-down conversion ratio is required, the use of a transformer can allow better converter optimization. By proper choice of the transformer turns ratio, the voltage or current stresses imposed on the transistors and diodes can be minimized, leading to improved efficiency and lower cost.

Multiple dc outputs can also be obtained in an inexpensive manner, by adding multiple secondary windings and converter secondary-side circuits. The secondary turns ratios are chosen to obtain the desired output voltages. Usually only one output voltage can be regulated via control of the converter duty cycle, so wider tolerances must be allowed for the auxiliary output voltages. *Cross regulation* is a measure of the variation in an auxiliary output voltage, given that the main output voltage is perfectly regulated [18–20].

A physical multiple-winding transformer having turns ratio $n_1:n_2:n_3:...$ is illustrated in Fig. 6.16(a). A simple equivalent circuit is illustrated in Fig. 6.16(b), which is sufficient for understanding the operation of most transformer-isolated converters. The model assumes perfect coupling between windings and neglects losses; more accurate models are discussed in a later chapter. The ideal transformer obeys the relationships

$$\frac{v_1(t)}{n_1} = \frac{v_2(t)}{n_2} = \frac{v_3(t)}{n_3} = ...$$
$$0 = n_1 i_1'(t) + n_2 i_2(t) + n_3 i_3(t) + ... \tag{6.16}$$

In parallel with the ideal transformer is an inductance L_M, called the *magnetizing inductance*, referred to the transformer primary in the figure.

Physical transformers must contain a magnetizing inductance. For example, suppose we disconnect all windings except for the primary winding. We are then left with a single winding on a magnetic core—an inductor. Indeed, the equivalent circuit of Fig. 6.16(b) predicts this behavior, via the magnetizing inductance.

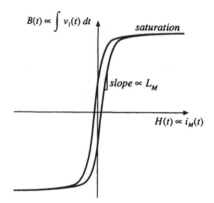

Fig. 6.17 *B–H* characteristics of transformer core.

The magnetizing current $i_M(t)$ is proportional to the magnetic field $H(t)$ inside the transformer core. The physical *B–H* characteristics of the transformer core material, illustrated in Fig. 6.17, govern the magnetizing current behavior. For example, if the magnetizing current $i_M(t)$ becomes too large, then the magnitude of the magnetic field $H(t)$ causes the core to saturate. The magnetizing inductance then becomes very small in value, effectively shorting out the transformer.

The presence of the magnetizing inductance explains why transformers do not work in dc circuits: at dc, the magnetizing inductance has zero impedance, and shorts out the windings. In a well-designed transformer, the impedance of the magnetizing inductance is large in magnitude over the intended range of operating frequencies, such that the magnetizing current $i_M(t)$ has much smaller magnitude than $i_1(t)$. Then $i_1'(t) \approx i_1(t)$, and the transformer behaves nearly as an ideal transformer. It should be emphasized that the magnetizing current $i_M(t)$ and the primary winding current $i_1(t)$ are independent quantities.

The magnetizing inductance must obey all of the usual rules for inductors. In the model of Fig. 6.16(b), the primary winding voltage $v_1(t)$ is applied across L_M, and hence

$$v_1(t) = L_M \frac{di_M(t)}{dt} \tag{6.17}$$

Integration leads to

$$i_M(t) - i_M(0) = \frac{1}{L_M} \int_0^t v_1(\tau) d\tau \tag{6.18}$$

So the magnetizing current is determined by the integral of the applied winding voltage. The principle of inductor volt-second balance also applies: when the converter operates in steady-state, the dc component of voltage applied to the magnetizing inductance must be zero:

$$0 = \frac{1}{T_s} \int_0^{T_s} v_1(t) dt \tag{6.19}$$

Since the magnetizing current is proportional to the integral of the applied winding voltage, it is important that the dc component of this voltage be zero. Otherwise, during each switching period there will be a net increase in magnetizing current, eventually leading to excessively large currents and transformer saturation.

The operation of converters containing transformers may be understood by inserting the model of Fig. 6.16(b) in place of the transformer in the converter circuit. Analysis then proceeds as described in the previous chapters, treating the magnetizing inductance as any other inductor of the converter.

Practical transformers must also contain leakage inductance. A small part of the flux linking a winding may not link the other windings. In the two-winding transformer, this phenomenon may be modeled with small inductors in series with the windings. In most isolated converters, leakage inductance is a nonideality that leads to switching loss, increased peak transistor voltage, and that degrades cross-regulation, but otherwise has no influence on basic converter operation.

There are several ways of incorporating transformer isolation into a dc–dc converter. The full-bridge, half-bridge, forward, and push-pull converters are commonly used isolated versions of the buck converter. Similar isolated variants of the boost converter are known. The flyback converter is an isolated version of the buck–boost converter. These isolated converters, as well as isolated versions of the SEPIC and the Ćuk converter, are discussed in this section.

6.3.1 Full-Bridge and Half-Bridge Isolated Buck Converters

The full-bridge transformer-isolated buck converter is sketched in Fig. 6.18(a). A version containing a center-tapped secondary winding is shown; this circuit is commonly used in converters producing low output voltages. The two halves of the center-tapped secondary winding may be viewed as separate windings, and hence we can treat this circuit element as a three-winding transformer having turns ratio $1{:}n{:}n$. When the transformer is replaced by the equivalent circuit model of Fig. 6.16(b), the circuit of Fig.

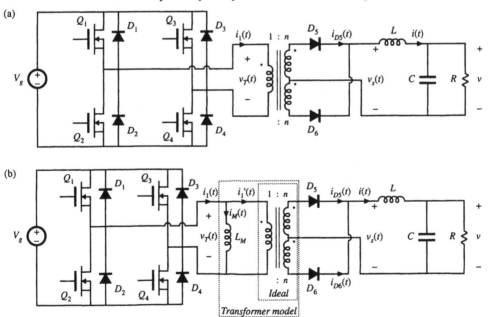

Fig. 6.18 Full-bridge transformer-isolated buck converter: (a) schematic diagram, (b) replacement of transformer with equivalent circuit model.

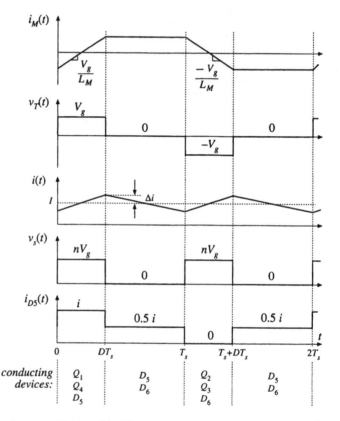

Fig. 6.19 Waveforms of the full-bridge transformer-isolated buck converter.

6.18(b) is obtained. Typical waveforms are illustrated in Fig. 6.19. The output portion of the converter is similar to the nonisolated buck converter—compare the $v_s(t)$ and $i(t)$ waveforms of Fig. 6.19 with Figs. 2.1(b) and 2.10.

During the first subinterval $0 < t < DT_s$, transistors Q_1 and Q_4 conduct, and the transformer primary voltage is $v_T = V_g$. This positive voltage causes the magnetizing current $i_M(t)$ to increase with a slope of V_g/L_M. The voltage appearing across each half of the center-tapped secondary winding is nV_g, with the polarity mark at positive potential. Diode D_5 is therefore forward-biased, and D_6 is reverse-biased. The voltage $v_s(t)$ is then equal to nV_g, and the output filter inductor current $i(t)$ flows through diode D_5.

Several transistor control schemes are possible for the second subinterval $DT_s < t < T_s$. In the most common scheme, all four transistors are switched off, and hence the transformer voltage is $v_T = 0$. Alternatively, transistors Q_2 and Q_4 could conduct, or transistors Q_1 and Q_3 could conduct. In any event, diodes D_5 and D_6 are both forward-biased during this subinterval; each diode conducts approximately one-half of the output filter inductor current.

Actually, the diode currents i_{D5} and i_{D6} during the second subinterval are functions of both the output inductor current and the transformer magnetizing current. In the ideal case (no magnetizing current), the transformer causes $i_{D5}(t)$ and $i_{D6}(t)$ to be equal in magnitude since, if $i_1'(t) = 0$, then $ni_{D5}(t) = ni_{D6}(t)$. But the sum of the two diode currents is equal to the output inductor current:

$$i_{D5}(t) + i_{D6}(t) = i(t) \tag{6.20}$$

Therefore, it must be true that $i_{D5} = i_{D6} = 0.5i$ during the second subinterval. In practice, the diode currents differ slightly from this result, because of the nonzero magnetizing current.

The ideal transformer currents in Fig. 6.18(b) obey

$$i_1'(t) - n i_{D5}(t) + n i_{D6}(t) = 0 \tag{6.21}$$

The node equation at the primary of the ideal transformer is

$$i_1(t) = i_M(t) + i_1'(t) \tag{6.22}$$

Elimination of $i_1'(t)$ from Eqs. (6.21) and (6.22) leads to

$$i_1(t) - n i_{D5}(t) + n i_{D6}(t) = i_M(t) \tag{6.23}$$

Equations (6.23) and (6.20) describe, in the general case, the transformer winding currents during the second subinterval. According to Eq. (6.23), the magnetizing current $i_M(t)$ may flow through the primary winding, through one of the secondary windings, or it may divide between all three of these windings. How the division occurs depends on the $i–v$ characteristics of the conducting transistors and diodes, and on the transformer leakage inductances. In the case where $i_1 = 0$, the solution to Eqs. (6.20) and (6.23) is

$$\begin{aligned} i_{D5}(t) &= \tfrac{1}{2} i(t) - \frac{1}{2n} i_M(t) \\ i_{D6}(t) &= \tfrac{1}{2} i(t) + \frac{1}{2n} i_M(t) \end{aligned} \tag{6.24}$$

Provided that $i_M \ll ni$, then i_{D5} and i_{D6} are each approximately $0.5i$.

The next switching period, $T_s < t < 2T_s$, proceeds in a similar manner, except that the transformer is excited with voltage of the opposite polarity. During $T_s < t < (T_s + DT_s)$, transistors Q_2 and Q_3 and diode D_6 conduct. The applied transformer primary voltage is $v_T = -V_g$, which causes the magnetizing current to decrease with slope $-V_g/L_M$. The voltage $v_s(t)$ is equal to nV_g, and the output inductor current $i(t)$ flows through diode D_6. Diodes D_5 and D_6 again both conduct during $(T_s + DT_s) < t < 2T_s$, with operation similar to subinterval 2 described previously. It can be seen that the switching ripple in the output filter elements has frequency $f_s = 1/T_s$. However, the transformer waveforms have frequency $0.5f_s$.

By application of the principle of inductor volt-second balance to the magnetizing inductance, the average value of the transformer voltage $v_T(t)$ must be zero when the converter operates in steady state. During the first switching period, positive volt-seconds are applied to the transformer, approximately equal to

$$\left[V_g - \left(Q_1 \text{ and } Q_4 \text{ forward voltage drops} \right) \right] \left(Q_1 \text{ and } Q_4 \text{ conduction time} \right) \tag{6.25}$$

During the next switching period, negative volt-seconds are applied to the transformer, given by

$$-\left[V_g - \left(Q_2 \text{ and } Q_3 \text{ forward voltage drops} \right) \right] \left(Q_2 \text{ and } Q_3 \text{ conduction time} \right) \tag{6.26}$$

The net volt-seconds, that is, the sum of Eqs. (6.25) and (6.26), should equal zero. While the full bridge scheme causes this to be approximately true, in practice there exist imbalances such as small differences in the transistor forward voltage drops or in the transistor switching times, so that $\langle v_T \rangle$ is small but non-zero. In consequence, during every two switching periods there is a net increase in the magnitude of the magnetizing current. This increase can cause the transistor forward voltage drops to change such that small imbalances are compensated. However, if the imbalances are too large, then the magnetizing current becomes large enough to saturate the transformer.

Transformer saturation under steady-state conditions can be avoided by placing a capacitor in series with the transformer primary. Imbalances then induce a dc voltage component across the capacitor, rather than across the transformer primary. Another solution is the use of current-programmed control, discussed in a later chapter. The series capacitor is omitted when current-programmed control is used.

By application of the principle of volt-second balance to the output filter inductor L, the dc load voltage must be equal to the dc component of $v_s(t)$:

$$V = \langle v_s \rangle \tag{6.27}$$

By inspection of the $v_s(t)$ waveform in Fig. 6.19, $\langle v_s \rangle = nDV_g$. Hence,

$$V = nDV_g \tag{6.28}$$

So as in the buck converter, the output voltage can be controlled by variation of the transistor duty cycle D. An additional increase or decrease of the voltage can be obtained via the physical transformer turns ratio n. Equation (6.28) is valid for operation in the continuous conduction mode; as in the nonisolated buck converter, the full-bridge and half-bridge converters can operate in discontinuous conduction mode at light load. The converter can operate over essentially the entire range of duty cycles $0 \le D < 1$.

Transistors Q_1 and Q_2 must not conduct simultaneously; doing so would short out the dc source V_g, causing a *shoot-through* current spike. This transistor *cross-conduction* condition can lead to low efficiency and transistor failure. Cross conduction can be prevented by introduction of delay between the turn-off of one transistor and the turn-on of the next transistor. Diodes D_1 to D_4 ensure that the peak transistor voltage is limited to the dc input voltage V_g, and also provide a conduction path for the transformer magnetizing current at light load. Details of the switching transitions of the full-bridge circuit are discussed further in a later chapter, in conjunction with zero-voltage switching phenomena.

The full-bridge configuration is typically used in switching power supplies at power levels of approximately 750 W and greater. It is usually not used at lower power levels because of its high parts count—four transistors and their associated drive circuits are required. The utilization of the transformer is good, leading to small transformer size. In particular, the utilization of the transformer core is very good, since the transformer magnetizing current can be both positive and negative. Hence, the entire core B–H loop can be used. However, in practice, the flux swing is usually limited by core loss. The transformer primary winding is effectively utilized. But the center-tapped secondary winding is not, since each half of the center-tapped winding transmits power only during alternate switching periods. Also, the secondary winding currents during subinterval 2 lead to winding power loss, but not to transmittal of energy to the load. Design of the transformer of the full-bridge configuration is discussed in detail in a later chapter.

The half-bridge transformer-isolated buck converter is illustrated in Fig. 6.20. Typical waveforms are illustrated in Fig. 6.21. This circuit is similar to the full-bridge of Fig. 6.18(a), except transistors Q_3 and Q_4, and their antiparallel diodes, have been replaced with large-value capacitors C_a and C_b. By volt-second balance of the transformer magnetizing inductance, the dc voltage across capacitor C_b is equal to the dc component of the voltage across transistor Q_2, or $0.5V_g$. The transformer primary voltage

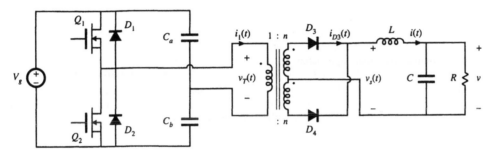

Fig. 6.20   Half-bridge transformer-isolated buck converter.

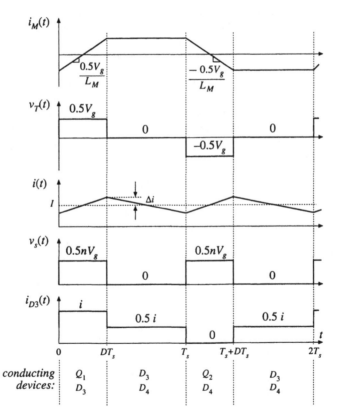

Fig. 6.21 Waveforms of the half-bridge transformer-isolated buck converter.

$v_T(t)$ is then $0.5V_g$ when transistor Q_1 conducts, and $-0.5V_g$ when transistor Q_2 conducts. The magnitude of $v_T(t)$ is half as large as in the full-bridge configuration, with the result that the output voltage is reduced by a factor of 0.5:

$$V = 0.5nDV_g \tag{6.29}$$

The factor of 0.5 can be compensated for by doubling the transformer turns ratio n. However, this causes the transistor currents to double.

So the half-bridge configuration needs only two transistors rather than four, but these two transistors must handle currents that are twice as large as those of the full-bridge circuit. In consequence, the half-bridge configuration finds application at lower power levels, for which transistors with sufficient current rating are readily available, and where low parts count is important. Utilization of the transformer core and windings is essentially the same as in the full-bridge, and the peak transistor voltage is clamped to the dc input voltage V_g by diodes D_1 and D_2. It is possible to omit capacitor C_a if desired. The current-programmed mode generally does not work with half-bridge converters.

6.3.2   Forward Converter

The forward converter is illustrated in Fig. 6.22. This transformer-isolated converter is based on the buck converter. It requires a single transistor, and hence finds application at power levels lower than those commonly encountered in the full-bridge and half-bridge configurations. Its nonpulsating output current, shared with other buck-derived converters, makes the forward converter well suited for applications involving high output currents. The maximum transistor duty cycle is limited in value; for the common choice $n_1 = n_2$, the duty cycle is limited to the range $0 \leq D < 0.5$.

The transformer magnetizing current is reset to zero while the transistor is in the off-state. How this occurs can be understood by replacing the three-winding transformer in Fig. 6.22 with the equivalent circuit of Fig. 6.16(b). The resulting circuit is illustrated in Fig. 6.23, and typical waveforms are given in Fig. 6.24. The magnetizing inductance L_M, in conjunction with diode D_1, must operate in the discontinuous conduction mode. The output inductor L, in conjunction with diode D_3, may operate in either continuous or discontinuous conduction mode. The waveforms of Fig. 6.24 are sketched for continuous mode operation of inductor L. During each switching period, three subintervals then occur as illustrated in Fig. 6.25.

During subinterval 1, transistor Q_1 conducts and the circuit of Fig. 6.25(a) is obtained. Diode D_2 becomes forward-biased, while diodes D_1 and D_3 are reverse-biased. Voltage V_g is applied to the transformer primary winding, and hence the transformer magnetizing current $i_M(t)$ increases with a slope of V_g/L_M as illustrated in Fig. 6.24. The voltage across diode D_3 is equal to V_g, multiplied by the turns ratio n_3/n_1.

The second subinterval begins when transistor Q_1 is switched off. The circuit of Fig. 6.25(b) is

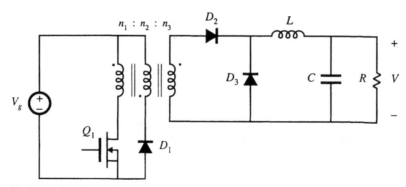

Fig. 6.22   Single-transistor forward converter.

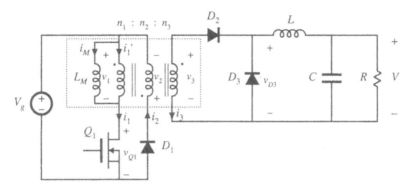

Fig. 6.23 Forward converter, with transformer equivalent circuit model.

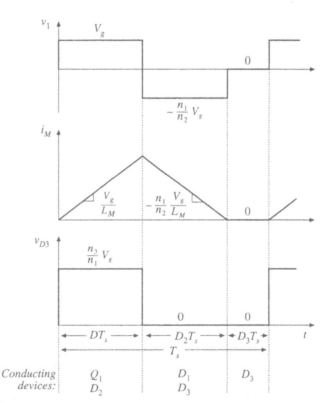

Fig. 6.24 Waveforms of the forward converter.

then obtained. The transformer magnetizing current $i_M(t)$ at this instant is positive, and must continue to flow. Since transistor Q_1 is off, the equivalent circuit model predicts that the magnetizing current must flow into the primary of the ideal transformer. It can be seen that $n_1 i_M$ ampere-turns flow out of the polarity mark of the primary winding. Hence, according to Eq. (6.16), an equal number of total ampere-turns must flow into the polarity marks of the other windings. Diode D_2 prevents current from flowing into the

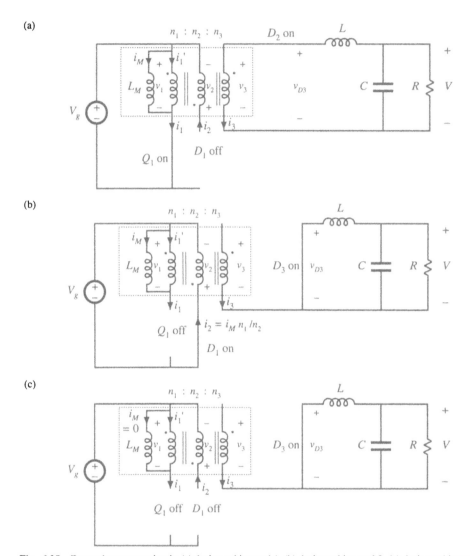

Fig. 6.25 Forward converter circuit: (a) during subinterval 1, (b) during subinterval 2, (c) during subinterval 3.

polarity mark of winding 3. Hence, the current $i_M n_1/n_2$ must flow into the polarity mark of winding 2. So diode D_1 becomes forward-biased, while diode D_2 is reverse-biased. Voltage V_g is applied to winding 2, and hence the voltage across the magnetizing inductance is $-V_g n_1/n_2$, referred to winding 1. This negative voltage causes the magnetizing current to decrease, with a slope of $-V_g n_1/n_2 L_M$. Since diode D_2 is reverse-biased, diode D_3 must turn on to conduct the output inductor current $i(t)$.

When the magnetizing current reaches zero, diode D_1 becomes reverse-biased. Subinterval 3 then begins, and the circuit of Fig. 6.25(c) is obtained. Elements Q_1, D_1, and D_2 operate in the off state, and the magnetizing current remains at zero for the balance of the switching period.

By application of the principle of inductor volt-second balance to the transformer magnetizing

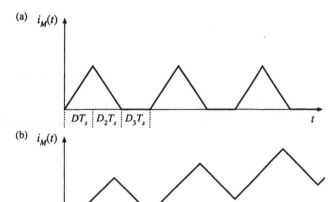

Fig. 6.26 Magnetizing current waveform, forward converter: (a) DCM, $D < 0.5$; (b) CCM, $D > 0.5$.

inductance, the primary winding voltage $v_1(t)$ must have zero average. Referring to Fig. 6.24, the average of $v_1(t)$ is given by

$$\langle v_1 \rangle = D(V_g) + D_2(-V_g n_1/n_2) + D_3(0) = 0 \tag{6.30}$$

Solution for the duty cycle D_2 yields

$$D_2 = \frac{n_2}{n_1} D \tag{6.31}$$

Note that the duty cycle D_3 cannot be negative. But since $D + D_2 + D_3 = 1$, we can write

$$D_3 = 1 - D - D_2 \geq 0 \tag{6.32}$$

Substitution of Eq. (6.31) into Eq. (6.32) leads to

$$D_3 = 1 - D\left(1 + \frac{n_2}{n_1}\right) \geq 0 \tag{6.33}$$

Solution for D then yields

$$D \leq \frac{1}{1 + \frac{n_2}{n_1}} \tag{6.34}$$

So the maximum duty cycle is limited. For the common choice $n_1 = n_2$, the limit becomes

$$D \leq \frac{1}{2} \tag{6.35}$$

If this limit is violated, then the transistor off-time is insufficient to reset the transformer magnetizing current to zero before the end of the switching period. Transformer saturation may then occur.

The transformer magnetizing current waveform $i_M(t)$ is illustrated in Fig. 6.26, for the typical case where $n_1 = n_2$. Figure 6.26(a) illustrates operation with $D < 0.5$. The magnetizing inductance, in conjunction with diode D_1, operates in the discontinuous conduction mode, and $i_M(t)$ is reset to zero before the end of each switching period. Figure 6.26(b) illustrates what happens when the transistor duty cycle D is greater than 0.5. There is then no third subinterval, and the magnetizing inductance operates in continuous conduction mode. Furthermore, subinterval 2 is not long enough to reset the magnetizing current to zero. Hence, there is a net increase of $i_M(t)$ over each switching period. Eventually, the magnetizing current will become large enough the saturate the transformer.

The converter output voltage can be found by application of the principle of inductor volt-second balance to inductor L. The voltage across inductor L must have zero dc component, and therefore the dc output voltage V is equal to the dc component of diode D_3 voltage $v_{D3}(t)$. The waveform $v_{D3}(t)$ is illustrated in Fig. 6.24. It has an average value of

$$\langle v_{D3} \rangle = V = \frac{n_3}{n_1} D V_g \tag{6.36}$$

This is the solution of the forward converter in the continuous conduction mode. The solution is subject to the constraint given in Eq. (6.34).

It can be seen from Eq. (6.34) that the maximum duty cycle could be increased by decreasing the turns ratio n_2/n_1. This would cause $i_M(t)$ to decrease more quickly during subinterval 2, resetting the transformer faster. Unfortunately, this also increases the voltage stress applied to transistor Q_1. The maximum voltage applied to transistor Q_1 occurs during subinterval 2; solution of the circuit of Fig. 6.25(b) for this voltage yields

$$\max\left(v_{Q1}\right) = V_g\left(1 + \frac{n_1}{n_2}\right) \tag{6.37}$$

For the common choice $n_1 = n_2$, the voltage applied to the transistor during subinterval 2 is $2V_g$. In practice, a somewhat higher voltage is observed, due to ringing associated with the transformer leakage inductance. So decreasing the turns ratio n_2/n_1 allows increase of the maximum transistor duty cycle, at the expense of increased transistor blocking voltage.

A two-transistor version of the forward converter is illustrated in Fig. 6.27. Transistors Q_1 and Q_2 are controlled by the same gate drive signal, such that they both conduct during subinterval 1, and are off during subintervals 2 and 3. The secondary side of the converter is identical to the single-transistor forward converter; diode D_3 conducts during subinterval 1, while diode D_4 conducts during subintervals 2 and 3. During subinterval 2, the magnetizing current $i_M(t)$ forward-biases diodes D_1 and D_2. The trans-

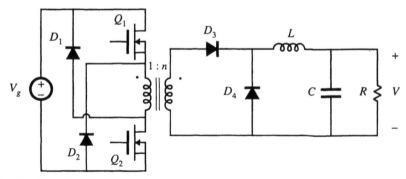

Fig. 6.27 Two-transistor forward converter.

former primary winding is then connected to V_g, with polarity opposite that of subinterval 1. The magnetizing current then decreases, with slope $- V_g/L_M$. When the magnetizing current reaches zero, diodes D_1 and D_2 become reverse-biased. The magnetizing current then remains at zero for the balance of the switching period. So operation of the two-transistor forward converter is similar to the single-transistor forward converter, in which $n_1 = n_2$. The duty cycle is limited to $D < 0.5$. This converter has the advantage that the transistor peak blocking voltage is limited to V_g, and is clamped by diodes D_1 and D_2. Typical power levels of the two-transistor forward converter are similar to those of the half-bridge configuration.

The utilization of the transformer of the forward converter is quite good. Since the transformer magnetizing current cannot be negative, only half of the core *B–H* loop can be used. This would seemingly imply that the transformer cores of forward converters should be twice as large as those of full- or half-bridge converters. However, in modern high-frequency converters, the flux swing is constrained by core loss rather than by the core material saturation flux density. In consequence, the utilization of the transformer core of the forward converter can be as good as in the full- or half-bridge configurations. Utilization of the primary and secondary windings of the transformer is better than in the full-bridge, half-bridge, or push-pull configurations, since the forward converter requires no center-tapped windings. During subinterval 1, all of the available winding copper is used to transmit power to the load. Essentially no unnecessary current flows during subintervals 2 and 3. Typically, the magnetizing current is small compared to the reflected load current, and has negligible effect on the transformer utilization. So the transformer core and windings are effectively utilized in modern forward converters.

6.3.3 Push-Pull Isolated Buck Converter

The push-pull isolated buck converter is illustrated in Fig. 6.28. The secondary-side circuit is identical with the full- and half-bridge converters, with identical waveforms. The primary-side circuit contains a center-tapped winding. Transistor Q_1 conducts for time DT_s during the first switching period. Transistor Q_2 conducts for an identical length of time during the next switching period, such that volt-second balance is maintained across the transformer primary winding. Converter waveforms are illustrated in Fig. 6.29. This converter can operate oven the entire range of duty cycles $0 \le D < 1$. Its conversion ratio is given by

$$V = nDV_g \tag{6.38}$$

This converter is sometimes used in conjunction with low input voltages. It tends to exhibit low primary-

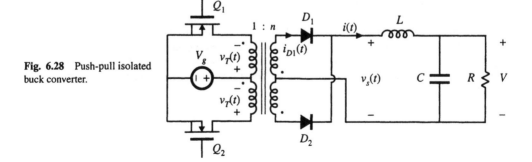

Fig. 6.28 Push-pull isolated buck converter.

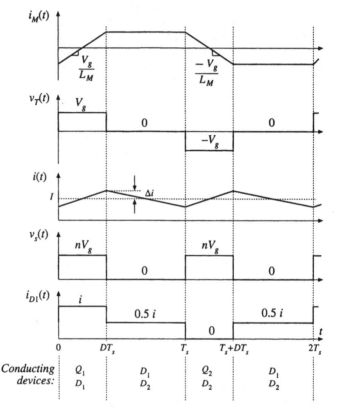

Fig. 6.29 Waveforms of the push-pull isolated buck converter.

side conduction losses, since at any given instant only one transistor is connected in series with the dc source V_g. The ability to operate with transistor duty cycles approaching unity also allows the turns ratio n to be minimized, reducing the transistor currents.

The push-pull configuration is prone to transformer saturation problems. Since it cannot be guaranteed that the forward voltage drops and conduction times of transistors Q_1 and Q_2 are exactly equal, small imbalances can cause the dc component of voltage applied to the transformer primary to be nonzero. In consequence, during every two switching periods there is a net increase in the magnitude of the magnetizing current. If this imbalance continues, then the magnetizing current can eventually become large enough to saturate the transformer.

Current-programmed control can be employed to mitigate the transformer saturation problems. Operation of the push-pull converter using only duty cycle control is not recommended.

Utilization of the transformer core material and secondary winding is similar to that for the full-bridge converter. The flux and magnetizing current can be both positive and negative, and therefore the entire *B–H* loop can be used, if desired. Since the primary and secondary windings are both center-tapped, their utilization is suboptimal.

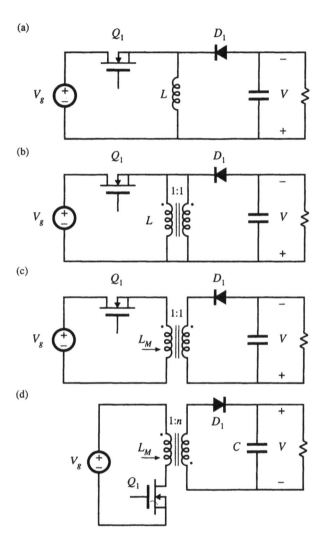

Fig. 6.30 Derivation of the fly-back converter: (a) buck–boost converter; (b) inductor *L* is wound with two parallel wires; (c) inductor windings are isolated, leading to the flyback converter; (d) with a 1:*n* turns ratio and positive output.

6.3.4 Flyback Converter

The flyback convérter is based on the buck–boost converter. Its derivation is illustrated in Fig. 6.30. Figure 6.30(a) depicts the basic buck-boost converter, with the switch realized using a MOSFET and diode. In Fig. 6.30(b), the inductor winding is constructed using two wires, with a 1:1 turns ratio. The basic function of the inductor is unchanged, and the parallel windings are equivalent to a single winding constructed of larger wire. In Fig. 6.30(c), the connections between the two windings are broken. One winding is used while the transistor Q_1 conducts, while the other winding is used when diode D_1 conducts. The total current in the two windings is unchanged from the circuit of Fig. 6.30(b); however, the current is now distributed between the windings differently. The magnetic fields inside the inductor in both cases

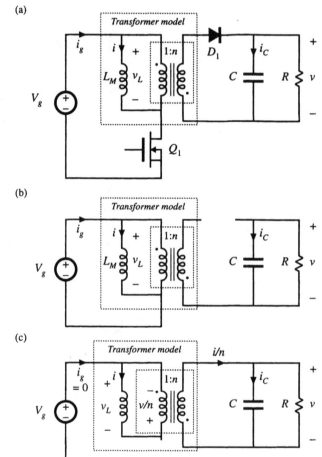

Fig. 6.31 Flyback converter circuit:
(a) with transformer equivalent circuit
model, (b) during subinterval 1,
(c) during subinterval 2.

are identical. Although the two-winding magnetic device is represented using the same symbol as the transformer, a more descriptive name is "two-winding inductor." This device is sometimes also called a *flyback transformer*. Unlike the ideal transformer, current does not flow simultaneously in both windings of the flyback transformer. Figure 6.30(d) illustrates the usual configuration of the flyback converter. The MOSFET source is connected to the primary-side ground, simplifying the gate drive circuit. The transformer polarity marks are reversed, to obtain a positive output voltage. A $1:n$ turns ratio is introduced; this allows better converter optimization.

The flyback converter may be analyzed by insertion of the model of Fig. 6.16(b) in place of the flyback transformer. The circuit of Fig. 6.31(a) is then obtained. The magnetizing inductance L_M functions in the same manner as inductor L of the original buck-boost converter of Fig. 6.30(a). When transistor Q_1 conducts, energy from the dc source V_g is stored in L_M. When diode D_1 conducts, this stored energy is transferred to the load, with the inductor voltage and current scaled according to the $1:n$ turns ratio.

During subinterval 1, while transistor Q_1 conducts, the converter circuit model reduces to Fig.

6.31(b). The inductor voltage v_L, capacitor current i_C, and dc source current i_g are given by

$$\begin{aligned} v_L &= V_g \\ i_C &= -\frac{v}{R} \\ i_g &= i \end{aligned}$$ (6.39)

With the assumption that the converter operates in the continuous conduction mode, with small inductor current ripple and small capacitor voltage ripple, the magnetizing current i and output capacitor voltage v can be approximated by their dc components, I and V, respectively. Equation (6.39) then becomes

$$\begin{aligned} v_L &= V_g \\ i_C &= -\frac{V}{R} \\ i_g &= I \end{aligned}$$ (6.40)

During the second subinterval, the transistor is in the off-state, and the diode conducts. The equivalent circuit of Fig. 6.31(c) is obtained. The primary-side magnetizing inductance voltage v_L, the capacitor current i_C, and the dc source current i_g for this subinterval are:

$$\begin{aligned} v_L &= -\frac{v}{n} \\ i_C &= \frac{i}{n} - \frac{v}{R} \\ i_g &= 0 \end{aligned}$$ (6.41)

It is important to consistently define $v_L(t)$ on the same side of the transformer for all subintervals. Upon making the small-ripple approximation, one obtains

$$\begin{aligned} v_L &= -\frac{V}{n} \\ i_C &= \frac{I}{n} - \frac{V}{R} \\ i_g &= 0 \end{aligned}$$ (6.42)

The $v_L(t)$, $i_C(t)$, and $i_g(t)$ waveforms are sketched in Fig. 6.32 for continuous conduction mode operation.

Application of the principle of volt-second balance to the primary-side magnetizing inductance yields

$$\langle v_L \rangle = D\left(V_g\right) + D'\left(-\frac{V}{n}\right) = 0$$ (6.43)

Solution for the conversion ratio then leads to

$$M(D) = \frac{V}{V_g} = n\frac{D}{D'}$$ (6.44)

So the conversion ratio of the flyback converter is similar to that of the buck-boost converter, but contains an added factor of n.

Application of the principle of charge balance to the output capacitor C leads to

$$\langle i_C \rangle = D\left(-\frac{V}{R}\right) + D'\left(\frac{I}{n} - \frac{V}{R}\right) = 0$$ (6.45)

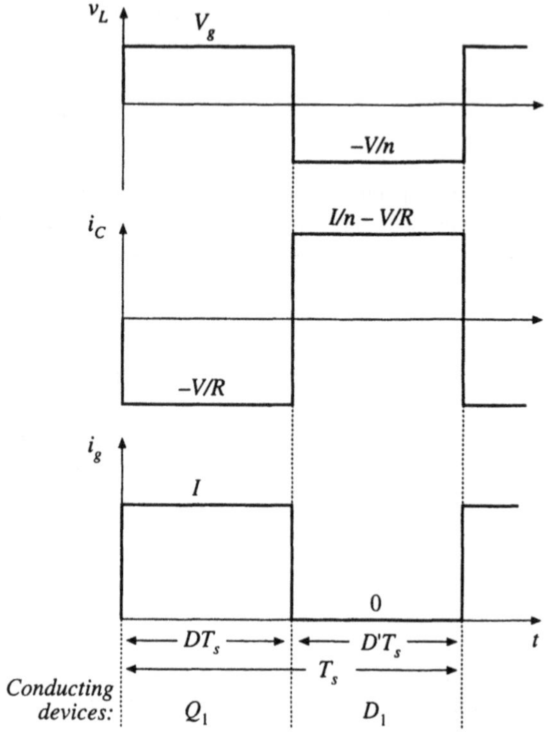

Fig. 6.32 Flyback converter waveforms, continuous conduction mode.

Solution for *I* yields

$$I = \frac{nV}{D'R} \tag{6.46}$$

This is the dc component of the magnetizing current, referred to the primary. The dc component of the source current i_g is

$$I_g = \langle i_g \rangle = D(I) + D'(0) \tag{6.47}$$

An equivalent circuit that models the dc components of the flyback converter waveforms can now be constructed. Circuits corresponding to the inductor loop equation (6.43) and to node equations (6.45) and (6.47) are illustrated in Fig. 6.33(a). By replacing the dependent sources with ideal dc transformers, one obtains Fig. 6.33(b). This is the dc equivalent circuit of the flyback converter. It contains a 1:*D* buck-type conversion ratio, followed by a *D'*:1 boost-type conversion ratio, and an added factor of 1:*n* arising from the flyback transformer turns ratio. By use of the method developed in Chapter 3, the model can be refined to account for losses and to predict the converter efficiency. The flyback converter can also be operated in the discontinuous conduction mode; analysis is left as a homework problem. The results are similar to the DCM buck-boost converter results tabulated in Chapter 5, but are generalized to account for the turns ratio 1:*n*.

The flyback converter is commonly used at the 50 to 100 W power range, as well as in high-voltage power supplies for televisions and computer monitors. It has the advantage of very low parts

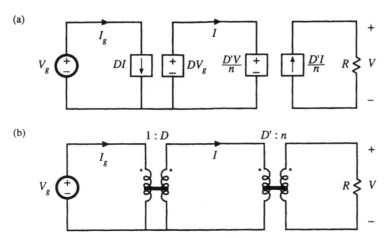

Fig. 6.33 Flyback converter equivalent circuit model, CCM: (a) circuits corresponding to Eqs. (6.43), (6.45), and (6.47); (b) equivalent circuit containing ideal dc transformers.

count. Multiple outputs can be obtained using a minimum number of parts: each additional output requires only an additional winding, diode, and capacitor. However, in comparison with the full-bridge, half-bridge, or two-transistor forward converters, the flyback converter has the disadvantages of high transistor voltage stress and poor cross-regulation. The peak transistor voltage is equal to the dc input voltage V_g plus the reflected load voltage V/n; in practice, additional voltage is observed due to ringing associated with the transformer leakage inductance. Rigorous comparison of the utilization of the flyback transformer with the transformers of buck-derived circuits is difficult because of the different functions performed by these elements. The magnetizing current of the flyback transformer is unipolar, and hence no more than half of the core material B–H loop can be utilized. The magnetizing current must contain a significant dc component. Yet, the size of the flyback transformer is quite small in designs intended to operate in the discontinuous conduction mode. However, DCM operation leads to increased peak currents in the transistor, diode, and filter capacitors. Continuous conduction mode designs require larger values of L_M, and hence larger flyback transformers, but the peak currents in the power stage elements are lower.

6.3.5   Boost-Derived Isolated Converters

Transformer-isolated boost converters can be derived by inversion of the source and load of buck-derived isolated converters. A number of configurations are known, and two of these are briefly discussed here. These converters find some employment in high-voltage power supplies, as well as in low-harmonic rectifier applications.

A full-bridge configuration is diagrammed in Fig. 6.34, and waveforms for the continuous conduction mode are illustrated in Fig. 6.35. The circuit topologies during the first and second subintervals are equivalent to those of the basic nonisolated boost converter, and when the turns ratio is 1:1, the inductor current $i(t)$ and output current $i_o(t)$ waveforms are identical to the inductor current and diode current waveforms of the nonisolated boost converter.

During subinterval 1, all four transistors operate in the on state. This connects the inductor L across the dc input source V_g, and causes diodes D_1 and D_2 to be reverse-biased. The inductor current $i(t)$

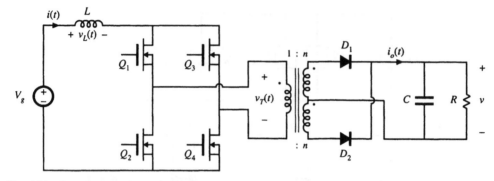

Fig. 6.34 Full-bridge transformer-isolated boost converter.

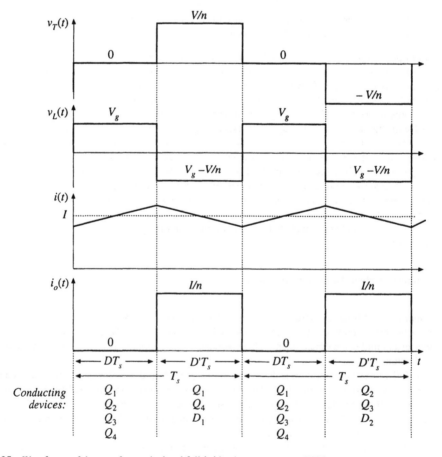

Fig. 6.35 Waveforms of the transformer-isolated full-bridge boost converter, CCM.

increases with slope V_g/L, and energy is transferred from the dc source V_g to inductor L. During the second subinterval, transistors Q_2 and Q_3 operate in the off state, so that inductor L is connected via transistors Q_1 and Q_4 through the transformer and diode D_1 to the dc output. The next switching period is similar, except that during subinterval 2, transistors Q_1 and Q_4 operate in the off state, and inductor L is connected via transistors Q_2 and Q_3 through the transformer and diode D_2 to the dc output. If the transistor off-times and the diode forward drops are identical, then the average transformer voltage is zero, and the net volt-seconds applied to the transformer magnetizing inductance over two switching periods is zero.

Application of the principle of inductor volt-second balance to the inductor voltage waveform $v_L(t)$ yields

$$\langle v_L \rangle = D\left(V_g\right) + D'\left(V_g - \frac{V}{n}\right) = 0 \tag{6.48}$$

Solution for the conversion ratio $M(D)$ then leads to

$$M(D) = \frac{V}{V_g} = \frac{n}{D'} \tag{6.49}$$

This result is similar to the boost converter $M(D)$, with an added factor of n due to the transformer turns ratio.

The transistors must block the reflected load voltage $V/n = V_g/D'$. In practice, additional voltage

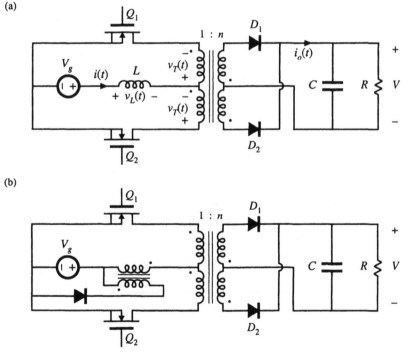

Fig. 6.36 Push-pull isolated converters: (a) based on the boost converter, (b) based on the Watkins-Johnson converter.

is observed due to ringing associated with the transformer leakage inductance. Because the instantaneous transistor current is limited by inductor L, saturation of the transformer due to small imbalances in the semiconductor forward voltage drops or conduction times is not catastrophic. Indeed, control schemes are known in which the transformer is purposely operated in saturation during subinterval 1 [13, 15].

A push-pull configuration is depicted in Fig. 6.36(a). This configuration requires only two transistors, each of which must block voltage $2V/n$. Operation is otherwise similar to that of the full-bridge. During subinterval 1, both transistors conduct. During subinterval 2, one of the transistors operates in the off state, and energy is transferred from the inductor through the transformer and one of the diodes to the output. Transistors conduct during subinterval 2 during alternate switching periods, such that transformer volt-second balance is maintained. A similar push-pull version of the Watkins-Johnson converter, converter 6 of Fig. 6.14, is illustrated in Fig. 6.36(b).

6.3.6   Isolated Versions of the SEPIC and the Ćuk Converter

The artifice used to obtain isolation in the flyback converter can also be applied to the SEPIC and inverse-SEPIC. Referring to Fig. 6.37(a), inductor L_2 can be realized using two windings, leading to the isolated SEPIC of Fig. 6.37(b). An equivalent circuit is given in Fig. 6.37(c). It can be seen that the mag-

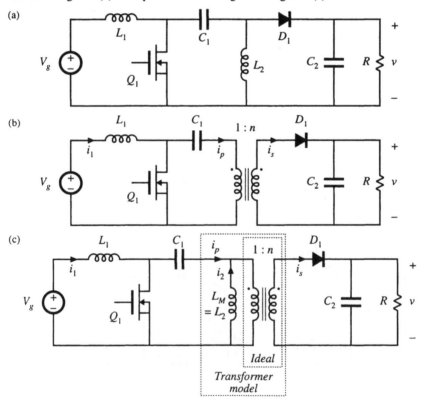

Fig. 6.37 Obtaining isolation in the SEPIC: (a) basic nonisolated converter, (b) isolated SEPIC, (c) with transformer equivalent circuit model.

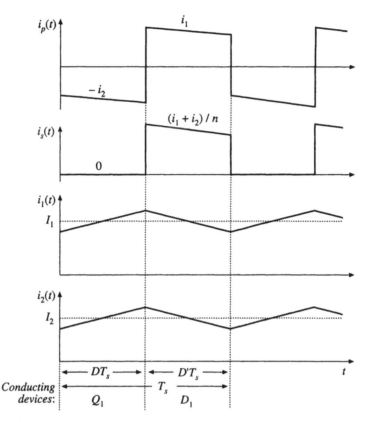

Fig. 6.38 Waveforms of the isolated SEPIC, continuous conduction mode.

netizing inductance performs the energy-storage function of the original inductor L_2. In addition, the ideal transformer provides isolation and a turns ratio.

Typical primary and secondary winding current waveforms $i_p(t)$ and $i_s(t)$ are portrayed in Fig. 6.38, for the continuous conduction mode. The magnetic device must function as both a flyback transformer and also a conventional two-winding transformer. During subinterval 1, while transistor Q_1 conducts, the magnetizing current flows through the primary winding, and the secondary winding current is zero. During subinterval 2, while diode D_1 conducts, the magnetizing current flows through the secondary winding to the load. In addition, the input inductor current i_1 flows through the primary winding. This induces an additional component of secondary current i_1/n, which also flows to the load. So design of the SEPIC transformer is somewhat unusual, and the rms winding currents are larger than those of the flyback transformer.

By application of the principle of volt-second balance to inductors L_1 and L_M, the conversion ratio can be shown to be

$$M(D) = \frac{V}{V_g} = \frac{nD}{D'} \tag{6.50}$$

Ideally, the transistor must block voltage V_g/D'. In practice, additional voltage is observed due to ringing associated with the transformer leakage inductance.

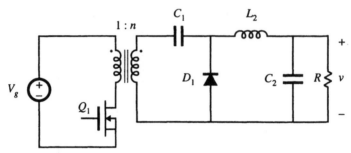

Fig. 6.39 Isolated inverse-SEPIC.

An isolated version of the inverse-SEPIC is shown in Fig. 6.39. Operation and design of the transformer is similar to that of the SEPIC.

Isolation in the Ćuk converter is obtained in a different manner [18]. The basic nonisolated Ćuk converter is illustrated in Fig. 6.40(a). In Fig. 6.40(b), capacitor C_1 is split into two series capacitors C_{1a} and C_{1b}. A transformer can now be inserted between these capacitors, as indicated in Fig. 6.40(c). The polarity marks have been reversed, so that a positive output voltage is obtained. Having capacitors in series with the transformer primary and secondary windings ensures that no dc voltage is applied to the transformer. The transformer functions in a conventional manner, with small magnetizing current and negligible energy storage within the magnetizing inductance.

Utilization of the transformer of the Ćuk converter is quite good. The magnetizing current can

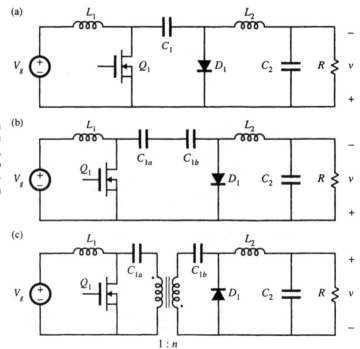

Fig. 6.40 Obtaining isolation in the Ćuk converter: (a) basic nonisolated Ćuk converter, (b) splitting capacitor C_1 into two series capacitors, (c) insertion of transformer between capacitors.

be both positive and negative, and hence the entire core *B–H* loop can be utilized if desired. There are no center-tapped windings, and all of the copper is effectively utilized. The transistor must block voltage V_g/D', plus some additional voltage due to ringing associated with the transformer leakage inductance. The conversion ratio is identical to that of the isolated SEPIC, Eq. (6.50).

The isolated SEPIC and Ćuk converter find application as switching power supplies, typically at power levels of several hundred watts. They are also now finding use as ac–dc low-harmonic rectifiers.

6.4   CONVERTER EVALUATION AND DESIGN

There is no ultimate converter perfectly suited for all possible applications. For a given application, with given specifications, trade studies should be performed to select a converter topology. Several approaches that meet the specifications should be considered, and for each approach important quantities such as worst-case transistor voltage, worst-case transistor rms current, transformer size, etc., should be computed. This type of quantitative comparison can lead to selection of the best approach, while avoiding the personal biases of the engineer.

6.4.1   Switch Stress and Utilization

Often, the largest single cost in a converter is the cost of the active semiconductor devices. Also, the conduction and switching losses associated with the semiconductor devices often dominates the other converter losses. This suggests evaluating candidate converter approaches by comparing the voltage and current stresses imposed on the active semiconductor devices. Minimization of the total switch stresses leads to minimization of the total silicon area required to realize the power devices of the converter.

So it is useful to compare the *total active switch stress* and *active switch utilization* of candidate converter approaches. In a good design, the voltages and currents imposed on the semiconductor devices is minimized, while the load power is maximized. If a converter contains k active semiconductor devices, the total active switch stress S can be defined as

$$S = \sum_{j=1}^{k} V_j I_j \tag{6.51}$$

where V_j is the peak voltage applied to semiconductor switch j, and I_j is the rms current applied to switch j. Peak rather than rms current is sometimes used, with qualitatively similar results. If the converter load power is P_{load}, then the active switch utilization U can be defined as

$$U = \frac{P_{load}}{S} \tag{6.52}$$

The switch utilization is less than one in transformer-isolated converters, and is a quantity to be maximized.

For example, consider the transistor utilization in the CCM flyback converter of Fig. 6.30(d). The peak transistor voltage occurs during subinterval 2, and is equal to the dc input voltage V_g plus the reflected load voltage V/n:

$$V_{Q1,pk} = V_g + \frac{V}{n} = \frac{V_g}{D'} \tag{6.53}$$

The transistor current waveform coincides with the input current waveform $i_g(t)$, which is sketched in Fig. 6.32. The rms value of this waveform is

$$I_{Q1,rms} = I\sqrt{D} = \frac{P_{load}}{V_g\sqrt{D}}$$

(6.54)

So the total active switch stress is

$$S = V_{Q1,pk}\,I_{Q1,rms} = \left(V_g + \frac{V}{n}\right)\!\left(I\sqrt{D}\right)$$

(6.55)

The load power P_{load} can be expressed in terms of V and I by solution of the equivalent circuit model, Fig. 6.33(b). The result is

$$P_{load} = D'V\frac{I}{n}$$

(6.56)

Use of Eq. (6.44) to eliminate V_g from Eq. (6.55), and evaluation of Eq. (6.52), leads to

$$U = D'\sqrt{D}$$

(6.57)

The transistor utilization U tends to zero at $D = 0$ and at $D = 1$, and reaches a maximum of $U = 0.385$ at $D = 1/3$.

For given values of V_g, V, and the load power, the designer can arbitrarily choose the duty cycle D. The turns ratio is then chosen to satisfy Eq. (6.44), as follows:

$$n = \frac{V}{V_g}\frac{D'}{D}$$

(6.58)

At low duty cycle, the transistor rms current becomes large because the transformer turns ratio must be large. At a duty cycle approaching one, the transistor peak voltage is large. So the choice $D = 1/3$ is a good one, which minimizes the product of peak transistor voltage and rms transistor current. In practice, the converter must be optimized to meet a number of different criteria, so a somewhat different duty cycle may be chosen. Also, the converter must usually be designed to operate with some given range of load powers and input voltages; this can lead to a different choice of D, as well as to reduced switch utilization.

For a simple comparison between converters, the switch utilizations of a number of isolated and nonisolated converters are collected in Table 6.1. For simplicity, the formulas assume that the converter is designed to function at a single operating point, that is, with no variations in V_g, V, or P_{load}.

It can be seen that the nonisolated buck and boost converters operate most efficiently when their conversion ratios $M(D)$ are near one. In the case of the boost converter, the switch utilization is greater than one for $D < 0.382$, and approaches infinity as D tends to zero. The reason for this is that, at $D = 0$, the transistor is always off and hence its rms current is zero. But at $D = 0$, $V = V_g$, so the output power is nonzero. All of the load power flows through the diode rather than the transistor. Of course, if it is desired that $V = V_g$, then it would be best to eliminate the boost converter, and directly connect the load to the input voltage. But it is nonetheless true that if the output voltage V is not too much greater than V_g, then a large amount of power can be controlled by a relatively small transistor. Similar arguments apply to the buck converter: all of the load power must flow through the transistor and hence $U \le 1$, yet converter efficiency and cost per watt are optimized when the output voltage V is not too much smaller than the input voltage.

Table 6.1 Active switch utilizations of some common dc–dc converters, single operating point

Converter	$U(D)$	max $U(D)$	max $U(D)$ occurs at $D =$
Buck	$\sqrt{D}$	1	1
Boost	$\dfrac{D'}{\sqrt{D}}$	∞	0
Buck–boost, flyback, nonisolated SEPIC, isolated SEPIC, nonisolated Ćuk, isolated Ćuk	$D'\sqrt{D}$	$\dfrac{2}{3\sqrt{3}} = 0.385$	$\dfrac{1}{3}$
Forward, $n_1 = n_2$	$\dfrac{1}{2}\sqrt{D}$	$\dfrac{1}{2\sqrt{2}} = 0.353$	$\dfrac{1}{2}$
Other isolated buck-derived converters (full–bridge, half–bridge, push–pull)	$\dfrac{\sqrt{D}}{2\sqrt{2}}$	$\dfrac{1}{2\sqrt{2}} = 0.353$	1
Isolated boost-derived converters (full–bridge, push–pull)	$\dfrac{D'}{2\sqrt{1+D}}$	$\dfrac{1}{2}$	0

Incorporation of an isolation transformer leads to reduced switch utilization. In general, transformer-isolated buck-derived converters should be designed to operate at as large a duty cycle as other considerations will allow. Even so, the switch utilization is reduced to $U \leq 0.353$, meaning that the switch stress is increased by a factor of approximately 2.8 as compared with the nonisolated buck converter at $D = 1$. On the other hand, the transformer turns ratio can be chosen to match the load voltage to the input voltage and better optimize the converter. For example, in a full-bridge buck-derived converter operating with $V_g = 500$ V and $V = 5$ V, the turns ratio could be chosen to be nearly 100:1, leading to a duty cycle close to one and switch utilization of approximately 0.35. To obtain a 1 kW output power, the total transistor stress would be 1 kW/0.35 = 2.86 kVA. By comparison, the nonisolated buck converter would operate with a duty cycle of 0.01 and a switch utilization of 0.1. Its total switch stress would be 1 kW/0.1 = 10 kVA; transistors with larger rated currents and lower on-resistances would be needed. Similar arguments apply to the transformer-isolated boost-derived converters: these converters are better optimized when they operate at low duty cycles.

The nonisolated buck-boost, nonisolated SEPIC, nonisolated Ćuk converter, and the isolated SEPIC, flyback, and Ćuk converters have similar switch utilizations. In all of these converters, $U \leq 0.385$, which is approximately the same as in the isolated buck-derived converters. So the nonisolated versions of these converters tend to have lower switch utilizations than the buck or boost converters; however, isolation can be obtained with no additional penalty in switch stress. Switch utilization of a single-operating-point design is maximized when the turns ratio is chosen such that $D = 1/3$.

The cost of the active semiconductor devices of a converter approach can be estimated using the converter switch utilization, as follows:

$$\begin{pmatrix} \text{semiconductor cost} \\ \text{per kW output power} \end{pmatrix} = \frac{\begin{pmatrix} \text{semiconductor device cost} \\ \text{per rated kVA} \end{pmatrix}}{\begin{pmatrix} \text{voltage} \\ \text{derating} \\ \text{factor} \end{pmatrix}\begin{pmatrix} \text{current} \\ \text{derating} \\ \text{factor} \end{pmatrix}\begin{pmatrix} \text{converter} \\ \text{switch} \\ \text{utilization} \end{pmatrix}} \tag{6.59}$$

The semiconductor device cost per rated kVA is equal to the cost of a semiconductor device, divided by the products of its maximum voltage rating and its maximum rms current capability, expressed in $/kVA. This figure depends on a variety of factors, including the device type, packaging, voltage and power levels, and market volume. A typical U.S. value in 2000 is less than $1 /kVA. Voltage and current derating is required to obtain reliable operation of the semiconductor devices. A typical *design guideline* is that the worst-case peak transistor voltage (including transients, voltage spikes due to ringing, and all other anticipated events) should not exceed 75% of the rated transistor voltage, leading to a voltage derating factor of (0.75). Hence, the cost of the active semiconductor switches in a 2000 isolated dc–dc converter is typically in the range $1 to $10 per kW of output power for medium to high-power applications.

6.4.2 Design Using Computer Spreadsheet

Computer spreadsheets are a useful tool for performing converter trade studies and designs. Given specifications regarding the desired output voltage V, the ranges of the input voltage V_g and the load power P_{load}, the desired output voltage ripple Δv, the switching frequency f_s, etc., various design options can be explored. The transformer turns ratio and the inductor current ripple Δi can be taken as design variables, chosen by the engineer. The range of duty cycle variations and the inductor and capacitor component values can then be computed. Worst-case values of the currents and voltages applied to the various power-stage elements can also be evaluated, as well as the sizes of the magnetic elements. By investigating several choices of the design variables, a good compromise between the worst-case voltage stresses and current stresses can be found.

    A short spreadsheet example is given in Table 6.2. The converter operates from a dc voltage derived by rectifying a 230 V ± 20% ac source voltage. The converter dc input voltage V_g is therefore $230\sqrt{2}$ V ± 20%. The load voltage is a regulated 15 V dc, with switching ripple Δv no greater than 0.1 V. The load power can vary over the range 20 W to 200 W. It is desired to operate with a switching frequency of f_s = 100 kHz. These values are entered as specifications, at the top of the spreadsheet. The design of a forward converter, Fig. 6.22, and of a flyback converter, Fig. 6.30(d), to meet these specifications is investigated in the spreadsheet. Continuous conduction mode designs are investigated: the inductor current ripple Δi is chosen small enough that the converter operates in CCM at full load power. Depending on the choice of Δi, the converter may operate in either CCM or DCM at minimum load power.

    For the single-transistor forward converter, the turns ratios n_2/n_1 and n_3/n_1, as well as the inductor current ripple Δi, can be taken as design variables. For this example, the reset-winding turns ratio n_2/n_1 is chosen to be one, and hence the duty cycle is limited to $D < 0.5$ as given by Eq. (6.35). The maximum duty cycle is computed first. The output voltage of the forward converter, in continuous conduction mode, is given by Eq. (6.36). Solution for the duty cycle D leads to

$$D = \frac{n_1}{n_3}\frac{V}{V_g} \qquad (6.60)$$

The maximum value of D occurs at minimum V_g and at full load, and is given in Table 6.2. The minimum CCM value of D, occurring at maximum V_g, is also listed.

    The value of the inductance L is computed next. The magnitude of the inductor current ripple Δi can be computed in a manner similar to that used for the nonisolated buck converter to obtain Eq. (2.15). The result is

$$\Delta i = \frac{D'VT_s}{2L} \qquad (6.61)$$

Table 6.2 Spreadsheet design example

Specifications			
Maximum input voltage V_g	390 V		
Minimum input voltage V_g	260 V		
Output voltage V	15 V		
Maximum load power P_{load}	200 W		
Minimum load power P_{load}	20 W		
Switching frequency f_s	100 kHz		
Maximum output ripple Δv	0.1 V		

Forward converter design, CCM		*Flyback converter design, CCM*	
Design variables		*Design variables*	
Reset winding turns ratio n_2/n_1	1	Turns ratio n_2/n_1	0.125
Turns ratio n_3/n_1	0.125	Inductor current ripple Δi	3 A ref to sec
Inductor current ripple Δi	2A ref to sec		
Results		*Results*	
Maximum duty cycle D	0.462	Maximum duty cycle D	0.316
Minimum D, at full load	0.308	Minimum D, at full load	0.235
Minimum D, at minimum load	0.251	Minimum D, at minimum load	0.179
Inductance L	26 µH	Inductance L	19 µH ref to sec
Capacitance C	25 µF	Capacitance C	210 µF
Worst-case stresses		*Worst-case stresses*	
Peak transistor voltage v_{Q1}	780 V	Peak transistor voltage v_{Q1}	510 V
Rms transistor current	1.13 A	Rms transistor current	1.38 A
Transistor utilization U	0.226	Transistor utilization U	0.284
Peak diode voltage v_{D2}	49 V	Peak diode voltage v_{D1}	64 V
Rms diode current i_{D2}	9.1 A	Rms diode current i_{D1}	16.3 A
Peak diode voltage v_{D3}	49 V	Peak diode current i_{D1}	22.2 A
Rms diode current i_{D3}	11.1 A		
Rms output capacitor current i_C	1.15 A	Rms output capacitor current i_C	9.1 A

The worst-case maximum ripple occurs in CCM at minimum duty cycle. Solution for L yields

$$L = \frac{D'VT_s}{2\Delta i} \tag{6.62}$$

This equation is used to select L such that the worst-case ripple is equal to the specified value of Δi. The required value of L is listed in Table 6.2. The required value of C that leads to the specified voltage ripple Δv is also computed, using Eq. (2.60). Since Eq. (2.60) neglects capacitor esr, a larger value of C may be required in practice.

   If the converter operates in the discontinuous conduction mode at light load, then the controller must reduce the duty cycle D to maintain the required output voltage V. The conversion ratio $M(D, K)$ of the DCM forward converter can be found analytically, using the method developed in the previous chapter. Alternatively, the nonisolated buck converter solution, Eq. (5.29), can be applied directly if all element values are referred to the transformer secondary side. Hence, the output voltage in DCM is given by

$$V = \frac{n_3}{n_1} V_g \frac{2}{1 + \sqrt{1 + \frac{4K}{D^2}}} \tag{6.63}$$

with $K = 2L/RT_s$, and $R = V^2/P_{load}$. Solution for the duty cycle D yields

$$D = \frac{2\sqrt{K}}{\sqrt{\left(\frac{2n_3 V_g}{n_1 V} - 1\right)^2 - 1}} \tag{6.64}$$

The actual duty cycle is the smaller of Eqs. (6.60) and (6.64). The minimum duty cycle occurs at minimum load power and maximum V_g, and is given in Table 6.2.

Worst-case component stresses can now be evaluated. The peak transistor voltage is given by Eq. (6.37). The rms transistor current is calculated with the help of Appendix 1. With the assumption that the transformer magnetizing current can be neglected, the transistor current is equal to the reflected inductor current $i(t)n_3/n_1$ during subinterval 1, and is equal to zero during subintervals 2 and 3. The rms transistor current is therefore

$$I_{Q1,rms} = \frac{n_3}{n_1} \sqrt{D} \sqrt{I^2 + \frac{(\Delta i)^2}{3}} \approx \frac{n_3}{n_1} \sqrt{D} \, I \tag{6.65}$$

where $I = P_{load}/V$. The worst-case value of $I_{Q1\text{-}rms}$ occurs at maximum load power and at maximum duty cycle. Expressions for the worst-case stresses in the diodes and output capacitor, as well as for the flyback converter, are found in a similar manner. Their derivation is left as an exercise for the student.

The designs of Table 6.2 are good ones which illustrate the tradeoffs inherent in selection of an isolated converter topology, although some additional design optimization is possible and is left as a homework problem. Both designs utilize a turns ratio of 8:1. The rms transistor current is 22% higher in the flyback converter. This current could be reduced, at the expense of increased transistor voltage. The flyback converter imposes only 510 V on the transistor. A transistor rated at 800 V or 1000 V could be used, with an adequate voltage derating factor and some margin for voltage ringing due to transformer leakage inductance. The 780 V imposed on the transistor of the forward converter is 53% higher than in the flyback converter. Power MOSFETs with voltage ratings greater than 1000 V are not available in 1997; hence, when voltage ringing due to transformer leakage inductance is accounted for, this design will have an inadequate voltage design margin. This problem could be overcome by changing the reset winding turns ratio n_2/n_1, or by using a two-transistor forward converter. It can be concluded that the transformer reset mechanism of the flyback converter is better than that of the conventional forward converter.

Because of the pulsating nature of the secondary-side currents in the flyback converter, the rms and peak secondary currents are significantly higher than in the forward converter. The flyback converter diode must conduct an rms current that is 47% greater than that of forward converter diode D_3, and 80% greater than the current in forward converter diode D_2. The secondary winding of the flyback transformer must also conduct this current. Furthermore, the output capacitor of the flyback converter must be rated to conduct an rms current of 9.1 A. This capacitor will be much more expensive than its counterpart in the forward converter. It can be concluded that the nonpulsating output current property of the forward converter is superior to the pulsating output current of the flyback. For these reasons, flyback converters and other converters having pulsating output currents are usually avoided when the application calls for a high-current output.

6.5 SUMMARY OF KEY POINTS

1. The boost converter can be viewed as an inverse buck converter, while the buck-boost and Ćuk converters arise from cascade connections of buck and boost converters. The properties of these converters are consistent with their origins. Ac outputs can be obtained by differential connection of the load. An infinite number of converters are possible, and several are listed in this chapter.

2. For understanding the operation of most converters containing transformers, the transformer can be modeled as a magnetizing inductance in parallel with an ideal transformer. The magnetizing inductance must obey all of the usual rules for inductors, including the principle of volt-second balance.

3. The steady-state behavior of transformer-isolated converters may be understood by first replacing the transformer with the magnetizing-inductance-plus-ideal-transformer equivalent circuit. The techniques developed in the previous chapters can then be applied, including use of inductor volt-second balance and capacitor charge balance to find dc currents and voltages, use of equivalent circuits to model losses and efficiency, and analysis of the discontinuous conduction mode.

4. In the full–bridge, half–bridge, and push–pull isolated versions of the buck and/or boost converters, the transformer frequency is twice the output ripple frequency. The transformer is reset while it transfers energy: the applied voltage polarity alternates on successive switching periods.

5. In the conventional forward converter, the transformer is reset while the transistor is off. The transformer magnetizing inductance operates in the discontinuous conduction mode, and the maximum duty cycle is limited.

6. The flyback converter is based on the buck–boost converter. The flyback transformer is actually a two-winding inductor, which stores and transfers energy.

7. The transformer turns ratio is an extra degree-of-freedom which the designer can choose to optimize the converter design. Use of a computer spreadsheet is an effective way to determine how the choice of turns ratio affects the component voltage and current stresses.

8. Total active switch stress, and active switch utilization, are two simplified figures-of-merit which can be used to compare the various converter circuits.

REFERENCES

[1] S. ĆUK, "Modeling, Analysis, and Design of Switching Converters," Ph.D. thesis, California Institute of Technology, November 1976.

[2] S. ĆUK and R. D. MIDDLEBROOK, "A New Optimum Topology Switching Dc-to-Dc Converter," *IEEE Power Electronics Specialists Conference*, 1977 Record, pp. 160-179, June 1977.

[3] E. LANDSMAN, "A Unifying Derivation of Switching Dc-Dc Converter Topologies," *IEEE Power Electronics Specialists Conference*, 1979 Record, pp. 239-243, June 1979.

[4] R. TYMERSKI and V. VORPERIAN, "Generation, Classification, and Analysis of Switched-Mode Dc-to-Dc Converters by the Use of Converter Cells," *Proceedings International Telecommunications Energy Conference*, pp. 181-195, October 1986.

[5] S. ĆUK and R. ERICKSON, "A Conceptually New High-Frequency Switched-Mode Amplifier Technique Eliminates Current Ripple," *Proceedings Fifth National Solid-State Power Conversion Conference* (Powercon 5), pp. G3.1-G3.22, May 1978.

[6] F. BARZEGAR and S. ĆUK, "A New Switched-Mode Amplifier Produces Clean Three-Phase Power," *Proceedings Ninth International Solid-State Power Conversion Conference* (Powercon 9), pp. E3.1-E3.15, July 1982.

[7] K. D. T. NGO, S. ĆUK, and R. D. MIDDLEBROOK, "A New Flyback Dc-to-Three-Phase Converter with Sinusoidal Outputs," *IEEE Power Electronics Specialists Conference*, 1983 Record, pp. 377-388.

[8] R. W. ERICKSON, "Synthesis of Switched-Mode Converters," *IEEE Power Electronics Specialists Conference*, 1983 Record, pp. 9-22, June 1983.

[9] D. MAKSIMOVIĆ and S. ĆUK, "General Properties and Synthesis of PWM Dc-Dc Converters," *IEEE Power Electronics Specialists Conference*, 1989 Record, pp. 515-525, June 1989.

[10] M. S. MAKOWSKI, "On Topological Assumptions on PWM Converters—A Reexamination," *IEEE Power Electronics Specialists Conference*, 1993 Record, pp. 141-147, June 1993.

[11] B. ISRAELSEN, J. MARTIN, C. REEVE, and V. SCOWN, "A 2.5 kV High Reliability TWT Power Supply: Design Techniques for High Efficiency and Low Ripple," *IEEE Power Electronics Specialists Conference*, 1977 Record, pp. 109-130, June 1977.

[12] R. SEVERNS, "A New Current-Fed Converter Topology," *IEEE Power Electronics Specialists Conference*, 1979 Record, pp. 277-283, June 1979.

[13] V. J. THOTTUVELIL, T. G. WILSON, and H. A. OWEN, "Analysis and Design of a Push-Pull Current-Fed Converter," *IEEE Power Electronics Specialists Conference*, 1981 Record, pp. 192-203, June 1981.

[14] R. REDL AND N. SOKAL, "Push–Pull Current-Fed Multiple-Output Dc–Dc Power Converter with Only One Inductor and with 0–100% Switch Duty Ratio," *IEEE Power Electronics Specialists Conference*, 1980 Record, pp. 341-345, June 1982.

[15] P. W. CLARKE, "Converter Regulation by Controlled Conduction Overlap," U. S. Patent 3,938,024, February 10, 1976.

[16] R. P. MASSEY and E. C. SNYDER, "High-Voltage Single-Ended Dc-Dc Converter," *IEEE Power Electronics Specialists Conference*, 1977 Record, pp. 156-159, June 1977.

[17] D. MAKSIMOVIĆ and S. ĆUK, "Switching Converters with Wide Dc Conversion Range," *IEEE Transactions on Power Electronics*, Vol. 6, No. 1, pp. 151-157, January 1991.

[18] R. D. MIDDLEBROOK and S. ĆUK, "Isolation and Multiple Outputs of a New Optimum Topology Switching Dc-to-Dc Converter," *IEEE Power Electronics Specialists Conference*, 1978 Record, pp. 256-264, June 13-15, 1978.

[19] T. G. WILSON, "Cross Regulation in an Energy-Storage Dc-to-Dc Converter with Two Regulated Outputs," *IEEE Power Electronics Specialists Conference*, 1977 Record, pp. 190-199, June 1977.

[20] H. MATSUO, "Comparison of Multiple-Output Dc-Dc Converters Using Cross Regulation," *IEEE Power Electronics Specialists Conference*, 1979 Record, pp. 169-185, June 1979.

[21] K. HARADA, T. NABESHIMA, and K. HISANAGA, "State-Space Analysis of the Cross-Regulation," *IEEE Power Electronics Specialists Conference*, 1979 Record, pp. 186-192, June 1979.

[22]   J. N. PARK and T. R. ZALOUM, "A Dual Mode Forward/Flyback Converter," *IEEE Power Electronics Specialists Conference*, 1982 Record, pp. 3-13, June 1982.

[23]   S. ĆUK, "General Topological Properties of Switching Structures," *IEEE Power Electronics Specialists Conference*, 1979 Record, pp. 109-130, June 1979.

[24]   R. SEVERNS and G. BLOOM, *Modern Dc–to–Dc Switchmode Power Converter Circuits*, New York: Van Nostrand Reinhold, 1985.

[25]   N. MOHAN, T. UNDELAND, and W. ROBBINS, *Power Electronics: Converters, Applications, and Design*, 2nd edit., New York: John Wiley & Sons, 1995.

[26]   J. KASSAKIAN, M. SCHLECHT, and G. VERGHESE, *Principles of Power Electronics*, Reading, MA: Addison-Wesley, 1991.

[27]   D. MITCHELL, *Dc–Dc Switching Regulator Analysis*, New York: McGraw-Hill, 1988.

[28]   K. KIT SUM, *Switch Mode Power Conversion: Basic Theory and Design*, New York: Marcel Dekker, 1984.

[29]   R. E. TARTER, *Solid-State Power Conversion Handbook*, New York: John Wiley & Sons, 1993.

[30]   Q. CHEN, F. C. LEE, and M. M. JOVANOVIĆ, "DC Analysis and Design of Multiple-Output Forward Converters with Weighted Voltage-Mode Control," *IEEE Applied Power Electronics Conference*, 1993 Record, pp. 449-455, March 1993.

PROBLEMS

6.1   Tapped-inductor boost converter. The boost converter is sometimes modified as illustrated in Fig. 6.41, to obtain a larger conversion ratio than would otherwise occur. The inductor winding contains a total of $(n_1 + n_2)$ turns. The transistor is connected to a tap placed n_1 turns from the left side of the inductor, as shown. The tapped inductor can be viewed as a two-winding $(n_1 : n_2)$ transformer, in which the two windings are connected in series. The inductance of the entire $(n_1 + n_2)$ turn winding is L.

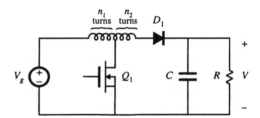

Fig. 6.41 Tapped-inductor boost converter, Problem 6.1

(a)   Sketch an equivalent circuit model for the tapped inductor, which includes a magnetizing inductance and an ideal transformer. Label the values of the magnetizing inductance and turns ratio.

(b)   Determine an analytical expression for the conversion ratio $M = V/V_g$. You may assume that the transistor, diode, tapped inductor, and capacitor are lossless. You may also assume that the converter operates in continuous conduction mode.

(c)   Sketch $M(D)$ vs. D for $n_1 = n_2$, and compare to the nontapped $(n_2 = 0)$ case.

6.2   Analysis of the DCM flyback converter. The flyback converter of Fig. 6.30(d) operates in the discontinuous conduction mode.

(a)   Model the flyback transformer as a magnetizing inductance in parallel with an ideal transformer, and sketch the converter circuits during the three subintervals.

(b) Derive the conditions for operation in discontinuous conduction mode.

(c) Solve the converter: derive expressions for the steady-state output voltage V and subinterval 2 (diode conduction interval) duty cycle D_2.

6.3 Analysis of the isolated inverse–SEPIC of Fig. 6.39. You may assume that the converter operates in the continuous conduction mode, and that all inductor current ripples and capacitor voltage ripples are small.

(a) Derive expressions for the dc components of the magnetizing current, inductor current, and capacitor voltages.

(b) Derive analytical expressions for the rms values of the primary and secondary winding currents. Note that these quantities do not simply scale by the turns ratio.

6.4 The two–transistor flyback converter. The converter of Fig. 6.42 is sometimes used when the dc input voltage is high. Transistors Q_1 and Q_2 are driven with the same gating signal, such that they turn on and off simultaneously with the same duty cycle D. Diodes D_1 and D_2 ensure that the off state voltages of the transistors do not exceed V_g. The converter operates in discontinuous conduction mode. The magnetizing inductance, referred to the primary side, is L_M.

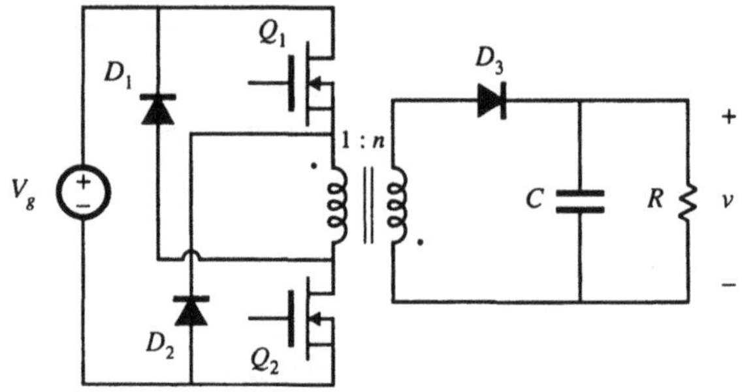

Fig. 6.42 Two-transistor flyback converter, Problem 6.4.

(a) Determine an analytical expression for the steady-state output voltage V.

(b) Over what range of duty cycles does the transformer reset properly? Explain.

6.5 A nonideal flyback converter. The flyback converter shown in Fig. 6.30(d) operates in the continuous conduction mode. The MOSFET has on-resistance R_{on}, and the diode has a constant forward voltage drop V_D. The flyback transformer has primary winding resistance R_p and secondary winding resistance R_s.

(a) Derive a complete steady-state equivalent circuit model, which is valid in the continuous conduction mode, and which correctly models the loss elements listed above as well as the converter input and output ports. Sketch your equivalent circuit.

(b) Derive an analytical expression for the converter efficiency.

6.6 A low-voltage computer power supply with synchronous rectification. The trend in digital integrated circuits is towards lower power supply voltages. It is difficult to construct a high-efficiency low-voltage power supply, because the conduction loss arising in the secondary-side diodes becomes very large. The objective of this problem is to estimate how the efficiency of a forward converter varies as the output voltage is reduced, and to investigate the use of synchronous rectifiers.

The forward converter of Fig. 6.22 operates from a dc input of $V_g = 325$ V, and supplies 20 A to its dc load. Consider three cases: (*i*) $V = 5$ V, (*ii*) $V = 3.3$ V, and (*iii*) $V = 1.5$ V. For each case, the turns ratio n_3/n_1 is chosen such that the converter produces the required output voltage at a transistor duty cycle of $D = 0.4$. The MOSFET has on-resistance $R_{on} = 5$ Ω. The secondary-side schottky diodes have forward voltage drops of $V_F = 0.5$ V. All other elements can be considered ideal.

(a)   Derive an equivalent circuit for the forward converter, which models the semiconductor conduction losses described above.

(b)   Solve your model for cases (*i*), (*ii*), and (*iii*) described above. For each case, determine numerical values of the turns ratio n_3/n_1 and for the efficiency η.

(c)   The secondary-side Schottky diodes are replaced by MOSFETs operating as synchronous rectifiers. The MOSFETs each have an on-resistance of 4 mΩ. Determine the new numerical values of the turns ratio n_3/n_1 and the efficiency η, for cases (*i*), (*ii*), and (*iii*).

6.7   Rotation of switching cells. A network containing switches and reactive elements has terminals a, b, and c, as illustrated in Fig. 6.43(a). You are given that the relationship between the terminal voltages is $V_{bc}/V_{ac} = \mu(D)$.

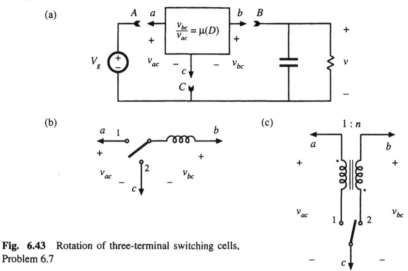

Fig. 6.43 Rotation of three-terminal switching cells, Problem 6.7

(a)   Derive expressions for the source-to-load conversion ratio $V/V_g = M(D)$, in terms of $\mu(D)$, for the following three connection schemes:

      (*i*) a-A b-B c-C

      (*ii*) a-B b-C c-A

      (*iii*) a-C b-A c-B

(b)   Consider the three-terminal network of Fig. 6.43(b). Determine $\mu(D)$ for this network. Plug your answer into your results from part (a), to verify that the buck, boost, and buck–boost converters are generated.

(c)   Consider the three-terminal network of Fig. 6.43(c). Determine $\mu(D)$ for this network. Plug your answer into your results from part (a). What converters are generated?

6.8   Transformer-isolated current-sense circuit. It is often required that the current flowing in a power transistor be sensed. A noninductive resistor R placed in series with the transistor will produce a voltage $v(t)$ that is proportional to the transistor drain current $i_D(t)$. Use of a transformer allows isolation between the power transistor and the control circuit. The transformer turns ratio also allows reduction of the current and power loss and increase of the voltage of the resistor. This problem is concerned with design of the transformer-isolated current-sense circuit of Fig. 6.44.

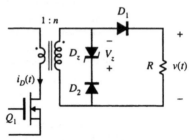

Fig. 6.44 Transformer-isolated circuit for sensing the transistor switch current, Problem 6.8

The transformer has a single-turn primary and an n-turn secondary winding. The transistor switches on and off with duty cycle D and switching frequency f_s. While the transistor conducts, its current is essentially constant and is equal to I. Diodes D_1 and D_2 are conventional silicon diodes having forward voltage drop V_D. Diode D_Z is a zener diode, which can be modeled as a voltage source of value V_Z, with the polarity indicated in the figure. For a proper design, the circuit elements should be chosen such that the transformer magnetizing current, in conjunction with diode D_2, operates in discontinuous conduction mode. In a good design, the magnetizing current is much smaller than the transistor current. Three subintervals occur during each switching period: subinterval 1, in which Q_1 and D_1 conduct; subinterval 2, in which D_2 and D_Z conduct; subinterval 3, in which Q_1, D_1 and D_2 are off.

(a) Sketch the current sense circuit, replacing the transformer and zener diode by their equivalent circuits.

(b) Sketch the waveforms of the transistor current $i_D(t)$, the transformer magnetizing current $i_M(t)$, the primary winding voltage, and the voltage $v(t)$. Label salient features.

(c) Determine the conditions on the zener voltage V_Z that ensure that the transformer magnetizing current is reset to zero before the end of the switching period.

(d) You are given the following specifications:

> Switching frequency $\qquad f_s = 100$ kHz
>
> Transistor duty cycle $\qquad D \leq 0.75$
>
> Transistor peak current $\qquad$ max $i_D(t) \leq 25$ A

The output voltage $v(t)$ should equal 5 V when the transistor current is 25 A. To avoid saturating the transformer core, the volt-seconds applied to the single-turn primary winding while the transistor conducts should be no greater than 2 volt-μsec. The silicon diode forward voltage drops are $V_D = 0.7$ V.
Design the circuit: select values of R, n, and V_Z.

6.9   Optimal reset of the forward converter transformer. As illustrated in Fig. 6.45, it is possible to reset the transformer of the forward converter using a voltage source other than the dc input V_g; several such schemes appear in the literature. By optimally choosing the value of the reset voltage V_r, the peak voltage stresses imposed on transistor Q_1 and diode D_2 can be reduced. The maximum duty cycle can also be increased, leading to a lower transformer turns ratio and lower transistor current. The resulting improvement in converter cost and efficiency can be significant when the dc input voltage varies over a wide range.

(a) As a function of V_g, the transistor duty cycle D, and the transformer turns ratios, what is the minimum value of V_r that causes the transformer magnetizing current to be reset to zero by the end of the switching period?

(b) For your choice of V_r from part (a), what is the peak voltage imposed on transistor Q_1?

This converter is to be used in a universal-input off-line application, with the following specifications. The input voltage V_g can vary between 127 and 380 V. The load voltage is regulated by variation of the

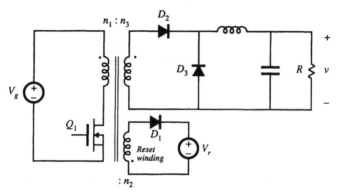

Fig. 6.45　Forward converter with auxiliary reset winding, Problem 6.9

duty cycle, and is equal to 12 V. The load power is 480 W.

(c)　Choose the turns ratio n_3/n_1 such that the total active switch stress is minimized. For your choice of n_3/n_1, over what range will the duty cycle vary? What is the peak transistor current?

(d)　Compare your design of Part (c) with the conventional scheme in which $n_1 = n_2$ and $V_r = V_g$. Compare the worst-case peak transistor voltage and peak transistor current.

(e)　Suggest a way to implement the voltage source V_r. Give a schematic of the power-stage components of your implementation. Use a few sentences to describe the control-circuit functions required by your implementation, if any.

6.10　Design of a multiple-output dc-dc flyback converter. For this problem, you may neglect all losses and transformer leakage inductances. It is desired that the three-output flyback converter shown in Fig. 6.46 operates in the discontinuous conduction mode, with a switching frequency of f_s = 100 kHz. The nominal operating conditions are given in the diagram, and you may that there are no variations in the input voltage or the load currents. Select D_3 = 0.1 (the duty cycle of subinterval 3, in which all semiconductors are off). The objective of this problem is to find a good steady-state design, in which the semiconductor peak blocking voltages and peak currents are reasonably low.

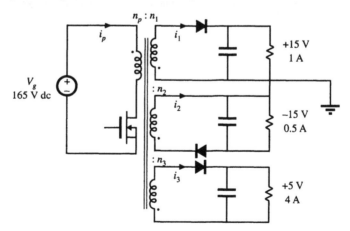

Fig. 6.46　Three-output flyback converter design, Problem 6.10.

(a)   It is possible to find a design in which the transistor peak blocking voltage is less than 300 V, and the peak diode blocking voltages are all less than 35 V, under steady-state conditions. Design the converter such that this is true. Specify: (*i*) the transistor duty cycle D, (*ii*) the magnetizing inductance L_M, referred to the primary, (*iii*) the turns ratios n_1/n_p and n_3/n_p.

(b)   For your design of part (a), determine the rms currents of the four windings. Note that they don't simply scale by the turns ratios.

6.11   Spreadsheet design.

(a)   Develop the analytical expressions for the "Results" and "Worst-case stresses" of the forward converter spreadsheet design example of Table 6.2.

(b)   Enter the formulas you developed in part (a) into a computer spreadsheet, and verify that your computed values agree with those of Table 6.2.

(c)   It is desired to reduce the forward converter peak transistor voltage to a value no greater than 650 V. Modify the design numbers to accomplish this, and briefly discuss the effect on the other component stresses.

(d)   For these specifications, what is the largest possible value of the transistor utilization of the CCM forward converter? How should the spreadsheet design variables be chosen to attain the maximum transistor utilization?

6.12   Spreadsheet design of an isolated Ćuk converter. The isolated Ćuk converter of Fig. 6.40(c) is to be designed to meet the specifications listed in Table 6.2. The converter is to be designed such that it operates in continuous conduction mode at full load.

(a)   Develop analytical expressions for the following quantities:
- The maximum and minimum duty cycles, for CCM operation
- The peak voltages and rms currents of both semiconductor devices
- The ripple magnitudes of the capacitor voltages and inductor currents
- The rms capacitor currents
- The transistor utilization U

(b)   Enter the formulas you developed in part (a) into a computer spreadsheet. What are the design variables?

(c)   For the specifications listed in Table 6.2, select the design variables to attain what you believe is the best design. Compare the performance of your design with the flyback and forward converter designs of Table 6.2.

Part II

Converter Dynamics and Control

7

AC Equivalent Circuit Modeling

7.1   INTRODUCTION

Converter systems invariably require feedback. For example, in a typical dc–dc converter application, the output voltage $v(t)$ must be kept constant, regardless of changes in the input voltage $v_g(t)$ or in the effective load resistance R. This is accomplished by building a circuit that varies the converter control input [i.e., the duty cycle $d(t)$] in such a way that the output voltage $v(t)$ is regulated to be equal to a desired reference value v_{ref}. In inverter systems, a feedback loop causes the output voltage to follow a sinusoidal reference voltage. In modern low-harmonic rectifier systems, a control system causes the converter input current to be proportional to the input voltage, such that the input port presents a resistive load to the ac source. So feedback is commonly employed.

     A typical dc–dc system incorporating a buck converter and feedback loop block diagram is illustrated in Fig. 7.1. It is desired to design this feedback system in such a way that the output voltage is accurately regulated, and is insensitive to disturbances in $v_g(t)$ or in the load current. In addition, the feedback system should be stable, and properties such as transient overshoot, settling time, and steady-state regulation should meet specifications. The ac modeling and design of converters and their control systems such as Fig. 7.1 is the subject of Part II of this book.

     To design the system of Fig. 7.1, we need a dynamic model of the switching converter. How do variations in the power input voltage, the load current, or the duty cycle affect the output voltage? What are the small-signal transfer functions? To answer these questions, we will extend the steady-state models developed in Chapters 2 and 3 to include the dynamics introduced by the inductors and capacitors of the converter. Dynamics of converters operating in the continuous conduction mode can be modeled using techniques quite similar to those of Chapters 2 and 3; the resulting ac equivalent circuits bear a strong resemblance to the dc equivalent circuits derived in Chapter 3.

     Modeling is the representation of physical phenomena by mathematical means. In engineering,

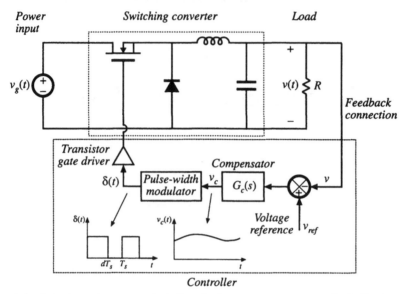

Fig. 7.1   A simple dc–dc regulator system, including a buck converter power stage and a feedback network.

it is desired to model the important dominant behavior of a system, while neglecting other insignificant phenomena. Simplified terminal equations of the component elements are used, and many aspects of the system response are neglected altogether, that is, they are "unmodeled." The resulting simplified model yields physical insight into the system behavior, which aids the engineer in designing the system to operate in a given specified manner. Thus, the modeling process involves use of approximations to neglect small but complicating phenomena, in an attempt to understand what is most important. Once this basic insight is gained, it may be desirable to carefully refine the model, by accounting for some of the previously ignored phenomena. It is a fact of life that real, physical systems are complex, and their detailed analysis can easily lead to an intractable and useless mathematical mess. Approximate models are an important tool for gaining understanding and physical insight.

As discussed in Chapter 2, the switching ripple is small in a well-designed converter operating in continuous conduction mode (CCM). Hence, we should ignore the switching ripple, and model only the underlying ac variations in the converter waveforms. For example, suppose that some ac variation is introduced into the converter duty cycle $d(t)$, such that

$$d(t) = D + D_m \cos \omega_m t \tag{7.1}$$

where D and D_m are constants, $|D_m| \ll D$, and the modulation frequency ω_m is much smaller than the converter switching frequency $\omega_s = 2\pi f_s$. The resulting transistor gate drive signal is illustrated in Fig. 7.2(a), and a typical converter output voltage waveform $v(t)$ is illustrated in Fig. 7.2(b). The spectrum of $v(t)$ is illustrated in Fig. 7.3. This spectrum contains components at the switching frequency as well as its harmonics and sidebands; these components are small in magnitude if the switching ripple is small. In addition, the spectrum contains a low-frequency component at the modulation frequency ω_m. The magnitude and phase of this component depend not only on the duty cycle variation, but also on the frequency response of the converter. If we neglect the switching ripple, then this low-frequency compo-

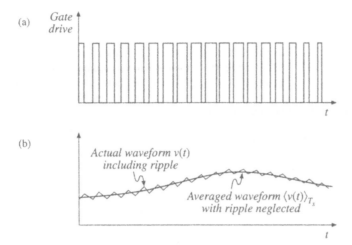

Fig.   7.2   Ac variation of the converter signals: (a) transistor gate drive signal, in which the duty cycle varies slowly, and (b) the resulting converter output voltage waveform. Both the actual waveform $v(t)$ (including high frequency switching ripple) and its averaged, low-frequency component, $\langle v(t) \rangle_{T_s}$, are illustrated.

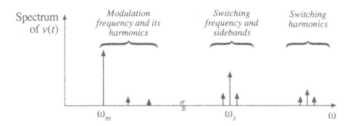

Fig.   7.3   Spectrum of the output voltage waveform $v(t)$ of Fig. 7.2.

nent remains [also illustrated in Fig. 7.2(b)]. The objective of our ac modeling efforts is to predict this low-frequency component.

A simple method for deriving the small-signal model of CCM converters is explained in Section 7.2. The switching ripples in the inductor current and capacitor voltage waveforms are removed by averaging over one switching period. Hence, the low-frequency components of the inductor and capacitor waveforms are modeled by equations of the form

$$L \frac{d\langle i_L(t) \rangle_{T_s}}{dt} = \langle v_L(t) \rangle_{T_s}$$
$$C \frac{d\langle v_C(t) \rangle_{T_s}}{dt} = \langle i_C(t) \rangle_{T_s}$$

$$(7.2)$$

where $\langle x(t) \rangle_{T_s}$ denotes the average of $x(t)$ over an interval of length T_s:

$$\left\langle x(t) \right\rangle_{T_s} = \frac{1}{T_s} \int_{t}^{t+T_s} x(\tau)d\tau \tag{7.3}$$

So we will employ the basic approximation of removing the high-frequency switching ripple by averaging over one switching period. Yet the average value is allowed to vary from one switching period to the next, such that low-frequency variations are modeled. In effect, the "moving average" of Eq. (7.3) constitutes low-pass filtering of the waveform. A few of the numerous references on averaged modeling of switching converters are listed at the end of this chapter [1–20].

Note that the principles of inductor volt-second balance and capacitor charge balance predict that the right-hand sides of Eqs. (7.2) are zero when the converter operates in equilibrium. Equations (7.2) describe how the inductor currents and capacitor voltages change when nonzero average inductor voltage and capacitor current are applied over a switching period.

The averaged inductor voltage and capacitor currents of Eq. (7.2) are, in general, nonlinear functions of the signals in the converter, and hence Eqs. (7.2) constitute a set of nonlinear differential equations. Indeed, the spectrum in Fig. 7.3 also contains harmonics of the modulation frequency ω_m. In most converters, these harmonics become significant in magnitude as the modulation frequency ω_m approaches the switching frequency ω_s, or as the modulation amplitude D_m approaches the quiescent duty cycle D. Nonlinear elements are not uncommon in electrical engineering; indeed, all semiconductor devices exhibit nonlinear behavior. To obtain a linear model that is easier to analyze, we usually construct a small-signal model that has been linearized about a quiescent operating point, in which the harmonics of the modulation or excitation frequency are neglected. As an example, Fig. 7.4 illustrates linearization of the familiar diode i–v characteristic shown in Fig. 7.4(b). Suppose that the diode current $i(t)$ has a quiescent (dc) value I and a signal component $\hat{i}(t)$. As a result, the voltage $v(t)$ across the diode has a quiescent value V and a signal component $\hat{v}(t)$. If the signal components are small compared to the quiescent values,

$$\left| \hat{v} \right| \ll \left| V \right|, \quad \left| \hat{i} \right| \ll \left| I \right| \tag{7.4}$$

then the relationship between $\hat{v}(t)$ and $\hat{i}(t)$ is approximately linear, $\hat{v}(t) = r_D \hat{i}(t)$. The conductance $1/r_D$

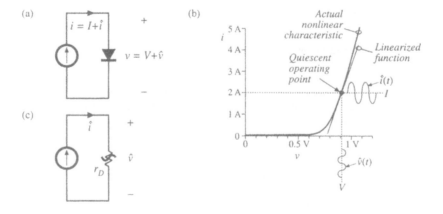

Fig.   7.4 Small-signal equivalent circuit modeling of the diode: (a) a nonlinear diode conducting current i; (b) linearization of the diode characteristic around a quiescent operating point; (c) a linearized small-signal model.

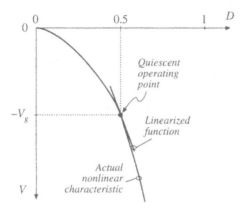

Fig. 7.5 Linearization of the static control-to-output characteristic of the buck-boost converter about the quiescent operating point $D = 0.5$.

represents the slope of the diode characteristic, evaluated at the quiescent operating point. The small-signal equivalent circuit model of Fig. 7.4(c) describes the diode behavior for small variations around the quiescent operating point.

An example of a nonlinear converter characteristic is the dependence of the steady-state output voltage V of the buck-boost converter on the duty cycle D, illustrated in Fig. 7.5. Suppose that the converter operates with some dc output voltage, say, $V = -V_g$, corresponding to a quiescent duty cycle of $D = 0.5$. Duty cycle variations $\hat{d}$ about this quiescent value will excite variations $\hat{v}$ in the output voltage. If the magnitude of the duty cycle variation is sufficiently small, then we can compute the resulting output voltage variations by linearizing the curve. The slope of the linearized characteristic in Fig. 7.5 is chosen to be equal to the slope of the actual nonlinear characteristic at the quiescent operating point; this slope is the dc control-to-output gain of the converter. The linearized and nonlinear characteristics are approximately equal in value provided that the duty cycle variations $\hat{d}$ are sufficiently small.

Although it illustrates the process of small-signal linearization, the buck-boost example of Fig. 7.5 is oversimplified. The inductors and capacitors of the converter cause the gain to exhibit a frequency response. To correctly predict the poles and zeroes of the small-signal transfer functions, we must linearize the converter averaged differential equations, Eqs. (7.2). This is done in Section 7.2. A small-signal ac equivalent circuit can then be constructed using the methods developed in Chapter 3. The resulting small-signal model of the buck-boost converter is illustrated in Fig. 7.6; this model can be solved using conventional circuit analysis techniques, to find the small-signal transfer functions, output impedance, and other frequency-dependent properties. In systems such as Fig. 7.1, the equivalent circuit model can be inserted in place of the converter. When small-signal models of the other system elements (such as the

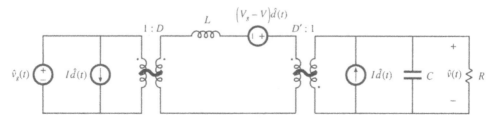

Fig. 7.6 Small-signal ac equivalent circuit model of the buck-boost converter.

pulse-width modulator) are inserted, then a complete linearized system model is obtained. This model can be analyzed using standard linear techniques, such as the Laplace transform, to gain insight into the behavior and properties of the system.

Two well-known variants of the ac modeling method, state-space averaging and circuit averaging, are explained in Sections 7.3 and 7.4. An extension of circuit averaging, known as *averaged switch modeling*, is also discussed in Section 7.4. Since the switches are the only elements that introduce switching harmonics, equivalent circuit models can be derived by averaging only the switch waveforms. The converter models suitable for analysis or simulation are obtained simply by replacing the switches with the averaged switch model. The averaged switch modeling technique can be extended to other modes of operation such as the discontinuous conduction mode, as well as to current programmed control and to resonant converters. In Section 7.5, it is shown that the small-signal model of any dc–dc pulse-width modulated CCM converter can be written in a standard form. Called the *canonical model*, this equivalent circuit describes the basic physical functions that any of these converters must perform. A simple model of the pulse-width modulator circuit is described in Section 7.6.

These models are useless if you don't know how to apply them. So in Chapter 8, the frequency response of converters is explored, in a design-oriented and detailed manner. Small-signal transfer functions of the basic converters are tabulated. Bode plots of converter transfer functions and impedances are derived in a simple, approximate manner, which allows insight to be gained into the origins of the frequency response of complex converter systems.

These results are used to design converter control systems in Chapter 9 and input filters in Chapter 10. The modeling techniques are extended in Chapters 11 and 12 to cover the discontinuous conduction mode and the current programmed mode.

7.2   THE BASIC AC MODELING APPROACH

Let us derive a small-signal ac model of the buck-boost converter of Fig. 7.7. The analysis begins as usual, by determining the voltage and current waveforms of the inductor and capacitor. When the switch is in position 1, the circuit of Fig. 7.8(a) is obtained. The inductor voltage and capacitor current are:

$$v_L(t) = L \frac{di(t)}{dt} = v_g(t) \tag{7.5}$$

$$i_C(t) = C \frac{dv(t)}{dt} = -\frac{v(t)}{R} \tag{7.6}$$

We now make the small-ripple approximation. But rather than replacing $v_g(t)$ and $v(t)$ with their dc components V_g and V as in Chapter 2, we now replace them with their low-frequency averaged values $\langle v_g(t)\rangle_{T_s}$ and $\langle v(t)\rangle_{T_s}$, defined by Eq. (7.3). Equations (7.5) and (7.6) then become

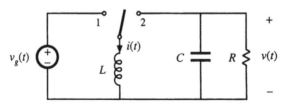

Fig. 7.7   Buck-boost converter example.

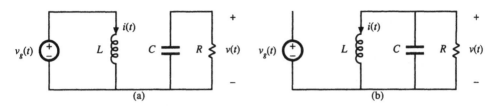

Fig. 7.8 Buck-boost converter circuit: (a) when the switch is in position 1, (b) when the switch is in position 2.

$$v_L(t) = L\frac{di(t)}{dt} \approx \left\langle v_g(t)\right\rangle_{T_s} \tag{7.7}$$

$$i_C(t) = C\frac{dv(t)}{dt} \approx -\frac{\left\langle v(t)\right\rangle_{T_s}}{R} \tag{7.8}$$

Hence, during the first subinterval, the inductor current $i(t)$ and the capacitor voltage $v(t)$ change with the essentially constant slopes given by Eqs. (7.7) and (7.8). With the switch in position 2, the circuit of Fig. 7.8(b) is obtained. Its inductor voltage and capacitor current are:

$$v_L(t) = L\frac{di(t)}{dt} = v(t) \tag{7.9}$$

$$i_C(t) = C\frac{dv(t)}{dt} = -i(t) - \frac{v(t)}{R} \tag{7.10}$$

Use of the small-ripple approximation, to replace $i(t)$ and $v(t)$ with their averaged values, yields

$$v_L(t) = L\frac{di(t)}{dt} \approx \left\langle v(t)\right\rangle_{T_s} \tag{7.11}$$

$$i_C(t) = C\frac{dv(t)}{dt} \approx -\left\langle i(t)\right\rangle_{T_s} - \frac{\left\langle v(t)\right\rangle_{T_s}}{R} \tag{7.12}$$

During the second subinterval, the inductor current and capacitor voltage change with the essentially constant slopes given by Eqs. (7.11) and (7.12).

7.2.1   Averaging the Inductor Waveforms

The inductor voltage and current waveforms are sketched in Fig. 7.9. The low-frequency average of the inductor voltage is found by evaluation of Eq. (7.3)—the inductor voltage during the first and second subintervals, given by Eqs. (7.7) and (7.11), are averaged:

$$\left\langle v_L(t)\right\rangle_{T_s} = \frac{1}{T_s}\int_t^{t+T_s} v_L(\tau)d\tau \approx d(t)\left\langle v_g(t)\right\rangle_{T_s} + d'(t)\left\langle v(t)\right\rangle_{T_s} \tag{7.13}$$

where $d'(t) = 1 - d(t)$. The right-hand side of Eq. (7.13) contains no switching harmonics, and models

(a)

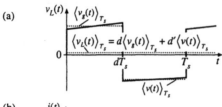

(b)

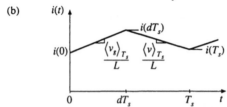

Fig. 7.9 Buck-boost converter waveforms: (a) inductor voltage, (b) inductor current.

only the low-frequency components of the inductor voltage waveform. Insertion of this equation into Eq. (7.2) leads to

$$L \frac{d\langle i(t)\rangle_{T_s}}{dt} = d(t)\langle v_g(t)\rangle_{T_s} + d'(t)\langle v(t)\rangle_{T_s} \tag{7.14}$$

This equation describes how the low-frequency components of the inductor current vary with time.

7.2.2   Discussion of the Averaging Approximation

In steady-state, the actual inductor current waveform $i(t)$ is periodic with period equal to the switching period T_s: $i(t + T_s) = i(t)$. During transients, there is a net change in $i(t)$ over one switching period. This net change in inductor current is correctly predicted by use of the average inductor voltage. We can show that this is true, based on the inductor equation

$$L \frac{di(t)}{dt} = v_L(t) \tag{7.15}$$

Divide by L, and integrate both sides from t to $t + T_s$:

$$\int_{t}^{t+T_s} di = \frac{1}{L} \int_{t}^{t+T_s} v_L(\tau)d\tau \tag{7.16}$$

The left-hand side of Eq. (7.16) is $i(t + T_s) - i(t)$, while the right-hand side can be expressed in terms of the definition of $\langle v_L(t)\rangle_{T_s}$, Eq. (7.3), by multiplying and dividing by T_s to obtain

$$i(t + T_s) - i(t) = \frac{1}{L} T_s \langle v_L(t)\rangle_{T_s} \tag{7.17}$$

The left-hand side of Eq. (7.17) is the net change in inductor current over one complete switching period. Equation (7.17) states that this change is exactly equal to the switching period T_s multiplied by the average slope $\langle v_L(t)\rangle_{T_s}/L$.

Equation (7.17) can be rearranged to obtain

$$L\frac{i(t+T_s)-i(t)}{T_s} = \langle v_L(t)\rangle_{T_s} \tag{7.18}$$

Let us now find the derivative of $\langle i(t)\rangle_{T_s}$:

$$\frac{d\langle i(t)\rangle_{T_s}}{dt} = \frac{d}{dt}\left(\frac{1}{T_s}\int_t^{t+T_s} i(\tau)d\tau\right) = \frac{i(t+T_s)-i(t)}{T_s} \tag{7.19}$$

Substitution of Eq. (7.19) into (7.18) leads to

$$L\frac{d\langle i(t)\rangle_{T_s}}{dt} = \langle v_L(t)\rangle_{T_s} \tag{7.20}$$

which coincides with Eq. (7.2).

Let us next compute how the inductor current changes over one switching period in our buck-boost example. The inductor current waveform is sketched in Fig. 7.9(b). Assume that the inductor current begins at some arbitrary value $i(0)$. During the first subinterval, the inductor current changes with the essentially constant value given by Eq. (7.7). The value at the end of the first subinterval is

$$\underbrace{i(dT_s)}_{} \quad = \quad \underbrace{i(0)}_{} \quad + \quad \underbrace{(dT_s)}_{} \quad \underbrace{\left(\frac{\langle v_g(t)\rangle_{T_s}}{L}\right)}_{} \tag{7.21}$$

(final value) = (initial value) + (length of interval) (average slope)

During the second subinterval, the inductor current changes with the essentially constant value given by Eq. (7.11). Hence, the value at the end of the second subinterval is

$$\underbrace{i(T_s)}_{} \quad = \quad \underbrace{i(dT_s)}_{} \quad + \quad \underbrace{(d'T_s)}_{} \quad \underbrace{\left(\frac{\langle v(t)\rangle_{T_s}}{L}\right)}_{} \tag{7.22}$$

(final value) = (initial value) + (length of interval) (average slope)

By substitution of Eq. (7.21) into Eq. (7.22), we can express $i(T_s)$ in terms of $i(0)$,

$$i(T_s) = i(0) + \frac{T_s}{L}\underbrace{\left(d(t)\langle v_g(t)\rangle_{T_s} + d'(t)\langle v(t)\rangle_{T_s}\right)}_{\langle v_L(t)\rangle_{T_s}} \tag{7.23}$$

Equations (7.21) to (7.23) are illustrated in Fig. 7.10. Equation (7.23) expresses the final value $i(T_s)$ directly in terms of $i(0)$, without the intermediate step of calculating $i(DT_s)$. This equation can be interpreted in the same manner as Eqs. (7.21) and (7.22): the final value $i(T_s)$ is equal to the initial value $i(0)$, plus the length of the interval T_s multiplied by the average slope $\langle v_L(t)\rangle_{T_s}/L$. But note that the interval length is chosen to coincide with the switching period, such that the switching ripple is effectively

Fig.   7.10   Use of the average slope to predict how the inductor current waveform changes over one switching period. The actual waveform $i(t)$ and its low-frequency component $\langle i(t)\rangle_{T_s}$ are illustrated.

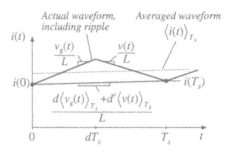

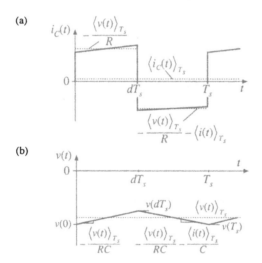

(a)

(b)

Fig.   7.11   Buck-boost converter waveforms: (a) capacitor current, (b) capacitor voltage.

removed. Also, the use of the average slope leads to correct prediction of the final value $i(T_s)$. It can be easily verified that, when Eq. (7.23) is inserted into Eq. (7.19), the previous result (7.14) is obtained.

7.2.3   Averaging the Capacitor Waveforms

A similar procedure leads to the capacitor dynamic equation. The capacitor voltage and current waveforms are sketched in Fig. 7.11. The average capacitor current can be found by averaging Eqs. (7.8) and (7.12); the result is

$$\langle i_C(t)\rangle_{T_s} = d(t)\left(-\frac{\langle v(t)\rangle_{T_s}}{R}\right) + d'(t)\left(-\langle i(t)\rangle_{T_s} - \frac{\langle v(t)\rangle_{T_s}}{R}\right) \tag{7.24}$$

Upon inserting this equation into Eq. (7.2) and collecting terms, one obtains

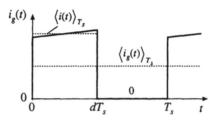

Fig. 7.12 Buck-boost converter waveforms: input source current $i_g(t)$.

$$C \frac{d\langle v(t)\rangle_{T_s}}{dt} = -d'(t)\langle i(t)\rangle_{T_s} - \frac{\langle v(t)\rangle_{T_s}}{R} \tag{7.25}$$

This is the basic averaged equation which describes dc and low-frequency ac variations in the capacitor voltage.

7.2.4 The Average Input Current

In Chapter 3, it was found to be necessary to write an additional equation that models the dc component of the converter input current. This allowed the input port of the converter to be modeled by the dc equivalent circuit. A similar procedure must be followed here, so that low-frequency variations at the converter input port are modeled by the ac equivalent circuit.

For the buck-boost converter example, the current $i_g(t)$ drawn by the converter from the input source is equal to the inductor current $i(t)$ during the first subinterval, and zero during the second subinterval. By neglecting the inductor current ripple and replacing $i(t)$ with its averaged value $\langle i(t)\rangle_{T_s}$, we can express the input current as follows:

$$i_g(t) = \begin{cases} \langle i(t)\rangle_{T_s} & \text{during subinterval 1} \\ 0 & \text{during subinterval 2} \end{cases} \tag{7.26}$$

The input current waveform is illustrated in Fig. 7.12. Upon averaging over one switching period, one obtains

$$\langle i_g(t)\rangle_{T_s} = d(t)\langle i(t)\rangle_{T_s} \tag{7.27}$$

This is the basic averaged equation which describes dc and low-frequency ac variations in the converter input current.

7.2.5 Perturbation and Linearization

The buck-boost converter averaged equations, Eqs. (7.14), (7.25), and (7.27), are collected below:

$$L\frac{d\langle i(t)\rangle_{T_s}}{dt} = d(t)\langle v_g(t)\rangle_{T_s} + d'(t)\langle v(t)\rangle_{T_s}$$

$$C\frac{d\langle v(t)\rangle_{T_s}}{dt} = -d'(t)\langle i(t)\rangle_{T_s} - \frac{\langle v(t)\rangle_{T_s}}{R} \tag{7.28}$$

$$\langle i_g(t)\rangle_{T_s} = d(t)\langle i(t)\rangle_{T_s}$$

These equations are nonlinear because they involve the multiplication of time-varying quantities. For example, the capacitor current depends on the product of the control input $d'(t)$ and the low-frequency component of the inductor current, $\langle i(t)\rangle_{T_s}$. Multiplication of time-varying signals generates harmonics, and is a nonlinear process. Most of the techniques of ac circuit analysis, such as the Laplace transform and other frequency-domain methods, are not useful for nonlinear systems. So we need to linearize Eqs. (7.28) by constructing a small-signal model.

Suppose that we drive the converter at some steady-state, or quiescent, duty ratio $d(t) = D$, with quiescent input voltage $v_g(t) = V_g$. We know from our steady-state analysis of Chapters 2 and 3 that, after any transients have subsided, the inductor current $\langle i(t)\rangle_{T_s}$, the capacitor voltage $\langle v(t)\rangle_{T_s}$, and the input current $\langle i_g(t)\rangle_{T_s}$ will reach the quiescent values I, V, and I_g, respectively, where

$$V = -\frac{D}{D'}V_g$$

$$I = -\frac{V}{D'R} \tag{7.29}$$

$$I_g = DI$$

Equations (7.29) are derived as usual via the principles of inductor volt-second and capacitor charge balance. They could also be derived from Eqs. (7.28) by noting that, in steady state, the derivatives must equal zero.

To construct a small-signal ac model at a quiescent operating point (I, V), one assumes that the input voltage $v_g(t)$ and the duty cycle $d(t)$ are equal to some given quiescent values V_g and D, plus some superimposed small ac variations $\hat{v}_g(t)$ and $\hat{d}(t)$. Hence, we have

$$\langle v_g(t)\rangle_{T_s} = V_g + \hat{v}_g(t)$$

$$d(t) = D + \hat{d}(t) \tag{7.30}$$

In response to these inputs, and after any transients have subsided, the averaged inductor current $\langle i(t)\rangle_{T_s}$, the averaged capacitor voltage $\langle v(t)\rangle_{T_s}$, and the averaged input current $\langle i_g(t)\rangle_{T_s}$ waveforms will be equal to the corresponding quiescent values I, V, and I_g, plus some superimposed small ac variations $\hat{i}(t)$, $\hat{v}(t)$, and $\hat{i}_g(t)$:

$$\langle i(t)\rangle_{T_s} = I + \hat{i}(t)$$

$$\langle v(t)\rangle_{T_s} = V + \hat{v}(t) \tag{7.31}$$

$$\langle i_g(t)\rangle_{T_s} = I_g + \hat{i}_g(t)$$

With the assumptions that the ac variations are small in magnitude compared to the dc quiescent values, or

$$\left|\hat{v}_g(t)\right| \ll \left|V_g\right|$$
$$\left|\hat{d}(t)\right| \ll \left|D\right|$$
$$\left|\hat{i}(t)\right| \ll \left|I\right| \qquad (7.32)$$
$$\left|\hat{v}(t)\right| \ll \left|V\right|$$
$$\left|\hat{i}_g(t)\right| \ll \left|I_g\right|$$

then the nonlinear equations (7.28) can be linearized. This is done by inserting Eqs. (7.30) and (7.31) into Eq. (7.28). For the inductor equation, one obtains

$$L \frac{d\left(I + \hat{i}(t)\right)}{dt} = \left(D + \hat{d}(t)\right)\left(V_g + \hat{v}_g(t)\right) + \left(D' - \hat{d}(t)\right)\left(V + \hat{v}(t)\right) \qquad (7.33)$$

It should be noted that the complement of the duty cycle is given by

$$d'(t) = \left(1 - d(t)\right) = 1 - \left(D + \hat{d}(t)\right) = D' - \hat{d}(t) \qquad (7.34)$$

where $D' = 1 - D$. The minus sign arises in the expression for $d'(t)$ because a $d(t)$ variation that causes $d(t)$ to increase will cause $d'(t)$ to decrease.

By multiplying out Eq. (7.33) and collecting terms, one obtains

$$L\left(\frac{dI}{dt} + \frac{d\hat{i}(t)}{dt}\right) = \underbrace{\left(DV_g + D'V\right)}_{\text{Dc terms}} + \underbrace{\left(D\hat{v}_g(t) + D'\hat{v}(t) + \left(V_g - V\right)\hat{d}(t)\right)}_{\substack{1^{st} \text{ order ac terms} \\ \text{(linear)}}} + \underbrace{\hat{d}(t)\left(\hat{v}_g(t) - \hat{v}(t)\right)}_{\substack{2^{nd} \text{ order ac terms} \\ \text{(nonlinear)}}} \qquad (7.35)$$

The derivative of I is zero, since I is by definition a dc (constant) term. For the purposes of deriving a small-signal ac model, the dc terms can be considered known constant quantities. On the right-hand side of Eq. (7.35), three types of terms arise:

Dc terms: These terms contain dc quantities only.

First-order ac terms: Each of these terms contains a single ac quantity, usually multiplied by a constant coefficient such as a dc term. These terms are linear functions of the ac variations.

Second-order ac terms: These terms contain the products of ac quantities. Hence they are nonlinear, because they involve the multiplication of time-varying signals.

It is desired to neglect the nonlinear ac terms. Provided that the small-signal assumption, Eq. (7.32), is satisfied, then each of the second-order nonlinear terms is much smaller in magnitude that one or more of the linear first-order ac terms. For example, the second-order ac term $\hat{d}(t)\hat{v}_g(t)$ is much smaller in magnitude than the first-order ac term $D\hat{v}_g(t)$ whenever $\left|\hat{d}(t)\right| \ll D$. So we can neglect the second-order terms. Also, by definition [or by use of Eq. (7.29)], the dc terms on the right-hand side of the equation are equal to the dc terms on the left-hand side, or zero.

We are left with the first-order ac terms on both sides of the equation. Hence,

$$L \frac{d\hat{i}(t)}{dt} = D\hat{v}_g(t) + D'\hat{v}(t) + \left(V_g - V\right)\hat{d}(t) \qquad (7.36)$$

This is the desired result: the small-signal linearized equation that describes variations in the inductor current.

The capacitor equation can be linearized in a similar manner. Insertion of Eqs. (7.30) and (7.31) into the capacitor equation of Eq. (7.28) yields

$$C\frac{d\left(V + \hat{v}(t)\right)}{dt} = -\left(D' - \hat{d}(t)\right)\left(I + \hat{i}(t)\right) - \frac{\left(V + \hat{v}(t)\right)}{R} \tag{7.37}$$

Upon multiplying out Eq. (7.37) and collecting terms, one obtains

$$C\left(\frac{dV}{dt} + \frac{d\hat{v}(t)}{dt}\right) = \underbrace{\left(-D'I - \frac{V}{R}\right)}_{\text{Dc terms}} + \underbrace{\left(-D'\hat{i}(t) - \frac{\hat{v}(t)}{R} + I\hat{d}(t)\right)}_{\substack{1^{st}\text{ order ac terms} \\ \text{(linear)}}} + \underbrace{\hat{d}(t)\hat{i}(t)}_{\substack{2^{nd}\text{ order ac term} \\ \text{(nonlinear)}}} \tag{7.38}$$

By neglecting the second-order terms, and noting that the dc terms on both sides of the equation are equal, we again obtain a linearized first-order equation, containing only the first-order ac terms of Eq. (7.38):

$$C\frac{d\hat{v}(t)}{dt} = -D'\hat{i}(t) - \frac{\hat{v}(t)}{R} + I\hat{d}(t) \tag{7.39}$$

This is the desired small-signal linearized equation that describes variations in the capacitor voltage.

Finally, the equation of the average input current is also linearized. Insertion of Eqs. (7.30) and (7.31) into the input current equation of Eq. (7.28) yields

$$I_g + \hat{i}_g(t) = \left(D + \hat{d}(t)\right)\left(I + \hat{i}(t)\right) \tag{7.40}$$

By collecting terms, we obtain

$$\underbrace{I_g}_{\text{Dc term}} + \underbrace{\hat{i}_g(t)}_{1^{st}\text{ order ac term}} = \underbrace{\left(DI\right)}_{\text{Dc term}} + \underbrace{\left(D\hat{i}(t) + I\hat{d}(t)\right)}_{\substack{1^{st}\text{ order ac terms} \\ \text{(linear)}}} + \underbrace{\hat{d}(t)\hat{i}(t)}_{\substack{2^{nd}\text{ order ac term} \\ \text{(nonlinear)}}} \tag{7.41}$$

We again neglect the second-order nonlinear terms. The dc terms on both sides of the equation are equal. The remaining first-order linear ac terms are

$$\hat{i}_g(t) = D\hat{i}(t) + I\hat{d}(t) \tag{7.42}$$

This is the linearized small-signal equation that describes the low-frequency ac components of the converter input current.

In summary, the nonlinear averaged equations of a switching converter can be linearized about a quiescent operating point. The converter independent inputs are expressed as constant (dc) values, plus small ac variations. In response, the converter averaged waveforms assume similar forms. Insertion of Eqs. (7.30) and (7.31) into the converter averaged nonlinear equations yields dc terms, linear ac terms, and nonlinear terms. If the ac variations are sufficiently small in magnitude, then the nonlinear terms are much smaller than the linear ac terms, and so can be neglected. The remaining linear ac terms comprise the small-signal ac model of the converter.

7.2.6 Construction of the Small-Signal Equivalent Circuit Model

Equations (7.36), (7.39), and (7.42) are the small-signal ac description of the ideal buck-boost converter, and are collected below:

$$L\frac{d\hat{i}(t)}{dt} = D\hat{v}_g(t) + D'\hat{v}(t) + \left(V_g - V\right)\hat{d}(t)$$

$$C\frac{d\hat{v}(t)}{dt} = -D'\hat{i}(t) - \frac{\hat{v}(t)}{R} + I\hat{d}(t) \tag{7.43}$$

$$\hat{i}_g(t) = D\hat{i}(t) + I\hat{d}(t)$$

In Chapter 3, we collected the averaged dc equations of a converter, and reconstructed an equivalent circuit that modeled the dc properties of the converter. We can use the same procedure here, to construct averaged small-signal ac models of converters.

The inductor equation of (7.43), or Eq. (7.36), describes the voltages around a loop containing the inductor. Indeed, this equation was derived by finding the inductor voltage via loop analysis, then averaging, perturbing, and linearizing. So the equation represents the voltages around a loop of the small-signal model, which contains the inductor. The loop current is the small-signal ac inductor current $\hat{i}(t)$. As illustrated in Fig. 7.13, the term $L\,d\hat{i}(t)/dt$ represents the voltage across the inductor L in the small-signal model. This voltage is equal to three other voltage terms. $D\hat{v}_g(t)$ and $D'\hat{v}(t)$ represent dependent sources as shown. These terms will be combined into ideal transformers. The term $(V_g - V)\hat{d}(t)$ is driven by the control input $\hat{d}(t)$, and is represented by an independent source as shown.

The capacitor equation of (7.43), or Eq. (7.39), describes the currents flowing into a node attached to the capacitor. This equation was derived by finding the capacitor current via node analysis, then averaging, perturbing, and linearizing. Hence, this equation describes the currents flowing into a node of the small-signal model, attached to the capacitor. As illustrated in Fig. 7.14, the term $C\,d\hat{v}(t)/dt$ represents the current flowing through capacitor C in the small-signal model. The capacitor voltage is $\hat{v}(t)$. According to the equation, this current is equal to three other terms. The term $-D'\hat{i}(t)$ represents a dependent source, which will eventually be combined into an ideal transformer. The term $-\hat{v}(t)/R$ is rec-

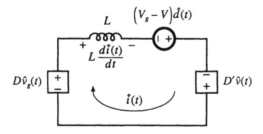

Fig. 7.13 Circuit equivalent to the small-signal ac inductor loop equation of Eq. (7.43) or (7.36).

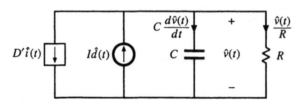

Fig. 7.14   Circuit equivalent to the small-signal ac capacitor node equation of Eq. (7.43) or (7.39).

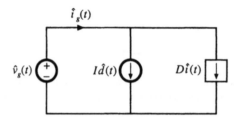

Fig. 7.15   Circuit equivalent to the small-signal ac input source current equation of Eq. (7.43) or (7.42).

ognized as the current flowing through the load resistor in the small-signal model. The resistor is connected in parallel with the capacitor, such that the ac voltage across the resistor R is $\hat{v}(t)$ as expected. The term $I\hat{d}(t)$ is driven by the control input $\hat{d}(t)$, and is represented by an independent source as shown.

Finally, the input current equation of (7.43), or Eq. (7.42), describes the small-signal ac current $\hat{i}_g(t)$ drawn by the converter out of the input voltage source $\hat{v}_g(t)$. This is a node equation which states that $\hat{i}_g(t)$ is equal to the currents in two branches, as illustrated in Fig. 7.15. The first branch, corresponding to the $D\hat{i}(t)$ term, is dependent on the ac inductor current $\hat{i}(t)$. Hence, we represent this term using a dependent current source; this source will eventually be incorporated into an ideal transformer. The second branch, corresponding to the $I\hat{d}(t)$ term, is driven by the control input $\hat{d}(t)$, and is represented by an independent source as shown.

The circuits of Figs. 7.13, 7.14, and 7.15 are collected in Fig. 7.16(a). As discussed in Chapter 3, the dependent sources can be combined into effective ideal transformers, as illustrated in Fig. 7.16(b). The sinusoid superimposed on the transformer symbol indicates that the transformer is ideal, and is part of the averaged small-signal ac model. So the effective dc transformer property of CCM dc-dc converters also influences small-signal ac variations in the converter signals.

The equivalent circuit of Fig. 7.16(b) can now be solved using techniques of conventional linear circuit analysis, to find the converter transfer functions, input and output impedances, etc. This is done in detail in the next chapter. Also, the model can be refined by inclusion of losses and other nonidealities—an example is given in Section 7.2.9.

7.2.7   Discussion of the Perturbation and Linearization Step

In the perturbation and linearization step, it is assumed that an averaged voltage or current consists of a constant (dc) component and a small-signal ac variation around the dc component. In Section 7.2.5, the

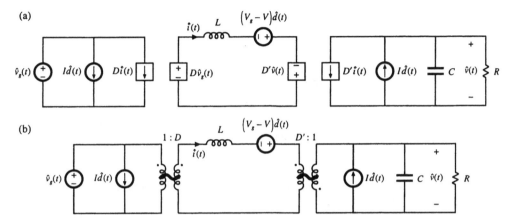

Fig. 7.16 Buck-boost converter small-signal ac equivalent circuit: (a) the circuits of Figs. 7.13 to 7.15, collected together; (b) combination of dependent sources into effective ideal transformer, leading to the final model.

linearization step was completed by neglecting nonlinear terms that correspond to products of the small-signal ac variations. In general, the linearization step amounts to taking the Taylor expansion of a nonlinear relation and retaining only the constant and linear terms. For example, the large-signal averaged equation for the inductor current in Eq. (7.28) can be written as:

$$L \frac{d\langle i(t)\rangle_{T_s}}{dt} = d(t) \langle v_g(t)\rangle_{T_s} + d'(t) \langle v(t)\rangle_{T_s} = f_1\left(\langle v_g(t)\rangle_{T_s}, \langle v(t)\rangle_{T_s}, d(t)\right) \tag{7.44}$$

Let us expand this expression in a three-dimensional Taylor series, about the quiescent operating point (V_g, V, D):

$$L\left(\frac{dI}{dt} + \frac{d\hat{i}(t)}{dt}\right) = f_1\left(V_g, V, D\right) + \hat{v}_g(t) \left.\frac{\partial f_1\left(v_g, V, D\right)}{\partial v_g}\right|_{v_g = V_g}$$

$$+ \hat{v}(t) \left.\frac{\partial f_1\left(V_g, v, D\right)}{\partial v}\right|_{v = V} + \hat{d}(t) \left.\frac{\partial f_1\left(V_g, V, d\right)}{\partial d}\right|_{d = D} \tag{7.45}$$

+ higher-order nonlinear terms

For simplicity of notation, the angle brackets denoting average values are dropped in the above equation. The derivative of I is zero, since I is by definition a dc (constant) term. Equating the dc terms on both sides of Eq. (7.45) gives:

$$0 = f_1\left(V_g, V, D\right) \tag{7.46}$$

which is the volt-second balance relationship for the inductor. The coefficients with the linear terms on the right-hand side of Eq. (7.45) are found as follows:

$$\left.\frac{\partial f_1(v_g, V, D)}{\partial v_g}\right|_{v_g = V_g} = D \tag{7.47}$$

$$\left.\frac{\partial f_1(V_g, v, D)}{\partial v}\right|_{v = V} = D' \tag{7.48}$$

$$\left.\frac{\partial f_1(V_g, V, d)}{\partial d}\right|_{d = D} = V_g - V \tag{7.49}$$

Using (7.47), (7.48) and (7.49), neglecting higher-order nonlinear terms, and equating the linear ac terms on both sides of Eq. (7.45) gives:

$$L\frac{d\hat{\imath}(t)}{dt} = D\hat{v}_g(t) + D'\hat{v}(t) + \left(V_g - V\right)\hat{d}(t) \tag{7.50}$$

which is identical to Eq. (7.36) derived in Section 7.2.5. In conclusion, the linearization step can always be accomplished using the Taylor expansion.

7.2.8   Results for Several Basic Converters

The equivalent circuit models for the buck, boost, and buck-boost converters operating in the continuous conduction mode are summarized in Fig. 7.17. The buck and boost converter models contain ideal transformers having turns ratios equal to the converter conversion ratio. The buck-boost converter contains ideal transformers having buck and boost conversion ratios; this is consistent with the derivation of Section 6.1.2 of the buck-boost converter as a cascade connection of buck and boost converters. These models can be solved to find the converter transfer functions, input and output impedances, inductor current variations, etc. By insertion of appropriate turns ratios, the equivalent circuits of Fig. 7.17 can be adapted to model the transformer-isolated versions of the buck, boost, and buck-boost converters, including the forward, flyback, and other converters.

7.2.9   Example: A Nonideal Flyback Converter

To illustrate that the techniques of the previous section are useful for modeling a variety of converter phenomena, let us next derive a small-signal ac equivalent circuit of a converter containing transformer isolation and resistive losses. An isolated flyback converter is illustrated in Fig. 7.18. The flyback transformer has magnetizing inductance L, referred to the primary winding, and turns ratio $1{:}n$. MOSFET Q_1 has on-resistance R_{on}. Other loss elements, as well as the transformer leakage inductances and the switching losses, are negligible. The ac modeling of this converter begins in a manner similar to the dc converter analysis of Section 6.3.4. The flyback transformer is replaced by an equivalent circuit consisting of the magnetizing inductance L in parallel with an ideal transformer, as illustrated in Fig. 7.19(a).

During the first subinterval, when MOSFET Q_1 conducts, diode D_1 is off. The circuit then

(a)

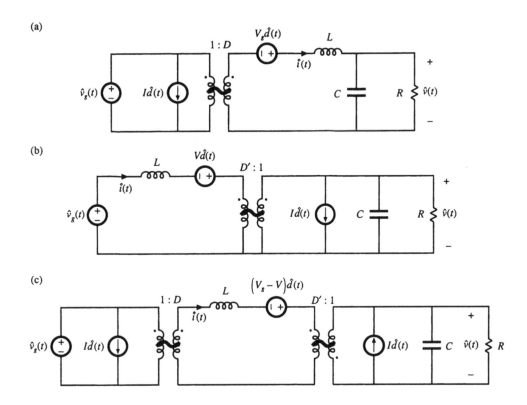

(b)

(c)

Fig. 7.17 Averaged small-signal ac models for several basic converters operating in continuous conduction mode: (a) buck, (b) boost, (c) buck-boost.

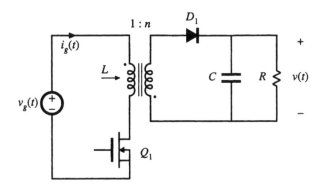

Fig. 7.18 Flyback converter example.

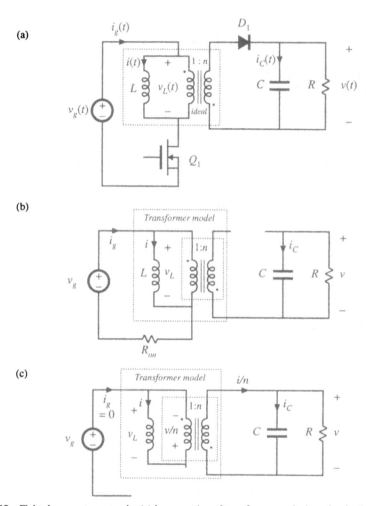

Fig. 7.19 Flyback converter example: (a) incorporation of transformer equivalent circuit, (b) circuit during sub-interval 1, (c) circuit during subinterval 2.

reduces to Fig. 7.19(b). The inductor voltage $v_L(t)$, capacitor current $i_C(t)$, and converter input current $i_g(t)$ are:

$$v_L(t) = v_g(t) - i(t)R_{on}$$
$$i_C(t) = -\frac{v(t)}{R} \qquad (7.51)$$
$$i_g(t) = i(t)$$

We next make the small ripple approximation, replacing the voltages and currents with their average values as defined by Eq. (7.3), to obtain

$$v_L(t) = \langle v_g(t) \rangle_{T_s} - \langle i(t) \rangle_{T_s} R_{on}$$
$$i_C(t) = -\frac{\langle v(t) \rangle_{T_s}}{R} \tag{7.52}$$
$$i_g(t) = \langle i(t) \rangle_{T_s}$$

During the second subinterval, MOSFET Q_1 is off, diode D_1 conducts, and the circuit of Fig. 7.19(c) is obtained. Analysis of this circuit shows that the inductor voltage, capacitor current, and input current are given by

$$v_L(t) = -\frac{v(t)}{n}$$
$$i_C(t) = \frac{i(t)}{n} - \frac{v(t)}{R} \tag{7.53}$$
$$i_g(t) = 0$$

The small-ripple approximation leads to

$$v_L(t) = -\frac{\langle v(t) \rangle_{T_s}}{n}$$
$$i_C(t) = \frac{\langle i(t) \rangle_{T_s}}{n} - \frac{\langle v(t) \rangle_{T_s}}{R} \tag{7.54}$$
$$i_g(t) = 0$$

The inductor voltage and current waveforms are sketched in Fig. 7.20. The average inductor voltage can now be found by averaging the waveform of Fig. 7.20(a) over one switching period. The result is

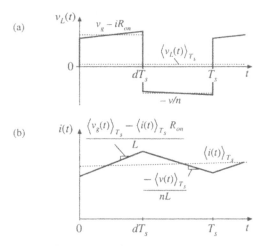

Fig.   7.20   Inductor waveforms for the flyback example: (a) inductor voltage, (b) inductor current.

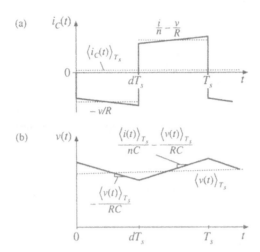

(a)

(b)

Fig. 7.21 Capacitor waveforms for the flyback example: (a) capacitor current, (b) capacitor voltage.

$$\left\langle v_L(t)\right\rangle_{T_s} = d(t)\left(\left\langle v_g(t)\right\rangle_{T_s} - \left\langle i(t)\right\rangle_{T_s} R_{on}\right) + d'(t)\left(\frac{-\left\langle v(t)\right\rangle_{T_s}}{n}\right) \tag{7.55}$$

By inserting this result into Eq. (7.20), we obtain the averaged inductor equation,

$$L\frac{d\left\langle i(t)\right\rangle_{T_s}}{dt} = d(t)\left\langle v_g(t)\right\rangle_{T_s} - d(t)\left\langle i(t)\right\rangle_{T_s} R_{on} - d'(t)\frac{\left\langle v(t)\right\rangle_{T_s}}{n} \tag{7.56}$$

The capacitor waveforms are constructed in Fig. 7.21. The average capacitor current is

$$\left\langle i_C(t)\right\rangle_{T_s} = d(t)\left(\frac{-\left\langle v(t)\right\rangle_{T_s}}{R}\right) + d'(t)\left(\frac{\left\langle i(t)\right\rangle_{T_s}}{n} - \frac{\left\langle v(t)\right\rangle_{T_s}}{R}\right) \tag{7.57}$$

This leads to the averaged capacitor equation

$$C\frac{d\left\langle v(t)\right\rangle_{T_s}}{dt} = d'(t)\frac{\left\langle i(t)\right\rangle_{T_s}}{n} - \frac{\left\langle v(t)\right\rangle_{T_s}}{R} \tag{7.58}$$

The converter input current $i_g(t)$ is sketched in Fig. 7.22. Its average is

$$\left\langle i_g(t)\right\rangle_{T_s} = d(t)\left\langle i(t)\right\rangle_{T_s} \tag{7.59}$$

The averaged converter equations (7.56), (7.58) and (7.59) are collected below:

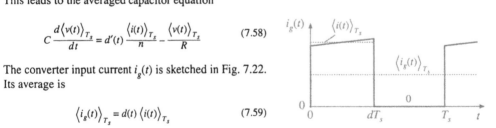

Fig. 7.22 Input source current waveform, flyback example.

$$L \frac{d\langle i(t)\rangle_{T_s}}{dt} = d(t) \langle v_g(t)\rangle_{T_s} - d(t) \langle i(t)\rangle_{T_s} R_{on} - d'(t) \frac{\langle v(t)\rangle_{T_s}}{n}$$

$$C \frac{d\langle v(t)\rangle_{T_s}}{dt} = d'(t) \frac{\langle i(t)\rangle_{T_s}}{n} - \frac{\langle v(t)\rangle_{T_s}}{R} \qquad (7.60)$$

$$\langle i_g(t)\rangle_{T_s} = d(t) \langle i(t)\rangle_{T_s}$$

This is a nonlinear set of differential equations, and hence the next step is to perturb and linearize, to construct the converter small-signal ac equations. We assume that the converter input voltage $v_g(t)$ and duty cycle $d(t)$ can be expressed as quiescent values plus small ac variations, as follows:

$$\langle v_g(t)\rangle_{T_s} = V_g + \hat{v}_g(t)$$

$$d(t) = D + \hat{d}(t) \qquad (7.61)$$

In response to these inputs, and after all transients have decayed, the average converter waveforms can also be expressed as quiescent values plus small ac variations:

$$\langle i(t)\rangle_{T_s} = I + \hat{i}(t)$$

$$\langle v(t)\rangle_{T_s} = V + \hat{v}(t) \qquad (7.62)$$

$$\langle i_g(t)\rangle_{T_s} = I_g + \hat{i}_g(t)$$

With these substitutions, the large-signal averaged inductor equation becomes

$$L \frac{d\left(I + \hat{i}(t)\right)}{dt} = \left(D + \hat{d}(t)\right)\left(V_g + \hat{v}_g(t)\right) - \left(D' - \hat{d}(t)\right)\frac{\left(V + \hat{v}(t)\right)}{n} - \left(D + \hat{d}(t)\right)\left(I + \hat{i}(t)\right)R_{on} \qquad (7.63)$$

Upon multiplying this expression out and collecting terms, we obtain

$$L\left(\frac{dI}{dt} + \frac{d\hat{i}(t)}{dt}\right) = \underbrace{\left(DV_g - D'\frac{V}{n} - DR_{on}I\right)}_{\text{Dc terms}} + \underbrace{\left(D\hat{v}_g(t) - D'\frac{\hat{v}(t)}{n} + \left(V_g + \frac{V}{n} - IR_{on}\right)\hat{d}(t) - DR_{on}\hat{i}(t)\right)}_{1^{st}\text{ order ac terms (linear)}}$$

$$+ \underbrace{\left(\hat{d}(t)\hat{v}_g(t) + \hat{d}(t)\frac{\hat{v}(t)}{n} - \hat{d}(t)\hat{i}(t)R_{on}\right)}_{2^{nd}\text{ order ac terms (nonlinear)}} \qquad (7.64)$$

As usual, this equation contains three types of terms. The dc terms contain no time-varying quantities. The first-order ac terms are linear functions of the ac variations in the circuit, while the second-order ac terms are functions of the products of the ac variations. If the small-signal assumptions of Eq. (7.32) are satisfied, then the second-order terms are much smaller in magnitude that the first-order terms, and hence can be neglected. The dc terms must satisfy

$$0 = DV_g - D'\frac{V}{n} - DR_{on}I \tag{7.65}$$

This result could also be derived by applying the principle of inductor volt-second balance to the steady-state inductor voltage waveform. The first-order ac terms must satisfy

$$L\frac{d\hat{i}(t)}{dt} = D\hat{v}_g(t) - D'\frac{\hat{v}(t)}{n} + \left(V_g + \frac{V}{n} - IR_{on}\right)\hat{d}(t) - DR_{on}\hat{i}(t) \tag{7.66}$$

This is the linearized equation that describes ac variations in the inductor current.

Upon substitution of Eqs. (7.61) and (7.62) into the averaged capacitor equation (7.60), one obtains

$$C\frac{d\left(V + \hat{v}(t)\right)}{dt} = \left(D' - \hat{d}(t)\right)\frac{\left(I + \hat{i}(t)\right)}{n} - \frac{\left(V + \hat{v}(t)\right)}{R} \tag{7.67}$$

By collecting terms, we obtain

$$C\left(\frac{dV}{dt} + \frac{d\hat{v}(t)}{dt}\right) = \underbrace{\left(\frac{D'I}{n} - \frac{V}{R}\right)}_{\text{Dc terms}} + \underbrace{\left(\frac{D'\hat{i}(t)}{n} - \frac{\hat{v}(t)}{R} - \frac{I\hat{d}(t)}{n}\right)}_{\substack{1^{st}\text{ order ac terms}\\(\text{linear})}} - \underbrace{\frac{\hat{d}(t)\hat{i}(t)}{n}}_{\substack{2^{nd}\text{ order ac term}\\(\text{nonlinear})}} \tag{7.68}$$

We neglect the second-order terms. The dc terms of Eq. (7.68) must satisfy

$$0 = \left(\frac{D'I}{n} - \frac{V}{R}\right) \tag{7.69}$$

This result could also be obtained by use of the principle of capacitor charge balance on the steady-state capacitor current waveform. The first-order ac terms of Eq. (7.68) lead to the small-signal ac capacitor equation

$$C\frac{d\hat{v}(t)}{dt} = \frac{D'\hat{i}(t)}{n} - \frac{\hat{v}(t)}{R} - \frac{I\hat{d}(t)}{n} \tag{7.70}$$

Substitution of Eqs. (7.61) and (7.62) into the averaged input current equation (7.60) leads to

$$I_g + \hat{i}_g(t) = \left(D + \hat{d}(t)\right)\left(I + \hat{i}(t)\right) \tag{7.71}$$

Upon collecting terms, we obtain

$$\underbrace{I_g}_{\text{Dc term}} + \underbrace{\hat{i}_g(t)}_{1^{st}\text{ order ac term}} = \underbrace{\left(DI\right)}_{\text{Dc term}} + \underbrace{\left(D\hat{i}(t) + I\hat{d}(t)\right)}_{\substack{1^{st}\text{ order ac terms}\\(\text{linear})}} + \underbrace{\hat{d}(t)\hat{i}(t)}_{\substack{2^{nd}\text{ order ac term}\\(\text{nonlinear})}} \tag{7.72}$$

The dc terms must satisfy

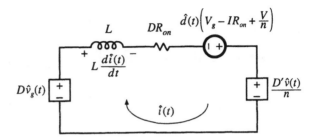

Fig. 7.23 Circuit equivalent to the small-signal ac inductor loop equation, Eq. (7.76) or (7.66).

$$I_g = DI \tag{7.73}$$

We neglect the second-order nonlinear terms of Eq. (7.72), leaving the following linearized ac equation:

$$\hat{i}_g(t) = D\hat{i}(t) + I\hat{d}(t) \tag{7.74}$$

This result models the low-frequency ac variations in the converter input current.

The equations of the quiescent values, Eqs. (7.65), (7.69), and (7.73) are collected below:

$$
\begin{aligned}
0 &= DV_g - D'\frac{V}{n} - DR_{on}I \\
0 &= \left(\frac{D'I}{n} - \frac{V}{R}\right) \\
I_g &= DI
\end{aligned}
\tag{7.75}
$$

For given quiescent values of the input voltage V_g and duty cycle D, this system of equations can be evaluated to find the quiescent output voltage V, inductor current I, and input current dc component I_g. The results are then inserted into the small-signal ac equations.

The small-signal ac equations, Eqs. (7.66), (7.70), and (7.74), are summarized below:

$$
\begin{aligned}
L\frac{d\hat{i}(t)}{dt} &= D\hat{v}_g(t) - D'\frac{\hat{v}(t)}{n} + \left(V_g + \frac{V}{n} - IR_{on}\right)\hat{d}(t) - DR_{on}\hat{i}(t) \\
C\frac{d\hat{v}(t)}{dt} &= \frac{D'\hat{i}(t)}{n} - \frac{\hat{v}(t)}{R} - \frac{I\hat{d}(t)}{n} \\
\hat{i}_g(t) &= D\hat{i}(t) + I\hat{d}(t)
\end{aligned}
\tag{7.76}
$$

The final step is to construct an equivalent circuit that corresponds to these equations.

The inductor equation was derived by first writing loop equations, to find the applied inductor voltage during each subinterval. These equations were then averaged, perturbed, and linearized, to obtain Eq. (7.66). So this equation describes the small-signal ac voltages around a loop containing the inductor. The loop current is the ac inductor current $\hat{i}(t)$. The quantity $Ld\hat{i}(t)/dt$ is the low-frequency ac voltage across the inductor. The four terms on the right-hand side of the equation are the voltages across the four other elements in the loop. The terms $D\hat{v}_g(t)$ and $-D'\hat{v}(t)/n$ are dependent on voltages elsewhere in the converter, and hence are represented as dependent sources in Fig. 7.23. The third term is driven by the duty cycle variations $\hat{d}(t)$ and hence is represented as an independent source. The fourth term, $-DR_{on}\hat{i}(t)$, is a voltage that is proportional to the loop current $\hat{i}(t)$. Hence this term obeys Ohm's law, with effective resistance DR_{on} as shown in the figure. So the influence of the MOSFET on-resistance on the converter

Fig. 7.24 Circuit equivalent to the small-signal ac capacitor node equation, Eq. (7.76) or (7.70).

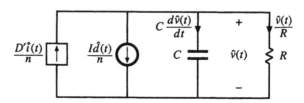

Fig. 7.25 Circuit equivalent to the small-signal ac input source current equation, Eq. (7.76) or (7.74).

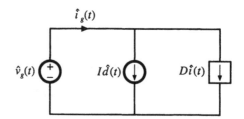

small-signal transfer functions is modeled by an effective resistance of value DR_{on}.

Small-signal capacitor equation (7.70) leads to the equivalent circuit of Fig. 7.24. The equation constitutes a node equation of the equivalent circuit model. It states that the capacitor current $C d\hat{v}(t)/dt$ is equal to three other currents. The current $D'\hat{i}(t)/n$ depends on a current elsewhere in the model, and hence is represented by a dependent current source. The term $-\hat{v}(t)/R$ is the ac component of the load current, which we model with a load resistance R connected in parallel with the capacitor. The last term is driven by the duty cycle variations $\hat{d}(t)$, and is modeled by an independent source.

The input port equation, Eq. (7.74), also constitutes a node equation. It describes the small-signal ac current $\hat{i}_g(t)$, drawn by the converter out of the input voltage source $\hat{v}_g(t)$. There are two other terms in the equation. The term $D\hat{i}(t)$ is dependent on the inductor current ac variation $\hat{i}(t)$, and is represented with a dependent source. The term $I\hat{d}(t)$ is driven by the control variations, and is modeled by an independent source. The equivalent circuit for the input port is illustrated in Fig. 7.25.

The circuits of Figs. 7.23, 7.24, and 7.25 are combined in Fig. 7.26. The dependent sources can be replaced by ideal transformers, leading to the equivalent circuit of Fig. 7.27. This is the desired result: an equivalent circuit that models the low-frequency small-signal variations in the converter waveforms. It can now be solved, using conventional linear circuit analysis techniques, to find the converter transfer functions, output impedance, and other ac quantities of interest.

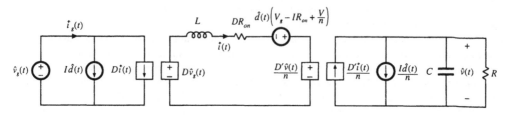

Fig. 7.26 The equivalent circuits of Figs. 7.23 to 7.25, collected together.

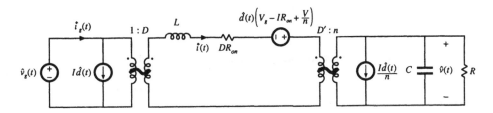

Fig. 7.27 Small-signal ac equivalent circuit model of the flyback converter.

7.3 STATE-SPACE AVERAGING

A number of ac converter modeling techniques have appeared in the literature, including the current-injected approach, circuit averaging, and the state-space averaging method. Although the proponents of a given method may prefer to express the end result in a specific form, the end results of nearly all methods are equivalent. And everybody will agree that averaging and small-signal linearization are the key steps in modeling PWM converters.

The state-space averaging approach [1, 2] is described in this section. The state-space description of dynamical systems is a mainstay of modern control theory; the state-space averaging method makes use of this description to derive the small-signal averaged equations of PWM switching converters. The state-space averaging method is otherwise identical to the procedure derived in Section 7.2. Indeed, the procedure of Section 7.2 amounts to state-space averaging, but without the formality of writing the equations in matrix form. A benefit of the state-space averaging procedure is the generality of its result: a small-signal averaged model can always be obtained, provided that the state equations of the original converter can be written.

Section 7.3.1 summarizes how to write the state equations of a network. The basic results of state-space averaging are described in Section 7.3.2, and a short derivation is given in Section 7.3.3. Section 7.3.4 contains an example, in which the state-space averaging method is used to derive the quiescent dc and small-signal ac equations of a buck-boost converter.

7.3.1 The State Equations of a Network

The state-space description is a canonical form for writing the differential equations that describe a system. For a linear network, the derivatives of the *state variables* are expressed as linear combinations of the system independent inputs and the state variables themselves. The physical state variables of a system are usually associated with the storage of energy, and for a typical converter circuit, the physical state variables are the independent inductor currents and capacitor voltages. Other typical state variables include the position and velocity of a motor shaft. At a given point in time, the values of the state variables depend on the previous history of the system, rather than on the present values of the system inputs. To solve the differential equations of the system, the initial values of the state variables must be specified. So if we know the *state* of a system, that is, the values of all of the state variables, at a given time t_0, and if we additionally know the system inputs, then we can in principle solve the system state equations to find the system waveforms at any future time.

The state equations of a system can be written in the compact matrix form of Eq. (7.77):

$$\mathbf{K}\frac{d\mathbf{x}(t)}{dt} = \mathbf{A}\mathbf{x}(t) + \mathbf{B}\mathbf{u}(t)$$

$$\mathbf{y}(t) = \mathbf{C}\mathbf{x}(t) + \mathbf{E}\mathbf{u}(t)$$

(7.77)

Here, the state vector $\mathbf{x}(t)$ is a vector containing all of the state variables, that is, the inductor currents, capacitor voltages, etc. The input vector $\mathbf{u}(t)$ contains the independent inputs to the system, such as the input voltage source $v_g(t)$. The derivative of the state vector is a vector whose elements are equal to the derivatives of the corresponding elements of the state vector:

$$\mathbf{x}(t) = \begin{vmatrix} x_1(t) \\ x_2(t) \\ \vdots \end{vmatrix}, \qquad \frac{d\mathbf{x}(t)}{dt} = \begin{vmatrix} \dfrac{dx_1(t)}{dt} \\ \dfrac{dx_2(t)}{dt} \\ \vdots \end{vmatrix}$$

(7.78)

In the standard form of Eq. (7.77), $\mathbf{K}$ is a matrix containing the values of capacitance, inductance, and mutual inductance (if any), such that $\mathbf{K}d\mathbf{x}(t)/dt$ is a vector containing the inductor winding voltages and capacitor currents. In other physical systems, $\mathbf{K}$ may contain other quantities such as moment of inertia or mass. Equation (7.77) states that the inductor voltages and capacitor currents of the system can be expressed as linear combinations of the state variables and the independent inputs. The matrices $\mathbf{A}$ and $\mathbf{B}$ contain constants of proportionality.

It may also be desired to compute other circuit waveforms that do not coincide with the elements of the state vector $\mathbf{x}(t)$ or the input vector $\mathbf{u}(t)$. These other signals are, in general, dependent waveforms that can be expressed as linear combinations of the elements of the state vector and input vector. The vector $\mathbf{y}(t)$ is usually called the *output vector*. We are free to place any dependent signal in this vector, regardless of whether the signal is actually a physical output. The converter input current $i_g(t)$ is often chosen to be an element of $\mathbf{y}(t)$. In the state equations (7.77), the elements of $\mathbf{y}(t)$ are expressed as a linear combination of the elements of the $\mathbf{x}(t)$ and $\mathbf{u}(t)$ vectors. The matrices $\mathbf{C}$ and $\mathbf{E}$ contain constants of proportionality.

As an example, let us write the state equations of the circuit of Fig. 7.28. This circuit contains two capacitors and an inductor, and hence the physical state variables are the independent capacitor voltages $v_1(t)$ and $v_2(t)$, as well as the inductor current $i(t)$. So we can define the state vector as

$$\mathbf{x}(t) = \begin{vmatrix} v_1(t) \\ v_2(t) \\ i(t) \end{vmatrix}$$

(7.79)

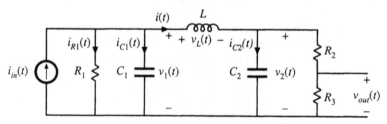

Fig. 7.28 Circuit example.

Since there are no coupled inductors, the matrix $\mathbf{K}$ is diagonal, and simply contains the values of capacitance and inductance:

$$\mathbf{K} = \begin{bmatrix} C_1 & 0 & 0 \\ 0 & C_2 & 0 \\ 0 & 0 & L \end{bmatrix} \tag{7.80}$$

The circuit has one independent input, the current source $i_{in}(t)$. Hence we should define the input vector as

$$\mathbf{u}(t) = \begin{bmatrix} i_{in}(t) \end{bmatrix} \tag{7.81}$$

We are free to place any dependent signal in vector $\mathbf{y}(t)$. Suppose that we are interested in also computing the voltage $v_{out}(t)$ and the current $i_{R1}(t)$. We can therefore define $\mathbf{y}(t)$ as

$$\mathbf{y}(t) = \begin{bmatrix} v_{out}(t) \\ i_{R1}(t) \end{bmatrix} \tag{7.82}$$

To write the state equations in the canonical form of Eq. (7.77), we need to express the inductor voltages and capacitor currents as linear combinations of the elements of $\mathbf{x}(t)$ and $\mathbf{u}(t)$, that is, as linear combinations of $v_1(t)$, $v_2(t)$, $i(t)$, and $i_{in}(t)$.

The capacitor current $i_{C1}(t)$ is given by the node equation

$$i_{C1}(t) = C_1 \frac{dv_1(t)}{dt} = i_{in}(t) - \frac{v_1(t)}{R_1} - i(t) \tag{7.83}$$

This equation will become the top row of the matrix equation (7.77). The capacitor current $i_{C2}(t)$ is given by the node equation,

$$i_{C2}(t) = C_2 \frac{dv_2(t)}{dt} = i(t) - \frac{v_2(t)}{R_2 + R_3} \tag{7.84}$$

Note that we have been careful to express this current as a linear combination of the elements of $\mathbf{x}(t)$ and $\mathbf{u}(t)$ alone. The inductor voltage is given by the loop equation,

$$v_L(t) = L \frac{di(t)}{dt} = v_1(t) - v_2(t) \tag{7.85}$$

Equations (7.83) to (7.85) can be written in the following matrix form:

$$\underbrace{\begin{bmatrix} C_1 & 0 & 0 \\ 0 & C_2 & 0 \\ 0 & 0 & L \end{bmatrix}}_{\mathbf{K}} \underbrace{\begin{bmatrix} \dfrac{dv_1(t)}{dt} \\ \dfrac{dv_2(t)}{dt} \\ \dfrac{di(t)}{dt} \end{bmatrix}}_{\dfrac{d\mathbf{x}(t)}{dt}} = \underbrace{\begin{bmatrix} -\dfrac{1}{R_1} & 0 & -1 \\ 0 & -\dfrac{1}{R_2 + R_3} & 1 \\ 1 & -1 & 0 \end{bmatrix}}_{\mathbf{A}} \underbrace{\begin{bmatrix} v_1(t) \\ v_2(t) \\ i(t) \end{bmatrix}}_{\mathbf{x}(t)} + \underbrace{\begin{bmatrix} 1 \\ 0 \\ 0 \end{bmatrix}}_{\mathbf{B}} \underbrace{\begin{bmatrix} i_{in}(t) \end{bmatrix}}_{\mathbf{u}(t)} \tag{7.86}$$

Matrices **A** and **B** are now known.

It is also necessary to express the elements of $\mathbf{y}(t)$ as linear combinations of the elements of $\mathbf{x}(t)$ and $\mathbf{u}(t)$. By solution of the circuit of Fig. 7.28, $v_{out}(t)$ can be written in terms of $v_2(t)$ as

$$v_{out}(t) = v_2(t) \frac{R_3}{R_2 + R_3} \tag{7.87}$$

Also, $i_{R1}(t)$ can be expressed in terms of $v_1(t)$ as

$$i_{R1}(t) = \frac{v_1(t)}{R_1} \tag{7.88}$$

By collecting Eqs. (7.87) and (7.88) into the standard matrix form of Eq. (7.77), we obtain

$$\underbrace{\begin{bmatrix} v_{out}(t) \\ i_{R1}(t) \end{bmatrix}}_{\mathbf{y}(t)} = \underbrace{\begin{bmatrix} 0 & \dfrac{R_3}{R_2 + R_3} & 0 \\ \dfrac{1}{R_1} & 0 & 0 \end{bmatrix}}_{\mathbf{C}} \underbrace{\begin{bmatrix} v_1(t) \\ v_2(t) \\ i(t) \end{bmatrix}}_{\mathbf{x}(t)} + \underbrace{\begin{bmatrix} 0 \\ 0 \end{bmatrix}}_{\mathbf{E}} \underbrace{\begin{bmatrix} i_{in}(t) \end{bmatrix}}_{\mathbf{u}(t)} \tag{7.89}$$

We can now identify the matrices **C** and **E** as shown above.

It should be recognized that, starting in Chapter 2, we have always begun the analysis of converters by writing their state equations. We are now simply writing these equations in matrix form.

7.3.2   The Basic State-Space Averaged Model

Consider now that we are given a PWM converter, operating in the continuous conduction mode. The converter circuit contains independent states that form the state vector $\mathbf{x}(t)$, and the converter is driven by independent sources that form the input vector $\mathbf{u}(t)$. During the first subinterval, when the switches are in position 1, the converter reduces to a linear circuit that can be described by the following state equations:

$$\mathbf{K} \frac{d\mathbf{x}(t)}{dt} = \mathbf{A}_1 \mathbf{x}(t) + \mathbf{B}_1 \mathbf{u}(t)$$
$$\mathbf{y}(t) = \mathbf{C}_1 \mathbf{x}(t) + \mathbf{E}_1 \mathbf{u}(t) \tag{7.90}$$

During the second subinterval, with the switches in position 2, the converter reduces to another linear circuit whose state equations are

$$\mathbf{K} \frac{d\mathbf{x}(t)}{dt} = \mathbf{A}_2 \mathbf{x}(t) + \mathbf{B}_2 \mathbf{u}(t)$$
$$\mathbf{y}(t) = \mathbf{C}_2 \mathbf{x}(t) + \mathbf{E}_2 \mathbf{u}(t) \tag{7.91}$$

During the two subintervals, the circuit elements are connected differently; therefore, the respective state equation matrices $\mathbf{A}_1$, $\mathbf{B}_1$, $\mathbf{C}_1$, $\mathbf{E}_1$ and $\mathbf{A}_2$, $\mathbf{B}_2$, $\mathbf{C}_2$, $\mathbf{E}_2$ may also differ. Given these state equations, the result of state-space averaging is the state equations of the equilibrium and small-signal ac models.

Provided that the natural frequencies of the converter, as well as the frequencies of variations of the converter inputs, are much slower than the switching frequency, then the state-space averaged model that describes the converter in equilibrium is

$$0 = \mathbf{AX} + \mathbf{BU}$$
$$\mathbf{Y} = \mathbf{CX} + \mathbf{EU}$$

(7.92)

where the averaged matrices are

$$\mathbf{A} = D\mathbf{A}_1 + D'\mathbf{A}_2$$
$$\mathbf{B} = D\mathbf{B}_1 + D'\mathbf{B}_2$$
$$\mathbf{C} = D\mathbf{C}_1 + D'\mathbf{C}_2$$
$$\mathbf{E} = D\mathbf{E}_1 + D'\mathbf{E}_2$$

(7.93)

The equilibrium dc components are

$\mathbf{X}$ = equilibrium (dc) state vector
$\mathbf{U}$ = equilibrium (dc) input vector
$\mathbf{Y}$ = equilibrium (dc) output vector
D = equilibrium (dc) duty cycle

(7.94)

Quantities defined in Eq. (7.94) represent the equilibrium values of the averaged vectors. Equation (7.92) can be solved to find the equilibrium state and output vectors:

$$\mathbf{X} = -\mathbf{A}^{-1}\mathbf{BU}$$
$$\mathbf{Y} = \left(-\mathbf{CA}^{-1}\mathbf{B} + \mathbf{E} \right)\mathbf{U}$$

(7.95)

The state equations of the small-signal ac model are

$$\mathbf{K}\frac{d\hat{\mathbf{x}}(t)}{dt} = \mathbf{A}\hat{\mathbf{x}}(t) + \mathbf{B}\hat{\mathbf{u}}(t) + \left\{ \left(\mathbf{A}_1 - \mathbf{A}_2\right)\mathbf{X} + \left(\mathbf{B}_1 - \mathbf{B}_2\right)\mathbf{U} \right\}\hat{d}(t)$$
$$\hat{\mathbf{y}}(t) = \mathbf{C}\hat{\mathbf{x}}(t) + \mathbf{E}\hat{\mathbf{u}}(t) + \left\{ \left(\mathbf{C}_1 - \mathbf{C}_2\right)\mathbf{X} + \left(\mathbf{E}_1 - \mathbf{E}_2\right)\mathbf{U} \right\}\hat{d}(t)$$

(7.96)

The quantities $\hat{\mathbf{x}}(t)$, $\hat{\mathbf{u}}(t)$, $\hat{\mathbf{y}}(t)$, and $\hat{d}(t)$ in Eq. (7.96) are small ac variations about the equilibrium solution, or quiescent operating point, defined by Eqs. (7.92) to (7.95).

So if we can write the converter state equations, Eqs. (7.90) and (7.91), then we can always find the averaged dc and small-signal ac models, by evaluation of Eqs. (7.92) to (7.96).

7.3.3 Discussion of the State-Space Averaging Result

As in Sections 7.1 and 7.2, the low-frequency components of the inductor currents and capacitor voltages are modeled by averaging over an interval of length T_s. Hence, we can define the average of the state vector $\mathbf{x}(t)$ as

$$\left\langle \mathbf{x}(t) \right\rangle_{T_s} = \frac{1}{T_s}\int_t^{t+T_s} \mathbf{x}(\tau)d\tau$$

(7.97)

The low-frequency components of the input and output vectors are modeled in a similar manner. By averaging the inductor voltages and capacitor currents, one then obtains the following low-frequency state equation:

$$\mathbf{K}\frac{d\langle \mathbf{x}(t)\rangle_{T_s}}{dt} = \Big(d(t)\,\mathbf{A}_1 + d'(t)\,\mathbf{A}_2\Big)\langle \mathbf{x}(t)\rangle_{T_s} + \Big(d(t)\,\mathbf{B}_1 + d'(t)\,\mathbf{B}_2\Big)\langle \mathbf{u}(t)\rangle_{T_s} \tag{7.98}$$

This result is equivalent to Eq. (7.2).

For example, let us consider how the elements of the state vector $\mathbf{x}(t)$ change over one switching period. During the first subinterval, with the switches in position 1, the converter state equations are given by Eq. (7.90). Therefore, the elements of $\mathbf{x}(t)$ change with the slopes $\mathbf{K}^{-1}(\mathbf{A}_1\mathbf{x}(t) + \mathbf{B}_1\mathbf{u}(t))$. If we make the small ripple approximation, that $\mathbf{x}(t)$ and $\mathbf{u}(t)$ do not change much over one switching period, then the slopes are essentially constant and are approximately equal to

$$\frac{d\mathbf{x}(t)}{dt} = \mathbf{K}^{-1}\Big(\mathbf{A}_1\langle \mathbf{x}(t)\rangle_{T_s} + \mathbf{B}_1\langle \mathbf{u}(t)\rangle_{T_s}\Big) \tag{7.99}$$

This assumption coincides with the requirements for small switching ripple in all elements of $\mathbf{x}(t)$ and that variations in $\mathbf{u}(t)$ be slow compared to the switching frequency. If we assume that the state vector is initially equal to $\mathbf{x}(0)$, then we can write

$$\underbrace{\mathbf{x}(dT_s)}_{\substack{\text{final}\\\text{value}}} = \underbrace{\mathbf{x}(0)}_{\substack{\text{initial}\\\text{value}}} + \underbrace{(dT_s)}_{\substack{\text{interval}\\\text{length}}}\ \underbrace{\mathbf{K}^{-1}\Big(\mathbf{A}_1\langle \mathbf{x}(t)\rangle_{T_s} + \mathbf{B}_1\langle \mathbf{u}(t)\rangle_{T_s}\Big)}_{\text{slope}} \tag{7.100}$$

Similar arguments apply during the second subinterval. With the switch in position 2, the state equations are given by Eq. (7.91). With the assumption of small ripple during this subinterval, the state vector now changes with slope

$$\frac{d\mathbf{x}(t)}{dt} = \mathbf{K}^{-1}\Big(\mathbf{A}_2\langle \mathbf{x}(t)\rangle_{T_s} + \mathbf{B}_2\langle \mathbf{u}(t)\rangle_{T_s}\Big) \tag{7.101}$$

The state vector at the end of the switching period is

$$\underbrace{\mathbf{x}(T_s)}_{\substack{\text{final}\\\text{value}}} = \underbrace{\mathbf{x}(dT_s)}_{\substack{\text{initial}\\\text{value}}} + \underbrace{(d'T_s)}_{\substack{\text{interval}\\\text{length}}}\ \underbrace{\mathbf{K}^{-1}\Big(\mathbf{A}_2\langle \mathbf{x}(t)\rangle_{T_s} + \mathbf{B}_2\langle \mathbf{u}(t)\rangle_{T_s}\Big)}_{\text{slope}} \tag{7.102}$$

Substitution of Eq. (7.100) into Eq. (7.102) allows us to determine $\mathbf{x}(T_s)$ in terms of $\mathbf{x}(0)$:

$$\mathbf{x}(T_s) = \mathbf{x}(0) + dT_s\mathbf{K}^{-1}\Big(\mathbf{A}_1\langle \mathbf{x}(t)\rangle_{T_s} + \mathbf{B}_1\langle \mathbf{u}(t)\rangle_{T_s}\Big) + d'T_s\mathbf{K}^{-1}\Big(\mathbf{A}_2\langle \mathbf{x}(t)\rangle_{T_s} + \mathbf{B}_2\langle \mathbf{u}(t)\rangle_{T_s}\Big) \tag{7.103}$$

Upon collecting terms, one obtains

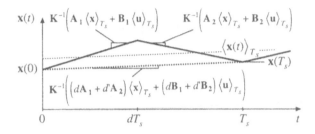

Fig. 7.29 How an element of the state vector, and its average, evolve over one switching period.

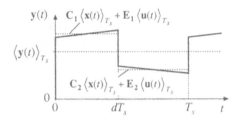

Fig. 7.30 Averaging an element of the output vector $y(t)$.

$$\mathbf{x}(T_s) = \mathbf{x}(0) + T_s\mathbf{K}^{-1}\Big(d(t)\mathbf{A}_1 + d'(t)\mathbf{A}_2\Big)\big\langle\mathbf{x}(t)\big\rangle_{T_s} + T_s\mathbf{K}^{-1}\Big(d(t)\mathbf{B}_1 + d'(t)\mathbf{B}_2\Big)\big\langle\mathbf{u}(t)\big\rangle_{T_s} \qquad (7.104)$$

Next, we approximate the derivative of $\langle\mathbf{x}(t)\rangle_{T_s}$ using the net change over one switching period:

$$\frac{d\big\langle\mathbf{x}(t)\big\rangle_{T_s}}{dt} \approx \frac{\mathbf{x}(T_s) - \mathbf{x}(0)}{T_s} \qquad (7.105)$$

Substitution of Eq. (7.104) into (7.105) leads to

$$\mathbf{K}\frac{d\big\langle\mathbf{x}(t)\big\rangle_{T_s}}{dt} = \Big(d(t)\,\mathbf{A}_1 + d'(t)\,\mathbf{A}_2\Big)\big\langle\mathbf{x}(t)\big\rangle_{T_s} + \Big(d(t)\,\mathbf{B}_1 + d'(t)\,\mathbf{B}_2\Big)\big\langle\mathbf{u}(t)\big\rangle_{T_s} \qquad (7.106)$$

which is identical to Eq. (7.99). This is the basic averaged model which describes the converter dynamics. It is nonlinear because the control input $d(t)$ is multiplied by $\langle\mathbf{x}(t)\rangle_{T_s}$ and $\langle\mathbf{u}(t)\rangle_{T_s}$. Variation of a typical element of $\mathbf{x}(t)$ and its average are illustrated in Fig. 7.29.

It is also desired to find the low-frequency components of the output vector $\mathbf{y}(t)$ by averaging. The vector $\mathbf{y}(t)$ is described by Eq. (7.90) for the first subinterval, and by Eq. (7.91) for the second subinterval. Hence, the elements of $\mathbf{y}(t)$ may be discontinuous at the switching transitions, as illustrated in Fig. 7.30. We can again remove the switching harmonics by averaging over one switching period; the result is

$$\left\langle \mathbf{y}(t) \right\rangle_{T_s} = d(t) \left(\mathbf{C_1} \left\langle \mathbf{x}(t) \right\rangle_{T_s} + \mathbf{E_1} \left\langle \mathbf{u}(t) \right\rangle_{T_s} \right) + d'(t) \left(\mathbf{C_2} \left\langle \mathbf{x}(t) \right\rangle_{T_s} + \mathbf{E_2} \left\langle \mathbf{u}(t) \right\rangle_{T_s} \right) \qquad (7.107)$$

Rearrangement of terms yields

$$\left\langle \mathbf{y}(t) \right\rangle_{T_s} = \left(d(t)\, \mathbf{C_1} + d'(t)\, \mathbf{C_2} \right) \left\langle \mathbf{x}(t) \right\rangle_{T_s} + \left(d(t)\, \mathbf{E_1} + d'(t)\, \mathbf{E_2} \right) \left\langle \mathbf{u}(t) \right\rangle_{T_s} \qquad (7.108)$$

This is again a nonlinear equation.

The averaged state equations, (7.106) and (7.108), are collected below:

$$\mathbf{K} \frac{d\left\langle \mathbf{x}(t) \right\rangle_{T_s}}{dt} = \left(d(t)\, \mathbf{A_1} + d'(t)\, \mathbf{A_2} \right) \left\langle \mathbf{x}(t) \right\rangle_{T_s} + \left(d(t)\, \mathbf{B_1} + d'(t)\, \mathbf{B_2} \right) \left\langle \mathbf{u}(t) \right\rangle_{T_s}$$
$$\left\langle \mathbf{y}(t) \right\rangle_{T_s} = \left(d(t)\, \mathbf{C_1} + d'(t)\, \mathbf{C_2} \right) \left\langle \mathbf{x}(t) \right\rangle_{T_s} + \left(d(t)\, \mathbf{E_1} + d'(t)\, \mathbf{E_2} \right) \left\langle \mathbf{u}(t) \right\rangle_{T_s} \qquad (7.109)$$

The next step is the linearization of these equations about a quiescent operating point, to construct a small-signal ac model. When dc inputs $d(t) = D$ and $\mathbf{u}(t) = \mathbf{U}$ are applied, converter operates in equilibrium when the derivatives of all of the elements of $\left\langle \mathbf{x}(t) \right\rangle_{T_s}$ are zero. Hence, by setting the derivative of $\left\langle \mathbf{x}(t) \right\rangle_{T_s}$ to zero in Eq. (7.109), we can define the converter quiescent operating point as the solution of

$$0 = \mathbf{AX} + \mathbf{BU}$$
$$\mathbf{Y} = \mathbf{CX} + \mathbf{EU} \qquad (7.110)$$

where definitions (7.93) and (7.94) have been used. We now perturb and linearize the converter waveforms about this quiescent operating point:

$$\left\langle \mathbf{x}(t) \right\rangle_{T_s} = \mathbf{X} + \hat{\mathbf{x}}(t)$$
$$\left\langle \mathbf{u}(t) \right\rangle_{T_s} = \mathbf{U} + \hat{\mathbf{u}}(t)$$
$$\left\langle \mathbf{y}(t) \right\rangle_{T_s} = \mathbf{Y} + \hat{\mathbf{y}}(t) \qquad (7.111)$$
$$d(t) = D + \hat{d}(t) \ \Rightarrow\ d'(t) = D' - \hat{d}(t)$$

Here, $\hat{\mathbf{u}}(t)$ and $\hat{d}(t)$ are small ac variations in the input vector and duty ratio. The vectors $\hat{\mathbf{x}}(t)$ and $\hat{\mathbf{y}}(t)$ are the resulting small ac variations in the state and output vectors. We must assume that these ac variations are much smaller than the quiescent values. In other words,

$$\| \mathbf{U} \| \gg \| \hat{\mathbf{u}}(t) \|$$
$$D \gg \left| \hat{d}(t) \right|$$
$$\| \mathbf{X} \| \gg \| \hat{\mathbf{x}}(t) \| \qquad (7.112)$$
$$\| \mathbf{Y} \| \gg \| \hat{\mathbf{y}}(t) \|$$

Here, $\| \mathbf{x} \|$ denotes a norm of the vector $\mathbf{x}$.

Substitution of Eq. (7.111) into Eq. (7.109) yields

$$\mathbf{K}\frac{d\left(\mathbf{X}+\hat{\mathbf{x}}(t)\right)}{dt}=\left(\left(D+\hat{d}(t)\right)\mathbf{A}_1+\left(D'-\hat{d}(t)\right)\mathbf{A}_2\right)\left(\mathbf{X}+\hat{\mathbf{x}}(t)\right)$$

$$+\left(\left(D+\hat{d}(t)\right)\mathbf{B}_1+\left(D'-\hat{d}(t)\right)\mathbf{B}_2\right)\left(\mathbf{U}+\hat{\mathbf{u}}(t)\right)$$

$$\left(\mathbf{Y}+\hat{\mathbf{y}}(t)\right)=\left(\left(D+\hat{d}(t)\right)\mathbf{C}_1+\left(D'-\hat{d}(t)\right)\mathbf{C}_2\right)\left(\mathbf{X}+\hat{\mathbf{x}}(t)\right)$$

$$+\left(\left(D+\hat{d}(t)\right)\mathbf{E}_1+\left(D'-\hat{d}(t)\right)\mathbf{E}_2\right)\left(\mathbf{U}+\hat{\mathbf{u}}(t)\right)$$

(7.113)

The derivative $d\mathbf{X}/dt$ is zero. By collecting terms, one obtains

$$\underbrace{\mathbf{K}\frac{d\hat{\mathbf{x}}(t)}{dt}}_{\text{first-order ac}}=\underbrace{\left(\mathbf{AX}+\mathbf{BU}\right)}_{\text{dc terms}}+\underbrace{\mathbf{A}\hat{\mathbf{x}}(t)+\mathbf{B}\hat{\mathbf{u}}(t)+\left\{\left(\mathbf{A}_1-\mathbf{A}_2\right)\mathbf{X}+\left(\mathbf{B}_1-\mathbf{B}_2\right)\mathbf{U}\right\}\hat{d}(t)}_{\text{first-order ac terms}}$$

$$+\underbrace{\left(\mathbf{A}_1-\mathbf{A}_2\right)\hat{\mathbf{x}}(t)\hat{d}(t)+\left(\mathbf{B}_1-\mathbf{B}_2\right)\hat{\mathbf{u}}(t)\hat{d}(t)}_{\text{second-order nonlinear terms}}$$

(7.114)

$$\underbrace{\left(\mathbf{Y}+\hat{\mathbf{y}}(t)\right)}_{\text{dc + 1}^{st}\text{ order ac}}=\underbrace{\left(\mathbf{CX}+\mathbf{EU}\right)}_{\text{dc terms}}+\underbrace{\mathbf{C}\hat{\mathbf{x}}(t)+\mathbf{E}\hat{\mathbf{u}}(t)+\left\{\left(\mathbf{C}_1-\mathbf{C}_2\right)\mathbf{X}+\left(\mathbf{E}_1-\mathbf{E}_2\right)\mathbf{U}\right\}\hat{d}(t)}_{\text{first-order ac terms}}$$

$$+\underbrace{\left(\mathbf{C}_1-\mathbf{C}_2\right)\hat{\mathbf{x}}(t)\hat{d}(t)+\left(\mathbf{E}_1-\mathbf{E}_2\right)\hat{\mathbf{u}}(t)\hat{d}(t)}_{\text{second-order nonlinear terms}}$$

Since the dc terms satisfy Eq. (7.110), they drop out of Eq. (7.114). Also, if the small-signal assumption (7.112) is satisfied, then the second-order nonlinear terms of Eq. (7.114) are small in magnitude compared to the first-order ac terms. We can therefore neglect the nonlinear terms, to obtain the following linearized ac model:

$$\mathbf{K}\frac{d\hat{\mathbf{x}}(t)}{dt}=\mathbf{A}\hat{\mathbf{x}}(t)+\mathbf{B}\hat{\mathbf{u}}(t)+\left\{\left(\mathbf{A}_1-\mathbf{A}_2\right)\mathbf{X}+\left(\mathbf{B}_1-\mathbf{B}_2\right)\mathbf{U}\right\}\hat{d}(t)$$

$$\hat{\mathbf{y}}(t)=\mathbf{C}\hat{\mathbf{x}}(t)+\mathbf{E}\hat{\mathbf{u}}(t)+\left\{\left(\mathbf{C}_1-\mathbf{C}_2\right)\mathbf{X}+\left(\mathbf{E}_1-\mathbf{E}_2\right)\mathbf{U}\right\}\hat{d}(t)$$

(7.115)

This is the desired result, which coincides with Eq. (7.95).

7.3.4   Example: State-Space Averaging of a Nonideal Buck-Boost Converter

Let us apply the state-space averaging method to model the buck-boost converter of Fig. 7.31. We will model the conduction loss of MOSFET Q_1 by on-resistance R_{on}, and the forward voltage drop of diode

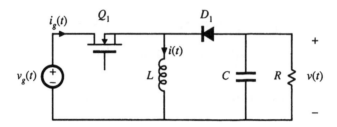

Fig. 7.31   Buck-boost converter example.

D_1 by an independent voltage source of value V_D. It is desired to obtain a complete equivalent circuit, which models both the input port and the output port of the converter.

The independent states of the converter are the inductor current $i(t)$ and the capacitor voltage $v(t)$. Therefore, we should define the state vector $\mathbf{x}(t)$ as

$$\mathbf{x}(t) = \begin{bmatrix} i(t) \\ v(t) \end{bmatrix} \tag{7.116}$$

The input voltage $v_g(t)$ is an independent source which should be placed in the input vector $\mathbf{u}(t)$. In addition, we have chosen to model the diode forward voltage drop with an independent voltage source of value V_D. So this voltage source should also be included in the input vector $\mathbf{u}(t)$. Therefore, let us define the input vector as

$$\mathbf{u}(t) = \begin{bmatrix} v_g(t) \\ V_D \end{bmatrix} \tag{7.117}$$

To model the converter input port, we need to find the converter input current $i_g(t)$. To calculate this dependent current, it should be included in the output vector $\mathbf{y}(t)$. Therefore, let us choose to define $\mathbf{y}(t)$ as

$$\mathbf{y}(t) = \begin{bmatrix} i_g(t) \end{bmatrix} \tag{7.118}$$

Note that it isn't necessary to include the output voltage $v(t)$ in the output vector $\mathbf{y}(t)$, since $v(t)$ is already included in the state vector $\mathbf{x}(t)$.

Next, let us write the state equations for each subinterval. When the switch is in position 1, the converter circuit of Fig. 7.32(a) is obtained. The inductor voltage, capacitor current, and converter input current are

$$
\begin{aligned}
L\frac{di(t)}{dt} &= v_g(t) - i(t) R_{on} \\
C\frac{dv(t)}{dt} &= -\frac{v(t)}{R} \\
i_g(t) &= i(t)
\end{aligned}
\tag{7.119}
$$

These equations can be written in the following state-space form:

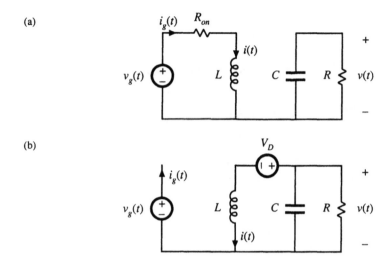

Fig. 7.32 Buck-boost converter circuit: (a) during subinterval 1, (b) during subinterval 2.

$$\underbrace{\begin{bmatrix} L & 0 \\ 0 & C \end{bmatrix}}_{\mathbf{K}} \underbrace{\frac{d}{dt}\begin{bmatrix} i(t) \\ v(t) \end{bmatrix}}_{\frac{d\mathbf{x}(t)}{dt}} = \underbrace{\begin{bmatrix} -R_{on} & 0 \\ 0 & -\frac{1}{R} \end{bmatrix}}_{\mathbf{A}_1} \underbrace{\begin{bmatrix} i(t) \\ v(t) \end{bmatrix}}_{\mathbf{x}(t)} + \underbrace{\begin{bmatrix} 1 & 0 \\ 0 & 0 \end{bmatrix}}_{\mathbf{B}_1} \underbrace{\begin{bmatrix} v_g(t) \\ V_D \end{bmatrix}}_{\mathbf{u}(t)}$$

$$\underbrace{\begin{bmatrix} i_g(t) \end{bmatrix}}_{\mathbf{y}(t)} = \underbrace{\begin{bmatrix} 1 & 0 \end{bmatrix}}_{\mathbf{C}_1} \underbrace{\begin{bmatrix} i(t) \\ v(t) \end{bmatrix}}_{\mathbf{x}(t)} + \underbrace{\begin{bmatrix} 0 & 0 \end{bmatrix}}_{\mathbf{E}_1} \underbrace{\begin{bmatrix} v_g(t) \\ V_D \end{bmatrix}}_{\mathbf{u}(t)}$$

(7.120)

So we have identified the state equation matrices $\mathbf{A}_1$, $\mathbf{B}_1$, $\mathbf{C}_1$, and $\mathbf{E}_1$.

With the switch in position 2, the converter circuit of Fig. 7.32(b) is obtained. For this subinterval, the inductor voltage, capacitor current, and converter input current are given by

$$L\frac{di(t)}{dt} = v(t) - V_D$$
$$C\frac{dv(t)}{dt} = -\frac{v(t)}{R} - i(t)$$
$$i_g(t) = 0$$

(7.121)

When written in state-space form, these equations become

$$
\underbrace{\begin{bmatrix} L & 0 \\ 0 & C \end{bmatrix}}_{\mathbf{K}} \underbrace{\frac{d}{dt}\begin{bmatrix} i(t) \\ v(t) \end{bmatrix}}_{\dfrac{d\mathbf{x}(t)}{dt}} = \underbrace{\begin{bmatrix} 0 & 1 \\ -1 & -\frac{1}{R} \end{bmatrix}}_{\mathbf{A_2}} \underbrace{\begin{bmatrix} i(t) \\ v(t) \end{bmatrix}}_{\mathbf{x}(t)} + \underbrace{\begin{bmatrix} 0 & -1 \\ 0 & 0 \end{bmatrix}}_{\mathbf{B_2}} \underbrace{\begin{bmatrix} v_g(t) \\ V_D \end{bmatrix}}_{\mathbf{u}(t)}
\tag{7.122}
$$

$$
\underbrace{\begin{bmatrix} i_g(t) \end{bmatrix}}_{\mathbf{y}(t)} = \underbrace{\begin{bmatrix} 0 & 0 \end{bmatrix}}_{\mathbf{C_2}} \underbrace{\begin{bmatrix} i(t) \\ v(t) \end{bmatrix}}_{\mathbf{x}(t)} + \underbrace{\begin{bmatrix} 0 & 0 \end{bmatrix}}_{\mathbf{E_2}} \underbrace{\begin{bmatrix} v_g(t) \\ V_D \end{bmatrix}}_{\mathbf{u}(t)}
$$

So we have also identified the subinterval 2 matrices $\mathbf{A_2}$, $\mathbf{B_2}$, $\mathbf{C_2}$, and $\mathbf{E_2}$.

The next step is to evaluate the state-space averaged equilibrium equations (7.92) to (7.94). The averaged matrix $\mathbf{A}$ is

$$
\mathbf{A} = D\mathbf{A}_1 + D'\mathbf{A}_2 = D\begin{bmatrix} -R_{on} & 0 \\ 0 & -\frac{1}{R} \end{bmatrix} + D'\begin{bmatrix} 0 & 1 \\ -1 & -\frac{1}{R} \end{bmatrix} = \begin{bmatrix} -DR_{on} & D' \\ -D' & -\frac{1}{R} \end{bmatrix}
\tag{7.123}
$$

In a similar manner, the averaged matrices $\mathbf{B}$, $\mathbf{C}$, and $\mathbf{E}$ are evaluated, with the following results:

$$
\begin{aligned}
\mathbf{B} &= D\mathbf{B}_1 + D'\mathbf{B}_2 = \begin{bmatrix} D & -D' \\ 0 & 0 \end{bmatrix} \\
\mathbf{C} &= D\mathbf{C}_1 + D'\mathbf{C}_2 = \begin{bmatrix} D & 0 \end{bmatrix} \\
\mathbf{E} &= D\mathbf{E}_1 + D'\mathbf{E}_2 = \begin{bmatrix} 0 & 0 \end{bmatrix}
\end{aligned}
\tag{7.124}
$$

The dc state equations (7.92) therefore become

$$
\begin{aligned}
\begin{bmatrix} 0 \\ 0 \end{bmatrix} &= \begin{bmatrix} -DR_{on} & D' \\ -D' & -\frac{1}{R} \end{bmatrix}\begin{bmatrix} I \\ V \end{bmatrix} + \begin{bmatrix} D & -D' \\ 0 & 0 \end{bmatrix}\begin{bmatrix} V_g \\ V_D \end{bmatrix} \\
\begin{bmatrix} I_g \end{bmatrix} &= \begin{bmatrix} D & 0 \end{bmatrix}\begin{bmatrix} I \\ V \end{bmatrix} + \begin{bmatrix} 0 & 0 \end{bmatrix}\begin{bmatrix} V_g \\ V_D \end{bmatrix}
\end{aligned}
\tag{7.125}
$$

Evaluation of Eq. (7.95) leads to the following solution for the equilibrium state and output vectors:

$$
\begin{aligned}
\begin{bmatrix} I \\ V \end{bmatrix} &= \left(\frac{1}{1 + \dfrac{D}{D'^2}\dfrac{R_{on}}{R}} \right) \begin{bmatrix} \dfrac{D}{D'^2 R} & \dfrac{1}{D' R} \\ -\dfrac{D}{D'} & 1 \end{bmatrix}\begin{bmatrix} V_g \\ V_D \end{bmatrix} \\
\begin{bmatrix} I_g \end{bmatrix} &= \left(\frac{1}{1 + \dfrac{D}{D'^2}\dfrac{R_{on}}{R}} \right) \begin{bmatrix} \dfrac{D^2}{D'^2 R} & \dfrac{D}{D' R} \end{bmatrix}\begin{bmatrix} V_g \\ V_D \end{bmatrix}
\end{aligned}
\tag{7.126}
$$

Alternatively, the steady-state equivalent circuit of Fig. 7.33 can be constructed as usual from

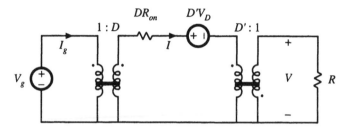

Fig. 7.33 Dc circuit model for the buck-boost converter example, equivalent to Eq. (7.125).

Eq. (7.125). The top row of Eq. (7.125) could have been obtained by application of the principle of inductor volt-second balance to the inductor voltage waveform. The second row of Eq. (7.125) could have been obtained by application of the principle of capacitor charge balance to the capacitor current waveform. The $i_g(t)$ equation expresses the dc component of the converter input current. By reconstructing circuits that are equivalent to these three equations, the dc model of Fig. 7.33 is obtained.

The small-signal model is found by evaluation of Eq. (7.95). The vector coefficients of $\hat{d}(t)$ in Eq. (7.95) are

$$\left(\mathbf{A}_1 - \mathbf{A}_2\right)\mathbf{X} + \left(\mathbf{B}_1 - \mathbf{B}_2\right)\mathbf{U} = \begin{bmatrix} -V - IR_{on} \\ I \end{bmatrix} + \begin{bmatrix} V_g + V_D \\ 0 \end{bmatrix} = \begin{bmatrix} V_g - V - IR_{on} + V_D \\ I \end{bmatrix} \tag{7.127}$$

$$\left(\mathbf{C}_1 - \mathbf{C}_2\right)\mathbf{X} + \left(\mathbf{E}_1 - \mathbf{E}_2\right)\mathbf{U} = [I]$$

The small-signal ac state equations (7.95) therefore become

$$\begin{bmatrix} L & 0 \\ 0 & C \end{bmatrix} \frac{d}{dt} \begin{bmatrix} \hat{i}(t) \\ \hat{v}(t) \end{bmatrix} = \begin{bmatrix} -DR_{on} & D' \\ -D' & -\frac{1}{R} \end{bmatrix} \begin{bmatrix} \hat{i}(t) \\ \hat{v}(t) \end{bmatrix} + \begin{bmatrix} D & -D' \\ 0 & 0 \end{bmatrix} \begin{bmatrix} \hat{v}_g(t) \\ 0 \end{bmatrix} + \begin{bmatrix} V_g - V - IR_{on} + V_D \\ I \end{bmatrix} \hat{d}(t) \tag{7.128}$$

$$\begin{bmatrix} \hat{i}_g(t) \end{bmatrix} = \begin{bmatrix} D & 0 \end{bmatrix} \begin{bmatrix} \hat{i}(t) \\ \hat{v}(t) \end{bmatrix} + \begin{bmatrix} 0 & 0 \end{bmatrix} \begin{bmatrix} \hat{v}_g(t) \\ 0 \end{bmatrix} + \begin{bmatrix} I \end{bmatrix} \hat{d}(t)$$

Note that, since the diode forward voltage drop is modeled as the constant value V_D, there are no ac variations in this source, and $\hat{v}_D(t)$ equals zero. Again, a circuit model equivalent to Eq. (7.128) can be constructed, in the usual manner. When written in scalar form, Eq. (7.128) becomes

$$L \frac{d\hat{i}(t)}{dt} = D' \hat{v}(t) - DR_{on} \hat{i}(t) + D \hat{v}_g(t) + \left(V_g - V - IR_{on} + V_D\right) \hat{d}(t)$$

$$C \frac{d\hat{v}(t)}{dt} = -D' \hat{i}(t) - \frac{\hat{v}(t)}{R} + I \hat{d}(t) \tag{7.129}$$

$$\hat{i}_g(t) = D \hat{i}(t) + I \hat{d}(t)$$

Circuits corresponding to these equations are listed in Fig. 7.34. These circuits can be combined into the complete small-signal ac equivalent circuit model of Fig. 7.35.

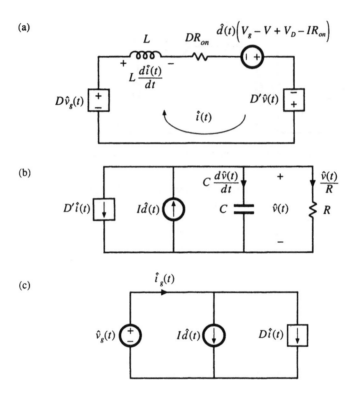

Fig. 7.34   Circuits equivalent to the small-signal converter equations: (a) inductor loop, (b) capacitor node, (c) input port.

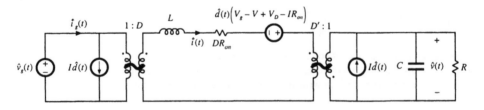

Fig. 7.35   Complete small-signal ac equivalent circuit model, nonideal buck-boost converter example.

7.4   CIRCUIT AVERAGING AND AVERAGED SWITCH MODELING

Circuit averaging is another well-known technique for derivation of converter equivalent circuits. Rather than averaging the converter state equations, with the circuit averaging technique we average the converter waveforms directly. All manipulations are performed on the circuit diagram, instead of on its equations, and hence the circuit averaging technique gives a more physical interpretation to the model. Since circuit averaging involves averaging and small-signal linearization, it is equivalent to state-space averaging. However, in many cases circuit averaging is easier to apply, and allows the small-signal ac model to be written almost by inspection. The circuit averaging technique can also be applied directly to a number of different types of converters and switch elements, including phase-controlled rectifiers, PWM converters operated in discontinuous conduction mode or with current programming, and quasi-resonant converters—these are described in later chapters. However, in other cases it may lead to involuted models that are less easy to analyze and understand. To overcome this problem, the circuit averaging and state-space averaging approaches can be combined. Circuit averaging was developed before state-space averaging, and is described in [4]. Because of its generality, there has been a recent resurgence of interest in circuit averaging of switch networks [13–20].

The key step in circuit averaging is to replace the converter switches with voltage and current sources, to obtain a time-invariant circuit topology. The waveforms of the voltage and current generators are defined to be identical to the switch waveforms of the original converter. Once a time-invariant circuit network is obtained, then the converter waveforms can be averaged over one switching period to remove the switching harmonics. Any nonlinear elements in the averaged circuit model can then be perturbed and linearized, leading to the small-signal ac model.

In Fig. 7.36, the switching elements are separated from the remainder of the converter. The converter therefore consists of a switch network containing the converter switching elements, and a time-invariant network, containing the reactive and other remaining elements. Figure 7.36 illustrates the simple case in which there are two single-pole single-throw (SPST) switches; the switches can then be represented using a two-port network. In more complicated systems containing multiple transistors or diodes, such as in polyphase converters, the switch network may contain more than two ports.

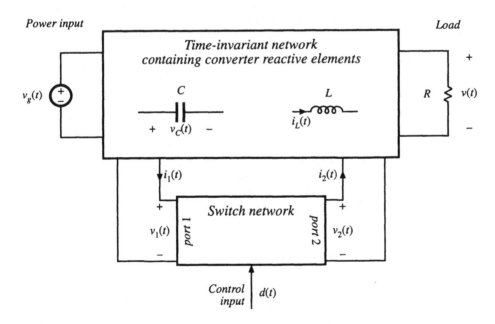

Fig. 7.36   A switching converter can be viewed as a switch network connected to a time-invariant network.

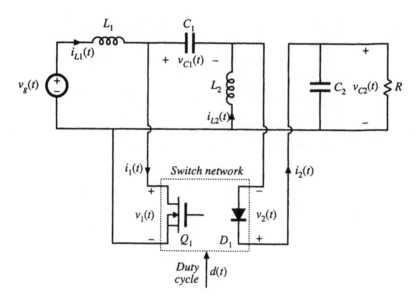

Fig. 7.37 Schematic of the SEPIC, arranged in the form of Fig. 7.36.

The central idea of the *averaged switch modeling* approach is to find an averaged circuit model for the switch network. The resulting averaged switch model can then be inserted into the converter circuit to obtain a complete averaged circuit model of the converter. An important advantage of the averaged switch modeling approach is that the same model can be used in many different converter configurations. It is not necessary to rederive an averaged circuit model for each particular converter. Furthermore, in many cases, the averaged switch model simplifies converter analysis and yields good intuitive understanding of the converter steady-state and dynamic properties.

The first step in the process of finding an averaged switch model for a switch network is to sketch the converter in the form of Fig. 7.36, in which a switch network containing only the converter switching elements is explicitly defined. The CCM SEPIC example shown in Fig. 7.37 is used to illustrate the process. There is usually more than one way to define the two ports of the switch network; a natural way to define the two-port switch network of the SEPIC is illustrated in Fig. 7.37. The switch network terminal quantities $v_1(t)$, $i_1(t)$, $v_2(t)$, and $i_2(t)$ are illustrated in Fig. 7.38 for CCM operation. Note that it is not necessary that the ports of the switch network be electrically connected within the switch network itself. Furthermore, there is no requirement that any of the terminal voltage or current waveforms of the switch network be nonpulsating.

7.4.1 Obtaining a Time-Invariant Circuit

The first step in the circuit averaging technique is to replace the switch network with voltage and current sources, such that the circuit connections do not vary in time. The switch network defined in the SEPIC is shown in Fig. 7.39(a). As with any two-port network, two of the four terminal voltages and

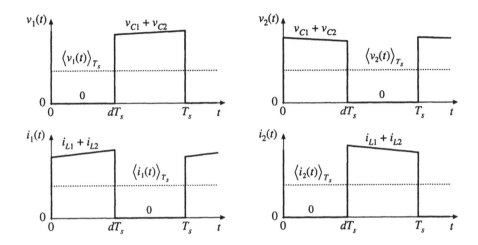

Fig.   7.38   Terminal switch network waveforms in the CCM SEPIC.

currents can be taken as independent inputs to the switch network. The remaining two voltages and/or currents are viewed as dependent outputs of the switch network. In general, the choice of independent inputs is arbitrary, as long as the inputs can indeed be independent in the given converter circuit. For CCM operation, one can choose one terminal current and one terminal voltage as the independent inputs. Let us select $i_1(t)$ and $v_2(t)$ as the switch network independent inputs. In addition, the duty cycle $d(t)$ is the independent control input.

In Fig. 7.39(b), the ports of the switch network are replaced by dependent voltage and current sources. The waveforms of these dependent sources are defined to be identical to the actual dependent outputs $v_1(t)$ and $i_2(t)$ given in Fig. 7.38. Since all waveforms in Fig. 7.39(b) match the waveforms of Figs. 7.39(a) and 7.38, the circuits are electrically equivalent. So far, no approximations have been made.

7.4.2   Circuit Averaging

The next step is determination of the average values of the switch network terminal waveforms in terms of the converter state variables (inductor currents and capacitor voltages) and the converter independent inputs (such as the input voltage and the transistor duty cycle). The basic assumption is made that the natural time constants of the converter network are much longer that the switching period T_s. This assumption coincides with the requirement for small switching ripple. One may average the waveforms over a time interval which is short compared to the system natural time constants, without significantly altering the system response. Hence, when the basic assumption is satisfied, it is a good approximation to average the converter waveforms over the switching period T_s. The resulting averaged model predicts the low-frequency behavior of the system, while neglecting the high-frequency switching harmonics. In the SEPIC example, by use of the usual small ripple approximation, the average values of the switch network terminal waveforms of Fig. 7.38 can be expressed in terms of the independent inputs and the state variables as follows:

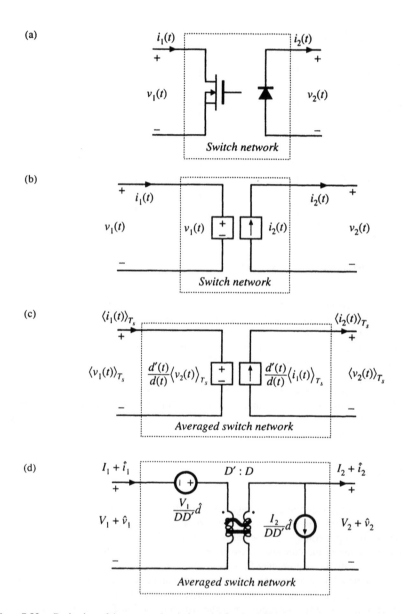

Fig.   7.39   Derivation of the averaged switch model for the CCM SEPIC: (a) switch network; (b) switch network where the switches are replaced with dependent sources whose waveforms match the switch terminal dependent waveforms; (c) large-signal, nonlinear averaged switch model obtained by averaging the switch network terminal waveforms in (b); (d) dc and ac small-signal averaged switch model.

$$\langle v_1(t) \rangle_{T_s} = d'(t) \left(\langle v_{C1}(t) \rangle_{T_s} + \langle v_{C2}(t) \rangle_{T_s} \right) \tag{7.130}$$

$$\langle i_1(t) \rangle_{T_s} = d(t) \left(\langle i_{L1}(t) \rangle_{T_s} + \langle i_{L2}(t) \rangle_{T_s} \right) \tag{7.131}$$

$$\langle v_2(t) \rangle_{T_s} = d(t) \left(\langle v_{C1}(t) \rangle_{T_s} + \langle v_{C2}(t) \rangle_{T_s} \right) \tag{7.132}$$

$$\langle i_2(t) \rangle_{T_s} = d'(t) \left(\langle i_{L1}(t) \rangle_{T_s} + \langle i_{L2}(t) \rangle_{T_s} \right) \tag{7.133}$$

We have selected $\langle i_1(t) \rangle_{T_s}$ and $\langle v_2(t) \rangle_{T_s}$ as the switch network independent inputs. The dependent outputs of the averaged switch network are then $\langle i_2(t) \rangle_{T_s}$ and $\langle v_1(t) \rangle_{T_s}$. The next step is to express, if possible, the switch network dependent outputs $\langle i_2(t) \rangle_{T_s}$ and $\langle v_1(t) \rangle_{T_s}$ as functions *solely* of the switch network independent inputs $\langle i_1(t) \rangle_{T_s}$, $\langle v_2(t) \rangle_{T_s}$, and the control input $d(t)$. In this step, the averaged switch outputs should not be written as functions of other converter signals such as $\langle v_g(t) \rangle_{T_s}$, $\langle v_{C1}(t) \rangle_{T_s}$, $\langle v_{C2}(t) \rangle_{T_s}$, $\langle i_{L1}(t) \rangle_{T_s}$, $\langle i_{L2}(t) \rangle_{T_s}$, etc.

We can use Eqs. (7.131) and (7.132) to write

$$\langle i_{L1}(t) \rangle_{T_s} + \langle i_{L2}(t) \rangle_{T_s} = \frac{\langle i_1(t) \rangle_{T_s}}{d(t)} \tag{7.134}$$

$$\langle v_{C1}(t) \rangle_{T_s} + \langle v_{C2}(t) \rangle_{T_s} = \frac{\langle v_2(t) \rangle_{T_s}}{d(t)} \tag{7.135}$$

Substitution of these expressions into Eqs. (7.130) and (7.133) leads to

$$\langle v_1(t) \rangle_{T_s} = \frac{d'(t)}{d(t)} \langle v_2(t) \rangle_{T_s} \tag{7.136}$$

$$\langle i_2(t) \rangle_{T_s} = \frac{d'(t)}{d(t)} \langle i_1(t) \rangle_{T_s} \tag{7.137}$$

The averaged equivalent circuit for the switch network, that corresponds to Eqs. (7.136) and (7.137), is illustrated in Fig. 7.39(c). Upon completing the averaging step, the switching harmonics have been removed from all converter waveforms, leaving only the dc and low-frequency ac components. This large-signal, nonlinear, time-invariant model is valid for frequencies sufficiently less than the switching frequency. Averaging the waveforms of Fig. 7.38 modifies only the switch network; the remainder of the converter circuit is unchanged. Therefore, the averaged circuit model of the converter is obtained simply by replacing the switch network with the averaged switch model. The switch network of Fig. 7.39(a) can be identified in any two-switch converter, such as buck, boost, buck-boost, SEPIC, or Ćuk. If the converter operates in continuous conduction mode, the derivation of the averaged switch model follows the same steps, and the result shown in Fig. 7.39(c) is the same as in the SEPIC example. This means that the model of Fig. 7.39(c) can be used as a general large-signal averaged switch model for all two-switch converters operating in CCM.

7.4.3   Perturbation and Linearization

The model of Fig. 7.39(c) is nonlinear, because the dependent generators given by Eqs. (7.136) and (7.137) are nonlinear functions of $d(t)$, $\langle i_2(t) \rangle_{T_s}$ and $\langle v_1(t) \rangle_{T_s}$. To construct a small-signal ac model, we perturb and linearize Eqs. (7.136) and (7.137) in the usual fashion. Let

$$
\begin{aligned}
d(t) &= D + \hat{d}(t) \\
\langle v_1(t) \rangle_{T_s} &= V_1 + \hat{v}_1(t) \\
\langle i_1(t) \rangle_{T_s} &= I_1 + \hat{i}_1(t) \\
\langle v_2(t) \rangle_{T_s} &= V_2 + \hat{v}_2(t) \\
\langle i_2(t) \rangle_{T_s} &= I_2 + \hat{i}_2(t)
\end{aligned}
\tag{7.138}
$$

With these substitutions, Eq. (7.136) becomes

$$
\left(D + \hat{d} \right)\left(V_1 + \hat{v}_1 \right) = \left(D' - \hat{d} \right)\left(V_2 + \hat{v}_2 \right)
\tag{7.139}
$$

It is desired to solve for the dependent quantity $V_1 + \hat{v}_1$. Equation (7.139) can be manipulated as follows:

$$
D\left(V_1 + \hat{v}_1 \right) = D'\left(V_2 + \hat{v}_2 \right) - \hat{d}\left(V_1 + V_2 \right) - \hat{d}\hat{v}_1 - \hat{d}\hat{v}_2
\tag{7.140}
$$

The terms $\hat{d}(t)\hat{v}_1(t)$ and $\hat{d}(t)\hat{v}_2(t)$ are nonlinear, and are small in magnitude provided that the ac variations are much smaller than the quiescent values [as in Eq. (7.32)]. When the small-signal assumption is satisfied, these terms can be neglected. Upon eliminating the nonlinear terms and solving for the switch network dependent output $V_1 + \hat{v}_1$, we obtain

$$
\begin{aligned}
\left(V_1 + \hat{v}_1 \right) &= \frac{D'}{D}\left(V_2 + \hat{v}_2 \right) - \hat{d}\left(\frac{V_1 + V_2}{D} \right) \\
&= \frac{D'}{D}\left(V_2 + \hat{v}_2 \right) - \hat{d}\left(\frac{V_1}{DD'} \right)
\end{aligned}
\tag{7.141}
$$

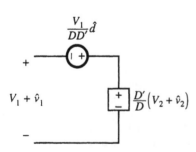

Fig. 7.40 Linearization of the dependent voltage source.

The term $(V_1/DD')\hat{d}(t)$ is driven by the control input $\hat{d}$, and hence can be represented by an independent voltage source as in Fig. 7.40. The term $(D'/D)(V_2 + \hat{v}_2(t))$ is equal to the constant value (D'/D) multiplied by the port 2 independent voltage $(V_2 + \hat{v}_2(t))$. This term is represented by a dependent voltage source in Fig. 7.40. This dependent source will become the primary winding of an ideal transformer.

In a similar manner, substitution of the relationships (7.138) into Eq. (7.137) leads to:

$$
\left(D + \hat{d} \right)\left(I_2 + \hat{i}_2 \right) = \left(D' - \hat{d} \right)\left(I_1 + \hat{i}_1 \right)
\tag{7.142}
$$

The terms $\hat{i}_1(t)\hat{d}(t)$ and $\hat{i}_2(t)\hat{d}(t)$ are nonlinear, and can be neglected when the small-signal assumption is satisfied. Elimination of the nonlinear terms, and solution for $I_2 + \hat{i}_2$, yields:

$$\left(I_2 + \hat{i}_2\right) = \frac{D'}{D}\left(I_1 + \hat{i}_1\right) - \hat{d}\left(\frac{I_1 + I_2}{D}\right)$$

$$= \frac{D'}{D}\left(I_1 + \hat{i}_1\right) - \hat{d}\left(\frac{I_2}{DD'}\right)$$

(7.143)

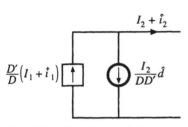

Fig. 7.41 Linearization of the dependent current source.

The term $(I_2/DD')\hat{d}(t)$ is driven by the control input $\hat{d}(t)$, and is represented by an independent current source in Fig. 7.41. The term $(D'/D)(I_1 + \hat{i}_1(t))$ is dependent on the port 1 current $(I_1 + \hat{i}_1(t))$. This term is modeled by a dependent current source in Fig. 7.41; this source will become the secondary winding of an ideal transformer. Equations (7.141) and (7.143) describe the averaged switch network model of Fig. 7.39(d). Note that the model contains both dc and small-signal ac terms: one equivalent circuit is used for both the dc and the small-signal ac models. The transformer symbol contains both a solid line (indicating that it is an ideal transformer capable of passing dc voltages and currents) and a sinusoidal line (which indicates that small-signal ac variations are modeled). The averaged switch model of Fig. 7.39(d) reveals that the switch network performs the functions of: (i) transformation of dc and small-signal ac voltage and current levels according to the $D':D$ conversion ratio, and (ii) introduction of ac voltage and current variations into the converter circuit, driven by the control input $d(t)$. When this model is inserted into Fig. 7.37, the dc and small-signal ac SEPIC model of Fig. 7.42 is obtained. This model can now be solved to determine the steady-state voltages and currents as well as the small-signal converter transfer functions.

The switch network of Fig. 7.39(a) can be identified in all two-switch converters, including buck, boost, SEPIC, Ćuk, etc. As illustrated Fig. 7.43, a complete averaged circuit model of the converter can be constructed simply by replacing the switch network with the averaged switch model. For exam-

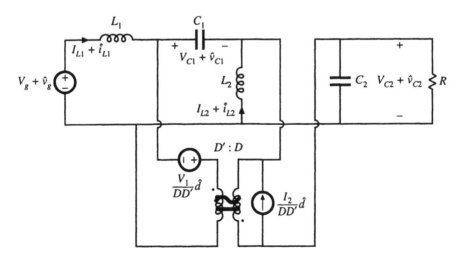

Fig. 7.42 A dc and small-signal ac averaged circuit model of the CCM SEPIC.

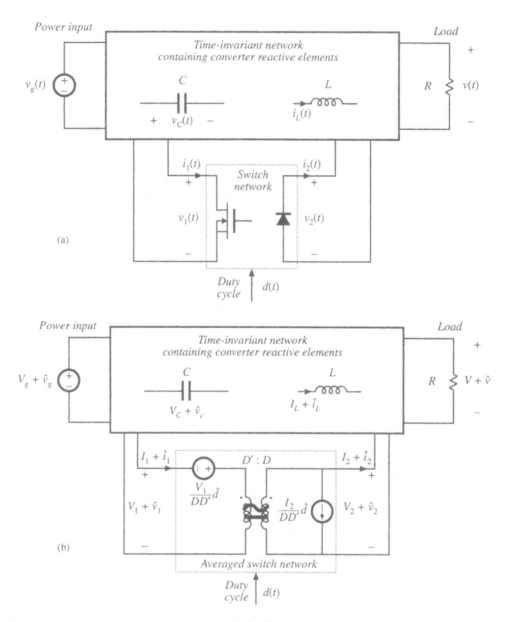

Fig. 7.43 Construction of an averaged circuit model for a two-switch converter operating in CCM: (a) the converter circuit with the general two-switch network identified; (b) dc and ac small-signal averaged circuit model obtained by replacing the switch network with the averaged model.

(a)

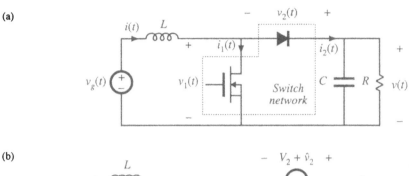

(b)

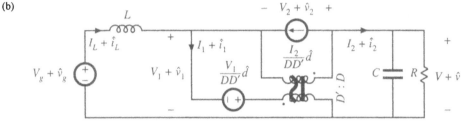

Fig. 7.44 Construction of an averaged circuit model for an ideal boost converter example: (a) converter circuit with the switch network of Fig. 7.39(a) identified; (b) a dc and small-signal ac averaged circuit model obtained by replacing the switch network with the model of Fig. 7.39(d).

ple, Fig. 7.44 shows an averaged circuit model of the boost converter obtained by identifying the switch network of Fig. 7.39(a) and replacing the switch network with the model of Fig. 7.39(d).

In summary, the circuit averaging method involves replacing the switch network with equivalent voltage and current sources, such that a time-invariant network is obtained. The converter waveforms are then averaged over one switching period to remove the switching harmonics. The large-signal model is perturbed and linearized about a quiescent operating point, to obtain a dc and a small-signal averaged switch model. Replacement of the switch network with the averaged switch model yields a complete averaged circuit model of the converter.

7.4.4 Switch Networks

So far, we have described derivation of the averaged switch model for the general two-switch network where the ports of the switch network coincide with the switch ports. No connections are assumed between the switches within the switch network itself. As a result, this switch network and its averaged model can be used to easily construct averaged circuit models of many two-switch converters, as illustrated in Fig. 7.43. It is important to note, however, that the definition of the switch network ports is not unique. Different definitions of the switch network lead to equivalent, but not identical, averaged switch models. The alternative forms of the averaged switch model may result in simpler circuit models, or models that provide better physical insight. Two alternative averaged switch models, better suited for analyses of boost and buck converters, are described in this section.

Consider the ideal boost converter of Fig. 7.45(a). The switch network contains the transistor

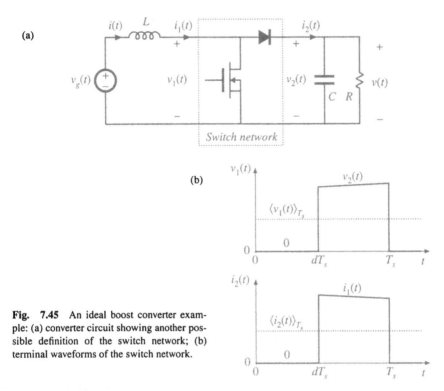

(a)

(b)

Fig. 7.45 An ideal boost converter example: (a) converter circuit showing another possible definition of the switch network; (b) terminal waveforms of the switch network.

and the diode, as in Fig. 7.44(a), but the switch network ports are defined differently. Let us proceed with the derivation of the corresponding averaged switch model. The switch network terminal waveforms are shown in Fig. 7.45(b). Since $i_1(t)$ and $v_2(t)$ coincide with the converter inductor current and capacitor voltage, it is convenient to choose these waveforms as the independent inputs to the switch network. The steps in the derivation of the averaged switch model are illustrated in Fig. 7.46.

First, we replace the switch network with dependent voltage and current generators as illustrated in Fig. 7.46(b). The voltage generator $v_1(t)$ models the dependent voltage waveform at the input port of the switch network, i.e., the transistor voltage. As illustrated in Fig. 7.45(b), $v_1(t)$ is zero when the transistor conducts, and is equal to $v_2(t)$ when the diode conducts:

$$v_1(t) = \begin{cases} 0, & 0 < t < dT_s \\ v_2(t), & dT_s < t < T_s \end{cases} \tag{7.144}$$

When $v_1(t)$ is defined in this manner, the inductor voltage waveform is unchanged. Likewise, $i_2(t)$ models the dependent current waveform at port 2 of the network, i.e., the diode current. As illustrated in Fig. 7.45(b), $i_2(t)$ is equal to zero when the transistor conducts, and is equal to $i_1(t)$ when the diode conducts:

$$i_2(t) = \begin{cases} 0, & 0 < t < dT_s \\ i_1(t), & dT_s < t < T_s \end{cases} \tag{7.145}$$

With $i_2(t)$ defined in this manner, the capacitor current waveform is unchanged. Therefore, the original converter circuit shown in Fig. 7.45(a), and the circuit obtained by replacing the switch network of Fig. 7.46(a) with the switch network of Fig. 7.46(b), are electrically identical. So far, no approximations have been made. Next, we remove the switching harmonics by averaging all signals over one switching period, as in Eq. (7.3). The results are

$$\begin{aligned} \langle v_1(t) \rangle_{T_s} &= d'(t) \langle v_2(t) \rangle_{T_s} \\ \langle i_2(t) \rangle_{T_s} &= d'(t) \langle i_1(t) \rangle_{T_s} \end{aligned} \tag{7.146}$$

Here we have assumed that the switching ripples of the inductor current and capacitor voltage are small, or at least linear functions of time. The averaged switch model of Fig. 7.46(c) is now obtained. This is a large-signal, nonlinear model, which can replace the switch network in the original converter circuit, for construction of a large-signal nonlinear circuit model of the converter. The switching harmonics have been removed from all converter waveforms, leaving only the dc and low-frequency ac components.

The model can be linearized by perturbing and linearizing the converter waveforms about a quiescent operating point, in the usual manner. Let

$$\begin{aligned} \langle v_g(t) \rangle_{T_s} &= V_g + \hat{v}_g(t) \\ d(t) &= D + \hat{d}(t) \quad \Rightarrow d'(t) = D' - \hat{d}(t) \\ \langle i(t) \rangle_{T_s} &= \langle i_1(t) \rangle_{T_s} = I + \hat{i}(t) \\ \langle v(t) \rangle_{T_s} &= \langle v_2(t) \rangle_{T_s} = V + \hat{v}(t) \\ \langle v_1(t) \rangle_{T_s} &= V_1 + \hat{v}_1(t) \\ \langle i_2(t) \rangle_{T_s} &= I_2 + \hat{i}_2(t) \end{aligned} \tag{7.147}$$

The nonlinear voltage generator at port 1 of the averaged switch network has value

$$\left(D' - \hat{d}(t) \right) \left(V + \hat{v}(t) \right) = D' \left(V + \hat{v}(t) \right) - V\hat{d}(t) - \hat{v}(t)\hat{d}(t) \tag{7.148}$$

The term $\hat{v}(t)\hat{d}(t)$ is nonlinear, and is small in magnitude provided that the ac variations are much smaller than the quiescent values [as in Eq. (7.32)]. When the small-signal assumption is satisfied, this term can be neglected. The term $V\hat{d}(t)$ is driven by the control input, and hence can be represented by an independent voltage source. The term $D'(V + \hat{v}(t))$ is equal to the constant value D' multiplied by the output voltage $(V + \hat{v}(t))$. This term is dependent on the output capacitor voltage; it is represented by a dependent voltage source. This dependent source will become the primary winding of an ideal transformer.

The nonlinear current generator at the port 2 of the averaged switch network is treated in a similar manner. Its current is

$$\left(D' - \hat{d}(t) \right) \left(I + \hat{i}(t) \right) = D' \left(I + \hat{i}(t) \right) - I\hat{d}(t) - \hat{i}(t)\hat{d}(t) \tag{7.149}$$

The term $\hat{i}(t)\hat{d}(t)$ is nonlinear, and can be neglected provided that the small-signal assumption is satisfied.

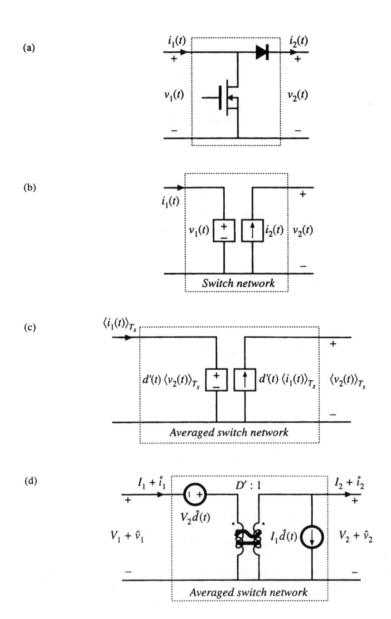

Fig. 7.46 Derivation of the averaged switch model for the CCM boost of Fig. 7.45: (a) switch network; (b) switch network where the switches are replaced by dependent sources whose waveforms match the switch terminal waveforms; (c) large-signal, nonlinear averaged switch model obtained by averaging the switch network terminal waveforms; (d) dc and ac small-signal averaged switch network model.

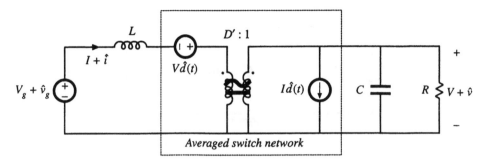

Fig. 7.47 Dc and small-signal ac averaged circuit model of the boost converter.

The term $I\hat{d}(t)$ is driven by the control input $\hat{d}(t)$, and is represented by an independent current source. The term $D'(I + \hat{i}(t))$ is dependent on the inductor current $(I + \hat{i}(t))$. This term is modeled by a dependent current source; this source will become the secondary winding of an ideal transformer.

Upon elimination of the nonlinear terms, and replacement of the dependent generators with an ideal $D'{:}1$ transformer, the combined dc and small-signal ac averaged switch model of Fig. 7.46(d) is obtained. Figure 7.47 shows the complete averaged circuit model of the boost converter.

It is interesting to compare the models of Fig. 7.44(b) and Fig. 7.47. The two averaged circuit models of the boost converter are equivalent—they result in the same steady-state solution, and the same converter transfer functions. However, since both ports of the switch network in Fig. 7.45(a) share the same reference ground, the resulting averaged circuit model in Fig. 7.47 is easier to solve, and gives better physical insight into steady-state operation and dynamics of the boost converter. The circuit model of Fig. 7.47 reveals that the switch network performs the functions of: (*i*) transformation of dc and small-signal ac voltage and current levels according to the $D'{:}1$ conversion ratio, and (*ii*) introduction of ac voltage and current variations into the converter circuit, driven by the control input $d(t)$. The model of Fig. 7.47 obtained using the circuit averaging approach is identical to the model of Fig. 7.17(b) obtained using the basic ac modeling technique of Section 7.2.

Next, we consider the CCM buck converter of Fig. 7.48, where the switch network ports are defined to share a common ground terminal. The derivation of the corresponding averaged switch model follows the same steps as in the SEPIC and the boost examples. Let us select $v_1(t)$ and $i_2(t)$ as the independent terminal variables of the two-port switch network, since these quantities coincide with the applied converter input voltage $v_g(t)$ and the inductor current $i(t)$, respectively. We then need to express the averaged dependent terminal waveforms $\langle i_1(t)\rangle_{T_s}$ and $\langle v_2(t)\rangle_{T_s}$ as functions of the control input $d(t)$ and of $\langle v_1(t)\rangle_{T_s}$ and $\langle i_2(t)\rangle_{T_s}$. Upon averaging the waveforms of Fig. 7.48(b), one obtains

$$\begin{aligned}\langle i_1(t)\rangle_{T_s} &= d(t)\,\langle i_2(t)\rangle_{T_s}\\ \langle v_2(t)\rangle_{T_s} &= d(t)\,\langle v_1(t)\rangle_{T_s}\end{aligned} \tag{7.150}$$

Perturbation and linearization of Eq. (7.150) then leads to

$$\begin{aligned}I_1 + \hat{i}_1(t) &= D\left(I_2 + \hat{i}_2(t)\right) + I_2\,\hat{d}(t)\\ V_2 + \hat{v}_2(t) &= D\left(V_1 + \hat{v}_1(t)\right) + V_1\,\hat{d}(t)\end{aligned} \tag{7.151}$$

An equivalent circuit corresponding to Eq. (7.151) is illustrated in Fig. 7.49(a). Replacement of the

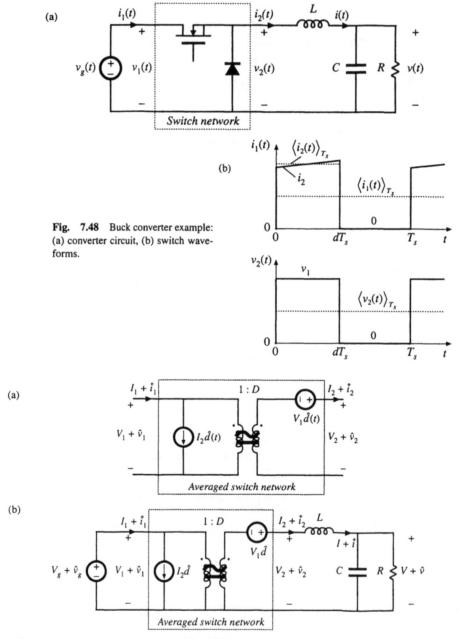

Fig. 7.48 Buck converter example: (a) converter circuit, (b) switch waveforms.

Fig. 7.49 Averaged switch modeling, buck converter example: (a) dc and small-signal ac averaged switch model; (b) Averaged circuit model of the buck converter obtained by replacement of the switch network by the averaged switch model.

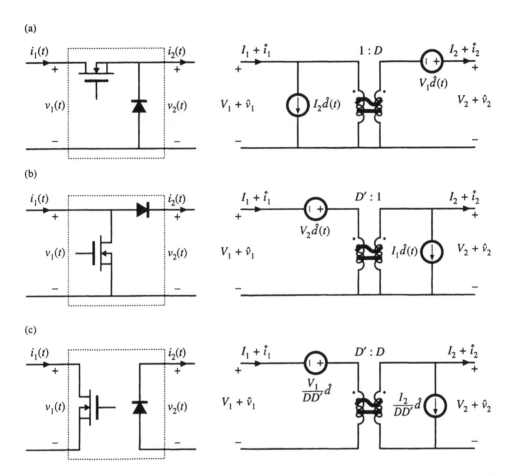

Fig. 7.50 Three basic switch networks, and their CCM dc and small-signal ac averaged switch models: (a) the buck switch network, (b) the boost switch network, and (c) the general two-switch network.

switch network in Fig. 7.48(a) with the averaged switch model of Fig. 7.49(a) leads to the converter averaged circuit model of Fig. 7.49(b). The circuit model of Fig. 7.49(b) reveals that the switch network performs the functions of: (*i*) transformation of dc and small-signal ac voltage and current levels according to the 1:*D* conversion ratio, and (*ii*) introduction of ac voltage and current variations into the converter circuit, driven by the control input *d(t)*. The model is easy to solve for both dc conversion ratio and small-signal frequency responses. It is identical to the model shown in Fig. 7.17(a).

The three basic switch networks—the buck switch network, the boost switch network, and the general two-switch network—together with the corresponding averaged switch models are shown in Fig. 7.50. Averaged switch models can be refined to include conduction and switching losses. These models can then be used to predict the voltages, currents, and efficiencies of nonideal converters. Two examples of averaged switch models that include losses are described in Sections 7.4.5 and 7.4.6.

7.4.5   Example: Averaged Switch Modeling of Conduction Losses

An averaged switch model can be refined to include switch conduction losses. Consider again the SEPIC of Fig. 7.37. Suppose that the transistor on-resistance is R_{on} and the diode forward voltage drop V_D are approximately constant. In this example, all other conduction or switching losses are neglected. Our objective is to derive an averaged switch model that includes conduction losses caused by the voltage drops across R_{on} and V_D. Let us define the switch network as in Fig. 7.39(a). The waveforms of the switch network terminal currents are the same as in Fig. 7.38, but the voltage waveforms are affected by the voltage drops across R_{on} and V_D as shown in Fig. 7.51. We select $i_1(t)$ and $v_2(t)$ as the switch network independent inputs, as in Section 7.4.1. The average values of $v_1(t)$ and $v_2(t)$ can be found as follows:

$$\langle v_1(t) \rangle_{T_s} = d(t)R_{on}\left(\langle i_{L1}(t) \rangle_{T_s} + \langle i_{L2}(t) \rangle_{T_s} \right) + d'(t)\left(\langle v_{C1}(t) \rangle_{T_s} + \langle v_{C2}(t) \rangle_{T_s} + V_D \right) \tag{7.152}$$

$$\langle v_2(t) \rangle_{T_s} = d(t)\left(\langle v_{C1}(t) \rangle_{T_s} + \langle v_{C2}(t) \rangle_{T_s} - R_{on}\left(\langle i_{L1}(t) \rangle_{T_s} + \langle i_{L2}(t) \rangle_{T_s} \right) \right) + d'(t)\left(-V_D \right) \tag{7.153}$$

Next, we proceed to eliminate $\langle i_{L1}(t) \rangle_{T_s}$, $\langle i_{L2}(t) \rangle_{T_s}$, $\langle v_{C1}(t) \rangle_{T_s}$, and $\langle v_{C2}(t) \rangle_{T_s}$, to write the above equations in terms of the averaged independent terminal currents and voltages of the switch network. By combining Eqs. (7.152) and (7.153), we obtain:

$$\langle v_{C1}(t) \rangle_{T_s} + \langle v_{C2}(t) \rangle_{T_s} = \langle v_1(t) \rangle_{T_s} + \langle v_2(t) \rangle_{T_s} \tag{7.154}$$

Since the current waveforms are the same as in Fig. 7.38, Eq. (7.134) can be used here:

$$\langle i_{L1}(t) \rangle_{T_s} + \langle i_{L2}(t) \rangle_{T_s} = \frac{\langle i_1(t) \rangle_{T_s}}{d(t)} \tag{7.155}$$

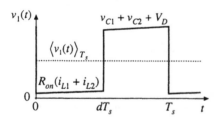

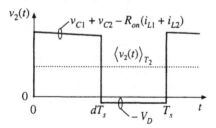

Fig. 7.51 The switch network terminal voltages $v_1(t)$ and $v_2(t)$ for the case when the transistor on-resistance is R_{on} and the diode forward voltage drop is V_D.

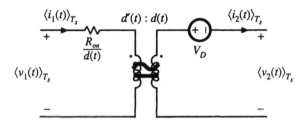

Fig. 7.52 Large-signal averaged switch model for the general two-switch network of Fig. 7.50. This model includes conduction losses due to the transistor on-resistance R_{on} and the diode forward voltage drop V_D.

Substitution of Eqs. (7.154) and (7.155) into Eq. (7.152) results in:

$$\langle v_1(t) \rangle_{T_s} = R_{on} \langle i_1(t) \rangle_{T_s} + d'(t) \left(\langle v_1(t) \rangle_{T_s} + \langle v_2(t) \rangle_{T_s} + V_D \right) \tag{7.156}$$

Equation (7.156) can be solved for the voltage $\langle v_1(t) \rangle_{T_s}$:

$$\langle v_1(t) \rangle_{T_s} = \frac{R_{on}}{d(t)} \langle i_1(t) \rangle_{T_s} + \frac{d'(t)}{d(t)} \left(\langle v_2(t) \rangle_{T_s} + V_D \right) \tag{7.157}$$

The expression for the averaged current $\langle i_2(t) \rangle_{T_s}$ is given by Eq. (7.137) derived in Section 7.4.2:

$$\langle i_2(t) \rangle_{T_s} = \frac{d'(t)}{d(t)} \langle i_1(t) \rangle_{T_s} \tag{7.158}$$

Equations (7.157) and (7.158) constitute the averaged terminal relations of the switch network. An equivalent circuit corresponding to these relationships is shown in Fig. 7.52. The generators that depend on the transistor duty cycle $d(t)$ are combined into an ideal transformer with the turns ratio $d'(t):d(t)$. This part of the model is the same as in the averaged switch model derived earlier for the switch network with ideal switches. The elements R_{on}/d and V_D model the conduction losses in the switch network. This is a large-signal, nonlinear model. If desired, this model can be perturbed and linearized in the usual manner, to obtain a small-signal ac switch model.

The model of Fig. 7.52 is also well suited for computer simulations. As an example of this application, consider the buck-boost converter in Fig 7.53(a). In this converter, the transistor on-resistance is R_{on} = 50 mΩ, while the diode forward voltage drop is V_D = 0.8 V. Resistor R_L = 100 mΩ models the copper loss of the inductor. All other losses are neglected. Figure 7.53(b) shows the averaged circuit model of the converter obtained by replacing the switch network with the averaged switch model of Fig. 7.52.

Let's investigate how the converter output voltage reaches its steady-state value, starting from zero initial conditions. A transient simulation can be used to generate converter waveforms during the start-up transient. It is instructive to compare the responses obtained by simulation of the converter switching circuit shown in Fig. 7.53(a) against the responses obtained by simulation of the averaged circuit model shown in Fig. 7.53(b). Details of how these simulations are performed can be found in Appendix B.1. Figure 7.54 shows the start-up transient waveforms of the inductor current and the output voltage. In the waveforms obtained by simulation of the averaged circuit model, the switching ripple is removed, but other features of the converter transient responses match very closely the responses

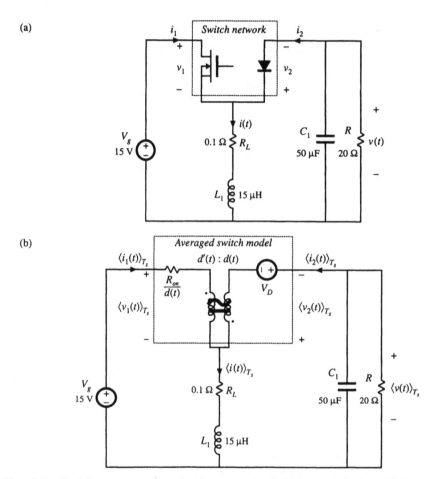

Fig. 7.53 Buck-boost converter example: (a) converter circuit; (b) averaged circuit model of the converter.

obtained from the switching circuit. Simulations of averaged circuit models can be used to predict converter steady-state and dynamic responses, as well as converter losses and efficiency.

7.4.6   Example: Averaged Switch Modeling of Switching Losses

Switching losses can also be modeled via averaged switch modeling. As an example, consider again the CCM buck converter of Fig. 7.48(a). Let us suppose that the transistor is ideal, and that the diode exhibits reverse recovery described in Section 4.3.2. The simplified switch waveforms are shown in Fig. 7.55. Initially, the diode conducts the inductor current and the transistor is in the off state. When the transistor turns on, a negative current flows through the diode so that the transistor current i_1 exceeds the inductor current. The time it takes to remove the charge Q_r stored within the diode is the reverse recovery time t_r.

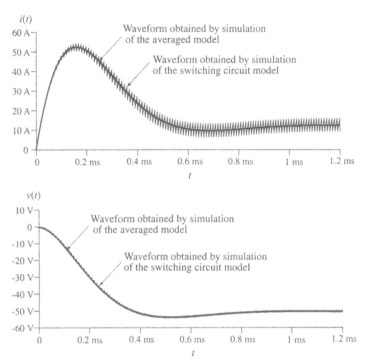

Fig. 7.54 Waveforms obtained by simulation of the switching converter circuit shown in Fig. 7.53(a) and by simulation of the averaged circuit model of Fig. 7.53(b)

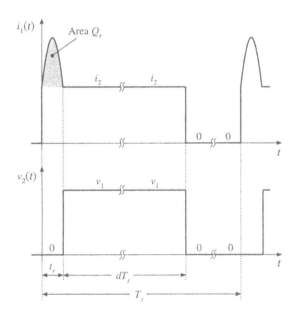

Fig. 7.55 Switch waveforms, buck converter switching loss example.

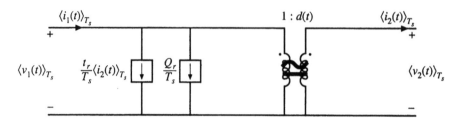

Fig. 7.56 Large-signal averaged switch model for the buck converter switching loss example.

It is assumed that the diode is "snappy," so that the voltage drop across the diode remains small during the reverse recovery time. After the diode reverse recovery is completed, the diode turns off, and the voltage v_2 across the diode quickly jumps to the input voltage $v_1 = v_g$. For this simple example, conduction losses and other switching losses are neglected.

Let us select $v_1(t)$ and $i_2(t)$ as the independent terminal variables of the two-port switch network, and derive expressions for the averaged dependent terminal waveforms $\langle i_1(t) \rangle_{T_s}$ and $\langle v_2(t) \rangle_{T_s}$. The average value of $i_1(t)$ is equal to the area under the $i_1(t)$ waveform, divided by the switching period T_s:

$$\langle i_1(t) \rangle_{T_s} = \frac{1}{T_s} \int_0^{T_s} i_1(t) \, dt = \frac{1}{T_s} \left(Q_r + t_r \langle i_2(t) \rangle_{T_s} + dT_s \langle i_2(t) \rangle_{T_s} \right)$$
$$= \frac{Q_r}{T_s} + \frac{t_r}{T_s} \langle i_2(t) \rangle_{T_s} + d \langle i_2(t) \rangle_{T_s} \tag{7.159}$$

The quantity $d(t)$ is the effective transistor duty cycle, defined in Fig. 7.55 as the transistor on-time minus the reverse recovery time, divided by the switching period. The average value of $v_2(t)$ is equal to:

$$\langle v_2(t) \rangle_{T_s} = d \langle v_1(t) \rangle_{T_s} \tag{7.160}$$

Equations (7.159) and (7.160) constitute the averaged terminal relations of the switch network. An equivalent circuit corresponding to these relationships is constructed in Fig. 7.56. The generators that depend on the effective transistor duty cycle $d(t)$ are combined into an ideal transformer. To complete the model, the recovered charge Q_r and the reverse recovery time t_r can be expressed as functions of the current $\langle i_2(t) \rangle_{T_s}$ [20]. This is a large-signal averaged switch model, which accounts for the switching loss of the idealized waveforms of Fig. 7.55. If desired, this model can be perturbed and linearized in the usual manner, to obtain a small-signal ac switch model.

The model of Fig. 7.56 has the following physical interpretation. The transistor operates with the effective duty cycle $d(t)$. This is the turns ratio of the ideal dc transformer, which models the first-order switch property of lossless transfer of power from the switch input to the switch output port. The additional current generators model the switching loss. Note that both generators consume power. The total switching loss is:

$$P_{sw} = \langle v_1(t) \rangle_{T_s} \left(\frac{Q_r}{T_s} + \frac{t_r}{T_s} \langle i_2(t) \rangle_{T_s} \right) \tag{7.161}$$

These generators also correctly model how the switching loss increases the average switch input current.

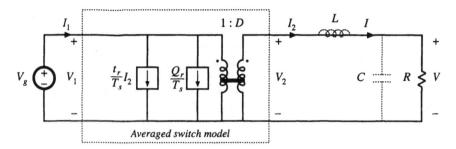

Fig. 7.57 Dc equivalent circuit model, buck converter switching loss example.

By inserting the switch model of Fig. 7.56 into the original converter circuit of Fig. 7.48(a), and by letting all waveforms be equal to their quiescent values, we obtain the steady-state model of Fig. 7.57. This model predicts that the steady-state output voltage is:

$$V = DV_g \tag{7.162}$$

To find the efficiency, we must compute the average input and output powers. The converter input power is

$$P_{in} = V_g I_1 = V_g \left(\frac{Q_r}{T_s} + \frac{t_r}{T_s} I_2 + DI_2 \right) \tag{7.163}$$

The average output power is

$$P_{out} = VI_2 = DV_g I_2 \tag{7.164}$$

Hence the converter efficiency is

$$\eta = \frac{P_{out}}{P_{in}} = \frac{1}{1 + \frac{Q_r}{DT_s I} + \frac{t_r}{DT_s}} \tag{7.165}$$

Beware, the efficiency is not simply equal to V/DV_g.

7.5   THE CANONICAL CIRCUIT MODEL

Having discussed several methods for deriving the ac equivalent circuit models of switching converters, let us now pause to interpret the results. All PWM CCM dc–dc converters perform similar basic functions. First, they transform the voltage and current levels, ideally with 100% efficiency. Second, they contain low-pass filtering of the waveforms. While necessary to remove the high-frequency switching ripple, this filtering also influences low-frequency voltage and current variations. Third, the converter waveforms can be controlled by variation of the duty cycle.

We expect that converters having similar physical properties should have qualitatively similar equivalent circuit models. Hence, we can define a *canonical circuit model* that correctly accounts for all of these basic properties [1-3]. The ac equivalent circuit of any CCM PWM dc–dc converter can be manipulated into this canonical form. This allows us to extract physical insight, and to compare the ac properties of converters. The canonical model is used in several later chapters, where it is desired to analyze converter phenomena in a general manner, without reference to a specific converter. So the canonical model allows us to define and discuss the physical ac properties of converters.

In this section, the canonical circuit model is developed, based on physical arguments. An example is given which illustrates how to manipulate a converter equivalent circuit into canonical form. Finally, the parameters of the canonical model are tabulated for several basic ideal converters.

7.5.1   Development of the Canonical Circuit Model

The physical elements of the canonical circuit model are collected, one at a time, in Fig. 7.58. The converter contains a power input port $v_g(t)$ and a control input port $d(t)$, as well as a power output port and load having voltage $v(t)$. As discussed in Chapter 3, the basic function of any CCM PWM dc–dc converter is the conversion of dc voltage and current levels, ideally with 100% efficiency. As illustrated in Fig. 7.58(a), we have modeled this property with an ideal dc transformer, having effective turns ratio $1:M(D)$ where M is the conversion ratio. This conversion ratio is a function of the quiescent duty cycle D. As discussed in Chapter 3, this model can be refined, if desired, by addition of resistors and other elements that model the converter losses.

Slow variations $v_g(t)$ in the power input induce ac variations $v(t)$ in the converter output voltage. As illustrated in Fig. 7.58(b), we expect these variations also to be transformed by the conversion ratio $M(D)$.

The converter must also contain reactive elements that filter the switching harmonics and transfer energy between the power input and power output ports. Since it is desired that the output switching ripple be small, the reactive elements should comprise a low-pass filter having a cutoff frequency well below the switching frequency. This low-pass characteristic also affects how ac line voltage variations influence the output voltage. So the model should contain an effective low-pass filter as illustrated in Fig. 7.58(c). This figure predicts that the line-to-output transfer function is

$$G_{vg}(s) = \frac{\hat{v}(s)}{\hat{v}_g(s)} = M(D)\, H_e(s) \tag{7.166}$$

where $H_e(s)$ is the transfer function of the effective low-pass filter loaded by resistance R. When the load is nonlinear, R is the incremental load resistance, evaluated at the quiescent operating point. The effective filter also influences other properties of the converter, such as the small-signal input and output impedances. It should be noted that the elemental values in the effective low-pass filter do not necessarily coincide with the physical element values in the converter. In general, the element values, transfer function, and terminal impedances of the effective low-pass filter can vary with quiescent operating point. Examples are given in the following subsections.

Control input variations, specifically, duty cycle variations $\hat{d}(t)$, also induce ac variations in the converter voltages and currents. Hence, the model should contain voltage and current sources driven by $\hat{d}(t)$. In the examples of the previous section, we have seen that both voltage sources and current sources appear, which are distributed around the circuit model. It is possible to manipulate the model such that all of the $\hat{d}(t)$ sources are pushed to the input side of the equivalent circuit. In the process, the sources may become frequency-dependent; an example is given in the next subsection. In general, the sources can be combined into a single voltage source $e(s)\hat{d}(s)$ and a single current source $j(s)\hat{d}(s)$ as shown in

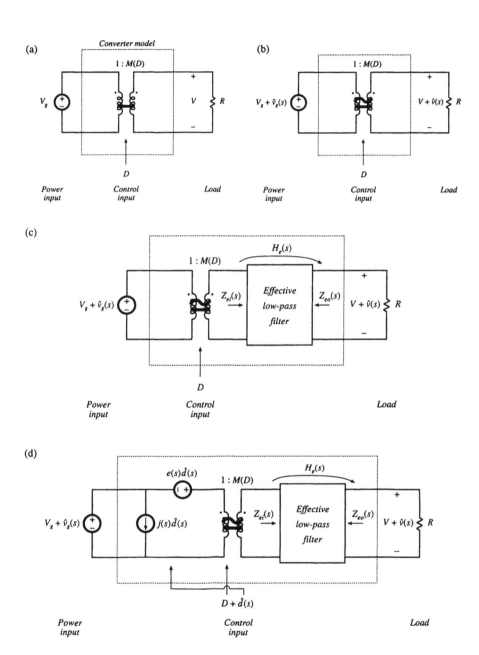

Fig. 7.58 Development of the canonical circuit model, based on physical arguments: (a) dc transformer model, (b) inclusion of ac variations, (c) reactive elements introduce effective low-pass filter, (d) inclusion of ac duty cycle variations.

Fig. 7.58(d). This model predicts that the small-signal control-to-output transfer function is

$$G_{vd}(s) = \frac{\hat{v}(s)}{\hat{d}(s)} = e(s)\, M(D)\, H_e(s) \tag{7.167}$$

This transfer function is found by setting the $\hat{v}_g(s)$ variations to zero, and solving for the dependence of $\hat{v}(s)$ on $\hat{d}(s)$. Figure 7.58(d) is the complete canonical circuit, which can model any PWM CCM dc–dc converter.

7.5.2   Example: Manipulation of the Buck-Boost Converter Model into Canonical Form

To illustrate the steps in the derivation of the canonical circuit model, let us manipulate the equivalent circuit of the buck-boost converter into canonical form. A small-signal ac equivalent circuit for the buck-boost converter is derived in Section 7.2. The result, Fig. 7.16(b), is reproduced in Fig. 7.59. To manipulate this network into canonical form, it is necessary to push all of the independent $d(t)$ generators to the left, while pushing the inductor to the right and combining the transformers.

The $(V_g - V)\hat{d}(t)$ voltage source is in series with the inductor, and hence the positions of these two elements can be interchanged. In Fig. 7.60(a), the voltage source is placed on the primary side of the $1{:}D$ ideal transformer; this requires dividing by the effective turns ratio D. The output-side $I\hat{d}(t)$ current source has also been moved to the primary side of the $D'{:}1$ transformer. This requires multiplying by the turns ratio $1/D'$. The polarity is also reversed, in accordance with the polarities of the $D'{:}1$ transformer windings.

Next, we need to move the $I\hat{d}(t)/D$ current source to the left of the inductor. This can be done using the artifice illustrated in Fig. 7.60(b). The ground connection of the current source is broken, and the source is connected to node A instead. A second, identical, current source is connected from node A to ground. The second source causes the current flowing into node A to be unchanged, such that the node equations of Figs. 7.60(a) and 7.60(b) are identical.

In Fig. 7.60(c), the parallel combination of the inductor and current source is converted into Thevenin equivalent form. The series combination of an inductor and voltage source are obtained.

In Fig. 7.60(d), the $I\hat{d}(t)/D$ current source is pushed to the primary side of the $1{:}D$ transformer. The magnitude of the current source is multiplied by the turns ratio D. In addition, the current source is pushed through the $(V_g - V)\hat{d}(t)/D$ voltage source, using the previously described artifice. The ground connection of the source is moved to node B, and an identical source is connected from node B to ground such that the circuit node equations are unchanged.

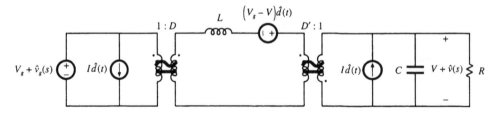

Fig.   7.59   Small-signal ac model of the buck-boost converter, before manipulation into canonical form.

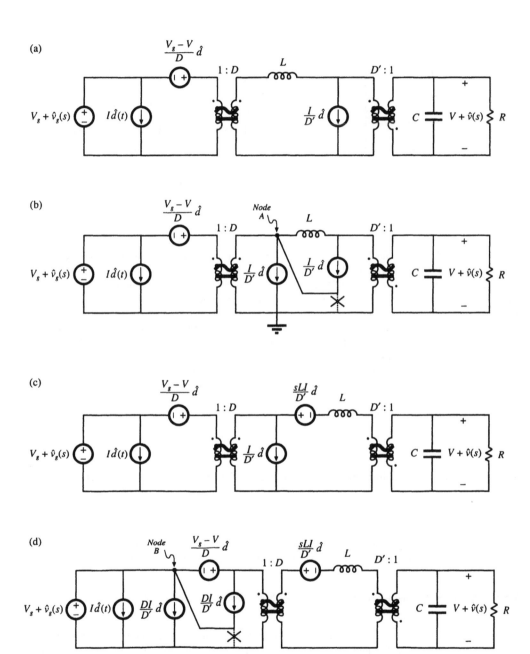

Fig. 7.60 Steps in the manipulation of the buck-boost ac model into canonical form.

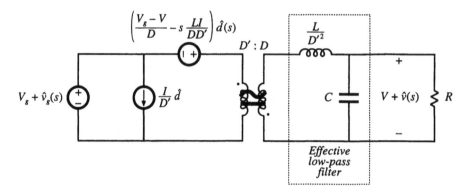

Fig. 7.61   The buck-boost converter model, in canonical form.

Figure 7.61 is the final form of the model. The inductor is moved to the secondary side of the $D':1$ transformer, by multiplying by the square of the turns ratio as shown. The $sLI\hat{d}(t)/D'$ voltage source is moved to the primary side of the $1:D$ transformer, by dividing by the turns ratio D. The voltage and current sources are combined as shown, and the two transformers are combined into a single $D':D$ transformer. The circuit is now in canonical form.

It can be seen that the inductance of the effective low-pass filter is not simply equal to the physical inductor value L, but rather is equal to L/D'^2. At different quiescent operating points, with different values of D', the value of the effective inductance will change. In consequence, the transfer function, input impedance, and output impedance of the effective low-pass filter will also vary with quiescent operating point. The reason for this variation is the transformation of the inductance value by the effective $D':1$ transformer.

It can also be seen from Fig. 7.61 that the coefficient of the $\hat{d}(t)$ voltage generator is

$$e(s) = \frac{V_g - V}{D} - s\frac{LI}{DD'} \tag{7.168}$$

This expression can be simplified by substitution of the dc relationships (7.29). The result is

$$e(s) = -\frac{V}{D^2}\left(1 - s\frac{DL}{D'^2R}\right) \tag{7.169}$$

When we pushed the output-side $I\hat{d}(t)$ current source through the inductor, we obtained a voltage source having a frequency dependence. In consequence, the $e(s)\hat{d}$ voltage generator is frequency-dependent.

7.5.3   Canonical Circuit Parameter Values for Some Common Converters

For ideal CCM PWM dc–dc converters containing a single inductor and capacitor, the effective low-pass filter of the canonical model should contain a single inductor and a single capacitor. The canonical model then reduces to the circuit of Fig. 7.62. It is assumed that the capacitor is connected directly across the load. The parameter values for the basic buck, boost, and buck-boost converters are collected in Table 7.1. Again, it should be pointed out that the effective inductance L_e depends not only on the physical

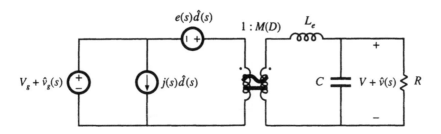

Fig. 7.62 The canonical model, for ideal CCM converters containing a single inductor and capacitor.

Table 7.1 Canonical model parameters for the ideal buck, boost and buck-boost converters

Converter	$M(D)$	L_e	$e(s)$	$j(s)$
Buck	D	L	$\dfrac{V}{D^2}$	$\dfrac{V}{R}$
Boost	$\dfrac{1}{D'}$	$\dfrac{L}{D'^2}$	$V\left(1-\dfrac{sL}{D'^2R}\right)$	$\dfrac{V}{D'^2R}$
Buck-boost	$-\dfrac{D}{D'}$	$\dfrac{L}{D'^2}$	$-\dfrac{V}{D^2}\left(1-\dfrac{sDL}{D'^2R}\right)$	$-\dfrac{V}{D'^2R}$

inductor value L, but also on the quiescent duty cycle D. Furthermore, the current flowing in the effective inductance L_e does not in general coincide with the physical inductor current $I + \hat{i}(t)$.

The model of Fig. 7.62 can be solved using conventional linear circuit analysis, to find quantities of interest such as the converter transfer functions, input impedance, and output impedance. Transformer isolated versions of the buck, boost, and buck-boost converters, such as the full bridge, forward, and flyback converters, can also be modeled using the equivalent circuit of Fig. 7.62 and the parameters of Table 7.1, provided that one correctly accounts for the transformer turns ratio.

7.6 MODELING THE PULSE-WIDTH MODULATOR

We have now achieved the goal, stated at the beginning of this chapter, of deriving a useful equivalent circuit model for the switching converter in Fig. 7.1. One detail remains: modeling the pulse-width modulator. The pulse-width modulator block shown in Fig. 7.1 produces a logic signal $\delta(t)$ that commands the converter power transistor to switch on and off. The logic signal $\delta(t)$ is periodic, with frequency f_s and duty cycle $d(t)$. The input to the pulse-width modulator is an analog control signal $v_c(t)$. The function of the pulse-width modulator is to produce a duty cycle $d(t)$ that is proportional to the analog control volt-

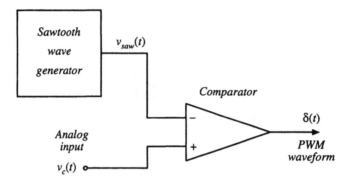

Fig. 7.63   A simple pulse-width modulator circuit.

age $v_c(t)$.

A schematic diagram of a simple pulse-width modulator circuit is given in Fig. 7.63. A sawtooth wave generator produces the voltage waveform $v_{saw}(t)$ illustrated in Fig. 7.64. The peak-to-peak amplitude of this waveform is V_M. The converter switching frequency f_s is determined by and equal to the frequency of $v_{saw}(t)$. An analog comparator compares the analog control voltage $v_c(t)$ to $v_{saw}(t)$. This comparator produces a logic-level output which is high whenever $v_c(t)$ is greater than $v_{saw}(t)$, and is otherwise low. Typical waveforms are illustrated in Fig. 7.64.

If the sawtooth waveform $v_{saw}(t)$ has minimum value zero, then the duty cycle will be zero whenever $v_c(t)$ is less than or equal to zero. The duty cycle will be $D = 1$ whenever $v_c(t)$ is greater than or equal to V_M. If, over a given switching period, $v_{saw}(t)$ varies linearly with t, then for $0 \leq v_c(t) \leq V_M$ the duty cycle d will be a linear function of v_c. Hence, we can write

$$d(t) = \frac{v_c(t)}{V_M} \quad \text{for } 0 \leq v_c(t) \leq V_M \tag{7.170}$$

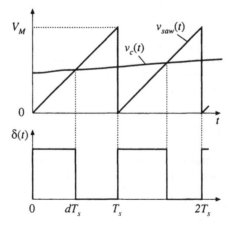

Fig. 7.64   Waveforms of the circuit of Fig. 7.63.

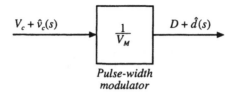

Pulse-width
modulator

Fig. 7.65 Pulse-width modulator block diagram.

This equation is the input-output characteristic of the pulse-width modulator [2,11].

To be consistent with the perturbed-and-linearized converter models of the previous sections, we can perturb Eq. (7.170). Let

$$v_c(t) = V_c + \hat{v}_c(t)$$
$$d(t) = D + \hat{d}(t)$$
(7.171)

Insertion of Eq. (7.171) into Eq. (7.170) leads to

$$D + \hat{d}(t) = \frac{V_c + \hat{v}_c(t)}{V_M}$$
(7.172)

A block diagram representing Eq. (7.172) is illustrated in Fig. 7.65. The pulse-width modulator has linear gain $1/V_M$. By equating like terms on both sides of Eq. (7.172), one obtains

$$D = \frac{V_c}{V_M}$$
$$\hat{d}(t) = \frac{\hat{v}_c(t)}{V_M}$$
(7.173)

So the quiescent value of the duty cycle is determined in practice by V_c.

The pulse-width modulator model of Fig. 7.65 is sufficiently accurate for nearly all applications. However, it should be pointed out that pulse-width modulators also introduce sampling of the waveform. Although the analog input signal $v_c(t)$ is a continuous function of time, there can be only one discrete value of the duty cycle during every switching period. Therefore, the pulse-width modulator samples the waveform, with sampling rate equal to the switching frequency f_s. Hence, a more accurate modulator block diagram is as in Fig. 7.66 [10]. In practice, this sampling restricts the useful frequencies of the ac variations to values much less than the switching frequency. The designer must ensure that the bandwidth of the control system be sufficiently less than the Nyquist rate $f_s/2$.

Significant high-frequency variations in the control signal $v_c(t)$ can also alter the behavior of the pulse-width modulator. A common example is when $v_c(t)$ contains switching ripple, introduced by the feedback loop. This phenomenon has been analyzed by several authors [10,19], and effects of inductor current ripple on the transfer functions of current-programmed converters are investigated in Chapter 12. But it is generally best to avoid the case where $v_c(t)$ contains significant components at the switching frequency or higher, since the pulse-width modulators of such systems exhibit poor noise immunity.

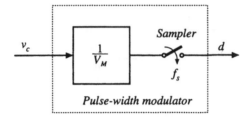

Fig. 7.66   A more accurate pulse-width modulator model, including sampling.

7.7   SUMMARY OF KEY POINTS

1.   The CCM converter analytical techniques of Chapters 2 and 3 can be extended to predict converter ac behavior. The key step is to average the converter waveforms over one switching period. This removes the switching harmonics, thereby exposing directly the desired dc and low-frequency ac components of the waveforms. In particular, expressions for the averaged inductor voltages, capacitor currents, and converter input current are usually found.

2.   Since switching converters are nonlinear systems, it is desirable to construct small-signal linearized models. This is accomplished by perturbing and linearizing the averaged model about a quiescent operating point.

3.   Ac equivalent circuits can be constructed, in the same manner used in Chapter 3 to construct dc equivalent circuits. If desired, the ac equivalent circuits may be refined to account for the effects of converter losses and other nonidealities.

4.   The state-space averaging method of Section 7.3 is essentially the same as the basic approach of Section 7.2, except that the formality of the state-space network description is used. The general results are listed in Section 7.3.2.

5.   The circuit averaging technique also yields equivalent results, but the derivation involves manipulation of circuits rather than equations. Switching elements are replaced by dependent voltage and current sources, whose waveforms are defined to be identical to the switch waveforms of the actual circuit. This leads to a circuit having a time-invariant topology. The waveforms are then averaged to remove the switching ripple, and perturbed and linearized about a quiescent operating point to obtain a small-signal model.

6.   When the switches are the only time-varying elements in the converter, then circuit averaging affects only the switch network. The converter model can then be derived by simply replacing the switch network with its averaged model. Dc and small-signal ac models of several common CCM switch networks are listed in Section 7.4.4. Conduction and switching losses can also be modeled using this approach.

7.   The canonical circuit describes the basic properties shared by all dc–dc PWM converters operating in the continuous conduction mode. At the heart of the model is the ideal $1{:}M(D)$ transformer, introduced in Chapter 3 to represent the basic dc–dc conversion function, and generalized here to include ac variations. The converter reactive elements introduce an effective low-pass filter into the network. The model also includes independent sources that represent the effect of duty cycle variations. The parameter values in the canonical models of several basic converters are tabulated for easy reference.

8.   The conventional pulse-width modulator circuit has linear gain, dependent on the slope of the sawtooth waveform, or equivalently on its peak-to-peak magnitude.

REFERENCES

[1] R. D. MIDDLEBROOK and SLOBODAN ĆUK, "A General Unified Approach to Modeling Switching-Converter Power Stages," *International Journal of Electronics*, Vol. 42, No. 6, pp. 521-550, June 1977.

[2] SLOBODAN ĆUK, "Modeling, Analysis, and Design of Switching Converters," Ph.D. thesis, California Institute of Technology, November 1976.

[3] R. D. MIDDLEBROOK AND SLOBODAN ĆUK, "Modeling and Analysis Methods for Dc-to-Dc Switching Converters," *Proceedings of the IEEE International Semiconductor Power Converter Conference*, 1977 Record, pp. 90-111, March 1977. Reprinted in *Advances in Switched-Mode Power Conversion*, Vol. 1, Irvine: Teslaco, 1983.

[4] G. W. WESTER and R. D. MIDDLEBROOK, "Low-Frequency Characterization of Switched Dc-Dc Converters," *IEEE Transactions an Aerospace and Electronic Systems*, Vol. AES-9, pp. 376-385, May 1973.

[5] DANIEL M. MITCHELL, *Dc–Dc Switching Regulator Analysis*, New York: McGraw-Hill, 1988.

[6] SETH R. SANDERS and GEORGE C. VERGESE, "Synthesis of Averaged Circuit Models for Switched Power Converters," *IEEE Transactions on Circuits and Systems*, Vol. 38, No. 8, pp. 905-915, August 1991.

[7] P. T. KREIN, J. BENTSMAN, R. M. BASS, and B. C. LESIEUTRE, "On the Use of Averaging for the Analysis of Power Electronic Systems," *IEEE Transactions on Power Electronics*, Vol. 5, No. 2, pp. 182-190, April 1990.

[8] B. LEHMAN and R. M. BASS, "Switching Frequency Dependent Averaged Models for PWM DC-DC Converters," *IEEE Transactions on Power Electronics*, Vol. 11, No. 1, pp. 89-98, January 1996.

[9] R. M. BASS and J. SUN, "Averaging Under Large-Ripple Conditions," *IEEE Power Electronics Specialists Conference*, 1998 Record, pp. 630-632, May 1998.

[10] ARTHUR R. BROWN and R. D. MIDDLEBROOK, "Sampled-Data Modeling of Switching Regulators," *IEEE Power Electronics Specialists Conference*, 1981 Record, pp. 349-369, June 1981.

[11] R. D. MIDDLEBROOK, "Predicting Modulator Phase Lag in PWM Converter Feedback Loops," *Proceedings of the Eighth National Solid-State Power Conversion Conference* (Powercon 8), April 1981.

[12] A. KISLOVSKI, R. REDL, AND N. SOKAL, *Dynamic Analysis of Switching-Mode DC/DC Converters*, New York: Van Nostrand Reinhold, 1994.

[13] R. TYMERSKI and V. VORPERIAN, "Generation, Classification and Analysis of Switched-Mode DC-to-DC Converters by the Use of Converter Cells," *Proceedings of the 1986 International Telecommunications Energy Conference* (INTELEC'86), pp. 181-195, October 1986.

[14] V. VORPERIAN, R. TYMERSKI, and F. C. LEE, "Equivalent Circuit Models for Resonant and PWM Switches," *IEEE Transactions on Power Electronics*, Vol. 4, No. 2, pp. 205-214, April 1989.

[15] V. VORPERIAN, "Simplified Analysis of PWM Converters Using the Model of the PWM Switch: Parts I and II," *IEEE Transactions on Aerospace and Electronic Systems*, Vol. AES-26, pp. 490-505, May 1990.

[16] S. FREELAND and R. D. MIDDLEBROOK, "A Unified Analysis of Converters with Resonant Switches," *IEEE Power Electronics Specialists Conference*, 1987 Record, pp. 20-30.

[17]   ARTHUR WITULSKI and ROBERT ERICKSON, "Extension of State-Space Averaging to Resonant Switches —and Beyond," *IEEE Transactions on Power Electronics*, Vol. 5, No. 1, pp. 98-109, January 1990.

[18]   D. MAKSIMOVIĆ and S. ĆUK, "A Unified Analysis of PWM Converters in Discontinuous Modes," *IEEE Transactions on Power Electronics*, Vol. 6, No. 3, pp. 476-490, July 1991.

[19]   D. J. SHORTT and F. C. LEE, "Extensions of the Discrete-Average Models for Converter Power Stages," *IEEE Power Electronics Specialists Conference*, 1983 Record, pp. 23-37, June 1983.

[20]   O. AL-NASEEM and R.W. ERICKSON, "Prediction of Switching Loss Variations by Averaged Switch Modeling," *IEEE Applied Power Electronics Conference*, 2000 Record, pp. 242-248, February 2000.

PROBLEMS

7.1   An ideal boost converter operates in the continuous conduction mode.

    (a)   Determine the nonlinear averaged equations of this converter.

    (b)   Now construct a small-signal ac model. Let

$$\left\langle v_g(t) \right\rangle_{T_s} = V_g + \hat{v}_g(t)$$
$$d(t) = D + \hat{d}(t)$$
$$\left\langle i(t) \right\rangle_{T_s} = I + \hat{i}(t)$$
$$\left\langle v(t) \right\rangle_{T_s} = V + \hat{v}(t)$$

where V_g, D, I, and V are steady-state dc values; $\hat{v}_g(t)$ and $\hat{d}(t)$ are small ac variations in the power and control inputs; and $\hat{i}(t)$ and $\hat{v}(t)$ are the resulting small ac variations in the inductor current and output voltage, respectively. Show that the following model results:

Large-signal dc components

$$0 = - D'V + V_g$$
$$0 = D'I - \frac{V}{R}$$

Small-signal ac components

$$L \frac{d\hat{i}(t)}{dt} = - D'\hat{v}(t) + V\hat{d}(t) + \hat{v}_g(t)$$
$$C \frac{d\hat{v}(t)}{dt} = D'\hat{i}(t) - I\hat{d}(t) - \frac{\hat{v}(t)}{R}$$

7.2   Construct an equivalent circuit that corresponds to the boost converter small-signal ac equations derived in Problem 7.1(b).

7.3   Manipulate your boost converter equivalent circuit of Problem 7.2 into canonical form. Explain each step in your derivation. Verify that the elements in your canonical model agree with Table 7.1.

7.4   The ideal current-fed bridge converter of Fig. 2.31 operates in the continuous conduction mode.

    (a)   Determine the nonlinear averaged equations of this converter.

    (b)   Perturb and linearize these equations, to determine the small-signal ac equations of the converter.

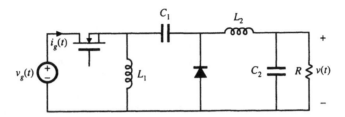

Fig. 7.67   Inverse SEPIC, Problem 7.7.

(c)   Construct a small-signal ac equivalent circuit model for this converter.

7.5   Construct a complete small-signal ac equivalent circuit model for the flyback converter shown in Fig. 7.18, operating in continuous conduction mode. The transformer contains magnetizing inductance L, referred to the primary. In addition, the transformer exhibits significant core loss, which can be modeled by a resistor R_C in parallel with the primary winding. All other elements are ideal. You may use any valid method to solve this problem. Your model should correctly predict variations in $i_g(t)$.

7.6   Modeling the Ćuk converter. You may use any valid method to solve this problem.

(a)   Derive the small-signal dynamic equations that model the ideal Ćuk converter.

(b)   Construct a complete small-signal equivalent circuit model for the Ćuk converter.

7.7   Modeling the inverse-SEPIC. You may use any valid method to solve this problem.

(a)   Derive the small-signal dynamic equations that model the converter shown in Fig. 7.67.

(b)   Construct a complete small-signal equivalent circuit model for the inverse-SEPIC.

7.8   Consider the nonideal buck converter of Fig. 7.68. The input voltage source $v_g(t)$ has internal resistance R_g. Other component nonidealities may be neglected.

(a)   Using the state-space averaging method, determine the small-signal ac equations that describe variations in i, v, and i_g, which occur owing to variations in the transistor duty cycle d and input voltage v_g.

(b)   . Construct an ac equivalent circuit model corresponding to your equations of part (a).

(c)   Solve your model to determine an expression for the small-signal control-to-output transfer function.

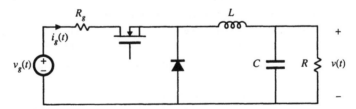

Fig. 7.68   Nonideal buck converter, Problem 7.8.

7.9   Use the circuit-averaging technique to derive the dc and small-signal ac equivalent circuit of the buck converter with input filter, illustrated in Fig. 2.32. All elements are ideal.

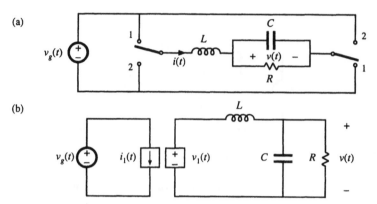

Fig. 7.69 Bridge inverter, Problem 7.11: (a) circuit, (b) large-signal averaged model.

7.10   A flyback converter operates in the continuous conduction mode. The MOSFET switch has on-resistance R_{on}, and the secondary-side diode has a constant forward voltage drop V_D. The flyback transformer has primary winding resistance R_p and secondary winding resistance R_s.

    (a)   Derive the small-signal ac equations for this converter.

    (b)   Derive a complete small-signal ac equivalent circuit model, which is valid in the continuous conduction mode and which correctly models the above losses, as well as the converter input and output ports.

7.11   Circuit averaging of the bridge inverter circuit of Fig. 7.69(a).

    (a)   Show that the converter of Fig. 7.69(a) can be written in the electrically identical form shown in Fig. 7.69(b). Sketch the waveforms $i_1(t)$ and $v_1(t)$.

    (b)   Use the circuit-averaging method to derive a large-signal averaged model for this converter.

    (c)   Perturb and linearize your circuit model of part (b), to obtain a single equivalent circuit that models dc and small-signal ac signals in the bridge inverter.

7.12   Use the circuit averaging method to derive an equivalent circuit that models dc and small-signal ac signals in the buck-boost converter. You may assume that the converter operates in the continuous conduction mode, and that all elements are ideal.

    (a)   Give a time-invariant electrically identical circuit, in which the switching elements are replaced by equivalent voltage and current sources. Define the waveforms of the sources.

    (b)   Derive a large-signal averaged model for this converter.

    (c)   Perturb and linearize your circuit model of part (b), to obtain a single equivalent circuit that models dc and small-signal ac signals in the buck-boost converter.

7.13   The two-output flyback converter of Fig. 7.70(a) operates in the continuous conduction mode. It may be assumed that the converter is lossless.

    (a)   Derive a small-signal ac equivalent circuit for this converter.

    (b)   Show that the small-signal ac equivalent circuit for this two-output converter can be written in the generalized canonical form of Fig. 7.70(b). Give analytical expressions for the generators $e(s)$ and $j(s)$.

7.14   A pulse-width modulator circuit is constructed in which the sawtooth-wave generator is replaced by a triangle-wave generator, as illustrated in Fig. 7.71(a). The triangle waveform is illustrated in Fig. 7.71(b).

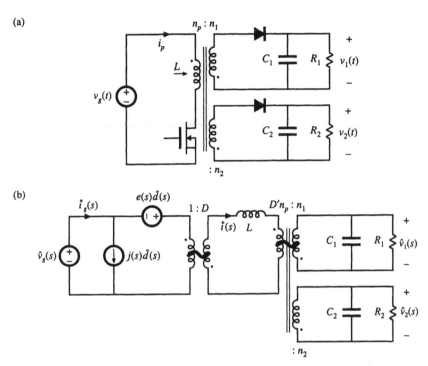

Fig. 7.70 Two-output flyback converter, Problem 7.13: (a) converter circuit, (b) small-signal ac equivalent circuit.

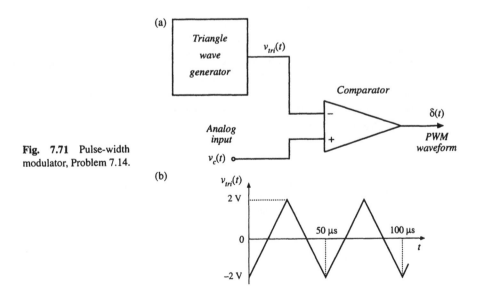

Fig. 7.71 Pulse-width modulator, Problem 7.14.

(a)   Determine the converter switching frequency, in Hz.

(b)   Determine the gain $d(t)/v_c(t)$ for this circuit.

(c)   Over what range of v_c is your answer to (b) valid?

7.15   Use the averaged switch modeling technique to derive an ac equivalent circuit model for the buck-boost converter of Fig. 7.31:

(a)   Replace the switches in Fig. 7.31 with the averaged switch model given in Fig. 7.50(c).

(b)   Compare your result with the model given in Fig. 7.16(b). Show that the two models predict the same small-signal line-to-output transfer function $G_{vg}(s) = \hat{v}/\hat{v}_g$.

7.16   Modify the CCM dc and small-signal ac averaged switch models of Fig. 7.50, to account for MOSFET on-resistance R_{on} and diode forward voltage drop V_D.

7.17   Use the averaged switch modeling technique to derive a dc and ac equivalent circuit model for the flyback converter of Fig. 7.18. You can neglect all losses and the transformer leakage inductances.

(a)   Define a switch network containing the transistor Q_1 and the diode D_1 as in Fig. 7.39(a). Derive a large-signal averaged switch model of the switch network. The model should account for the transformer turns ratio n.

(b)   Perturb and linearize the model you derived in part (a) to obtain the dc and ac small-signal averaged switch model. Verify that for $n = 1$ your model reduces to the model shown in Fig. 7.39(d).

(c)   Using the averaged switch model you derived in part (b), sketch a complete dc and small-signal ac model of the flyback converter. Solve the model for the steady-state conversion ratio $M(D) = V/V_g$.

(d)   The averaged switch models you derived in parts (a) and (b) could be used in other converters having an isolation transformer. Which ones?

7.18   In the flyback converter of Fig. 7.18, the transistor on-resistance is R_{on}, and the diode forward voltage drop is V_D. Other losses and the transformer leakage inductances can be neglected. Derive a dc and small-signal ac averaged switch model for the switch network containing the transistor Q_1 and the diode D_1. The model should account for the on-resistance R_{on}, the diode forward voltage drop V_D, and the transformer turns ratio n.

7.19   In the boost converter of Fig. 7.72(a), the $v_1(t)$ and $i_2(t)$ waveforms of Fig. 7.72(b) are observed. During the transistor turn-on transition, a reverse current flows through the diode which removes the diode stored charge. As illustrated in Fig. 7.72(b), the reverse current spike has area $-Q_r$ and duration t_r. The inductor winding has resistance R_L. You may neglect all losses other than the switching loss due to the diode stored charge and the conduction loss due to the inductor winding resistance.

(a)   Derive an averaged switch model for the boost switch network in Fig. 7.72(a).

(b)   Use your result of part (a) to sketch a dc equivalent circuit model for the boost converter.

(c)   The diode stored charge can be expressed as a function of the current I_1 as:

$$Q_r = k_q \sqrt{I_1}$$

while the reverse recovery time t_r is approximately constant. Given $V_g = 100$ V, $D = 0.5$, $f_s = 100$ kHz, $k_q = 100$ nC/A$^{1/2}$, $t_r = 100$ ns, $R_L = 0.1$ Ω, use a dc sweep simulation to plot the converter efficiency as a function of the load current I_{LOAD} in the range:

$$1\ \text{A} \le I_{LOAD} \le 10\ \text{A}$$

(a)

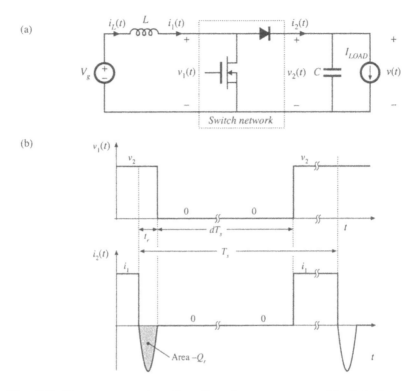

(b)

Fig. 7.72 Boost converter and waveforms illustrating reverse recovery of the diode. Averaged switch modeling in this converter is addressed in Problem 7.19.

8

Converter Transfer Functions

The engineering design process is comprised of several major steps:

1. *Specifications and other design goals* are defined.

2. *A circuit is proposed.* This is a creative process that draws on the physical insight and experience of the engineer.

3. *The circuit is modeled.* The converter power stage is modeled as described in Chapter 7. Components and other portions of the system are modeled as appropriate, often with vendor-supplied data.

4. *Design-oriented analysis* of the circuit is performed. This involves development of equations that allow element values to be chosen such that specifications and design goals are met. In addition, it may be necessary for the engineer to gain additional understanding and physical insight into the circuit behavior, so that the design can be improved by adding elements to the circuit or by changing circuit connections.

5. *Model verification.* Predictions of the model are compared to a laboratory prototype, under nominal operating conditions. The model is refined as necessary, so that the model predictions agree with laboratory measurements.

6. *Worst-case analysis* (or other reliability and production yield analysis) of the circuit is performed. This involves quantitative evaluation of the model performance, to judge whether specifications are met under all conditions. Computer simulation is well-suited to this task.

7. *Iteration.* The above steps are repeated to improve the design until the worst-case behavior meets specifications, or until the reliability and production yield are acceptably high.

This chapter covers techniques of design-oriented analysis, measurement of experimental transfer functions, and computer simulation, as needed in steps 4, 5, and 6.

Sections 8.1 to 8.3 discuss techniques for analysis and construction of the Bode plots of the converter transfer functions, input impedance, and output impedance predicted by the equivalent circuit

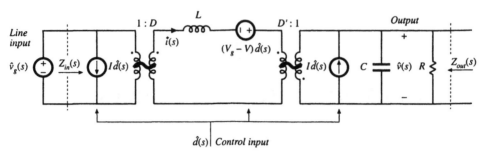

Fig. 8.1   Small-signal equivalent circuit model of the buck-boost converter, as derived in Chapter 7.

models of Chapter 7. For example, the small-signal equivalent circuit model of the buck-boost converter is illustrated in Fig. 7.17(c). This model is reproduced in Fig. 8.1, with the important inputs and terminal impedances identified. The line-to-output transfer function $G_{vg}(s)$ is found by setting duty cycle variations $\hat{d}(s)$ to zero, and then solving for the transfer function from $\hat{v}_g(s)$ to $\hat{v}(s)$:

$$G_{vg}(s) = \left. \frac{\hat{v}(s)}{\hat{v}_g(s)} \right|_{\hat{d}(s) = 0} \tag{8.1}$$

This transfer function describes how variations or disturbances in the applied input voltage $v_g(t)$ lead to disturbances in the output voltage $v(t)$. It is important in design of an output voltage regulator. For example, in an off-line power supply, the converter input voltage $v_g(t)$ contains undesired even harmonics of the ac power line voltage. The transfer function $G_{vg}(s)$ is used to determine the effect of these harmonics on the converter output voltage $v(t)$.

The control-to-output transfer function $G_{vd}(s)$ is found by setting the input voltage variations $\hat{v}_g(s)$ to zero, and then solving the equivalent circuit model for $\hat{v}(s)$ as a function of $\hat{d}(s)$:

$$G_{vd}(s) = \left. \frac{\hat{v}(s)}{\hat{d}(s)} \right|_{\hat{v}_g(s) = 0} \tag{8.2}$$

This transfer function describes how control input variations $\hat{d}(s)$ influence the output voltage $\hat{v}(s)$. In an output voltage regulator system, $G_{vd}(s)$ is a key component of the loop gain and has a significant effect on regulator performance.

The output impedance $Z_{out}(s)$ is found under the conditions that $\hat{v}_g(s)$ and $\hat{d}(s)$ variations are set to zero. $Z_{out}(s)$ describes how variations in the load current affect the output voltage. This quantity is also important in voltage regulator design. It may be appropriate to define $Z_{out}(s)$ either including or not including the load resistance R.

The converter input impedance $Z_{in}(s)$ plays a significant role when an electromagnetic interference (EMI) filter is added at the converter power input. The relative magnitudes of Z_{in} and the EMI filter output impedance influence whether the EMI filter disrupts the transfer function $G_{vd}(s)$. Design of input EMI filters is the subject of Chapter 10.

An objective of this chapter is the construction of Bode plots of the important transfer functions and terminal impedances of switching converters. For example, Fig. 8.2 illustrates the magnitude and phase plots of $G_{vd}(s)$ for the buck-boost converter model of Fig. 8.1. Rules for construction of magnitude and phase asymptotes are reviewed in Section 8.1, including two types of features that often appear in

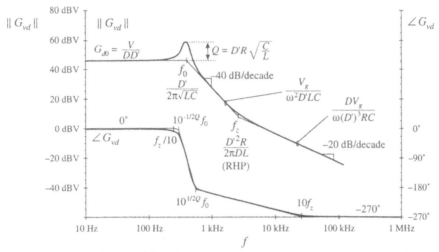

Fig. 8.2 Bode plot of control-to-output transfer function predicted by the model of Fig. 8.1, with analytical expressions for the important features.

converter transfer functions: resonances and right half-plane zeroes. Bode diagrams of the small-signal transfer functions of the buck-boost converter are derived in detail in Section 8.2, and the transfer functions of the basic buck, boost, and buck-boost converters are tabulated. The physical origins of the right half-plane zero are also described.

A difficulty usually encountered in circuit analysis (step 5 of the above list) is the complexity of the circuit model: practical circuits may contains hundreds of elements, and hence their analysis may leads to complicated derivations, intractable equations, and lots of algebra mistakes. *Design-oriented analysis*[1] is a collection of tools and techniques that can alleviate these problems. Some tools for approaching the design of a complicated converter system are described in this chapter. Writing the transfer functions in normalized form directly exposes the important features of the response. Analytical expressions for these features, as well as for the asymptotes, lead to simple equations that are useful in design. Well-separated roots of transfer function polynomials can be approximated in a simple way. Section 8.3 describes a graphical method for constructing Bode plots of transfer functions and impedances, essentially by inspection. This method can: (1) reduce the amount of algebra and associated algebra mistakes; (2) lead to greater insight into circuit behavior, which can be applied to design the circuit; and (3) lead to the insight necessary to make suitable approximations that render the equations tractable.

Experimental measurement of transfer functions and impedances (needed in step 4, model verification) is discussed in Section 8.5. Use of computer simulation to plot converter transfer functions (as needed in step 6, worst-case analysis) is covered in Appendix B.

8.1   REVIEW OF BODE PLOTS

A Bode plot is a plot of the magnitude and phase of a transfer function or other complex-valued quantity, vs. frequency. Magnitude in decibels, and phase in degrees, are plotted vs. frequency, using semilogarithmic axes. The magnitude plot is effectively a log-log plot, since the magnitude is expressed in decibels and the frequency axis is logarithmic.

The magnitude of a dimensionless quantity G can be expressed in decibels as follows:

$$|G|_{dB} = 20 \log_{10}(|G|) \qquad (8.3)$$

Decibel values of some simple magnitudes are listed in Table 8.1. Care must be used when the magnitude is not dimensionless. Since it is not proper to take the logarithm of a quantity having dimensions, the magnitude must first be normalized. For example, to express the magnitude of an impedance Z in decibels, we should normalize by dividing by a base impedance R_{base}:

Table 8.1 Expressing magnitudes in decibels

Actual magnitude	Magnitude in dB
1/2	– 6 dB
1	0 dB
2	6 dB
5 = 10/2	20 dB – 6 dB = 14 dB
10	20 dB
1000 = 10^3	3·20 dB = 60 dB

$$|Z|_{dB} = 20 \log_{10}\left(\frac{|Z|}{R_{base}}\right) \qquad (8.4)$$

The value of R_{base} is arbitrary, but we need to tell others what value we have used. So if $\| Z \|$ is 5 Ω, and we choose $R_{base} = 10\ \Omega$, then we can say that $\| Z \|_{dB} = 20 \log_{10}(5\ \Omega/10\Omega) = -6\text{dB}$ with respect to 10 Ω. A common choice is $R_{base} = 1\Omega$; decibel impedances expressed with $R_{base} = 1\ \Omega$ are said to be expressed in dBΩ. So 5 Ω is equivalent to 14 dBΩ. Current switching harmonics at the input port of a converter are often expressed in dBμA, or dB using a base current of 1 μA: 60 dBμA is equivalent to 1000 μA, or 1 mA.

The magnitude Bode plots of functions equal to powers of f are linear. For example, suppose that the magnitude of a dimensionless quantity $G(f)$ is

$$|G| = \left(\frac{f}{f_0}\right)^n \qquad (8.5)$$

where f_0 and n are constants. The magnitude in decibels is

$$|G|_{dB} = 20 \log_{10}\left(\frac{f}{f_0}\right)^n = 20n \log_{10}\left(\frac{f}{f_0}\right) \qquad (8.6)$$

This equation is plotted in Fig. 8.3, for several values of n. The magnitudes have value $1 \Rightarrow 0$ dB at frequency $f = f_0$. They are linear functions of $\log_{10}(f)$. The slope is the change in $\| G \|_{dB}$ arising from a unit change in $\log_{10}(f)$; a unit increase in $\log_{10}(f)$ corresponds to a factor of 10, or decade, increase in f. From Eq. (8.6), a decade increase in f leads to an increase in $\| G \|_{dB}$ of $20n$ dB. Hence, the slope is $20n$ dB per decade. Equivalently, we can say that the slope is $20n \log_{10}(2) \approx 6n$ dB per octave, where an octave is a factor of 2 change in frequency. In practice, the magnitudes of most frequency-dependent functions can usually be approximated over a limited range of frequencies by functions of the form (8.5); over this range of frequencies, the magnitude Bode plot is approximately linear with slope $20n$ dB/decade.

A simple transfer function whose magnitude is of the form (8.5) is the *pole at the origin*:

$$G(s) = \frac{1}{\left(\frac{s}{\omega_0}\right)} \qquad (8.7)$$

The magnitude is

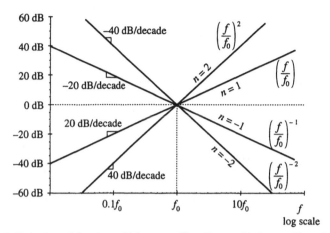

Fig. 8.3 Magnitude Bode plots of functions which vary as f^n are linear, with slope n dB per decade.

$$|G(j\omega)| = \frac{1}{\left|\frac{j\omega}{\omega_0}\right|} = \frac{1}{\left(\frac{\omega}{\omega_0}\right)} \tag{8.8}$$

If we define $f = \omega/2\pi$ and $f_0 = \omega_0/2\pi$, then Eq. (8.8) becomes

$$|G| = \left(\frac{f}{f_0}\right)^{-1} \tag{8.9}$$

which is of the form of Eq. (8.5) with $n = -1$. As illustrated in Fig. 8.3, the magnitude Bode plot of the pole at the origin (8.7) has a -20 dB per decade slope, and passes through 0 dB at frequency $f = f_0$.

8.1.1 Single Pole Response

Consider the simple R-C low-pass filter illustrated in Fig. 8.4. The transfer function is given by the voltage divider ratio

$$G(s) = \frac{v_2(s)}{v_1(s)} = \frac{\frac{1}{sC}}{\frac{1}{sC} + R} \tag{8.10}$$

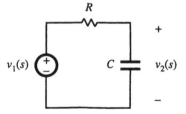

Fig. 8.4 Simple R–C low-pass filter example.

This transfer function is a ratio of voltages, and hence is dimensionless. By multiplying the numerator and denominator by sC, we can express the transfer function as a rational fraction:

$$G(s) = \frac{1}{1 + sRC} \tag{8.11}$$

The transfer function now coincides with the following standard normalized form for a single pole:

$$G(s) = \frac{1}{\left(1 + \frac{s}{\omega_0}\right)} \tag{8.12}$$

The parameter $\omega_0 = 2\pi f_0$ is found by equating the coefficients of s in the denominators of Eqs. (8.11) and (8.12). The result is

$$\omega_0 = \frac{1}{RC} \tag{8.13}$$

Since R and C are real positive quantities, ω_0 is also real and positive. The denominator of Eq. (8.12) contains a root at $s = -\omega_0$, and hence $G(s)$ contains a real pole in the left half of the complex plane.

To find the magnitude and phase of the transfer function, we let $s = j\omega$, where j is the square root of -1. We then find the magnitude and phase of the resulting complex-valued function. With $s = j\omega$, Eq. (8.12) becomes

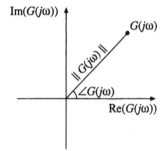

$$G(j\omega) = \frac{1}{\left(1 + j\frac{\omega}{\omega_0}\right)} = \frac{1 - j\frac{\omega}{\omega_0}}{1 + \left(\frac{\omega}{\omega_0}\right)^2} \tag{8.14}$$

The complex-valued $G(j\omega)$ is illustrated in Fig. 8.5, for one value of ω. The magnitude is

Fig. 8.5 Magnitude and phase of the complex-valued function $G(j\omega)$.

$$\begin{aligned} \left| G(j\omega) \right| &= \sqrt{\left[\text{Re}\,(G(j\omega)) \right]^2 + \left[\text{Im}\,(G(j\omega)) \right]^2} \\ &= \frac{1}{\sqrt{1 + \left(\frac{\omega}{\omega_0}\right)^2}} \end{aligned} \tag{8.15}$$

Here, we have assumed that ω_0 is real. In decibels, the magnitude is

$$\left| G(j\omega) \right|_{dB} = -20 \log_{10}\left(\sqrt{1 + \left(\frac{\omega}{\omega_0}\right)^2} \right) \text{ dB} \tag{8.16}$$

The easy way to sketch the magnitude Bode plot of G is to investigate the asymptotic behavior for large and small frequency.

For small frequency, $\omega \ll \omega_0$ and $f \ll f_0$, it is true that

$$\left(\frac{\omega}{\omega_0}\right) \ll 1 \tag{8.17}$$

The $(\omega/\omega_0)^2$ term of Eq. (8.15) is therefore much smaller than 1, and hence Eq. (8.15) becomes

$$\left| G(j\omega) \right| \approx \frac{1}{\sqrt{1}} = 1 \tag{8.18}$$

In decibels, the magnitude is approximately

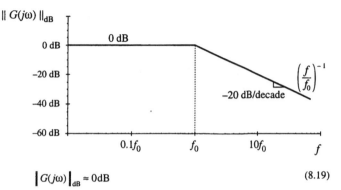
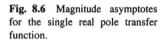

Fig. 8.6 Magnitude asymptotes for the single real pole transfer function.

$$\left| G(j\omega) \right|_{dB} \approx 0dB \tag{8.19}$$

Thus, as illustrated in Fig. 8.6, at low frequency $\| G(j\omega) \|_{dB}$ is asymptotic to 0 dB.

At high frequency, $\omega \gg \omega_0$ and $f \gg f_0$. In this case, it is true that

$$\left(\frac{\omega}{\omega_0} \right) \gg 1 \tag{8.20}$$

We can then say that

$$1 + \left(\frac{\omega}{\omega_0} \right)^2 \approx \left(\frac{\omega}{\omega_0} \right)^2 \tag{8.21}$$

Hence, Eq. (8.15) now becomes

$$\left| G(j\omega) \right| \approx \frac{1}{\sqrt{\left(\frac{\omega}{\omega_0} \right)^2}} = \left(\frac{f}{f_0} \right)^{-1} \tag{8.22}$$

This expression coincides with Eq. (8.5), with $n = -1$. So at high frequency, $\| G(j\omega) \|_{dB}$ has slope -20 dB per decade, as illustrated in Fig. 8.6. Thus, the asymptotes of $\| G(j\omega) \|$ are equal to 1 at low frequency, and $(f/f_0)^{-1}$ at high frequency. The asymptotes intersect at f_0. The actual magnitude tends toward these asymptotes at very low frequency and very high frequency. In the vicinity of the corner frequency f_0, the actual curve deviates somewhat from the asymptotes.

The deviation of the exact curve from the asymptotes can be found by simply evaluating Eq. (8.15). At the corner frequency $f = f_0$, Eq. (8.15) becomes

$$\left| G(j\omega_0) \right| = \frac{1}{\sqrt{1 + \left(\frac{\omega_0}{\omega_0} \right)^2}} = \frac{1}{\sqrt{2}} \tag{8.23}$$

In decibels, the magnitude is

$$\left| G(j\omega_0) \right|_{dB} = -20 \log_{10} \left(\sqrt{1 + \left(\frac{\omega_0}{\omega_0} \right)^2} \right) \approx -3 \text{ dB} \tag{8.24}$$

So the actual curve deviates from the asymptotes by -3 dB at the corner frequency, as illustrated in Fig. 8.7. Similar arguments show that the actual curve deviates from the asymptotes by -1 dB at $f = f_0/2$

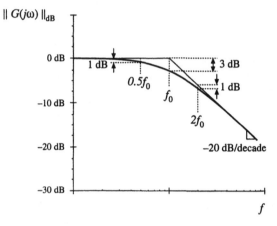

$\| G(j\omega) \|_{dB}$

Fig. 8.7 Deviation of the actual curve from the asymptotes, real pole.

0 dB

1 dB

3 dB

$0.5f_0$

1 dB

f_0

−10 dB

$2f_0$

−20 dB

−20 dB/decade

−30 dB

f

and at $f = 2f_0$.

The phase of $G(j\omega)$ is

$$\angle G(j\omega) = \tan^{-1}\left(\frac{\mathrm{Im}\left(G(j\omega)\right)}{\mathrm{Re}\left(G(j\omega)\right)}\right) \qquad (8.25)$$

Insertion of the real and imaginary parts of Eq. (8.14) into Eq. (8.25) leads to

$$\angle G(j\omega) = -\tan^{-1}\left(\frac{\omega}{\omega_0}\right) \qquad (8.26)$$

This function is plotted in Fig. 8.8. It tends to 0° at low frequency, and to −90° at high frequency. At the corner frequency $f = f_0$, the phase is −45°.

Since the high-frequency and low-frequency phase asymptotes do not intersect, we need a third asymptote to approximate the phase in the vicinity of the corner frequency f_0. One way to do this is illus-

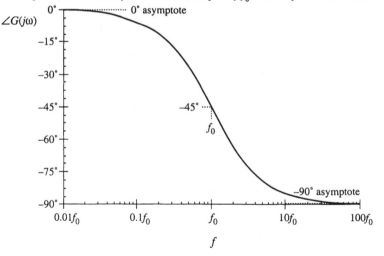

$\angle G(j\omega)$

0°   0° asymptote

−15°

Fig. 8.8 Exact phase plot, single real pole.

−30°

−45°   −45°

f_0

−60°

−75°

−90° asymptote

−90°

$0.01f_0$ $0.1f_0$ f_0 $10f_0$ $100f_0$

f

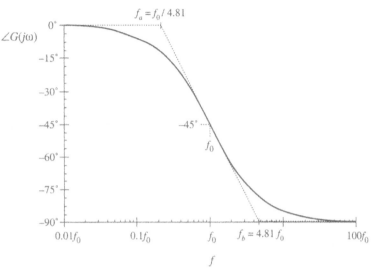

Fig. 8.9 One choice for the midfrequency phase asymptote, which correctly predicts the actual slope at $f = f_0$.

trated in Fig. 8.9, where the slope of the asymptote is chosen to be identical to the slope of the actual curve at $f = f_0$. It can be shown that, with this choice, the asymptote intersection frequencies f_a and f_b are given by

$$f_a = f_0 e^{-\pi/2} \approx \frac{f_0}{4.81}$$

$$f_b = f_0 e^{\pi/2} \approx 4.81\, f_0 \tag{8.27}$$

A simpler choice, which better approximates the actual curve, is

$$f_a = \frac{f_0}{10}$$

$$f_b = 10 f_0 \tag{8.28}$$

This asymptote is compared to the actual curve in Fig. 8.10. The pole causes the phase to change over a frequency span of approximately two decades, centered at the corner frequency. The slope of the asymptote in this frequency span is $-45°$ per decade. At the break frequencies f_a and f_b, the actual phase deviates from the asymptotes by $\tan^{-1}(0.1) = 5.7°$.

The magnitude and phase asymptotes for the single-pole response are summarized in Fig. 8.11.

It is good practice to consistently express single-pole transfer functions in the normalized form of Eq. (8.12). Both terms in the denominator of Eq. (8.12) are dimensionless, and the coefficient of s^0 is unity. Equation (8.12) is easy to interpret, because of its normalized form. At low frequencies, where the (s/ω_0) term is small in magnitude, the transfer function is approximately equal to 1. At high frequencies, where the (s/ω_0) term has magnitude much greater than 1, the transfer function is approximately $(s/\omega_0)^{-1}$. This leads to a magnitude of $(f/f_0)^{-1}$. The corner frequency is $f_0 = \omega_0/2\pi$. So the transfer function is written directly in terms of its salient features, that is, its asymptotes and its corner frequency.

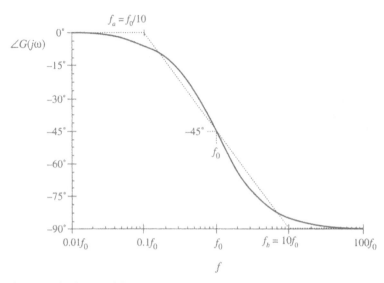

Fig. 8.10 A simpler choice for the midfrequency phase asymptote, which better approximates the curve over the entire frequency range.

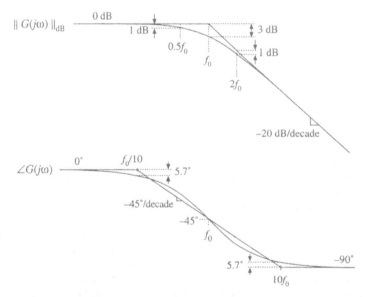

Fig. 8.11 Summary of the magnitude and phase Bode plot for the single real pole.

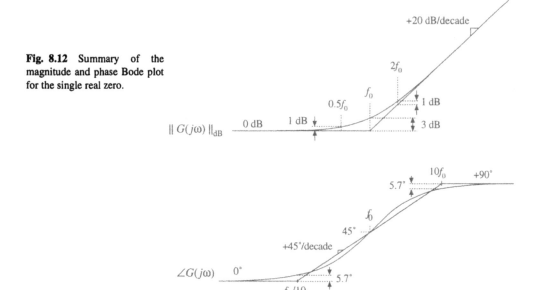

Fig. 8.12 Summary of the magnitude and phase Bode plot for the single real zero.

8.1.2 Single Zero Response

A single zero response contains a root in the numerator of the transfer function, and can be written in the following normalized form:

$$G(s) = \left(1 + \frac{s}{\omega_0}\right) \tag{8.29}$$

This transfer function has magnitude

$$\left|G(j\omega)\right| = \sqrt{1 + \left(\frac{\omega}{\omega_0}\right)^2} \tag{8.30}$$

At low frequency, $f \ll f_0 = \omega_0/2\pi$, the transfer function magnitude tends to $1 \Rightarrow 0$ dB. At high frequency, $f \gg f_0$, the transfer function magnitude tends to (f/f_0). As illustrated in Fig. 8.12, the high-frequency asymptote has slope +20 dB/decade.

The phase is given by

$$\angle G(j\omega) = \tan^{-1}\left(\frac{\omega}{\omega_0}\right) \tag{8.31}$$

With the exception of a minus sign, the phase is identical to Eq. (8.26). Hence, suitable asymptotes are as illustrated in Fig. 8.12. The phase tends to 0° at low frequency, and to +90° at high frequency. Over the interval $f_0/10 < f < 10f_0$, the phase asymptote has a slope of +45°/decade.

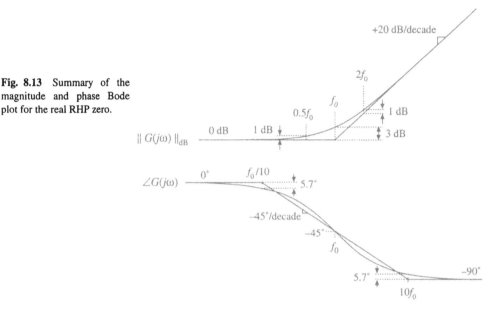

Fig. 8.13 Summary of the magnitude and phase Bode plot for the real RHP zero.

8.1.3   Right Half-Plane Zero

Right half-plane zeroes are often encountered in the small-signal transfer functions of switching converters. These terms have the following normalized form:

$$G(s) = \left(1 - \frac{s}{\omega_0}\right) \tag{8.32}$$

The root of Eq. (8.32) is positive, and hence lies in the right half of the complex s-plane. The right half-plane zero is also sometimes called a nonminimum phase zero. Its normalized form, Eq. (8.32), resembles the normalized form of the (left half-plane) zero of Eq. (8.29), with the exception of a minus sign in the coefficient of s. The minus sign causes a phase reversal at high frequency.

The transfer function has magnitude

$$\left| G(j\omega) \right| = \sqrt{1 + \left(\frac{\omega}{\omega_0}\right)^2} \tag{8.33}$$

This expression is identical to Eq. (8.30). Hence, it is impossible to distinguish a right half-plane zero from a left half-plane zero by the magnitude alone. The phase is given by

$$\angle G(j\omega) = -\tan^{-1}\left(\frac{\omega}{\omega_0}\right) \tag{8.34}$$

This coincides with the expression for the phase of the single pole, Eq. (8.26). So the right half-plane zero exhibits the magnitude response of the left half-plane zero, but the phase response of the pole. Magnitude and phase asymptotes are summarized in Fig. 8.13.

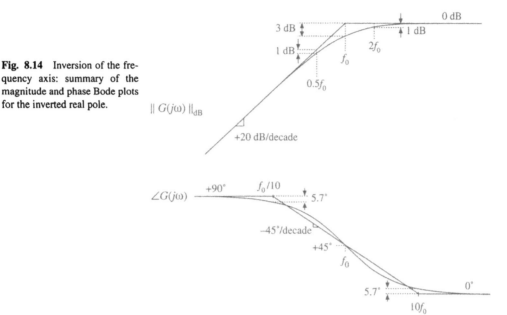

Fig. 8.14 Inversion of the frequency axis: summary of the magnitude and phase Bode plots for the inverted real pole.

8.1.4 Frequency Inversion

Two other forms arise, from inversion of the frequency axis. The inverted pole has the transfer function

$$G(s) = \frac{1}{\left(1 + \frac{\omega_0}{s}\right)} \tag{8.35}$$

As illustrated in Fig. 8.14, the inverted pole has a high-frequency gain of 1, and a low frequency asymptote having a +20 dB/decade slope. This form is useful for describing the gain of high-pass filters, and of other transfer functions where it is desired to emphasize the high frequency gain, with attenuation of low frequencies. Equation (8.35) is equivalent to

$$G(s) = \frac{\left(\frac{s}{\omega_0}\right)}{\left(1 + \frac{s}{\omega_0}\right)} \tag{8.36}$$

However, Eq. (8.35) more directly emphasizes that the high frequency gain is 1.

The inverted zero has the form

$$G(s) = \left(1 + \frac{\omega_0}{s}\right) \tag{8.37}$$

As illustrated in Fig. 8.15, the inverted zero has a high-frequency gain asymptote equal to 1, and a low-frequency asymptote having a slope equal to −20 dB/decade. An example of the use of this type of trans-

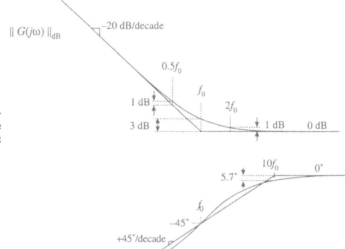

Fig. 8.15 Inversion of the frequency axis: summary of the
magnitude and phase Bode plot
for the inverted real zero.

fer function is the proportional-plus-integral controller, discussed in connection with feedback loop
design in the next chapter. Equation (8.37) is equivalent to

$$G(s) = \frac{\left(1 + \frac{s}{\omega_0}\right)}{\left(\frac{s}{\omega_0}\right)} \tag{8.38}$$

However, Eq. (8.37) is the preferred form when it is desired to emphasize the value of the high-frequency
gain asymptote.

    The use of frequency inversion is illustrated by example in the next section.

8.1.5   Combinations

The Bode diagram of a transfer function containing several pole, zero, and gain terms, can be constructed
by simple addition. At any given frequency, the magnitude (in decibels) of the composite transfer function is equal to the sum of the decibel magnitudes of the individual terms. Likewise, at a given frequency
the phase of the composite transfer function is equal to the sum of the phases of the individual terms.

    For example, suppose that we have already constructed the Bode diagrams of two complex-val-
ued functions of ω, $G_1(\omega)$ and $G_2(\omega)$. These functions have magnitudes $R_1(\omega)$ and $R_2(\omega)$, and phases
$\theta_1(\omega)$ and $\theta_2(\omega)$, respectively. It is desired to construct the Bode diagram of the product $G_3(\omega) = G_1(\omega)G_2(\omega)$. Let $G_3(\omega)$ have magnitude $R_3(\omega)$, and phase $\theta_3(\omega)$. To find this magnitude and phase, we
can express $G_1(\omega)$, $G_2(\omega)$, and $G_3(\omega)$ in polar form:

$$G_1(\omega) = R_1(\omega)\, e^{j\theta_1(\omega)}$$
$$G_2(\omega) = R_2(\omega)\, e^{j\theta_2(\omega)} \qquad\qquad (8.39)$$
$$G_3(\omega) = R_3(\omega)\, e^{j\theta_3(\omega)}$$

The product $G_3(\omega)$ can then be expressed as

$$G_3(\omega) = G_1(\omega)G_2(\omega) = R_1(\omega)e^{j\theta_1(\omega)}\,R_2(\omega)e^{j\theta_2(\omega)} \qquad\qquad (8.40)$$

Simplification leads to

$$G_3(\omega) = \left(R_1(\omega)R_2(\omega)\right) e^{j(\theta_1(\omega)+\theta_2(\omega))} \qquad\qquad (8.41)$$

Hence, the composite phase is

$$\theta_3(\omega) = \theta_1(\omega) + \theta_2(\omega) \qquad\qquad (8.42)$$

The total magnitude is

$$R_3(\omega) = R_1(\omega)R_2(\omega) \qquad\qquad (8.43)$$

When expressed in decibels, Eq. (8.43) becomes

$$\left|R_3(\omega)\right|_{dB} = \left|R_1(\omega)\right|_{dB} + \left|R_2(\omega)\right|_{dB} \qquad\qquad (8.44)$$

So the composite phase is the sum of the individual phases, and when expressed in decibels, the composite magnitude is the sum of the individual magnitudes. The composite magnitude slope, in dB per decade, is therefore also the sum of the individual slopes in dB per decade.

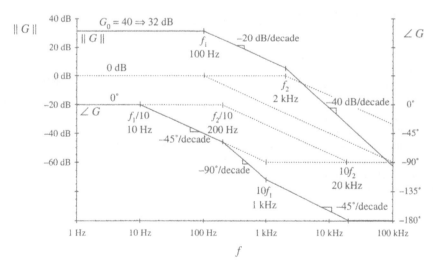

Fig. 8.16 Construction of magnitude and phase asymptotes for the transfer function of Eq.(8.45). Dashed line:

For example, consider construction of the Bode plot of the following transfer function:

$$G(s) = \frac{G_0}{\left(1 + \frac{s}{\omega_1}\right)\left(1 + \frac{s}{\omega_2}\right)} \tag{8.45}$$

where $G_0 = 40 \Rightarrow 32$ dB, $f_1 = \omega_1/2\pi = 100$ Hz, $f_2 = \omega_2/2\pi = 2$ kHz. This transfer function contains three terms: the gain G_0, and the poles at frequencies f_1 and f_2. The asymptotes for each of these terms are illustrated in Fig. 8.16. The gain G_0 is a positive real number, and therefore contributes zero phase shift with the gain 32 dB. The poles at 100 Hz and 2 kHz each contribute asymptotes as in Fig. 8.11.

At frequencies less than 100 Hz, the G_0 term contributes a gain magnitude of 32 dB, while the two poles each contribute magnitude asymptotes of 0 dB. So the low-frequency composite magnitude asymptote is 32 dB + 0 dB + 0 dB = 32 dB. For frequencies between 100 Hz and 2 kHz, the G_0 gain again contributes 32 dB, and the pole at 2 kHz continues to contribute a 0 dB magnitude asymptote. However, the pole at 100 Hz now contributes a magnitude asymptote that decreases with a −20 dB per decade slope. The composite magnitude asymptote therefore also decreases with a −20 dB per decade slope, as illustrated in Fig. 8.16. For frequencies greater than 2 kHz, the poles at 100 Hz and 2 kHz each contribute decreasing asymptotes having slopes of −20 dB/decade. The composite asymptote therefore decreases with a slope of −20 dB/decade −20 dB/decade = −40 dB/decade, as illustrated.

The composite phase asymptote is also constructed in Fig. 8.16. Below 10 Hz, all terms contribute 0° asymptotes. For frequencies between $f_1/10 = 10$ Hz, and $f_2/10 = 200$ Hz, the pole at f_1 contributes a decreasing phase asymptote having a slope of −45°/decade. Between 200 Hz and $10f_1 = 1$ kHz, both poles contribute decreasing asymptotes with −45°/decade slopes; the composite slope is therefore −90°/decade. Between 1 kHz and $10f_2 = 20$ kHz, the pole at f_1 contributes a constant −90° phase asymptote, while the pole at f_2 contributes a decreasing asymptote with −45°/decade slope. The composite slope is then −45°/decade. For frequencies greater than 20 kHz, both poles contribute constant −90° asymptotes, leading to a composite phase asymptote of −180°.

As a second example, consider the transfer function $A(s)$ represented by the magnitude and phase asymptotes of Fig. 8.17. Let us write the transfer function that corresponds to these asymptotes. The dc asymptote is A_0. At corner frequency f_1, the asymptote slope increases from 0 dB/decade to +20 dB/decade. Hence, there must be a zero at frequency f_1. At frequency f_2, the asymptote slope decreases from +20 dB/decade to 0 dB/decade. Therefore the transfer function contains a pole at frequency f_2. So we can express the transfer function as

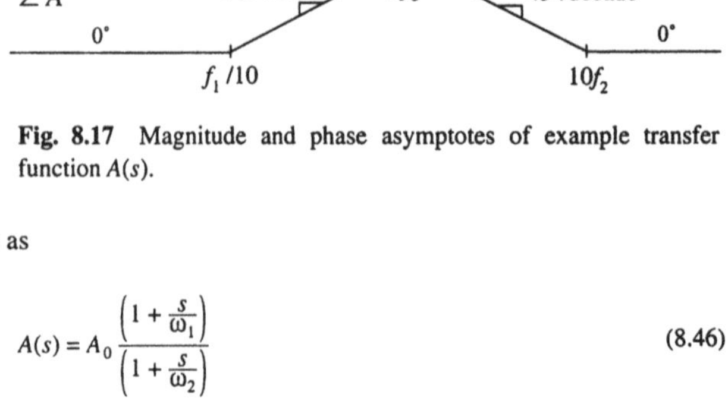

Fig. 8.17 Magnitude and phase asymptotes of example transfer function $A(s)$.

$$A(s) = A_0 \frac{\left(1 + \frac{s}{\omega_1}\right)}{\left(1 + \frac{s}{\omega_2}\right)} \tag{8.46}$$

where ω_1 and ω_2 are equal to $2\pi f_1$ and $2\pi f_2$, respectively.

We can use Eq. (8.46) to derive analytical expressions for the asymptotes. For $f < f_1$, and letting $s = j\omega$, we can see that the (s/ω_1) and (s/ω_2) terms each have magnitude less than 1. The asymptote is derived by neglecting these terms. Hence, the low-frequency magnitude asymptote is

$$\left| A_0 \frac{\left(1 + \frac{\cancel{s}}{\cancel{\omega_1}}\right)}{\left(1 + \frac{\cancel{s}}{\cancel{\omega_2}}\right)} \right|_{s = j\omega} = A_0 \frac{1}{1} = A_0 \tag{8.47}$$

For $f_1 < f < f_2$, the numerator term (s/ω_1) has magnitude greater than 1, while the denominator term (s/ω_2) has magnitude less than 1. The asymptote is derived by neglecting the smaller terms:

$$\left| A_0 \frac{\left(\cancel{1} + \frac{s}{\omega_1}\right)}{\left(1 + \frac{\cancel{s}}{\cancel{\omega_2}}\right)} \right|_{s = j\omega} = A_0 \frac{\left| \frac{s}{\omega_1} \right|_{s=j\omega}}{1} = A_0 \frac{\omega}{\omega_1} = A_0 \frac{f}{f_1} \tag{8.48}$$

This is the expression for the midfrequency magnitude asymptote of $A(s)$. For $f > f_2$, the (s/ω_1) and (s/ω_2) terms each have magnitude greater than 1. The expression for the high-frequency asymptote is therefore:

$$\left| A_0 \frac{\left(\cancel{1} + \frac{s}{\omega_1}\right)}{\left(\cancel{1} + \frac{s}{\omega_2}\right)} \right|_{s = j\omega} = A_0 \frac{\left| \frac{s}{\omega_1} \right|_{s=j\omega}}{\left| \frac{s}{\omega_2} \right|_{s=j\omega}} = A_0 \frac{\omega_2}{\omega_1} = A_0 \frac{f_2}{f_1} \tag{8.49}$$

We can conclude that the high-frequency gain is

$$A_\infty = A_0 \frac{f_2}{f_1} \tag{8.50}$$

Thus, we can derive analytical expressions for the asymptotes.

The transfer function $A(s)$ can also be written in a second form, using inverted poles and zeroes. Suppose that $A(s)$ represents the transfer function of a high-frequency amplifier, whose dc gain is not important. We are then interested in expressing $A(s)$ directly in terms of the high-frequency gain A_∞. We can view the transfer function as having an inverted pole at frequency f_2, which introduces attenuation at frequencies less than f_2. In addition, there is an inverted zero at $f = f_1$. So $A(s)$ could also be written

$$A(s) = A_\infty \frac{\left(1 + \frac{\omega_1}{s}\right)}{\left(1 + \frac{\omega_2}{s}\right)} \tag{8.51}$$

It can be verified that Eqs. (8.51) and (8.46) are equivalent.

8.1.6   Quadratic Pole Response: Resonance

Consider next the transfer function $G(s)$ of the two-pole low-pass filter of Fig. 8.18. The buck converter contains a filter of this type. When manipulated into canonical form, the models of the boost and buck-boost also contain similar filters. One can show that the transfer function of this network is

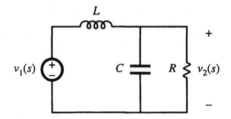

Fig. 8.18   Two-pole low-pass filter example.

$$G(s) = \frac{v_2(s)}{v_1(s)} = \frac{1}{1 + s\frac{L}{R} + s^2 LC} \qquad (8.52)$$

This transfer function contains a second-order denominator polynomial, and is of the form

$$G(s) = \frac{1}{1 + a_1 s + a_2 s^2} \qquad (8.53)$$

with $a_1 = L/R$ and $a_2 = LC$.

To construct the Bode plot of this transfer function, we might try to factor the denominator into its two roots:

$$G(s) = \frac{1}{\left(1 - \frac{s}{s_1}\right)\left(1 - \frac{s}{s_2}\right)} \qquad (8.54)$$

Use of the quadratic formula leads to the following expressions for the roots:

$$s_1 = -\frac{a_1}{2a_2}\left[1 - \sqrt{1 - \frac{4a_2}{a_1^2}}\right] \qquad (8.55)$$

$$s_2 = -\frac{a_1}{2a_2}\left[1 + \sqrt{1 - \frac{4a_2}{a_1^2}}\right] \qquad (8.56)$$

If $4a_2 \le a_1^2$, then the roots are real. Each real pole then exhibits a Bode diagram as derived in Section 8.1.1, and the composite Bode diagram can be constructed as described in Section 8.1.5 (but a better approach is described in Section 8.1.7).

If $4a_2 > a_1^2$, then the roots (8.55) and (8.56) are complex. In Section 8.1.1, the assumption was made that ω_0 is real; hence, the results of that section cannot be applied to this case. We need to do some additional work, to determine the magnitude and phase for the case when the roots are complex.

The transfer functions of Eqs. (8.52) and (8.53) can be written in the following standard normalized form:

$$G(s) = \frac{1}{1 + 2\zeta \frac{s}{\omega_0} + \left(\frac{s}{\omega_0}\right)^2} \qquad (8.57)$$

If the coefficients a_1 and a_2 are real and positive, then the parameters ζ and ω_0 are also real and positive. The parameter ω_0 is again the angular corner frequency, and we can define $f_0 = \omega_0/2\pi$. The parameter ζ is called the *damping factor*: ζ controls the shape of the transfer function in the vicinity of $f = f_0$. An alternative standard normalized form is

$$G(s) = \frac{1}{1 + \frac{s}{Q\omega_0} + \left(\frac{s}{\omega_0}\right)^2}$$ (8.58)

where

$$Q = \frac{1}{2\zeta}$$ (8.59)

The parameter Q is called the *quality factor* of the circuit, and is a measure of the dissipation in the system. A more general definition of Q, for sinusoidal excitation of a passive element or network, is

$$Q = 2\pi \frac{\text{(peak stored energy)}}{\text{(energy dissipated per cycle)}}$$ (8.60)

For a second-order passive system, Eqs. (8.59) and (8.60) are equivalent. We will see that the Q-factor has a very simple interpretation in the magnitude Bode diagrams of second-order transfer functions.

Analytical expressions for the parameters Q and ω_0 can be found by equating like powers of s in the original transfer function, Eq. (8.52), and in the normalized form, Eq. (8.58). The result is

$$f_0 = \frac{\omega_0}{2\pi} = \frac{1}{2\pi\sqrt{LC}}$$
$$Q = R\sqrt{\frac{C}{L}}$$ (8.61)

The roots s_1 and s_2 of Eqs. (8.55) and (8.56) are real when $Q \le 0.5$, and are complex when $Q > 0.5$.

The magnitude of G is

$$\left| G(j\omega) \right| = \frac{1}{\sqrt{\left(1 - \left(\frac{\omega}{\omega_0}\right)^2\right)^2 + \frac{1}{Q^2}\left(\frac{\omega}{\omega_0}\right)^2}}$$ (8.62)

Asymptotes of $\| G \|$ are illustrated in Fig. 8.19. At low frequencies, $(\omega/\omega_0) \ll 1$, and hence

$$\left| G \right| \to 1 \quad \text{for } \omega \ll \omega_0$$ (8.63)

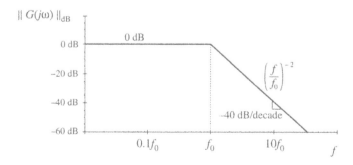

Fig. 8.19   Magnitude asymptotes for the two-pole transfer function.

At high frequencies where $(\omega/\omega_0) \gg 1$, the $(\omega/\omega_0)^4$ term dominates the expression inside the radical of Eq. (8.62). Hence, the high-frequency asymptote is

$$|G| \rightarrow \left(\frac{f}{f_0}\right)^{-2} \quad \text{for } \omega \gg \omega_0 \qquad (8.64)$$

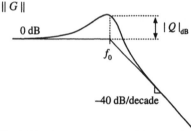

This expression coincides with Eq. (8.5), with $n = -2$. Therefore, the high-frequency asymptote has slope –40 dB/ decade. The asymptotes intersect at $f = f_0$, and are independent of Q.

The parameter Q affects the deviation of the actual curve from the asymptotes, in the neighborhood of the corner frequency f_0. The exact magnitude at $f = f_0$ is found by substitution of $\omega = \omega_0$ into Eq. (8.62):

Fig. 8.20 Important features of the magnitude Bode plot, for the two-pole transfer function.

$$|G(j\omega_0)| = Q \qquad (8.65)$$

So the exact transfer function has magnitude Q at the corner frequency f_0. In decibels, Eq. (8.65) is

$$|G(j\omega_0)|_{dB} = |Q|_{dB} \qquad (8.66)$$

So if, for example, $Q = 2 \Rightarrow 6$ dB, then the actual curve deviates from the asymptotes by 6 dB at the corner frequency $f = f_0$. Salient features of the magnitude Bode plot of the second-order transfer function are summarized in Fig. 8.20.

The phase of G is

$$\angle G(j\omega) = -\tan^{-1}\left[\frac{\frac{1}{Q}\left(\frac{\omega}{\omega_0}\right)}{1 - \left(\frac{\omega}{\omega_0}\right)^2}\right] \qquad (8.67)$$

The phase tends to 0° at low frequency, and to –180° at high frequency. At $f = f_0$, the phase is –90°. As illustrated in Fig. 8.21, increasing the value of Q causes a sharper phase change between the 0° and –180° asymptotes. We again need a midfrequency asymptote, to approximate the phase transition in the

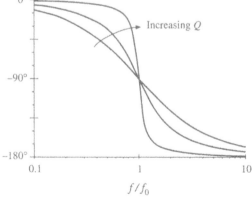

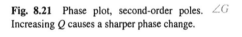

Fig. 8.21 Phase plot, second-order poles. $\angle G$ Increasing Q causes a sharper phase change.

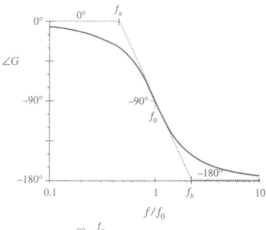

Fig. 8.22 One choice for the midfrequency phase asymptote of the two-pole response, which correctly predicts the actual slope at $f = f_0$.

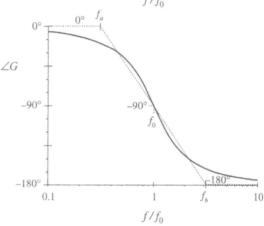

Fig. 8.23 A simpler choice for the midfrequency phase asymptote, which better approximates the curve over the entire frequency range and is consistent with the asymptote used for real poles.

vicinity of the corner frequency f_0, as illustrated in Fig. 8.22. As in the case of the real single pole, we could choose the slope of this asymptote to be identical to the slope of the actual curve at $f = f_0$. It can be shown that this choice leads to the following asymptote break frequencies:

$$f_a = \left(e^{\pi/2}\right)^{-\frac{1}{2Q}} f_0$$
$$f_b = \left(e^{\pi/2}\right)^{\frac{1}{2Q}} f_0$$

(8.68)

A better choice, which is consistent with the approximation (8.28) used for the real single pole, is

$$f_a = 10^{-1/2Q} f_0$$
$$f_b = 10^{1/2Q} f_0$$

(8.69)

With this choice, the midfrequency asymptote has slope $-180Q$ degrees per decade. The phase asymptotes are summarized in Fig. 8.23. With $Q = 0.5$, the phase changes from $0°$ to $-180°$ over a frequency span of approximately two decades, centered at the corner frequency f_0. Increasing the Q causes this frequency span to decrease rapidly.

Second-order response magnitude and phase curves are plotted in Figs. 8.24 and 8.25.

Fig. 8.24 Exact magnitude curves, two-pole response, for several values of Q.

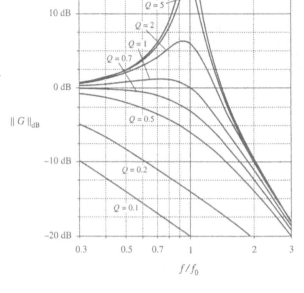

Fig. 8.25 Exact phase curves, two-pole response, for several values of Q.

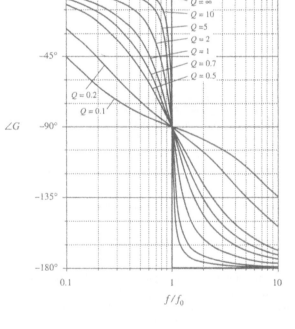

8.1.7 The Low-Q Approximation

As mentioned in Section 8.1.6, when the roots of second-order denominator polynomial of Eq. (8.53) are real, then we can factor the denominator, and construct the Bode diagram using the asymptotes for real poles. We would then use the following normalized form:

$$G(s) = \frac{1}{\left(1 + \frac{s}{\omega_1}\right)\left(1 + \frac{s}{\omega_2}\right)} \tag{8.70}$$

This is a particularly desirable approach when the corner frequencies ω_1 and ω_2 are well separated in value.

The difficulty in this procedure lies in the complexity of the quadratic formula used to find the corner frequencies. Expressing the corner frequencies ω_1 and ω_2 in terms of the circuit elements R, L, C, etc., invariably leads to complicated and unilluminating expressions, especially when the circuit contains many elements. Even in the case of the simple circuit of Fig. 8.18, whose transfer function is given by Eq. (8.52), the conventional quadratic formula leads to the following complicated formula for the corner frequencies:

$$\omega_1, \omega_2 = \frac{\frac{L}{R} \pm \sqrt{\left(\frac{L}{R}\right)^2 - 4LC}}{2LC} \tag{8.71}$$

This equation yields essentially no insight regarding how the corner frequencies depend on the element values. For example, it can be shown that when the corner frequencies are well separated in value, they can be expressed with high accuracy by the much simpler relations

$$\omega_1 \approx \frac{R}{L}, \quad \omega_2 \approx \frac{1}{RC} \tag{8.72}$$

In this case, ω_1 is essentially independent of the value of C, and ω_2 is essentially independent of L, yet Eq. (8.71) apparently predicts that both corner frequencies are dependent on all element values. The simple expressions of Eq. (8.72) are far preferable to Eq. (8.71), and can be easily derived using the low-Q approximation [2].

Let us assume that the transfer function has been expressed in the standard normalized form of Eq. (8.58), reproduced below:

$$G(s) = \frac{1}{1 + \frac{s}{Q\omega_0} + \left(\frac{s}{\omega_0}\right)^2} \tag{8.73}$$

For $Q \leq 0.5$, let us use the quadratic formula to write the real roots of the denominator polynomial of Eq. (8.73) as

$$\omega_1 = \frac{\omega_0}{Q} \frac{1 - \sqrt{1 - 4Q^2}}{2} \tag{8.74}$$

$$\omega_2 = \frac{\omega_0}{Q} \frac{1 + \sqrt{1 - 4Q^2}}{2} \tag{8.75}$$

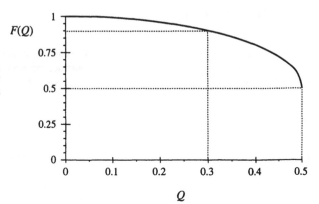

Fig. 8.26 $F(Q)$ vs. Q, as given by Eq. (8.77). The approximation $F(Q) \approx 1$ is within 10% of the exact value for $Q < 3$.

The corner frequency ω_2 can be expressed

$$\omega_2 = \frac{\omega_0}{Q} F(Q) \tag{8.76}$$

where $F(Q)$ is defined as [2]:

$$F(Q) = \frac{1}{2}\left(1 + \sqrt{1 - 4Q^2}\right) \tag{8.77}$$

Note that, when $Q \ll 0.5$, then $4Q^2 \ll 1$ and $F(Q)$ is approximately equal to 1. We then obtain

$$\omega_2 \approx \frac{\omega_0}{Q} \quad \text{for } Q \ll \frac{1}{2} \tag{8.78}$$

The function $F(Q)$ is plotted in Fig. 8.26. It can be seen that $F(Q)$ approaches 1 very rapidly as Q decreases below 0.5.

To derive a similar approximation for ω_1, we can multiply and divide Eq. (8.74) by $F(Q)$, Eq. (8.77). Upon simplification of the numerator, we obtain

$$\omega_1 = \frac{Q\omega_0}{F(Q)} \tag{8.79}$$

Again, $F(Q)$ tends to 1 for small Q. Hence, ω_1 can be approximated as

$$\omega_1 \approx Q\omega_0 \quad \text{for } Q \ll \frac{1}{2} \tag{8.80}$$

Magnitude asymptotes for the low-Q case are summarized in Fig. 8.27. For $Q < 0.5$, the two poles at ω_0 split into real poles. One real pole occurs at corner frequency $\omega_1 < \omega_0$, while the other occurs at corner frequency $\omega_2 > \omega_0$. The corner frequencies are easily approximated, using Eqs. (8.78) and (8.80).

For the filter circuit of Fig. 8.18, the parameters Q and ω_0 are given by Eq. (8.61). For the case when $Q \ll 0.5$, we can derive the following analytical expressions for the corner frequencies, using Eqs. (8.78) and (8.80):

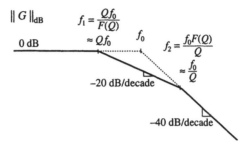

Fig. 8.27 Magnitude asymptotes predicted by the low-Q approximation. Real poles occur at frequencies Qf_0 and f_0/Q.

$$\omega_1 \approx Q\omega_0 = R\sqrt{\frac{C}{L}}\,\frac{1}{\sqrt{LC}} = \frac{R}{L}$$

$$\omega_2 \approx \frac{\omega_0}{Q} = \frac{1}{\sqrt{LC}}\,\frac{1}{R\sqrt{\frac{C}{L}}} = \frac{1}{RC}$$

(8.81)

So the low-Q approximation allows us to derive simple design-oriented analytical expressions for the corner frequencies.

8.1.8 Approximate Roots of an Arbitrary-Degree Polynomial

The low-Q approximation can be generalized, to find approximate analytical expressions for the roots of the nth-order polynomial

$$P(s) = 1 + a_1 s + a_2 s^2 + \cdots + a_n s^n$$

(8.82)

It is desired to factor the polynomial $P(s)$ into the form

$$P(s) = \left(1 + \tau_1 s\right)\left(1 + \tau_2 s\right) \cdots \left(1 + \tau_n s\right)$$

(8.83)

In a real circuit, the coefficients a_1, ..., a_n are real, while the time constants τ_1, ..., τ_n may be either real or complex. Very often, some or all of the time constants are well separated in value, and depend in a very simple way on the circuit element values. In such cases, simple approximate analytical expressions for the time constants can be derived.

The time constants τ_1, ..., τ_n can be related to the original coefficients a_1, ..., a_n by multiplying out Eq. (8.83). The result is

$$
\begin{aligned}
a_1 &= \tau_1 + \tau_2 + \cdots + \tau_n \\
a_2 &= \tau_1\left(\tau_2 + \cdots + \tau_n\right) + \tau_2\left(\tau_3 + \cdots + \tau_n\right) + \cdots \\
a_3 &= \tau_1\tau_2\left(\tau_3 + \cdots + \tau_n\right) + \tau_2\tau_3\left(\tau_4 + \cdots + \tau_n\right) + \cdots \\
&\vdots \\
a_n &= \tau_1\tau_2\tau_3\cdots\tau_n
\end{aligned}
$$

(8.84)

General solution of this system of equations amounts to exact factoring of the arbitrary degree polynomial, a hopeless task. Nonetheless, Eq. (8.84) does suggest a way to approximate the roots.

Suppose that all of the time constants τ_1, ..., τ_n are real and well separated in value. We can further assume, without loss of generality, that the time constants are arranged in decreasing order of magnitude:

$$|\tau_1| \gg |\tau_2| \gg \cdots \gg |\tau_n|$$ (8.85)

When the inequalities of Eq. (8.85) are satisfied, then the expressions for a_1, ..., a_n of Eq. (8.84) are each dominated by their first terms:

$$
\begin{aligned}
a_1 &\approx \tau_1 \\
a_2 &\approx \tau_1 \tau_2 \\
a_3 &\approx \tau_1 \tau_2 \tau_3 \\
&\vdots \\
a_n &= \tau_1 \tau_2 \tau_3 \cdots \tau_n
\end{aligned}
$$ (8.86)

These expressions can now be solved for the time constants, with the result

$$
\begin{aligned}
\tau_1 &\approx a_1 \\
\tau_2 &\approx \frac{a_2}{a_1} \\
\tau_3 &\approx \frac{a_3}{a_2} \\
&\vdots \\
\tau_n &\approx \frac{a_n}{a_{n-1}}
\end{aligned}
$$ (8.87)

Hence, if

$$|a_1| \gg \left|\frac{a_2}{a_1}\right| \gg \left|\frac{a_3}{a_2}\right| \gg \cdots \gg \left|\frac{a_n}{a_{n-1}}\right|$$ (8.88)

then the polynomial $P(s)$ given by Eq. (8.82) has the approximate factorization

$$P(s) \approx \left(1 + a_1 s\right)\left(1 + \frac{a_2}{a_1} s\right)\left(1 + \frac{a_3}{a_2} s\right)\cdots\left(1 + \frac{a_n}{a_{n-1}} s\right)$$ (8.89)

Note that if the original coefficients in Eq. (8.82) are simple functions of the circuit elements, then the approximate roots given by Eq. (8.89) are similar simple functions of the circuit elements. So approximate analytical expressions for the roots can be obtained. Numerical values are substituted into Eq. (8.88) to justify the approximation.

In the case where two of the roots are not well separated, then one of the inequalities of Eq. (8.88) is violated. We can then leave the corresponding terms in quadratic form. For example, suppose that inequality k is not satisfied:

$$|a_1| \gg \left|\frac{a_2}{a_1}\right| \gg \cdots \gg \left|\frac{a_k}{a_{k-1}}\right| \not\gg \left|\frac{a_{k+1}}{a_k}\right| \gg \cdots \gg \left|\frac{a_n}{a_{n-1}}\right|$$ (8.90)

Then an approximate factorization is

$$P(s) \approx \left(1 + a_1 s\right)\left(1 + \frac{a_2}{a_1} s\right) \cdots \left(1 + \frac{a_k}{a_{k-1}} s + \frac{a_{k+1}}{a_{k-1}} s^2\right) \cdots \left(1 + \frac{a_n}{a_{n-1}} s\right) \qquad (8.91)$$

The conditions for accuracy of this approximation are

$$\left|a_1\right| \gg \left|\frac{a_2}{a_1}\right| \gg \cdots \gg \left|\frac{a_k}{a_{k-1}}\right| \gg \left|\frac{a_{k-2} a_{k+1}}{a_{k-1}^2}\right| \gg \left|\frac{a_{k+2}}{a_{k+1}}\right| \gg \cdots \gg \left|\frac{a_n}{a_{n-1}}\right| \qquad (8.92)$$

Complex conjugate roots can be approximated in this manner.

When the first inequality of Eq. (8.88) is violated, that is,

$$\left|a_1\right| \not\gg \left|\frac{a_2}{a_1}\right| \gg \left|\frac{a_3}{a_2}\right| \gg \cdots \gg \left|\frac{a_n}{a_{n-1}}\right| \qquad (8.93)$$

then the first two roots should be left in quadratic form:

$$P(s) \approx \left(1 + a_1 s + a_2 s^2\right)\left(1 + \frac{a_3}{a_2} s\right) \cdots \left(1 + \frac{a_n}{a_{n-1}} s\right) \qquad (8.94)$$

This approximation is justified provided that

$$\left|\frac{a_2^2}{a_3}\right| \gg \left|a_1\right| \gg \left|\frac{a_3}{a_2}\right| \gg \left|\frac{a_4}{a_3}\right| \gg \cdots \gg \left|\frac{a_n}{a_{n-1}}\right| \qquad (8.95)$$

If none of the above approximations is justified, then there are three or more roots that are close in magnitude. One must then resort to cubic or higher-order forms.

As an example, consider the damped EMI filter illustrated in Fig. 8.28. Filters such as this are typically placed at the power input of a converter, to attenuate the switching harmonics present in the converter input current. By circuit analysis, on can show that this filter exhibits the following transfer function:

$$G(s) = \frac{i_g(s)}{i_c(s)} = \frac{1 + s \dfrac{L_1 + L_2}{R}}{1 + s \dfrac{L_1 + L_2}{R} + s^2 L_1 C + s^3 \dfrac{L_1 L_2 C}{R}} \qquad (8.96)$$

This transfer function contains a third-order denominator, with the following coefficients:

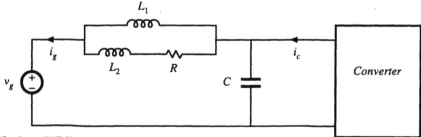

Fig. 8.28 Input EMI filter example.

$$a_1 = \frac{L_1 + L_2}{R}$$

$$a_2 = L_1 C \qquad\qquad (8.97)$$

$$a_3 = \frac{L_1 L_2 C}{R}$$

It is desired to factor the denominator, to obtain analytical expressions for the poles. The correct way to do this depends on the numerical values of R, L_1, L_2, and C. When the roots are real and well separated, then Eq. (8.89) predicts that the denominator can be factored as follows:

$$\left(1 + s\frac{L_1 + L_2}{R}\right)\left(1 + sRC\frac{L_1}{L_1 + L_2}\right)\left(1 + s\frac{L_2}{R}\right) \qquad\qquad (8.98)$$

According to Eq. (8.88), this approximation is justified provided that

$$\frac{L_1 + L_2}{R} \gg RC\frac{L_1}{L_1 + L_2} \gg \frac{L_2}{R} \qquad\qquad (8.99)$$

These inequalities cannot be satisfied unless $L_1 \gg L_2$. When $L_1 \gg L_2$, then Eq. (8.99) can be further simplified to

$$\frac{L_1}{R} \gg RC \gg \frac{L_2}{R} \qquad\qquad (8.100)$$

The approximate factorization, Eq. (8.98), can then be further simplified to

$$\left(1 + s\frac{L_1}{R}\right)\left(1 + sRC\right)\left(1 + s\frac{L_2}{R}\right) \qquad\qquad (8.101)$$

Thus, in this case the transfer function contains three well separated real poles. Equations (8.98) and (8.101) represent approximate analytical factorizations of the denominator of Eq. (8.96). Although numerical values must be substituted into Eqs. (8.99) or (8.100) to justify the approximation, we can nonetheless express Eqs. (8.98) and (8.101) as analytical functions of L_1, L_2, R, and C. Equations (8.98) and (8.101) are design-oriented, because they yield insight into how the element values can be chosen such that given specified pole frequencies are obtained.

When the second inequality of Eq. (8.99) is violated,

$$\frac{L_1 + L_2}{R} \gg RC\frac{L_1}{L_1 + L_2} \not\gg \frac{L_2}{R} \qquad\qquad (8.102)$$

then the second and third roots should be left in quadratic form:

$$\left(1 + s\frac{L_1 + L_2}{R}\right)\left(1 + sRC\frac{L_1}{L_1 + L_2} + s^2 L_1 \| L_2 C\right) \qquad\qquad (8.103)$$

This expression follows from Eq. (8.91), with $k = 2$. Equation (8.92) predicts that this approximation is justified provided that

$$\frac{L_1 + L_2}{R} \gg RC \frac{L_1}{L_1 + L_2} \gg \frac{L_1 \| L_2}{L_1 + L_2} RC \qquad (8.104)$$

In application of Eq. (8.92), we take a_0 to be equal to 1. The inequalities of Eq. (8.104) can be simplified to obtain

$$L_1 \gg L_2, \quad \text{and} \quad \frac{L_1}{R} \gg RC \qquad (8.105)$$

Note that it is no longer required that $RC \gg L_2/R$. Equation (8.105) implies that factorization (8.103) can be further simplified to

$$\left(1 + s\frac{L_1}{R}\right)\left(1 + sRC + s^2 L_2 C\right) \qquad (8.106)$$

Thus, for this case, the transfer function contains a low-frequency pole that is well separated from a high-frequency quadratic pole pair. Again, the factored result (8.106) is expressed as an analytical function of the element values, and consequently is design-oriented.

In the case where the first inequality of Eq. (8.99) is violated:

$$\frac{L_1 + L_2}{R} \not\gg RC \frac{L_1}{L_1 + L_2} \gg \frac{L_2}{R} \qquad (8.107)$$

then the first and second roots should be left in quadratic form:

$$\left(1 + s\frac{L_1 + L_2}{R} + s^2 L_1 C\right)\left(1 + s\frac{L_2}{R}\right) \qquad (8.108)$$

This expression follows directly from Eq. (8.94). Equation (8.95) predicts that this approximation is justified provided that

$$\frac{L_1 RC}{L_2} \gg \frac{L_1 + L_2}{R} \gg \frac{L_2}{R} \qquad (8.109)$$

that is,

$$L_1 \gg L_2, \quad \text{and} \quad RC \gg \frac{L_2}{R} \qquad (8.110)$$

For this case, the transfer function contains a low-frequency quadratic pole pair that is well separated from a high-frequency real pole. If none of the above approximations are justified, then all three of the roots are similar in magnitude. We must then find other means of dealing with the original cubic polynomial. Design of input filters, including the filter of Fig. 8.28, is covered in Chapter 10.

8.2 ANALYSIS OF CONVERTER TRANSFER FUNCTIONS

Let us next derive analytical expressions for the poles, zeroes, and asymptote gains in the transfer functions of the basic converters.

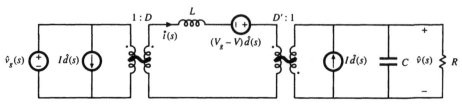

Fig. 8.29   Buck-boost converter equivalent circuit derived in Section 7.2.

8.2.1   Example: Transfer Functions of the Buck-Boost Converter

The small-signal equivalent circuit model of the buck-boost converter is derived in Section 7.2, with the result [Fig. 7.16(b)] repeated in Fig. 8.29. Let us derive and plot the control-to-output and line-to-output transfer functions for this circuit.

The converter contains two independent ac inputs: the control input $\hat{d}(s)$ and the line input $\hat{v}_g(s)$. The ac output voltage variations $\hat{v}(s)$ can be expressed as the superposition of terms arising from these two inputs:

$$\hat{v}(s) = G_{vd}(s)\hat{d}(s) + G_{vg}(s)\,\hat{v}_g(s) \tag{8.111}$$

Hence, the transfer functions $G_{vd}(s)$ and $G_{vg}(s)$ can be defined as

$$G_{vd}(s) = \left.\frac{\hat{v}(s)}{\hat{d}(s)}\right|_{\hat{v}_g(s)=0} \quad \text{and} \quad G_{vg}(s) = \left.\frac{\hat{v}(s)}{\hat{v}_g(s)}\right|_{\hat{d}(s)=0} \tag{8.112}$$

To find the line-to-output transfer function $G_{vg}(s)$, we set the $\hat{d}$ sources to zero as in Fig. 8.30(a). We can then push the $v_g(s)$ source and the inductor through the transformers, to obtain the circuit of Fig. 8.30(b). The transfer function $G_{vg}(s)$ is found using the voltage divider formula:

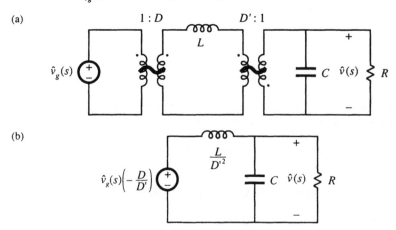

Fig. 8.30   Manipulation of buck-boost equivalent circuit to find the line-to-output transfer function $G_{vg}(s)$: (a) set $\hat{d}$ sources to zero; (b) push inductor and $\hat{v}_g$ source through transformers.

$$G_{vg}(s) = \frac{\hat{v}(s)}{\hat{v}_g(s)}\bigg|_{\hat{d}(s)=0} = -\frac{D}{D'}\frac{\left(R \parallel \frac{1}{sC}\right)}{\frac{sL}{D'^2} + \left(R \parallel \frac{1}{sC}\right)} \tag{8.113}$$

We next expand the parallel combination, and express as a rational fraction:

$$G_{vg}(s) = \left(-\frac{D}{D'}\right)\frac{\left(\frac{R}{1+sRC}\right)}{\frac{sL}{D'^2} + \left(\frac{R}{1+sRC}\right)} \tag{8.114}$$

$$= \left(-\frac{D}{D'}\right)\frac{R}{R + \frac{sL}{D'^2} + \frac{s^2 RLC}{D'^2}}$$

We aren't done yet—the next step is to manipulate the expression into normalized form, such that the coefficients of s^0 in the numerator and denominator polynomials are equal to one. This can be accomplished by dividing the numerator and denominator by R:

$$G_{vg}(s) = \frac{\hat{v}(s)}{\hat{v}_g(s)}\bigg|_{\hat{d}(s)=0} = \left(-\frac{D}{D'}\right)\frac{1}{1 + s\frac{L}{D'^2 R} + s^2 \frac{LC}{D'^2}} \tag{8.115}$$

Thus, the line-to-output transfer function contains a dc gain G_{g0} and a quadratic pole pair:

$$G_{vg}(s) = G_{g0}\frac{1}{1 + \frac{s}{Q\omega_0} + \left(\frac{s}{\omega_0}\right)^2} \tag{8.116}$$

Analytical expressions for the salient features of the line-to-output transfer function are found by equating like terms in Eqs. (8.115) and (8.116). The dc gain is

$$G_{g0} = -\frac{D}{D'} \tag{8.117}$$

By equating the coefficients of s^2 in the denominators of Eqs. (8.115) and (8.116), we obtain

$$\frac{1}{\omega_0^2} = \frac{LC}{D'^2} \tag{8.118}$$

Hence, the angular corner frequency is

$$\omega_0 = \frac{D'}{\sqrt{LC}} \tag{8.119}$$

By equating coefficients of s in the denominators of Eqs. (8.115) and (8.116), we obtain

$$\frac{1}{Q\omega_0} = \frac{L}{D'^2 R} \tag{8.120}$$

Elimination of ω_0 using Eq. (8.119) and solution for Q leads to

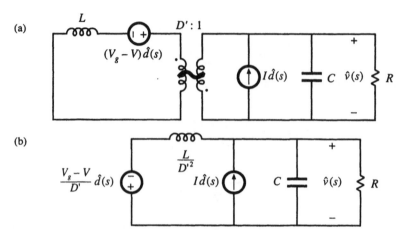

Fig. 8.31 Manipulation of buck-boost equivalent circuit to find the control-to-output transfer function $G_{vd}(s)$: (a) set $\hat{v}_g$ source to zero; (b) push inductor and voltage source through transformer.

$$Q = D'R \sqrt{\frac{C}{L}} \qquad (8.121)$$

Equations (8.117), (8.119), and (8.121) are the desired results in the analysis of the line-to-output transfer function. These expressions are useful not only in analysis situations, where it is desired to find numerical values of the salient features G_{g0}, ω_0, and Q, but also in design situations, where it is desired to select numerical values for R, L, and C such that given values of the salient features are obtained.

Derivation of the control-to-output transfer function $G_{vd}(s)$ is complicated by the presence in Fig. 8.29 of three generators that depend on $\hat{d}(s)$. One good way to find $G_{vd}(s)$ is to manipulate the circuit model as in the derivation of the canonical model, Fig. 7.60. Another approach, used here, employs the principle of superposition. First, we set the $\hat{v}_g$ source to zero. This shorts the input to the 1:D transformer, and we are left with the circuit illustrated in Fig. 8.31(a). Next, we push the inductor and $\hat{d}$ voltage source through the D':1 transformer, as in Fig. 8.31(b).

Figure 8.31(b) contains a $\hat{d}$-dependent voltage source and a $\hat{d}$-dependent current source. The transfer function $G_{vd}(s)$ can therefore be expressed as a superposition of terms arising from these two sources. When the current source is set to zero (i.e., open-circuited), the circuit of Fig. 8.32(a) is obtained. The output $\hat{v}(s)$ can then be expressed as

$$\frac{\hat{v}(s)}{\hat{d}(s)} = \left(-\frac{V_g - V}{D'}\right) \frac{\left(R \parallel \frac{1}{sC}\right)}{\frac{sL}{D'^2} + \left(R \parallel \frac{1}{sC}\right)} \qquad (8.122)$$

When the voltage source is set to zero (i.e., short-circuited), Fig. 8.31(b) reduces to the circuit illustrated in Fig. 8.32(b). The output $\hat{v}(s)$ can then be expressed as

$$\frac{\hat{v}(s)}{\hat{d}(s)} = I \left(\frac{sL}{D'^2} \parallel R \parallel \frac{1}{sC}\right) \qquad (8.123)$$

The transfer function $G_{vd}(s)$ is the sum of Eqs. (8.122) and (8.123):

(a)

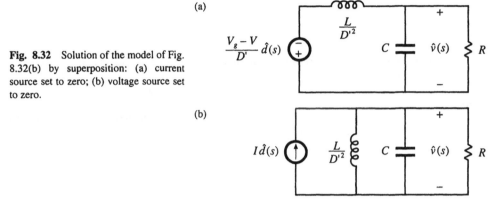

Fig. 8.32 Solution of the model of Fig. 8.32(b) by superposition: (a) current source set to zero; (b) voltage source set to zero.

(b)

$$G_{vd}(s) = \left(-\frac{V_g - V}{D'}\right) \frac{\left(R \parallel \frac{1}{sC}\right)}{\frac{sL}{D'^2} + \left(R \parallel \frac{1}{sC}\right)} + I\left(\frac{sL}{D'^2} \parallel R \parallel \frac{1}{sC}\right) \tag{8.124}$$

By algebraic manipulation, one can reduce this expression to

$$G_{vd}(s) = \frac{\hat{v}(s)}{\hat{d}(s)}\bigg|_{\hat{v}_g(s) = 0} = \left(-\frac{V_g - V}{D'}\right) \frac{\left(1 - s\frac{LI}{D'\left(V_g - V\right)}\right)}{\left(1 + s\frac{L}{D'^2 R} + s^2 \frac{LC}{D'^2}\right)} \tag{8.125}$$

This equation is of the form

$$G_{vd}(s) = G_{d0} \frac{\left(1 - \frac{s}{\omega_z}\right)}{\left(1 + \frac{s}{Q\omega_0} + \left(\frac{s}{\omega_0}\right)^2\right)} \tag{8.126}$$

The denominators of Eq. (8.125) and (8.115) are identical, and hence $G_{vd}(s)$ and $G_{vg}(s)$ share the same ω_0 and Q, given by Eqs. (8.119) and (8.121). The dc gain is

$$G_{d0} = -\frac{V_g - V}{D'} = -\frac{V_g}{D'^2} = \frac{V}{DD'} \tag{8.127}$$

The angular frequency of the zero is found by equating coefficients of s in the numerators of Eqs. (8.125) and (8.126). One obtains

$$\omega_z = \frac{D'\left(V_g - V\right)}{LI} = \frac{D'^2 R}{DL} \quad \text{(RHP)} \tag{8.128}$$

This zero lies in the right half-plane. Equations (8.127) and (8.128) have been simplified by use of the dc relationships

$$V = -\frac{D}{D'} V_g$$
$$I = -\frac{V}{D'R}$$

(8.129)

Equations (8.119), (8.121), (8.127), and (8.128) constitute the results of the analysis of the control-to-output transfer function: analytical expressions for the salient features ω_0, Q, G_{d0}, and ω_z. These expressions can be used to choose the element values such that given desired values of the salient features are obtained.

Having found analytical expressions for the salient features of the transfer functions, we can now plug in numerical values and construct the Bode plot. Suppose that we are given the following values:

$$D = 0.6$$
$$R = 10\ \Omega$$
$$V_g = 30\ \text{V}$$
$$L = 160\ \mu\text{H}$$
$$C = 160\ \mu\text{F}$$

(8.130)

We can evaluate Eqs. (8.117), (8.119), (8.121), (8.127), and (8.128), to determine numerical values of the salient features of the transfer functions. The results are:

$$\left| G_{g0} \right| = \frac{D}{D'} = 1.5 \Rightarrow 3.5\ \text{dB}$$
$$\left| G_{d0} \right| = \frac{|V|}{DD'} = 187.5\ \text{V} \Rightarrow 45.5\ \text{dBV}$$
$$f_0 = \frac{\omega_0}{2\pi} = \frac{D'}{2\pi\sqrt{LC}} = 400\ \text{Hz}$$
$$Q = D'R\sqrt{\frac{C}{L}} = 4 \Rightarrow 12\ \text{dB}$$
$$f_z = \frac{\omega_z}{2\pi} = \frac{D'^2 R}{2\pi DL} = 2.65\ \text{kHz}$$

(8.131)

The Bode plot of the magnitude and phase of G_{vd} is constructed in Fig. 8.33. The transfer function contains a dc gain of 45.5 dBV, resonant poles at 400 Hz having a Q of 4 $\Rightarrow$ 12 dB, and a right half-plane zero at 2.65 kHz. The resonant poles contribute –180° to the high-frequency phase asymptote, while the right half-plane zero contributes –90°. In addition, the inverting characteristic of the buck-boost converter leads to a 180° phase reversal, not included in Fig. 8.33.

The Bode plot of the magnitude and phase of the line-to-output transfer function G_{vg} is constructed in Fig. 8.34. This transfer function contains the same resonant poles at 400 Hz, but is missing the right half-plane zero. The dc gain G_{g0} is equal to the conversion ratio $M(D)$ of the converter. Again, the 180° phase reversal, caused by the inverting characteristic of the buck-boost converter, is not included in Fig. 8.34.

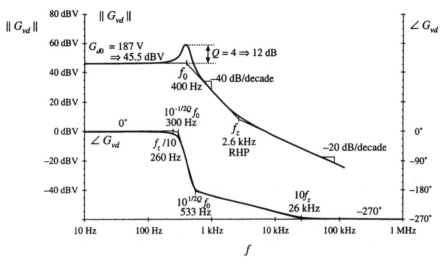

Fig. 8.33 Bode plot of the control-to-output transfer function G_{vd}, buck-boost converter example. Phase reversal owing to output voltage inversion is not included.

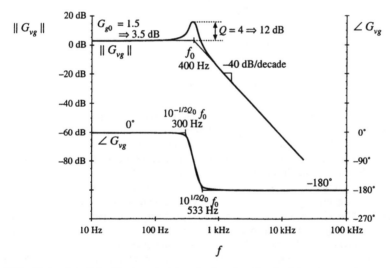

Fig. 8.34 Bode plot of the line-to-output transfer function G_{vg}, buck-boost converter example. Phase reversal owing to output voltage reversal is not included.

Table 8.2 Salient features of the small-signal CCM transfer functions of some basic dc–dc converters

Converter	G_{g0}	G_{d0}	ω_0	Q	ω_z
Buck	D	$\dfrac{V}{D}$	$\dfrac{1}{\sqrt{LC}}$	$R\sqrt{\dfrac{C}{L}}$	∞
Boost	$\dfrac{1}{D'}$	$\dfrac{V}{D'}$	$\dfrac{D'}{\sqrt{LC}}$	$D'R\sqrt{\dfrac{C}{L}}$	$\dfrac{D'^2R}{L}$
Buck-boost	$-\dfrac{D}{D'}$	$\dfrac{V}{DD'}$	$\dfrac{D'}{\sqrt{LC}}$	$D'R\sqrt{\dfrac{C}{L}}$	$\dfrac{D'^2R}{DL}$

8.2.2 Transfer Functions of Some Basic CCM Converters

The salient features of the line-to-output and control-to-output transfer functions of the basic buck, boost, and buck-boost converters are summarized in Table 8.2. In each case, the control-to-output transfer function is of the form

$$G_{vd}(s) = G_{d0} \frac{\left(1 - \dfrac{s}{\omega_z}\right)}{\left(1 + \dfrac{s}{Q\omega_0} + \left(\dfrac{s}{\omega_0}\right)^2\right)} \tag{8.132}$$

and the line-to-output transfer function is of the form

$$G_{vg}(s) = G_{g0} \frac{1}{1 + \dfrac{s}{Q\omega_0} + \left(\dfrac{s}{\omega_0}\right)^2} \tag{8.133}$$

The boost and buck-boost converters exhibit control-to-output transfer functions containing two poles and a right half-plane zero. The buck converter $G_{vg}(s)$ exhibits two poles but no zero. The line-to-output transfer functions of all three ideal converters contain two poles and no zeroes.

These results can be easily adapted to transformer-isolated versions of the buck, boost, and buck-boost converters. The transformer has negligible effect on the transfer functions $G_{vg}(s)$ and $G_{vd}(s)$, other than introduction of a turns ratio. For example, when the transformer of the bridge topology is driven symmetrically, its magnetizing inductance does not contribute dynamics to the converter small-signal transfer functions. Likewise, when the transformer magnetizing inductance of the forward converter is reset by the input voltage v_g, as in Fig. 6.23 or 6.28, then it also contributes negligible dynamics. In all transformer-isolated converters based on the buck, boost, and buck-boost converters, the line-to-output transfer function $G_{vg}(s)$ should be multiplied by the transformer turns ratio; the transfer functions (8.132) and (8.133) and the parameters listed in Table 8.2 can otherwise be directly applied.

8.2.3 Physical Origins of the Right Half-Plane Zero in Converters

Figure 8.35 contains a block diagram that illustrates the behavior of the right half-plane zero. At low frequencies, the gain (s/ω_z) has negligible magnitude, and hence $u_{out} \approx u_{in}$. At high frequencies, where the

Fig. 8.35 Block diagram having a right half-plane zero transfer function, as in Eq. (8.32), with $\omega_0 = \omega_z$.

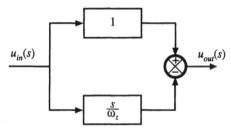

magnitude of the gain (s/ω_z) is much greater than 1, $u_{out} \approx -(s/\omega_z)u_{in}$. The negative sign causes a phase reversal at high frequency. The implication for the transient response is that the output initially tends in the opposite direction of the final value.

We have seen that the control-to-output transfer functions of the boost and buck-boost converters, Fig. 8.36, exhibit RHP zeroes. Typical transient response waveforms for a step change in duty cycle are illustrated in Fig. 8.37. For this example, the converter initially operates in equilibrium, at $d = 0.4$ and $d' = 0.6$. Equilibrium inductor current $i_L(t)$, diode current $i_D(t)$, and output voltage $v(t)$ waveforms are illustrated. The average diode current is

$$\langle i_D \rangle_{T_s} = d' \langle i_L \rangle_{T_s} \tag{8.134}$$

By capacitor charge balance, this average diode current is equal to the dc load current when the converter operates in equilibrium. At time $t = t_1$, the duty cycle is increased to 0.6. In consequence, d' decreases to 0.4. The average diode current, given by Eq. (8.134), therefore decreases, and the output capacitor begins to discharge. The output voltage magnitude initially decreases as illustrated.

(a)

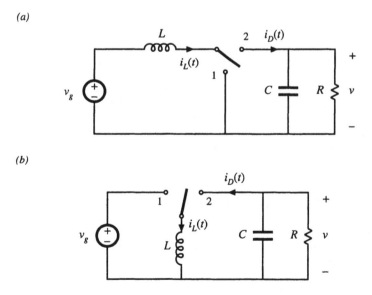

(b)

Fig. 8.36 Two basic converters whose CCM control-to-output transfer functions exhibit RHP zeroes: (a) boost, (b) buck-boost.

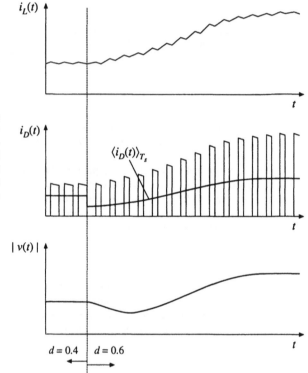

Fig. 8.37 Waveforms of the converters of Fig. 8.36, for a step response in duty cycle. The average diode current and output voltage initially decrease, as predicted by the RHP zero. Eventually, the inductor current increases, causing the average diode current and the output voltage to increase.

The increased duty cycle causes the inductor current to slowly increase, and hence the average diode current eventually exceeds its original $d = 0.4$ equilibrium value. The output voltage eventually increases in magnitude, to the new equilibrium value corresponding to $d = 0.6$.

The presence of a right half-plane zero tends to destabilize wide-bandwidth feedback loops, because during a transient the output initially changes in the wrong direction. The phase margin test for feedback loop stability is discussed in the next chapter; when a RHP zero is present, it is difficult to obtain an adequate phase margin in conventional single-loop feedback systems having wide bandwidth. Prediction of the right half-plane zero, and the consequent explanation of why the feedback loops controlling CCM boost and buck-boost converters tend to oscillate, was one of the early successes of averaged converter modeling.

8.3   GRAPHICAL CONSTRUCTION OF IMPEDANCES AND TRANSFER FUNCTIONS

Often, we can draw approximate Bode diagrams by inspection, without large amounts of messy algebra and the inevitable associated algebra mistakes. A great deal of insight can be gained into the operation of the circuit using this method. It becomes clear which components dominate the circuit response at various frequencies, and so suitable approximations become obvious. Analytical expressions for the approximate corner frequencies and asymptotes can be obtained directly. Impedances and transfer functions of quite complicated networks can be constructed. Thus insight can be gained, so that the design engineer can modify the circuit to obtain a desired frequency response.

The graphical construction method, also known as "doing algebra on the graph," involves use of a few simple rules for combining the magnitude Bode plots of impedances and transfer functions.

8.3.1 Series Impedances: Addition of Asymptotes

A series connection represents the addition of impedances. If the Bode diagrams of the individual impedance magnitudes are known, then the asymptotes of the series combination are found by simply taking the largest of the individual impedance asymptotes. In many cases, the result is exact. In other cases, such as when the individual asymptotes have the same slope, then the result is an approximation; nonetheless, the accuracy of the approximation can be quite good.

Consider the series-connected R–C network of Fig. 8.38. It is desired to construct the magnitude asymptotes of the total series impedance $Z(s)$, where

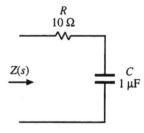

Fig. 8.38 Series R–C network example.

$$Z(s) = R + \frac{1}{sC} \tag{8.135}$$

Let us first sketch the magnitudes of the individual impedances. The 10 Ω resistor has an impedance magnitude of 10 $\Omega \Rightarrow$ 20 dBΩ. This value is independent of frequency, and is given in Fig. 8.39. The capacitor has an impedance magnitude of $1/\omega C$. This quantity varies inversely with ω, and hence its magnitude Bode plot is a line with slope −20 dB/decade. The line passes through 1 $\Omega \Rightarrow$ 0 dBΩ at the angular frequency ω where

$$\frac{1}{\omega C} = 1 \ \Omega \tag{8.136}$$

that is, at

$$\omega = \frac{1}{(1 \ \Omega)C} = \frac{1}{(1 \ \Omega)(10^{-6} \ \mathrm{F})} = 10^6 \ \mathrm{rad/sec} \tag{8.137}$$

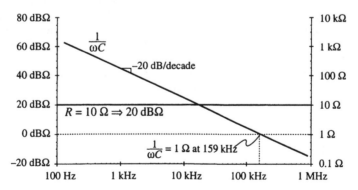

Fig. 8.39 Impedance magnitudes of the individual elements in the network of Fig. 8.38.

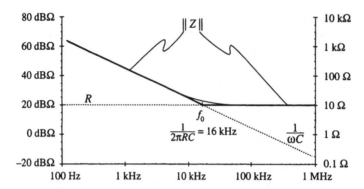

Fig. 8.40 Construction of the composite asymptotes of $\| Z \|$. The asymptotes of the series combination can be approximated by simply selecting the larger of the individual resistor and capacitor asymptotes.

In terms of frequency f, this occurs at

$$f = \frac{\omega}{2\pi} = \frac{10^6}{2\pi} = 159 \text{ kHz} \tag{8.138}$$

So the capacitor impedance magnitude is a line with slope -20 dB/dec, and which passes through 0 dBΩ at 159 kHz, as shown in Fig. 8.39. It should be noted that, for simplicity, the asymptotes in Fig. 8.39 have been labeled R and $1/\omega C$. But to draw the Bode plot, we must actually plot dBΩ; for example, $20 \log_{10} (R/1 \ \Omega)$ and $20 \log_{10} ((1/\omega C)/1 \ \Omega)$.

Let us now construct the magnitude of $Z(s)$, given by Eq. (8.135). The magnitude of Z can be approximated as follows:

$$\left| Z(j\omega) \right| = \left| R + \frac{1}{j\omega C} \right| \approx \begin{cases} R & \text{for } R \gg 1/\omega C \\ \frac{1}{\omega C} & \text{for } R \ll 1/\omega C \end{cases} \tag{8.139}$$

The asymptotes of the series combination are simply the larger of the individual resistor and capacitor asymptotes, as illustrated by the heavy lines in Fig. 8.40. For this example, these are in fact the exact asymptotes of $\| Z \|$. In the limiting case of zero frequency (dc), then the capacitor tends to an open circuit. The series combination is then dominated by the capacitor, and the exact function tends asymptotically to the capacitor impedance magnitude. In the limiting case of infinite frequency, then the capacitor tends to a short circuit, and the total impedance becomes simply R. So the R and $1/\omega C$ lines are the exact asymptotes for this example.

The corner frequency f_0, where the asymptotes intersect, can now be easily deduced. At angular frequency $\omega_0 = 2\pi f_0$, the two asymptotes are equal in value:

$$\frac{1}{\omega_0 C} = R \tag{8.140}$$

Solution for ω_0 and f_0 leads to:

$$\omega_0 = \frac{1}{RC} = \frac{1}{(10 \ \Omega)(10^{-6} \ \text{F})} = 10^5 \ \text{rad/sec}$$

$$f_0 = \frac{\omega_0}{2\pi} = \frac{1}{2\pi RC} = 16 \ \text{kHz} \tag{8.141}$$

So if we can write analytical expressions for the asymptotes, then we can equate the expressions to find analytical expressions for the corner frequencies where the asymptotes intersect.

The deviation of the exact curve from the asymptotes follows all of the usual rules. The slope of the asymptotes changes by +20 dB/decade at the corner frequency f_0 (i.e., from −20 dBΩ/decade to 0 dBΩ/decade), and hence there is a zero at $f = f_0$. So the exact curve deviates from the asymptotes by +3 dBΩ at $f = f_0$, and by +1 dBΩ at $f = 2f_0$ and at $f = f_0/2$.

8.3.2   Series Resonant Circuit Example

As a second example, let us construct the magnitude asymptotes for the series R–L–C circuit of Fig. 8.41. The series impedance $Z(s)$ is

$$Z(s) = R + sL + \frac{1}{sC} \tag{8.142}$$

The magnitudes of the individual resistor, inductor, and capacitor asymptotes are plotted in Fig. 8.42, for the values

$$R = 1 \ \text{k}\Omega$$
$$L = 1 \ \text{mH} \tag{8.143}$$
$$C = 0.1 \ \mu\text{F}$$

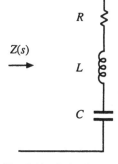

The series impedance $Z(s)$ is dominated by the capacitor at low frequency, by the resistor at mid frequencies, and by the inductor at high frequencies, as illustrated by the bold line in Fig. 8.42. The impedance $Z(s)$ contains a

Fig. 8.41 Series R–L–C network example.

zero at angular frequency ω_1, where the capacitor and resistor asymptotes intersect. By equating the expressions for the resistor and capacitor asymptotes, we can find ω_1:

$$R = \frac{1}{\omega_1 C} \quad \Rightarrow \quad \omega_1 = \frac{1}{RC} \tag{8.144}$$

Fig. 8.42 Graphical construction of $\| Z \|$ of the series R–L–C network of Fig. 8.41, for the element values specified by Eq. (8.143).

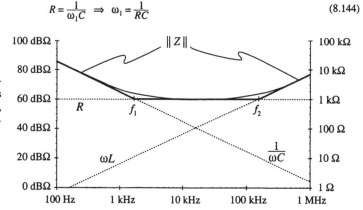

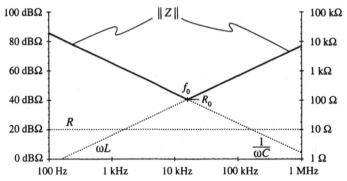

Fig. 8.43 Graphical construction of impedance asymptotes for the series R–L–C network example, with R decreased to 10 Ω.

A second zero occurs at angular frequency ω_2, where the inductor and resistor asymptotes intersect. Upon equating the expressions for the resistor and inductor asymptotes at ω_2, we obtain the following:

$$R = \omega_2 L \quad \Rightarrow \quad \omega_2 = \frac{R}{L} \tag{8.145}$$

So simple expressions for all important features of the magnitude Bode plot of $Z(s)$ can be obtained directly. It should be noted that Eqs. (8.144) and (8.145) are approximate, rather than exact, expressions for the corner frequencies ω_1 and ω_2. Equations (8.144) and (8.145) coincide with the results obtained via the low-Q approximation of Section 8.1.7.

Next, suppose that the value of R is decreased to 10 Ω. As R is reduced in value, the approximate corner frequencies ω_1 and ω_2 move closer together until, at $R = 100$ Ω, they are both 100 krad/sec. Reducing R further in value causes the asymptotes to become independent of the value of R, as illustrated in Fig. 8.43 for $R = 10$ Ω. The $\| Z \|$ asymptotes now switch directly from ωL to $1/\omega C$.

So now there are two zeroes at $\omega = \omega_0$. At corner frequency ω_0, the inductor and capacitor asymptotes are equal in value. Hence,

$$\omega_0 L = \frac{1}{\omega_0 C} = R_0 \tag{8.146}$$

Solution for the angular corner frequency ω_0 leads to

$$\omega_0 = \frac{1}{\sqrt{LC}} \tag{8.147}$$

At $\omega = \omega_0$, the inductor and capacitor impedances both have magnitude R_0, called the characteristic impedance.

Since there are two zeroes at $\omega = \omega_0$, there is a possibility that the two poles could be complex conjugates, and that peaking could occur in the vicinity of $\omega = \omega_0$. So let us investigate what the actual curve does at $\omega = \omega_0$. The actual value of the series impedance $Z(j\omega_0)$ is

$$Z(j\omega_0) = R + j\omega_0 L + \frac{1}{j\omega_0 C} \tag{8.148}$$

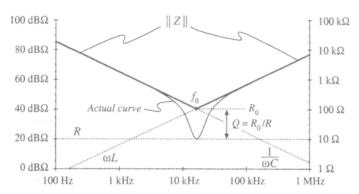

Fig. 8.44 Actual impedance magnitude (solid line) for the series resonant R–L–C example. The inductor and capacitor impedances cancel out at $f = f_0$, and hence $Z(j\omega_0) = R$.

Substitution of Eq. (8.146) into Eq. (8.148) leads to

$$Z(j\omega_0) = R + jR_0 + \frac{R_0}{j} = R + jR_0 - jR_0 = R \tag{8.149}$$

At $\omega = \omega_0$, the inductor and capacitor impedances are equal in magnitude but opposite in phase. Hence, they exactly cancel out in the series impedance, and we are left with $Z(j\omega_0) = R$, as illustrated in Fig. 8.44. The actual curve in the vicinity of the resonance at $\omega = \omega_0$ can deviate significantly from the asymptotes, because its value is determined by R rather than ωL or $1/\omega C$.

We know from Section 8.1.6 that the deviation of the actual curve from the asymptotes at $\omega = \omega_0$ is equal to Q. From Fig. 8.44, one can see that

$$\left|Q\right|_{dB} = \left|R_0\right|_{dB\Omega} - \left|R\right|_{dB\Omega} \tag{8.150}$$

or,

$$Q = \frac{R_0}{R} \tag{8.151}$$

Equations (8.146) to (8.151) are exact results for the series resonant circuit.

The practice of adding asymptotes by simply selecting the larger asymptote can be applied to transfer functions as well as impedances. For example, suppose that we have already constructed the magnitude asymptotes of two transfer functions, G_1 and G_2, and we wish to find the asymptotes of $G = G_1 + G_2$. At each frequency, the asymptote for G can be approximated by simply selecting the larger of the asymptotes for G_1 and G_2:

$$G = G_1 + G_2 \approx \begin{cases} G_1, & \left|G_1\right| \gg \left|G_2\right| \\ G_2, & \left|G_2\right| \gg \left|G_1\right| \end{cases} \tag{8.152}$$

Corner frequencies can be found by equating expressions for asymptotes as illustrated in the preceding examples. In the next chapter, we will see that this approach yields a simple and powerful method for determining the closed-loop transfer functions of feedback systems.

8.3.3   Parallel Impedances: Inverse Addition of Asymptotes

A parallel combination represents inverse addition of impedances:

$$Z_{par} = \cfrac{1}{\cfrac{1}{Z_1} + \cfrac{1}{Z_2} + \cdots} \qquad (8.153)$$

If the asymptotes of the individual impedances Z_1, Z_2, ..., are known, then the asymptotes of the parallel combination Z_{par} can be found by simply selecting the smallest individual impedance asymptote. This is true because the smallest impedance will have the largest inverse, and will dominate the inverse sum. As in the case of the series impedances, this procedure will often yield the exact asymptotes of Z_{par}.

Let us construct the magnitude asymptotes for the parallel R–L–C network of Fig. 8.45, using the following element values:

$$\begin{aligned} R &= 10\ \Omega \\ L &= 1\ \text{mH} \\ C &= 0.1\ \mu\text{F} \end{aligned} \qquad (8.154)$$

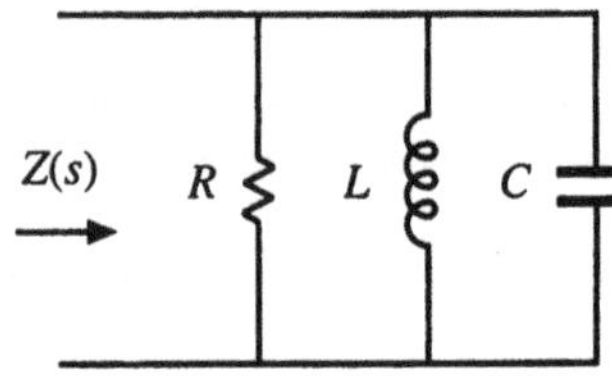

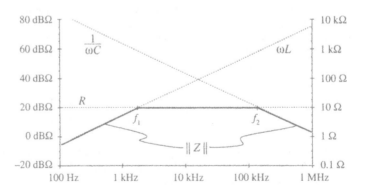

Fig. 8.45 Parallel R–L–C network example.

Impedance magnitudes of the individual elements are illustrated in Fig. 8.46. The asymptotes for the total parallel impedance Z are approximated by simply selecting the smallest individual element impedance, as shown by the heavy line in Fig. 8.46. So the parallel impedance is dominated by the inductor at low frequency, by the resistor at mid frequencies, and by the capacitor at high frequency. Approximate expressions for the angular corner frequencies are again found by equating asymptotes:

$$\begin{aligned} \text{at } \omega = \omega_1, \quad R &= \omega_1 L \;\Rightarrow\; \omega_1 = \frac{R}{L} \\ \text{at } \omega = \omega_2, \quad R &= \frac{1}{\omega_2 C} \;\Rightarrow\; \omega_2 = \frac{1}{RC} \end{aligned} \qquad (8.155)$$

These expressions could have been obtained by conventional analysis, combined with the low-Q approximation of Section 8.1.7.

Fig. 8.46 Construction of the composite asymptotes of $\| Z \|$, for the parallel R–L–C example. The asymptotes of the parallel combination can be approximated by simply selecting the smallest of the individual resistor, inductor, and capacitor asymptotes.

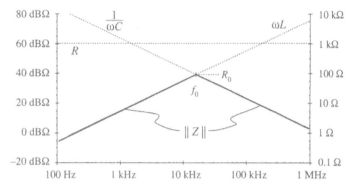

Fig. 8.47 Graphical construction of impedance asymptotes for the parallel R–L–C example, with R increased to 1 kΩ.

8.3.4 Parallel Resonant Circuit Example

Figure 8.47 illustrates what happens when the value of R in the parallel R–L–C network is increased to 1 kΩ. The asymptotes for $\parallel Z \parallel$ then become independent of R, and change directly from ωL to $1/\omega C$ at angular frequency ω_0. The corner frequency ω_0 is now the frequency where the inductor and capacitor asymptotes have equal value:

$$\omega_0 L = \frac{1}{\omega_0 C} = R_0 \tag{8.156}$$

which implies that

$$\omega_0 = \frac{1}{\sqrt{LC}} \tag{8.157}$$

At $\omega = \omega_0$, the slope of the asymptotes of $\parallel Z \parallel$ changes from +20 dB/decade to −20 dB/decade, and hence there are two poles. We should investigate whether peaking occurs, by determining the exact value of $\parallel Z \parallel$ at $\omega = \omega_0$, as follows:

$$\tag{8.158}$$

$$Z(j\omega_0) = R \parallel j\omega_0 L \parallel \frac{1}{j\omega_0 C} = \frac{1}{\frac{1}{R} + \frac{1}{j\omega_0 L} + j\omega_0 C}$$

Substitution of Eq. (8.156) into (8.158) yields

$$Z(j\omega_0) = \frac{1}{\frac{1}{R} + \frac{1}{jR_0} + \frac{j}{R_0}} = \frac{1}{\frac{1}{R} - \frac{j}{R_0} + \frac{j}{R_0}} = R \tag{8.159}$$

So at $\omega = \omega_0$, the impedances of the inductor and capacitor again cancel out, and we are left with $Z(j\omega_0) = R$. The values of L and C determine the values of the asymptotes, but R determines the value of the actual curve at $\omega = \omega_0$.

The actual curve is illustrated in 8.48. The deviation of the actual curve from the asymptotes at $\omega = \omega_0$ is

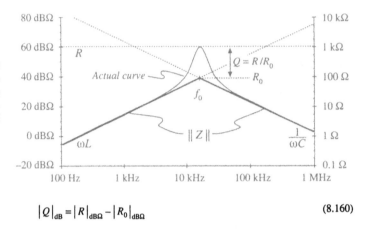

Fig. 8.48 Actual impedance magnitude (solid line) for the parallel *R–L–C* example. The inductor and capacitor impedances cancel out at $f = f_0$, and hence $Z(j\omega_0) = R$.

$$|Q|_{dB} = |R|_{dB\Omega} - |R_0|_{dB\Omega} \qquad (8.160)$$

or,

$$Q = \frac{R}{R_0} \qquad (8.161)$$

Equations (8.156) to (8.161) are exact results for the parallel resonant circuit.

The graphical construction method for impedance magnitudes is well known, and *reactance paper* can be purchased commercially. As illustrated in Fig. 8.49, the magnitudes of the impedances of various inductances, capacitances, and resistances are plotted on semilogarithmic axes. Asymptotes for the impedances of *R–L–C* networks can be sketched directly on these axes, and numerical values of corner frequencies can then be graphically determined.

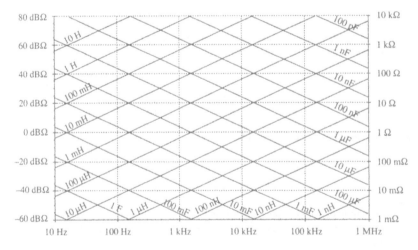

Fig. 8.49 "Reactance paper": an aid for graphical construction of impedances, with the magnitudes of various inductive, capacitive, and resistive impedances preplotted.

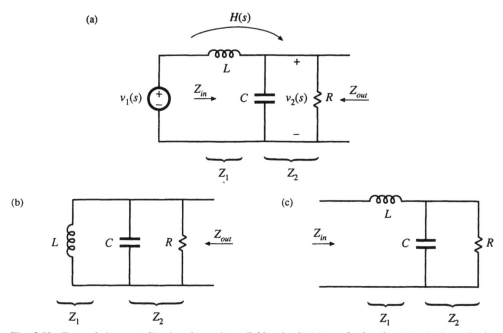

Fig. 8.50 Two-pole low-pass filter based on voltage divider circuit: (a) transfer function $H(s)$, (b) determination of Z_{out}, by setting independent sources to zero, (c) determination of $Z_{in}(s)$.

8.3.5 Voltage Divider Transfer Functions: Division of Asymptotes

Usually, we can express transfer functions in terms of impedances—for example, as the ratio of two impedances. If we can construct these impedances as described in the previous sections, then we can divide to construct the transfer function. In this section, construction of the transfer function $H(s)$ of the two-pole R–L–C low-pass filter (Fig. 8.50) is discussed in detail. A filter of this form appears in the canonical model for two-pole converters, and the results of this section are applied in the converter examples of the next section.

The familiar voltage divider formula shows that the transfer function of this circuit can be expressed as the ratio of impedances Z_2/Z_{in}, where $Z_{in} = Z_1 + Z_2$ is the network input impedance:

$$\frac{\hat{v}_2(s)}{\hat{v}_1(s)} = \frac{Z_2}{Z_1 + Z_2} = \frac{Z_2}{Z_{in}} \tag{8.162}$$

For this example, $Z_1(s) = sL$, and $Z_2(s)$ is the parallel combination of R and $1/sC$. Hence, we can find the transfer function asymptotes by constructing the asymptotes of Z_2 and of the series combination represented by Z_{in}, and then dividing. Another approach, which is easier to apply in this example, is to multiply the numerator and denominator of Eq. (8.162) by Z_1:

(a)

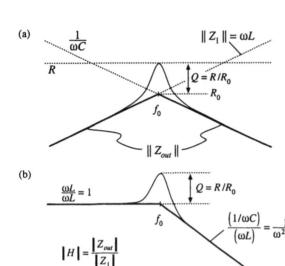

(b)

Fig. 8.51 Graphical construction of H and Z_{out} of the voltage divider circuit: (a) output impedance Z_{out}; (b) transfer function H.

$$\frac{\hat{v}_2(s)}{\hat{v}_1(s)} = \frac{Z_2 Z_1}{Z_1 + Z_2} \frac{1}{Z_1} = \frac{Z_{out}}{Z_1} \tag{8.163}$$

where $Z_{out} = Z_1 \| Z_2$ is the output impedance of the voltage divider. So another way to construct the voltage divider transfer function is to first construct the asymptotes for Z_1 and for the parallel combination represented by Z_{out}, and then divide. This method is useful when the parallel combination $Z_1 \| Z_2$ is easier to construct than the series combination $Z_1 + Z_2$. It often gives a different approximate result, which may be more (or sometimes less) accurate than the result obtained using Z_{in}.

The output impedance Z_{out} in Fig. 8.50(b) is

$$Z_{out}(s) = R \| \frac{1}{sC} \| sL \tag{8.164}$$

The impedance of the parallel R–L–C network is constructed in Section 8.3.3, and is illustrated in Fig. 8.51(a) for the high-Q case.

According to Eq. (8.163), the voltage divider transfer function magnitude is $\| H \| = \| Z_{out} \| / \| Z_1 \|$. This quantity is constructed in Fig. 8.51(b). For $\omega < \omega_0$, the asymptote of $\| Z_{out} \|$ coincides with $\| Z_1 \|$: both are equal to ωL. Hence, the ratio is $\| Z_{out} \| / \| Z_1 \| = 1$. For $\omega > \omega_0$, the asymptote of $\| Z_{out} \|$ is $1/\omega C$, while $\| Z_1 \|$ is equal to ωL. The ratio then becomes $\| Z_{out} \| / \| Z_1 \| = 1/\omega^2 LC$, and hence the high-

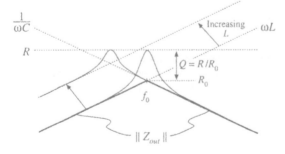

Fig. 8.52 Effect of increasing L on the output impedance asymptotes, corner frequency, and Q-factor.

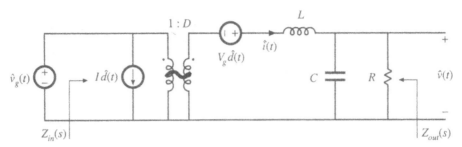

Fig. 8.53 Small-signal model of the buck converter, with input impedance $Z_{in}(s)$ and output impedance $Z_{out}(s)$ explicitly defined.

frequency asymptote has a -40 dB/decade slope. At $\omega = \omega_0$, $\| Z_{out} \|$ has exact value R, while $\| Z_1 \|$ has exact value R_0. The ratio is then $\| H(j\omega_0) \| = \| Z_{out}(j\omega_0) \| / \| Z_1(j\omega_0) \| = R/R_0 = Q$. So the filter transfer function H has the same ω_0 and Q as the impedance Z_{out}.

It now becomes obvious how variations in element values affect the salient features of the transfer function and output impedance. For example, the effect of increasing L is illustrated in Fig. 8.52. This causes the angular resonant frequency ω_0 to be reduced, and also reduces the Q-factor.

8.4 GRAPHICAL CONSTRUCTION OF CONVERTER TRANSFER FUNCTIONS

The small-signal equivalent circuit model of the buck converter, derived in Chapter 7, is reproduced in Fig. 8.53. Let us construct the transfer functions and terminal impedances of this converter, using the graphical approach of the previous section.

The output impedance $Z_{out}(s)$ is found with the $\hat{d}(s)$ and $\hat{v}_g(s)$ sources set to zero; the circuit of Fig. 8.54(a) is then obtained. This model coincides with the parallel R–L–C circuit analyzed in Sections 8.3.3 and 8.3.4. As illustrated in Fig. 8.54(b), the output impedance is dominated by the inductor at low frequency, and by the capacitor at high frequency. At the resonant frequency f_0, given by

$$f_0 = \frac{1}{2\pi\sqrt{LC}} \tag{8.165}$$

the output impedance is equal to the load resistance R. The Q-factor of the circuit is equal to

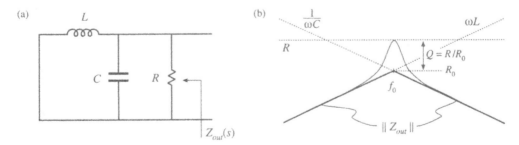

Fig. 8.54 Construction of buck converter output impedance $Z_{out}(s)$: (a) circuit model; (b) impedance asymptotes.

(a)

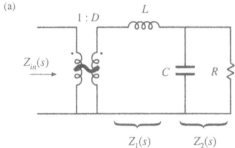

Fig. 8.55 Construction of the input impedance $Z_{in}(s)$ for the buck converter: (a) circuit model; (b) the individual resistor, inductor, and capacitor impedance magnitudes; (c) construction of the impedance magnitudes $\| Z_1 \|$ and $\| Z_2 \|$; (d) construction of $\| Z_{out} \|$; (e) final result $\| Z_{in} \|$.

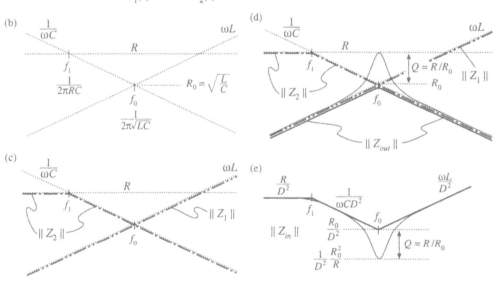

$$Q = \frac{R}{R_0} \tag{8.166}$$

where

$$R_0 = \omega_0 L = \frac{1}{\omega_0 C} = \sqrt{\frac{L}{C}} \tag{8.167}$$

Thus, the circuit is lightly damped (high Q) at light load, where the value of R is large.

The converter input impedance $Z_{in}(s)$ is also found with the $\hat{d}(s)$ and $\hat{v}_g(s)$ sources set to zero, as illustrated in Fig. 8.55(a). The input impedance is referred to the primary side of the $1{:}D$ transformer, and is equal to

$$Z_{in}(s) = \frac{1}{D^2}\left[Z_1(s) + Z_2(s)\right] \tag{8.168}$$

where

$$Z_1(s) = sL \tag{8.169}$$

and

$$Z_2(s) = R \parallel \frac{1}{sC} \tag{8.170}$$

We begin construction of the impedance asymptotes corresponding to Eqs. (8.168) to (8.170) by constructing the individual resistor, capacitor, and inductor impedances as in Fig. 8.55(b). The impedances in Fig. 8.55 are constructed for the case $R > R_0$. As illustrated in Fig. 8.55(c), $\parallel Z_1 \parallel$ coincides with the inductor reactance ωL. The impedance $\parallel Z_2 \parallel$ is asymptotic to resistance R at low frequencies, and to the capacitor reactance $1/\omega C$ at high frequency. The resistor and capacitor asymptotes intersect at corner frequency f_1, given by

$$f_1 = \frac{1}{2\pi RC} \tag{8.171}$$

According to Eq. (8.168), the input impedance $Z_{in}(s)$ is equal to the series combination of $Z_1(s)$ and $Z_2(s)$, divided by the square of the turns ratio D. The asymptotes for the series combination $[Z_1(s) + Z_2(s)]$ are found by selecting the larger of the $\parallel Z_1 \parallel$ and $\parallel Z_2 \parallel$ asymptotes. The $\parallel Z_1 \parallel$ and $\parallel Z_2 \parallel$ asymptotes intersect at frequency f_0, given by Eq. (8.165). It can be seen from Fig. 8.55(c) that the series combination is dominated by Z_2 for $f < f_0$, and by Z_1 for $f > f_0$. Upon scaling the $[Z_1(s) + Z_2(s)]$ asymptotes by the factor $1/D^2$, the input impedance asymptotes of Fig. 8.55(e) are obtained.

The zeroes of $Z_{in}(s)$, at frequency f_0, have the same Q-factor as the poles of $Z_{out}(s)$ [Eq. (8.166)]. One way to see that this is true is to note that the output impedance can be expressed as

$$Z_{out}(s) = \frac{Z_1(s)Z_2(s)}{Z_1(s) + Z_2(s)} = \frac{Z_1(s)Z_2(s)}{D^2 Z_{in}(s)} \tag{8.172}$$

Hence, we can relate $Z_{out}(s)$ to $Z_{in}(s)$ as follows:

$$Z_{in}(s) = \frac{1}{D^2} \frac{Z_1(s)Z_2(s)}{Z_{out}(s)} \tag{8.173}$$

The impedances $\parallel Z_1 \parallel$, $\parallel Z_2 \parallel$, and $\parallel Z_{out} \parallel$ are illustrated in Fig. 8.55(d). At the resonant frequency $f = f_0$, impedance Z_1 has magnitude R_0 and impedance Z_2 has magnitude approximately equal to R_0. The output impedance Z_{out} has magnitude R. Hence, Eq. (8.173) predicts that the input impedance has the magnitude

$$|Z_{in}| \approx \frac{1}{D^2} \frac{R_0 R_0}{R} \quad \text{at } f = f_0 \tag{8.174}$$

At $f = f_0$, the asymptotes of the input impedance have magnitude R_0/D^2. The deviation from the asymptotes is therefore equal to $Q = R/R_0$, as illustrated in Fig. 8.55(e).

The control-to-output transfer function $G_{vd}(s)$ is found with the $\hat{v}_g(s)$ source set to zero, as in Fig. 8.56(a). This circuit coincides with the voltage divider analyzed in Section 8.3.5. Hence, $G_{vd}(s)$ can be expressed as

$$G_{vd}(s) = V_g \frac{Z_{out}(s)}{Z_1(s)} \tag{8.175}$$

Fig. 8.56 Construction of the control-to-output transfer function $G_{vd}(s)$ for the buck converter: (a) circuit model; (b) relevant impedance asymptotes; (c) transfer function $\| G_{vd}(s) \|$.

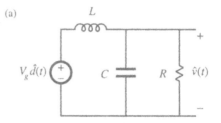

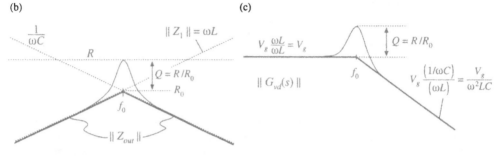

(b)   (c)

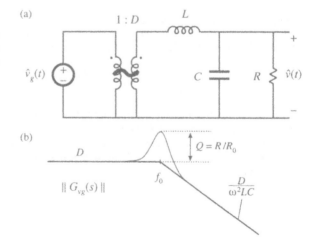

The quantities $\| Z_{out} \|$ and $\| Z_1 \|$ are constructed in Fig. 8.56(b). According to Eq. (8.175), we can construct $\| G_{vd}(s) \|$ by finding the ratio of $\| Z_{out} \|$ and $\| Z_1 \|$, and then scaling the result by V_g. For $f < f_0$, $\| Z_{out} \|$ and $\| Z_1 \|$ are both equal to ωL and hence $\| Z_{out} \| / \| Z_1 \|$ is equal to 1. As illustrated in Fig. 8.56(c), the low-frequency asymptote of $\| G_{vd}(s) \|$ has value V_g. For $f > f_0$, $\| Z_{out} \|$ has asymptote $1/\omega C$, and $\| Z_1 \|$ is equal to ωL. Hence, $\| Z_{out} \| / \| Z_1 \|$ has asymptote $1/\omega^2 LC$, and the high-frequency asymptote of $\| G_{vd}(s) \|$ is equal to $V_g/\omega^2 LC$. The Q-factor of the two poles at $f = f_0$ is again equal to R/R_0.

The line-to-output transfer function $G_{vg}(s)$ is found with the $\hat{d}(s)$ sources set to zero, as in Fig. 8.57(a). This circuit contains the same voltage divider as in Fig. 8.56, and additionally contains the 1:D transformer. The transfer function $G_{vg}(s)$ can be expressed as

Fig. 8.57 The line-to-output transfer function $G_{vg}(s)$ for the buck converter: (a) circuit model; (b) magnitude asymptotes.

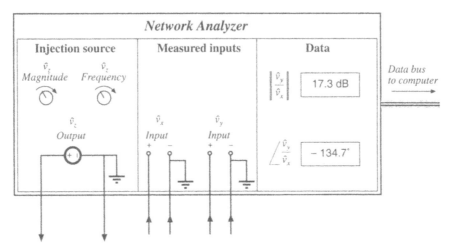

Fig. 8.58 Key features and functions of a network analyzer: sinusoidal source of controllable amplitude and frequency, two inputs, and determination of relative magnitude and phase of the input components at the injection frequency.

$$G_{vg}(s) = D \frac{Z_{out}(s)}{Z_i(s)} \tag{8.176}$$

This expression is similar to Eq. (8.175), except for the scaling factor of D. Therefore, the line-to-output transfer function of Fig. 8.57(b) has the same shape as the control-to-output transfer function $G_{vd}(s)$.

8.5 MEASUREMENT OF AC TRANSFER FUNCTIONS AND IMPEDANCES

It is good engineering practice to measure the transfer functions of prototype converters and converter systems. Such an exercise can verify that the system has been correctly modeled and designed. Also, it is often useful to characterize individual circuit elements through measurement of their terminal impedances.

Small-signal ac magnitude and phase measurements can be made using an instrument known as a network analyzer, or frequency response analyzer. The key inputs and outputs of a basic network analyzer are illustrated in Fig. 8.58. The network analyzer provides a sinusoidal output voltage $\hat{v}_z$ of controllable amplitude and frequency. This signal can be injected into the system to be measured, at any desired location. The network analyzer also has two (or more) inputs, $\hat{v}_x$ and $\hat{v}_y$. The return electrodes of $\hat{v}_z$, $\hat{v}_y$, and $\hat{v}_x$ are internally connected to earth ground. The network analyzer performs the function of a narrowband tracking voltmeter: it measures the components of $\hat{v}_x$ and $\hat{v}_y$ at the injection frequency, and displays the magnitude and phase of the quantity $\hat{v}_y/\hat{v}_x$. The narrowband tracking voltmeter feature is essential for switching converter measurements; otherwise, switching ripple and noise corrupt the desired sinusoidal signals and make accurate measurements impossible [3]. Modern network analyzers can automatically sweep the frequency of the injection source $\hat{v}_z$ to generate magnitude and phase Bode plots of the transfer function $\hat{v}_y/\hat{v}_x$.

A typical test setup for measuring the transfer function of an amplifier is illustrated in Fig. 8.59. A potentiometer, connected between a dc supply voltage V_{CC} and ground, is used to bias the

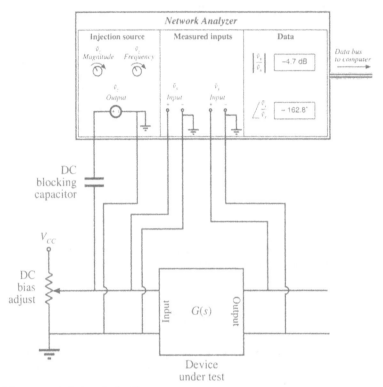

Fig. 8.59   Measurement of a transfer function.

amplifier input to attain the correct quiescent operating point. The injection source voltage $\hat{v}_z$ is coupled to the amplifier input terminals via a dc blocking capacitor. This blocking capacitor prevents the injection voltage source from upsetting the dc bias. The network analyzer inputs $\hat{v}_x$ and $\hat{v}_y$ are connected to the input and output terminals of the amplifier. Hence, the measured transfer function is

$$\frac{\hat{v}_y(s)}{\hat{v}_x(s)} = G(s) \tag{8.177}$$

Note that the blocking capacitance, bias potentiometer, and $\hat{v}_z$ amplitude have no effect on the measured transfer function
    An impedance

$$Z(s) = \frac{\hat{v}(s)}{\hat{i}(s)} \tag{8.178}$$

can be measured by treating the impedance as a transfer function from current to voltage. For example, measurement of the output impedance of an amplifier is illustrated in Fig. 8.60. The quiescent operating condition is again established by a potentiometer which biases the amplifier input. The injection source $\hat{v}_z$ is coupled to the amplifier output through a dc blocking capacitor. The injection source voltage $\hat{v}_z$ excites a current $\hat{i}_{out}$ in impedance Z_s. This current flows into the output of the amplifier, and excites a

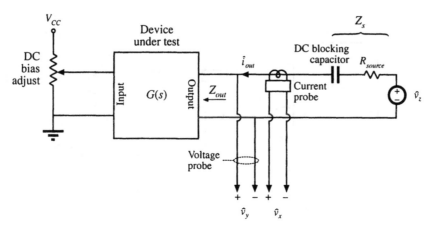

Fig. 8.60 Measurement of the output impedance of a circuit.

voltage across the amplifier output impedance:

$$Z_{out}(s) = \frac{\hat{v}_y(s)}{\hat{i}_{out}(s)} \Bigg|_{\substack{amplifier \\ ac\ input\ = 0}} \tag{8.179}$$

A current probe is used to measure $\hat{i}_{out}$. The current probe produces a voltage proportional to $\hat{i}_{out}$; this voltage is connected to the network analyzer input $\hat{v}_x$. A voltage probe is used to measure the amplifier output voltage $\hat{v}_y$. The network analyzer displays the transfer function $\hat{v}_y/\hat{v}_x$, which is proportional to Z_{out}. Note that the value of Z_s and the amplitude of $\hat{v}_z$ do not affect the measurement of Z_{out}.

In power applications, it is sometimes necessary to measure impedances that are very small in magnitude. Grounding problems[4] cause the test setup of Fig. 8.60 to fail in such cases. The reason is illustrated in Fig. 8.61(a). Since the return connections of the injection source $\hat{v}_z$ and the analyzer input $\hat{v}_y$ are both connected to earth ground, the injected current $\hat{i}_{out}$ can return to the source through the return connections of either the injection source or the voltage probe. In practice, $\hat{i}_{out}$ divides between the two paths according to their relative impedances. Hence, a significant current $(1 - k)\,\hat{i}_{out}$ flows through the return connection of the voltage probe. If the voltage probe return connection has some total contact and wiring impedance Z_{probe}, then the current induces a voltage drop $(1 - k)\hat{i}_{out}Z_{probe}$ in the voltage probe wiring, as illustrated in Fig. 8.61(a). Hence, the network analyzer does not correctly measure the voltage drop across the impedance Z. If the internal ground connections of the network analyzer have negligible impedance, then the network analyzer will display the following impedance:

$$Z + (1 - k)Z_{probe} = Z + Z_{probe}\|Z_{rz} \tag{8.180}$$

Here, Z_{rz} is the impedance of the injection source return connection. So to obtain an accurate measurement, the following condition must be satisfied:

$$|Z| \gg \left| \left(Z_{probe}\|Z_{rz}\right) \right| \tag{8.181}$$

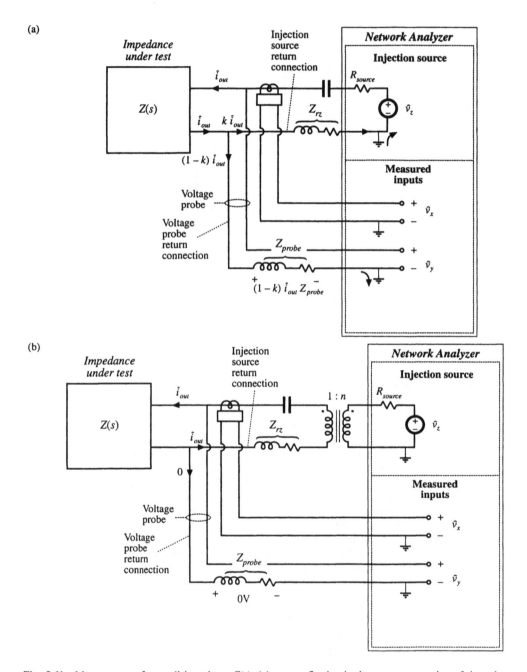

Fig. 8.61 Measurement of a small impedance $Z(s)$: (a) current flowing in the return connection of the voltage probe induces a voltage drop that corrupts the measurement; (b) an improved experiment, incorporating isolation of the injection source.

A typical lower limit on $\| Z \|$ is a few tens or hundreds of milliohms.

An improved test setup for measurement of small impedances is illustrated in Fig. 8.61(b). An isolation transformer is inserted between the injection source and the dc blocking capacitor. The return connections of the voltage probe and injection source are no longer in parallel, and the injected current $\hat{i}_{out}$ must now return entirely through the injection source return connection. An added benefit is that the transformer turns ratio n can be increased, to better match the injection source impedance to the imped- ance under test. Note that the impedances of the transformer, of the blocking capacitor, and of the probe and injection source return connections, do not affect the measurement. Much smaller impedances can therefore be measured using this improved approach.

8.6   SUMMARY OF KEY POINTS

1. The magnitude Bode diagrams of functions which vary as $(f/f_0)^n$ have slopes equal to $20n$ dB per decade, and pass through 0 dB at $f = f_0$.

2. It is good practice to express transfer functions in normalized pole-zero form; this form directly exposes expressions for the salient features of the response, that is, the corner frequencies, reference gain, etc.

3. The right half-plane zero exhibits the magnitude response of the left half-plane zero, but the phase response of the pole.

4. Poles and zeroes can be expressed in frequency-inverted form, when it is desirable to refer the gain to a high-frequency asymptote.

5. A two-pole response can be written in the standard normalized form of Eq. (8.58). When $Q > 0.5$, the poles are complex conjugates. The magnitude response then exhibits peaking in the vicinity of the corner fre- quency, with an exact value of Q at $f = f_0$. High Q also causes the phase to change sharply near the corner frequency.

6. When Q is less than 0.5, the two pole response can be plotted as two real poles. The low-Q approximation predicts that the two poles occur at frequencies f_0/Q and Qf_0. These frequencies are within 10% of the exact values for $Q \leq 0.3$.

7. The low-Q approximation can be extended to find approximate roots of an arbitrary degree polynomial. Approximate analytical expressions for the salient features can be derived. Numerical values are used to justify the approximations.

8. Salient features of the transfer functions of the buck, boost, and buck-boost converters are tabulated in Section 8.2.2. The line-to-output transfer functions of these converters contain two poles. Their control-to- output transfer functions contain two poles, and may additionally contain a right half-plane zero.

9. Approximate magnitude asymptotes of impedances and transfer functions can be easily derived by graphi- cal construction. This approach is a useful supplement to conventional analysis, because it yields physical insight into the circuit behavior, and because it exposes suitable approximations. Several examples, includ- ing the impedances of basic series and parallel resonant circuits and the transfer function $H(s)$ of the volt- age divider circuit, are worked in Section 8.3. The input impedance, output impedance, and transfer functions of the buck converter are constructed in Section 8.4, and physical origins of the asymptotes, cor- ner frequencies, and Q-factor are found.

10. Measurement of transfer functions and impedances using a network analyzer is discussed in Section 8.5. Careful attention to ground connections is important when measuring small impedances.

REFERENCES

[1]   R.D. MIDDLEBROOK, "Low Entropy Expressions: The Key to Design-Oriented Analysis," *IEEE Frontiers in Education Conference*, 1991 Proceedings, pp. 399-403, Sept. 1991.

[2]   R. D. MIDDLEBROOK, "Methods of Design-Oriented Analysis: The Quadratic Equation Revisited," *IEEE Frontiers in Education Conference*, 1992 Proceedings, pp. 95-102, Nov. 1991.

[3]   F. BARZEGAR, S. ĆUK, and R. D. MIDDLEBROOK, "Using Small Computers to Model and Measure Magnitude and Phase of Regulator Transfer Functions and Loop Gain," *Proceedings of Powercon 8*, April 1981. Also in *Advances in Switched-Mode Power Conversion*, Irvine: Teslaco, Vol. 1, pp. 251-278, 1981.

[4]   H. W. OTT, *Noise Reduction Techniques in Electronic Systems*, 2nd edit., New York: John Wiley & Sons, 1988, Chapter 3.

PROBLEMS

8.1   Express the gains represented by the asymptotes of Figs. 8.62(a) to (c) in factored pole-zero form. You may assume that all poles and zeroes have negative real parts.

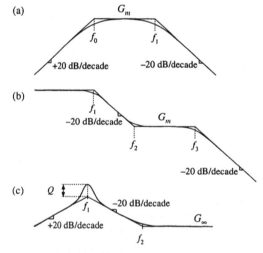

Fig. 8.62   Gain asymptotes for Problem 8.1.

8.2   Express the gains represented by the asymptotes of Figs. 8.63(a) to (c) in factored pole-zero form. You may assume that all poles and zeroes have negative real parts.

8.3   Derive analytical expressions for the low-frequency asymptotes of the magnitude Bode plots shown in Fig. 8.63(a) to (c).

8.4   Derive analytical expressions for the three magnitude asymptotes of Fig. 8.16.

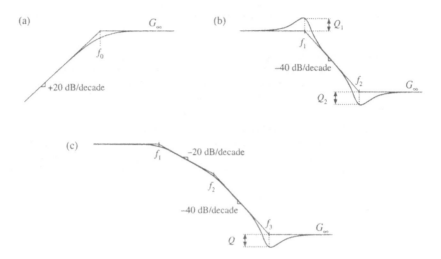

Fig. 8.63 Gain asymptotes for Problems 8.2 and 8.3.

8.5 An experimentally measured transfer function. Figure 8.64 contains experimentally measured magnitude and phase data for the gain function $A(s)$ of a certain amplifier. The object of this problem is to find an expression for $A(s)$. Overlay asymptotes as appropriate on the magnitude and phase data, and hence deduce numerical values for the gain asymptotes and corner frequencies of $A(s)$. Your magnitude and

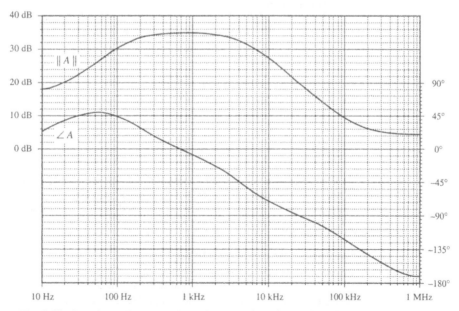

Fig. 8.64 Experimentally-measured magnitude and phase data, Problem 8.5.

phase asymptotes must, of course, follow all of the rules: magnitude slopes must be multiples of ±20 dB per decade, phase slopes for real poles must be multiples of ±45° per decade, etc. The phase and magnitude asymptotes must be consistent with each other.

It is suggested that you start by guessing $A(s)$ based on the magnitude data. Then construct the phase asymptotes for your guess, and compare them with the given data. If there are discrepancies, then modify your guess accordingly and redo your magnitude and phase asymptotes. You should turn in: (1) your analytical expression for $A(s)$, with numerical values given, and (2) a copy of Fig. 8.64, with your magnitude and phase asymptotes superimposed and with all break frequencies and slopes clearly labeled.

8.6   An experimentally-measured impedance. Figure 8.65 contains experimentally measured magnitude and phase data for the driving-point impedance $Z(s)$ of a passive network. The object of this problem is the find an expression for $Z(s)$. Overlay asymptotes as appropriate on the magnitude and phase data, and hence deduce numerical values for the salient features of the impedance function. You should turn in: (1) your analytical expression for $Z(s)$, with numerical values given, and (2) a copy of Fig. 8.65, with your magnitude and phase asymptotes superimposed and with all salient features and asymptote slopes clearly labeled.

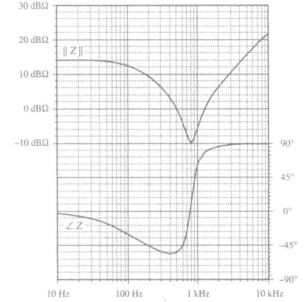

Fig. 8.65 Impedance magnitude and phase data, Problem 8.6.

8.7   In Section 7.2.9, the small-signal ac model of a nonideal flyback converter is derived, with the result illustrated in Fig. 7.27. Construct a Bode plot of the magnitude and phase of the converter output impedance $Z_{out}(s)$. Give both analytical expressions and numerical values for all important features in your plot. Note: $Z_{out}(s)$ includes the load resistance R. The element values are: $D = 0.4$, $n = 0.2$, $R = 6\ \Omega$, $L = 600\ \mu H$, $C = 100\ \mu F$, $R_{on} = 5\ \Omega$.

8.8   For the nonideal flyback converter modeled in Section 7.2.9:

(a)   Derive analytical expressions for the control-to-output and line-to-output transfer functions $G_{vd}(s)$ and $G_{vg}(s)$. Express your results in standard normalized form.

(b)   Derive analytical expressions for the salient features of these transfer functions.

(c)   Construct the magnitude and phase Bode plots of the control-to-output transfer function, using the following values: $n = 2$, $V_g = 48$ V, $D = 0.3$, $R = 5$ Ω, $L = 250$ μH, $C = 100$ μF, $R_{on} = 1.2$ Ω. Label the numerical values of the constant asymptotes, all corner frequencies, the Q-factor, and asymptote slopes.

8.9   Magnitude Bode diagram of an R–L–C filter circuit. For the filter circuit of Fig. 8.66, construct the Bode plots for the magnitudes of the Thevenin-equivalent output impedance Z_{out} and the transfer function $H(s)$ = v_2/v_1. Plot your results on semilog graph paper. Give approximate analytical expressions and numerical values for the important corner frequencies and asymptotes. Do all of the elements significantly affect Z_{out} and H?

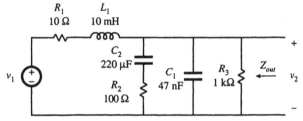

Fig. 8.66   Filter circuit of Problem 8.9.

8.10   Operational amplifier filter circuit. The op amp circuit shown in Fig. 8.67 is a practical realization of what is known as a *PID controller*, and is sometimes used to modify the loop gain of feedback circuits to improve their performance. Using semilog graph paper, sketch the Bode diagram of the magnitude of the transfer function $v_2(s)/v_1(s)$ of the circuit shown. Label all corner frequencies, flat asymptote gains, and asymptote slopes, as appropriate, giving both analytical expressions and numerical values. You may assume that the op amp is ideal.

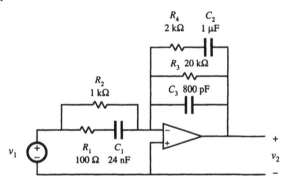

Fig. 8.67   Op-amp PID controller circuit, Problem 8.10.

8.11   Phase asymptotes. Construct the phase asymptotes for the transfer function $v_2(s)/v_1(s)$ of Problem 8.10. Label all break frequencies, flat asymptotes, and asymptote slopes.

8.12   Construct the Bode diagram for the magnitude of the output impedance Z_{out} of the network shown in Fig. 8.68. Give suitable analytical expressions for each asymptote, corner frequency, and Q-factor, as appropriate. Justify any approximations that you use.
The component values are:

$$L_1 = 100 \; \mu H \qquad\qquad L_2 = 16 \; mH$$
$$C_1 = 1000 \; \mu F \qquad\qquad C_2 = 10 \; \mu F$$
$$R_1 = 5 \; \Omega \qquad\qquad\quad R_2 = 50 \; \Omega$$

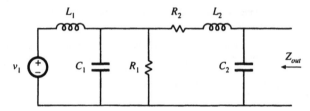

Fig. 8.68   Filter network of Problem 8.12.

8.13   The two section input filter in the circuit of Fig. 8.69 should be designed such that its output impedance $Z_{out}|_{v_g = 0}$ meets certain input filter design criteria, and hence it is desirable to construct the Bode plot for the magnitude of Z_s. Although this filter contains six reactive elements, $\| Z_s \|$ can nonetheless be constructed in a relatively straightforward manner using graphical construction techniques. The element values are:

$$L_1 = 32 \text{ mH} \qquad\qquad C_1 = 32 \text{ }\mu\text{F}$$
$$L_2 = 400 \text{ }\mu\text{H} \qquad\qquad C_2 = 6.8 \text{ }\mu\text{F}$$
$$L_3 = 800 \text{ }\mu\text{H} \qquad\qquad R_1 = 10 \text{ }\Omega$$
$$L_4 = 1 \text{ }\mu\text{H} \qquad\qquad R_2 = 1 \text{ }\Omega$$

(a)   Construct $\| Z_s \|$ using the "algebra on the graph" method. Give simple approximate analytical expressions for all asymptotes and corner frequencies.

(b)   It is desired that $\| Z_s \|$ be approximately equal to 5 Ω at 500 Hz and 2.5 Ω at 1 kHz. Suggest a simple way to accomplish this by changing the value of one component.

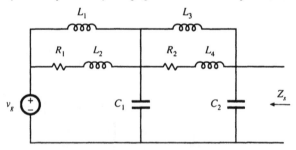

Fig. 8.69   Input filter circuit of Problem 8.13.

8.14   Construct the Bode plot of the magnitude of the output impedance of the filter illustrated in Fig. Fig. 8.70. Give approximate analytical expressions for each corner frequency. No credit will be given for computer-generated plots.

8.15   A certain open-loop buck-boost converter contains an input filter. Its small-signal ac model is shown in Fig. 8.71, and the element values are specified below. Construct the Bode plot for the magnitude of the converter output impedance $\| Z_{out}(s) \|$. Label the values of all important corner frequencies and asymptotes.

$$D = 0.6 \qquad\qquad L_f = 150 \text{ }\mu\text{H}$$
$$R = 6 \text{ }\Omega \qquad\qquad C_f = 16 \text{ }\mu\text{F}$$
$$C = 0.33 \text{ }\mu\text{F} \qquad\qquad C_b = 2200 \text{ }\mu\text{F}$$
$$L = 25 \text{ }\mu\text{H} \qquad\qquad R_f = 1 \text{ }\Omega$$

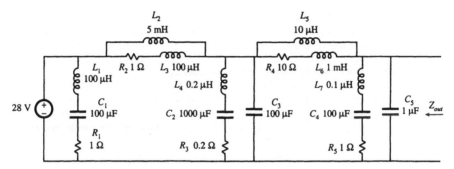

Fig. 8.70   Input filter circuit of Problem 8.14.

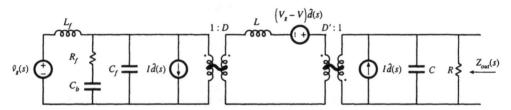

Fig. 8.71   Small-signal model of a buck-boost converter with input filter, Problem 8.15.

8.16   The small-signal equations of the Watkins-Johnson converter operating in continuous conduction mode are:

$$L\frac{d\hat{i}(t)}{dt} = -D\hat{v}(t) + (2V_g - V)\hat{d}(t) + (D - D')\hat{v}_g(t)$$

$$C\frac{d\hat{v}(t)}{dt} = D\hat{i}(t) - \frac{\hat{v}(t)}{R}$$

$$\hat{i}_g(t) = (D - D')\hat{i}(t) + 2I\hat{d}(t)$$

(a)   Derive analytical expressions for the line-to-output transfer function $G_{vg}(s)$ and the control-to-output transfer function $G_{vd}(s)$.

(b)   Derive analytical expressions for the salient features (dc gains, corner frequencies, and Q-factors) of the transfer functions $G_{vg}(s)$ and $G_{vd}(s)$. Express your results as functions of V_g, D, R, L, and C.

(c)   The converter operates at $V_g = 28$ V, $D = 0.25$, $R = 28$ Ω, $C = 100$ μF, $L = 400$ μF. Sketch the Bode diagram of the magnitude and phase of $G_{vd}(s)$. Label salient features.

8.17   The element values in the buck converter of Fig. 7.68 are:

$$V_g = 120 \text{ V} \qquad\qquad D = 0.6$$
$$R = 10 \text{ Ω} \qquad\qquad R_g = 2 \text{ Ω}$$
$$L = 550 \text{ μH} \qquad\qquad C = 100 \text{ μF}$$

(a)   Determine an analytical expression for the control-to-output transfer function $G_{vg}(s)$ of this converter.

(b)   Find analytical expressions for the salient features of $G_{vg}(s)$.

(c)   Construct magnitude and phase asymptotes for G_{vg}. Label the numerical values of all slopes and other important features.

8.18   Loss mechanisms in capacitors, such as dielectric loss and contact and foil resistance, can be modeled electrically using an *equivalent series resistance* (esr). Capacitors whose dielectric materials exhibit a high dielectric constant, such as electrolytic capacitors, tantalum capacitors, and some types of multi-layer ceramic capacitors, typically exhibit relatively high esr.

A buck converter contains a 1.6 mH inductor, and operates with a quiescent duty cycle of 0.5. Its output capacitor can be modeled as a 16 μF capacitor in series with a 0.2 Ω esr. The load resistance is 10 Ω. The converter operates in continuous conduction mode. The quiescent input voltage is $V_g = 120$ V.

(a)   Determine an analytical expression for the control-to-output transfer function $G_{vg}(s)$ of this converter.

(b)   Find analytical expressions for the salient features of $G_{vg}(s)$.

(c)   Construct magnitude and phase asymptotes for G_{vg}. Label the numerical values of all slopes and other important features.

8.19   The *LCC* resonant inverter circuit contains the following transfer function:

$$H(s) = \frac{sC_1R}{1 + sR(C_1 + C_2) + s^2LC_1 + s^3LC_1C_2R}$$

(a)   When C_1 is sufficiently large, this transfer function can be expressed as an inverted pole and a quadratic pole pair. Derive analytical expressions for the corner frequencies and Q-factor in this case, and sketch typical magnitude asymptotes. Determine analytical conditions for validity of your approximation.

(b)   When C_2 is sufficiently large, the transfer function can be also expressed as an inverted pole and a quadratic pole pair. Derive analytical expressions for the corner frequencies and Q-factor in this case, and sketch typical magnitude asymptotes. Determine analytical conditions for validity of your approximation in this case.

(c)   When $C_1 = C_2$ and when the quadratic poles have sufficiently high Q, then the transfer function can again be expressed as an inverted pole and a quadratic pole pair. Derive analytical expressions for the corner frequencies and Q-factor in this case, and sketch typical magnitude asymptotes. Determine analytical conditions for validity of your approximation in this case.

8.20   A two-section *L–C* filter has the following transfer function:

$$G(s) = \frac{1}{1 + s\left(\dfrac{L_1 + L_2}{R}\right) + s^2\left(L_1\left(C_1 + C_2\right) + L_2C_2\right) + s^3\left(\dfrac{L_1L_2C_1}{R}\right) + s^4\left(L_1L_2C_1C_2\right)}$$

The element values are:

$$R = 50 \text{ m}\Omega$$

$$C_1 = 680 \ \mu\text{F} \qquad\qquad C_2 = 4.7 \ \mu\text{F}$$

$$L_1 = 500 \ \mu\text{H} \qquad\qquad L_2 = 50 \ \mu\text{H}$$

(a)   Factor $G(s)$ into approximate real and quadratic poles, as appropriate. Give analytical expressions for the salient features. Justify your approximation using the numerical element values.

(b)   Construct the magnitude and phase asymptotes of $G(s)$.

(c)   It is desired to reduce the Q to 2, without significantly changing the corner frequencies or other features of the response. It is possible to do this by changing only two element values. Specify how to accomplish this.

8.21   The boost converter of Fig. 8.72 operates in the continuous conduction mode, with quiescent duty cycle

$D = 0.6$. On semi-log axes, construct the magnitude and phase Bode plots of

(a)   the control-to-output transfer function $G_{vd}(s)$,

(b)   the line-to-output transfer function $G_{vg}(s)$,

(c)   the output impedance $Z_{out}(s)$, and

(d)   the input impedance $Z_{in}(s)$.

On each plot, label the corner frequencies and asymptotes.

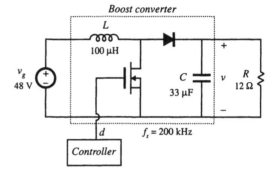

Fig. 8.72 Boost converter of Problem 8.21.

8.22   The forward converter of Fig. 8.73 operates in the continuous conduction mode, with the quiescent values $V_g = 380$ V and $V = 28$ V. The transformer turns ratio is $n_1/n_3 = 4.5$. On semi-log axes, construct the magnitude and phase Bode plots of

(a)   the control-to-output transfer function $G_{vd}(s)$, and

(b)   the line-to-output transfer function $G_{vg}(s)$.

On each plot, label the corner frequencies and asymptotes. *Hint:* other than introduction of the turns ratio n_1/n_3, the transformer does not significantly affect the small-signal behavior of the forward converter.

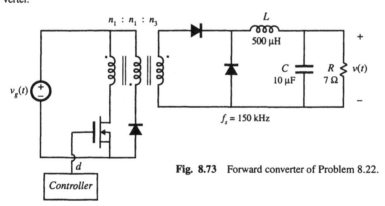

Fig. 8.73 Forward converter of Problem 8.22.

8.23   The boost converter of Fig. 8.74 operates in the continuous conduction mode, with the following quiescent values: $V_g = 120$ V, $V = 300$ V. It is desired to control the converter input current waveform, and hence it is necessary to determine the small-signal transfer function

$$G_{id}(s) = \left. \frac{\hat{i}_g(s)}{\hat{d}(s)} \right|_{\hat{v}_g(s) = 0}$$

(a)   Derive an analytical expression for $G_{id}(s)$. Express all poles and zeroes in normalized standard form, and give analytical expressions for the corner frequencies, Q-factor, and dc gain.

(b)   On semi-log axes, construct the Bode plot for the magnitude and phase of $G_{id}(s)$.

Fig. 8.74 Boost converter of Problem 8.23.

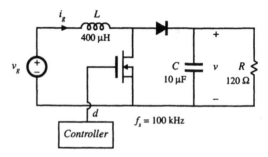

8.24   The buck-boost converter of Fig. 8.75 operates in the continuous conduction mode, with the following quiescent values: $V_g = 48$ V, $V = -24$ V. On semi-log axes, construct the magnitude and phase Bode plots of:

(a)   the control-to-output transfer function $G_{vd}(s)$, and

(b)   the output impedance $Z_{out}(s)$.

On each plot, label the corner frequencies and asymptotes as appropriate.

Fig. 8.75 Buck-boost converter of Problem 8.24.

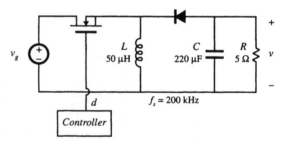

9

Controller Design

9.1 INTRODUCTION

In all switching converters, the output voltage $v(t)$ is a function of the input line voltage $v_g(t)$, the duty cycle $d(t)$, and the load current $i_{load}(t)$, as well as the converter circuit element values. In a dc–dc converter application, it is desired to obtain a constant output voltage $v(t) = V$, in spite of disturbances in $v_g(t)$ and $i_{load}(t)$, and in spite of variations in the converter circuit element values. The sources of these disturbances and variations are many, and a typical situation is illustrated in Fig. 9.1. The input voltage $v_g(t)$ of an off-line power supply may typically contain periodic variations at the second harmonic of the ac power system frequency (100 Hz or 120 Hz), produced by a rectifier circuit. The magnitude of $v_g(t)$ may also vary when neighboring power system loads are switched on or off. The load current $i_{load}(t)$ may contain variations of significant amplitude, and a typical power supply specification is that the output voltage must remain within a specified range (for example, 3.3 V ± 0.05 V) when the load current takes a step change from, for example, full rated load current to 50% of the rated current, and vice versa. The values of the circuit elements are constructed to a certain tolerance, and so in high-volume manufacturing of a converter, converters are constructed whose output voltages lie in some distribution. It is desired that essentially all of this distribution fall within the specified range; however, this is not practical to achieve without the use of negative feedback. Similar considerations apply to inverter applications, except that the output voltage is ac.

So we cannot expect to simply set the dc–dc converter duty cycle to a single value, and obtain a given constant output voltage under all conditions. The idea behind the use of negative feedback is to build a circuit that automatically adjusts the duty cycle as necessary, to obtain the desired output voltage with high accuracy, regardless of disturbances in $v_g(t)$ or $i_{load}(t)$ or variations in component values. This is

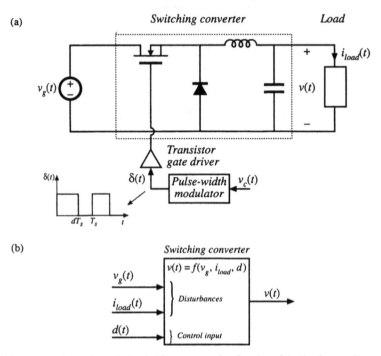

Fig. 9.1 The output voltage of a typical switching converter is a function of the line input voltage v_g, the duty cycle d, and the load current i_{load}: (a) open-loop buck converter; (b) functional diagram illustrating dependence of v on the independent quantities v_g, d, and i_{load}.

a useful thing to do whenever there are variations and unknowns that otherwise prevent the system from attaining the desired performance.

A block diagram of a feedback system is shown in Fig. 9.2. The output voltage $v(t)$ is measured, using a "sensor" with gain $H(s)$. In a dc voltage regulator or dc–ac inverter, the sensor circuit is usually a voltage divider, comprised of precision resistors. The sensor output signal $H(s)v(s)$ is compared with a reference input voltage $v_{ref}(s)$. The objective is to make $H(s)v(s)$ equal to $v_{ref}(s)$, so that $v(s)$ accurately follows $v_{ref}(s)$ regardless of disturbances or component variations in the compensator, pulse-width modulator, gate driver, or converter power stage.

The difference between the reference input $v_{ref}(s)$ and the sensor output $H(s)v(s)$ is called the error signal $v_e(s)$. If the feedback system works perfectly, then $v_{ref}(s) = H(s)v(s)$, and hence the error signal is zero. In practice, the error signal is usually nonzero but nonetheless small. Obtaining a small error is one of the objectives in adding a compensator network $G_c(s)$ as shown in Fig. 9.2. Note that the output voltage $v(s)$ is equal to the error signal $v_e(s)$, multiplied by the gains of the compensator, pulse-width modulator, and converter power stage. If the compensator gain $G_c(s)$ is large enough in magnitude, then a small error signal can produce the required output voltage $v(t) = V$ for a dc regulator (*Q:* how should H and v_{ref} then be chosen?). So a large compensator gain leads to a small error, and therefore the output follows the reference input with good accuracy. This is the key idea behind feedback systems.

The averaged small-signal converter models derived in Chapter 7 are used in the following sections to find the effects of feedback on the small-signal transfer functions of the regulator. The loop gain $T(s)$ is defined as the product of the small-signal gains in the forward and feedback paths of the feedback

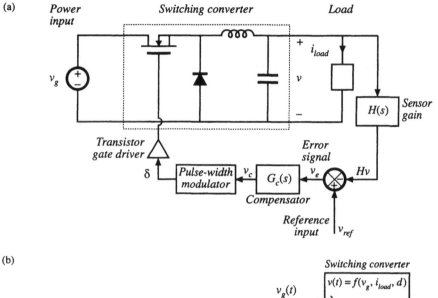

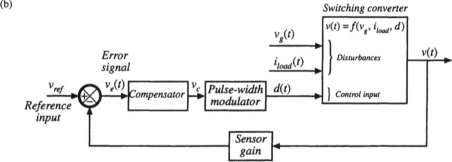

Fig. 9.2 Feedback loop for regulation of the output voltage: (a) buck converter, with feedback loop block diagram; (b) functional block diagram of the feedback system.

loop. It is found that the transfer function from a disturbance to the output is multiplied by the factor $1/(1 + T(s))$. Hence, when the loop gain T is large in magnitude, then the influence of disturbances on the output voltage is small. A large loop gain also causes the output voltage $v(s)$ to be nearly equal to $v_{ref}(s)/H(s)$, with very little dependence on the gains in the forward path of the feedback loop. So the loop gain magnitude $\| T \|$ is a measure of how well the feedback system works. All of these gains can be easily constructed using the algebra-on-the-graph method; this allows easy evaluation of important closed-loop performance measures, such as the output voltage ripple resulting from 120 Hz rectification ripple in $v_g(t)$ or the closed-loop output impedance.

Stability is another important issue in feedback systems. Adding a feedback loop can cause an otherwise well-behaved circuit to exhibit oscillations, ringing and overshoot, and other undesirable behavior. An in-depth treatment of stability is beyond the scope of this book; however, the simple phase margin criterion for assessing stability is used here. When the phase margin of the loop gain T is positive, then the feedback system is stable. Moreover, increasing the phase margin causes the system transient response to be better behaved, with less overshoot and ringing. The relation between phase margin and closed-loop response is quantified in Section 9.4.

An example is given in Section 9.5, in which a compensator network is designed for a dc regulator system. The compensator network is designed to attain adequate phase margin and good rejection of expected disturbances. Lead compensators and P–D controllers are used to improve the phase margin and extend the bandwidth of the feedback loop. This leads to better rejection of high-frequency disturbances. Lag compensators and P–I controllers are used to increase the low-frequency loop gain. This leads to better rejection of low-frequency disturbances and very small steady-state error. More complicated compensators can achieve the advantages of both approaches.

Injection methods for experimental measurement of loop gain are introduced in Section 9.6. The use of voltage or current injection solves the problem of establishing the correct quiescent operating point in high-gain systems. Conditions for obtaining an accurate measurement are exposed. The injection method also allows measurement of the loop gains of unstable systems.

9.2   EFFECT OF NEGATIVE FEEDBACK ON THE NETWORK TRANSFER FUNCTIONS

We have seen how to derive the small-signal ac transfer functions of a switching converter. For example, the equivalent circuit model of the buck converter can be written as in Fig. 9.3. This equivalent circuit contains three independent inputs: the control input variations $\hat{d}$, the power input voltage variations $\hat{v}_g$, and the load current variations $\hat{i}_{load}$. The output voltage variation $\hat{v}$ can therefore be expressed as a linear combination of the three independent inputs, as follows:

$$\hat{v}(s) = G_{vd}(s)\hat{d}(s) + G_{vg}(s)\hat{v}_g(s) - Z_{out}(s)\hat{i}_{load}(s) \qquad (9.1)$$

where

$$G_{vd}(s) = \left. \frac{\hat{v}(s)}{\hat{d}(s)} \right|_{\substack{\hat{v}_g = 0 \\ \hat{i}_{load} = 0}} \qquad \text{converter control-to-output transfer function} \qquad (9.1a)$$

$$G_{vg}(s) = \left. \frac{\hat{v}(s)}{\hat{v}_g(s)} \right|_{\substack{\hat{d} = 0 \\ \hat{i}_{load} = 0}} \qquad \text{converter line-to-output transfer function} \qquad (9.1b)$$

$$Z_{out}(s) = -\left. \frac{\hat{v}(s)}{\hat{i}_{load}(s)} \right|_{\substack{\hat{d} = 0 \\ \hat{v}_g = 0}} \qquad \text{converter output impedance} \qquad (9.1c)$$

The Bode diagrams of these quantities are constructed in Chapter 8. Equation (9.1) describes how distur-

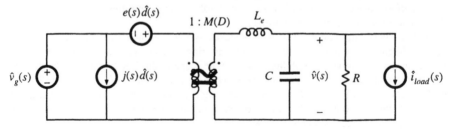

Fig. 9.3   Small-signal converter model, which represents variations in v_g, d, and i_{load}.

bances v_g and i_{load} propagate to the output v, through the transfer function $G_{vg}(s)$ and the output imped-ance $Z_{out}(s)$. If the disturbances v_g and i_{load} are known to have some maximum worst-case amplitude, then Eq. (9.1) can be used to compute the resulting worst-case open-loop variation in v.

As described previously, the feedback loop of Fig. 9.2 can be used to reduce the influences of v_g and i_{load} on the output v. To analyze this system, let us perturb and linearize its averaged signals about their quiescent operating points. Both the power stage and the control block diagram are perturbed and linearized:

$$v_{ref}(t) = V_{ref} + \hat{v}_{ref}(t)$$
$$v_e(t) = V_e + \hat{v}_e(t)$$

(9.2)

etc.

In a dc regulator system, the reference input is constant, so $v_{ref}(t) = 0$. In a switching amplifier or dc–ac inverter, the reference input may contain an ac variation. In Fig. 9.4(a), the converter model of Fig. 9.3 is combined with the perturbed and linearized control circuit block diagram. This is equivalent to the reduced block diagram of Fig. 9.4(b), in which the converter model has been replaced by blocks representing Eq. (9.1).

Solution of Fig. 9.4(b) for the output voltage variation v yields

$$\hat{v} = \hat{v}_{ref} \frac{G_c G_{vd}/V_M}{1 + HG_c G_{vd}/V_M} + \hat{v}_g \frac{G_{vg}}{1 + HG_c G_{vd}/V_M} - \hat{i}_{load} \frac{Z_{out}}{1 + HG_c G_{vd}/V_M}$$

(9.3)

which can be written in the form

$$\hat{v} = \hat{v}_{ref} \frac{1}{H} \frac{T}{1 + T} + \hat{v}_g \frac{G_{vg}}{1 + T} - \hat{i}_{load} \frac{Z_{out}}{1 + T}$$

(9.4)

with

$$T(s) = H(s)G_c(s)G_{vd}(s)/V_M = \text{"loop gain"}$$

Equation (9.4) is a general result. The loop gain $T(s)$ is defined in general as the product of the gains around the forward and feedback paths of the loop. This equation shows how the addition of a feedback loop modifies the transfer functions and performance of the system, as described in detail below.

9.2.1   Feedback Reduces the Transfer Functions from Disturbances to the Output

The transfer function from v_g to v in the open-loop buck converter of Fig. 9.3 is $G_{vg}(s)$, as given in Eq. (9.1). When feedback is added, this transfer function becomes

$$\left. \frac{\hat{v}(s)}{\hat{v}_g(s)} \right|_{\substack{\hat{v}_{ref}=0 \\ i_{load}=0}} = \frac{G_{vg}(s)}{1 + T(s)}$$

(9.5)

from Eq. (9.4). So this transfer function is reduced via feedback by the factor $1/(1 + T(s))$. If the loop gain $T(s)$ is large in magnitude, then the reduction can be substantial. Hence, the output voltage variation v resulting from a given v_g variation is attenuated by the feedback loop.

(a)

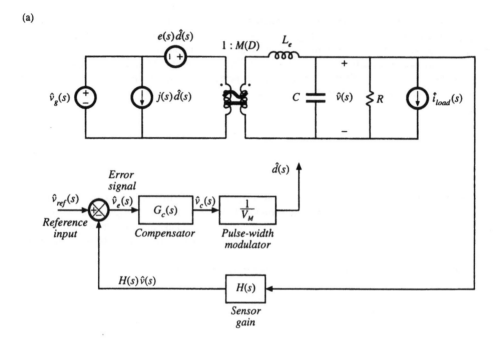

(b)

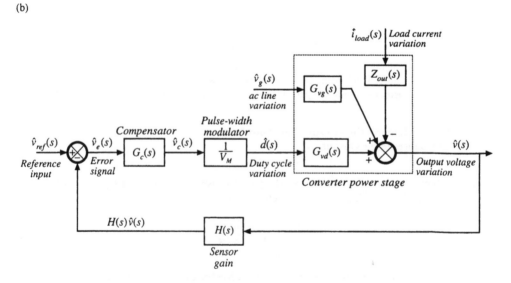

Fig. 9.4 Voltage regulator system small-signal model: (a) with converter equivalent circuit; (b) complete block diagram.

Equation (9.4) also predicts that the converter output impedance is reduced, from $Z_{out}(s)$ to

$$\left.\frac{\hat{v}(s)}{-\hat{i}_{load}(s)}\right|_{\substack{\hat{v}_{ref}=0 \\ \hat{v}_g=0}} = \frac{Z_{out}(s)}{1 + T(s)} \tag{9.6}$$

So the feedback loop also reduces the converter output impedance by a factor of $1/(1 + T(s))$, and the influence of load current variations on the output voltage is reduced.

9.2.2   Feedback Causes the Transfer Function from the Reference Input to the Output to be Insensitive to Variations in the Gains in the Forward Path of the Loop

According to Eq. (9.4), the closed-loop transfer function from v_{ref} to v is

$$\left.\frac{\hat{v}(s)}{\hat{v}_{ref}(s)}\right|_{\substack{\hat{v}_g=0 \\ i_{load}=0}} = \frac{1}{H(s)}\frac{T(s)}{1 + T(s)} \tag{9.7}$$

If the loop gain is large in magnitude, that is, $\| T \| \gg 1$, then $(1 + T) \approx T$ and $T/(1 + T) \approx T/T = 1$. The transfer function then becomes

$$\frac{\hat{v}(s)}{\hat{v}_{ref}(s)} \approx \frac{1}{H(s)} \tag{9.8}$$

which is independent of $G_c(s)$, V_M, and $G_{vd}(s)$. So provided that the loop gain is large in magnitude, then variations in $G_c(s)$, V_M, and $G_{vd}(s)$ have negligible effect on the output voltage. Of course, in the dc regulator application, v_{ref} is constant and $\hat{v}_{ref} = 0$. But Eq. (9.8) applies equally well to the dc values. For example, if the system is linear, then we can write

$$\frac{V}{V_{ref}} = \frac{1}{H(0)}\frac{T(0)}{1 + T(0)} \approx \frac{1}{H(0)} \tag{9.9}$$

So to make the dc output voltage V accurately follow the dc reference V_{ref}, we need only ensure that the dc sensor gain $H(0)$ and dc reference V_{ref} are well-known and accurate, and that $T(0)$ is large. Precision resistors are normally used to realize H, but components with tightly-controlled values need not be used in G_c, the pulse-width modulator, or the power stage. The sensitivity of the output voltage to the gains in the forward path is reduced, while the sensitivity of v to the feedback gain H and the reference input v_{ref} is increased.

9.3   CONSTRUCTION OF THE IMPORTANT QUANTITIES 1/(1 + T) AND T/(1 + T) AND THE CLOSED-LOOP TRANSFER FUNCTIONS

The transfer functions in Eqs. (9.4) to (9.9) can be easily constructed using the algebra-on-the-graph method [4]. Let us assume that we have analyzed the blocks in our feedback system, and have plotted the Bode diagram of $\| T(s) \|$. To use a concrete example, suppose that the result is given in Fig. 9.5, for which $T(s)$ is

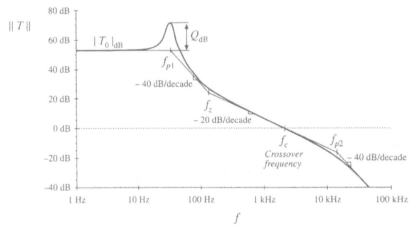

Fig. 9.5 Magnitude of the loop gain example, Eq. (9.10).

$$T(s) = T_0 \frac{\left(1 + \frac{s}{\omega_z}\right)}{\left(1 + \frac{s}{Q\omega_{p1}} + \left(\frac{s}{\omega_{p1}}\right)^2\right)\left(1 + \frac{s}{\omega_{p2}}\right)} \tag{9.10}$$

This example appears somewhat complicated. But the loop gains of practical voltage regulators are often even more complex, and may contain four, five, or more poles. Evaluation of Eqs. (9.5) to (9.7), to determine the closed-loop transfer functions, requires quite a bit of work. The loop gain T must be added to 1, and the resulting numerator and denominator must be refactored. Using this approach, it is difficult to obtain physical insight into the relationship between the closed-loop transfer functions and the loop gain. In consequence, design of the feedback loop to meet specifications is difficult.

Using the algebra-on-the-graph method, the closed-loop transfer functions can be constructed by inspection, and hence the relation between these transfer functions and the loop gain becomes obvious. Let us first investigate how to plot $\| T/(1 + T) \|$. It can be seen from Fig. 9.5 that there is a frequency f_c, called the "crossover frequency," where $\| T \| = 1$. At frequencies less than f_c, $\| T \| > 1$; indeed, $\| T \| \gg 1$ for $f \ll f_c$. Hence, at low frequency, $(1 + T) \approx T$, and $T/(1 + T) \approx T/T = 1$. At frequencies greater than f_c, $\| T \| < 1$, and $\| T \| \ll 1$ for $f \gg f_c$. So at high frequency, $(1 + T) \approx 1$ and $T/(1 + T) \approx T/1 = T$. So we have

$$\frac{T}{1 + T} \approx \begin{cases} 1 & \text{for } |T| \gg 1 \\ T & \text{for } |T| \ll 1 \end{cases} \tag{9.11}$$

The asymptotes corresponding to Eq. (9.11) are relatively easy to construct. The low-frequency asymptote, for $f < f_c$, is 1 or 0 dB. The high-frequency asymptotes, for $f > f_c$, follow T. The result is shown in Fig. 9.6.

So at low frequency, where $\| T \|$ is large, the reference-to-output transfer function is

$$\frac{\hat{v}(s)}{\hat{v}_{ref}(s)} = \frac{1}{H(s)} \frac{T(s)}{1 + T(s)} \approx \frac{1}{H(s)} \tag{9.12}$$

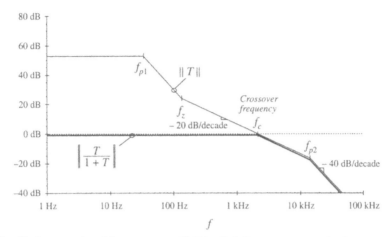

Fig. 9.6   Graphical construction of the asymptotes of $\| T/(1 + T) \|$. Exact curves are omitted.

This is the desired behavior, and the feedback loop works well at frequencies where $\| T \|$ is large. At high frequency $(f \gg f_c)$ where $\| T \|$ is small, the reference-to-output transfer function is

$$\frac{\hat{v}(s)}{\hat{v}_{ref}(s)} = \frac{1}{H(s)} \frac{T(s)}{1 + T(s)} \approx \frac{T(s)}{H(s)} = \frac{G_c(s)G_{vd}(s)}{V_M} \tag{9.13}$$

This is not the desired behavior; in fact, this is the gain with the feedback connection removed $(H \to 0)$. At high frequencies, the feedback loop is unable to reject the disturbance because the bandwidth of T is limited. The reference-to-output transfer function can be constructed on the graph by multiplying the $T/(1 + T)$ asymptotes of Fig. 9.6 by $1/H$.

We can plot the asymptotes of $\| 1/(1 + T) \|$ using similar arguments. At low frequencies where $\| T \| \gg 1$, then $(1 + T) \approx T$, and hence $1/(1 + T) \approx 1/T$. At high frequencies where $\| T \| \ll 1$, then $(1 + T) \approx 1$ and $1/(1 + T) \approx 1$. So we have

$$\frac{1}{1 + T(s)} \approx \begin{cases} \dfrac{1}{T(s)} & \text{for } | T | \gg 1 \\ 1 & \text{for } | T | \ll 1 \end{cases} \tag{9.14}$$

The asymptotes for the $T(s)$ example of Fig. 9.5 are plotted in Fig. 9.7.

At low frequencies where $\| T \|$ is large, the disturbance transfer function from v_g to v is

$$\frac{\hat{v}(s)}{\hat{v}_g(s)} = \frac{G_{vg}(s)}{1 + T(s)} \approx \frac{G_{vg}(s)}{T(s)} \tag{9.15}$$

Again, $G_{vg}(s)$ is the original transfer function, with no feedback. The closed-loop transfer function has magnitude reduced by the factor $1/\| T \|$. So if, for example, we want to reduce this transfer function by a factor of 20 at 120 Hz, then we need a loop gain $\| T \|$ of at least $20 \Rightarrow 26$ dB at 120 Hz. The disturbance transfer function from v_g to v can be constructed on the graph, by multiplying the asymptotes of Fig. 9.7 by the asymptotes for $G_{vg}(s)$.

Similar arguments apply to the output impedance. The closed-loop output impedance at low fre-

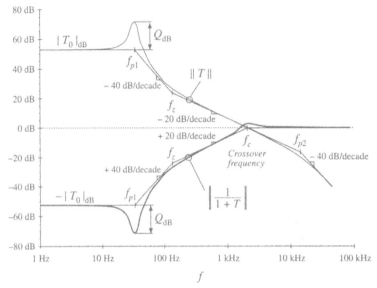

Fig. 9.7   Graphical construction of $\| 1/(1 + T) \|$.

quencies is

$$\frac{\hat{v}(s)}{-\hat{i}_{load}(s)} = \frac{Z_{out}(s)}{1 + T(s)} \approx \frac{Z_{out}(s)}{T(s)} \tag{9.16}$$

The output impedance is also reduced in magnitude by a factor of $1/\| T \|$ at frequencies below the crossover frequency.

At high frequencies $(f > f_c)$ where $\| T \|$ is small, then $1/(1 + T) \approx 1$, and

$$\frac{\hat{v}(s)}{\hat{v}_g(s)} = \frac{G_{vg}(s)}{1 + T(s)} \approx G_{vg}(s)$$

$$\frac{\hat{v}(s)}{-\hat{i}_{load}(s)} = \frac{Z_{out}(s)}{1 + T(s)} \approx Z_{out}(s) \tag{9.17}$$

This is the same as the original disturbance transfer function and output impedance. So the feedback loop has essentially no effect on the disturbance transfer functions at frequencies above the crossover frequency.

9.4   STABILITY

It is well known that adding a feedback loop can cause an otherwise stable system to become unstable. Even though the transfer functions of the original converter, Eq. (9.1), as well as of the loop gain $T(s)$, contain no right half-plane poles, it is possible for the closed-loop transfer functions of Eq. (9.4) to contain right half-plane poles. The feedback loop then fails to regulate the system at the desired quiescent operating point, and oscillations are usually observed. It is important to avoid this situation. And even when the feedback system is stable, it is possible for the transient response to exhibit undesirable ringing and overshoot. The stability problem is discussed in this section, and a method for ensuring that the feedback system is stable and well-behaved is explained.

When feedback destabilizes the system, the denominator $(1 + T(s))$ terms in Eq. (9.4) contain roots in the right half-plane (i.e., with positive real parts). If $T(s)$ is a rational fraction, that is, the ratio $N(s)/D(s)$ of two polynomial functions $N(s)$ and $D(s)$, then we can write

$$\frac{T(s)}{1 + T(s)} = \frac{\dfrac{N(s)}{D(s)}}{1 + \dfrac{N(s)}{D(s)}} = \frac{N(s)}{N(s) + D(s)}$$

$$\frac{1}{1 + T(s)} = \frac{1}{1 + \dfrac{N(s)}{D(s)}} = \frac{D(s)}{N(s) + D(s)}$$

(9.18)

So $T(s)/(1 + T(s))$ and $1/(1+T(s))$ contain the same poles, given by the roots of the polynomial $(N(s) + D(s))$. A brute-force test for stability is to evaluate $(N(s) + D(s))$, and factor the result to see whether any of the roots have positive real parts. However, for all but very simple loop gains, this involves a great deal of work. A simpler method is given by the Nyquist stability theorem, in which the number of right half-plane roots of $(N(s) + D(s))$ can be determined by testing $T(s)$ [1,2]. This theorem is not discussed here. However, a special case of the theorem known as the phase margin test is sufficient for designing most voltage regulators, and is discussed in this section.

9.4.1 The Phase Margin Test

The crossover frequency f_c is defined as the frequency where the magnitude of the loop gain is unity:

$$\left| T\left(j2\pi f_c\right) \right| = 1 \Rightarrow 0 \text{ dB}$$

(9.19)

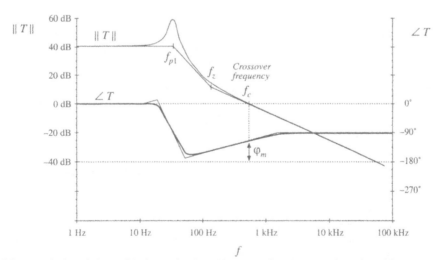

Fig. 9.8 Magnitude and phase of the loop gain of a stable system. The phase margin φ_m is positive.

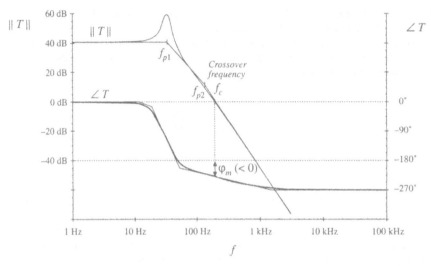

Fig. 9.9   Magnitude and phase of the loop gain of an unstable system. The phase margin φ_m is negative.

To compute the phase margin φ_m, the phase of the loop gain T is evaluated at the crossover frequency, and 180° is added. Hence,

$$\varphi_m = 180° + \angle\, T\!\left(j2\pi f_c\right) \tag{9.20}$$

If there is exactly one crossover frequency, and if the loop gain $T(s)$ contains no right half-plane poles, then the quantities $1/(1 + T)$ and $T/(1 + T)$ contain no right half-plane poles when the phase margin defined in Eq. (9.20) is positive. Thus, using a simple test on $T(s)$, we can determine the stability of $T/(1 + T)$ and $1/(1 + T)$. This is an easy-to-use design tool—we simple ensure that the phase of T is greater than –180° at the crossover frequency.

When there are multiple crossover frequencies, the phase margin test may be ambiguous. Also, when T contains right half-plane poles (i.e., the original open-loop system is unstable), then the phase margin test cannot be used. In either case, the more general Nyquist stability theorem must be employed.

The loop gain of a typical stable system is shown in Fig. 9.8. It can be seen that $\angle T(j2\pi f_c) = -112°$. Hence, $\varphi_m = 180° - 112° = +68°$. Since the phase margin is positive, $T/(1 + T)$ and $1/(1 + T)$ contain no right half-plane poles, and the feedback system is stable.

The loop gain of an unstable system is sketched in Fig. 9.9. For this example, $\angle T(j2\pi f_c) = -230°$. The phase margin is $\varphi_m = 180° - 230° = -50°$. The negative phase margin implies that $T/(1 + T)$ and $1/(1 + T)$ each contain at least one right half-plane pole.

9.4.2   The Relationship Between Phase Margin and Closed-Loop Damping Factor

How much phase margin is necessary? Is a worst-case phase margin of 1° satisfactory? Of course, good designs should have adequate design margins, but there is another important reason why additional phase margin is needed. A small phase margin (in T) causes the closed-loop transfer functions $T/(1 + T)$ and $1/(1 + T)$ to exhibit resonant poles with high Q in the vicinity of the crossover frequency. The system transient response exhibits overshoot and ringing. As the phase margin is reduced these characteristics

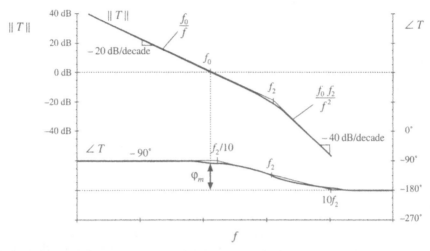

Fig. 9.10 Magnitude and phase asymptotes for the loop gain T of Eq. (9.21).

become worse (higher Q, longer ringing) until, for $\varphi_m \leq 0°$, the system becomes unstable.

Let us consider a loop gain $T(s)$ which is well-approximated, in the vicinity of the crossover frequency, by the following function:

$$T(s) = \frac{1}{\left(\frac{s}{\omega_0}\right)\left(1 + \frac{s}{\omega_2}\right)} \tag{9.21}$$

Magnitude and phase asymptotes are plotted in Fig. 9.10. This function is a good approximation near the crossover frequency for many common loop gains, in which $\| T \|$ approaches unity gain with a -20 dB/decade slope, with an additional pole at frequency $f_2 = \omega_2/2\pi$. Any additional poles and zeroes are assumed to be sufficiently far above or below the crossover frequency, such that they have negligible effect on the system transfer functions near the crossover frequency.

Note that, as $f_2 \to \infty$, the phase margin φ_m approaches 90°. As $f_2 \to 0$, $\varphi_m \to 0°$. So as f_2 is reduced, the phase margin is also reduced. Let's investigate how this affects the closed-loop response via $T/(1 + T)$. We can write

$$\frac{T(s)}{1 + T(s)} = \frac{1}{1 + \frac{1}{T(s)}} = \frac{1}{1 + \frac{s}{\omega_0} + \frac{s^2}{\omega_0\omega_2}} \tag{9.22}$$

using Eq. (9.21). By putting this into the standard normalized quadratic form, one obtains

$$\frac{T(s)}{1 + T(s)} = \frac{1}{1 + \frac{s}{Q\omega_c} + \left(\frac{s}{\omega_c}\right)^2} \tag{9.23}$$

where

$$\omega_c = \sqrt{\omega_0\omega_2} = 2\pi f_c$$

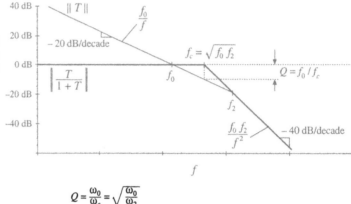

Fig. 9.11 Construction of magnitude asymptotes of the closed-loop transfer function $T/(1 + T)$, for the low-Q case.

$$Q = \frac{\omega_0}{\omega_c} = \sqrt{\frac{\omega_0}{\omega_2}}$$

So the closed-loop response contains quadratic poles at f_c, the geometric mean of f_0 and f_2. These poles have a low Q-factor when $f_0 \ll f_2$. In this case, we can use the low-Q approximation to estimate their frequencies:

$$\begin{aligned} Q\omega_c &= \omega_0 \\ \frac{\omega_c}{Q} &= \omega_2 \end{aligned}$$

(9.24)

Magnitude asymptotes are plotted in Fig. 9.11 for this case. It can be seen that these asymptotes conform to the rules of Section 9.3 for constructing $T/(1 + T)$ by the algebra-on-the-graph method.

　　Next consider the high-Q case. When the pole frequency f_2 is reduced, reducing the phase margin, then the Q-factor given by Eq. (9.23) is increased. For $Q > 0.5$, resonant poles occur at frequency f_c. The magnitude Bode plot for the case $f_2 < f_0$ is given in Fig. 9.12. The frequency f_c continues to be the geometric mean of f_2 and f_0, and f_c now coincides with the crossover (unity-gain) frequency of the $\| T \|$ asymptotes. The exact value of the closed-loop gain $T/(1 + T)$ at frequency f_c is equal to $Q = f_0/f_c$. As shown in Fig. 9.12, this is identical to the value of the low-frequency –20 dB/decade asymptote (f_0/f), evaluated at frequency f_c. It can be seen that the Q-factor becomes very large as the pole frequency f_2 is reduced.

　　The asymptotes of Fig. 9.12 also follow the algebra-on-the-graph rules of Section 9.3, but the deviation of the exact curve from the asymptotes is not predicted by the algebra-on-the-graph method.

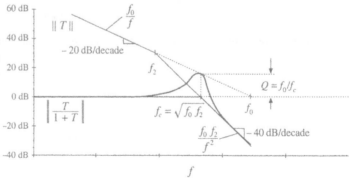

Fig. 9.12 Construction of magnitude asymptotes of the closed-loop transfer function $T/(1 + T)$, for the high-Q case.

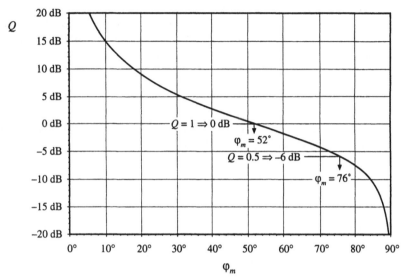

Fig. 9.13 Relationship between loop-gain phase margin φ_m and closed-loop peaking factor Q.

These two poles with Q-factor appear in both $T/(1 + T)$ and $1/(1 + T)$. We need an easy way to predict the Q-factor. We can obtain such a relationship by finding the frequency at which the magnitude of T is exactly equal to unity. We then evaluate the exact phase of T at this frequency, and compute the phase margin. This phase margin is a function of the ratio f_0/f_2, or Q^2. We can then solve to find Q as a function of the phase margin. The result is

$$Q = \frac{\sqrt{\cos \varphi_m}}{\sin \varphi_m}$$

$$\varphi_m = \tan^{-1}\sqrt{\frac{1 + \sqrt{1 + 4Q^4}}{2Q^4}}$$

(9.25)

This function is plotted in Fig. 9.13, with Q expressed in dB. It can be seen that obtaining real poles ($Q < 0.5$) requires a phase margin of at least 76°. To obtain $Q = 1$, a phase margin of 52° is needed. The system with a phase margin of 1° exhibits a closed-loop response with very high Q! With a small phase margin, $T(j\omega)$ is very nearly equal to –1 in the vicinity of the crossover frequency. The denominator $(1 + T)$ then becomes very small, causing the closed-loop transfer functions to exhibit a peaked response at frequencies near the crossover frequency f_c.

    Figure 9.13 is the result for the simple loop gain defined by Eq. (9.21). However, this loop gain is a good approximation for many other loop gains that are encountered in practice, in which $\| T \|$ approaches unity gain with a –20 dB/decade slope, with an additional pole at frequency f_2. If all other poles and zeroes of $T(s)$ are sufficiently far above or below the crossover frequency, then they have negligible effect on the system transfer functions near the crossover frequency, and Fig. 9.13 gives a good approximation for the relationship between φ_m and Q.

    Another common case is the one in which $\| T \|$ approaches unity gain with a –40 dB/decade slope, with an additional zero at frequency f_2. As f_2 is increased, the phase margin is decreased and Q is increased. It can be shown that the relation between φ_m and Q is exactly the same, Eq. (9.25).

    A case where Fig. 9.13 fails is when the loop gain $T(s)$ three or more poles at or near the crossover frequency. The closed-loop response then also contains three or more poles near the crossover frequency, and these poles cannot be completely characterized by a single Q-factor. Additional work is required to find the behavior of the exact $T/(1+T)$ and $1/(1+T)$ near the crossover frequency, but nonetheless it can be said that a small phase margin leads to a peaked closed-loop response.

9.4.3   Transient Response vs. Damping Factor

One can solve for the unit-step response of the $T/(1+T)$ transfer function, by multiplying Eq. (9.23) by $1/s$ and then taking the inverse Laplace transform. The result for $Q > 0.5$ is

$$\hat{v}(t) = 1 + \frac{2Qe^{-\omega_c t/2Q}}{\sqrt{4Q^2-1}}\,\sin\left[\frac{\sqrt{4Q^2-1}}{2Q}\,\omega_c t + \tan^{-1}\!\left(\sqrt{4Q^2-1}\right)\right] \tag{9.26}$$

For $Q < 0.5$, the result is

$$\hat{v}(t) = 1 - \frac{\omega_2}{\omega_2-\omega_1}\,e^{-\omega_1 t} - \frac{\omega_1}{\omega_1-\omega_2}\,e^{-\omega_2 t} \tag{9.27}$$

with

$$\omega_1, \omega_2 = \frac{\omega_c}{2Q}\left(1 \pm \sqrt{1-4Q^2}\right)$$

These equations are plotted in Fig. 9.14 for various values of Q.

According to Eq. (9.23), when $f_2 > 4f_0$, the Q-factor is less than 0.5, and the closed-loop

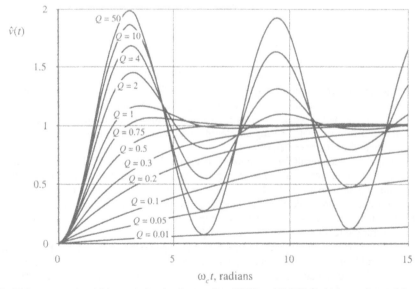

Fig. 9.14 Unit-step response of the second-order system, Eqs. (9.26) and (9.27), for various values of Q.

response contains a low-frequency and a high-frequency real pole. The transient response in this case, Eq. (9.27), contains decaying-exponential functions of time, of the form

$$A e^{(pole)t} \tag{9.28}$$

This is called the "overdamped" case. With very low Q, the low-frequency pole leads to a slow step response.

For $f_2 = 4f_0$, the Q-factor is equal to 0.5. The closed-loop response contains two real poles at frequency $2f_0$. This is called the "critically damped" case. The transient response is faster than in the overdamped case, because the lowest-frequency pole is at a higher frequency. This is the fastest response that does not exhibit overshoot. At $\omega_c t = \pi$ radians ($t = 1/2f_c$), the voltage has reached 82% of its final value. At $\omega_c t = 2\pi$ radians ($t = 1/f_c$), the voltage has reached 98.6% of its final value.

For $f_2 < 4f_0$, the Q-factor is greater than 0.5. The closed-loop response contains complex poles, and the transient response exhibits sinusoidal-type waveforms with decaying amplitude, Eq. (9.26). The rise time of the step response is faster than in the critically-damped case, but the waveforms exhibit overshoot. The peak value of $v(t)$ is

$$\text{peak } \hat{v}(t) = 1 + e^{-\pi/\sqrt{4Q^2 - 1}} \tag{9.29}$$

This is called the "underdamped" case. A Q-factor of 1 leads to an overshoot of 16.3%, while a Q-factor of 2 leads to a 44.4% overshoot. Large Q-factors lead to overshoots approaching 100%.

The exact transient response of the feedback loop may differ from the plots of Fig. 9.14, because of additional poles and zeroes in T, and because of differences in initial conditions. Nonetheless, Fig. 9.14 illustrates how high-Q poles lead to overshoot and ringing. In most power applications, overshoot is unacceptable. For example, in a 3.3 V computer power supply, the voltage must not be allowed to overshoot to 5 or 6 volts when the supply is turned on—this would likely destroy all of the integrated circuits in the computer! So the Q-factor must be sufficiently low, often 0.5 or less, corresponding to a phase margin of at least 76°.

9.5 REGULATOR DESIGN

Let's now consider how to design a regulator system, to meet specifications or design goals regarding rejection of disturbances, transient response, and stability. Typical dc regulator designs are defined using specifications such as the following:

1. *Effect of load current variations on the output voltage regulation.* The output voltage must remain within a specified range when the load current varies in a prescribed way. This amounts to a limit on the maximum magnitude of the closed-loop output impedance of Eq. (9.6), repeated below

$$\left. \frac{\hat{v}(s)}{-\hat{i}_{load}(s)} \right|_{\substack{\hat{v}_{ref}=0 \\ \hat{v}_g=0}} = \frac{Z_{out}(s)}{1 + T(s)} \tag{9.30}$$

If, over some frequency range, the open-loop output impedance Z_{out} has magnitude that exceeds the limit, then the loop gain T must be sufficiently large in magnitude over the same frequency range, such that the magnitude of the closed-loop output impedance given in Eq. (9.30) is less than the given limit.

2. *Effect of input voltage variations (for example, at the second harmonic of the ac line frequency) on the output voltage regulation.* Specific maximum limits are usually placed on the amplitude of variations in the

output voltage at the second harmonic of the ac line frequency (120 Hz or 100 Hz). If we know the magnitude of the rectification voltage ripple which appears at the converter input (as $\hat{v}_g$), then we can calculate the resulting output voltage ripple (in $\hat{v}$) using the closed loop line-to-output transfer function of Eq. (9.5), repeated below

$$\left. \frac{\hat{v}(s)}{\hat{v}_g(s)} \right|_{\substack{\hat{v}_{ref}=0 \\ i_{load}=0}} = \frac{G_{vg}(s)}{1+T(s)} \tag{9.31}$$

The output voltage ripple can be reduced by increasing the magnitude of the loop gain at the ripple frequency. In a typical good design, $\| T \|$ is 20 dB or more at 120 Hz, so that the transfer function of Eq. (9.31) is at least an order of magnitude smaller than the open-loop line-to-output transfer function $\| G_{vg} \|$.

3. *Transient response time.* When a specified large disturbance occurs, such as a large step change in load current or input voltage, the output voltage may undergo a transient. During this transient, the output voltage typically deviates from its specified allowable range. Eventually, the feedback loop operates to return the output voltage within tolerance. The time required to do so is the transient response time; typically, the response time can be shortened by increasing the feedback loop crossover frequency.

4. *Overshoot and ringing.* As discussed in Section 9.4.3, the amount of overshoot and ringing allowed in the transient response may be limited. Such a specification implies that the phase margin must be sufficiently large.

Each of these requirements imposes constraints on the loop gain $T(s)$. Therefore, the design of the control system involves modifying the loop gain. As illustrated in Fig. 9.2, a compensator network is added for this purpose. Several well-known strategies for design of the compensator transfer function $G_c(s)$ are discussed below.

9.5.1   Lead (PD) compensator

This type of compensator transfer function is used to improve the phase margin. A zero is added to the loop gain, at a frequency f_z sufficiently far below the crossover frequency f_c, such that the phase margin of $T(s)$ is increased by the desired amount. The lead compensator is also called a *proportional-plus-derivative*, or *PD*, controller—at high frequencies, the zero causes the compensator to differentiate the error signal. It often finds application in systems originally containing a two-pole response. By use of this type of compensator, the bandwidth of the feedback loop (i.e., the crossover frequency f_c) can be

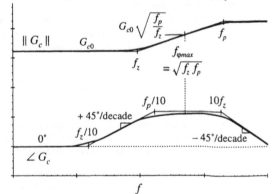

Fig. 9.15 Magnitude and phase asymptotes of the *PD* compensator transfer function G_c of Eq. (9.32).

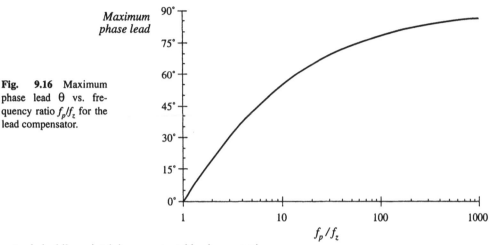

Fig. 9.16 Maximum phase lead θ vs. frequency ratio f_p/f_z for the lead compensator.

extended while maintaining an acceptable phase margin.

A side effect of the zero is that it causes the compensator gain to increase with frequency, with a +20 dB/decade slope. So steps must be taken to ensure that $\| T \|$ remains equal to unity at the desired crossover frequency. Also, since the gain of any practical amplifier must tend to zero at high frequency, the compensator transfer function $G_c(s)$ must contain high frequency poles. These poles also have the beneficial effect of attenuating high-frequency noise. Of particular concern are the switching frequency harmonics present in the output voltage and feedback signals. If the compensator gain at the switching frequency is too great, then these switching harmonics are amplified by the compensator, and can disrupt the operation of the pulse-width modulator (see Section 7.6). So the compensator network should contain poles at a frequency less than the switching frequency. These considerations typically restrict the crossover frequency f_c to be less than approximately 10% of the converter switching frequency f_s. In addition, the circuit designer must take care not to exceed the gain-bandwidth limits of available operational amplifiers.

The transfer function of the lead compensator therefore contains a low-frequency zero and several high-frequency poles. A simplified example containing a single high-frequency pole is given in Eq. (9.32) and illustrated in Fig. 9.15.

$$G_c(s) = G_{c0} \frac{\left(1 + \frac{s}{\omega_z}\right)}{\left(1 + \frac{s}{\omega_p}\right)} \tag{9.32}$$

The maximum phase occurs at a frequency $f_{\varphi max}$ given by the geometrical mean of the pole and zero frequencies:

$$f_{\varphi max} = \sqrt{f_z f_p} \tag{9.33}$$

To obtain the maximum improvement in phase margin, we should design our compensator so that the frequency $f_{\varphi max}$ coincides with the loop gain crossover frequency f_c. The value of the phase at this frequency can be shown to be

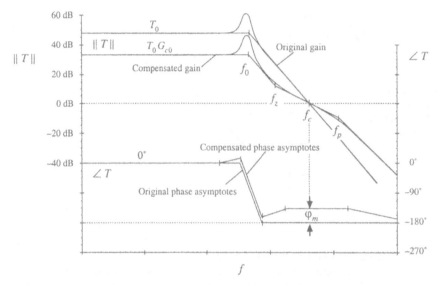

Fig. 9.17 Compensation of a loop gain containing two poles, using a lead (*PD*) compensator. The phase margin φ_m is improved.

$$\angle G_c\!\left(f_{\varphi max}\right) = \tan^{-1}\!\left(\frac{\sqrt{\dfrac{f_p}{f_z}} - \sqrt{\dfrac{f_z}{f_p}}}{2}\right) \tag{9.34}$$

This equation is plotted in Fig. 9.16. Equation (9.34) can be inverted to obtain

$$\frac{f_p}{f_z} = \frac{1 + \sin\left(\theta\right)}{1 - \sin\left(\theta\right)} \tag{9.35}$$

where $\theta = \angle G_c(f_{\varphi max})$. Equations (9.34) and (9.32) imply that, to optimally obtain a compensator phase lead of θ at frequency f_c, the pole and zero frequencies should be chosen as follows:

$$f_z = f_c \sqrt{\frac{1 - \sin\left(\theta\right)}{1 + \sin\left(\theta\right)}}$$
$$f_p = f_c \sqrt{\frac{1 + \sin\left(\theta\right)}{1 - \sin\left(\theta\right)}} \tag{9.36}$$

When it is desired to avoid changing the crossover frequency, the magnitude of the compensator gain is chosen to be unity at the loop gain crossover frequency f_c. This requires that G_{c0} be chosen according to the following formula:

$$G_{c0} = \sqrt{\frac{f_z}{f_p}} \tag{9.37}$$

It can be seen that G_{c0} is less than unity, and therefore the lead compensator reduces the dc gain of the feedback loop. Other choices of G_{c0} can be selected when it is desired to shift the crossover frequency f_c; for example, increasing the value of G_{c0} causes the crossover frequency to increase. If the frequencies f_p and f_z are chosen as in Eq. (9.36), then $f_{\varphi max}$ of Eq. (9.32) will coincide with the new crossover frequency f_c.

The Bode diagram of a typical loop gain $T(s)$ containing two poles is illustrated in Fig. 9.17. The phase margin of the original $T(s)$ is small, since the crossover frequency f_c is substantially greater than the pole frequency f_0. The result of adding a lead compensator is also illustrated. The lead compensator of this example is designed to maintain the same crossover frequency but improve the phase margin.

9.5.2 Lag (*PI*) Compensator

This type of compensator is used to increase the low-frequency loop gain, such that the output is better regulated at dc and at frequencies well below the loop crossover frequency. As given in Eq. (9.38) and illustrated in Fig. 9.18, an inverted zero is added to the loop gain, at frequency f_L.

$$G_c(s) = G_{c\infty}\left(1 + \frac{\omega_L}{s}\right) \tag{9.38}$$

If f_L is sufficiently lower than the loop crossover frequency f_c, then the phase margin is unchanged. This type of compensator is also called a *proportional-plus-integral*, or *PI*, controller—at low frequencies, the inverted zero causes the compensator to integrate the error signal.

To the extent that the compensator gain can be made arbitrarily large at dc, the dc loop gain $T(0)$ becomes arbitrarily large. This causes the dc component of the error signal to approach zero. In consequence, the steady-state output voltage is perfectly regulated, and the disturbance-to-output transfer functions approach zero at dc. Such behavior is easily obtained in practice, with the compensator of Eq. (9.38) realized using a conventional operational amplifier.

Although the *PI* compensator is useful in nearly all types of feedback systems, it is an especially simple and effective approach for systems originally containing a single pole. For the example of Fig. 9.19, the original uncompensated loop gain is of the form

Fig. 9.18 Magnitude and phase asymptotes of the *PI* compensator transfer function G_c of Eq. (9.38).

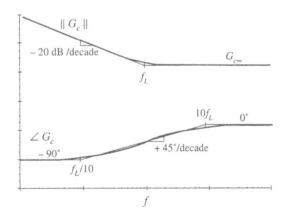

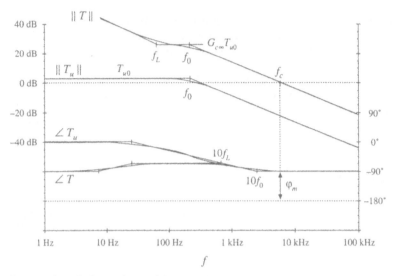

Fig. 9.19 Compensation of a loop gain containing a single pole, using a lag (*PI*) compensator. The loop gain magnitude is increased.

$$T_u(s) = \frac{T_{u0}}{\left(1 + \frac{s}{\omega_0}\right)} \tag{9.39}$$

The compensator transfer function of Eq. (9.38) is used, so that the compensated loop gain is $T(s) = T_u(s)G_c(s)$. Magnitude and phase asymptotes of $T(s)$ are also illustrated in Fig. 9.19. The compensator high-frequency gain $G_{c\infty}$ is chosen to obtain the desired crossover frequency f_c. If we approximate the compensated loop gain by its high-frequency asymptote, then at high frequencies we can write

$$|T| \approx \frac{T_{u0}G_{c\infty}}{\left(\frac{f}{f_0}\right)} \tag{9.40}$$

At the crossover frequency $f = f_c$, the loop gain has unity magnitude. Equation (9.40) predicts that the crossover frequency is

$$f_c \approx T_{u0}G_{c\infty}f_0 \tag{9.41}$$

Hence, to obtain a desired crossover frequency f_c, we should choose the compensator gain $G_{c\infty}$ as follows:

$$G_{c\infty} = \frac{f_c}{T_{u0}f_0} \tag{9.42}$$

The corner frequency f_L is then chosen to be sufficiently less than f_c, such that an adequate phase margin is maintained.

Magnitude asymptotes of the quantity $1/(1 + T(s))$ are constructed in Fig. 9.20. At frequencies

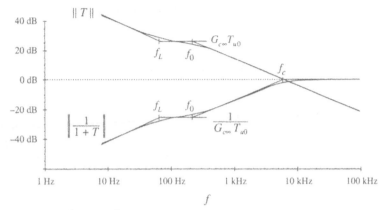

Fig. 9.20 Construction of $\| 1/(1 + T) \|$ for the *PI*-compensated example of Fig. 9.19.

less than f_L, the *PI* compensator improves the rejection of disturbances. At dc, where the magnitude of G_c approaches infinity, the magnitude of $1/(1 + T)$ tends to zero. Hence, the closed-loop disturbance-to-output transfer functions, such as Eqs. (9.30) and (9.31), tend to zero at dc.

9.5.3 Combined (*PID*) Compensator

The advantages of the lead and lag compensators can be combined, to obtain both wide bandwidth and zero steady-state error. At low frequencies, the compensator integrates the error signal, leading to large low-frequency loop gain and accurate regulation of the low-frequency components of the output voltage. At high frequency (in the vicinity of the crossover frequency), the compensator introduces phase lead into the loop gain, improving the phase margin. Such a compensator is sometimes called a *PID* controller.

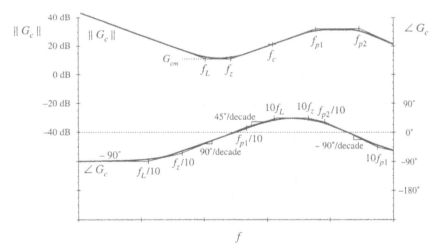

Fig. 9.21 Magnitude and phase asymptotes of the combined (*PID*) compensator transfer function G_c of Eq. (9.43).

A typical Bode diagram of a practical version of this compensator is illustrated in Fig. 9.21. The compensator has transfer function

$$G_c(s) = G_{cm} \frac{\left(1 + \frac{\omega_L}{s}\right)\left(1 + \frac{s}{\omega_z}\right)}{\left(1 + \frac{s}{\omega_{p1}}\right)\left(1 + \frac{s}{\omega_{p2}}\right)} \tag{9.43}$$

The inverted zero at frequency f_L functions in the same manner as the *PI* compensator. The zero at frequency f_z adds phase lead in the vicinity of the crossover frequency, as in the *PD* compensator. The high-frequency poles at frequencies f_{p1} and f_{p2} must be present in practical compensators, to cause the gain to roll off at high frequencies and to prevent the switching ripple from disrupting the operation of the pulse-width modulator. The loop gain crossover frequency f_c is chosen to be greater than f_L and f_z, but less than f_{p1} and f_{p2}.

9.5.4   Design Example

To illustrate the design of *PI* and *PD* compensators, let us consider the design of a combined *PID* compensator for the dc–dc buck converter system of Fig. 9.22. The input voltage $v_g(t)$ for this system has nominal value 28 V. It is desired to supply a regulated 15 V to a 5 A load. The load is modeled here with a 3 Ω resistor. An accurate 5 V reference is available.

The first step is to select the feedback gain $H(s)$. The gain H is chosen such that the regulator produces a regulated 15 V dc output. Let us assume that we will succeed in designing a good feedback system, which causes the output voltage to accurately follow the reference voltage. This is accomplished via a large loop gain T, which leads to a small error voltage: $v_e \approx 0$. Hence, $Hv \approx v_{ref}$. So we should choose

$$H = \frac{V_{ref}}{V} = \frac{5}{15} = \frac{1}{3} \tag{9.44}$$

The quiescent duty cycle is given by the steady-state solution of the converter:

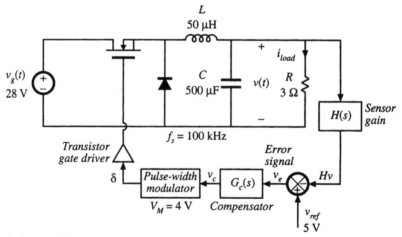

Fig. 9.22   Design example.

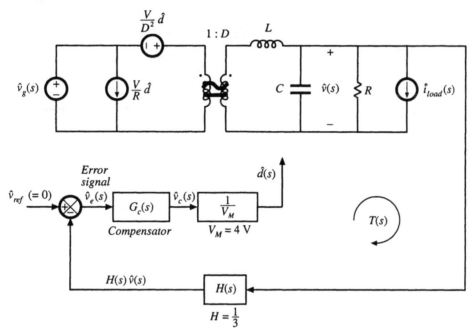

Fig. 9.23 System small-signal ac model, design example.

$$D = \frac{V}{V_g} = \frac{15}{28} = 0.536 \tag{9.45}$$

The quiescent value of the control voltage, V_c, must satisfy Eq. (7.173). Hence,

$$V_c = DV_M = 2.14 \text{ V} \tag{9.46}$$

Thus, the quiescent conditions of the system are known. It remains to design the compensator gain $G_c(s)$.

A small-signal ac model of the regulator system is illustrated in Fig. 9.23. The buck converter ac model is represented in canonical form. Disturbances in the input voltage and in the load current are modeled. For generality, reference voltage variations $\hat{v}_{ref}$ are included in the diagram; in a dc voltage regulator, these variations are normally zero.

The open-loop converter transfer functions are discussed in the previous chapters. The open-loop control-to-output transfer function is

$$G_{vd}(s) = \frac{V}{D} \frac{1}{1 + s\frac{L}{R} + s^2 LC} \tag{9.47}$$

The open-loop control-to-output transfer function contains two poles, and can be written in the following normalized form:

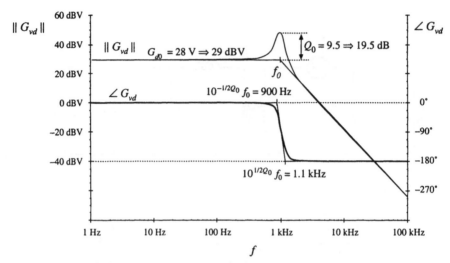

Fig. 9.24   Converter small-signal control-to-output transfer function G_{vd}, design example.

$$G_{vd}(s) = G_{d0} \frac{1}{1 + \dfrac{s}{Q_0\omega_0} + \left(\dfrac{s}{\omega_0}\right)^2} \tag{9.48}$$

By equating like coefficients in Eqs. (9.47) and (9.48), one finds that the dc gain, corner frequency, and Q-factor are given by

$$\begin{aligned}
G_{d0} &= \frac{V}{D} = 28 \text{ V} \\
f_0 &= \frac{\omega_0}{2\pi} = \frac{1}{2\pi\sqrt{LC}} = 1 \text{ kHz} \\
Q_0 &= R\sqrt{\frac{C}{L}} = 9.5 \Rightarrow 19.5 \text{ dB}
\end{aligned} \tag{9.49}$$

In practice, parasitic loss elements, such as the capacitor equivalent series resistance (*esr*), would cause a lower Q-factor to be observed. Figure 9.24 contains a Bode diagram of $G_{vd}(s)$.

The open-loop line-to-output transfer function is

$$G_{vg}(s) = D \frac{1}{1 + s\dfrac{L}{R} + s^2 LC} \tag{9.50}$$

This transfer function contains the same poles as in $G_{vd}(s)$, and can be written in the normalized form

$$G_{vg}(s) = G_{g0} \frac{1}{1 + \dfrac{s}{Q_0\omega_0} + \left(\dfrac{s}{\omega_0}\right)^2} \tag{9.51}$$

with $G_{g0} = D$. The open-loop output impedance of the buck converter is

$$Z_{out}(s) = R \parallel \frac{1}{sC} \parallel sL = \frac{sL}{1 + s\dfrac{L}{R} + s^2 LC} \tag{9.52}$$

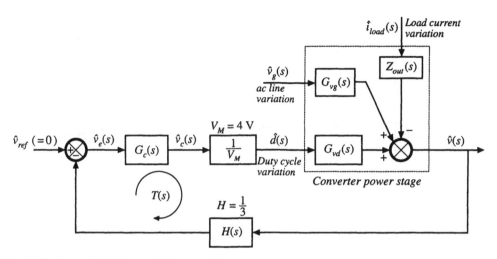

Fig. 9.25 System block diagram, design example.

Use of these equations to represent the converter in block-diagram form leads to the complete system block diagram of Fig. 9.25. The loop gain of the system is

$$T(s) = G_c(s) \left(\frac{1}{V_M}\right) G_{vd}(s) H(s) \tag{9.53}$$

Substitution of Eq. (9.48) into (9.53) leads to

$$T(s) = \frac{G_c(s)H(s)}{V_M} \frac{V}{D} \frac{1}{\left(1 + \frac{s}{Q_0 \omega_0} + \left(\frac{s}{\omega_0}\right)^2\right)} \tag{9.54}$$

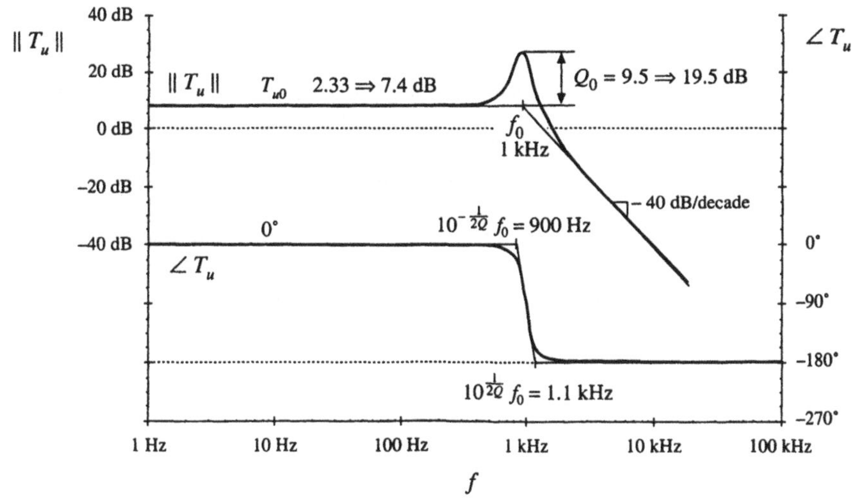

Fig. 9.26 Uncompensated loop gain T_u, design example.

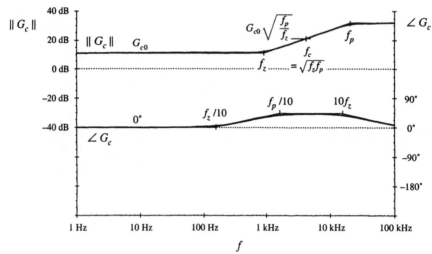

Fig. 9.27   *PD* compensator transfer function G_c, design example.

The closed-loop disturbance-to-output transfer functions are given by Eqs. (9.5) and (9.6).

The uncompensated loop gain $T_u(s)$, with unity compensator gain, is sketched in Fig. 9.26. With $G_c(s) = 1$, Eq. (9.54) can be written

$$T_u(s) = T_{u0} \frac{1}{1 + \frac{s}{Q_0\omega_0} + \left(\frac{s}{\omega_0}\right)^2} \tag{9.55}$$

where the dc gain is

$$T_{u0} = \frac{HV}{DV_M} = 2.33 \Rightarrow 7.4 \text{ dB} \tag{9.56}$$

The uncompensated loop gain has a crossover frequency of approximately 1.8 kHz, with a phase margin of less than five degrees.

Let us design a compensator, to attain a crossover frequency of $f_c = 5$ kHz, or one twentieth of the switching frequency. From Fig. 9.26, the uncompensated loop gain has a magnitude at 5 kHz of approximately $T_{u0} (f_0/f_c)^2 = 0.093 \Rightarrow -20.6$ dB. So to obtain unity loop gain at 5 kHz, our compensator should have a 5 kHz gain of +20.6 dB. In addition, the compensator should improve the phase margin, since the phase of the uncompensated loop gain is nearly $-180°$ at 5 kHz. So a lead (*PD*) compensator is needed. Let us (somewhat arbitrarily) choose to design for a phase margin of 52°. According to Fig. 9.13, this choice leads to closed-loop poles having a Q-factor of 1. The unit step response, Fig. 9.14, then exhibits a peak overshoot of 16%. Evaluation of Eq. (9.36), with $f_c = 5$ kHz and $\theta = 52°$, leads to the following compensator pole and zero frequencies:

$$f_z = (5 \text{ kHz}) \sqrt{\frac{1 - \sin(52°)}{1 + \sin(52°)}} = 1.7 \text{ kHz}$$
$$f_p = (5 \text{ kHz}) \sqrt{\frac{1 + \sin(52°)}{1 - \sin(52°)}} = 14.5 \text{ kHz} \tag{9.57}$$

To obtain a compensator gain of 20.6 dB $\Rightarrow$ 10.7 at 5 kHz, the low-frequency compensator gain must be

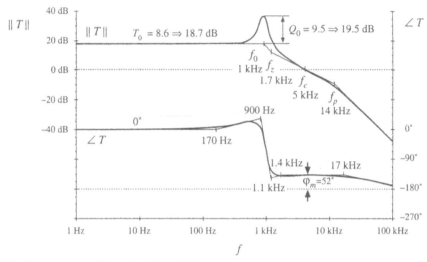

Fig. 9.28 The compensated loop gain of Eq. (9.59).

$$G_{c0} = \left(\frac{f_c}{f_0}\right)^2 \frac{1}{T_{u0}} \sqrt{\frac{f_z}{f_p}} = 3.7 \Rightarrow 11.3 \text{ dB} \qquad (9.58)$$

A Bode diagram of the *PD* compensator magnitude and phase is sketched in Fig. 9.27.

With this *PD* controller, the loop gain becomes

$$T(s) = T_{u0} G_{c0} \frac{\left(1 + \frac{s}{\omega_z}\right)}{\left(1 + \frac{s}{\omega_p}\right)\left(1 + \frac{s}{Q_0 \omega_0} + \left(\frac{s}{\omega_0}\right)^2\right)} \qquad (9.59)$$

The compensated loop gain is sketched in Fig. 9.28. It can be seen that the phase of $T(s)$ is approximately equal to 52° over the frequency range of 1.4 kHz to 17 kHz. Hence variations in component values, which cause the crossover frequency to deviate somewhat from 5 kHz, should have little impact on the phase margin. In addition, it can be seen from Fig. 9.28 that the loop gain has a dc magnitude of $T_{u0} G_{c0}$ ⇒ 18.7 dB.

Asymptotes of the quantity $1/(1 + T)$ are constructed in Fig. 9.29. This quantity has a dc asymptote of –18.7 dB. Therefore, at frequencies less than 1 kHz, the feedback loop attenuates output voltage disturbances by 18.7 dB. For example, suppose that the input voltage $v_g(t)$ contains a 100 Hz variation of amplitude 1 V. With no feedback loop, this disturbance would propagate to the output according to the open-loop transfer function $G_{vg}(s)$, given in Eq. (9.51). At 100 Hz, this transfer function has a gain essentially equal to the dc asymptote $D = 0.536$. Therefore, with no feedback loop, a 100 Hz variation of amplitude 0.536 V would be observed at the output. In the presence of feedback, the closed-loop line-to-output transfer function of Eq. (9.5) is obtained; for our example, this attenuates the 100 Hz variation by an additional factor of 18.7 dB ⇒ 8.6. The 100 Hz output voltage variation now has magnitude 0.536/8.6 = 0.062 V.

The low-frequency regulation can be further improved by addition of an inverted zero, as discussed in Section 9.5.2. A *PID* controller, as in Section 9.5.3, is then obtained. The compensator transfer

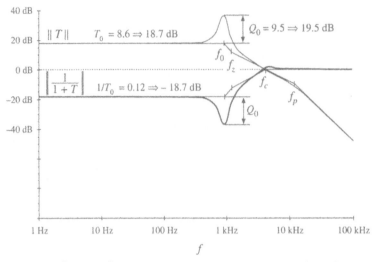

Fig. 9.29   Construction of $\| 1/(1 + T) \|$ for the *PD*-compensated design example of Fig. 9.28.

function becomes

$$G_c(s) = G_{cm} \frac{\left(1 + \frac{s}{\omega_z}\right)\left(1 + \frac{\omega_L}{s}\right)}{\left(1 + \frac{s}{\omega_p}\right)} \tag{9.60}$$

The Bode diagram of this compensator gain is illustrated in Fig. 9.30. The pole and zero frequencies f_z and f_p are unchanged, and are given by Eq. (9.57). The midband gain G_{cm} is chosen to be the same as the previous G_{c0}, Eq. (9.58). Hence, for frequencies greater than f_L, the magnitude of the loop gain is

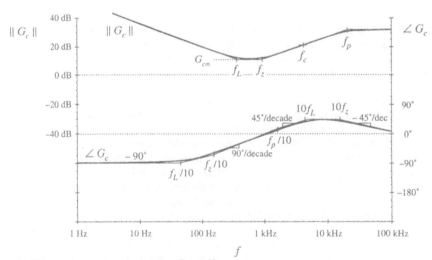

Fig. 9.30   *PID* compensator transfer function, Eq. (9.60).

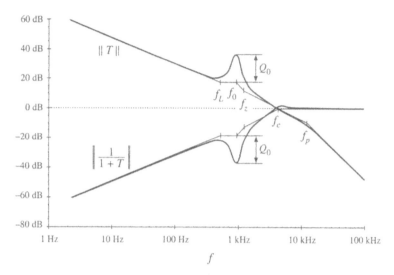

Fig. 9.31 Construction of $\| T \|$ and $\| 1/(1 + T) \|$ with the *PID*-compensator of Fig. 9.30.

unchanged by the inverted zero. The loop continues to exhibit a crossover frequency of 5 kHz.

So that the inverted zero does not significantly degrade the phase margin, let us (somewhat arbitrarily) choose f_L to be one-tenth of the crossover frequency, or 500 Hz. The inverted zero will then increase the loop gain at frequencies below 500 Hz, improving the low-frequency regulation of the output voltage. The loop gain of Fig. 9.31 is obtained. The magnitude of the quantity $1/(1 + T)$ is also constructed. It can be seen that the inverted zero at 500 Hz causes the magnitude of $1/(1 + T)$ at 100 Hz to be reduced by a factor of approximately (100 Hz)/(500 Hz) = 1/5. The total attenuation of $1/(1 + T)$ at 100 Hz is −32.7dB. A 1 V, 100 Hz variation in $v_g(t)$ would now induce a 12 mV variation in $v(t)$. Further improvements could be obtained by increasing f_L; however, this would require redesign of the *PD* portion of the compensator to maintain an adequate phase margin.

The line-to-output transfer function is constructed in Fig. 9.32. Both the open-loop transfer function $G_{vg}(s)$, Eq. (9.51), and the closed-loop transfer function $G_{vg}(s)/(1 + T(s))$, are constructed using the algebra-on-the-graph method. The two transfer functions coincide at frequencies greater than the crossover frequency. At frequencies less than the crossover frequency f_c, the closed-loop transfer function is reduced by a factor of $T(s)$. It can be seen that the poles of $G_{vg}(s)$ are cancelled by zeroes of $1/(1 + T)$. Hence the closed-loop line-to-output transfer function is approximately

$$\frac{G_{vg}(s)}{\left(1 + T(s)\right)} \approx \frac{D}{T_{u0}G_{cm}} \frac{1}{\left(1 + \frac{\omega_L}{s}\right)\left(1 + \frac{s}{\omega_z}\right)\left(1 + \frac{s}{\omega_c}\right)} \tag{9.61}$$

So the algebra-on-the-graph method allows simple approximate disturbance-to-output closed-loop transfer functions to be written. Armed with such an analytical expression, the system designer can easily compute the output disturbances, and can gain the insight required to shape the loop gain $T(s)$ such that system specifications are met. Computer simulations can then be used to judge whether the specifications are met under all operating conditions, and over expected ranges of component parameter values. Results of computer simulations of the design example described in this section can be found in Appendix B, Section B.2.2.

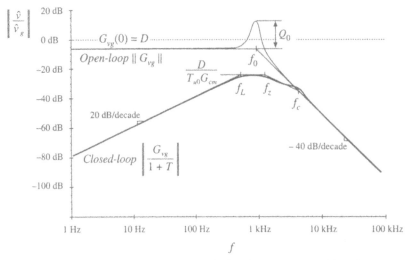

Fig. 9.32   Comparison of open-loop line-to-output transfer function G_{vg} and closed-loop line-to-output transfer function of Eq. (9.61).

9.6   MEASUREMENT OF LOOP GAINS

It is good engineering practice to measure the loop gains of prototype feedback systems. The objective of such an exercise is to verify that the system has been correctly modeled. If so, then provided that a good controller design has been implemented, then the system behavior will meet expectations regarding transient overshoot (and phase margin), rejection of disturbances, dc output voltage regulation, etc. Unfortunately, there are reasons why practical system prototypes are likely to differ from theoretical models. Phenomena may occur that were not accounted for in the original model, and that significantly influence the system behavior. Noise and electromagnetic interference (EMI) can be present, which cause the system transfer functions to deviate in unexpected ways.

So let us consider the measurement of the loop gain $T(s)$ of the feedback system of Fig. 9.33.

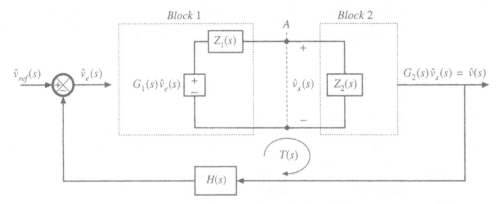

Fig. 9.33   It is desired to determine the loop gain $T(s)$ experimentally, by making measurements at point A.

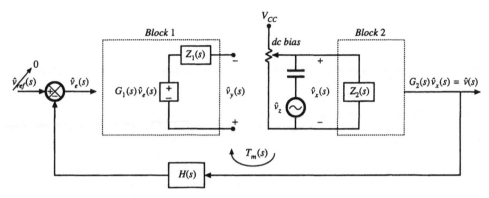

Fig. 9.34 Measurement of loop gain by breaking the loop.

We will make measurements at some point A, where two blocks of the network are connected electrically. In Fig. 9.33, the output port of block 1 is represented by a Thevenin-equivalent network, composed of the dependent voltage source $G_1\hat{v}_e$ and output impedance Z_1. Block 1 is loaded by the input impedance Z_2 of block 2. The remainder of the feedback system is represented by a block diagram as shown. The loop gain of the system is

$$T(s) = G_1(s)\left(\frac{Z_2(s)}{Z_1(s) + Z_2(s)}\right)G_2(s)H(s) \tag{9.62}$$

Measurement of this loop gain presents several challenges not present in other frequency response measurements.

In principle, one could break the loop at point A, and attempt to measure $T(s)$ using the transfer function measurement method of the previous chapter. As illustrated in Fig. 9.34, a dc supply voltage V_{CC} and potentiometer would be used, to establish a dc bias in the voltage v_x, such that all of the elements of the network operate at the correct quiescent point. Ac voltage variations in $v_z(t)$ are coupled into the injection point via a dc blocking capacitor. Any other independent ac inputs to the system are disabled. A network analyzer is used to measure the relative magnitudes and phases of the ac components of the voltages $v_y(t)$ and $v_x(t)$:

$$T_m(s) = \frac{\hat{v}_y(s)}{\hat{v}_x(s)}\bigg|_{\substack{\hat{v}_{ref}=0 \\ \hat{v}_g=0}} \tag{9.63}$$

The measured gain $T_m(s)$ differs from the actual gain $T(s)$ because, by breaking the connection between blocks 1 and 2 at the measurement point, we have removed the loading of block 2 on block 1. Solution of Fig. 9.34 for the measured gain $T_m(s)$ leads to

$$T_m(s) = G_1(s)G_2(s)H(s) \tag{9.64}$$

Equations (9.62) and (9.64) can be combined to express $T_m(s)$ in terms of $T(s)$:

$$T_m(s) = T(s)\left(1 + \frac{Z_1(s)}{Z_2(s)}\right) \tag{9.65}$$

Hence,

$$T_m(s) \approx T(s) \quad \text{provided that} \quad |Z_2| \gg |Z_1| \tag{9.66}$$

So to obtain an accurate measurement, we need to find an injection point where loading is negligible over the range of frequencies to be measured.

Other difficulties are encountered when using the method of Fig. 9.34. The most serious problem is adjustment of the dc bias using a potentiometer. The dc loop gain is typically very large, especially when a *PI* controller is used. A small change in the dc component of $v_x(t)$ can therefore lead to very large changes in the dc biases of some elements in the system. So it is difficult to establish the correct dc conditions in the circuit. The dc gains may drift during the experiment, making the problem even worse, and saturation of the error amplifier is a common complaint. Also, we have seen that the gains of the converter can be a function of the quiescent operating point; significant deviation from the correct operating point can cause the measured gain to differ from the loop gain of actual operating conditions.

9.6.1   Voltage Injection

An approach that avoids the dc biasing problem [3] is illustrated in Fig. 9.35. The voltage source $v_z(t)$ is injected between blocks 1 and 2, without breaking the feedback loop. Ac variations in $v_z(t)$ again excite variations in the feedback system, but dc bias conditions are determined by the circuit. Indeed, if $v_z(t)$ contains no dc component, then the biasing circuits of the system itself establish the quiescent operating point. Hence, the loop gain measurement is made at the actual system operating point.

The injection source is modeled in Fig. 9.35 by a Thevenin equivalent network, containing an independent voltage source with source impedance $Z_s(s)$. The magnitudes of v_z and Z_s are irrelevant in the determination of the loop gain. However, the injection of v_z does disrupt the loading of block 2 on block 1. Hence, a suitable injection point must be found, where the loading effect is negligible.

To measure the loop gain by voltage injection, we connect a network analyzer to measure the transfer function from $\hat{v}_x$ to $\hat{v}_y$. The system independent ac inputs are set to zero, and the network analyzer sweeps the injection voltage $\hat{v}_z(t)$ over the intended frequency range. The measured gain is

$$T_v(s) = \left. \frac{\hat{v}_y(s)}{\hat{v}_x(s)} \right|_{\substack{\hat{v}_{ref} = 0 \\ \hat{v}_g = 0}} \tag{9.67}$$

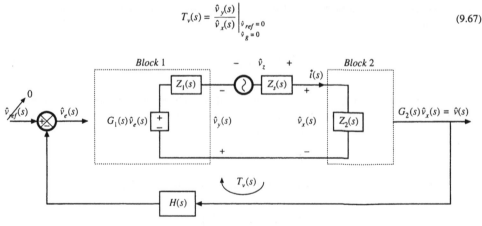

Fig. 9.35   Measurement of loop gain by voltage injection.

Let us solve Fig. 9.35, to compare the measured gain $T_v(s)$ with the actual loop gain $T(s)$ given by (9.62). The error signal is

$$\hat{v}_e(s) = -H(s)G_2(s)\hat{v}_x(s) \tag{9.68}$$

The voltage $\hat{v}_y$ can be written

$$-\hat{v}_y(s) = G_1(s)\hat{v}_e(s) - \hat{i}(s)Z_1(s) \tag{9.69}$$

where $\hat{i}(s)Z_1(s)$ is the voltage drop across the source impedance Z_1. Substitution of Eq. (9.68) into (9.69) leads to

$$-\hat{v}_y(s) = -\hat{v}_x(s)G_2(s)H(s)G_1(s) - \hat{i}(s)Z_1(s) \tag{9.70}$$

But $\hat{i}(s)$ is

$$\hat{i}(s) = \frac{\hat{v}_x(s)}{Z_2(s)} \tag{9.71}$$

Therefore, Eq. (9.70) becomes

$$\hat{v}_y(s) = \hat{v}_x(s)\left(G_1(s)G_2(s)H(s) + \frac{Z_1(s)}{Z_2(s)}\right) \tag{9.72}$$

Substitution of Eq. (9.72) into (9.67) leads to the following expression for the measured gain $T_v(s)$:

$$T_v(s) = G_1(s)G_2(s)H(s) + \frac{Z_1(s)}{Z_2(s)} \tag{9.73}$$

Equations (9.62) and (9.73) can be combined to determine the measured gain $T_v(s)$ in terms of the actual loop gain $T(s)$:

$$T_v(s) = T(s)\left(1 + \frac{Z_1(s)}{Z_2(s)}\right) + \frac{Z_1(s)}{Z_2(s)} \tag{9.74}$$

Thus, $T_v(s)$ can be expressed as the sum of two terms. The first term is proportional to the actual loop gain $T(s)$, and is approximately equal to $T(s)$ whenever $\| Z_1 \| \ll \| Z_2 \|$. The second term is not proportional to $T(s)$, and limits the minimum $T(s)$ that can be measured with the voltage injection technique. If Z_1/Z_2 is much smaller in magnitude than $T(s)$, then the second term can be ignored, and $T_v(s) \approx T(s)$. At frequencies where $T(s)$ is smaller in magnitude than Z_1/Z_2, the measured data must be discarded. Thus,

$$T_v(s) \approx T(s) \tag{9.75}$$

provided

$$(i) \quad \left| Z_1(s) \right| \ll \left| Z_2(s) \right|$$

and

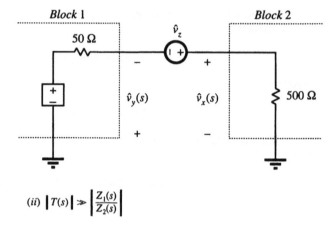

Fig. 9.36 Voltage injection example.

$$(ii) \; |T(s)| \gg \left| \frac{Z_1(s)}{Z_2(s)} \right|$$

Again, note that the value of the injection source impedance Z_s is irrelevant.

As an example, consider voltage injection at the output of an operational amplifier, having a 50 Ω output impedance, which drives a 500 Ω effective load. The system in the vicinity of the injection point is illustrated in Fig. 9.36. So $Z_1(s) = 50$ Ω and $Z_2(s) = 500$ Ω. The ratio Z_1/Z_2 is 0.1, or -20 dB. Let us further suppose that the actual loop gain $T(s)$ contains poles at 10 Hz and 100 kHz, with a dc gain of 80 dB. The actual loop gain magnitude is illustrated in Fig. 9.37.

Voltage injection would result in measurement of $T_v(s)$ given in Eq. (9.74). Note that

$$\left(1 + \frac{Z_1(s)}{Z_2(s)}\right) = 1.1 \Rightarrow 0.83 \text{ dB} \tag{9.76}$$

Hence, for large $\| T \|$, the measured $\| T_v \|$ deviates from the actual loop gain by less than 1 dB. However, at high frequency where $\| T \|$ is less than -20 dB, the measured gain differs significantly. Apparently,

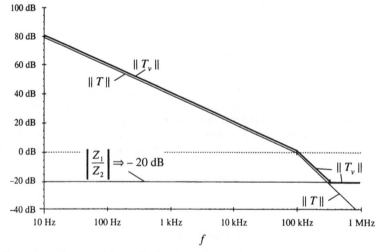

Fig. 9.37 Comparison of measured loop gain T_v and actual loop gain T, voltage injection example. The measured gain deviates at high frequency.

$T_v(s)$ contains two high-frequency zeroes that are not present in $T(s)$. Depending on the Q-factor of these zeroes, the phase of T_v at the crossover frequency could be influenced. To ensure that the phase margin is correctly measured, it is important that Z_1/Z_2 be sufficiently small in magnitude.

9.6.2 Current Injection

The results of the preceding paragraphs can also be obtained in dual form, where the loop gain is measured by current injection [3]. As illustrated in Fig. 9.38, we can model block 1 and the analyzer injection source by their Norton equivalents, and use current probes to measure $\hat{i}_x$ and $\hat{i}_y$. The gain measured by current injection is

$$T_i(s) = \frac{\hat{i}_y(s)}{\hat{i}_x(s)}\bigg|_{\substack{\hat{v}_{ref}=0 \\ \hat{v}_g=0}} \tag{9.77}$$

It can be shown that

$$T_i(s) = T(s)\left(1 + \frac{Z_2(s)}{Z_1(s)}\right) + \frac{Z_2(s)}{Z_1(s)} \tag{9.78}$$

Hence,

$T_i(s) \approx T(s)$ provided

$$(i) \quad \left|Z_2(s)\right| \ll \left|Z_1(s)\right|, \text{ and} \tag{9.79}$$

$$(ii) \quad \left|T(s)\right| \gg \left|\frac{Z_2(s)}{Z_1(s)}\right|$$

So to obtain an accurate measurement of the loop gain by current injection, we must find a point in the network where block 2 has sufficiently small input impedance. Again, note that the injection source impedance Z_s does not affect the measurement. In fact, we can realize i_z by use of a Thevenin-equivalent source, as illustrated in Fig. 9.39. The network analyzer injection source is represented by voltage source

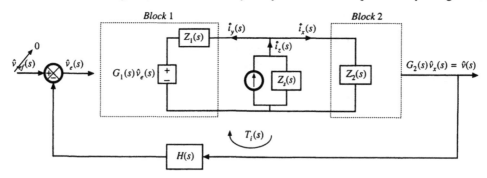

Fig. 9.38 Measurement of loop gain by current injection.

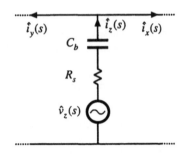

Fig. 9.39   Current injection using Thevenin-equivalent source.

$\hat{v}_z$ and output resistance R_s. A series capacitor, C_b, is inserted to avoid disrupting the dc bias at the injection point.

9.6.3   Measurement of Unstable Systems

When the prototype feedback system is unstable, we are even more eager to measure the loop gain—to find out what went wrong. But measurements cannot be made while the system oscillates. We need to stabilize the system, yet measure the original unstable loop gain. It is possible to do this by recognizing that the injection source impedance Z_s does not influence the measured loop gain [3]. As illustrated in Fig. 9.40, we can even add additional resistance R_{ext}, effectively increasing the source impedance Z_s. The measured loop gain $T_v(s)$ is unaffected.

Adding series impedance generally lowers the loop gain of a system, leading to a lower crossover frequency and a more positive phase margin. Hence, it is usually possible to add a resistor R_{ext} that is sufficiently large to stabilize the system. The gain $T_v(s)$, Eq. (9.67), continues to be approximately equal to the original unstable loop gain, according to Eq. (9.75). To avoid disturbing the dc bias conditions, it may be necessary to bypass R_{ext} with inductor L_{ext}. If the inductance value is sufficiently large, then it will not influence the stability of the modified system.

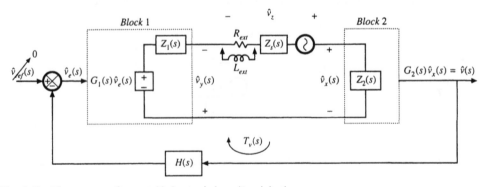

Fig. 9.40   Measurement of an unstable loop gain by voltage injection.

9.7   SUMMARY OF KEY POINTS

1. Negative feedback causes the system output to closely follow the reference input, according to the gain $1/H(s)$. The influence on the output of disturbances and variation of gains in the forward path is reduced.

2. The loop gain $T(s)$ is equal to the products of the gains in the forward and feedback paths. The loop gain is a measure of how well the feedback system works: a large loop gain leads to better regulation of the output. The crossover frequency f_c is the frequency at which the loop gain T has unity magnitude, and is a measure of the bandwidth of the control system.

3. The introduction of feedback causes the transfer functions from disturbances to the output to be multiplied by the factor $1/(1 + T(s))$. At frequencies where T is large in magnitude (i.e., below the crossover frequency), this factor is approximately equal to $1/T(s)$. Hence, the influence of low-frequency disturbances on the output is reduced by a factor of $1/T(s)$. At frequencies where T is small in magnitude (i.e., above the crossover frequency), the factor is approximately equal to 1. The feedback loop then has no effect. Closed-loop disturbance-to-output transfer functions, such as the line-to-output transfer function or the output impedance, can easily be constructed using the algebra-on-the-graph method.

4. Stability can be assessed using the phase margin test. The phase of T is evaluated at the crossover frequency, and the stability of the important closed-loop quantities $T/(1 + T)$ and $1/(1 + T)$ is then deduced. Inadequate phase margin leads to ringing and overshoot in the system transient response, and peaking in the closed-loop transfer functions.

5. Compensators are added in the forward paths of feedback loops to shape the loop gain, such that desired performance is obtained. Lead compensators, or *PD* controllers, are added to improve the phase margin and extend the control system bandwidth. *PI* controllers are used to increase the low-frequency loop gain, to improve the rejection of low-frequency disturbances and reduce the steady-state error.

6. Loop gains can be experimentally measured by use of voltage or current injection. This approach avoids the problem of establishing the correct quiescent operating conditions in the system, a common difficulty in systems having a large dc loop gain. An injection point must be found where interstage loading is not significant. Unstable loop gains can also be measured.

REFERENCES

[1]   B. KUO, *Automatic Control Systems*, New York: Prentice-Hall, Inc.

[2]   J. D'AZZO and C. HOUPIS, *Linear Control System Analysis and Design: Conventional and Modern*, New York: McGraw-Hill, 1995.

[3]   R. D. MIDDLEBROOK, "Measurement of Loop Gain in Feedback Systems," *International Journal of Electronics*, Vol. 38, No. 4, pp. 485-512, 1975.

[4]   R. D. MIDDLEBROOK, "Design-Oriented Analysis of Feedback Amplifiers," *Proceedings National Electronics Conference*, Vol. XX, October 1964, pp. 234-238.

PROBLEMS

9.1   Derive both forms of Eq. (9.25).

9.2   The flyback converter system of Fig. 9.41 contains a feedback loop for regulation of the main output voltage v_1. An auxiliary output produces voltage v_2. The dc input voltage v_g lies in the range $280\text{ V} \le v_g \le 380\text{ V}$. The compensator network has transfer function

$$G_c(s) = G_{c\infty}\left(1 + \frac{\omega_1}{s}\right)$$

where $G_{c\infty} = 0.05$, and $f_1 = \omega_1/2\pi = 400$ Hz.

(a)   What is the steady-state value of the error voltage $v_e(t)$? Explain your reasoning.

(b)   Determine the steady-state value of the main output voltage v_1.

(c)   Estimate the steady-state value of the auxiliary output voltage v_2.

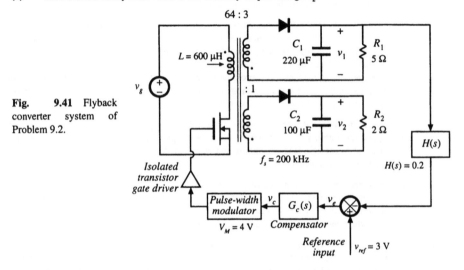

Fig.   9.41 Flyback converter   system   of Problem 9.2.

9.3   In the boost converter system of Fig. 9.42, all elements are ideal. The compensator has gain $G_c(s) = 250/s$.

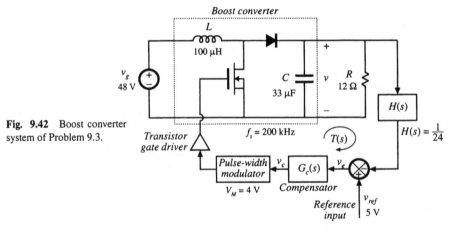

Fig. 9.42 Boost converter system of Problem 9.3.

(a)   Construct the Bode plot of the loop gain $T(s)$ magnitude and phase. Label values of all corner frequencies and Q-factors, as appropriate.

(b)   Determine the crossover frequency and phase margin.

(c)   Construct the Bode diagram of the magnitude of $1/(1 + T)$, using the algebra-on-the-graph method. Label values of all corner frequencies and Q-factors, as appropriate.

(d)   Construct the Bode diagram of the magnitude of the closed-loop line-to-output transfer function. Label values of all corner frequencies and Q-factors, as appropriate.

9.4   A certain inverter system has the following loop gain

$$T(s) = T_0 \frac{\left(1 + \frac{s}{\omega_z}\right)}{\left(1 + \frac{s}{\omega_1}\right)\left(1 + \frac{s}{\omega_2}\right)\left(1 + \frac{s}{\omega_3}\right)}$$

and the following open-loop line-to-output transfer function

$$G_{vg}(s) = G_{g0} \frac{1}{\left(1 + \frac{s}{\omega_1}\right)\left(1 + \frac{s}{\omega_3}\right)}$$

where

$T_0 = 100$	$\omega_1 = 500$ rad/sec
$\omega_2 = 1000$ rad/sec	$\omega_3 = 24000$ rad/sec
$\omega_z = 4000$ rad/sec	$G_{g0} = 0.5$

The gain of the feedback connection is $H(s) = 0.1$.

(a)   Sketch the magnitude and phase asymptotes of the loop gain $T(s)$. Determine numerical values of the crossover frequency in Hz and phase margin in degrees.

(b)   Construct the magnitude asymptotes of the closed-loop line-to-output transfer function. Label important features.

(c)   Construct the magnitude asymptotes of the closed-loop transfer function from the reference voltage to the output voltage. Label important features.

9.5   The forward converter system of Fig. 9.43(a) is constructed with the element values shown. The quiescent value of the input voltage is $V_g = 380$ V. The transformer has turns ratio $n_1/n_3 = 4.5$. The duty cycle produced by the pulse-width modulator is restricted to the range $0 \le d(t) \le 0.5$. Within this range, $d(t)$ follows the control voltage $v_c(t)$ according to

$$d(t) = \frac{1}{2} \frac{v_c(t)}{V_M}$$

with $V_M = 3$ V.

(a)   Determine the quiescent values of: the duty cycle D, the output voltage V, and the control voltage V_c.

(b)   The op-amp circuit and feedback connection can be modeled using the block diagram illustrated in Fig. 9.43(b), with $H(s) = R_2/(R_1 + R_2)$. Determine the transfer functions $G_c(s)$ and $G_r(s)$.

(c)   Sketch a block diagram which models the small-signal ac variations of the complete system, and determine the transfer function of each block. *Hint*: the transformer magnetizing inductance has negligible influence on the converter dynamics, and can be ignored. The small-signal models of the forward and buck converters are similar.

(d)   Construct a Bode plot of the loop gain magnitude and phase. What is the crossover frequency? What is the phase margin?

(e)   Construct the Bode plot of the closed-loop line-to-output transfer function magnitude

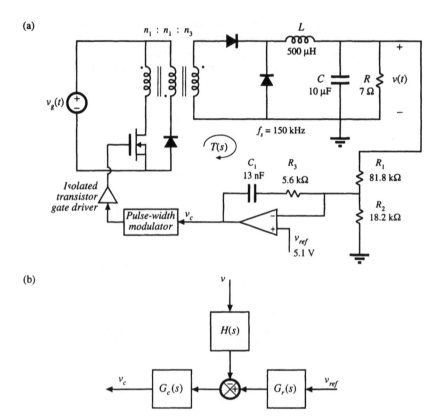

Fig. 9.43 Forward converter system of Problem 9.5: (a) system diagram, (b) modeling the op amp circuit using a block diagram.

$$\left| \frac{\hat{v}}{\hat{v}_g} \right|$$

Label important features. What is the gain at 120 Hz? At what frequency do disturbances in v_g have the greatest influence on the output voltage?

9.6   In the voltage regulator system of Fig. 9.43, described in Problem 9.5, the input voltage $v_g(t)$ contains a 120 Hz variation of peak amplitude 10 V.

(a)   What is the amplitude of the resulting 120 Hz variation in $v(t)$?

(b)   Modify the compensator network such that the 120 Hz output voltage variation has peak amplitude less than 25 mV. Your modification should leave the dc output voltage unchanged, and should result in a crossover frequency no greater than 10 kHz.

9.7   Design of a boost converter with current feedback and a *PI* compensator. In some applications, it is desired to control the converter input terminal current waveform. The boost converter system of Fig. 9.44 contains a feedback loop which causes the converter input current $i_g(t)$ to be proportional to a reference voltage $v_{ref}(t)$. The feedback connection is a current sense circuit having gain $H(s) = 0.2$ volts per ampere. A conventional pulse width modulator circuit (Fig. 7.63) is employed, having a sawtooth wave-

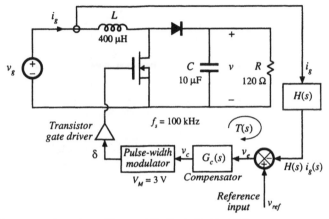

Fig. 9.44 Boost converter system with current feedback, Problem 9.7.

form with peak-peak amplitude of $V_M = 3$ V. The quiescent values of the inputs are: $V_g = 120$ V, $V_{ref} = 2$ V. All elements are ideal.

(a)   Determine the quiescent values D, V, and I_g.

(b)   Determine the small-signal transfer function

$$G_{id}(s) = \frac{\hat{i}_g(s)}{\hat{d}(s)}$$

(c)   Sketch the magnitude and phase asymptotes of the uncompensated ($G_c(s) = 1$) loop gain.

(d)   It is desired to obtain a loop gain magnitude of at least 35 dB at 120 Hz, while maintaining a phase margin of at least 72°. The crossover frequency should be no greater than $f_s/10 = 10$ kHz. Design a *PI* compensator that accomplishes this. Sketch the magnitude and phase asymptotes of the resulting loop gain, and label important features.

(e)   For your design of part (d), sketch the magnitude of the closed-loop transfer function

$$\frac{\hat{i}_g(s)}{\hat{v}_{ref}(s)}$$

Label important features.

9.8   Design of a buck regulator to meet closed-loop output impedance specifications. The buck converter with control system illustrated inFig. 9.45 is to be designed to meet the following specifications. The closed-loop output impedance should be less than 0.2 Ω over the entire frequency range 0 to 20 kHz. To ensure that the transient response is well-behaved, the poles of the closed-loop transfer functions, in the vicinity of the crossover frequency, should have Q-factors no greater than unity. The quiescent load current I_{LOAD} can vary from 5 A to 50 A, and the above specifications must be met for every value of I_{LOAD} in this range. For simplicity, you may assume that the input voltage vg does not vary. The loop gain crossover frequency f_c may be chosen to be no greater than $f_s/10$, or 10 kHz. You may also assume that all elements are ideal. The pulse-width modulator circuit obeys Eq. (7.173).

(a)   What is the intended dc output voltage V? Over what range does the effective load resistance R_{LOAD} vary?

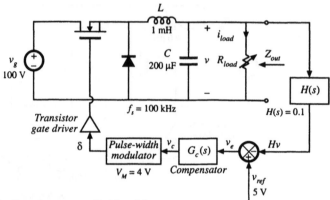

Fig. 9.45   Buck regulator system, Problem 9.8.

(b)   Construct the magnitude asymptotes of the open-loop output impedance $Z_{out}(s)$. Over what range of frequencies is the output impedance specification not met? Hence, deduce how large the minimum loop gain $T(s)$ must be in magnitude, such that the closed-loop output impedance meets the specification. Choose a suitable crossover frequency f_c.

(c)   Design a compensator network $G_c(s)$ such that all specifications are met. Additionally, the dc loop gain $T(s)$ should be at least 20 dB. Specify the following:

   (i)   Your choice for the transfer function $G_c(s)$

   (ii)   The worst-case closed-loop Q

   (iii) Bode plots of the loop gain $T(s)$ and the closed-loop output impedance, for load currents of 5 A and 50 A. What effect does variation of R_{LOAD} have on the closed-loop behavior of your design?

(d)   Design a circuit using resistors, capacitors, and an op amp to realize your compensator transfer function $G_c(s)$.

9.9   Design of a buck-boost voltage regulator. The buck-boost converter of Fig. 9.46 operates in the continuous conduction mode, with the element values shown. The nominal input voltage is $V_g = 48$ V, and it is desired to regulate the output voltage at −15 V. Design the best compensator that you can, which has high crossover frequency (but no greater than 10% of the switching frequency), large loop gain over the bandwidth of the feedback loop, and phase margin of at least 52°.

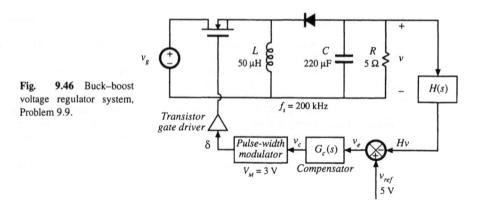

Fig.   9.46   Buck–boost voltage   regulator   system, Problem 9.9.

(a) Specify the required value of *H*. Sketch Bode plots of the uncompensated loop gain magnitude and phase, as well as the magnitude and phase of your proposed compensator transfer function $G_c(s)$. Label the important features of your plots.

(b) Construct Bode diagrams of the magnitude and phase of your compensated loop gain $T(s)$, and also of the magnitude of the quantities $T/(1 + T)$ and $1/(1 + T)$.

(c) Discuss your design. What prevents you from further increasing the crossover frequency? How large is the loop gain at 120 Hz? Can you obtain more loop gain at 120 Hz?

9.10 The loop gain of a certain feedback system is measured, using voltage injection at a point in the forward path of the loop as illustrated in Fig. 9.47(a). The data in Fig. 9.47(b) is obtained. What is $T(s)$? Specify $T(s)$ in factored pole-zero form, and give numerical values for all important features. Over what range of frequencies does the measurement give valid results?

(a)

(b)

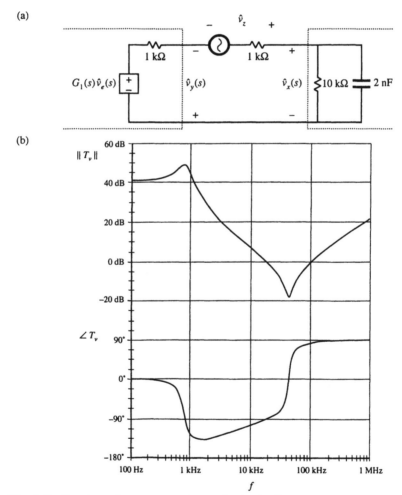

Fig. 9.47 Experimental measurement of loop gain, Problem 9.10: (a) measurement via voltage injection, (b) measured data.

10

Input Filter Design

10.1   INTRODUCTION

10.1.1   Conducted EMI

It is nearly always required that a filter be added at the power input of a switching converter. By attenuating the switching harmonics that are present in the converter input current waveform, the input filter allows compliance with regulations that limit *conducted electromagnetic interference* (EMI). The input filter can also protect the converter and its load from transients that appear in the input voltage $v_g(t)$, thereby improving the system reliability.

A simple buck converter example is illustrated in Fig. 10.1. The converter injects the pulsating current $i_g(t)$ of Fig. 10.1(b) into the power source $v_g(t)$. The Fourier series of $i_g(t)$ contains harmonics at multiples of the switching frequency f_s, as follows:

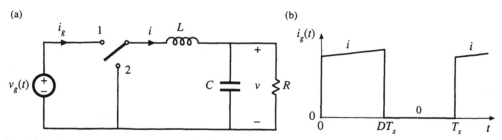

Fig. 10.1   Buck converter example: (a) circuit of power stage, (b) pulsating input current waveform.

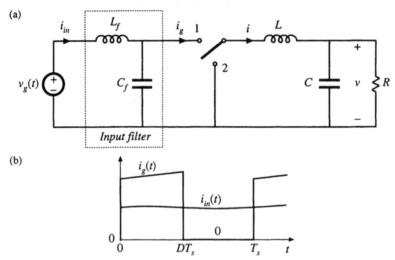

Fig. 10.2 Addition of a simple *L-C* low-pass filter to the power input terminals of the buck converter: (a) circuit, (b) input current waveforms.

$$i_g(t) = DI + \sum_{k=1}^{\infty} \frac{2I}{k\pi} \sin\left(k\pi D\right) \cos\left(k\omega t\right) \tag{10.1}$$

In practice, the magnitudes of the higher-order harmonics can also be significantly affected by the current spike caused by diode reverse recovery, and also by the finite slopes of the switching transitions. The large high-frequency current harmonics of $i_g(t)$ can interfere with television and radio reception, and can disrupt the operation of nearby electronic equipment. In consequence, regulations and standards exist that limit the amplitudes of the harmonic currents injected by a switching converter into its power source [1-8]. As an example, if the dc inductor current i of Fig. 10.2 has a magnitude of several Amperes, then the fundamental component ($n = 1$) has an rms amplitude in the vicinity of one Ampere. Regulations may require attenuation of this current to a value typically in the range 10 μA to 100 μA.

To meet limits on conducted EMI, it is necessary to add an input filter to the converter. Figure 10.2 illustrates a simple single-section *L-C* low-pass filter, added to the input of the converter of Fig. 10.1. This filter attenuates the current harmonics produced by the switching converter, and thereby smooths the current waveform drawn from the power source. If the filter has transfer function $H(s) = i_{in}/i_g$, then the input current Fourier series becomes

$$i_{in}(t) = H(0)DI + \sum_{k=1}^{\infty} \left| H(kj\omega) \right| \frac{2I}{k\pi} \sin\left(k\pi D\right) \cos\left(k\omega t + \angle H(kj\omega)\right) \tag{10.2}$$

In other words, the amplitude of each current harmonic at angular frequency $k\omega$ is attenuated by the filter transfer function at the harmonic frequency, $\| H(kj\omega) \|$. Typical requirements effectively limit the current harmonics to have amplitudes less than 100 μA, and hence input filters are often required to attenuate the current amplitudes by 80 dB or more.

To improve the reliability of the system, input filters are sometimes required to operate normally when transients or periodic disturbances are applied to the power input. Such *conducted susceptibility* specifications force the designer to damp the input filter resonances, so that input disturbances do not excite excessive currents or voltages within the filter or converter.

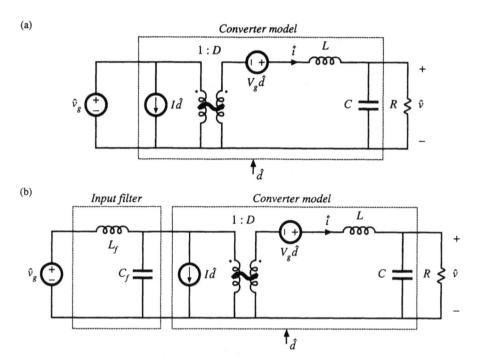

Fig. 10.3 Small-signal equivalent circuit models of the buck converter: (a) basic converter model, (b) with addition of input filter.

10.1.2 The Input Filter Design Problem

The situation faced by the design engineer is typically as follows. A switching regulator has been designed, which meets performance specifications. The regulator was properly designed as discussed in Chapter 9, using a small-signal model of the converter power stage such as the equivalent circuit of Fig. 10.3(a). In consequence, the transient response is well damped and sufficiently fast, with adequate phase margin at all expected operating points. The output impedance is sufficiently small over a wide frequency range. The line-to-output transfer function $G_{vg}(s)$, or *audiosusceptibility*, is sufficiently small, so that the output voltage remains regulated in spite of variations in $\hat{v}_g(t)$.

Having developed a good design that meets the above goals regarding dynamic response, the problem of conducted EMI is then addressed. A low-pass filter having attenuation sufficient to meet conducted EMI specifications is constructed and added to the converter input. A new problem then arises: the input filter changes the dynamics of the converter. The transient response is modified, and the control system may even become unstable. The output impedance may become large over some frequency range, possibly exhibiting resonances. The audiosusceptibility may be degraded.

The problem is that the input filter affects the dynamics of the converter, often in a manner that degrades regulator performance. For example, when a single-section L-C input filter is added to a buck converter as in Fig. 10.2(a), the small-signal equivalent circuit model is modified as shown in Fig. 10.3(b). The input filter elements affect all transfer functions of the converter, including the control-to-

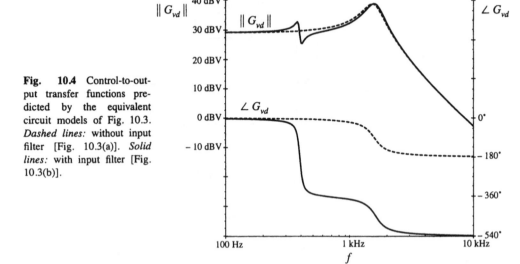

Fig. 10.4 Control-to-output transfer functions predicted by the equivalent circuit models of Fig. 10.3. *Dashed lines:* without input filter [Fig. 10.3(a)]. *Solid lines:* with input filter [Fig. 10.3(b)].

output transfer function $G_{vd}(s)$, the line-to-output transfer function $G_{vg}(s)$, and the converter output impedance $Z_{out}(s)$. Moreover, the influence of the input filter on these transfer functions can be quite severe.

As an illustration, let's examine how the control-to-output transfer function $G_{vd}(s)$ of the buck converter of Fig. 10.1 is altered when a simple L-C input filter is added as in Fig. 10.2. For this example, the element values are chosen to be: $D = 0.5$, $L = 100\ \mu\mathrm{H}$, $C = 100\ \mu\mathrm{F}$, $R = 3\ \Omega$, $L_f = 330\ \mu\mathrm{H}$, $C_f = 470\ \mu\mathrm{F}$. Figure 10.4 contains the Bode plot of the magnitude and phase of the control-to-output transfer function $G_{vd}(s)$. The dashed lines are the magnitude and phase before the input filter was added, generated by solution of the model of Fig. 10.3(a). The complex poles of the converter output filter cause the phase to approach $-180°$ at high frequency. Usually, this is the model used to design the regulator feedback loop and to evaluate the phase margin (*see* Chapter 9). The solid lines of Fig. 10.4 show the magnitude and phase after addition of the input filter, generated by solution of the model of Fig. 10.3(b). The magnitude exhibits a "glitch" at the resonant frequency of the input filter, and an additional $-360°$ of phase shift is introduced into the phase. It can be shown that $G_{vd}(s)$ now contains an additional complex pole pair and a complex right half-plane zero pair, associated with the input filter dynamics. If the crossover frequency of the regulator feedback loop is near to or greater than the resonant frequency of the input filter, then the loop phase margin will become negative and instability will result. Such behavior is typical; consequently, input filters are notorious for destabilizing switching regulator systems.

This chapter shows how to mitigate the stability problem, by introducing damping into the input filter and by designing the input filter such that its output impedance is sufficiently small [9-21]. The result of these measures is that the effect of the input filter on the control-to-output transfer function becomes negligible, and hence the converter dynamics are much better behaved. Although analysis of the fourth-order system of Fig. 10.3(b) is potentially quite complex, the approach used here simplifies the problem through use of impedance inequalities involving the converter input impedance and the filter output impedance [9,10]. These inequalities are based on Middlebrook's extra element theorem of Appendix C. This approach allows the engineer to gain the insight needed to effectively design the input filter. Optimization of the damping networks of input filters, and design of multiple-section filters, is also discussed.

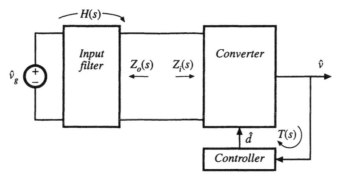

Fig. 10.5 Addition of an input filter to a switching voltage regulator system.

10.2 EFFECT OF AN INPUT FILTER ON CONVERTER TRANSFER FUNCTIONS

The control-to-output transfer function $G_{vd}(s)$ is defined as follows:

$$G_{vd}(s) = \left. \frac{\hat{v}(s)}{\hat{d}(s)} \right|_{\hat{v}_g(s) = 0} \tag{10.3}$$

The control-to-output transfer functions of basic CCM converters with no input filters are listed in Section 8.2.2.

Addition of an input filter to a switching regulator leads to the system illustrated in Fig. 10.5. To determine the control-to-output transfer function in the presence of the input filter, we set $\hat{v}_g(s)$ to zero and solve for $\hat{v}(s)/\hat{d}(s)$ according to Eq. (10.3). The input filter can then be represented simply by its output impedance $Z_o(s)$ as illustrated in Fig. 10.6. Thus, the input filter can be treated as an extra element having impedance $Z_o(s)$. In Appendix C, Section C.4.3, Middlebrook's extra element theorem is employed to determine how addition of the input filter modifies the control-to-output transfer function. It is found that the modified control-to-output transfer function can be expressed as follows [9]:

$$G_{vd}(s) = \left(\left. G_{vd}(s) \right|_{Z_o(s) = 0} \right) \frac{\left(1 + \dfrac{Z_o(s)}{Z_N(s)} \right)}{\left(1 + \dfrac{Z_o(s)}{Z_D(s)} \right)} \tag{10.4}$$

Fig. 10.6 Determination of the control-to-output transfer function $G_{vd}(s)$ for the system of Fig. 10.5.

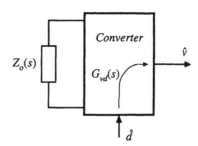

Table 10.1   Input filter design criteria for basic converters

Converter	$Z_N(s)$	$Z_D(s)$	$Z_e(s)$
Buck	$-\dfrac{R}{D^2}$	$\dfrac{R}{D^2}\dfrac{\left(1+s\dfrac{L}{R}+s^2LC\right)}{\left(1+sRC\right)}$	$\dfrac{sL}{D^2}$
Boost	$-D'^2R\left(1-\dfrac{sL}{D'^2R}\right)$	$D'^2R\dfrac{\left(1+s\dfrac{L}{D'^2R}+s^2\dfrac{LC}{D'^2}\right)}{\left(1+sRC\right)}$	sL
Buck–boost	$-\dfrac{D'^2R}{D^2}\left(1-\dfrac{sDL}{D'^2R}\right)$	$\dfrac{D'^2R}{D^2}\dfrac{\left(1+s\dfrac{L}{D'^2R}+s^2\dfrac{LC}{D'^2}\right)}{\left(1+sRC\right)}$	$\dfrac{sL}{D^2}$

where

$$G_{vd}(s)\big|_{Z_o(s)=0} \tag{10.5}$$

is the original control-to-output transfer function with no input filter. The quantity $Z_D(s)$ is equal to the converter input impedance $Z_i(s)$ under the condition that $\hat{d}(s)$ is equal to zero:

$$Z_D(s) = Z_i(s)\big|_{\hat{d}(s)=0} \tag{10.6}$$

The quantity $Z_N(s)$ is equal to the converter input impedance $Z_i(s)$ under the condition that the feedback controller of Fig. 10.5 operates ideally; in other words, the controller varies $\hat{d}(s)$ as necessary to maintain $\hat{v}(s)$ equal to zero:

$$Z_N(s) = Z_i(s)\big|_{\hat{v}(s)\xrightarrow{null}0} \tag{10.7}$$

In terms of the canonical circuit model parameters described in Section 7.5, $Z_N(s)$ can be shown to be

$$Z_N(s) = -\frac{e(s)}{j(s)} \tag{10.8}$$

Expressions for $Z_N(s)$ and $Z_D(s)$ for the basic buck, boost, and buck-boost converters are listed in Table 10.1.

10.2.1   Discussion

Equation (10.4) relates the power stage control-to-output transfer function $G_{vd}(s)$ to the output impedance $Z_o(s)$ of the input filter, and also to the quantities $Z_N(s)$ and $Z_D(s)$ measured at the power input port of the converter. The quantity $Z_D(s)$ coincides with the open-loop input impedance of the converter.

As described above, the quantity $Z_N(s)$ is equal to the input port impedance of the converter

(a)

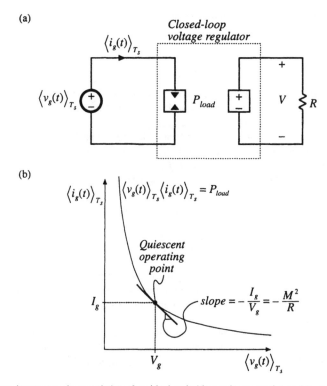

(b)

Fig. 10.7 Power input port characteristics of an ideal switching voltage regulator: (a) equivalent circuit model, including dependent power sink, (b) constant power characteristic of input port.

power stage, under the conditions that $\hat{d}(s)$ is varied as necessary to null $\hat{v}(s)$ to zero. This is, in fact, the function performed by an ideal controller: it varies the duty cycle as necessary to maintain zero error of the output voltage. Therefore, $Z_N(s)$ coincides with the impedance that would be measured at the converter power input terminals, if an ideal feedback loop perfectly regulated the converter output voltage. Of course, Eq. (10.4) is valid in general, regardless of whether a control system is present.

Figure 10.7 illustrates the large-signal behavior of a feedback loop that perfectly regulates the converter output voltage. Regardless of the applied input voltage $v_g(t)$, the output voltage is maintained equal to the desired value V. The load power is therefore constant, and equal to $P_{load} = V^2/R$. In the idealized case of a lossless converter, the power flowing into the converter input terminals will also be equal to P_{load}, regardless of the value of $v_g(t)$. Hence, the power input terminal of the converter obeys the equation

$$\left\langle v_g(t) \right\rangle_{T_s} \left\langle i_g(t) \right\rangle_{T_s} = P_{load} \tag{10.9}$$

This characteristic is illustrated in Fig. 10.7(b), and is represented in Fig. 10.7(a) by the dependent power sink symbol. The properties of power sources and power sinks are discussed in detail in Chapter 11.

Figure 10.7(b) also illustrates linearization of the constant input power characteristic, about a quiescent operating point. The resulting line has negative slope; therefore, the incremental (small signal) input resistance of the ideal voltage regulator is negative. For example, increasing the voltage $\left\langle v_g(t) \right\rangle_{T_s}$

causes the current $\langle i_g(t) \rangle_{T_s}$ to decrease, such that the power remains constant. This incremental resistance has the value [9,14]:

$$-\frac{R}{M^2} \tag{10.10}$$

where R is the output load resistance, and M is the conversion ratio V/V_g. For each of the converters listed in Table 10.1, the dc asymptote of $Z_N(s)$ coincides with the negative incremental resistance given by the equation above.

Practical control systems exhibit a limited bandwidth, determined by the crossover frequency f_c of the feedback loop. Therefore, we would expect the closed-loop regulator input impedance to be approximately equal to $Z_N(s)$ at low frequency $(f \ll f_c)$ where the loop gain is large and the regulator works well. At frequencies above the bandwidth of the regulator $(f \gg f_c)$, we expect the converter input impedance to follow the open-loop value $Z_D(s)$. For closed-loop conditions, it can be shown that the regulator input impedance $Z_i(s)$ is, in fact, described by the following equation:

$$\frac{1}{Z_i(s)} = \frac{1}{Z_N(s)} \frac{T(s)}{1 + T(s)} + \frac{1}{Z_D(s)} \frac{1}{1 + T(s)} \tag{10.11}$$

where $T(s)$ is the controller loop gain. Thus, the regulator input impedance follows the negative resistance of $Z_N(s)$ at low frequency where the magnitude of the loop gain is large [and hence $T/(1 + T) \approx 1$, $1/(1 + T) \approx 0$], and reverts to the (positive) open-loop impedance $Z_D(s)$ at high frequency where $\| T \|$ is small [i.e., where $T/(1 + T) \approx 0$, $1/(1 + T) \approx 1$].

When an undamped or lightly damped input filter is connected to the regulator input port, the input filter can interact with the negative resistance characteristic of Z_N to form a *negative resistance oscillator*. This further explains why addition of an input filter tends to lead to instabilities.

10.2.2 Impedance Inequalities

Equation (10.4) reveals that addition of the input filter causes the control-to-output transfer function to be modified by the factor

$$\frac{\left(1 + \dfrac{Z_o(s)}{Z_N(s)}\right)}{\left(1 + \dfrac{Z_o(s)}{Z_D(s)}\right)} \tag{10.12}$$

called the *correction factor*. When the following inequalities are satisfied,

$$\begin{aligned} |Z_o| &\ll |Z_N|, \quad \text{and} \\ |Z_o| &\ll |Z_D| \end{aligned} \tag{10.13}$$

then the correction factor has a magnitude of approximately unity, and the input filter does not substantially alter the control-to-output transfer function [9,10]. These inequalities limit the maximum allowable output impedance of the input filter, and constitute useful filter design criteria. One can sketch the Bode plots of $\| Z_N(j\omega) \|$ and $\| Z_D(j\omega) \|$, and compare with the Bode plot of $\| Z_o(j\omega) \|$. This allows the engineer to gain the insight necessary to design an input filter that satisfies Eq. (10.13).

A similar analysis shows that the converter output impedance is not substantially affected by the input filter when the following inequalities are satisfied:

$$|Z_o| \ll |Z_e|, \text{ and}$$
$$|Z_o| \ll |Z_D| \qquad (10.14)$$

where $Z_D(s)$ is again as given in Table 10.1. The quantity $Z_e(s)$ is equal to the converter input impedance $Z_i(s)$ under the conditions that the converter output is shorted:

$$Z_e = Z_i|_{\hat{v}=0} \qquad (10.15)$$

Expressions for $Z_e(s)$ for basic converters are also listed in Table 10.1.

Similar impedance inequalities can be derived for the case of current-programmed converters [12,13], or converters operating in the discontinuous conduction mode. In [12], impedance inequalities nearly identical to the above equations were shown to guarantee that the input filter does not degrade transient response and stability in the current-programmed case. Feedforward of the converter input voltage was suggested in [16].

10.3 BUCK CONVERTER EXAMPLE

Let us again consider the example of a simple buck converter with *L-C* input filter, as illustrated in Fig. 10.8(a). Upon replacing the converter with its small-signal model, we obtain the equivalent circuit of Fig. 10.8(b). Let's evaluate Eq. (10.4) for this example, to find how the input filter modifies the control-to-output transfer function of the converter.

10.3.1 Effect of Undamped Input Filter

The quantities $Z_N(s)$ and $Z_D(s)$ can be read from Table 10.1, or can be derived using Eqs. (10.6) and (10.7) as further described in Appendix C. The quantity $Z_D(s)$ is given by Eq. (10.6). Upon setting $\hat{d}(s)$ to zero, the converter small signal model reduces to the circuit of Fig. 10.9(a). It can be seen that $Z_D(s)$ is equal to the input impedance of the *R-L-C* filter, divided by the square of the turns ratio:

$$Z_D(s) = \frac{1}{D^2}\left(sL + R\|\frac{1}{sC}\right) \qquad (10.16)$$

Construction of asymptotes for this impedance is treated in Section 8.4, with the results for the numerical values of this example given in Fig. 10.10. The load resistance dominates the impedance at low frequency, leading to a dc asymptote of $R/D^2 = 12 \ \Omega$. For the high-Q case shown, $\| Z_D(j\omega) \|$ follows the output capacitor asymptote, reflected through the square of the effective turns ratio, at intermediate frequencies. A series resonance occurs at the output filter resonant frequency f_0, given by

$$f_0 = \frac{1}{2\pi\sqrt{LC}} \qquad (10.17)$$

For the element values listed in Fig. 10.8(a), the resonant frequency is $f_0 = 1.6$ kHz. The values of the asymptotes at the resonant frequency f_0 are given by the characteristic impedance R_0, referred to the

(a)

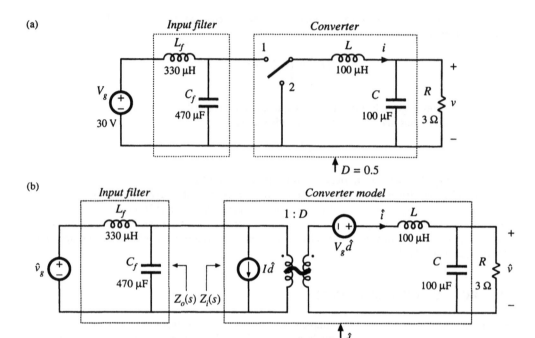

Fig. 10.8 Buck converter example: (a) converter circuit, (b) small-signal model.

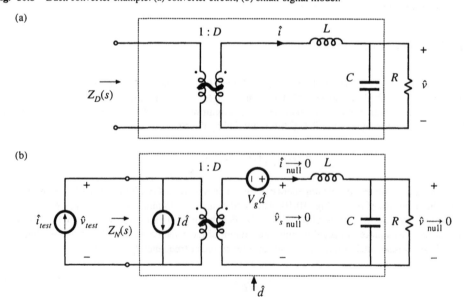

Fig. 10.9 Determination of the quantities $Z_N(s)$ and $Z_D(s)$ for the circuit of Fig. 10.8(b): (a) determination of $Z_D(s)$, (b) determination of $Z_N(s)$.

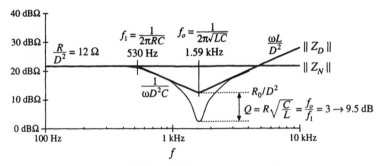

Fig. 10.10 Construction of $\| Z_N(j\omega) \|$ and $\| Z_D(j\omega) \|$, buck converter example.

transformer primary:

$$\frac{R_0}{D^2} = \frac{1}{D^2} \sqrt{\frac{L}{C}} \tag{10.18}$$

For the element values given in Fig. 10.8(a), this expression is equal to 4 Ω. The Q-factor is given by

$$Q = \frac{R}{R_0} = R \sqrt{\frac{C}{L}} \tag{10.19}$$

This expression yields a numerical value of $Q = 3$. The value of $\| Z_D(j\omega) \|$ at the resonant frequency 1.6 kHz is therefore equal to $(4\ \Omega)/(3) = 1.33\ \Omega$. At high frequency, $\| Z_D(j\omega) \|$ follows the reflected inductor asymptote.

The quantity $Z_N(s)$ is given by Eq. (10.7). This impedance is equal to the converter input impedance $Z_i(s)$, under the conditions that $\hat{d}(s)$ is varied to maintain the output voltage $\hat{v}(s)$ at zero. Figure 10.9(b) illustrates the derivation of an expression for $Z_N(s)$. A test current source $\hat{i}_{test}(s)$ is injected at the converter input port. The impedance $Z_N(s)$ can be viewed as the transfer function from $\hat{i}_{test}(s)$ to $\hat{v}_{test}(s)$:

$$Z_N(s) = \frac{\hat{v}_{test}(s)}{\hat{i}_{test}(s)} \bigg|_{\hat{v} \xrightarrow{null} 0} \tag{10.20}$$

The null condition $\hat{v}(s) \xrightarrow{null} 0$ greatly simplifies analysis of the circuit of Fig. 10.9(b). Since the voltage $\hat{v}(s)$ is zero, the currents through the capacitor and load impedances are also zero. This further implies that the inductor current $\hat{i}(s)$ and transformer winding currents are zero, and hence the voltage across the inductor is also zero. Finally, the voltage $\hat{v}_s(s)$, equal to the output voltage plus the inductor voltage, is zero.

Since the currents in the windings of the transformer model are zero, the current $i_{test}(s)$ is equal to the independent source current $I\hat{d}(s)$:

$$\hat{i}_{test}(s) = I\hat{d}(s) \tag{10.21}$$

Because $\hat{v}_s(s)$ is equal to zero, the voltage applied to the secondary of the transformer model is equal to the independent source voltage $- V_g \hat{d}(s)$. Upon dividing by the turns ratio D, we obtain $\hat{v}_{test}(s)$:

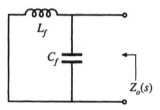

Fig. 10.11   Determination of the filter output impedance $Z_o(s)$.

$$\hat{v}_{test}(s) = -\frac{V_g \hat{d}(s)}{D} \tag{10.22}$$

Insertion of Eqs. (10.21) and (10.22) into Eq. (10.20) leads to the following result:

$$Z_N(s) = \frac{\left(-\dfrac{V_g \hat{d}(s)}{D}\right)}{\left(I\hat{d}(s)\right)} = -\frac{R}{D^2} \tag{10.23}$$

The steady-state relationship $I = DV_g/R$ has been used to simplify the above result. This equation coincides with the expression listed in Table 10.1. The Bode diagram of $\| Z_N(j\omega) \|$ is constructed in Fig. 10.10; this plot coincides with the dc asymptote of $\| Z_D(j\omega) \|$.

Next, let us construct the Bode diagram of the filter output impedance $Z_o(s)$. When the independent source $\hat{v}_g(s)$ is set to zero, the input filter network reduces to the circuit of Fig. 10.11. It can be seen that $Z_o(s)$ is given by the parallel combination of the inductor L_f and the capacitor C_f:

$$Z_o(s) = sL_f \| \frac{1}{sC_f} \tag{10.24}$$

Construction of the Bode diagram of this parallel resonant circuit is discussed in Section 8.3.4. As illustrated in Fig. 10.12, the magnitude $\| Z_o(j\omega) \|$ is dominated by the inductor impedance at low frequency, and by the capacitor impedance at high frequency. The inductor and capacitor asymptotes intersect at the filter resonant frequency:

Fig. 10.12   Magnitude plot of the output impedance of the input filter of Fig. 10.11. Since the filter is not damped, the Q-factor is very large.

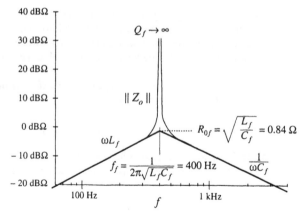

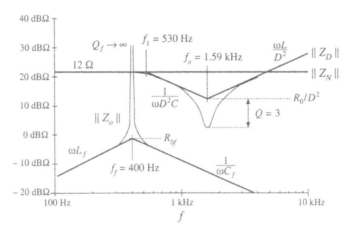

Fig. 10.13 Impedance design criteria $\| Z_N(j\omega) \|$ and $\| Z_D(j\omega) \|$ from Fig. 10.10, with the filter output impedance $\| Z_o(j\omega) \|$ of Fig. 10.12 superimposed. The design criteria of Eq. (10.13) are not satisfied at the input filter resonance.

$$f_f = \frac{1}{2\pi\sqrt{L_f C_f}} \tag{10.25}$$

For the given values, the input filter resonant frequency is $f_f = 400$ Hz. This filter has characteristic impedance

$$R_{0f} = \sqrt{\frac{L_f}{C_f}} \tag{10.26}$$

equal to 0.84 Ω. Since the input filter is undamped, its Q-factor is ideally infinite. In practice, parasitic elements such as inductor loss and capacitor equivalent series resistance limit the value of Q_f. Nonetheless, the impedance $\| Z_o(j\omega) \|$ is very large in the vicinity of the filter resonant frequency f_f.

The Bode plot of the filter output impedance $\| Z_o(j\omega) \|$ is overlaid on the $\| Z_N(j\omega) \|$ and $\| Z_D(j\omega) \|$ plots in Fig. 10.13, for the element values listed in Fig. 10.8(a). We can now determine whether the impedance inequalities (10.13) are satisfied. Note the design-oriented nature of Fig. 10.13: since analytical expressions are given for each impedance asymptote, the designer can easily adjust the component values to satisfy Eq. (10.13). For example, the values of L_f and C_f should be chosen to ensure that the asymptotes of $\| Z_o(j\omega) \|$ lie below the worst-case value of R/D^2, as well as the other asymptotes of $\| Z_D(j\omega) \|$.

It should also be apparent that it is a bad idea to choose the input and output filter resonant frequencies f_0 and f_f to be equal, because it would then be more difficult to satisfy the inequalities of Eq. (10.13). Instead, the resonant frequencies f_0 and f_f should be well separated in value.

Since the input filter is undamped, it is impossible to satisfy the impedance inequalities (10.13) in the vicinity of the input filter resonant frequency f_f. Regardless of the choice of element values, the input filter changes the control-to-output transfer function $G_{vd}(s)$ in the vicinity of frequency f_f. Figures 10.14 and 10.15 illustrate the resulting correction factor [Eq. (10.12)] and the modified control-to-output transfer function [Eq. (10.4)], respectively. At frequencies well below the input filter resonant frequency, impedance inequalities (10.13) are well satisfied. The correction factor tends to the value $1\angle 0°$, and the

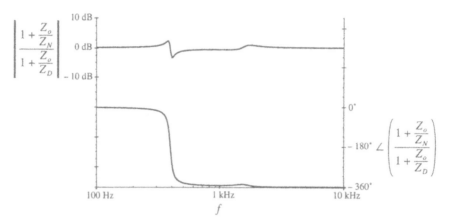

Fig. 10.14   Magnitude of the correction factor, Eq. (10.12), for the buck converter example of Fig. 10.8.

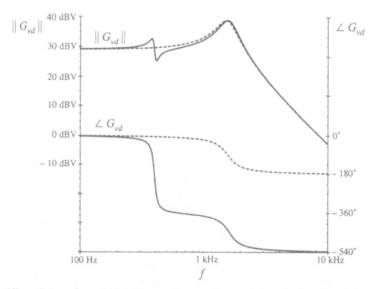

Fig. 10.15   Effect of the undamped input filter on the control-to-output transfer function of the buck converter example. *Dashed lines:* without input filter. *Solid lines:* with undamped input filter.

control-to-output transfer function $G_{vd}(s)$ is essentially unchanged. In the vicinity of the resonant frequency f_f, the correction factor contains a pair of complex poles, and also a pair of right half-plane complex zeroes. These cause a "glitch" in the magnitude plot of the correction factor, and they contribute $360°$ of lag to the phase of the correction factor. The glitch and its phase lag can be seen in the Bode plot of $G_{vd}(s)$. At high frequency, the correction factor tends to a value of approximately $1\angle - 360°$; consequently, the high-frequency magnitude of G_{vd} is unchanged. However, when the $- 360°$ contributed by the correction factor is added to the $- 180°$ contributed at high frequency by the two poles of the original $G_{vd}(s)$, a high-frequency phase asymptote of $- 540°$ is obtained. If the crossover frequency of the converter feedback loop is placed near to or greater than the input filter resonant frequency f_f, then a negative phase margin is inevitable. This explains why addition of an input filter often leads to instabilities and oscillations in switching regulators.

10.3.2 Damping the Input Filter

Let's damp the resonance of the input filter, so that impedance inequalities (10.13) are satisfied at all frequencies.

One approach to damping the filter is to add resistor R_f in parallel with capacitor C_f as illustrated in Fig. 10.16(a). The output impedance of this network is identical to the parallel resonant impedance analyzed in Section 8.3.4. The maximum value of the output impedance occurs at the resonant frequency f_f, and is equal in value to the resistance R_f. Hence, to satisfy impedance inequalities (10.13), we should choose R_f to be much less than the $\| Z_N(j\omega) \|$ and $\| Z_D(j\omega) \|$ asymptotes. The condition $R_f \ll \| Z_N(j\omega) \|$ can be expressed as:

$$R_f \ll \frac{R}{D^2} \tag{10.27}$$

Unfortunately, this raises a new problem: the power dissipation in R_f. The dc input voltage V_g is applied across resistor R_f, and therefore R_f dissipates power equal to V_g^2/R_f. Equation (10.27) implies that this power loss is greater than the load power! Therefore, the circuit of Fig. 10.16(a) is not a practical solution.

One solution to the power loss problem is to place R_f in parallel with L_f as illustrated in Fig. 10.16(b). The value of R_f in Fig. 10.16(b) is also chosen according to Eq. (10.27). Since the dc voltage across inductor L_f is zero, there is now no dc power loss in resistor R_f. The problem with this circuit is that its transfer function contains a high-frequency zero. Addition of R_f degrades the slope of the high-frequency asymptote, from -40 dB/decade to -20 dB/decade. The circuit of Fig. 10.16(b) is effectively a single-pole R-C low-pass filter, with no attenuation provided by inductor L_f.

One practical solution is illustrated in Fig. 10.17 [10]. Dc blocking capacitor C_b is added in series with resistor R_f. Since no dc current can flow through resistor R_f, its dc power loss is eliminated. The value of C_b is chosen to be very large such that, at the filter resonant frequency f_f, the impedance of the R_f-C_b branch is dominated by resistor R_f. When C_b is sufficiently large, then the output impedance of this network reduces to the output impedances of the filters of Fig. 10.16. The impedance asymptotes for the case of large C_b are illustrated in Fig. 10.17(b).

(a)　　　　　　　　　　　　　　　　　(b)

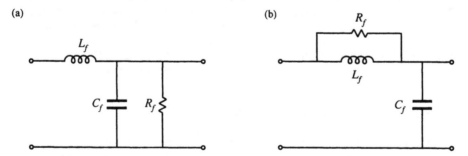

Fig. 10.16 Two attempts to damp the input filter: (a) addition of damping resistance R_f across C_f, (b) addition of damping resistance R_f in parallel with L_f.

(a)   (b)

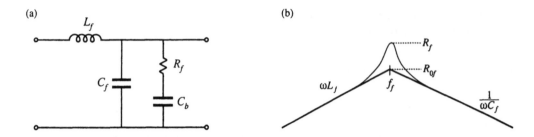

Fig. 10.17 A practical method to damping the input filter, including damping resistance R_f and dc blocking capacitor C_b: (a) circuit, (b) output impedance asymptotes.

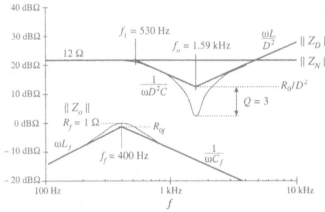

Fig. 10.18 Impedance design criteria $\| Z_N(j\omega) \|$ and $\| Z_D(j\omega) \|$ from Fig. 10.10, with the filter output impedance $\| Z_o(j\omega) \|$ of Fig. 10.17(b) superimposed. The design criteria of Eq. (10.13) are well satisfied.

The low-frequency asymptotes of $\| Z_N(j\omega) \|$ and $\| Z_D(j\omega) \|$ in Fig. 10.10 are equal to $R/D^2 = 12\ \Omega$. The choice $R_f = 1\ \Omega$ therefore satisfies impedance inequalities (10.13) very well. The choice $C_b = 4700\ \mu F$ leads to $1/2\pi f_f C_b = 0.084\ \Omega$, which is much smaller than R_f. The resulting magnitude $\| Z_o(j\omega) \|$ is compared with $\| Z_N(j\omega) \|$ and $\| Z_D(j\omega) \|$ in Fig. 10.18. It can be seen that the chosen values of R_f and C_b lead to adequate damping, and impedance inequalities (10.13) are now well satisfied.

Figure 10.19 illustrates how addition of the damped input filter modifies the magnitude and phase of the control-to-output transfer function. There is now very little change in $G_{vd}(s)$, and we would expect that the performance of the converter feedback loop is unaffected by the input filter.

10.4   DESIGN OF A DAMPED INPUT FILTER

As illustrated by the example of the previous section, design of an input filter requires not only that the filter impedance asymptotes satisfy impedance inequalities, but also that the filter be adequately damped. Damping of the input filter is also necessary to prevent transients and disturbances in $v_g(t)$ from exciting filter resonances. Other design constraints include attaining the desired filter attenuation, and minimizing

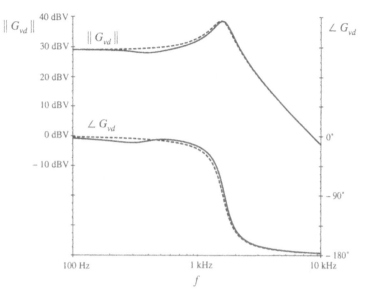

Fig. 10.19 Effect of the damped input filter on the control-to-output transfer function of the buck converter example. *Dashed lines:* without input filter. *Solid lines:* with damped input filter.

the size of the reactive elements. Although a large number of classical filter design techniques are well known, these techniques do not address the problems of limiting the maximum output impedance and damping filter resonances.

The value of the blocking capacitor C_b used to damp the input filter in Section 10.3.2 is ten times larger than the value of C_f, and hence its size and cost are of practical concern. Optimization of an input filter design therefore includes minimization of the size of the elements used in the damping networks.

Several practical approaches to damping the single-section L-C low-pass filter are illustrated in Fig. 10.20 [10,11,17]. Figure 10.20(a) contains the R_f-C_b damping branch considered in the previous section. In Fig. 10.20(b), the damping resistor R_f is placed in parallel with the filter inductor L_f, and a high-frequency blocking inductor L_b is placed in series with R_f. Inductor L_b causes the filter transfer function to roll off with a high-frequency slope of -40 dB/decade. In Fig. 10.20(c), the damping resistor R_f is placed in series with the filter inductor L_f, and the dc current is bypassed by inductor L_b. In each case, it is desired to obtain a given amount of damping [i.e., to cause the peak value of the filter output impedance to be no greater than a given value that satisfies the impedance inequalities (10.13)], while minimizing the value of C_b or L_b. This problem can be formulated in an alternate but equivalent form: for a given choice of C_b or L_b, find the value of R_f that minimizes the peak output impedance [10]. The solutions to this optimization problem, for the three filter networks of Fig. 21, are summarized in this section. In each case, the quantities R_{0f} and f_f are defined by Eqs. (10.25) and (10.26).

Consider the filter of Fig. 10.20(b), with fixed values of L_f, C_f, and L_b. Figure 10.21 contains Bode plots of the filter output impedance $\| Z_o(j\omega) \|$ for several values of damping resistance R_f. For the limiting case $R_f = \infty$, the circuit reduces to the original undamped filter with infinite Q_f. In the limiting case $R_f = 0$, the filter is also undamped, but the resonant frequency is increased because L_b becomes connected in parallel with L_f. Between these two extremes, there must exist an optimum value of R_f that causes the peak filter output impedance to be minimized. It can be shown [10,17] that all magnitude plots must pass through a common point, and therefore the optimum attains its peak at this point. This fact has been used to derive the design equations of optimally-damped L-C filter sections.

(a)

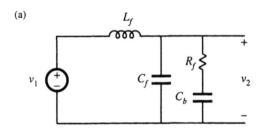

(b)

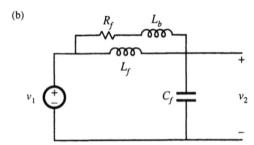

Fig. 10.20 Several practical approaches to damping the single-section input filter: (a) R_f-C_b parallel damping, (b) R_f-L_b parallel damping, (c) R_f-L_b series damping.

(c)

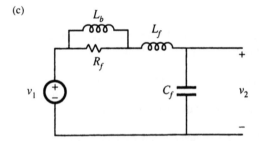

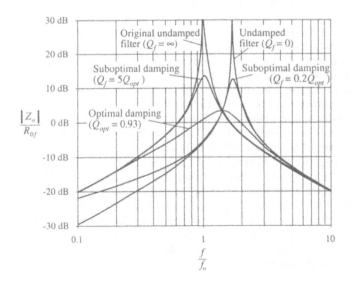

Fig. 10.21 Comparison of output impedance curves for optimal parallel R_f-L_b damping with undamped and several suboptimal designs. For this example, $n = L_b/L = 0.516$

10.4.1 R_f-C_b Parallel Damping

Optimization of the filter network of Fig. 10.20(a) and Section 10.3.2 was described in [10]. The high-frequency attenuation of this filter is not affected by the choice of C_b, and the high-frequency asymptote is identical to that of the original undamped filter. The sole tradeoff in design of the damping elements for this filter is in the size of the blocking capacitor C_b vs. the damping achieved.

For this filter, let us define the quantity n as the ratio of the blocking capacitance C_b to the filter capacitance C_f:

$$n = \frac{C_b}{C_f} \tag{10.28}$$

For the optimum design, the peak filter output impedance occurs at the frequency

$$f_m = f_f \sqrt{\frac{2}{2+n}} \tag{10.29}$$

The value of the peak output impedance for the optimum design is

$$\left| Z_o \right|_{mm} = R_{0f} \frac{\sqrt{2(2+n)}}{n} \tag{10.30}$$

The value of damping resistance that leads to optimum damping is described by

$$Q_{opt} = \frac{R_f}{R_{0f}} = \sqrt{\frac{(2+n)(4+3n)}{2n^2(4+n)}} \tag{10.31}$$

The above equations allow choice of the damping values R_f and C_b.

For example, let's redesign the damping network of Section 10.3.2, to achieve the same peak output impedance $\| Z_o(j\omega) \|_{mm} = 1\ \Omega$, while minimizing the value of the blocking capacitance C_b. From Section 10.3.2, the other parameter values are $R_{0f} = 0.84\ \Omega$, $C_f = 470\ \mu F$, and $L_f = 330\ \mu H$. First, we solve Eq. (10.30) to find the required value of n:

$$n = \frac{R_{0f}^2}{\left| Z_o \right|_{mm}^2} \left(1 + \sqrt{1 + 4 \frac{\left| Z_o \right|_{mm}^2}{R_{0f}^2}} \right) \tag{10.32}$$

Evaluation of this expression with the given numerical values leads to $n = 2.5$. The blocking capacitor is therefore required to have a value of $nC_f = 1200\ \mu F$. This is one-quarter of the value employed in Section 10.3.2. The value of R_f is then found by evaluation of Eq. (10.31), leading to

$$R_f = R_{0f} \sqrt{\frac{(2+n)(4+3n)}{2n^2(4+n)}} = 0.67\ \Omega \tag{10.33}$$

The output impedance of this filter design is compared with the output impedances of the original undamped filter of Section 10.3.1, and of the suboptimal design of Section 10.3.2, in Fig. 10.22. It can be

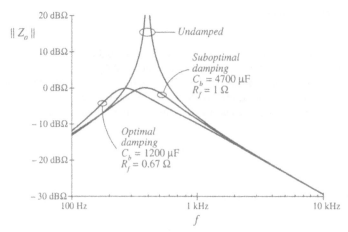

Fig. 10.22   Comparison of the output impedances of the design with optimal parallel R_f-C_b damping, the suboptimal design of Section 10.3.2, and the original undamped filter.

seen that the optimally damped filter does indeed achieve the desired peak output impedance of 1 Ω, at the slightly lower peak frequency given by Eq. (10.29)

The R_f-C_b parallel damping approach finds significant application in dc–dc converters. Since a series resistor is placed in series with C_b, C_b can be realized using capacitor types having substantial equivalent series resistance, such as electrolytic and tantalum types. However, in some applications, the R_f-L_b approaches of the next subsections can lead to smaller designs. Also, the large blocking capacitor value may be undesirable in applications having an ac input.

10.4.2 R_f-L_b Parallel Damping

Figure 10.20(b) illustrates the placement of damping resistor R_f in parallel with inductor L_f. Inductor L_b causes the filter to exhibit a two-pole attenuation characteristic at high frequency. To allow R_f to damp the filter, inductor L_b should have an impedance magnitude that is sufficiently smaller than R_f at the filter resonant frequency f_f. Optimization of this damping network is described in [17].

With this approach, inductor L_b can be physically much smaller than L_f. Since R_f is typically much greater than the dc resistance of L_f, essentially none of the dc current flows through L_b. Furthermore, R_f could be realized as the equivalent series resistance of L_b at the filter resonant frequency f_f. Hence, this is a very simple, low-cost approach to damping the input filter.

The disadvantage of this approach is the fact that the high-frequency attenuation of the filter is degraded: the high-frequency asymptote of the filter transfer function is increased from $1/\omega^2 L_f C_f$ to $1/\omega^2 (L_f \| L_b) C_f$. Furthermore, since the need for damping limits the maximum value of L_b, significant loss of high-frequency attenuation is unavoidable. To compensate, the value of L_f must be increased. Therefore, a tradeoff occurs between damping and degradation of high-frequency attenuation, as illustrated in Fig. 10.23. For example, limiting the degradation of high-frequency attenuation to 6 dB leads to an optimum peak filter output impedance $\| Z_o \|_{mm}$ of $\sqrt{6}$ times the original characteristic impedance R_{0f}. Additional damping leads to further degradation of the high-frequency attenuation.

The optimally damped design (i.e., the choice of R_f that minimizes the peak output impedance

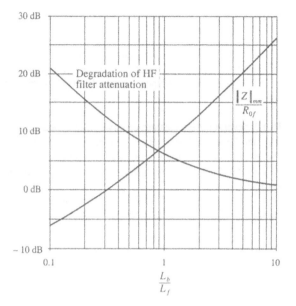

Fig. 10.23 Performance attained via optimal design procedure, parallel R_f-L_b circuit of 10.20(b). Optimum peak filter output impedance $\| Z_o \|_{mm}$ and increase of filter high-frequency gain, vs. $n = L_b/L$.

$\| Z_o \|$ for a given choice of L_b) is described by the following equations:

$$Q_{opt} = \frac{R_f}{R_{0f}} = \sqrt{\frac{n(3 + 4n)(1 + 2n)}{2(1 + 4n)}}$$

(10.34)

where

$$n = \frac{L_b}{L_f}$$

(10.35)

The peak filter output impedance occurs at frequency

$$f_m = f_f \sqrt{\frac{1 + 2n}{2n}}$$

(10.36)

and has the value

$$\left| Z_o \right|_{mm} = R_{0f} \sqrt{2n(1 + 2n)}$$

(10.37)

The attenuation of the filter high-frequency asymptote is degraded by the factor

$$\frac{L_f}{L_f \| L_b} = 1 + \frac{1}{n}$$

(10.38)

So, given an undamped L_f-C_f filter having corner frequency f_f, and characteristic impedance R_{0f}, and given a requirement for the maximum allowable output impedance $\| Z_o \|_{mm}$, one can solve Eq. (10.37) for the required value of n. One can then determine the required numerical values of L_b and R_f.

10.4.3 R_f-L_b Series Damping

Figure 10.20(c) illustrates the placement of damping resistor R_f in series with inductor L_f. Inductor L_b provides a dc bypass to avoid significant power dissipation in R_f. To allow R_f to damp the filter, inductor L_b should have an impedance magnitude that is sufficiently greater than R_f at the filter resonant frequency.

Although this circuit is theoretically equivalent to the parallel damping R_f-L_b case of Section 10.4.2, several differences are observed in practical designs. Both inductors must carry the full dc current, and hence both have significant size. The filter high-frequency attenuation is not affected by the choice of L_b, and the high-frequency asymptote is identical to that of the original undamped filter. The tradeoff in design of this filter does not involve high-frequency attenuation; rather, the issue is damping vs. bypass inductor size.

Design equations similar to those of the previous sections can be derived for this case. The optimum peak filter output impedance occurs at frequency

$$f_m = f_f \sqrt{\frac{2+n}{2(1+n)}} \tag{10.39}$$

and has the value

$$|Z_o|_{mm} = R_{0f} \frac{\sqrt{2(1+n)(2+n)}}{n} \tag{10.40}$$

The value of damping resistance that leads to optimum damping is described by

$$Q_{opt} = \frac{R_{0f}}{R_f} = \left(\frac{1+n}{n}\right) \sqrt{\frac{2(1+n)(4+n)}{(2+n)(4+3n)}} \tag{10.41}$$

For this case, the peak output impedance cannot be reduced below $\sqrt{2}\,R_{0f}$ via damping. Nonetheless, it is possible to further reduce the filter output impedance by redesign of L_f and C_f, to reduce the value of R_{0f}.

10.4.4   Cascading Filter Sections

A cascade connection of multiple L-C filter sections can achieve a given high-frequency attenuation with less volume and weight than a single-section L-C filter. The increased cutoff frequency of the multiple-section filter allows use of smaller inductance and capacitance values. Damping of each L-C section is usually required, which implies that damping of each section should be optimized. Unfortunately, the results of the previous sections are restricted to single-section filters. Interactions between cascaded L-C sections can lead to additional resonances and increased filter output impedance.

It is nonetheless possible to design cascaded filter sections such that interaction between L-C sections is negligible. In the approach described below, the filter output impedance is approximately equal to the output impedance of the last section, and resonances caused by interactions between stages are avoided. Although the resulting filter may not be "optimal" in any sense, insight can be gained that allows intelligent design of multiple-section filters with economical damping of each section.

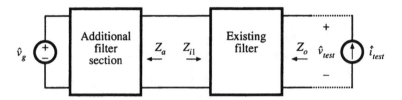

Fig. 10.24 Addition of a filter section at the input of an existing filter.

Consider the addition of a filter section to the input of an existing filter, as in Fig. 10.24. Let us assume that the existing filter has been correctly designed to meet the output impedance design criteria of Eq. (10.13): under the conditions $Z_a(s) = 0$ and $\hat{v}_g(s) = 0$, $\| Z_o \|$ is sufficiently small. It is desired to add a damped filter section that does not significantly increase $\| Z_o \|$.

Middlebrook's extra element theorem of Appendix C can again be invoked, to express how addition of the filter section modifies $Z_o(s)$:

$$\text{modified } Z_o(s) = \left[Z_o(s)\right]_{Z_a(s) = 0} \frac{\left(1 + \dfrac{Z_a(s)}{Z_{N1}(s)}\right)}{\left(1 + \dfrac{Z_a(s)}{Z_{D1}(s)}\right)} \tag{10.42}$$

where

$$Z_{N1}(s) = Z_{i1}(s)\Big|_{\hat{v}_{test}(s) \xrightarrow{\text{null}} 0} \tag{10.43}$$

is the impedance at the input port of the existing filter, with its output port short-circuited. Note that, in this particular case, nulling $\hat{v}_{test}(s)$ is the same as shorting the filter output port because the short-circuit current flows through the $\hat{i}_{test}$ source. The quantity

$$Z_{D1}(s) = Z_{i1}(s)\Big|_{\hat{i}_{test}(s) = 0} \tag{10.44}$$

is the impedance at the input port of the existing filter, with its output port open-circuited. Hence, the additional filter section does not significantly alter Z_o provided that

$$\begin{aligned} |Z_a| &\ll |Z_{N1}| \quad \text{and} \\ |Z_a| &\ll |Z_{D1}| \end{aligned} \tag{10.45}$$

Bode plots of the quantities Z_{N1} and Z_{D1} can be constructed either analytically or by computer simulation, to obtain limits of Z_a. When $\| Z_a \|$ satisfies Eq. (10.45), then the "correction factor" $(1 + Z_a/Z_{N1})/(1 + Z_a/Z_{D1})$ is approximately equal to 1, and the modified Z_o is approximately equal to the original Z_o.

To satisfy the design criteria (10.45), it is advantageous to select the resonant frequencies of Z_a to differ from the resonant frequencies of Z_{D1}. In other words, we should stagger-tune the filter sections. This minimizes the interactions between filter sections, and can allow use of smaller reactive element values.

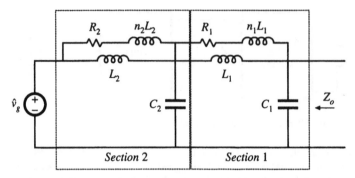

Fig. 10.25   Two-section input filter example, employing R_f-L_b parallel damping in each section.

10.4.5   Example: Two Stage Input Filter

As an example, let us consider the design of a two-stage filter using R_f-L_b parallel damping in each section as illustrated in Fig. 10.25 [17]. It is desired to achieve the same attenuation as the single-section filters designed in Sections 10.3.2 and 10.4.1, and to filter the input current of the same buck converter example of Fig. 10.8. These filters exhibit an attenuation of 80 dB at 250 kHz, and satisfy the design inequalities of Eq. (10.13) with the $\| Z_N \|$ and $\| Z_D \|$ impedances of Fig. 10.10. Hence, let's design the filter of Fig. 10.25 to attain 80 dB of attenuation at 250 kHz.

   As described in the previous section and below, it is advantageous to stagger-tune the filter sections so that interaction between filter sections is reduced. We will find that the cutoff frequency of filter section 1 should be chosen to be smaller than the cutoff frequency of section 2. In consequence, the attenuation of section 1 will be greater than that of section 2. Let us (somewhat arbitrarily) design to obtain 45 dB of attenuation from section 1, and 35 dB of attenuation from section 2 (so that the total is the specified 80 dB). Let us also select $n_1 = n_2 = n = L_b/L_f = 0.5$ for each section; as illustrated in Fig. 10.23, this choice leads to a good compromise between damping of the filter resonance and degradation of high frequency filter attenuation. Equation (10.38) and Fig. 10.23 predict that the R_f-L_b damping network will degrade the high frequency attenuation by a factor of $(1 + 1/n) = 3$, or 9.5 dB. Hence, the section 1 undamped resonant frequency f_{f1} should be chosen to yield 45 dB + 9.5 dB = 54.5 dB $\Rightarrow$ 533 of attenuation at 250 kHz. Since section 1 exhibits a two-pole ($-$ 40 dB/decade) roll-off at high frequencies, f_{f1} should be chosen as follows:

$$f_{f1} = \frac{(250 \text{ kHz})}{\sqrt{533}} = 10.8 \text{ kHz} \tag{10.46}$$

Note that this frequency is well above the 1.6 kHz resonant frequency f_0 of the buck converter output filter. Consequently, the output impedance $\| Z_o \|$ can be as large as 3 Ω, and still be well below the $\| Z_N(j\omega) \|$ and $\| Z_D(j\omega) \|$ plots of Fig. 10.10.

   Solution of Eq. (10.37) for the required section 1 characteristic impedance that leads to a peak output impedance of 3 Ω with $n = 0.5$ leads to

$$R_{0f1} = \frac{|Z_o|_{mm}}{\sqrt{2n(1 + 2n)}} = \frac{3\,\Omega}{\sqrt{2(0.5)(1 + 2(0.5))}} = 2.12\,\Omega \tag{10.47}$$

The filter inductance and capacitance values are therefore

$$L_1 = \frac{R_{0f1}}{2\pi f_{f1}} = 31.2\,\mu H$$

$$C_1 = \frac{1}{2\pi f_{f1}R_{0f1}} = 6.9\,\mu F \tag{10.48}$$

The section 1 damping network inductance is

$$n_1 L_1 = 15.6\,\mu H \tag{10.49}$$

The section 1 damping resistance is found from Eq. (10.34):

$$R_1 = Q_{opt}R_{0f1} = R_{0f1}\sqrt{\frac{n(3 + 4n)(1 + 2n)}{2(1 + 4n)}} = 1.9\,\Omega \tag{10.50}$$

The peak output impedance will occur at the frequency given by Eq. (10.36), 15.3 kHz. The quantities $\| Z_{N1}(j\omega) \|$ and $\| Z_{D1}(j\omega) \|$ for filter section 1 can now be constructed analytically or plotted by computer simulation. $\| Z_{N1}(j\omega) \|$ is the section 1 input impedance Z_{i1} with the output of section 1 shorted, and is given by the parallel combination of the sL_1 and the $(R_1 + sn_1L_1)$ branches. $\| Z_{D1}(j\omega) \|$ is the section 1 input impedance Z_{i1} with the output of section 1 open-circuited, and is given by the series combination of $Z_{N1}(s)$ with the capacitor impedance $1/sC_1$. Figure 10.26 contains plots of $\| Z_{N1}(j\omega) \|$ and $\| Z_{D1}(j\omega) \|$ for filter section 1, generated using Spice.

One way to approach design of filter section 2 is as follows. To avoid significantly modifying the overall filter output impedance Z_o, the section 2 output impedance $\| Z_a(j\omega) \|$ must be sufficiently less than $\| Z_{N1}(j\omega) \|$ and $\| Z_{D1}(j\omega) \|$. It can be seen from Fig. 10.26 that, with respect to $\| Z_{D1}(j\omega) \|$, this is most difficult to accomplish when the peak frequencies of sections 1 and 2 coincide. It is most difficult to satisfy the $\| Z_{N1}(j\omega) \|$ design criterion when the peak frequency of sections 2 is lower than the peak frequency of section 1. Therefore, the best choice is to stagger-tune the filter sections, with the resonant frequency of section 1 being lower than the peak frequency of section 2. This implies that section 1 will produce more high-frequency attenuation than section 2. For this reason, we have chosen to achieve 45 dB of attenuation with section 1, and 35 dB of attenuation from section 2.

The section 2 undamped resonant frequency f_{f2} should be chosen in the same manner used in Eq. (10.46) for section 1. We have chosen to select $n_2 = n = L_b/L_f = 0.5$ for section 2; this again means that the R_f-L_b damping network will degrade the high frequency attenuation by a factor of $(1 + 1/n) = 3$, or 9.5 dB. Hence, the section 2 undamped resonant frequency f_{f2} should be chosen to yield 35 dB + 9.5 dB = 44.5 dB $\Rightarrow$ 169 of attenuation at 250 kHz. Since section 2 exhibits a two-pole (− 40 dB/decade) roll-off at high frequencies, f_{f2} should be chosen as follows:

$$f_{f2} = \frac{(250\text{ kHz})}{\sqrt{169}} = 19.25\text{ kHz} \tag{10.51}$$

The output impedance of section 2 will peak at the frequency 27.2 kHz, as given by Eq. (10.36). Hence, the peak frequencies of sections 1 and 2 differ by almost a factor of 2.

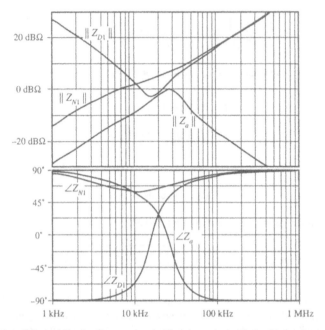

Fig. 10.26   Bode plot of Z_{N1} and Z_{D1} for filter section 1. Also shown is the Bode plot for the output impedance Z_a of filter section 2.

Figure 10.26 shows that, at 27.2 kHz, $\| Z_{D1}(j\omega) \|$ has a magnitude of roughly 3 dBΩ, and that $\| Z_{N1}(j\omega) \|$ is approximately 7 dBΩ. Hence, let us design section 2 to have a peak output impedance of 0 dBΩ $\Rightarrow$ 1 Ω. Solution of Eq. (10.37) for the required section 2 characteristic impedance leads to

$$R_{0f2} = \frac{|Z_a|_{mm}}{\sqrt{2n(1+2n)}} = \frac{1\ \Omega}{\sqrt{2(0.5)(1+2(0.5))}} = 0.71\ \Omega \tag{10.52}$$

The section 2 element values are therefore

$$L_2 = \frac{R_{0f2}}{2\pi f_{f2}} = 5.8\ \mu H$$

$$C_2 = \frac{1}{2\pi f_{f2} R_{0f2}} = 11.7\ \mu F$$

$$n_2 L_2 = 2.9\ \mu H \tag{10.53}$$

$$R_2 = Q_{opt} R_{0f2} = R_{0f2} \sqrt{\frac{n(3+4n)(1+2n)}{2(1+4n)}} = 0.65\ \Omega$$

A Bode plot of the resulting Z_a is overlaid on Fig. 10.26. It can be seen that $\| Z_a(j\omega) \|$ is less than, but very close to, $\| Z_{D1}(j\omega) \|$ between the peak frequencies of 15 kHz and 27 kHz. The impedance inequalities (10.45) are satisfied somewhat better below 15 kHz, and are satisfied very well at high frequency.

The resulting filter output impedance $\| Z_o(j\omega) \|$ is plotted in Fig. 10.27, for section 1 alone and for the complete cascaded two-section filter. It can be seen that the peak output impedance is approxi-

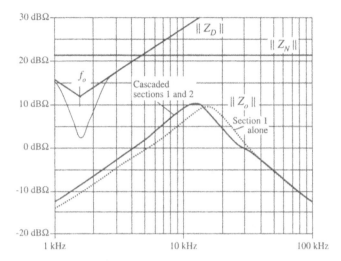

Fig. 10.27 Comparison of the impedance design criteria $\| Z_N(j\omega) \|$ and $\| Z_D(j\omega) \|$, Eq. (10.13), with the filter output impedance $\| Z_o(j\omega) \|$. *Solid line:* $\| Z_o(j\omega) \|$ of cascaded design. *Dashed line:* $\| Z_o(j\omega) \|$ of section 1 alone.

mately 10 dBΩ, or roughly 3 Ω. The impedance design criteria (10.13) are also shown, and it can be seen that the filter meets these design criteria. Note the absence of resonances in $\| Z_o(j\omega) \|$.

The effect of stage 2 on $\| Z_o(j\omega) \|$ is very small above 40 kHz [where inequalities (10.45) are very well satisfied], and has moderate-to-small effect at lower frequencies. It is interesting that, above approximately 12 kHz, the addition of stage 2 actually *decreases* $\| Z_o(j\omega) \|$. The reason for this can be seen from Fig. C.8 of Appendix C: when the phase difference between $\angle Z_a(j\omega)$ and $\angle Z_{D1}(j\omega)$ is not too large ($\leq 90°$), then the $1/(1 + Z_a/Z_{D1})$ term decreases the magnitude of the resulting $\| Z_o(j\omega) \|$. As can be seen from the phase plot of Fig. 10.26, this is indeed what happens. So allowing $\| Z_a(j\omega) \|$ to be similar in magnitude to $\| Z_{D1}(j\omega) \|$ above 12 kHz was an acceptable design choice.

The resulting filter transfer function is illustrated in Fig. 10.28. It can be seen that it does indeed attain the goal of 80 dB attenuation at 250 kHz.

Figure 10.29 compares the single stage design of Section 10.4.1 to the two-stage design of this section. Both designs attain 80 dB attenuation at 250 kHz, and both designs meet the impedance design criteria of Eq. (10.13). However, the single-stage approach requires much larger filter elements.

10.5 SUMMARY OF KEY POINTS

1. Switching converters usually require input filters, to reduce conducted electromagnetic interference and possibly also to meet requirements concerning conducted susceptibility.

2. Addition of an input filter to a converter alters the control-to-output and other transfer functions of the converter. Design of the converter control system must account for the effects of the input filter.

3. If the input filter is not damped, then it typically introduces complex poles and RHP zeroes into the converter control-to-output transfer function, at the resonant frequencies of the input filter. If these resonant frequencies are lower than the crossover frequency of the controller loop gain, then the phase margin will become negative and the regulator will be unstable.

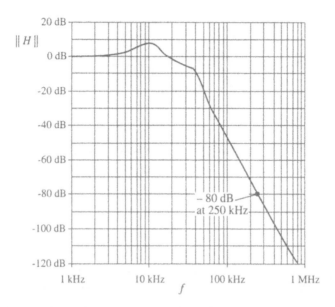

Fig. 10.28   Input filter transfer function, cascaded two-section design.

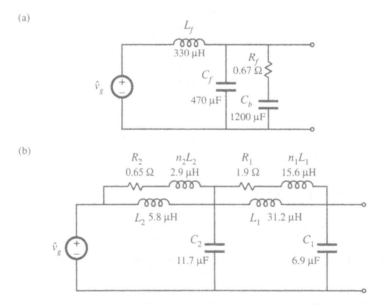

Fig. 10.29   Comparison of single-section (a) and two section (b) input filter designs. Both designs meet the design criteria (10.13), and both exhibit 80 dB of attenuation at 250 kHz.

4. The input filter can be designed so that it does not significantly change the converter control-to-output and other transfer functions. Impedance inequalities (10.13) give simple design criteria that guarantee this. To meet these design criteria, the resonances of the input filter must be sufficiently damped.

5. Optimization of the damping networks of single-section filters can yield significant savings in filter element size. Equations for optimizing three different filter sections are listed.

6. Substantial savings in filter element size can be realized via cascading filter sections. The design of noninteracting cascaded filter sections can be achieved by an approach similar to the original input filter design method. Impedance inequalities (10.45) give design criteria that guarantee that interactions are not substantial.

REFERENCES

[1]   M. NAVE, *Power Line Filter Design for Switched Mode Power Supplies*, New York: Van Nostrand Reinhold, 1991.

[2]   *Design Guide for Electromagnetic Interference (EMI) Reduction in Power Supplies*, MIL-HDBK-241B, U.S. Department of Defense, April 1 1981.

[3]   C. MARSHAM, *The Guide to the EMC Directive 89/336/EEC*, New York: IEEE Press, 1992.

[4]   P. DEGAUQUE and J. HAMELIN, *Electromagnetic Compatibility*, Oxford: Oxford University Press, 1993.

[5]   R. REDL, "Power Electronics and Electromagnetic Compatibility," *IEEE Power Electronics Specialists Conference*, 1996 Record, pp. 15-21.

[6]   P. R. WILLCOCK, J. A. FERREIRA, J. D. VAN WYK, "An Experimental Approach to Investigate the Generation and Propagation of Conducted EMI in Converters," *IEEE Power Electronics Specialists Conference*, 1998 Record, pp. 1140-1146.

[7]   L. ROSSETTO, S. BUSO and G. SPIAZZI, "Conducted EMI Issues in a 600W Single-Phase Boost PFC Design," *IEEE Transactions on Industry Applications*, Vol. 36, No. 2, pp. 578-585, March/April 2000.

[8]   F. DOS REIS, J. SEBASTIAN and J. UCEDA, "Determination of EMI Emissions in Power Factor Preregulators by Design," *IEEE Power Electronics Specialists Conference*, 1994 Record, pp. 1117-1126.

[9]   R. D. MIDDLEBROOK, "Input Filter Considerations in Design and Application of Switching Regulators," *IEEE Industry Applications Society Annual Meeting*, 1976 Record, pp. 366-382.

[10] R. D. MIDDLEBROOK, "Design Techniques for Preventing Input Filter Oscillations in Switched-Mode Regulators," *Proceedings of Powercon 5*, pp. A3.1 – A3.16, May 1978.

[11] T. PHELPS and W. TATE, "Optimizing Passive Input Filter Design," *Proceedings of Powercon 6*, pp. G1.1-G1.10, May 1979.

[12] Y. JANG and R. ERICKSON, "Physical Origins of Input Filter Oscillations in Current Programmed Converters," *IEEE Transactions on Power Electronics*, Vol 7, No. 4, pp. 725-733, October 1992.

[13] S. ERICH and W. POLIVKA, "Input Filter Design for Current-Programmed Regulators," *IEEE Applied Power Electronics Conference*, 1990 Proceedings, pp. 781-791, March 1990.

[14]   N. SOKAL, "System Oscillations Caused by Negative Input Resistance at the Power Input Port of a Switching Mode Regulator, Amplifier, Dc/Dc Converter, or Dc/Ac Inverter," *IEEE Power Electronics Specialists Conference*, 1973 Record, pp. 138-140.

[15]   A. KISLOVSKI, R. REDL, and N. SOKAL, *Dynamic Analysis of Switching-Mode Dc/Dc Converters*, New York: Van Nostrand Reinhold, Chapter 10, 1991.

[16]   S. KELKAR and F. LEE, "A Novel Input Filter Compensation Scheme for Switching Regulators," *IEEE Power Electronics Specialists Conference*, 1982 Record, pp. 260-271.

[17]   R. ERICKSON, "Optimal Single-Resistor Damping of Input Filters," *IEEE Applied Power Electronics Conference*, 1999 Proceedings, pp. 1073-1097, March 1999.

[18]   M. FLOREZ-LIZARRAGA and A. F. WITULSKI, "Input Filter Design for Multiple-Module Dc Power Systems," *IEEE Transactions on Power Electronics*, Vol 11, No. 3, pp. 472-479, May 1996.

[19]   V. VLATKOVIĆ, D. BOROJEVIĆ and F. LEE, "Input Filter Design for Power Factor Correction Circuits," *IEEE Transactions on Power Electronics*, Vol 11, No.1, pp. 199-205, January 1996.

[20]   F. YUAN, D. Y. CHEN, Y. WU and Y. CHEN, "A Procedure for Designing EMI Filters for Ac Line Applications," *IEEE Transactions on Power Electronics*, Vol 11, No. 1, pp. 170-181, January 1996.

[21]   G. SPIAZZI and J. POMILIO, "Interaction Between EMI Filter and Power Factor Preregulators with Average Current Control: Analysis and Design Considerations," *IEEE Transactions on Industrial Electronics*, Vol. 46, No. 3, pp. 577-584, June 1999.

PROBLEMS

10.1   It is required to design an input filter for the flyback converter of Fig. 10.30. The maximum allowed amplitude of switching harmonics of $i_{in}(t)$ is 10 μA rms. Calculate the required attenuation of the filter at the switching frequency.

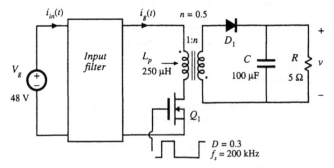

Fig. 10.30   Flyback converter, Problems 10.1, 10.4, 10.6, 10.8, and 10.10.

10.2   In the boost converter of Fig. 10.31, the input filter is designed so that the maximum amplitude of switching harmonics of $i_{in}(t)$ is not greater than 10 μA rms. Find the required attenuation of the filter at the switching frequency.

10.3   Derive the expressions for Z_N and Z_D in Table 10.1.

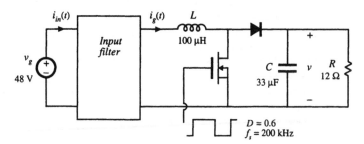

Fig. 10.31   Boost converter, Problems 10.2, 10.5, 10.7, and 10.9.

10.4   The input filter for the flyback converter of Fig. 10.30 is designed using a single L_f-C_f section. The filter is damped using a resistor R_f in series with a very large blocking capacitor C_b.

    (a)   Sketch a small-signal model of the flyback converter. Derive expressions for $Z_N(s)$ and $Z_D(s)$ using your model. Sketch the magnitude Bode plots of Z_N and Z_D, and label all salient features.

    (b)   Design the input filter, i.e., select the values of L_f, C_f, and R_f, so that: (*i*) the filter attenuation at the switching frequency is at least 100 dB, and (*ii*) the magnitude of the filter output impedance $Z_o(s)$ satisfies the conditions $\| Z_o(j\omega) \| < 0.3\| Z_D(j\omega) \|$, $\| Z_o(j\omega) \| < 0.3\| Z_N(j\omega) \|$, for all frequencies.

    (c)   Use Spice simulations to verify that the filter designed in part (b) meets the specifications.

    (d)   Using Spice simulations, plot the converter control-to-output magnitude and phase responses without the input filter, and with the filter designed in part (b). Comment on the changes introduced by the filter.

10.5   It is required to design the input filter for the boost converter of Fig. 10.31 using a single L_f-C_f section. The filter is damped using a resistor R_f in series with a very large blocking capacitor C_b.

    (a)   Sketch the magnitude Bode plots of $Z_N(s)$ and $Z_D(s)$ for the boost converter, and label all salient features.

    (b)   Design the input filter, i.e., select the values of L_f, C_f, and R_f, so that: (*i*) the filter attenuation at the switching frequency is at least 80 dB, and (*ii*) the magnitude of the filter output impedance $Z_o(s)$ satisfies the conditions $\| Z_o(j\omega) \| < 0.2\| Z_D(j\omega) \|$, $\| Z_o(j\omega) \| < 0.2\| Z_N(j\omega) \|$, for all frequencies.

    (c)   Use Spice simulations to verify that the filter designed in part (b) meets the specifications.

    (d)   Using Spice simulations, plot the converter control-to-output magnitude and phase responses without the input filter, and with the filter designed in part (b). Comment on the changes in the control-to-output responses introduced by the filter.

10.6   Repeat the filter design of Problem 10.4 using the optimum filter damping approach described in Section 10.4.1. Find the values of L_f, C_f, R_f, and C_b.

10.7   Repeat the filter design of Problem 10.5 using the optimum filter damping approach of Section 10.4.1. Find the values of L_f, C_f, R_f, and C_b.

10.8   Repeat the filter design of Problem 10.4 using the optimum R_f-L_b parallel damping approach described in Section 10.4.2. Find the values of L_f, C_f, R_f, and L_b.

10.9   Repeat the filter design of Problem 10.5 using the optimum R_f-L_b parallel damping approach described in Section 10.4.2. Find the values of L_f, C_f, R_f, and L_b.

10.10   It is required to design the input filter for the flyback converter of Fig. 10.30 using two filter sections. Each filter section is damped using a resistor in series with a blocking capacitor.

   (a)   Design the input filter, i.e., select values of all circuit parameters, so that (*i*) the filter attenuation at the switching frequency is at least 100 dB, and (*ii*) the magnitude of the filter output impedance $Z_o(s)$ satisfies the conditions $\| Z_o(j\omega) \| < 0.3 \| Z_D(j\omega) \|$, $\| Z_o(j\omega) \| < 0.3 \| Z_N(j\omega) \|$, for all frequencies.

   (b)   Use Spice simulations to verify that the filter designed in part (a) meets the specifications.

   (c)   Using Spice simulations, plot the converter control-to-output magnitude and phase responses without the input filter, and with the filter designed in part (b). Comment on the changes introduced by the filter.

11

AC and DC Equivalent Circuit Modeling of the Discontinuous Conduction Mode

So far, we have derived equivalent circuit models for dc–dc pulse-width modulation (PWM) converters operating in the continuous conduction mode. As illustrated in Fig. 11.1, the basic dc conversion property is modeled by an effective dc transformer, having a turns ratio equal to the conversion ratio $M(D)$. This model predicts that the converter has a voltage-source output characteristic, such that the output voltage is essentially independent of the load current or load resistance R. We have also seen how to refine this model, to predict losses and efficiency, converter dynamics, and small-signal ac transfer functions. We found that the transfer functions of the buck converter contain two low-frequency poles, owing to the converter filter inductor and capacitor. The control-to-output transfer functions of the boost and buck-boost converters additionally contain a right half-plane zero. Finally, we have seen how to utilize these results in the design of converter control systems.

What are the basic dc and small-signal ac equivalent circuits of converters operating in the discontinuous conduction mode (DCM)? It was found in Chapter 5 that, in DCM, the output voltage becomes load-dependent: the conversion ratio $M(D, K)$ is a function of the dimensionless parameter $K = 2L/RT_s$, which in turn is a function of the load resistance R. So the converter no longer has a voltage-source output characteristic, and hence the dc transformer model is less appropriate. In this chapter, the averaged switch modeling [1-8] approach is employed, to derive equivalent circuits of the DCM switch network.

In Section 11.1, it is shown that the *loss-free resistor* model [9-11] is the averaged switch model of the DCM switch network. This equivalent circuit represents the steady-state and large-signal dynamic characteristics of the DCM switch network, in a clear and simple manner. In the discontinuous conduction mode, the average transistor voltage and current obey Ohm's law, and hence the transistor is modeled by an effective resistor R_e. The average diode voltage and current obey a power source characteristic, with power equal to the power effectively dissipated in R_e. Therefore, the diode is modeled with a *dependent power source*.

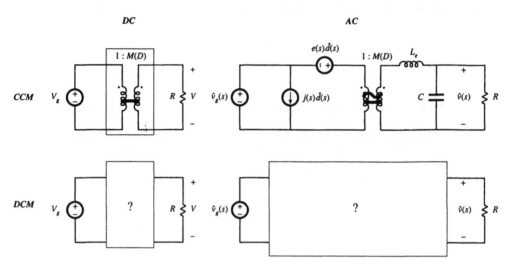

Fig. 11.1 The objective of this chapter is the derivation of large-signal dc and small-signal ac equivalent circuit models for converters operating in the discontinuous conduction mode.

Since most converters operate in discontinuous conduction mode at some operating points, small-signal ac DCM models are needed, to prove that the control systems of such converters are correctly designed. In Section 11.2, a small-signal model of the DCM switch network is derived by linearization of the loss-free resistor model. The transfer functions of DCM converters are quite different from their respective CCM transfer functions. The basic DCM buck, boost, and buck-boost converters essentially exhibit simple single-pole transfer functions [12, 13], in which the second pole and the RHP zero (in the case of boost and buck-boost converters) are at high frequencies. So the basic converters operating in DCM are easy to control; for this reason, converters are sometimes purposely operated in DCM for all loads. The transfer functions of higher order converters such as the DCM Ćuk or SEPIC are considerably more complicated; but again, one pole is shifted to high frequency, where it has negligible practical effect. This chapter concludes, in Section 11.3, with a discussion of a more detailed analysis used to predict high-frequency dynamics of DCM converters. The more detailed analysis predicts that the high-frequency pole of DCM converters occurs at frequencies near or exceeding the switching frequency [2-6]. The RHP zero, in the case of DCM buck-boost and boost converters, also occurs at high frequencies. This is why, in practice, the high-frequency dynamics can usually be neglected in DCM.

11.1   DCM AVERAGED SWITCH MODEL

Consider the buck-boost converter of Fig. 11.2. Let us follow the averaged switch modeling approach of Section 7.4, to derive an equivalent circuit that models the averaged terminal waveforms of the switch network. The general two-switch network and its terminal quantities $v_1(t)$, $i_1(t)$, $v_2(t)$, and $i_2(t)$ are defined as illustrated in Fig. 11.2, consistent with Fig. 7.39(a). The inductor and switch network voltage and current waveforms are illustrated in Fig. 11.3, for DCM operation.

The inductor current is equal to zero at the beginning of each switching period. During the first subinterval, while the transistor conducts, the inductor current increases with a slope of $v_g(t)/L$. At the

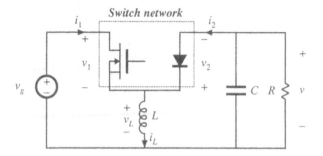

Fig. 11.2 Buck-boost converter example, with switch network terminal quantities identified.

end of the first subinterval, the inductor current $i_L(t)$ attains the peak value given by

$$i_{pk} = \frac{v_g}{L} d_1 T_s \tag{11.1}$$

During the second subinterval, while the diode conducts, the inductor current decreases with a slope equal to $v(t)/L$. The second subinterval ends when the diode becomes reverse-biased, at time $t = (d_1 + d_2)T_s$. The inductor current then remains at zero for the balance of the switching period. The inductor voltage is zero during the third subinterval.

A DCM averaged switch model can be derived with reference to the waveforms of Fig. 11.3. Following the approach of Section 7.4.2, let us find the average values of the switch network terminal waveforms $v_1(t)$, $v_2(t)$, $i_1(t)$, and $i_2(t)$ in terms of the converter state variables (inductor currents and capacitor voltages), the input voltage $v_g(t)$, and the subinterval lengths d_1 and d_2.

The average switch network input voltage $\langle v_1(t) \rangle_{T_s}$, or the average transistor voltage, is found by averaging the $v_1(t)$ waveform of Fig. 11.3:

$$\langle v_1(t) \rangle_{T_s} = d_1(t) \cdot 0 + d_2(t) \left(\langle v_g(t) \rangle_{T_s} - \langle v(t) \rangle_{T_s} \right) + d_3(t) \langle v_g(t) \rangle_{T_s} \tag{11.2}$$

Use of the identity $d_3(t) = 1 - d_1(t) - d_2(t)$ yields

$$\langle v_1(t) \rangle_{T_s} = \left(1 - d_1(t) \right) \langle v_g(t) \rangle_{T_s} - d_2(t) \langle v(t) \rangle_{T_s} \tag{11.3}$$

Similar analysis leads to the following expression for the average diode voltage:

$$\langle v_2(t) \rangle_{T_s} = d_1(t) \left(\langle v_g(t) \rangle_{T_s} - \langle v(t) \rangle_{T_s} \right) + d_2(t) \cdot 0 + d_3(t) \left(- \langle v(t) \rangle_{T_s} \right)$$
$$= d_1(t) \langle v_g(t) \rangle_{T_s} - \left(1 - d_2(t) \right) \langle v(t) \rangle_{T_s} \tag{11.4}$$

The average switch network input current $\langle i_1(t) \rangle_{T_s}$ is found by integrating the $i_1(t)$ waveform of Fig. 11.3 over one switching period:

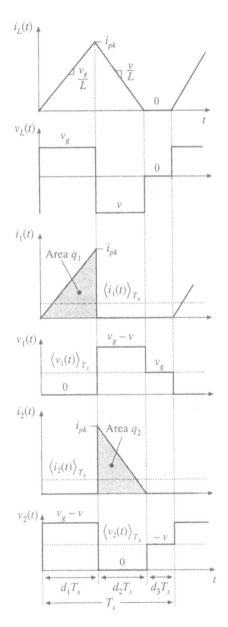

Fig. 11.3 Inductor and switch network voltage and current waveforms.

$$\langle i_1(t)\rangle_{T_s} = \frac{1}{T_s}\int_t^{t+T_s} i_1(t)dt = \frac{q_1}{T_s} \qquad (11.5)$$

The integral q_1 is equal to the area under the $i_1(t)$ waveform during the first subinterval. This area is easily evaluated using the triangle area formula:

$$q_1 = \int_t^{t+T_s} i_1(t)dt = \frac{1}{2}\left(d_1 T_s\right)\left(i_{pk}\right) \qquad (11.6)$$

Substitution of Eqs. (11.1) and (11.6) into Eq. (11.5) gives:

$$\langle i_1(t)\rangle_{T_s} = \frac{d_1^2(t)T_s}{2L}\langle v_g(t)\rangle_{T_s} \qquad (11.7)$$

Note that $\langle i_1(t)\rangle_{T_s}$ is not equal to $d_1\langle i_L(t)\rangle_{T_s}$. Since the inductor current ripple is not small, it is necessary to sketch the actual input current waveform, including the large switching ripple, and then correctly compute the average as in Eqs. (11.5) to (11.7).

The average diode current $\langle i_2(t)\rangle_{T_s}$ is found in a manner similar to that used above for $\langle i_1(t)\rangle_{T_s}$:

$$\langle i_2(t)\rangle_{T_s} = \frac{1}{T_s}\int_t^{t+T_s} i_2(t)dt = \frac{q_2}{T_s} \qquad (11.8)$$

The integral q_2 is equal to the area under the $i_2(t)$ waveform during the second subinterval. This area is evaluated using the triangle area formula:

$$q_2 = \int_t^{t+T_s} i_2(t)dt = \frac{1}{2}\left(d_2 T_s\right)\left(i_{pk}\right) \qquad (11.9)$$

Substitution of Eqs. (11.1) and (11.9) into Eq. (11.8) leads to:

$$\langle i_2(t)\rangle_{T_s} = \frac{d_1(t)d_2(t)T_s}{2L}\langle v_g(t)\rangle_{T_s} \qquad (11.10)$$

Equations (11.3), (11.4), (11.7) and (11.10) constitute the averaged terminal equations of the switch network in the DCM buck-boost converter. In these equations, it remains to express the subinterval length d_2 in terms of the switch duty cycle $d_1 = d$, and the converter averaged waveforms. One approach to finding the subinterval length d_2 is by solving the inductor current waveform. In the buck-boost converter, the diode switches off when the inductor current reaches zero, at the end of the second subinterval. As a result, $i_L(T_s) = i_L(0) = 0$. There is no net change in inductor current over one complete switching period, and no net volt-seconds are applied to the inductor over any complete switching period that starts at the time when the transistor is turned on. Therefore, the average inductor voltage computed over this period is zero,

$$\left\langle v_L(t) \right\rangle_{T_s} = d_1 \left\langle v_g(t) \right\rangle_{T_s} + d_2 \left\langle v(t) \right\rangle_{T_s} + d_3 \cdot 0 = 0 \tag{11.11}$$

even when the converter is not in equilibrium. This equation can be used to find the length of the second subinterval:

$$d_2(t) = -d_1(t) \frac{\left\langle v_g(t) \right\rangle_{T_s}}{\left\langle v(t) \right\rangle_{T_s}} \tag{11.12}$$

Substitution of Eq. (11.12) into Eqs. (11.3), (11.4), (11.7) and (11.10), allows us to obtain simple expressions for the averaged terminal waveforms of the switch network in the discontinuous conduction mode:

$$\left\langle v_1(t) \right\rangle_{T_s} = \left\langle v_g(t) \right\rangle_{T_s} \tag{11.13}$$

$$\left\langle v_2(t) \right\rangle_{T_s} = -\left\langle v(t) \right\rangle_{T_s} \tag{11.14}$$

$$\left\langle i_1(t) \right\rangle_{T_s} = \frac{d_1^2(t)T_s}{2L} \left\langle v_1(t) \right\rangle_{T_s} \tag{11.15}$$

$$\left\langle i_2(t) \right\rangle_{T_s} = \frac{d_1^2(t)T_s}{2L} \frac{\left\langle v_1(t) \right\rangle_{T_s}^2}{\left\langle v_2(t) \right\rangle_{T_s}} \tag{11.16}$$

Let us next construct an equivalent circuit corresponding to the averaged switch network equations (11.15) and (11.16). The switch network input port is modeled by Eq. (11.15). This equation states that the average input current $\left\langle i_1(t) \right\rangle_{T_s}$ is proportional to the applied input voltage $\left\langle v_1(t) \right\rangle_{T_s}$. In other words, the low-frequency components of the switch network input port obey Ohm's law:

$$\left\langle i_1(t) \right\rangle_{T_s} = \frac{\left\langle v_1(t) \right\rangle_{T_s}}{R_e(d_1)} \tag{11.17}$$

where the effective resistance R_e is

$$R_e(d_1) = \frac{2L}{d_1^2 T_s} \tag{11.18}$$

An equivalent circuit is illustrated in Fig. 11.4. During the first subinterval, the slope of the input current waveform $i_1(t)$ is proportional to the input voltage $\left\langle v_g(t) \right\rangle_{T_s} = \left\langle v_1(t) \right\rangle_{T_s}$, as illustrated in Fig. 11.3. As a result, the peak current i_{pk}, the total charge q_1, and the average input current $\left\langle i_1(t) \right\rangle_{T_s}$, are also proportional to $\left\langle v_1(t) \right\rangle_{T_s}$. Of course, there is no physical resistor inside the converter. Indeed, if the converter elements are ideal, then no heat is generated inside the converter. Rather, the power apparently consumed by R_e is transferred to the switch network output port.

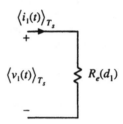

Fig. 11.4 Equivalent circuit that models the average waveforms of the switch input (transistor) port.

The switch network output (diode) port is modeled by Eq. (11.16), or

$$\langle i_2(t)\rangle_{T_s}\langle v_2(t)\rangle_{T_s} = \frac{\langle v_1(t)\rangle_{T_s}^2}{R_e(d_1)} = \langle p(t)\rangle_{T_s} \tag{11.19}$$

Note that $\langle v_1(t)\rangle_{T_s}^2/R_e$ is the average power $\langle p(t)\rangle_{T_s}$ apparently consumed by the effective resistor $R_e(d_1)$. Equation (11.19) states that this power flows out of the switch network output port. So the switch network consumes no net power—its average input and output powers are equal.

Equation (11.19) can also be derived by consideration of the inductor stored energy. During the first subinterval, the inductor current increases from 0 to i_{pk}. In the process, the inductor stores the following energy:

$$\frac{1}{2} L i_{pk}^2 = \frac{\langle v_1\rangle_{T_s}^2 d_1^2 T_s^2}{2L} = \frac{\langle v_1\rangle_{T_s}^2}{R_e(d_1)} T_s \tag{11.20}$$

Here, i_{pk} has been expressed in terms of $\langle v_1(t)\rangle_{T_s}$ using Eqs. (11.1) and (11.13). This energy is transferred from the source v_g, through the switch network input terminals (i.e., through the transistor), to the inductor. During the second subinterval, the inductor releases all of its stored energy through the switch network output terminals (i.e., through the diode), to the output. The average output power can therefore be expressed as the energy transferred per cycle, divided by the switching period:

$$\langle p(t)\rangle_{T_s} = \left(\frac{\langle v_1\rangle_{T_s}^2}{R_e(d_1)} T_s\right)\left(\frac{1}{T_s}\right) = \frac{\langle v_1\rangle_{T_s}^2}{R_e(d_1)} \tag{11.21}$$

This power is transferred to the load, and hence

$$\langle v\rangle_{T_s}\langle i_2\rangle_{T_s} = \langle v_2\rangle_{T_s}\langle i_2\rangle_{T_s} = \langle p(t)\rangle_{T_s} = \frac{\langle v_1\rangle_{T_s}^2}{R_e(d_1)} \tag{11.22}$$

This result coincides with Eq. (11.19).

The average power $\langle p(t)\rangle_{T_s}$ is independent of the load characteristics, and is determined solely by the effective resistance R_e and the applied switch network input terminal voltage or current. In other words, the switch network output port behaves as a source of power, equal to the power apparently consumed by the effective resistance R_e. This behavior is represented schematically by the dependent power source symbol illustrated in Fig. 11.5. In any lossless two-port network, when the voltage and current at one port are independent of the characteristics of the external network connected to the second port, then the second port must exhibit a dependent power source characteristic [10]. This situation arises in a num-

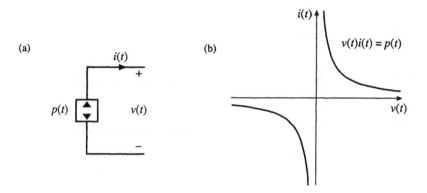

Fig. 11.5 The dependent power source: (a) schematic symbol, (b) *i–v* characteristic.

ber of common power-processing applications, including switch networks operating in the discontinuous conduction mode.

The power source characteristic illustrated in Fig. 11.5(b) is symmetrical with respect to voltage and current; in consequence, the power source exhibits several unique properties. Similar to the voltage source, the ideal power source must not be short-circuited; otherwise, infinite current occurs. And similar to the current source, the ideal power source must not be open-circuited, to avoid infinite terminal voltage. The power source must be connected to a load capable of absorbing the power $p(t)$, and the operating point is defined by the intersection of the load and power source *i–v* characteristics.

As illustrated in Fig. 11.6(a), series- and parallel-connected power sources can be combined

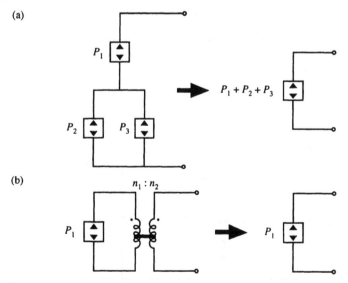

Fig. 11.6 Circuit manipulations of power source elements: (a) combination of series- and parallel- connected power sources into a single equivalent power source, (b) invariance of the power source to reflection through an ideal transformer of arbitrary turns ratio.

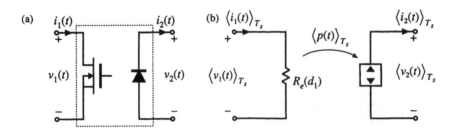

Fig. 11.7 (a) the general two-switch network, and (b) the corresponding averaged switch model in the discontinuous conduction mode: the average transistor waveforms obey Ohm's law, while the average diode waveforms behave as a dependent power source.

into a single power source, equal to the sum of the powers of the individual sources. Fig. 11.6(b) illustrates how reflection of a power source through a transformer, having an arbitrary turns ratio, leaves the power source unchanged. Power sources are also invariant to duality transformations.

The averaged large-signal model of the general two-switch network in DCM is illustrated in Fig. 11.7(b). The input port behaves effectively as resistance R_e. The instantaneous power apparently consumed by R_e is transferred to the output port, and the output port behaves as a dependent power source. This lossless two-port network is called the *loss-free resistor* model (LFR) [9]. The loss-free resistor represents the basic power conversion properties of DCM switch networks [11]. It can be shown that the loss-free resistor models the averaged properties of DCM switch networks not only in the buck-boost converter, but also in other PWM converters.

When the switch network of the DCM buck-boost converter is replaced by the averaged model of Fig. 11.7(b), the converter equivalent circuit of Fig. 11.8 is obtained. Upon setting all averaged waveforms to their quiescent values, and letting the inductor and capacitor become a short-circuit and an open-circuit, respectively, we obtain the dc model of Fig. 11.9.

Systems containing power sources or loss-free resistors can usually be easily solved, by equating average source and load powers. For example, in the dc network of Fig. 11.9, the power flowing into the converter input terminals is

$$P = \frac{V_g^2}{R_e} \tag{11.23}$$

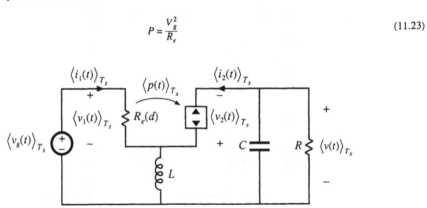

Fig. 11.8 Replacement of the switch network of the DCM buck-boost converter with the loss-free resistor model.

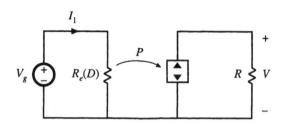

Fig. 11.9 Dc network example containing a loss-free resistor model.

The power flowing into the load resistor is

$$P = \frac{V^2}{R} \tag{11.24}$$

The loss-free resistor model states that these two powers must be equal:

$$P = \frac{V_g^2}{R_e} = \frac{V^2}{R} \tag{11.25}$$

Solution for the voltage conversion ratio $M = V/V_g$ yields

$$\frac{V}{V_g} = \pm \sqrt{\frac{R}{R_e}} \tag{11.26}$$

Equation (11.26) is a general result, valid for any converter that can be modeled by a loss-free resistor and that drives a resistive load. Other arguments must be used to determine the polarity of V/V_g. In the buck-boost converter shown in Fig. 11.2, the diode polarity indicates that V/V_g must be negative. The steady-state value of R_e is

$$R_e(D) = \frac{2L}{D^2 T_s} \tag{11.27}$$

where D is the quiescent transistor duty cycle. Substitution of Eq. (11.27) into (11.26) leads to

$$\frac{V}{V_g} = -\sqrt{\frac{D^2 T_s R}{2L}} = -\frac{D}{\sqrt{K}} \tag{11.28}$$

with $K = 2L/RT_s$. This equation coincides with the previous steady-state result given in Table 5.2.

Similar arguments apply when the waveforms contain ac components. For example, consider the network of Fig. 11.10, in which the voltages and currents are periodic functions of time. The rms val-

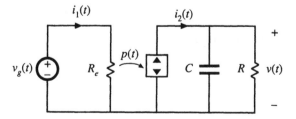

Fig. 11.10 Ac network example containing a loss-free resistor model.

ues of the waveforms can be determined by simply equating the average source and load powers. The average power flowing into the converter input port is

$$P_{av} = \frac{V_{g,rms}^2}{R_e} \tag{11.29}$$

where P_{av} is the average power consumed by the effective resistance R_e. No average power is consumed by capacitor C, and hence the average power P_{av} must flow entirely into the load resistor R:

$$P_{av} = \frac{V_{rms}^2}{R} \tag{11.30}$$

Upon equating Eqs. (11.29) and (11.30), we obtain

$$\frac{V_{rms}}{V_{g,rms}} = \sqrt{\frac{R}{R_e}} \tag{11.31}$$

Thus, the rms terminal voltages obey the same relationship as in the dc case.

Averaged equivalent circuits of the DCM buck, boost, and buck-boost converters, as well as the DCM Ćuk and SEPIC converters, are listed in Fig. 11.11. In each case, the averaged transistor waveforms obey Ohm's law, and are modeled by an effective resistance R_e. The averaged diode waveforms follow a power source characteristic, equal to the power effectively dissipated in R_e. For the buck, boost, and buck-boost converters, R_e is given by

$$R_e = \frac{2L}{d^2 T_s} \tag{11.32}$$

For the Ćuk and SEPIC converters, R_e is given by

$$R_e = \frac{2(L_1 \| L_2)}{d^2 T_s} \tag{11.33}$$

Here, d is the transistor duty cycle.

Steady-state conditions in the converters of Fig. 11.11 are found by letting the inductors and capacitors become short-circuits and open-circuits, respectively, and then solving the resulting dc circuits with $d(t) = D$. The buck-boost, Ćuk, and SEPIC then reduce to the circuit of Fig. 11.9. The buck and boost converters reduce to the circuits of Fig. 11.12. Equilibrium conversion ratios $M = V/V_g$ of these converters are summarized in Table 11.1, as functions of $R_e(D)$. It can be shown that these converters operate in the discontinuous conduction mode whenever the load current I is less than the critical current I_{crit}:

$$\begin{aligned} I &> I_{crit} \quad \text{for CCM} \\ I &< I_{crit} \quad \text{for DCM} \end{aligned} \tag{11.34}$$

For all of these converters, I_{crit} is given by

$$I_{crit} = \frac{1-D}{D} \frac{V_g}{R_e(D)} \tag{11.35}$$

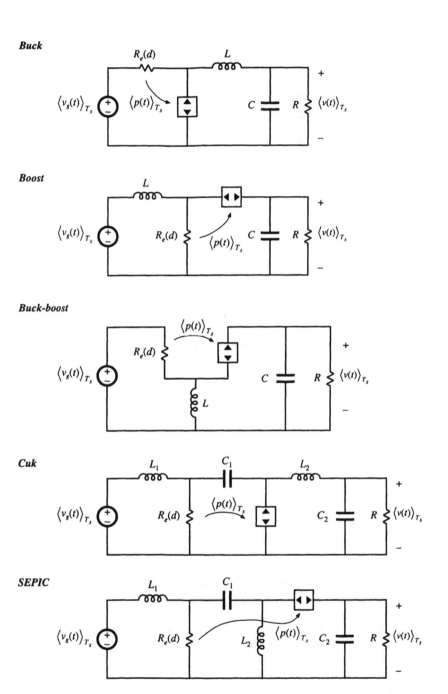

Fig. 11.11 Averaged large-signal equivalent circuits of five basic converters operating in the discontinuous conduction mode.

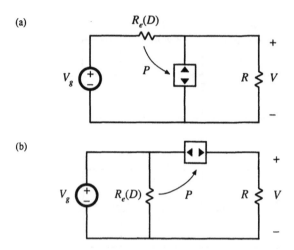

Fig. 11.12   Dc equivalent circuits representing the buck (a) and boost (b) converters operating in DCM.

Table 11.1   CCM and DCM conversion ratios of basic converters

Converter	M, CCM	M, DCM
Buck	D	$\dfrac{2}{1+\sqrt{1+4R_e/R}}$
Boost	$\dfrac{1}{1-D}$	$\dfrac{1+\sqrt{1+4R/R_e}}{2}$
Buck-boost, Ćuk	$\dfrac{-D}{1-D}$	$-\sqrt{\dfrac{R}{R_e}}$
SEPIC	$\dfrac{D}{1-D}$	$\sqrt{\dfrac{R}{R_e}}$

11.2   SMALL-SIGNAL AC MODELING OF THE DCM SWITCH NETWORK

The next step is construction of a small-signal equivalent circuit model for converters operating in the discontinuous conduction mode. In the large-signal ac equivalent circuits of Fig. 11.11, the averaged switch networks are nonlinear. Hence, construction of a small-signal ac model involves perturbation and linearization of the loss-free resistor network. The signals in the large-signal averaged DCM switch network model of Fig. 11.13(a) are perturbed about a quiescent operating point, as follows:

(a)

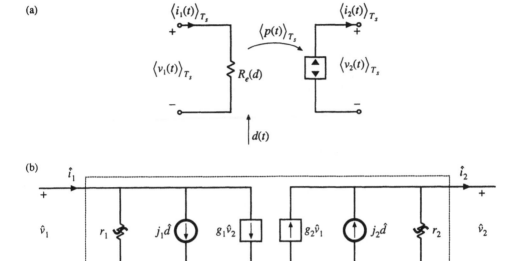

(b)

Fig. 11.13   Averaged models of the general two-switch network in a converter operating in DCM: (a) large-signal model, (b) small-signal model.

$$d(t) = D + \hat{d}(t)$$

$$\langle v_1(t) \rangle_{T_s} = V_1 + \hat{v}_1(t)$$

$$\langle i_1(t) \rangle_{T_s} = I_1 + \hat{i}_1(t) \qquad (11.36)$$

$$\langle v_2(t) \rangle_{T_s} = V_2 + \hat{v}_2(t)$$

$$\langle i_2(t) \rangle_{T_s} = I_2 + \hat{i}_2(t)$$

Here, D is the quiescent value of the transistor duty cycle, V_1 is the quiescent value of the applied average transistor voltage $\langle v_1(t) \rangle_{T_s}$, etc. The quantities $\hat{d}(t)$, $\hat{v}_1(t)$, etc., are small ac variations about the respective quiescent values. It is desired to linearize the average switch network terminal equations (11.15) and (11.16).

Equations (11.15) and (11.16) express the average terminal currents $\langle i_1(t) \rangle_{T_s}$ and $\langle i_2(t) \rangle_{T_s}$ as functions of the transistor duty cycle $d(t) = d_1(t)$ and the average terminal voltages $\langle v_1(t) \rangle_{T_s}$ and $\langle v_2(t) \rangle_{T_s}$. Upon perturbation and linearization of these equations, we will therefore find that $\hat{i}_1(t)$ and $\hat{i}_2(t)$ are expressed as linear functions of $\hat{d}(t)$, $\hat{v}_1(t)$, and $\hat{v}_2(t)$. So the small-signal switch network equations can be written in the following form:

$$\hat{i}_1 = \frac{\hat{v}_1}{r_1} + j_1\hat{d} + g_1\hat{v}_2$$

$$\hat{i}_2 = -\frac{\hat{v}_2}{r_2} + j_2\hat{d} + g_2\hat{v}_1 \qquad (11.37)$$

These equations describe the two-port equivalent circuit of Fig. 11.13(b).

The parameters r_1, j_1, and g_1 can be found by Taylor expansion of Eq. (11.15), as described in Section 7.2.7. The average transistor current $\langle i_1(t) \rangle_{T_s}$, Eq. (11.15), can be expressed in the following form:

$$\langle i_1(t) \rangle_{T_s} = \frac{\langle v_1(t) \rangle_{T_s}}{R_e(d(t))} = f_1\left(\langle v_1(t) \rangle_{T_s}, \langle v_2(t) \rangle_{T_s}, d(t) \right) \tag{11.38}$$

Let us expand this expression in a three-dimensional Taylor series, about the quiescent operating point (V_1, V_2, D):

$$I_1 + \hat{i}_1(t) = f_1\left(V_1, V_2, D \right) + \hat{v}_1(t) \left. \frac{\partial f_1\left(v_1, V_2, D \right)}{\partial v_1} \right|_{v_1 = V_1}$$

$$+ \hat{v}_2(t) \left. \frac{\partial f_1\left(V_1, v_2, D \right)}{\partial v_2} \right|_{v_2 = V_2} + \hat{d}(t) \left. \frac{\partial f_1\left(V_1, V_2, d \right)}{\partial d} \right|_{d = D} \tag{11.39}$$

+ higher–order nonlinear terms

For simplicity of notation, the angle brackets denoting average values are dropped in the above equation. The dc terms on both sides of Eq. (11.39) must be equal:

$$I_1 = f_1\left(V_1, V_2, D \right) = \frac{V_1}{R_e(D)} \tag{11.40}$$

As usual, we linearize the equation by discarding the higher-order nonlinear terms. The remaining first-order linear ac terms on both sides of Eq. (11.39) are equated:

$$\hat{i}_1(t) = \hat{v}_1(t) \frac{1}{r_1} + \hat{v}_2(t) g_1 + \hat{d}(t) j_1 \tag{11.41}$$

where

$$\frac{1}{r_1} = \left. \frac{\partial f_1\left(v_1, V_2, D \right)}{\partial v_1} \right|_{v_1 = V_1} = \frac{1}{R_e(D)} \tag{11.42}$$

$$g_1 = \left. \frac{\partial f_1\left(V_1, v_2, D \right)}{\partial v_2} \right|_{v_2 = V_2} = 0 \tag{11.43}$$

$$j_1 = \left. \frac{\partial f_1\left(V_1, V_2, d \right)}{\partial d} \right|_{d = D} = - \frac{V_1}{R_e^2(D)} \left. \frac{\partial R_e(d)}{\partial d} \right|_{d = D}$$

$$= \frac{2V_1}{DR_e(D)} \tag{11.44}$$

Thus, the small-signal input resistance r_1 is equal to the effective resistance R_e, evaluated at the quiescent operating point. This term describes how variations in $\langle v_1(t) \rangle_{T_s}$ affect $\langle i_1(t) \rangle_{T_s}$, via $R_e(D)$. The small-signal parameter g_1 is equal to zero, since the average transistor current $\langle i_1(t)\rangle_{T_s}$ is independent of the average diode voltage $\langle v_2(t)\rangle_{T_s}$. The small-signal gain j_1 describes how duty cycle variations, which affect the value of $R_e(d)$, lead to variations in $\langle i_1(t)\rangle_{T_s}$.

In a similar manner, $\langle i_2(t)\rangle_{T_s}$ from Eq. (11.16) can be expressed as

$$\langle i_2(t)\rangle_{T_s} = \frac{\langle v_1(t)\rangle^2_{T_s}}{R_e\big(d(t)\big)\langle v_2(t)\rangle_{T_s}} = f_2\Big(\langle v_1(t)\rangle_{T_s}, \langle v_2(t)\rangle_{T_s}, d(t)\Big) \tag{11.45}$$

Expansion of the function $f_2(v_1, v_2, d)$ in a three-dimensional Taylor series about the quiescent operating point leads to

$$I_2 + \hat{i}_2(t) = f_2\big(V_1, V_2, D\big) + \hat{v}_1(t)\left.\frac{\partial f_2\big(v_1, V_2, D\big)}{\partial v_1}\right|_{v_1 = V_1}$$

$$+ \hat{v}_2(t)\left.\frac{\partial f_2\big(V_1, v_2, D\big)}{\partial v_2}\right|_{v_2 = V_2} + \hat{d}(t)\left.\frac{\partial f_2\big(V_1, V_2, d\big)}{\partial d}\right|_{d = D} \tag{11.46}$$

$$+ \text{ higher–order nonlinear terms}$$

By equating the dc terms on both sides of Eq. (11.46), we obtain

$$I_2 = f_2\big(V_1, V_2, D\big) = \frac{V_1^2}{R_e(D)V_2} \tag{11.47}$$

The higher-order nonlinear terms are discarded, leaving the following first-order linear ac terms:

$$\hat{i}_2(t) = \hat{v}_2(t)\left(-\frac{1}{r_2}\right) + \hat{v}_1(t)g_2 + \hat{d}(t)j_2 \tag{11.48}$$

with

$$\frac{1}{r_2} = -\left.\frac{\partial f_2\big(V_1, v_2, D\big)}{\partial v_2}\right|_{v_2 = V_2} = \frac{1}{R} = \frac{1}{M^2 R_e(D)} \tag{11.49}$$

$$g_2 = \left.\frac{\partial f_2\big(v_1, V_2, D\big)}{\partial v_1}\right|_{v_1 = V_1} = \frac{2}{M R_e(D)} \tag{11.50}$$

$$j_2 = \left.\frac{\partial f_2\big(V_1, V_2, d\big)}{\partial d}\right|_{d = D} = -\frac{V_1^2}{R_e^2(D)V_2}\left.\frac{\partial R_e(d)}{\partial d}\right|_{d = D}$$

$$= \frac{2V_1}{DM R_e(D)} \tag{11.51}$$

The output resistance r_2 describes how variations in $\langle v_2(t)\rangle_{T_s}$ influence $\langle i_2(t)\rangle_{T_s}$. As illustrated in Fig. 11.14,

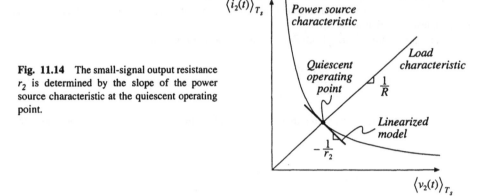

Fig. 11.14 The small-signal output resistance r_2 is determined by the slope of the power source characteristic at the quiescent operating point.

Table 11.2   Small-signal DCM switch model parameters

Switch network	g_1	j_1	r_1	g_2	j_2	r_2
General two-switch, Fig. 11.7(a)	0	$\dfrac{2V_1}{DR_e}$	R_e	$\dfrac{2}{MR_e}$	$\dfrac{2V_1}{DMR_e}$	M^2R_e
Buck, Fig. 11.16(a)	$\dfrac{1}{R_e}$	$\dfrac{2(1-M)V_1}{DR_e}$	R_e	$\dfrac{2-M}{MR_e}$	$\dfrac{2(1-M)V_1}{DMR_e}$	M^2R_e
Boost, Fig. 11.16(b)	$\dfrac{1}{(M-1)^2R_e}$	$\dfrac{2MV_1}{D(M-1)R_e}$	$\dfrac{(M-1)^2}{M^2}R_e$	$\dfrac{2M-1}{(M-1)^2R_e}$	$\dfrac{2V_1}{D(M-1)R_e}$	$(M-1)^2R_e$

r_2 is determined by the slope of the power source characteristic, evaluated at the quiescent operating point. For a linear resistive load, $r_2 = R$. For any type of load, it is true that $r_2 = M^2R_e(D)$. The parameters j_2 and g_2 describe how variations in the duty cycle $d(t)$ and in the average transistor voltage $\langle v_1(t)\rangle_{T_s}$ (which influence the average power $\langle p(t)\rangle_{T_s}$) lead to variations in the average diode current $\langle i_2(t)\rangle_{T_s}$. Values of the small-signal parameters in the DCM switch model of Fig. 11.13(b) are summarized in the top row of Table 11.2.

A small-signal model of the DCM buck-boost converter is obtained by replacing the transistor and diode of the converter with the switch model of Fig. 11.13(b). The result is illustrated in Fig. 11.15. This equivalent circuit can now be solved using conventional linear circuit analysis techniques, to determine the transfer functions and other small-signal quantities of interest.

The same small-signal switch model can be employed to model other DCM converters, by simply replacing the transistor and diode with ports 1 and 2, respectively, of the two-port model of Fig. 11.13(b). An alternative approach, which yields more convenient results in the analysis of the buck and boost converters, is to define the switch network as illustrated in Figs. 11.16(a) and 11.16(b), respectively. These switch networks can also be modeled using the two-port small-signal equivalent circuit of Fig. 11.16(c); however, new expressions for the parameters r_1, j_1, g_1, etc., must be derived. These expressions are again found by linearizing the equations of the averaged switch network terminal currents.

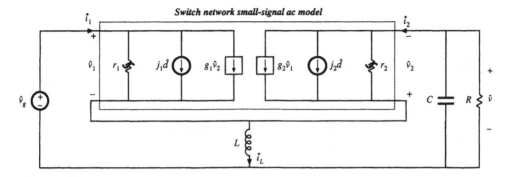

Fig. 11.15 Small-signal ac model of the DCM buck-boost converter obtained by insertion of the switch network two-port small-signal model into the original converter circuit.

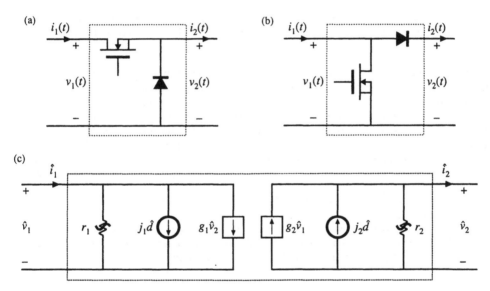

Fig. 11.16 A convenient way to model the switch networks of DCM buck and boost converters: (a) defined terminal quantities of the DCM buck switch network, (b) defined terminal quantities of the boost switch network, (c) two-port small-signal ac model. The model parameters are given in Table 11.2.

Table 11.2 lists the small-signal parameters for the buck switch network of Fig. 11.16(a) (middle row) and for the boost switch network of Fig. 11.16(b) (bottom row). Insertion of the small-signal two-port model into the DCM buck and boost converters leads to the equivalent circuits of Fig. 11.17.

The small-signal equivalent circuit models of Fig. 11.15 and Fig. 11.17 contain two dynamic elements: capacitor C and inductor L. Control-to-output transfer functions obtained by solving these equivalent circuit models have two poles. It has been shown [2-6] that one of the poles, due to the capacitor C, appears at a low frequency, while the other pole (and a RHP zero in the case of boost and buck-

(a)

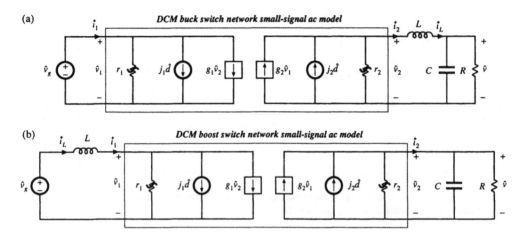

Fig. 11.17   Small-signal ac models of (a) the DCM buck converter, and (b) the DCM boost converter, obtained by replacing the switch networks defined in Fig. 11.16(a) and (b) with the small-signal switch model of Fig. 11.16(c).

boost converters) due to the inductor L, occurs at much higher frequency, close to the converter switching frequency. Therefore, in practice, the DCM buck, boost, and buck-boost converters exhibit essentially single-pole transfer functions, which are negligibly influenced by the inductor dynamics.

The small-signal equivalent circuit models have been derived in this section from the large-signal averaged switch network equations (11.15) and (11.16). These equations are based on Eq. (11.11), which states that the average inductor voltage, and therefore its small-signal ac voltage, is zero. This contradicts predictions of the resulting small-signal models in Figs. 11.15 and 11.17. As a result, we expect that the models derived in this section can be used to predict low-frequency dynamics, while predictions of the high-frequency dynamics due to the inductor L are of questionable validity. Equivalent circuit models that give more accurate predictions of high-frequency dynamics of DCM converters are discussed in Section 11.3.

A simple approximate way to determine the low-frequency small-signal transfer functions of the buck, boost, and buck-boost converters is to let the inductance L tend to zero. If L is shorted in the equivalent circuits of Figs. 11.15 and 11.17, the model in all three cases reduces to Fig. 11.18. This cir-

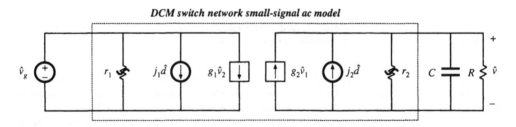

Fig. 11.18   Low-frequency ac model obtained by letting L approach zero. The buck, boost, or buck-boost converters can be modeled, by employing the appropriate parameters from Table 11.2.

cuit is relatively easy to solve.

The control-to-output transfer function $G_{vd}(s)$ is found by letting $\hat{v}_g = 0$ in Fig. 11.18. Solution for $\hat{v}$ then leads to

$$G_{vd}(s) = \left.\frac{\hat{v}}{\hat{d}}\right|_{\hat{v}_g = 0} = \frac{G_{d0}}{1 + \frac{s}{\omega_p}} \qquad (11.52)$$

with

$$\begin{aligned} G_{d0} &= j_2\left(R \| r_2\right) \\ \omega_p &= \frac{1}{\left(R \| r_2\right)C} \end{aligned} \qquad (11.53)$$

The line-to-output transfer function $G_{vg}(s)$ is found by letting $\hat{d} = 0$ in Fig. 11.18. One then obtains

$$G_{vg}(s) = \left.\frac{\hat{v}}{\hat{v}_g}\right|_{\hat{d} = 0} = \frac{G_{g0}}{1 + \frac{s}{\omega_p}} \qquad (11.54)$$

with

$$G_{g0} = g_2\left(R \| r_2\right) = M \qquad (11.55)$$

Expressions for G_{d0}, G_{g0}, and ω_p are listed in Table 11.3, for the DCM buck, boost, and buck-boost converters with resistive loads [12,13].

The ac modeling approach described in this section is both general and useful. The transistor and diode of a DCM converter can be simply replaced by the two-port network of Fig. 11.13(b), leading to the small-signal ac model. Alternatively, the switch network can be defined as in Fig. 11.16(a) or 11.16(b), and then modeled by the same two-port network, Fig. 11.16(c). The small-signal converter model can then be solved via conventional circuit analysis techniques, to obtain the small-signal transfer functions of the converter.

Table 11.3 Salient features of DCM converter small-signal transfer functions

Converter	G_{d0}	G_{g0}	ω_p
Buck	$\dfrac{2V}{D}\dfrac{1-M}{2-M}$	M	$\dfrac{2-M}{(1-M)RC}$
Boost	$\dfrac{2V}{D}\dfrac{M-1}{2M-1}$	M	$\dfrac{2M-1}{(M-1)RC}$
Buck–boost	$\dfrac{V}{D}$	M	$\dfrac{2}{RC}$

11.2.1   Example: Control-to-Output Frequency Response of a DCM Boost Converter

As a simple numerical example, let us find the small-signal control-to-output transfer function of a DCM boost converter having the following element and parameter values:

$$
\begin{aligned}
R &= 12\,\Omega \\
L &= 5\,\mu\text{H} \\
C &= 470\,\mu\text{F} \\
f_s &= 100\,\text{kHz}
\end{aligned}
\tag{11.56}
$$

The output voltage is regulated to be $V = 36$ V. It is desired to determine $G_{vd}(s)$ at the operating point where the load current is $I = 3$ A and the dc input voltage is $V_g = 24$ V.

The effective resistance $R_e(D)$ is found by solution of the dc equivalent circuit of Fig. 11.12(b). Since the load current I and the input and output voltages V and V_g are known, the power source value P is

$$
P = I\big(V - V_g\big) = \big(3\ \text{A}\big)\big(36\ \text{V} - 24\ \text{V}\big) = 36\ \text{W}
\tag{11.57}
$$

The effective resistance is therefore

$$
R_e = \frac{V_g^2}{P} = \frac{(24\ \text{V})^2}{36\ \text{W}} = 16\,\Omega
\tag{11.58}
$$

The steady-state duty cycle D can now be found using Eq. (11.32):

$$
D = \sqrt{\frac{2L}{R_e T_s}} = \sqrt{\frac{2(5\ \mu\text{H})}{(16\ \Omega)(10\ \mu\text{s})}} = 0.25
\tag{11.59}
$$

The expressions given in Table 11.3 for G_{d0} and ω_p of the boost converter can now be evaluated:

$$
G_{d0} = \frac{2V}{D}\frac{M-1}{2M-1} = \frac{2(36\ \text{V})}{(0.25)}\frac{\left(\dfrac{(36\ \text{V})}{(24\ \text{V})} - 1\right)}{\left(2\,\dfrac{(36\ \text{V})}{(24\ \text{V})} - 1\right)} = 72\ \text{V} \Rightarrow 37\ \text{dBV}
\tag{11.60}
$$

$$
f_p = \frac{\omega_p}{2\pi} = \frac{2M-1}{2\pi(M-1)RC} = \frac{\left(2\,\dfrac{(36\ \text{V})}{(24\ \text{V})} - 1\right)}{2\pi\left(\dfrac{(36\ \text{V})}{(24\ \text{V})} - 1\right)(12\ \Omega)(470\ \mu\text{F})} = 112\ \text{Hz}
$$

A Bode diagram of the control-to-output transfer function is constructed in Fig. 11.19. The solid lines illustrate the magnitude and phase predicted by the approximate single-pole model of Fig. 11.18. The dashed lines are the predictions of the more accurate model discussed in Section 11.3, which include a second pole at $f_2 = 64$ kHz and a RHP zero at $f_z = 127$ kHz, arising from the inductor dynamics. Since the switching frequency is 100 kHz, the accuracy of the model at these frequencies cannot be guaranteed. Nonetheless, in practice, the lagging phase asymptotes arising from the inductor dynamics can be

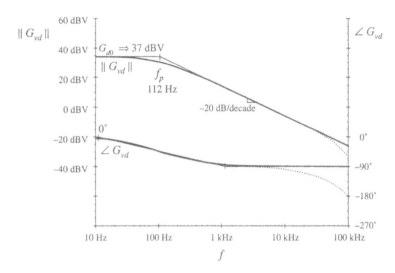

Fig. 11.19 Magnitude and phase of the control-to-output transfer function, DCM boost example. Solid lines: function and its asymptotes, approximate single-pole response predicted by the model of Fig. 11.18. Dashed lines: more accurate response that includes high-frequency inductor dynamics.

observed beginning at $f_2/10 = 6.4$ kHz.

11.2.2 Example: Control-to-Output Frequency Responses of a CCM/DCM SEPIC

As another example, consider the SEPIC of Fig. 11.20. According to Eq. (11.34), this converter operates in CCM if

$$\frac{V}{R} > \frac{1-D}{D}\frac{V_g}{R_e(D)} \tag{11.61}$$

where $R_e(D)$ is given by Eq. (11.33). Upon neglecting losses in the converter, one finds that the CCM conversion ratio is

$$\frac{V}{V_g} \approx \frac{D}{1-D} \tag{11.62}$$

When Eqs. (11.33) and (11.62) are substituted into Eq. (11.61), the condition for operation in CCM becomes:

$$R < \frac{2(L_1 \| L_2)}{(1-D)^2 T_s} = 46\,\Omega \tag{11.63}$$

The converter control-to-output frequency responses are generated using Spice ac simulations. Details of

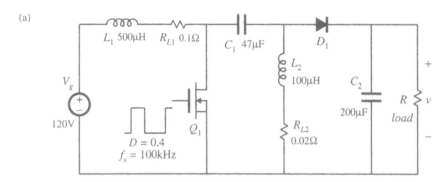

Fig. 11.20 SEPIC example.

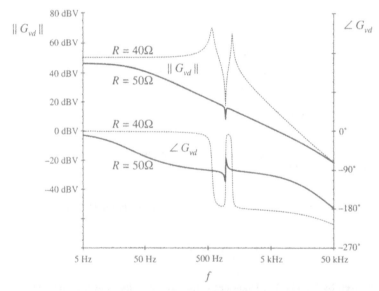

Fig. 11.21 Magnitude and phase of the control-to-output transfer function obtained by simulation of the SEPIC example shown in Fig. 11.20, for two values of the load resistance: $R = 50\Omega$ when the converter operates in DCM (solid lines), and $R = 40\Omega$ for which the converter operates in CCM (dashed lines).

the simulation setup are described in Appendix B, Section B.2.1. Figure 11.21 shows magnitude and phase responses of the control-to-output transfer function obtained for two different values of the load resistance: $R = 40\ \Omega$, for which the converter operates in CCM, and $R = 50\ \Omega$, for which the converter operates in DCM. For these two operating points, the quiescent (dc) voltages and currents in the circuit are nearly the same. Nevertheless, the frequency responses are qualitatively very different in the two operating modes. In CCM, the converter exhibits a fourth-order response with two pairs of high-Q complex-conjugate poles and a pair of complex-conjugate zeros. Another RHP (right-half plane) zero can be observed at frequencies approaching 50 kHz. In DCM, there is a dominant low-frequency pole followed by a pair of complex-conjugate poles and a pair of complex-conjugate zeros. The frequencies of the complex poles and zeros are very close in value. A high-frequency pole and a RHP zero contribute additional phase lag at higher frequencies.

11.3 HIGH-FREQUENCY DYNAMICS OF CONVERTERS IN DCM

As discussed in Section 11.2, transfer functions of converters operating in discontinuous conduction mode exhibit a dominant low-frequency pole. A pole and possibly a zero caused by inductor dynamics, are pushed to high frequencies. To correctly model the high-frequency dynamics of DCM converters, one must account for the fact that the ac voltage across the inductor is not zero. Equation (11.12) is employed in Section 11.1 to greatly simplify the equations of the DCM averaged switch model. Although this model gives good results at low frequencies, it cannot accurately predict high frequency inductor dynamics because it implies that the ac inductor voltage is zero.

A more accurate approach is employed in this section. The subinterval length d_2 is found by averaging the inductor current waveform $i_L(t)$ of Fig. 11.3 [4-6]:

$$\left\langle i_L(t) \right\rangle_{T_s} = \frac{1}{2} i_{pk} \big(d(t) + d_2(t) \big) = \frac{d(t)\big(d(t) + d_2(t) \big) T_s}{2L} \left\langle v_g(t) \right\rangle_{T_s} \tag{11.64}$$

Solution for $d_2(t)$ yields:

$$d_2(t) = \frac{2L \left\langle i_L(t) \right\rangle_{T_s}}{d(t) T_s \left\langle v_g(t) \right\rangle_{T_s}} - d(t) = \left(\frac{R_e(d) \left\langle i_L(t) \right\rangle_{T_s}}{\left\langle v_g(t) \right\rangle_{T_s}} - 1 \right) d(t) \tag{11.65}$$

Equation (11.65), together with Eqs. (11.3), (11.4), (11.7), and (11.10), constitutes a large-signal averaged model in DCM that can be used to investigate steady-state behavior, as well as low-frequency and high-frequency dynamics. Unfortunately, the model equations are more involved, and do not allow elimination of all converter voltages and currents in terms of the switch network average terminal waveforms.

Let us use this model to find predictions for the high-frequency pole caused by the inductor dynamics of DCM converters. Consider the buck-boost converter of Fig. 11.2 having the DCM waveforms shown in Fig. 11.3. The average transistor voltage $\left\langle v_1(t) \right\rangle_{T_s}$ and the average diode current $\left\langle i_2(t) \right\rangle_{T_s}$ are selected as the switch network dependent variables. Substitution of Eq. (11.65) into Eq. (11.3) yields

$$\left\langle v_1(t) \right\rangle_{T_s} = \big(1 - d(t) \big) \left\langle v_g(t) \right\rangle_{T_s} + d(t) \left\langle v(t) \right\rangle_{T_s} - \frac{R_e(d) \left\langle i_L(t) \right\rangle_{T_s} \left\langle v(t) \right\rangle_{T_s} d(t)}{\left\langle v_g(t) \right\rangle_{T_s}} \tag{11.66}$$

The averaged switch voltage $\left\langle v_1(t) \right\rangle_{T_s}$ in Eq. (11.66) is a nonlinear function of the switch duty cycle, the average inductor current, and the average input and output voltages:

$$\left\langle v_1(t) \right\rangle_{T_s} = \gamma_1 \left(\left\langle v_g(t) \right\rangle_{T_s}, \left\langle v(t) \right\rangle_{T_s}, \left\langle i_L(t) \right\rangle_{T_s}, d(t) \right) \tag{11.67}$$

A small-signal ac model can be obtained by Taylor expansion of Eq. (11.67). The small-signal ac component $\hat{v}_1$ of the average switch voltage can be found as:

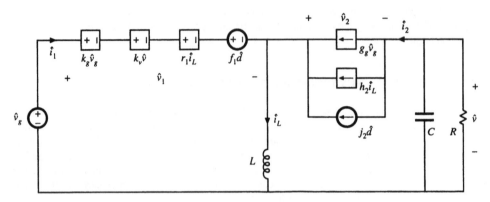

Fig. 11.22   A small-signal ac model of the DCM buck-boost converter.

$$\hat{v}_1(t) = \hat{v}_g(t)k_g + \hat{v}(t)k_v + \hat{i}_L r_1 + \hat{d}(t)f_1 \tag{11.68}$$

where the small-signal model parameters k_g, k_v, r_1, and f_1 are computed as partial derivatives of γ_1 evaluated at the quiescent operating point. In particular,

$$r_1 = \left. \frac{\partial \gamma_1 (V_g, V, i_L, D)}{\partial i_L} \right|_{i_L = I_L} = -\frac{V}{V_g} R_e D \tag{11.69}$$

Substitution of Eq. (11.65) into Eq. (11.10) yields

$$\left\langle i_2(t) \right\rangle_{T_s} = \left\langle i_L(t) \right\rangle_{T_s} - \frac{\left\langle v_g(t) \right\rangle_{T_s}}{R_e} = \gamma_2 \left(\left\langle v_g(t) \right\rangle_{T_s}, \left\langle i_L(t) \right\rangle_{T_s}, d(t) \right) \tag{11.70}$$

The small-signal ac component $\hat{i}_2$ of the average diode current can be found as:

$$\hat{i}_2(t) = \hat{v}_g(t)g_g + \hat{i}_L h_2 + \hat{d}(t)j_2 \tag{11.71}$$

where the small-signal model parameters g_g, h_2, and j_2 are computed as partial derivatives of γ_2 evaluated at the quiescent operating point. Figure 11.22 shows the small-signal ac model of the buck-boost converter, where the transistor and the diode switch are replaced by the sources specified by Eqs. (11.68) and (11.71), respectively. It can be shown that this model predicts essentially the same low-frequency dynamics as the model derived in Section 11.2.

To find the control-to-output transfer function, we set $\hat{v}_g = 0$. At high frequencies, the small-signal ac component of the capacitor voltage is very small, $\hat{v} \approx 0$. Therefore, the contribution of the dependent source $k_v \hat{v}$ can be neglected at high frequencies. Then, from the equivalent circuit model of Fig. 11.22, we have

$$sL\hat{i}_L + r_1 \hat{i}_L + f_1 \hat{d} = 0 \tag{11.72}$$

Equation (11.72) can be solved for the control-to-inductor current transfer function at high frequencies:

$$\frac{\hat{i}_L}{\hat{d}} = -\frac{f_1}{r_1}\frac{1}{1 + \frac{s}{\omega_2}} \tag{11.73}$$

where the pole frequency f_2 is given by

$$f_2 = \frac{\omega_2}{2\pi} = \frac{r_1}{2\pi L} \tag{11.74}$$

To simplify the expression for the pole frequency f_2, we use the steady-state relationship that follows from Eq. (11.12):

$$-\frac{V}{V_g} = \frac{D}{D_2} \tag{11.75}$$

Also, recall that the steady-state equivalent resistance $R_e(D)$ can be written as

$$R_e = \frac{2Lf_s}{D^2} \tag{11.76}$$

where f_s is the switching frequency. Upon substitution of Eqs. (11.69), (11.75) and (11.76) into Eq. (11.74) we get:

$$f_2 = \frac{f_s}{\pi D_2} \tag{11.77}$$

This is an expression for the frequency f_2 of the high-frequency pole that is caused by the inductor dynamics of the DCM buck-boost converter. It can be shown that Eq. (11.77) is a general result for the high-frequency pole, valid for all basic converters operating in DCM. Since $0 < D_2 < 1$, Eq. (11.77) implies that the high-frequency pole is always greater than approximately one third of the switching frequency.

Table 11.4 summarizes the expressions for the high-frequency pole ω_2 and the RHP zero ω_z caused by the inductor dynamics in control-to-output transfer functions $G_{vd}(s)$ of basic DCM converters [6]. The high-frequency pole and the RHP zero occur at frequencies close to or exceeding the switching frequency f_s. This is why, in practice, the high-frequency inductor dynamics can usually be neglected.

Table 11.4 High-frequency pole and RHP zero of the DCM converter control-to-output transfer function $G_{vd}(s)$

Converter	High-frequency pole ω_2	RHP zero ω_z
Buck	$\dfrac{2Mf_s}{D(1-M)}$	none
Boost	$\dfrac{2(M-1)f_s}{D}$	$\dfrac{2f_s}{D}$
Buck–boost	$\dfrac{2\lvert M\rvert f_s}{D}$	$\dfrac{2f_s}{D}$

11.4   SUMMARY OF KEY POINTS

1. In the discontinuous conduction mode, the average transistor voltage and current are proportional, and hence obey Ohm's law. An averaged equivalent circuit can be obtained by replacing the transistor with an effective resistor $R_e(d)$. The average diode voltage and current obey a power source characteristic, with power equal to the power effectively dissipated by R_e. In the averaged equivalent circuit, the diode is replaced with a dependent power source.

2. The two-port lossless network consisting of an effective resistor and power source, which results from averaging the transistor and diode waveforms of DCM converters, is called a loss-free resistor. This network models the basic power-processing functions of DCM converters, much in the same way that the ideal dc transformer models the basic functions of CCM converters.

3. The large-signal averaged model can be solved under equilibrium conditions to determine the quiescent values of the converter currents and voltages. Average power arguments can often be used.

4. A small-signal ac model for the DCM switch network can be derived by perturbing and linearizing the loss-free resistor network. The result has the form of a two-port y-parameter model. The model describes the small-signal variations in the transistor and diode currents, as functions of variations in the duty cycle and in the transistor and diode ac voltage variations.

5. To simplify the ac analysis of the DCM buck and boost converters, it is convenient to define two other forms of the small-signal switch model, corresponding to the switch networks of Figs. 11.16(a) and 11.16(b). These models are also y-parameter two-port models, but have different parameter values.

6. The inductor dynamics of the DCM buck, boost, and buck-boost converters occur at high frequency, above or just below the switching frequency. Hence, in most cases the high frequency inductor dynamics can be ignored. In the small-signal ac model, the inductance L is set to zero, and the remaining model is solved relatively easily for the low-frequency converter dynamics. The DCM buck, boost, and buck-boost converters exhibit transfer functions containing essentially a single low-frequency dominant pole.

7. To obtain a more accurate model of the inductor dynamics in DCM, it is necessary to write the equations of the averaged inductor waveforms in a way that does not assume that the average inductor voltage is zero.

REFERENCES

[1]   V. VORPERIAN, R. TYMERSKI, and F. C. LEE, "Equivalent Circuit Models for Resonant and PWM Switches," *IEEE Transactions on Power Electronics*, Vol. 4, No. 2, pp. 205-214, April 1989.

[2]   V. VORPERIAN, "Simplified Analysis of PWM Converters Using the Model of the PWM Switch," parts I and II, *IEEE Transactions on Aerospace and Electronic Systems*, Vol. 26, No. 3, May 1990, pp. 490-505.

[3]   D. MAKSIMOVIĆ and S. ĆUK, "A Unified Analysis of PWM Converters in Discontinuous Modes," *IEEE Transactions on Power Electronics*, Vol. 6, No. 3, pp. 476-490, July 1991.

[4]   J. SUN, D. M. MITCHELL, M. GREUEL, P. T. KREIN, and R. M. BASS, "Averaged Modelling of PWM Converters in Discontinuous Conduction Mode: a Reexamination," *IEEE Power Electronics Specialists Conference*, 1998 Record, pp. 615-622, June 1998.

[5]   S. BEN-YAAKOV and D. ADAR, "Average Models as Tools for Studying Dynamics of Switch Mode DC–DC Converters," *IEEE Power Electronics Specialists Conference*, 1994 Record, pp. 1369-1376, June 1994.

[6]   J. Sun, D. M. Mitchell, M. Greuel, P. T. Krein, and R. M. Bass, "Average Models for PWM Converters in Discontinuous Conduction Mode," *Proceedings of the 1998 International High Frequency Power Coversion Conference* (HFPC'98), pp. 61-72, November 1998.

[7]   A. Witulski and R. Erickson, "Extension of State-Space Averaging to Resonant Switches —and Beyond," *IEEE Transactions on Power Electronics*, Vol. 5, No. 1, pp. 98-109, January 1990.

[8]   S. Freeland and R. D. Middlebrook, "A Unified Analysis of Converters with Resonant Switches," *IEEE Power Electronics Specialists Conference*, 1987 Record, pp. 20-30, June 1987.

[9]   S. Singer, "Realization of Loss-Free Resistive Elements," *IEEE Transactions on Circuits and Systems*, Vol. CAS-36, No. 12, January 1990.

[10]   S. Singer and R.W. Erickson, "Power-Source Element and Its Properties," *IEE Proceedings—Circuits Devices and Systems*, Vol. 141, No. 3, pp. 220-226, June 1994.

[11]   S. Singer and R. Erickson, "Canonical Modeling of Power Processing Circuits Based on the POPI Concept," *IEEE Transactions on Power Electronics*, Vol. 7, No. 1, January 1992.

[12]   S. Ćuk and R. D. Middlebrook, "A General Unified Approach to Modeling Switching Dc-to-Dc Converters in Discontinuous Conduction Mode," *IEEE Power Electronics Specialists Conference*, 1977 Record, pp. 36-57.

[13]   S. Ćuk, "Modeling, Analysis, and Design of Switching Converters," Ph.D. Thesis, California Institute of Technology, November 1976.

Problems

11.1   Averaged switch modeling of a flyback converter. The converter of Fig. 11.23 operates in the discontinuous conduction mode. The two-winding inductor has a 1:n turns ratio and negligible leakage inductance, and can be modeled as an ideal transformer in parallel with primary-side magnetizing inductance L_p.

   (a)   Sketch the transistor and diode voltage and current waveforms, and derive expressions for their average values.

   (b)   Sketch an averaged model for the converter that includes a loss-free resistor network, and give an expression for $R_e(d)$.

   (c)   Solve your model to determine the voltage ratio V/V_g in the discontinuous conduction mode.

   (d)   Over what range of load current I is your answer of part (c) valid? Express the DCM boundary in the form $I < I_{crit}(D, R_e, V_g, n)$.

   (e)   Derive an expression for the small-signal control-to-output transfer function $G_{vd}(s)$. You may neglect inductor dynamics.

11.2   Averaged switch modeling of a nonisolated Watkins-Johnson converter. The converter of Fig. 11.24 operates in the discontinuous conduction mode. The two-winding inductor has a 1:1 turns ratio and negligible leakage inductance, and can be modeled as an ideal transformer in parallel with magnetizing inductance L.

   (a)   Sketch the transistor and diode voltage and current waveforms, and derive expressions for their average values.

   (b)   Sketch an averaged model for the converter that includes a loss-free resistor network, and give an expression for $R_e(d)$.

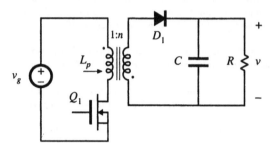

Fig. 11.23 Flyback converter, Problem 11.1.

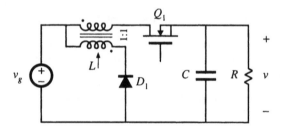

Fig. 11.24 Watkins-Johnson converter, Problem 11.2.

(c)   Solve your model to determine the converter conversion ratio $M(D) = V/V_g$ in the discontinuous conduction mode. Over what range of load currents is your expression valid?

11.3   Sketch the steady-state output characteristics of the buck-boost converter: plot the output voltage V vs. the load current I, for several values of duty cycle D. Include both CCM and DCM operation, and clearly label the boundary between modes.

11.4   In the network of Fig. 11.25, the power source waveform $p(t)$ is given by

$$p(t) = 1000 \cos^2 377t$$

The circuit operates in steady state. Determine the rms resistor voltage $V_{R,rms}$.

11.5   Verify the expressions for G_{d0} and ω_p given in Table 11.3.

11.6   A certain buck converter operates with an input voltage of $V_g = 28$ V and an output voltage of $V = 15$ V. The load resistance is $R = 10\Omega$. Other element and parameter values are: $L = 8\mu H$, $C = 220\mu F$, $f_s = 150kHz$.

    (a)   Determine the value of R_e.

    (b)   Determine the quiescent duty cycle D.

    (c)   Sketch a Bode plot of the control-to-output transfer function $G_{vd}(s)$. Label the values of all salient features. You may neglect inductor dynamics.

11.7   Using the approach of Section 11.3, determine the control-to-output transfer function $G_{vd}(s)$ of a boost converter. Do not make the approximation $L \approx 0$.

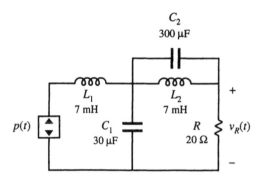

Fig. 11.25 Network with a power source, Problem 11.4.

(a) Derive analytical expressions for the dc gain G_{d0} and the RHP zero frequency ω_z, as functions of M, R_e, D, V_g, L, C, and R.

(b) With the assumption that C is sufficiently large and that L is sufficiently small, the poles of $G_{vd}(s)$ can be factored using the low-Q approximation. Do so, and express the two poles as functions of M, D, L, C, and R. Show that the low-frequency pole matches the expression in Table 11.3, and that the high-frequency pole is given by the expression in Table 11.4.

12

Current Programmed Control

So far, we have discussed duty ratio control of PWM converters, in which the converter output is controlled by direct choice of the duty ratio $d(t)$. We have therefore developed expressions and small-signal transfer functions that relate the converter waveforms and output voltage to the duty ratio.

Another control scheme, which finds wide application, is current programmed control [1–13], in which the converter output is controlled by choice of the peak transistor switch current peak($i_s(t)$). The control input signal is a current $i_c(t)$, and a simple control network switches the transistor on and off, such that the peak transistor current follows $i_c(t)$. The transistor duty cycle $d(t)$ is not directly controlled, but depends on $i_c(t)$ as well as on the converter inductor currents, capacitor voltages, and power input voltage. Converters controlled via current programming are said to operate in the *current programmed mode* (CPM).

The block diagram of a simple current programmed controller is illustrated in Fig. 12.1. Control signal $i_c(t)$ and switch current $i_s(t)$ waveforms are given in Fig. 12.2. A clock pulse at the Set input of a latch initiates the switching period, causing the latch output Q to be high and turning on the transistor. While the transistor conducts, its current $i_s(t)$ is equal to the inductor current $i_L(t)$; this current increases with some positive slope m_1 that depends on the value of inductance and the converter voltages. In more complicated converters, $i_s(t)$ may follow the sum of several inductor currents. Eventually, the switch current $i_s(t)$ becomes equal to the control signal $i_c(t)$. At this point, the controller turns the transistor switch off, and the inductor current decreases for the remainder of the switching period. The controller must measure the switch current $i_s(t)$ with some current sensor circuit, and compare $i_s(t)$ to $i_c(t)$ using an analog comparator. In practice, voltages proportional to $i_s(t)$ and $i_c(t)$ are compared, with constant of proportionality R_f. When $i_s(t) \geq i_c(t)$, the comparator resets the latch, turning the transistor off for the remainder of the switching period.

As usual, a feedback loop can be constructed for regulation of the output voltage. The output voltage $v(t)$ is compared to a reference voltage v_{ref}, to generate an error signal. This error signal is applied

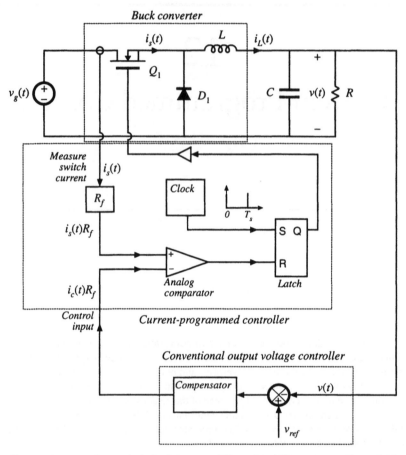

Fig. 12.1 Current-programmed control of a buck converter. The peak transistor current replaces the duty cycle as the control input.

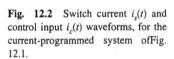

Fig. 12.2 Switch current $i_s(t)$ and control input $i_c(t)$ waveforms, for the current-programmed system of Fig. 12.1.

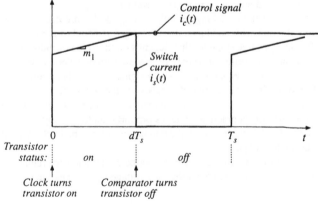

to the input of a compensation network, and the output of the compensator drives the control signal $i_c(t)R_f$. To design such a feedback system, we need to model how variations in the control signal $i_c(t)$ and in the line input voltage $v_g(t)$ affect the output voltage $v(t)$.

The chief advantage of the current programmed mode is its simpler dynamics. To first order, the small-signal control-to-output transfer function $\hat{v}(s)/\hat{i}_c(s)$ contains one less pole than $\hat{v}(s)/\hat{d}(s)$. Actually, this pole is moved to a high frequency, near the converter switching frequency. Nonetheless, simple robust wide-bandwidth output voltage control can usually be obtained, without the use of compensator lead networks. It is true that the current programmed controller requires a circuit for measurement of the switch current $i_s(t)$; however, in practice such a circuit is also required in duty ratio controlled systems, for protection of the transistor against excessive currents during transients and fault conditions. Current programmed control makes use of the available current sensor information during normal operation of the converter, to obtain simpler system dynamics. Transistor failures due to excessive switch current can then be prevented simply by limiting the maximum value of $i_c(t)$. This ensures that the transistor will turn off whenever the switch current becomes too large, on a cycle-by-cycle basis.

An added benefit is the reduction or elimination of transformer saturation problems in full-bridge or push–pull isolated converters. In these converters, small voltage imbalances induce a dc bias in the transformer magnetizing current; if sufficiently large, this dc bias can saturate the transformer. The dc current bias increases or decreases the transistor switch currents. In response, the current programmed controller alters the transistor duty cycles, such that transformer volt-second balance tends to be maintained. Current-programmed full-bridge isolated buck converters should be operated without a capacitor in series with the transformer primary winding; this capacitor tends to destabilize the system. For the same reason, current-programmed control of half-bridge isolated buck converters is generally avoided.

A disadvantage of current programmed control is its susceptibility to noise in the $i_s(t)$ or $i_c(t)$ signals. This noise can prematurely reset the latch, disrupting the operation of the controller. In particular, a small amount of filtering of the sensed switch current waveform is necessary, to remove the turn-on current spike caused by the diode stored charge. Addition of an artificial ramp to the current-programmed controller, as discussed in Section 12.1, can also improve the noise immunity of the circuit.

Commercial integrated circuits that implement current programmed control are widely available, and operation of converters in the current programmed mode is quite popular. In this chapter, converters operating in the current programmed mode are modeled. In Section 12.1, the stability of the current programmed controller and its inner switch-current-sensing loop is examined. It is found that this controller is unstable whenever converter steady-state duty cycle D is greater than 0.5. The current programmed controller can be stabilized by addition of an artificial ramp signal to the sensed switch current waveform. In Section 12.2, the system small-signal transfer functions are described, using a simple first-order model. The averaged terminal waveforms of the switch network can be described by a simple current source, in conjunction with a power source element. Perturbation and linearization leads to a simple small-signal model. Although this first-order model yields a great deal of insight into the control-to-output transfer function and converter output impedance, it does not predict the line-to-output transfer function $G_{vg}(s)$ of current-programmed buck converters. Hence, the model is refined in Section 12.3. Section 12.4 extends the modeling of current programmed converters to the discontinuous conduction mode.

12.1 OSCILLATION FOR $D > 0.5$

The current programmed controller of Fig. 12.1 is unstable whenever the steady-state duty cycle is greater than 0.5. To avoid this stability problem, the control scheme is usually modified, by addition of an artificial ramp to the sensed switch current waveform. In this section, the stability of the current programmed controller, with its inner switch-current-sensing loop, is analyzed. The effects of the addition of

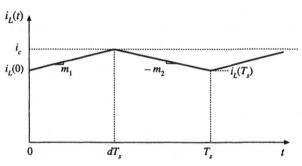

Fig. 12.3 Inductor current waveform of a current-programmed converter operating in the continuous conduction mode.

the artificial ramp are explained, using a simple first-order discrete-time analysis. Effects of the artificial ramp on controller noise susceptibility is also discussed.

Figure 12.3 illustrates a generic inductor current waveform of a switching converter operating in the continuous conduction mode. The inductor current changes with a slope m_1 during the first subinterval, and a slope $-m_2$ during the second subinterval. For the basic nonisolated converters, the slopes m_1 and $-m_2$ are given by

Buck converter

$$m_1 = \frac{v_g - v}{L} \qquad -m_2 = -\frac{v}{L}$$

Boost converter

$$m_1 = \frac{v_g}{L} \qquad -m_2 = \frac{v_g - v}{L} \qquad (12.1)$$

Buck–boost converter

$$m_1 = \frac{v_g}{L} \qquad -m_2 = \frac{v}{L}$$

With knowledge of the slopes m_1 and $-m_2$, we can determine the general relationships between $i_L(0)$, i_c, $i_L(T_s)$, and dT_s.

During the first subinterval, the inductor current $i_L(t)$ increases with slope m_1, until $i_L(t)$ reaches the control signal i_c. Hence,

$$i_L(dT_s) = i_c = i_L(0) + m_1 dT_s \qquad (12.2)$$

Solution for the duty cycle d leads to

$$d = \frac{i_c - i_L(0)}{m_1 T_s} \qquad (12.3)$$

In a similar manner, for the second subinterval we can write

$$\begin{aligned}
i_L(T_s) &= i_L(dT_s) - m_2 d'T_s \\
&= i_L(0) + m_1 dT_s - m_2 d'T_s
\end{aligned} \qquad (12.4)$$

In steady-state, $i_L(0) = i_L(T_s)$, $d = D$, $m_1 = M_1$, and $m_2 = M_2$. Insertion of these relationships into Eq. (12.4) yields

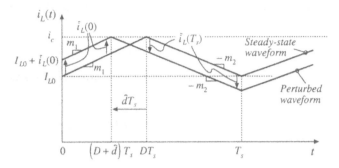

Fig. 12.4 Effect of initial perturbation $\hat{\imath}_L(0)$ on inductor current waveform.

$$0 = M_1 DT_s - M_2 D'T_s \tag{12.5}$$

Or,

$$\frac{M_2}{M_1} = \frac{D}{D'} \tag{12.6}$$

Steady-state Eq. (12.6) coincides with the requirement for steady-state volt-second balance on the inductor.

Consider now a small perturbation in $i_L(0)$:

$$i_L(0) = I_{L0} + \hat{\imath}_L(0) \tag{12.7}$$

I_{L0} is a steady-state value of $i_L(0)$, which satisfies Eqs. (12.4) and (12.5), while $\hat{\imath}_L(0)$ is a small perturbation such that

$$\left| \hat{\imath}_L(0) \right| \ll \left| I_{L0} \right| \tag{12.8}$$

It is desired to assess the stability of the current-programmed controller, by determining whether this small perturbation eventually decays to zero. To do so, let us solve for the perturbation after n switching periods, $\hat{\imath}_L(nT_s)$, and determine whether $\hat{\imath}_L(nT_s)$ tends to zero for large n.

The steady-state and perturbed inductor current waveforms are illustrated in Fig. 12.4. For clarity, the size of the inductor current perturbation $\hat{\imath}_L(0)$ is exaggerated. It is assumed that the converter operates near steady-state, such that the slopes m_1 and m_2 are essentially unchanged. Figure 12.4 is drawn for a positive $\hat{\imath}_L(0)$; the quantity $\hat{d}T_s$ is then negative. Since the slopes of the steady-state and perturbed waveforms are essentially equal over the interval $0 < t < (D + \hat{d})T_s$, the difference between the waveforms is equal to $\hat{\imath}_L(0)$ for this entire interval. Likewise, the difference between the two waveforms is a constant $\hat{\imath}_L(T_s)$ over the interval $DT_s < t < T_s$, since both waveforms then have the slope $- m_2$. Note that $\hat{\imath}_L(T_s)$ is a negative quantity, as sketched in Fig. 12.4. Hence, we can solve for $\hat{\imath}_L(T_s)$ in terms of $\hat{\imath}_L(0)$, by considering only the interval $(D + \hat{d})T_s < t < DT_s$ as illustrated in Fig. 12.5.

Fig. 12.5 Expanded view of the steady-state and perturbed inductor current waveforms, near the peak of $i_L(t)$.

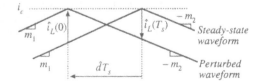

From Fig. 12.5, we can use the steady-state waveform to express $\hat{i}_L(0)$ as the slope m_1, multiplied by the interval length $-\hat{d}T_s$. Hence,

$$\hat{i}_L(0) = -m_1 \hat{d}T_s \tag{12.9}$$

Likewise, we can use the perturbed waveform to express $\hat{i}_L(T_s)$ as the slope $-m_2$, multiplied by the interval length $-\hat{d}T_s$:

$$\hat{i}_L(T_s) = m_2 \hat{d}T_s \tag{12.10}$$

Elimination of the intermediate variable $\hat{d}$ from Eqs. (12.9) and (12.10) leads to

$$\hat{i}_L(T_s) = \hat{i}_L(0)\left(-\frac{m_2}{m_1}\right) \tag{12.11}$$

If the converter operating point is sufficiently close to the quiescent operating point, then m_2/m_1 is given approximately by Eq. (12.6). Equation (12.11) then becomes

$$\hat{i}_L(T_s) = \hat{i}_L(0)\left(-\frac{D}{D'}\right) \tag{12.12}$$

A similar analysis can be performed during the next switching period, to show that

$$\hat{i}_L(2T_s) = \hat{i}_L(T_s)\left(-\frac{D}{D'}\right) = \hat{i}_L(0)\left(-\frac{D}{D'}\right)^2 \tag{12.13}$$

After n switching periods, the perturbation becomes

$$\hat{i}_L(nT_s) = \hat{i}_L((n-1)T_s)\left(-\frac{D}{D'}\right) = \hat{i}_L(0)\left(-\frac{D}{D'}\right)^n \tag{12.14}$$

Note that, as n tends to infinity, the perturbation $\hat{i}_L(nT_s)$ tends to zero provided that the characteristic value $-D/D'$ has magnitude less than one. Conversely, the perturbation $\hat{i}_L(nT_s)$ becomes large in magnitude when the characteristic value $\alpha = -D/D'$ has magnitude greater than one:

$$\left|\hat{i}_L(nT_s)\right| \rightarrow \begin{cases} 0 & \text{when } \left|-\dfrac{D}{D'}\right| < 1 \\[2mm] \infty & \text{when } \left|-\dfrac{D}{D'}\right| > 1 \end{cases} \tag{12.15}$$

Therefore, for stable operation of the current programmed controller, we need $|\alpha| = D/D' < 1$, or

$$D < 0.5 \tag{12.16}$$

As an example, consider the operation of the boost converter with the steady-state terminal voltages $V_g = 20$ V, $V = 50$ V. Since $V/V_g = 1/D'$, the boost converter should operate with $D = 0.6$. We therefore expect the current programmed controller to be unstable. The characteristic value will be

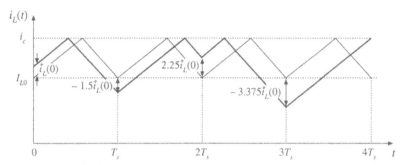

Fig. 12.6 Unstable oscillation for $D = 0.6$.

$$\alpha = -\frac{D}{D'} = \left(-\frac{0.6}{0.4}\right) = -1.5 \tag{12.17}$$

As given by Eq. (12.14), a perturbation in the inductor current will increase by a factor of -1.5 over every switching period. As illustrated in Fig. 12.6, the perturbation grows to $-1.5\hat{i}_L(0)$ after one switching period, to $+2.25\hat{i}_L(0)$ after two switching periods, and to $-3.375\hat{i}_L(0)$ after three switching periods. For the particular initial conditions illustrated in Fig. 12.6, this growing oscillation saturates the current programmed controller after three switching periods. The transistor remains on for the entire duration of the fourth switching period. The inductor current and controller waveforms may eventually become oscillatory and periodic in nature, with period equal to an integral number of switching periods. Alternatively, the waveforms may become chaotic. In either event, the controller does not operate as intended.

Figure 12.7 illustrates the inductor current waveforms when the output voltage is decreased to $V = 30$ V. The boost converter then operates with $D = 1/3$, and the characteristic value becomes

$$\alpha = -\frac{D}{D'} = \left(-\frac{1/3}{2/3}\right) = -0.5 \tag{12.18}$$

Perturbations now decrease in magnitude by a factor of 0.5 over each switching period. A disturbance in the inductor current becomes small in magnitude after a few switching periods.

The instability for $D > 0.5$ is a well-known problem of current programmed control, which is not dependent on the converter topology. The controller can be rendered stable for all duty cycles by addition of an artificial ramp to the sensed switch current waveform, as illustrated in Fig. 12.8. This arti-

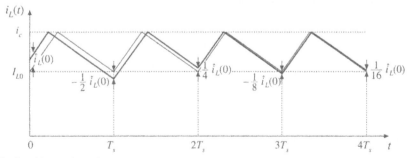

Fig. 12.7 A stable transient with $D = 1/3$.

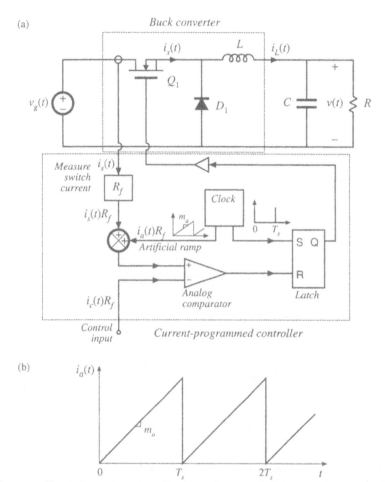

Fig. 12.8 Stabilization of the current programmed controller by addition of an artificial ramp to the measured switch current waveform: (a) block diagram, (b) artificial ramp waveform.

ficial ramp has the qualitative effect of reducing the gain of the inner switch-current-sensing discrete feedback loop. The artificial ramp has slope m_a as shown. The controller now switches the transistor off when

$$i_a(dT_s) + i_L(dT_s) = i_c \qquad (12.19)$$

where $i_a(t)$ is the artificial ramp waveform. Therefore, the transistor is switched off when the inductor current $i_L(t)$ is given by

$$i_L(dT_s) = i_c - i_a(dT_s) \qquad (12.20)$$

Figure 12.9 illustrates the analog comparison of the inductor current waveform $i_L(t)$ with the quantity $[i_c - i_a(t)]$.

Fig. 12.9 Addition of artificial ramp: the transistor is now switched off when $i_L(t) = i_c - i_a(t)$.

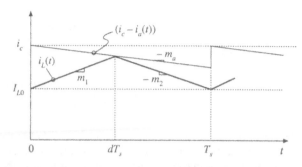

We can again determine the stability of the current programmed controller by analyzing the change in a perturbation of the inductor current waveform over a complete switching period. Figure 12.10 illustrates steady-state and perturbed inductor current waveforms, in the presence of the artificial ramp. Again, the magnitude of the perturbation $\hat{i}_L(0)$ is exaggerated. The perturbed waveform is sketched for a positive value of $\hat{i}_L(0)$; this causes $\hat{d}$, and usually also $\hat{i}_L(T_s)$, to be negative. If the perturbed waveforms are sufficiently close to the quiescent operating point, then the slopes m_1 and m_2 are essentially unchanged, and the relationship between $\hat{i}_L(0)$ and $\hat{i}_L(T_s)$ can be determined solely by consideration of the interval $(D + \hat{d})T_s < t < DT_s$. The perturbations $\hat{i}_L(0)$ and $\hat{i}_L(T_s)$ are expressed in terms of the slopes m_1, m_2, and m_a, and the interval length $-\hat{d}T_s$, as follows:

$$\hat{i}_L(0) = -\hat{d}T_s\left(m_1 + m_a\right) \tag{12.21}$$

$$\hat{i}_L(T_s) = -\hat{d}T_s\left(m_a - m_2\right) \tag{12.22}$$

Elimination of $\hat{d}$ yields

$$\hat{i}_L(T_s) = \hat{i}_L(0)\left(-\frac{m_2 - m_a}{m_1 + m_a}\right) \tag{12.23}$$

A similar analysis can be applied to the nth switching period, leading to

$$\hat{i}_L(nT_s) = \hat{i}_L((n-1)T_s)\left(-\frac{m_2 - m_a}{m_1 + m_a}\right) = \hat{i}_L(0)\left(-\frac{m_2 - m_a}{m_1 + m_a}\right)^n = \hat{i}_L(0)\,\alpha^n \tag{12.24}$$

The evolution of inductor current perturbations are now determined by the characteristic value

Fig. 12.10 Steady-state and perturbed inductor current waveforms, in the presence of an artificial ramp.

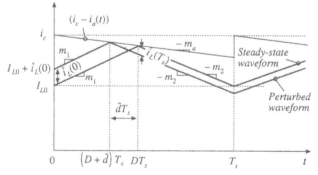

$$\alpha = -\frac{m_2 - m_a}{m_1 + m_a} \tag{12.25}$$

For large n, the perturbation magnitude tends to

$$\left| \hat{i}_L(nT_s) \right| \to \begin{cases} 0 & \text{when } |\alpha| < 1 \\ \infty & \text{when } |\alpha| > 1 \end{cases} \tag{12.26}$$

Therefore, for stability of the current programmed controller, we need to choose the slope of the artificial ramp m_a such that the characteristic value α has magnitude less than one. The artificial ramp gives us an additional degree of freedom, which we can use to stabilize the system for duty cycles greater than 0.5. Note that increasing the value of m_a causes the numerator of Eq. (12.25) to decrease, while the denominator increases. Therefore, the characteristic value α attains magnitude less than one for sufficiently large m_a.

In the conventional voltage regulator application, the output voltage $v(t)$ is well regulated by the converter control system, while the input voltage $v_g(t)$ is unknown. Equation (12.1) then predicts that the value of the slope m_2 is constant and known with a high degree of accuracy, for the buck and buck-boost converters. Therefore, let us use Eq. (12.6) to eliminate the slope m_1 from Eq. (12.25), and thereby express the characteristic value α as a function of the known slope m_2 and the steady-state duty cycle D:

$$\alpha = -\frac{1 - \dfrac{m_a}{m_2}}{\dfrac{D'}{D} + \dfrac{m_a}{m_2}} \tag{12.27}$$

One common choice of artificial ramp slope is

$$m_a = \tfrac{1}{2} m_2 \tag{12.28}$$

It can be verified, by substitution of Eq. (12.28) into (12.27), that this choice leads to $\alpha = -1$ at $D = 1$, and to $|\alpha| < 1$ for $0 \leq D < 1$. This is the minimum value of m_a that leads to stability for all duty cycles. We will see in Section 12.3 that this choice of m_a has the added benefit of causing the ideal line-to-output transfer function $G_{vg}(s)$ of the buck converter to become zero.

Another common choice of m_a is

$$m_a = m_2 \tag{12.29}$$

This causes the characteristic value α to become zero for all D. As a result, $\hat{i}_L(T_s)$ is zero for any $\hat{i}_L(0)$ that does not saturate the controller. The system removes any error after one switching period T_s. This behavior is known as *deadbeat control*, or *finite settling time*.

It should be noted that the above stability analysis employs a quasi-static approximation, in which the slopes m_1 and m_2 of the perturbed inductor current waveforms are assumed to be identical to the steady-state case. In the most general case, the stability and transient response of a complete system employing current programmed control must be assessed using a system-wide discrete time or sampled-data analysis. Nonetheless, in practice the above arguments are found to be sufficient for selection of the artificial ramp slope m_a.

Current-programmed controller circuits exhibit significant sensitivity to noise. The reason for this is illustrated in Fig. 12.11(a), in which the control signal $i_c(t)$ is perturbed by a small amount of noise (a)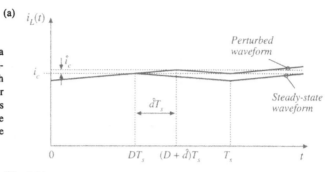

Fig. 12.11 When noise perturbs a controller signal such as i_c, a perturbation in the duty cycle results: (a) with no artificial ramp and small inductor current ripple, the perturbation $\hat{d}$ is large; (b) an artificial ramp reduces the controller gain, thereby reducing the perturbation $\hat{d}$.

(b)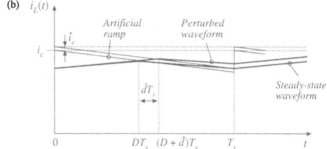

represented by $\hat{i}_c$. It can be seen that, when there is no artificial ramp and when the inductor current ripple is small, then a small perturbation in i_c leads to a large perturbation in the duty cycle: the controller has high gain. When noise is present in the controller circuit, then significant jitter in the duty cycle waveforms may be observed. A solution is to reduce the gain of the controller by introduction of an artificial ramp. As illustrated in Fig. 12.11(b), the same perturbation in i_c now leads to a reduced variation in the duty cycle. When the layout and grounding of the controller circuit introduce significant noise into the duty cycle waveform, it may be necessary to add an artificial ramp whose amplitude is substantially greater than the inductor current ripple.

12.2 A SIMPLE FIRST-ORDER MODEL

Once the current programmed controller has been constructed, and stabilized using an artificial ramp, then it is desired to design a feedback loop for regulation of the output voltage. As usual, this voltage feedback loop must be designed to meet specifications regarding line disturbance rejection, transient response, output impedance, etc. A block diagram of a typical system is illustrated in Fig. 12.12, containing an inner current programmed controller, with an outer voltage feedback loop.

To design the outer voltage feedback loop, an ac equivalent circuit model of the switching converter operating in the current programmed mode is needed. In Chapter 7, averaging was employed to develop small-signal ac equivalent circuit models for converters operating with duty ratio control. These models predict the circuit behavior in terms of variations d in the duty cycle. If we could find the relationship between the control signal $i_c(t)$ and the duty cycle $d(t)$ for the current programmed controller, then we could adapt the models of Chapter 7, to apply to the current programmed mode as well. In general, the duty cycle depends not only on $i_c(t)$, but also on the converter voltages and currents; hence, the current programmed controller incorporates multiple effective feedback loops as indicated in Fig. 12.12.

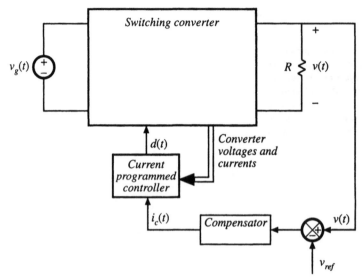

Fig. 12.12   Block diagram of a converter system incorporating current programmed control.

In this section, the averaging approach is extended, as described above, to treat current programmed converters. A simple first-order approximation is employed, in which it is assumed that the current programmed controller operates ideally, and hence causes the average inductor current $\langle i_L(t) \rangle_{T_s}$ to be identical to the control $i_c(t)$. This approximation is justified whenever the inductor current ripple and artificial ramp have negligible magnitudes. The inductor current then is no longer an independent state of the system, and no longer contributes a pole to the converter small-signal transfer functions.

This first-order model is derived in Section 12.2.1, using a simple algebraic approach. In Section 12.2.2, a simple physical interpretation is obtained via the averaged switch modeling technique. A more accurate, but more complicated, model is described in Section 12.3.

12.2.1   Simple Model via Algebraic Approach: Buck-Boost Example

The power stage of a simple buck-boost converter operating in the continuous conduction mode is illustrated in Fig. 12.13(a), and its inductor current waveform is given in Fig. 12.13(b). The small-signal averaged equations for this converter, under duty cycle control, were derived in Section 7.2. The result, Eq. (7.43), is reproduced below:

$$L\frac{d\hat{i}_L(t)}{dt} = D\hat{v}_g(t) + D'\hat{v}(t) + \left(V_g - V\right)\hat{d}(t)$$

$$C\frac{d\hat{v}(t)}{dt} = -D'\hat{i}_L - \frac{\hat{v}(t)}{R} + I_L\hat{d}(t) \tag{12.30}$$

$$\hat{i}_g(t) = D\hat{i}_L + I_L\hat{d}(t)$$

The Laplace transforms of these equations, with initial conditions set to zero, are

(a)

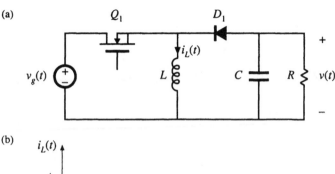

(b)

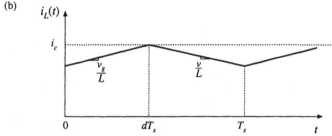

Fig. 12.13 Buck-boost converter example: (a) power stage, (b) inductor current waveform.

$$sL\hat{i}_L(s) = D\hat{v}_g(s) + D'\hat{v}(s) + \left(V_g - V\right)\hat{d}(s)$$

$$sC\hat{v}(s) = -D'\hat{i}_L(s) - \frac{\hat{v}(s)}{R} + I_L\hat{d}(s) \qquad (12.31)$$

$$\hat{i}_g(s) = D\hat{i}_L(s) + I_L\hat{d}(s)$$

We now make the assumption that the inductor current $\hat{i}_L(s)$ is identical to the programmed control current $\hat{i}_c(s)$. This is valid to the extent that the controller is stable, and that the magnitudes of the inductor current ripple and artificial ramp waveform are sufficiently small:

$$\hat{i}_L(s) \approx \hat{i}_c(s) \qquad (12.32)$$

This approximation, in conjunction with the inductor current equation of (12.31), can now be used to find the relationship between the control current $\hat{i}_c(s)$ and the duty cycle $\hat{d}(s)$, as follows:

$$sL\hat{i}_c(s) \approx D\hat{v}_g(s) + D'\hat{v}(s) + \left(V_g - V\right)\hat{d}(s) \qquad (12.33)$$

Solution for $\hat{d}(s)$ yields

$$\hat{d}(s) = \frac{sL\hat{i}_c(s) - D\hat{v}_g(s) - D'\hat{v}(s)}{\left(V_g - V\right)} \qquad (12.34)$$

This small-signal expression describes how the current programmed controller varies the duty cycle, in response to a given control input variation $\hat{i}_c(s)$. It can be seen that $\hat{d}(s)$ depends not only on $\hat{i}_c(s)$, but also on the converter output voltage and input voltage variations. Equation (12.34) can now be substituted into the second and third lines of Eq. (12.31), thereby eliminating $\hat{d}(s)$. One obtains

(a)

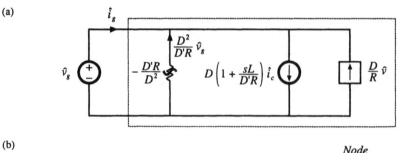

(b)

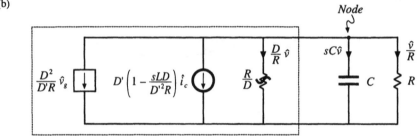

Fig. 12.14 Construction of CPM CCM buck–boost converter equivalent circuit: (a) input port model, corresponding to Eq. (12.38); (b) output port model, corresponding to Eq. (12.37).

$$sC\hat{v}(s) = -D'\hat{i}_c(s) - \frac{\hat{v}(s)}{R} + I_L \frac{sL\hat{i}_c(s) - D\hat{v}_g(s) - D'\hat{v}(s)}{\left(V_g - V\right)}$$

$$\hat{i}_g(s) = D\hat{i}_c(s) + I_L \frac{sL\hat{i}_c(s) - D\hat{v}_g(s) - D'\hat{v}(s)}{\left(V_g - V\right)}$$

(12.35)

These equations can be simplified by collecting terms, and by use of the steady-state relationships

$$V = -\frac{D}{D'} V_g$$

$$I_L = -\frac{V}{D'R} = \frac{D}{D'^2 R} V_g$$

(12.36)

Equation (12.35) then becomes

$$sC\hat{v}(s) = \left(\frac{sLD}{D'R} - D'\right) \hat{i}_c(s) - \left(\frac{D}{R} + \frac{1}{R}\right) \hat{v}(s) - \left(\frac{D^2}{D'R}\right) \hat{v}_g(s)$$

(12.37)

$$\hat{i}_g(s) = \left(\frac{sLD}{D'R} + D\right) \hat{i}_c(s) - \left(\frac{D}{R}\right) \hat{v}(s) - \left(\frac{D^2}{D'R}\right) \hat{v}_g(s)$$

(12.38)

These are the basic ac small-signal equations for the simplified first-order model of the current-programmed buck-boost converter. These equations can now be used to construct small-signal ac circuit models that represent the behavior of the converter input and output ports. In Eq. (12.37), the quantity $sC\hat{v}(s)$ is the output capacitor current. The $\hat{i}_c(s)$ term is represented in Fig. 12.14(b) by an independent

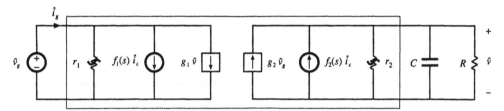

Fig. 12.15 Two-port equivalent circuit used to model the current-programmed CCM buck, boost, and buck–boost converters.

current source, while the $\hat{v}_g(s)$ term is represented by a dependent current source. $\hat{v}(s)/R$ is the current through the load resistor, and $\hat{v}(s)D/R$ is the current through an effective ac resistor of value R/D.

Equation (12.38) describes the current $\hat{i}_g(s)$ drawn by the converter input port, out of the source $\hat{v}_g(s)$. The $\hat{i}_c(s)$ term is again represented in Fig. 12.14(a) by an independent current source, and the $\hat{v}(s)$ term is represented by a dependent current source. The quantity $-\hat{v}_g(s)D^2/D'R$ is modeled by an effective ac resistor having the negative value $-D'R/D^2$.

Figures 12.14(a) and (b) can now be combined into the small-signal two-port model of Fig. 12.15. The current programmed buck and boost converters can also be modeled by a two-port equivalent circuit, of the same form. Table 12.1 lists the model parameters for the basic buck, boost, and buck-boost converters.

The two-port equivalent circuit can now be solved, to find the converter transfer functions and output impedance. The control-to-output transfer function is found by setting v_g to zero. Solution for the output voltage then leads to the transfer function $G_{vc}(s)$:

$$G_{vc}(s) = \frac{\hat{v}(s)}{\hat{i}_c(s)}\bigg|_{\hat{v}_g = 0} = f_2\left(r_2 \parallel R \parallel \frac{1}{sC}\right) \tag{12.39}$$

Substitution of the model parameters for the buck-boost converter yields

$$G_{vc}(s) = -R\,\frac{D'}{1+D}\,\frac{\left(1 - s\dfrac{DL}{D'^2 R}\right)}{\left(1 + s\dfrac{RC}{1+D}\right)} \tag{12.40}$$

Table 12.1 Current programmed mode small-signal equivalent circuit parameters, simple model

Converter	g_1	f_1	r_1	g_2	f_2	r_2
Buck	$\dfrac{D}{R}$	$D\left(1 + \dfrac{sL}{R}\right)$	$-\dfrac{R}{D^2}$	0	1	∞
Boost	0	1	∞	$\dfrac{1}{D'R}$	$D'\left(1 - \dfrac{sL}{D'^2 R}\right)$	R
Buck–boost	$-\dfrac{D}{R}$	$D\left(1 + \dfrac{sL}{D'R}\right)$	$-\dfrac{D'R}{D^2}$	$-\dfrac{D^2}{D'R}$	$-D'\left(1 - \dfrac{sDL}{D'^2 R}\right)$	$\dfrac{R}{D}$

It can be seen that this transfer function contains only one pole; the pole due to the inductor has been lost. The dc gain is now directly dependent on the load resistance R. In addition, the transfer function contains a right half-plane zero whose corner frequency is unchanged from the duty-cycle-controlled case. In general, introduction of current programming alters the transfer function poles and dc gain, but not the zeroes.

The line-to-output transfer function $G_{vg}(s)$ is found by setting the control input i_c to zero, and then solving for the output voltage. The result is

$$G_{vg}(s) = \left.\frac{\hat{v}(s)}{\hat{v}_g(s)}\right|_{i_c=0} = g_2\left(r_2 \parallel R \parallel \frac{1}{sC}\right) \tag{12.41}$$

Substitution of the parameters for the buck-boost converter leads to

$$G_{vg}(s) = -\frac{D^2}{1-D^2}\frac{1}{\left(1+s\frac{RC}{1+D}\right)} \tag{12.42}$$

Again, the inductor pole is lost. The output impedance is

$$Z_{out}(s) = r_2 \parallel R \parallel \frac{1}{sC} \tag{12.43}$$

For the buck-boost converter, one obtains

$$Z_{out}(s) = \frac{R}{1+D}\frac{1}{\left(1+s\frac{RC}{1+D}\right)} \tag{12.44}$$

12.2.2   Averaged Switch Modeling

Additional physical insight into the properties of current programmed converters can be obtained by use of the averaged switch modeling approach developed in Section 7.4. Consider the buck converter of Fig. 12.16. We can define the terminal voltages and currents of the switch network as shown. When the buck converter operates in the continuous conduction mode, the switch network average terminal waveforms are related as follows:

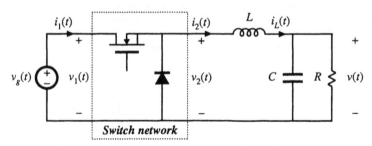

Fig. 12.16   Averaged switch modeling of a current-programmed converter: CCM buck example.

$$\langle v_2(t)\rangle_{T_s} = d(t)\,\langle v_1(t)\rangle_{T_s}$$
$$\langle i_1(t)\rangle_{T_s} = d(t)\,\langle i_2(t)\rangle_{T_s} \tag{12.45}$$

We again invoke the approximation in which the inductor current exactly follows the control current. In terms of the switch network terminal current i_2, we can therefore write

$$\langle i_2(t)\rangle_{T_s} \approx \langle i_c(t)\rangle_{T_s} \tag{12.46}$$

The duty cycle $d(t)$ can now be eliminated from Eq. (12.45), as follows:

$$\langle i_1(t)\rangle_{T_s} = d(t)\,\langle i_c(t)\rangle_{T_s} = \frac{\langle v_2(t)\rangle_{T_s}}{\langle v_1(t)\rangle_{T_s}}\,\langle i_c(t)\rangle_{T_s} \tag{12.47}$$

This equation can be written in the alternative form

$$\langle i_1(t)\rangle_{T_s}\,\langle v_1(t)\rangle_{T_s} = \langle i_c(t)\rangle_{T_s}\,\langle v_2(t)\rangle_{T_s} = \langle p(t)\rangle_{T_s} \tag{12.48}$$

Equations (12.46) and (12.48) are the desired result, which describes the average terminal relations of the CCM current-programmed buck switch network. Equation (12.46) states that the average terminal current $\langle i_2(t)\rangle_{T_s}$ is equal to the control current $\langle i_c(t)\rangle_{T_s}$. Equation (12.48) states that the input port of the switch network consumes average power $\langle p(t)\rangle_{T_s}$ equal to the average power flowing out of the switch output port. The averaged equivalent circuit of Fig. 12.17 is obtained.

Figure 12.17 describes the behavior of the current programmed buck converter switch network, in a simple and straightforward manner. The switch network output port behaves as a current source of value $\langle i_c(t)\rangle_{T_s}$. The input port follows a power sink characteristic, drawing power from the source v_g equal to the power supplied by the i_c current source. Properties of the power source and power sink elements are described in Chapters 11 and 18.

Similar arguments lead to the averaged switch models of the current programmed boost and buck-boost converters, illustrated in Fig. 12.18. In both cases, the switch network averaged terminal waveforms can be represented by a current source of value $\langle i_c(t)\rangle_{T_s}$, in conjunction with a dependent power source or power sink.

A small-signal ac model of the current-programmed buck converter can now be constructed by perturbation and linearization of the switch network averaged terminal waveforms. Let

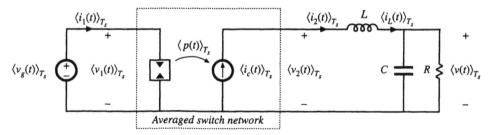

Fig. 12.17 Averaged model of CPM buck converter.

(a)

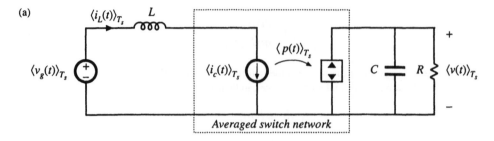

(b)

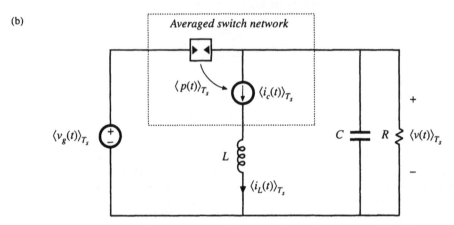

Fig. 12.18 Averaged models of CPM boost (a) and CPM buck–boost (b) converters, derived via averaged switch modeling.

$$\langle v_1(t) \rangle_{T_s} = V_1 + \hat{v}_1(t)$$
$$\langle i_1(t) \rangle_{T_s} = I_1 + \hat{i}_1(t)$$
$$\langle v_2(t) \rangle_{T_s} = V_2 + \hat{v}_2(t) \tag{12.49}$$
$$\langle i_2(t) \rangle_{T_s} = I_2 + \hat{i}_2(t)$$
$$\langle i_c(t) \rangle_{T_s} = I_c + \hat{i}_c(t)$$

Perturbation and linearization of the $\langle i_c(t) \rangle_{T_s}$ current source of Fig. 12.17 simply leads to a current source of value $\hat{i}_c(t)$. Perturbation of the power source characteristic, Eq. (12.48), leads to

$$\left(V_1 + \hat{v}_1(t) \right)\left(I_1 + \hat{i}_1(t) \right) = \left(I_c + \hat{i}_c(t) \right)\left(V_2 + \hat{v}_2(t) \right) \tag{12.50}$$

Upon equating the dc terms on both sides of this equation, we obtain

$$V_1 I_1 = I_c V_2 \quad \Rightarrow \quad I_1 = D I_c \tag{12.51}$$

The linear small-signal ac terms of Eq. (12.50) are

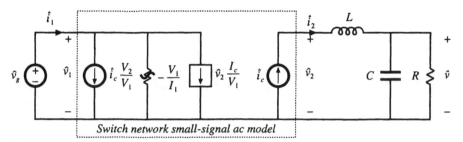

Switch network small-signal ac model

Fig. 12.19 Small-signal model of the CCM CPM buck converter, derived by perturbation and linearization of the switch network in Fig. 12.17.

$$\hat{v}_1(t)I_1 + V_1\hat{i}_1(t) = \hat{i}_c(t)V_2 + I_c\hat{v}_2(t) \tag{12.52}$$

Solution for the small-signal switch network input current $\hat{i}_1(t)$ yields

$$\hat{i}_1(t) = \hat{i}_c(t)\frac{V_2}{V_1} + \hat{v}_2(t)\frac{I_c}{V_1} - \hat{v}_1(t)\frac{I_1}{V_1} \tag{12.53}$$

The small-signal ac model of Fig. 12.19 can now be constructed. The switch network output port is again a current source, of value $\hat{i}_c(t)$. The switch network input port model is obtained by linearization of the power sink characteristic, as given by Eq. (12.53). The input port current $\hat{i}_1(t)$ is composed of three terms. The $\hat{i}_c(t)$ term is modeled by an independent current source, the $\hat{v}_2(t)$ term is modeled by a dependent current source, and the $\hat{v}_1(t)$ term is modeled by an effective ac resistor having the negative value $-V_1/I_1$. As illustrated in Fig. 12.20, this incremental resistance is determined by the slope of the power sink input port characteristic, evaluated at the quiescent operating point. The power sink leads to a negative incremental resistance because an increase in $\langle v_1(t)\rangle_{T_s}$ causes a decrease in $\langle i_1(t)\rangle_{T_s}$, such that constant $\langle p(t)\rangle_{T_s}$ is maintained.

The equivalent circuit of Fig. 12.19 can now be simplified by use of the dc relations $V_2 = DV_1$, $I_2 = V_2/R$, $I_1 = DI_2$, $I_2 = I_c$. Equation (12.53) then becomes

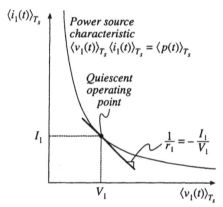

Fig. 12.20 Origin of the input port negative incremental resistance r_1: the slope of the power sink characteristic, evaluated at the quiescent operating point.

$$\hat{i}_1(t) = D\hat{i}_c(t) + \frac{D}{R}\hat{v}_2(t) - \frac{D^2}{R}\hat{v}_1(t) \tag{12.54}$$

Finally, we can eliminate the quantities $\hat{v}_1$ and $\hat{v}_2$ in favor of the converter terminal voltages $\hat{v}_g$ and $\hat{v}$, as follows. The quantity $\hat{v}_1$ is simply equal to $\hat{v}_g$. The quantity $\hat{v}_2$ is equal to the output voltage $\hat{v}$ plus the voltage across the inductor, $sL\hat{i}_c(s)$. Hence,

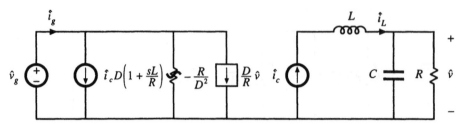

Fig. 12.21 Simplification of the CPM buck converter model of Fig. 12.19, with dependent source expressed in terms of the output voltage variations.

$$\hat{v}_2(s) = \hat{v}(s) + sL\hat{i}_c(s) \tag{12.55}$$

With these substitutions, Eq. (12.54) becomes

$$\hat{i}_1(s) = D\left(1 + s\frac{L}{R}\right)\hat{i}_c(s) + \frac{D}{R}\hat{v}(s) - \frac{D^2}{R}\hat{v}_g(s) \tag{12.56}$$

The equivalent circuit of Fig. 12.21 is now obtained. It can be verified that this equivalent circuit coincides with the model of Fig. 12.15 and the buck converter parameters of Table 12.1.

The approximate small-signal properties of the current programmed buck converter can now be explained. Since the inductor is in series with the current source $\hat{i}_c$, the inductor does not contribute to the control-to-output transfer function. The control-to-output transfer function is determined simply by the relation

$$G_{vc}(s) = \frac{\hat{v}(s)}{\hat{i}_c(s)}\bigg|_{\hat{v}_g = 0} = \left(R \parallel \frac{1}{sC}\right) \tag{12.57}$$

So current programming transforms the output characteristic of the buck converter into a current source. The power sink input characteristic of the current programmed buck converter leads to a negative incremental input resistance, as described above. Finally, Fig. 12.21 predicts that the buck converter line-to-output transfer function is zero:

$$G_{vg}(s) = \frac{\hat{v}(s)}{\hat{v}_g(s)}\bigg|_{\hat{i}_c = 0} = 0 \tag{12.58}$$

Disturbances in v_g do not influence the output voltage, since the inductor current depends only on i_c. The current programmed controller adjusts the duty cycle as necessary to maintain constant inductor current, regardless of variations in v_g. The more accurate models of Section 12.3 predict that $G_{vg}(s)$ is not zero, but is nonetheless small in magnitude.

Similar arguments lead to the boost converter small-signal equivalent circuit of Fig. 12.22. Derivation of this equivalent circuit is left as a homework problem. In the case of the boost converter, the switch network input port behaves as a current source, of value i_c, while the output port is a dependent power source, equal to the power apparently consumed by the current source i_c. In the small-signal model, the current source $\hat{i}_c$ appears in series with the inductor L, and hence the converter transfer functions cannot contain poles arising from the inductor. The switch network power source output characteristic leads to an ac resistance of value $r_2 = R$. The line-to-output transfer function $G_{vg}(s)$ is nonzero in the

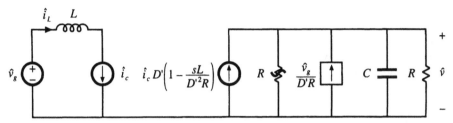

Fig. 12.22 Small-signal model of the CCM CPM boost converter, derived via averaged switch modeling and the approximation $i_L \approx i_c$.

boost converter, since the magnitude of the power source depends directly on the value of v_g. The control-to-output transfer function $G_{vc}(s)$ contains a right half-plane zero, identical to the right half-plane zero of the duty-cycle-controlled boost converter.

12.3   A MORE ACCURATE MODEL

The simple models discussed in the previous section yield much insight into the low-frequency behavior of current-programmed converters. Unfortunately, they do not always describe everything that we need to know. For example, the simple model of the buck converter predicts that the line-to-output transfer function $G_{vg}(s)$ is zero. While it is true that this transfer function is usually small in magnitude, the transfer function is not equal to zero. To predict the effect of input voltage disturbances on the output voltage, we need to compute the actual $G_{vg}(s)$.

  In this section, a more accurate analysis is performed which does not rely on the approximation $\langle i_L(t)\rangle_{T_s} \approx i_C(t)$. The analytical approach of [5,6] is combined with the controller model of [7]. A functional block diagram of the current programmed controller is constructed, which accounts for the presence of the artificial ramp and for the inductor current ripple. This block diagram is appended to the averaged converter models derived in Chapter 7, leading to a complete converter CPM model. Models for the CPM buck, boost, and buck–boost converters are listed, and the buck converter model is analyzed in detail.

12.3.1   Current Programmed Controller Model

Rather than using the approximation $\langle i_L(t)\rangle_{T_s} = \langle i_C(t)\rangle_{T_s}$, let us derive a more accurate expression relating the average inductor current $\langle i_L(t)\rangle_{T_s}$ to the control input $i_C(t)$. The inductor current waveform is illustrated in Fig. 12.23. It can be seen that the peak value of $i_L(t)$ differs from $i_C(t)$, by the magnitude of the artificial ramp waveform at time $t = dT_s$, that is, by $m_a dT_s$. The peak and average values of the inductor current waveform differ by the average value of the inductor current ripple. Under transient conditions, in which $i_L(0)$ is not equal to $i_L(T_s)$, the magnitudes of the inductor current ripples during the dT_s and $d'T_s$ subintervals are $m_1 dT_s/2$ and $m_2 d'T_s/2$, respectively. Hence, the average value of the inductor current ripple is $d(m_1 dT_s/2) + d'(m_2 d'T_s/2)$. We can express the average inductor current as

$$\langle i_L(t)\rangle_{T_s} = \langle i_C(t)\rangle_{T_s} - m_a dT_s - d\,\frac{m_1 dT_s}{2} - d'\,\frac{m_2 d'T_s}{2}$$
$$= \langle i_C(t)\rangle_{T_s} - m_a dT_s - m_1\,\frac{d^2 T_s}{2} - m_2\,\frac{d'^2 T_s}{2} \tag{12.59}$$

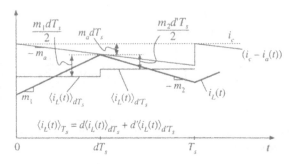

Fig. 12.23 Accurate determination of the relationship between the average inductor current $\langle i_L(t) \rangle_{T_s}$ and i_c.

This is the more accurate relationship which is employed in this section.

A small-signal current programmed controller model is found by perturbation and linearization of Eq. (12.59). Let

$$\begin{aligned}
\langle i_L(t) \rangle_{T_s} &= I_L + \hat{i}_L(t) \\
\langle i_c(t) \rangle_{T_s} &= I_c + \hat{i}_c(t) \\
d(t) &= D + \hat{d}(t) \\
m_1(t) &= M_1 + \hat{m}_1(t) \\
m_2(t) &= M_2 + \hat{m}_2(t)
\end{aligned} \tag{12.60}$$

Note that it is necessary to perturb the slopes m_1 and m_2, since the inductor current slope depends on the converter voltages according to Eq. (12.1). For the basic buck, boost, and buck–boost converters, the slope variations are given by

Buck converter

$$\hat{m}_1 = \frac{\hat{v}_g - \hat{v}}{L} \quad \hat{m}_2 = \frac{\hat{v}}{L}$$

Boost converter

$$\hat{m}_1 = \frac{\hat{v}_g}{L} \quad \hat{m}_2 = \frac{\hat{v} - \hat{v}_g}{L} \tag{12.61}$$

Buck–boost converter

$$\hat{m}_1 = \frac{\hat{v}_g}{L} \quad \hat{m}_2 = -\frac{\hat{v}}{L}$$

It is assumed that m_a does not vary: $\hat{m}_a = M_a$. Substitution of Eq. (12.60) into Eq. (12.59) leads to

$$\left(I_L + \hat{i}_L(t) \right) = \left(I_c + \hat{i}_c(t) \right) - M_a T_s \left(D + \hat{d}(t) \right) - \frac{T_s}{2} \left(M_1 + \hat{m}_1(t) \right) \left(D + \hat{d}(t) \right)^2 - \frac{T_s}{2} \left(M_2 + \hat{m}_2(t) \right) \left(D' - \hat{d}(t) \right)^2 \tag{12.62}$$

The first-order ac terms are

$$\hat{i}_L(t) = \hat{i}_c(t) - \left(M_a T_s + D M_1 T_s - D' M_2 T_s \right) \hat{d}(t) - \frac{D^2 T_s}{2} \hat{m}_1(t) - \frac{D'^2 T_s}{2} \hat{m}_2(t) \tag{12.63}$$

With use of the equilibrium relationship $D M_1 = D' M_2$, Eq. (12.63) can be further simplified:

Table 12.2 Current programmed controller gains for basic converters

Converter	F_g	F_v
Buck	$\dfrac{D^2 T_s}{2L}$	$\dfrac{\left(1 - 2D\right)T_s}{2L}$
Boost	$\dfrac{\left(2D - 1\right)T_s}{2L}$	$\dfrac{D'^2 T_s}{2L}$
Buck–boost	$\dfrac{D^2 T_s}{2L}$	$-\dfrac{D'^2 T_s}{2L}$

$$\hat{i}_L(t) = \hat{i}_c(t) - M_a T_s \hat{d}(t) - \frac{D^2 T_s}{2} \hat{m}_1(t) - \frac{D'^2 T_s}{2} \hat{m}_2(t) \tag{12.64}$$

Finally, solution for $\hat{d}(t)$ yields

$$\hat{d}(t) = \frac{1}{M_a T_s}\left[\hat{i}_c(t) - \hat{i}_L(t) - \frac{D^2 T_s}{2} \hat{m}_1(t) - \frac{D'^2 T_s}{2} \hat{m}_2(t)\right] \tag{12.65}$$

This is the actual relationship that the current programmed controller follows, to determine $\hat{d}(t)$ as a function of $\hat{i}_c(t)$, $\hat{i}_L(t)$, $\hat{m}_1(t)$, and $\hat{m}_2(t)$. Since the quantities $\hat{m}_1(t)$, and $\hat{m}_2(t)$ depend on $\hat{v}_g(t)$ and $\hat{v}(t)$, according to Eq. (12.61), we can express Eq. (12.65) in the following form:

$$\hat{d}(t) = F_m\left[\hat{i}_c(t) - \hat{i}_L(t) - F_g \hat{v}_g(t) - F_v \hat{v}(t)\right] \tag{12.66}$$

where $F_m = 1/M_a T_s$. Expressions for the gains F_g and F_v, for the basic buck, boost, and buck–boost converters, are listed in Table 12.2. A functional block diagram of the current programmed controller, corresponding to Eq. (12.66), is constructed in Fig. 12.24.

Current programmed converter models can now be obtained, by combining the controller block

Fig. 12.24 Functional block diagram of the current programmed controller.

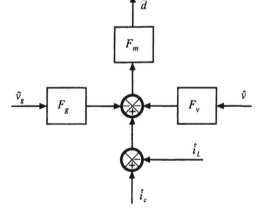

diagram of Fig. 12.24 with the averaged converter models derived in Chapter 7. Figure 12.25 illustrates the CPM converter models obtained by combination of Fig. 12.24 with the buck, boost, and buck-boost models of Fig. 7.17. For each converter, the current programmed controller contains effective feedback of the inductor current $\hat{i}_L(t)$ and the output voltage $\hat{v}(t)$, as well as effective feedforward of the input voltage $\hat{v}_g(t)$.

12.3.2 Solution of the CPM Transfer Functions

Next, let us solve the models of Fig. 12.25, to determine more accurate expressions for the control-to-output and line-to-output transfer functions of current-programmed buck, boost, and buck–boost converters. As discussed in Chapter 8, the converter output voltage $\hat{v}$ can be expressed as a function of the duty-cycle $\hat{d}$ and input voltage $\hat{v}_g$ variations, using the transfer functions $G_{vd}(s)$ and $G_{vg}(s)$:

$$\hat{v}(s) = G_{vd}(s)\hat{d}(s) + G_{vg}(s)\hat{v}_g(s) \qquad (12.67)$$

In a similar manner, the inductor current variation $\hat{i}$ can be expressed as a function of the duty-cycle $\hat{d}$ and input voltage $\hat{v}_g$ variations, by defining the transfer functions $G_{id}(s)$ and $G_{ig}(s)$:

$$\hat{i}_L(s) = G_{id}(s)\hat{d}(s) + G_{ig}(s)\hat{v}_g(s) \qquad (12.68)$$

where the transfer functions $G_{id}(s)$ and $G_{ig}(s)$ are given by:

(a) *Buck*

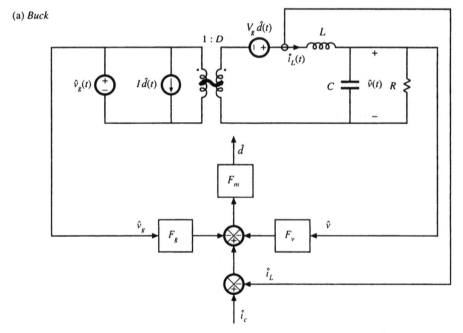

Fig. 12.25   More accurate models of current-programmed converters: (a) buck, (b) boost, (c) buck–boost.

(b) *Boost*

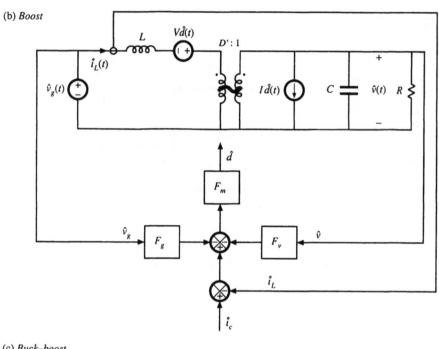

(c) *Buck–boost*

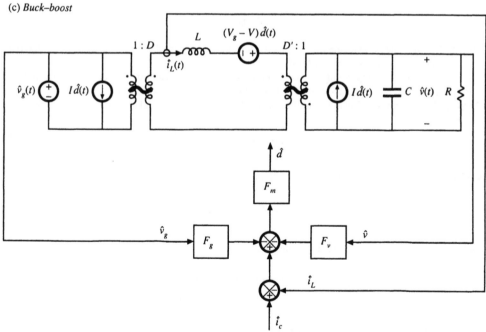

Fig. 12.25 Continued.

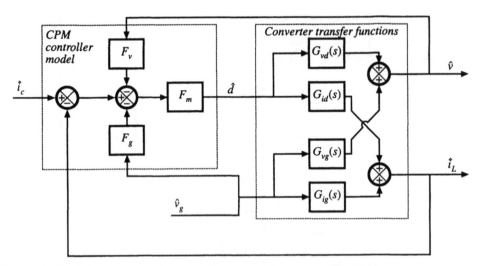

Fig. 12.26   Block diagram that models the current-programmed converters of Fig. 12.25.

$$G_{id}(s) = \left. \frac{\hat{i}_L(s)}{\hat{d}(s)} \right|_{\hat{v}_g(s) = 0}$$

$$G_{ig}(s) = \left. \frac{\hat{i}_L(s)}{\hat{v}_g(s)} \right|_{\hat{d}(s) = 0}$$

(12.69)

Figure 12.26 illustrates replacement of the converter circuit models of Fig. 12.25 with block diagrams that correspond to Eqs. (12.67) and (12.68).

The control-to-output and line-to-output transfer functions can now be found, by manipulation of the block diagram of Fig. 12.26, or by algebraic elimination of $\hat{d}$ and $\hat{i}_L$ from Eqs. (12.66), (12.67), and (12.68), and solution for $\hat{v}$. Substitution of Eq. (12.68) into Eq. (12.66) and solution for $\hat{d}$ leads to

$$\hat{d} = \frac{F_m}{\left(1 + F_m G_{id}\right)} \left[\hat{i}_c - \left(G_{ig} + F_g\right)\hat{v}_g - F_v \hat{v} \right]$$

(12.70)

By substituting this expression into Eq. (12.67), one obtains

$$\hat{v} = \frac{F_m G_{vd}}{\left(1 + F_m G_{id}\right)} \left[\hat{i}_c - \left(G_{ig} + F_g\right)\hat{v}_g - F_v \hat{v} \right] + G_{vg}\hat{v}_g$$

(12.71)

Solution of this equation for $\hat{v}$ leads to the desired result:

$$\hat{v} = \frac{F_m G_{vd}}{1 + F_m \left(G_{id} + F_v G_{vd}\right)} \hat{i}_c + \frac{G_{vg} - F_m F_g G_{vd} + F_m \left(G_{vg} G_{id} - G_{ig} G_{vd}\right)}{1 + F_m \left(G_{id} + F_v G_{vd}\right)} \hat{v}_g$$

(12.72)

Therefore, the current-programmed control-to-output transfer function is

$$G_{vc}(s) = \left.\frac{\hat{v}(s)}{\hat{i}_c(s)}\right|_{\hat{v}_g(s)=0} = \frac{F_m G_{vd}}{1 + F_m(G_{id} + F_v G_{vd})} \tag{12.73}$$

The current-programmed line-to-output transfer function is

$$G_{vg\text{-}cpm}(s) = \left.\frac{\hat{v}(s)}{\hat{v}_g(s)}\right|_{\hat{i}_c(s)=0} = \frac{G_{vg} - F_m F_g G_{vd} + F_m(G_{vg}G_{id} - G_{ig}G_{vd})}{1 + F_m(G_{id} + F_v G_{vd})} \tag{12.74}$$

Equations (12.73) and (12.74) are general expressions for the important transfer functions of single-inductor current-programmed converters operating in the continuous conduction mode.

12.3.3 Discussion

The controller model of Eq. (12.66) and Fig. 12.24 accounts for the differences between i_L and i_c that arise by two mechanisms: the inductor current ripple and the artificial ramp. The inductor current ripple causes the peak and average values of the inductor current to differ; this leads to a deviation between the average inductor current and i_c. Since the magnitude of the inductor current ripple is a function of the converter input and capacitor voltages, this mechanism introduces $\hat{v}_g$ and $\hat{v}$ dependencies into the controller small-signal block diagram. Thus, the F_g and F_v gain blocks of Fig. 12.24 model the small-signal effects of the inductor current ripple. For operation deep in continuous conduction mode ($2L/RT_s \gg 1$), the inductor current ripple is small. The F_g and F_v gain blocks can then be ignored, and the inductor current ripple has negligible effect on the current programmed controller gain.

The artificial ramp also causes the average inductor current to differ from i_c. This is modeled by the gain block F_m, which depends inversely on the artificial ramp slope M_a. With no artificial ramp, $M_a = 0$ and F_m tends to infinity. The current-programmed control systems of Fig. 12.25 then effectively have infinite loop gain. Since the duty cycle $\hat{d}$ is finite, the signal at the input to the F_m block ($\hat{d}/F_m$) must tend to zero. The block diagram then predicts that

$$\frac{\hat{d}}{F_m} = 0 = \hat{i}_c - \hat{i}_L - F_g \hat{v}_g - F_v \hat{v} \tag{12.75}$$

In the case of negligible inductor current ripple ($F_g \to 0$ and $F_v \to 0$), this equation further reduces to

$$0 = \hat{i}_c - \hat{i}_L \tag{12.76}$$

This coincides with the simple approximation employed in Section 12.2. Hence, the transfer functions predicted in this section reduce to the results of Section 12.2 when there is no artificial ramp and negligible inductor current ripple. In the limit when $F_m \to \infty$, $F_g \to 0$, and $F_v \to 0$, the control-to-output transfer function (12.73) reduces to

$$\lim_{\substack{F_m \to \infty \\ F_g \to 0 \\ F_v \to 0}} G_{vc}(s) = \frac{G_{vd}}{G_{id}} \tag{12.77}$$

the line-to-output transfer function reduces to

$$\lim_{\substack{F_m \to \infty \\ F_g \to 0 \\ F_v \to 0}} G_{vg\text{-}cpm}(s) = \frac{G_{vg}G_{id} - G_{ig}G_{vd}}{G_{id}}$$

(12.78)

It can be verified that Eqs. (12.77) and (12.78) are equivalent to the transfer functions derived in Section 12.2.

When an artificial ramp is present, then the gain F_m is reduced to a finite value. The current-programmed controller no longer perfectly regulates the inductor current i_L, and the terms on the right-hand side of Eq. (12.75) do not add to zero. In the extreme case of a very large artificial ramp (large M_a and hence small F_m), the current-programmed controller degenerates to duty-cycle control. The artificial ramp and analog comparator of Fig. 12.8 then function as a pulse-width modulator similar to Fig. 7.63, with small-signal gain F_m. For small F_m and for $F_g \to 0$, $F_v \to 0$, the control-to-output transfer function (12.73) reduces to

$$\lim_{\substack{\text{small } F_m \\ F_v \to 0 \\ F_g \to 0}} G_{vc}(s) = F_m G_{vd}(s)$$

(12.79)

which coincides with conventional duty cycle control. Likewise, Eq. (12.74) reduces to

$$\lim_{\substack{F_m \to \infty \\ F_g \to 0 \\ F_v \to 0}} G_{vg\text{-}cpm}(s) = G_{vg}$$

(12.80)

which is the line-to-output transfer function for conventional duty cycle control.

12.3.4   Current-Programmed Transfer Functions of the CCM Buck Converter

The control-to-output transfer function $G_{vd}(s)$ and line-to-output transfer function $G_{vg}(s)$ of the CCM buck converter with duty cycle control are tabulated in Chapter 8, by analysis of the equivalent circuit model in Fig. 7.17(a). The results are:

$$G_{vd}(s) = \frac{V}{D} \frac{1}{den(s)}$$

(12.81)

$$G_{vg}(s) = D \frac{1}{den(s)}$$

(12.82)

where the denominator polynomial is

$$den(s) = 1 + s \frac{L}{R} + s^2 LC$$

The inductor current transfer functions $G_{id}(s)$ and $G_{ig}(s)$ defined by Eqs. (12.68) and (' found by solution of the equivalent circuit model in Fig. 7.17(a), with the following re

$$G_{id}(s) = \frac{V}{DR} \frac{\left(1 + sRC\right)}{den(s)}$$

and

$$G_{ig}(s) = \frac{D}{R} \frac{(1 + sRC)}{den(s)} \tag{12.85}$$

where *den(s)* is again given by Eq. (12.83).

With no artificial ramp and negligible ripple, the control-to-output transfer function reduces to the ideal expression (12.77). Substitution of Eqs. (12.81) and (12.84) yields

$$\lim_{\substack{F_m \to \infty \\ F_g \to 0 \\ F_v \to 0}} G_{vc}(s) = \frac{G_{vd}(s)}{G_{id}(s)} = \frac{R}{1 + sRC} \tag{12.86}$$

Under the same conditions, the line-to-output transfer function reduces to the ideal expression (12.78). Substitution of Eqs. (12.81) to (12.85) leads to

$$\lim_{\substack{F_m \to \infty \\ F_g \to 0 \\ F_v \to 0}} G_{vg\text{-}cpm}(s) = \frac{G_{vg}(s)G_{id}(s) - G_{vd}(s)G_{ig}(s)}{G_{id}(s)} = 0 \tag{12.87}$$

Equations (12.86) and (12.87) coincide with the expressions derived in Section 12.2 for the CCM buck converter.

For arbitrary F_m, F_v, and F_g, the control-to-output transfer function is given by Eq. (12.73). Substitution of Eqs. (12.81) to (12.85) into Eq. (12.73) yields

$$G_{vc}(s) = \frac{F_m G_{vd}}{1 + F_m \left[G_{id} + F_v G_{vd} \right]} = \frac{F_m \left(\frac{V}{D} \frac{1}{den(s)} \right)}{1 + F_m \left[\left(\frac{V}{DR} \frac{1 + sRC}{den(s)} \right) + F_v \left(\frac{V}{D} \frac{1}{den(s)} \right) \right]} \tag{12.88}$$

Simplification leads to

$$G_{vc}(s) = \frac{F_m \frac{V}{D}}{den(s) + \frac{F_m V}{DR} \left(1 + sRC \right) + F_m F_v \frac{V}{D}} \tag{12.89}$$

Finally, the control-to-output transfer function can be written in the following normalized form:

$$G_{vc}(s) = \frac{G_{c0}}{1 + \frac{s}{Q_c \omega_c} + \left(\frac{s}{\omega_c} \right)^2} \tag{12.90}$$

where

$$G_{c0} = \frac{V}{D} \frac{F_m}{1 + \frac{F_m V}{DR} + \frac{F_m F_v V}{D}} \tag{12.91}$$

$$\omega_c = \frac{1}{\sqrt{LC}} \sqrt{1 + \frac{F_m V}{DR} + \frac{F_m F_v V}{D}} \tag{12.92}$$

$$Q_c = R \sqrt{\frac{C}{L}} \frac{\sqrt{1 + \frac{F_m V}{DR} + \frac{F_m F_v V}{D}}}{\left(1 + \frac{RCF_m V}{DL}\right)} \tag{12.93}$$

In the above equations, the salient features G_{c0}, ω_c, and Q_c are expressed as the duty-ratio-control value, multiplied by a factor that accounts for the effects of current-programmed control.

It can be seen from Eq. (12.93) that current programming tends to reduce the Q-factor of the poles. For large F_m, Q_c varies as $F_m^{-1/2}$; consequently, the poles become real and well-separated in magnitude. The low-Q approximation of Section 8.1.7 then predicts that the low-frequency pole becomes

$$Q_c \omega_c = \frac{R}{L} \frac{\left(1 + \frac{F_m V}{DR} + \frac{F_m F_v V}{D}\right)}{\left(1 + \frac{RCF_m V}{DL}\right)} \tag{12.94}$$

For large F_m and small F_v, this expression can be further approximated as

$$Q_c \omega_c \approx \frac{1}{RC} \tag{12.95}$$

which coincides with the low-frequency pole predicted by the simple model of Section 12.2. The low-Q approximation also predicts that the high-frequency pole becomes

$$\frac{\omega_c}{Q_c} = \frac{1}{RC} \left(1 + \frac{RCF_m V}{DL}\right) \tag{12.96}$$

For large F_m, this expression can be further approximated as

$$\frac{\omega_c}{Q_c} \approx \frac{F_m V}{DL} = f_s \frac{M_2}{DM_a} \tag{12.97}$$

The high-frequency pole is typically predicted to lie near to or greater than the switching frequency f_s. It should be pointed out that the converter switching and modulator sampling processes lead to discrete-time phenomena that affect the high-frequency behavior of the converter, and that are not predicted by the continuous-time averaged analysis employed here. Hence, the averaged model is valid only at frequencies sufficiently less than one-half of the switching frequency.

For arbitrary F_m, F_v, and F_g, the current-programmed line-to-output transfer function $G_{vg\text{-}cpm}(s)$ is given by Eq. (12.74). This equation is most easily evaluated by first finding the ideal transfer function, Eq. (12.78), and then using the result to simplify Eq. (12.74). In the case of the buck converter, Eq. (12.87) shows that the quantity $(G_{vg}G_{id} - G_{vd}G_{ig})$ is equal to zero. Hence, Eq. (12.74) becomes

$$G_{vg\text{-}cpm}(s) = \frac{G_{vg} - F_m F_g G_{vd} + F_m(0)}{1 + F_m\left(G_{id} + F_v G_{vd}\right)} \tag{12.98}$$

Substitution of Eqs. (12.81) to (12.85) into Eq. (12.98) yields

$$G_{vg-cpm}(s) = \frac{\dfrac{D}{den(s)} - F_m F_g \dfrac{V}{D} \dfrac{1}{den(s)}}{1 + F_m \left(\dfrac{V}{DR} \dfrac{1 + sRC}{den(s)} + F_v \dfrac{V}{D} \dfrac{1}{den(s)} \right)} \qquad (12.99)$$

Simplification leads to

$$G_{vg-cpm}(s) = \frac{\left(D - F_m F_g \dfrac{V}{D} \right)}{den(s) + \dfrac{F_m V}{DR} \left(1 + sRC \right) + F_m F_v \dfrac{V}{D}} \qquad (12.100)$$

Finally, the current-programmed line-to-output transfer function can be written in the following normalized form:

$$G_{vg-cpm}(s) = \frac{G_{g0}}{1 + \dfrac{s}{Q_c \omega_c} + \left(\dfrac{s}{\omega_c} \right)^2} \qquad (12.101)$$

where

$$G_{g0} = D \frac{\left(1 - \dfrac{F_m F_g V}{D^2} \right)}{\left(1 + \dfrac{F_m V}{DR} + \dfrac{F_m F_v V}{D} \right)} = D \frac{\left(1 - \dfrac{M_2}{2M_a} \right)}{\left(1 + \dfrac{F_m V}{DR} + \dfrac{F_m F_v V}{D} \right)} \qquad (12.102)$$

The quantities Q_c and ω_c are given by Eqs. (12.92) and (12.93).

Equation (12.102) shows how current programming reduces the dc gain of the buck converter line-to-output transfer function. For duty cycle control ($F_m \to 0$), G_{g0} is equal to D. Nonzero values of F_m reduce the numerator and increase the denominator of Eq. (12.102), which tends to reduce G_{g0}. We have already seen that, in the ideal case ($F_m \to \infty$, $F_g \to 0$, $F_v \to 0$), G_{g0} becomes zero. Equation (12.102) reveals that nonideal current-programmed buck converters can also exhibit zero G_{g0}, if the artificial ramp slope M_a is chosen equal to $0.5M_2$. The current programmed controller then prevents input line voltage variations from reaching the output. The mechanism that leads to this result is the effective feedforward of v_g, inherent in the current programmed controller via the $F_g \hat{v}_g$ term in Eq. (12.66). It can be seen from Fig. 12.26 that, when $F_g F_m G_{vd}(s) = G_{vg}(s)$, then the feedforward path from $\hat{v}_g$ through F_g induces variations in the output $\hat{v}$ that exactly cancel the $\hat{v}_g$-induced variations in the direct forward path of the converter through $G_{vg}(s)$. This cancellation occurs in the buck converter when $M_a = 0.5M_2$.

12.3.5 Results for Basic Converters

The transfer functions of the basic buck, boost, and buck-boost converters with current-programmed control are summarized in Tables 12.3 to 12.5. Control-to-output and line-to-output transfer functions for both the simple model of Section 12.2 and the more accurate model derived in this section are listed. For completeness, the transfer functions for duty cycle control are included. In each case, the salient features are expressed as the corresponding quantity with duty cycle control, multiplied by a factor that accounts for current-programmed control.

Table 12.3 Summary of results for the CPM buck converter

Simple model	Duty cycle controlled gains	
$\dfrac{\hat{v}}{\hat{i}_c} = \dfrac{R}{1 + sRC}$	$G_{vd}(s) = \dfrac{V}{D}\,\dfrac{1}{den(s)}$	$G_{id}(s) = \dfrac{V}{DR}\,\dfrac{1 + sRC}{den(s)}$
$\dfrac{\hat{v}}{\hat{v}_g} = 0$	$G_{vg}(s) = D\,\dfrac{1}{den(s)}$	$G_{ig}(s) = \dfrac{D}{R}\,\dfrac{1 + sRC}{den(s)}$
	$den(s) = 1 + s\dfrac{L}{R} + s^2 LC$	

More accurate model

$$\frac{\hat{v}}{\hat{i}_c} = G_{vc}(s) = G_{c0}\,\frac{1}{1 + \dfrac{s}{Q_c \omega_c} + \left(\dfrac{s}{\omega_c}\right)^2}$$

$$G_{c0} = \frac{V}{D}\,\frac{F_m}{\left(1 + \dfrac{F_m V}{DR} + \dfrac{F_m F_v V}{D}\right)}$$

$$\omega_c = \frac{1}{\sqrt{LC}}\sqrt{1 + \frac{F_m V}{DR} + \frac{F_m F_v V}{D}}$$

$$Q_c = R\sqrt{\frac{C}{L}}\,\frac{\sqrt{1 + \dfrac{F_m V}{DR} + \dfrac{F_m F_v V}{D}}}{\left(1 + \dfrac{RCF_m V}{DL}\right)}$$

$$\frac{\hat{v}}{\hat{v}_g} = G_{vg\text{-}cpm}(s) = G_{g0}\,\frac{1}{1 + \dfrac{s}{Q_c \omega_c} + \left(\dfrac{s}{\omega_c}\right)^2}$$

$$G_{g0} = D\,\frac{\left(1 - \dfrac{F_m F_g V}{D^2}\right)}{\left(1 + \dfrac{F_m V}{DR} + \dfrac{F_m F_v V}{D}\right)}$$

Table 12.4 Summary of results for the CPM boost converter

Simple model	Duty cycle controlled gains	
$\dfrac{\hat{v}}{\hat{i}_c} = \dfrac{D'R}{2}\,\dfrac{\left(1 - s\dfrac{L}{D'^2 R}\right)}{\left(1 + s\dfrac{RC}{2}\right)}$	$G_{vd}(s) = \dfrac{V}{D'}\,\dfrac{\left(1 - s\dfrac{L}{D'^2 R}\right)}{den(s)}$	$G_{id}(s) = \dfrac{2V}{D'^2 R}\,\dfrac{\left(1 + s\dfrac{RC}{2}\right)}{den(s)}$
$\dfrac{\hat{v}}{\hat{v}_g} = \dfrac{1}{2D'}\,\dfrac{1}{\left(1 + s\dfrac{RC}{2}\right)}$	$G_{vg}(s) = \dfrac{1}{D'}\,\dfrac{1}{den(s)}$	$G_{ig}(s) = \dfrac{1}{D'^2 R}\,\dfrac{\left(1 + sRC\right)}{den(s)}$
	$den(s) = 1 + s\dfrac{L}{D'^2 R} + s^2\dfrac{LC}{D'^2}$	

More accurate model

$$\frac{\hat{v}}{\hat{i}_c} = G_{vc}(s) = G_{c0}\,\frac{\left(1 - s\dfrac{L}{D'^2 R}\right)}{1 + \dfrac{s}{Q_c \omega_c} + \left(\dfrac{s}{\omega_c}\right)^2}$$

$$G_{c0} = \frac{V}{D'}\,\frac{F_m}{\left(1 + \dfrac{2F_m V}{D'^2 R} + \dfrac{F_m F_v V}{D'}\right)}$$

$$\omega_c = \frac{D'}{\sqrt{LC}}\sqrt{1 + \frac{2F_m V}{D'^2 R} + \frac{F_m F_v V}{D'}}$$

$$Q_c = D'R\sqrt{\frac{C}{L}}\,\frac{\sqrt{1 + \dfrac{2F_m V}{D'^2 R} + \dfrac{F_m F_v V}{D'}}}{\left(1 + RC\dfrac{F_m V}{L} - \dfrac{F_m F_v V}{D'}\right)}$$

$$\frac{\hat{v}}{\hat{v}_g} = G_{vg\text{-}cpm}(s) = G_{g0}\,\frac{\left(1 + \dfrac{s}{\omega_{gz}}\right)}{1 + \dfrac{s}{Q_c \omega_c} + \left(\dfrac{s}{\omega_c}\right)^2}$$

$$G_{g0} = \frac{1}{D'}\,\frac{\left(1 - F_m F_g V + \dfrac{F_m V}{D'^2 R}\right)}{\left(1 + \dfrac{2F_m V}{D'^2 R} + \dfrac{F_m F_v V}{D'}\right)}$$

$$\omega_{gz} = \frac{D'^3 R}{L}\,\frac{\left(1 - F_m F_g V + \dfrac{F_m V}{D'^2 R}\right)}{F_m F_g V}$$

Table 12.5 Summary of results for the CPM buck–boost converter

Simple model	Duty cycle controlled gains	

$$\frac{\hat{v}}{\hat{i}_c} = -\frac{D'R}{(1+D)} \frac{\left(1 - s\dfrac{DL}{D'^2R}\right)}{\left(1 + s\dfrac{RC}{1+D}\right)}$$

$$G_{vd}(s) = -\frac{|V|}{DD'} \frac{\left(1 - s\dfrac{DL}{D'^2R}\right)}{den(s)}$$

$$G_{id}(s) = -\frac{|V|(1+D)}{DD'^2R} \frac{\left(1 + s\dfrac{RC}{(1+D)}\right)}{den(s)}$$

$$\frac{\hat{v}}{\hat{v}_g} = -\frac{D^2}{1-D^2} \frac{1}{\left(1 + s\dfrac{RC}{1+D}\right)}$$

$$G_{vg}(s) = -\frac{D}{D'} \frac{1}{den(s)}$$

$$G_{ig}(s) = \frac{D}{D'^2R} \frac{(1 + sRC)}{den(s)}$$

$$den(s) = 1 + s\frac{L}{D'^2R} + s^2\frac{LC}{D'^2}$$

More accurate model

$$\frac{\hat{v}}{\hat{i}_c} = G_{vc}(s) = G_{c0} \frac{\left(1 - s\dfrac{DL}{D'^2R}\right)}{1 + \dfrac{s}{Q_c\omega_c} + \left(\dfrac{s}{\omega_c}\right)^2}$$

$$G_{c0} = -\frac{|V|}{DD'} \frac{F_m}{\left(1 + \dfrac{F_m|V|(1+D)}{DD'^2R} - \dfrac{F_mF_v|V|}{DD'}\right)}$$

$$\omega_c = \frac{D'}{\sqrt{LC}} \sqrt{1 + \frac{F_m|V|(1+D)}{DD'^2R} - \frac{F_mF_v|V|}{DD'}}$$

$$Q_c = D'R\sqrt{\frac{C}{L}} \frac{\sqrt{1 + \dfrac{F_m|V|(1+D)}{DD'^2R} - \dfrac{F_mF_v|V|}{DD'}}}{\left(1 + \dfrac{F_m|V|RC}{DL} + \dfrac{F_mF_v|V|}{D'}\right)}$$

$$\frac{\hat{v}}{\hat{v}_g} = G_{vg\text{-}cpm}(s) = G_{g0} \frac{\left(1 + \dfrac{s}{\omega_{gz}}\right)}{1 + \dfrac{s}{Q_c\omega_c} + \left(\dfrac{s}{\omega_c}\right)^2}$$

$$G_{g0} = -\frac{D}{D'}\left(1 + \frac{F_m|V|}{D'^2R} - \frac{F_mF_g|V|}{D^2}\right)$$

$$\omega_{gz} = \frac{DD'^2R}{|V|LF_mF_g}\left(1 + \frac{F_m|V|}{D'^2R} - \frac{F_mF_g|V|}{D^2}\right)$$

The two poles of the line-to-output transfer functions $G_{vg\text{-}cpm}$ and control-to-output transfer functions G_{vc} of all three converters typically exhibit low Q-factors in CPM. The low-Q approximation can be applied, as in Eqs. (12.94) to (12.97), to find the low-frequency pole. The line-to-output transfer functions of the boost and buck–boost converters exhibit two poles and one zero, with substantial dc gain.

12.3.6 Quantitative Effects of Current-Programmed Control on the Converter Transfer Functions

The frequency responses of a CCM buck converter, operating with current-programmed control and with duty cycle control, are compared in Appendix B, Section B.3.2. The buck converter of Fig. B.25 was simulated as described in Appendix B, and the resulting plots are reproduced here.

The magnitude and phase of the control-to-output transfer functions are illustrated in Fig. 12.27. It can be seen that, for duty cycle control, the transfer function $G_{vd}(s)$ exhibits a resonant two-pole response. The substantial damping introduced by current-programmed control leads to essentially a single-pole response in the current-programmed control-to-output transfer function $G_{vc}(s)$. A second pole appears in the vicinity of 100 kHz, which is near the 200 kHz switching frequency. Because of this effective single-pole response, it is relatively easy to design a controller that exhibits a well-behaved response,

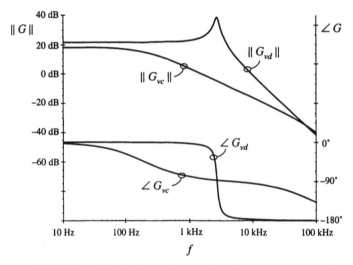

Fig. 12.27   Comparison of CPM control with duty-cycle control, for the control-to-output frequency response of the buck converter example.

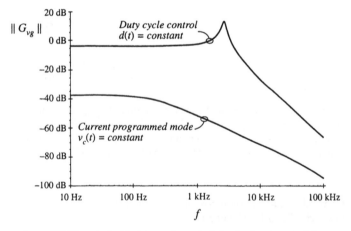

Fig. 12.28   Comparison of CPM control with duty-cycle control, for the line-to-output frequency response of the buck converter example.

having ample phase margin over a wide range of operating points. Proportional-plus-integral (*PI*) controllers are commonly used in current-programmed regulators.

The line-to-output transfer functions of the same example are compared in Fig. 12.28. The line-to-output transfer function $G_{vg}(s)$ for duty-cycle control is characterized by a dc asymptote approximately equal to the duty cycle $D = 0.676$. Resonant poles occur at the corner frequency of the L–C filter. The line-to-output transfer function $G_{vg-cpm}(s)$ with current-programmed control is significantly reduced, and exhibits more than 30 dB of additional attenuation over the frequencies of interest. It should again be

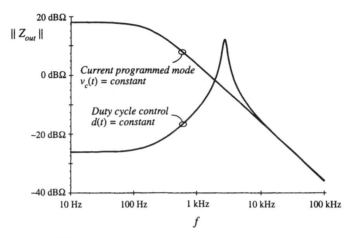

Fig. 12.29 Comparison of CPM control with duty-cycle control, for the output impedance of the buck converter example.

noted that the transfer function $G_{vg\text{-}cpm}(s)$ in Fig. 12.28 cannot be predicted by the simple models of Section 12.2; the more accurate model of Section 12.3 must be employed.

The effect of current-programmed control on the converter output impedance is illustrated in Fig. 12.29. The output impedance plotted in the figure includes the load resistance of 10 Ω. For duty-cycle control, the dc asymptote of the output impedance is dominated by the inductor winding resistance of 0.05 Ω. The inductor becomes significant in the vicinity of 200 Hz. Above the resonant frequency of the output filter, the output impedance is dominated by the output filter capacitor. For current-programmed control, the simple model of Section 12.2 predicts that the inductor branch of the circuit is driven by a current source; this effectively removes the influence of the inductor on the output impedance. The plot of Fig. 12.29 was generated using the more accurate model of this section; nonetheless, the output impedance is accurately predicted by the simple model. The dc asymptote is dominated by the load resistance, and the high-frequency asymptote follows the impedance of the output filter capacitor. It can be seen that current programming substantially increases the converter output impedance.

12.4 DISCONTINUOUS CONDUCTION MODE

Current-programmed converters operating in the discontinuous conduction mode can be described using the averaged switch modeling approaches of Sections 12.3 and 11.1. It is found in this section that the average transistor voltage and current follow a power sink characteristic, while the average diode voltage and current obey a power source characteristic. Perturbation and linearization of these characteristics leads to a small-signal equivalent circuit that models CPM DCM converters. The basic DCM CPM buck, boost, and buck–boost converters essentially exhibit single-pole transfer functions: the second pole and the right half-plane zero appear at frequencies near to or greater than the switching frequency, owing to the small value of L in DCM.

A DCM CPM buck–boost converter example is analyzed here. However, Eqs. (12.103) to (12.120) are written in general form, and apply equally well to DCM CPM buck and boost converters. The schematic of a buck-boost converter is illustrated in Fig. 12.30. The terminal waveforms of the switch network are defined as shown: $v_1(t)$ and $i_1(t)$ are the transistor waveforms, while $v_2(t)$ and $i_2(t)$ are

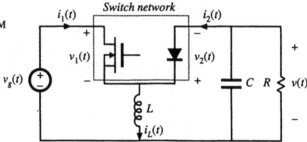

Fig. 12.30 Current-programmed DCM buck–boost converter example.

the diode waveforms. Figure 12.31 illustrates typical DCM waveforms, for current-programmed control with an artificial ramp having slope $-m_a$. The inductor current is zero at the beginning of each switching period. By solution of the transistor conduction subinterval, the programmed current i_{pk} can be related to the transistor duty cycle d_1 by:

$$
\begin{aligned}
i_c &= i_{pk} + m_a d_1 T_s \\
&= \left(m_1 + m_a \right) d_1 T_s
\end{aligned}
\tag{12.103}
$$

Solution for d_1 leads to

$$
d_1(t) = \frac{i_c(t)}{\left(m_1 + m_a \right) T_s}
\tag{12.104}
$$

The average transistor current is found by integrating the $i_1(t)$ waveform of Fig. 12.31 over one switching period:

$$
\left\langle i_1(t) \right\rangle_{T_s} = \frac{1}{T_s} \int_t^{t+T_s} i_1(\tau) d\tau = \frac{q_1}{T_s}
\tag{12.105}
$$

The total area q_1 is equal to one-half of the peak current i_{pk}, multiplied by the subinterval length $d_1 T_s$. Hence,

$$
\left\langle i_1(t) \right\rangle_{T_s} = \tfrac{1}{2} i_{pk}(t) d_1(t)
\tag{12.106}
$$

Elimination of i_{pk} and d_1, to express the average transistor current as a function of i_c, leads to

$$
\left\langle i_1(t) \right\rangle_{T_s} = \frac{\tfrac{1}{2} L i_c^2 f_s}{\left\langle v_1(t) \right\rangle_{T_s} \left(1 + \dfrac{m_a}{m_1} \right)^2}
\tag{12.107}
$$

Finally, Eq. (12.107) can be rearranged to obtain the averaged switch network input port relationship:

$$
\left\langle i_1(t) \right\rangle_{T_s} \left\langle v_1(t) \right\rangle_{T_s} = \frac{\tfrac{1}{2} L i_c^2 f_s}{\left(1 + \dfrac{m_a}{m_1} \right)^2} = \left\langle p(t) \right\rangle_{T_s}
\tag{12.108}
$$

Thus, the average transistor waveforms obey a power sink characteristic. When $m_a = 0$, then the average power $\langle p(t) \rangle_{T_s}$ is a function only of L, i_c, and f_s. The presence of the artificial ramp causes $\langle p(t) \rangle_{T_s}$ to additionally depend on the converter voltages, via m_1.

The power sink characteristic can also be explained via inductor energy arguments. During the first subinterval, the inductor current increases from 0 to i_{pk}. In the process, the inductor stores the following energy:

$$W = \tfrac{1}{2} L i_{pk}^2 \qquad (12.109)$$

The energy W is transferred from the power input v_g, through the switch network input port, to the inductor, once per switching period. This energy transfer process accounts for the power flow

$$\langle p(t) \rangle_{T_s} = W f_s = \tfrac{1}{2} L i_{pk}^2 f_s \qquad (12.110)$$

The switch network input port, that is, the transistor terminals, can therefore be modeled by a power sink element, as in Fig. 12.32.

The average switch network output port current, that is, the average diode current, is

$$\langle i_2(t) \rangle_{T_s} = \frac{1}{T_s} \int_t^{t+T_s} i_2(\tau) d\tau = \frac{q_2}{T_s} \qquad (12.111)$$

By inspection of Fig. 12.31, the area q_2 is given by

$$q_2 = \tfrac{1}{2} i_{pk} d_2 T_s \qquad (12.112)$$

The duty cycle d_2 is determined by the time required for the inductor current to return to zero, during the second subinterval. By arguments similar to those used to derive Eq. (11.12), the duty cycle d_2 can be found as follows:

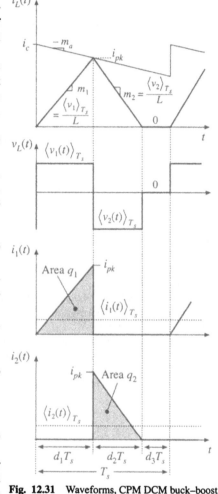

Fig. 12.31 Waveforms, CPM DCM buck–boost example.

$$d_2(t) = d_1(t) \frac{\langle v_1(t) \rangle_{T_s}}{\langle v_2(t) \rangle_{T_s}} \qquad (12.113)$$

Substitution of Eqs. (12.113), (12.112), and (12.110) into Eq. (12.111) yields

$$\langle i_2(t) \rangle_{T_s} = \frac{\langle p(t) \rangle_{T_s}}{\langle v_2(t) \rangle_{T_s}} \qquad (12.114)$$

The output port of the averaged switch network is therefore described by the relationship

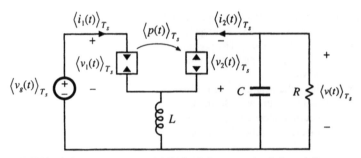

Fig. 12.32   CPM DCM buck–boost converter model, derived via averaged switch modeling.

$$\langle i_2(t)\rangle_{T_s} \langle v_2(t)\rangle_{T_s} = \frac{\frac{1}{2}Li_c^2(t)f_s}{\left(1+\frac{m_a}{m_1}\right)^2} = \langle p(t)\rangle_{T_s} \tag{12.115}$$

In the averaged model, the diode can be replaced by a power source of value $\langle p(t)\rangle_{T_s}$, equal to the power apparently consumed at the switch network input port. During the second subinterval, the inductor releases all of its stored energy through the diode, to the converter output. This results in an average power flow of value $\langle p(t)\rangle_{T_s}$.

A CPM DCM buck-boost averaged model is therefore as given in Fig. 12.32. In this model, the transistor is simply replaced by a power sink of value $\langle p(t)\rangle_{T_s}$, while the diode is replaced by a power source also of value $\langle p(t)\rangle_{T_s}$.

The steady-state equivalent circuit model of the CPM DCM buck-boost converter is obtained by letting the inductor and capacitor tend to short- and open-circuits, respectively. The model of Fig. 12.33 is obtained. The steady-state output voltage V can now be determined by equating the dc load power to the converter average power $\langle p(t)\rangle_{T_s}$. For a resistive load, one obtains

$$\frac{V^2}{R} = P \tag{12.116}$$

where the steady state value of $\langle p(t)\rangle_{T_s}$ is given by

$$P = \frac{\frac{1}{2}LI_c^2(t)f_s}{\left(1+\frac{M_a}{M_1}\right)^2} \tag{12.117}$$

and where I_c is the steady-state value of the control input $i_c(t)$. Solution for V yields the following result

Fig. 12.33   Steady-state model of the CPM DCM buck–boost converter.

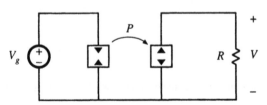

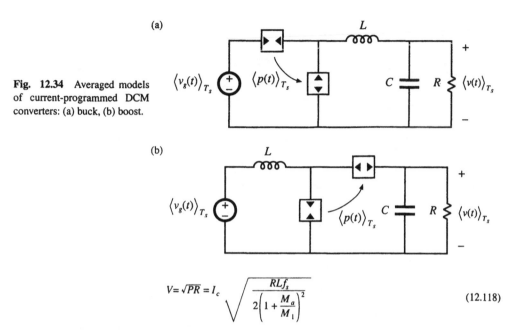

Fig. 12.34 Averaged models of current-programmed DCM converters: (a) buck, (b) boost.

$$V = \sqrt{PR} = I_c \sqrt{\frac{RLf_s}{2\left(1 + \frac{M_a}{M_1}\right)^2}} \qquad (12.118)$$

for the case of a resistive load.

Averaged models of the DCM CPM buck, boost, and other converters can be found in a similar manner. In each case, the average transistor waveforms are shown to follow a power sink characteristic, while the average diode waveforms follow a power source characteristic. The resulting equivalent circuits of the CPM DCM buck and boost converters are illustrated in Fig. 12.34. In each case, the average power is given by

Table 12.6 Steady-state DCM current-programmed characteristics of basic converters

Converter	M	I_{crit}	Stability range when $m_a = 0$
Buck	$\dfrac{P_{load} - P}{P_{load}}$	$\dfrac{1}{2}\left(I_c - M m_a T_s\right)$	$0 \leq M < \frac{2}{3}$
Boost	$\dfrac{P_{load}}{P_{load} - P}$	$\dfrac{\left(I_c - \dfrac{M-1}{M} m_a T_s\right)}{2M}$	$0 \leq D \leq 1$
Buck–boost	Depends on load characteristic: $P_{load} = P$	$\dfrac{\left(I_c - \dfrac{M}{M-1} m_a T_s\right)}{2(M-1)}$	$0 \leq D \leq 1$

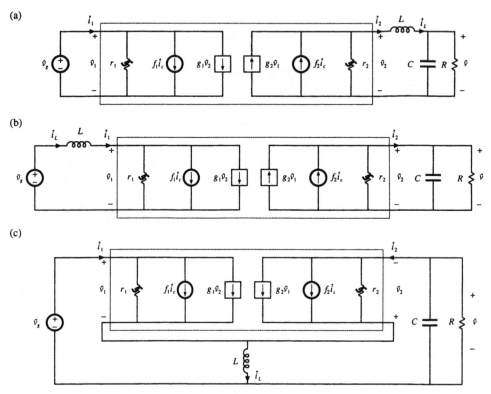

Fig. 12.35 Small-signal models of DCM CPM converters, derived by perturbation and linearization of Figs. 12.32 and 12.34: (a) buck, (b) boost, (c) buck–boost.

$$\langle p(t) \rangle_{T_s} = \frac{\frac{1}{2} L i_c^2(t) f_s}{\left(1 + \frac{m_a}{m_1}\right)^2} \tag{12.119}$$

with m_1 defined as in Eq. (12.1).

Steady-state characteristics of the DCM CPM buck, boost, and buck–boost converters are summarized in Table 12.6. In each case, the dc load power is $P_{load} = VI$ and P is given by Eq. (12.117). The conditions for operation of a current programmed converter in the discontinuous conduction mode can be expressed as follows:

$$\begin{aligned} |I| > |I_{crit}| \quad &\text{for CCM} \\ |I| < |I_{crit}| \quad &\text{for DCM} \end{aligned} \tag{12.120}$$

where I is the dc load current. The critical load current at the CCM–DCM boundary, I_{crit}, is expressed as a function of I_c and the voltage conversion ratio $M = V/V_g$ in Table 12.6.

In the discontinuous conduction mode, the inductor current is zero at the beginning and end of

Table 12.7 Current programmed DCM small-signal equivalent circuit parameters: input port

Converter	g_1	f_1	r_1
Buck	$\dfrac{1}{R}\left(\dfrac{M^2}{1-M}\right)\dfrac{\left(1-\dfrac{m_a}{m_1}\right)}{\left(1+\dfrac{m_a}{m_1}\right)}$	$2\dfrac{I_1}{I_c}$	$-R\left(\dfrac{1-M}{M^2}\right)\dfrac{\left(1+\dfrac{m_a}{m_1}\right)}{\left(1-\dfrac{m_a}{m_1}\right)}$
Boost	$-\dfrac{1}{R}\left(\dfrac{M}{M-1}\right)$	$2\dfrac{I}{I_c}$	$\dfrac{R}{M^2\left(\dfrac{2-M}{M-1}+\dfrac{2m_a/m_1}{1+\dfrac{m_a}{m_1}}\right)}$
Buck–boost	0	$2\dfrac{I_1}{I_c}$	$\dfrac{-R}{M^2}\dfrac{\left(1+\dfrac{m_a}{m_1}\right)}{\left(1-\dfrac{m_a}{m_1}\right)}$

Table 12.8 Current programmed DCM small-signal equivalent circuit parameters: output port

Converter	g_2	f_2	r_2
Buck	$\dfrac{1}{R}\left(\dfrac{M}{1-M}\right)\dfrac{\left(\dfrac{m_a}{m_1}(2-M)-M\right)}{\left(1+\dfrac{m_a}{m_1}\right)}$	$2\dfrac{I}{I_c}$	$R\dfrac{(1-M)\left(1+\dfrac{m_a}{m_1}\right)}{\left(1-2M+\dfrac{m_a}{m_1}\right)}$
Boost	$\dfrac{1}{R}\left(\dfrac{M}{M-1}\right)$	$2\dfrac{I_2}{I_c}$	$R\left(\dfrac{M-1}{M}\right)$
Buck–boost	$\dfrac{2M}{R}\dfrac{\left(\dfrac{m_a}{m_1}\right)}{\left(1+\dfrac{m_a}{m_1}\right)}$	$2\dfrac{I_2}{I_c}$	R

each switching period. As a result, the current programmed controller does not exhibit the type of instability described in Section 12.1. The current programmed controllers of DCM boost and buck–boost converters are stable for all duty cycles with no artificial ramp. However, the CPM DCM buck converter exhibits a different type of low-frequency instability when $M > 2/3$ and $m_a = 0$, that arises because the dc output characteristic is nonlinear and can exhibit two equilibrium points when the converter drives a resistive load. The stability range can be extended to $0 \leq D \leq 1$ by addition of an artificial ramp having slope $m_a > 0.086\,m_2$, or by addition of output voltage feedback.

Small-signal models of DCM CPM converters can be derived by perturbation and linearization of the averaged models of Figs. 12.32 and 12.34. The results are given in Fig. 12.35. Parameters of the small-signal models are listed in Tables 12.7 and 12.8.

The CPM DCM small-signal models of Fig. 12.35 are quite similar to the respective small-signal models of DCM duty-ratio controlled converters illustrated in Figs. 11.15 and 11.17. The sole differences are the parameter expressions of Tables 12.7 and 12.8. Transfer functions can be determined in a

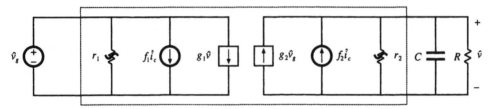

Fig. 12.36 Simplified small-signal model obtained by letting L become zero in Fig. 12.35 (a), (b), or (c).

similar manner. In particular, a simple approximate way to determine the low-frequency small-signal transfer functions of the CPM DCM buck, boost, and buck-boost converters is to simply let the inductance L tend to zero in the equivalent circuits of Fig. 12.35. This approximation is justified for frequencies sufficiently less than the converter switching frequency, because in the discontinuous conduction mode the value of L is small, and hence the pole and any RHP zero associated with L occur at frequencies near to or greater than the switching frequency. For all three converters, the equivalent circuit of Fig. 12.36 is obtained.

Figure 12.36 predicts that the control-to-output transfer function $G_{vc}(s)$ is

$$G_{vc}(s) = \left. \frac{\hat{v}}{\hat{i}_c} \right|_{\hat{v}_g = 0} = \frac{G_{c0}}{1 + \frac{s}{\omega_p}} \tag{12.121}$$

with

$$G_{c0} = f_2\left(R\|r_2\right)$$
$$\omega_p = \frac{1}{\left(R\|r_2\right)C}$$

The line-to-output transfer function is predicted to be

$$G_{vg}(s) = \left. \frac{\hat{v}}{\hat{v}_g} \right|_{\hat{i}_c = 0} = \frac{G_{g0}}{1 + \frac{s}{\omega_p}} \tag{12.122}$$

with

$$G_{g0} = g_2\left(R\|r_2\right)$$

If desired, more accurate expressions which account for inductor dynamics can be derived by solution of the models of Fig. 12.35.

12.5   SUMMARY OF KEY POINTS

1.   In current-programmed control, the peak switch current $i_s(t)$ follows the control input $i_c(t)$. This widely used control scheme has the advantage of a simpler control-to-output transfer function. The line-to-output transfer functions of current-programmed buck converters are also reduced.

2. The basic current-programmed controller is unstable when $D > 0.5$, regardless of the converter topology. The controller can be stabilized by addition of an artificial ramp having slope m_a. When $m_a > 0.5m_2$, then the controller is stable for all duty cycles.

3. The behavior of current-programmed converters can be modeled in a simple and intuitive manner by the first-order approximation $\langle i_L(t) \rangle_{T_s} \approx i_c(t)$. The averaged terminal waveforms of the switch network can then be modeled simply by a current source of value i_c, in conjunction with a power sink or power source element. Perturbation and linearization of these elements leads to the small-signal model. Alternatively, the small-signal converter equations derived in Chapter 7 can be adapted to cover the current programmed mode, using the simple approximation $i_L(t) \approx i_c(t)$.

4. The simple model predicts that one pole is eliminated from the converter line-to-output and control-to-output transfer functions. Current programming does not alter the transfer function zeroes. The dc gains become load-dependent.

5. The more accurate model of Section 12.3 correctly accounts for the difference between the average inductor current $\langle i_L(t) \rangle_{T_s}$ and the control input $i_c(t)$. This model predicts the nonzero line-to-output transfer function $G_{vg}(s)$ of the buck converter. The current-programmed controller behavior is modeled by a block diagram, which is appended to the small-signal converter models derived in Chapter 7. Analysis of the resulting multiloop feedback system then leads to the relevant transfer functions.

6. The more accurate model predicts that the inductor pole occurs at the crossover frequency f_c of the effective current feedback loop gain $T_i(s)$. The frequency f_c typically occurs in the vicinity of the converter switching frequency f_s. The more accurate model also predicts that the line-to-output transfer function $G_{vg}(s)$ of the buck converter is nulled when $m_a = 0.5m_2$.

7. Current programmed converters operating in the discontinuous conduction mode are modeled in Section 12.4. The averaged transistor waveforms can be modeled by a power sink, while the averaged diode waveforms are modeled by a power source. The power is controlled by $i_c(t)$. Perturbation and linearization of these averaged models, as usual, leads to small-signal equivalent circuits.

REFERENCES

[1] C. DEISCH, "Simple Switching Control Method Changes Power Converter into a Current Source," *IEEE Power Electronics Specialists Conference*, 1978 Record, pp. 300-306.

[2] A. CAPEL, G. FERRANTE, D. O'SULLIVAN, and A. WEINBERG, "Application of the Injected Current Model for the Dynamic Analysis of Switching Regulators with the New Concept of LC^3 Modulator," *IEEE Power Electronics Specialists Conference*, 1978 Record, pp. 135-147.

[3] S. HSU, A. BROWN, L. RENSINK, and R. D. MIDDLEBROOK, "Modeling and Analysis of Switching Dc-to-Dc Converters in Constant-Frequency Current Programmed Mode," *IEEE Power Electronics Specialists Conference*, 1979 Record, pp. 284-301.

[4] F. C. LEE and R. A. CARTER, "Investigations of Stability and Dynamic Performances of Switching Regulators Employing Current-Injected Control," *IEEE Power Electronics Specialists Conference*, 1981 Record, pp. 3-16.

[5] R. D. MIDDLEBROOK, "Topics in Multiple-Loop Regulators and Current-Mode Programming," *IEEE Power Electronics Specialists Conference*, 1985 Record, pp. 716-732.

[6] R. D. MIDDLEBROOK, "Modeling Current Programmed Buck and Boost Regulators," *IEEE Transactions on Power Electronics*, Vol. 4, No. 1, January 1989, pp. 36-52.

[7]   G. VERGHESE, C. BRUZOS, and K. MAHABIR, "Averaged and Sampled-Data Models for Current Mode Control: A Reexamination," *IEEE Power Electronics Specialists Conference*, 1989 Record, pp. 484-491.

[8]   D. M. MITCHELL, *Dc–Dc Switching Regulator Analysis*, New York: McGraw-Hill, 1988, Chapter 6.

[9]   A. KISLOVSKI, R. REDL, and N. SOKAL, *Dynamic Analysis of Switching-Mode DC/DC Converters*, New York: Van Nostrand Reinhold, 1994.

[10]   A. BROWN and R. D. MIDDLEBROOK, "Sampled-Data Modeling of Switching Regulators," *IEEE Power Electronics Specialists Conference*, 1981 Record, pp. 716-732.

[11]   R. RIDLEY, "A New Continuous-Time Model for Current-Mode Control," *IEEE Transactions on Power Electronics*, Vol. 6, No. 2, April 1991, pp. 271-280.

[12]   F. D. TAN and R. D. MIDDLEBROOK, "Unified Modeling and Measurement of Current-Programmed Converters," *IEEE Power Electronics Specialists Conference*, 1993 Record, pp. 380-387.

[13]   R. TYMERSKI, "Sampled-Data Modeling of Switched Circuits, Revisited," *IEEE Power Electronics Specialists Conference*, 1993 Record, pp. 395-401.

[14]   W. TANG, F. C. LEE, R. B. RIDLEY and I. COHEN, "Charge Control: Modeling, Analysis and Design," *IEEE Power Electronics Specialists Conference*, 1992 Record, pp. 503-511.

[15]   K. SMEDLEY and S. ĆUK, "One-Cycle Control of Switching Converters," *IEEE Power Electronics Specialists Conference*, 1991 Record, pp. 888-896.

PROBLEMS

12.1   A nonideal buck converter operates in the continuous conduction mode, with the values $V_g = 10$ V, $f_s = 100$ kHz, $L = 4$ μH, $C = 75$ μF, and $R = 0.25$ Ω. The desired full-load output is 5 V at 20 A. The power stage contains the following loss elements: MOSFET on-resistance $R_{on} = 0.1$ Ω, Schottky diode forward voltage drop $V_D = 0.5$ V, inductor winding resistance $R_L = 0.03$ Ω.

   (a)   Steady-state analysis: determine the converter steady-state duty cycle D, the inductor current ripple slopes m_1 and m_2, and the dimensionless parameter $K = 2L/RT_s$.

   (b)   Determine the small-signal equations for this converter, for duty cycle control.

   A current-programmed controller is now implemented for this converter. An artificial ramp is used, having a fixed slope $M_a = 0.5M_2$, where M_2 is the steady-state slope m_2 obtained with an output of 5 V at 20 A.

   (c)   Over what range of D is the current programmed controller stable? Is it stable at rated output? Note that the nonidealities affect the stability boundary.

   (d)   Determine the control-to-output transfer function $G_{vc}(s)$, using the simple approximation $\langle i_L(t) \rangle_{T_s} \approx i_c(t)$. Give analytical expressions for the corner frequency and dc gain. Sketch the Bode plot of $G_{vc}(s)$.

12.2   Use the averaged switch modeling approach to model the CCM boost converter with current-programmed control:

   (a)   Define the switch network terminal quantities as in Fig. 7.46(a). With the assumption that $\langle i_L(t) \rangle_{T_s} \approx i_c(t)$, determine expressions for the average values of the switch network terminal waveforms, and hence derive the equivalent circuit of Fig. 12.18(a).

(b) Perturb and linearize your model of part (a), to obtain the equivalent circuit of Fig. 12.22.

(c) Solve your model of part (b), to derive expressions for the control-to-output transfer function $G_{vc}(s)$ and the line-to-output transfer function $G_{vg}(s)$. Express your results in standard normalized form, and give analytical expressions for the corner frequencies and dc gains.

12.3 Use the averaged switch modeling approach to model the CCM Ćuk converter with current-programmed control. A Ćuk converter is diagrammed in Fig. 2.20.

(a) It is desired to model the switch network with an i_c current source and a dependent power source or sink, using the approach of Section 12.2.2. How should the switch network terminal voltages and currents be defined?

(b) Sketch the switch network terminal voltage and current waveforms. With the assumption that $\langle i_1(t) \rangle_{T_s} - \langle i_2(t) \rangle_{T_s} \approx i_c(t)$ (where i_1 and i_2 are the inductor currents defined in Fig. 2.20), determine expressions for the average values of the switch network terminal waveforms, and hence derive an equivalent circuit similar to the equivalent circuits of Fig. 12.18.

(c) Perturb and linearize your model of part (b), to obtain a small signal equivalent circuit similar to the model of Fig. 12.19. It is not necessary to solve your model.

12.4 The full-bridge converter of Fig. 6.19(a) operates with $V_g = 320$ V, and supplies 1000 W to a 42 V resistive load. Losses can be neglected, the duty cycle is 0.7, and the switching period T_s defined in Fig. 6.20 is 10 μsec. $L = 50$ μH and $C = 100$ μF. A current-programmed controller is employed, whose waveforms are referred to the secondary side of the transformer. In the following calculations, you may neglect the transformer magnetizing current.

(a) What is the minimum artificial ramp slope m_a that will stabilize the controller at the given operating point? Express your result in terms of m_2.

(b) An artificial ramp having the slope $m_a = m_2$ is employed. Sketch the Bode plot of the current loop gain $T_i(s)$, and label numerical values of the corner frequencies and dc gains. It is not necessary to re-derive the analytical expression for T_i. Determine the crossover frequency f_c.

(c) For $m_a = m_2$, sketch the Bode plots of the control-to-output transfer function $G_{vc}(s)$ and line-to-output transfer function $G_{vg}(s)$, and label numerical values of the corner frequencies and dc gains. It is not necessary to re-derive analytical expressions for these transfer functions.

12.5 In a CCM current-programmed buck converter, it is desired to minimize the line-to-output transfer function $G_{vg}(s)$ via the choice $m_a = 0.5m_2$. However, because of component tolerances, the value of inductance L can vary by ±10% from its nominal value of 100 μH. Hence, m_a is fixed in value while m_2 varies, and $m_a = 0.5m_2$ is obtained only at the nominal value of L. The switching frequency is 100 kHz, the output voltage is 15 V, the load current varies over the range 2 to 4 A, and the input voltage varies over the range 22 to 32 V. You may neglect losses. Determine the worst-case (maximum) value of the line-to-output dc gain $G_{vg}(0)$.

12.6 The nonideal flyback converter of Fig. 7.18 employs current-programmed control, with artificial ramp having slope m_a. MOSFET Q_1 exhibits on-resistance R_{on}. All current programmed controller waveforms are referred to the transformer primary side.

(a) Derive a block diagram which models the current-programmed controller, of form similar to Fig. 12.24. Give analytical expressions for the gains in your block diagram.

(b) Combine your result of part (a) with the converter small-signal model. Derive a new expression for the control-to-output transfer function $G_{vc}(s)$.

12.7 A buck converter operates with current-programmed control. The element values are:

$$V_g = 120 \text{ V} \qquad\qquad D = 0.6$$
$$R = 10 \ \Omega \qquad\qquad f_s = 100 \text{ kHz}$$
$$L = 550 \ \mu\text{H} \qquad\qquad C = 100 \ \mu\text{F}$$

An artificial ramp is employed, having slope 0.15 A/μsec.

(a)   Construct the magnitude and phase asymptotes of the control-to-output transfer function $G_{vd}(s)$ for duty-cycle control. On the same plot, construct the magnitude and phase asymptotes of the control-to-output transfer function $G_{vc}(s)$ for current-programmed control. Compare.

(b)   Construct the magnitude asymptotes of the line-to-output transfer function $G_{vg}(s)$ for duty-cycle control. On the same plot, construct the magnitude asymptotes of the line-to-output transfer function $G_{vg\text{-}cpm}(s)$ for current-programmed control. Compare.

12.8   A buck-boost converter operates in the discontinuous conduction mode. Its current-programmed controller has no compensating artificial ramp: $m_a = 0$.

(a)   Derive an expression for the control-to-output transfer function $G_{vc}(s)$, using the approximation $L \approx 0$. Give analytical expressions for the corner frequency and dc gain.

(b)   Repeat part (a), with the inductor included. Show that, provided the inductor is sufficiently small, then the inductor merely adds a high-frequency pole and zero to $G_{vc}(s)$, and the low-frequency pole derived in part (a) is essentially unchanged.

(c)   At the CCM-DCM boundary, what is the minimum value of the RHP zero frequency?

12.9   A current-programmed boost converter interfaces a 3 V battery to a small portable 5 V load. The converter operates in the discontinuous conduction mode, with constant transistor on-time t_{on} and variable off-time; the switching frequency can therefore vary and is used as the control variable. There is no artificial ramp, and the peak transistor current i_c is equal to a fixed value I_c; in practice, I_c is chosen to minimize the total loss.

(a)   Sketch the transistor and diode voltage and current waveforms. Determine expressions for the waveform average values, and hence derive a large-signal averaged equivalent circuit for this converter.

(b)   Perturb and linearize your model of part (a), to obtain a small-signal equivalent circuit. Note that the switching frequency f_s should be perturbed.

(c)   Solve your model of part (b), to derive an expression for the low-frequency control-to-output transfer function $G_{vf}(s) = \hat{v}(s)/\hat{f}_s(s)$. Express your results in standard normalized form, and give analytical expressions for the corner frequencies and dc gains. You may assume that L is small.

12.10   A current-programmed boost converter is employed in a low-harmonic rectifier system, in which the input voltage is a rectified sinusoid: $v_g(t) = V_M |\sin(\omega t)|$. The dc output voltage is $v(t) = V > V_M$. The capacitance C is large, such that the output voltage contains negligible ac variations. It is desired to control the converter such that the input current $i_g(t)$ is proportional to $v_g(t)$: $i_g(t) = v_g(t)/R_e$, where R_e is a constant, called the "emulated resistance." The averaged boost converter model of Fig. 12.18(a) suggests that this can be accomplished by simply letting $i_c(t)$ be proportional to $v_g(t)$, according to $i_c(t) = v_g(t)/R_e$. You may make the simplifying assumption that the converter always operates in the continuous conduction mode.

(a)   Solve the model of Fig. 12.18(a), subject to the assumptions listed above, to determine the power $\langle p(t) \rangle_{T_s}$. Find the average value of $\langle p(t) \rangle_{T_s}$, averaged over one cycle of the ac input $v_g(t)$.

(b)   An artificial ramp is necessary to stabilize the current-programmed controller at some operating points. What is the minimum value of m_a that ensures stability at all operating points along the input rectified sinusoid? Express your result as a function of V and L. Show your work.

(c)   The artificial ramp and inductor current ripple cause the average input current to differ from $i_c(t)$. Derive an algebraic expression for $\langle i_g(t) \rangle_{T_s}$, as a function of $i_c(t)$ and other quantities such as m_a, $v_g(t)$, V, L, and T_s. For this part, you may assume that the inductor dynamics are negligible. Show your work.

(d)   Substitute $v_g(t) = V_M |\sin(\omega t)|$ and $i_c(t) = v_g(t)/R_e$, into your result of part (c), to determine an expression for $i_g(t)$. How does $i_g(t)$ differ from a rectified sinusoid?

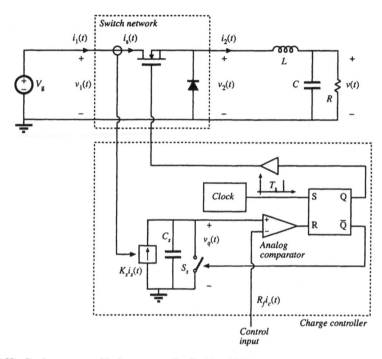

Fig. 12.37 Buck converter with charge controller, Problem 12.11.

12.11 Figure 12.37 shows a buck converter with a charge controller [14]. Operation of the charge controller is similar to operation of the current-programmed controller. At the beginning of each switching period, at time $t = 0$, a short clock pulse sets the SR latch. The logic high signal at the Q output of the latch turns the power MOSFET on. At the same time, the logic low signal at the $\bar{Q}$ output of the latch turns the switch S_s off. Current $K_s i_s$ proportional to the power MOSFET current charges the capacitor C_s. At $t = dT_s$, the capacitor voltage $v_q(t)$ reaches the control input voltage $R_f i_c$, the comparator output goes high and resets the latch. The logic low signal at the Q output of the latch turns the power MOSFET off. At the same time, the logic high signal at the $\bar{Q}$ output of the latch turns the switch S_s on, which quickly discharges the capacitor C_s to zero.

In this problem, the converter and controller parameters are: $V_g = 24$ V, $f_s = 1/T_s = 100$ kHz, $L = 60$ µH, $C = 100$ µF, $R = 3$ Ω, $K_s T_s/C_s = R_f = 1$ Ω. You can assume that the converter operates in continuous conduction mode.

(a) Find expressions for the average values of the switch network terminal waveforms, and hence derive a large-signal averaged switch model of the buck switch network with charge control. The control input to the model is the control current i_c. The averaged switch model should consist of a current source and a power source. The switch duty cycle d should not appear in the model.

(b) Using the averaged switch model derived in part (a), find an expression for the quiescent output voltage V as a function of V_g, I_c, and R. Given $I_c = 2$ A, find numerical values for V, I_1, I_2, and the duty cycle D. For this quiescent operating point, sketch the waveforms $i_1(t)$, $i_2(t)$, and $v_q(t)$ during one switching period.

(c) Perturb and linearize the averaged switch model from part (a) to derive a small-signal averaged switch model for the buck switch network with charge control. Find analytical expressions for all parameter values in terms of the converter parameters and the quiescent operating conditions. Sketch the complete small-signal model of the buck converter with the charge controller.

(d)   Solve the model obtained in part (c) to find the control-to-output transfer function $G_{vc}(s) = \hat{v}/\hat{i}_c$. At the quiescent operating point found in part (b), construct the Bode plot for the magnitude of G_{vc} and label all salient features of the magnitude response.

(e)   Comment on advantages charge control may have compared to duty-cycle control or current-programmed control.

12.12   Figure 12.38 shows a buck converter with a one-cycle controller [15]. Operation of the one-cycle controller is similar to operation of the current-programmed controller. At the beginning of each switching period, at time $t = 0$, a short clock pulse sets the SR latch. The logic high signal at the Q output of the latch turns the power MOSFET on. At the same time, the logic low signal at the $\bar{Q}$ output of the latch turns the switch S_s off. Current $G_s v_2(t)$ proportional to the voltage $v_2(t)$ charges the capacitor C_s. At $t = dT_s$, the capacitor voltage $v_s(t)$ reaches the control input voltage v_c, the comparator output goes high and resets the latch. The logic low signal at the Q output of the latch turns the power MOSFET off. At the same time, the logic high signal at the $\bar{Q}$ output of the latch turns the switch S_s on, which quickly discharges the capacitor C_s to zero.

In this problem, the converter and controller parameters are: $V_g = 24$ V, $f_s = 1/T_s = 100$ kHz, $L = 60$ μH, $C = 100$ μF, $R = 3$ Ω, $G_s T_s/C_s = 1$. You can assume that the converter operates in the continuous conduction mode.

(a)   Find expressions for the average values of the switch network terminal waveforms, and hence derive a large-signal averaged switch model of the buck switch network with one-cycle control. The control input to the model is the control voltage v_c. The switch duty cycle d should not appear in the model.

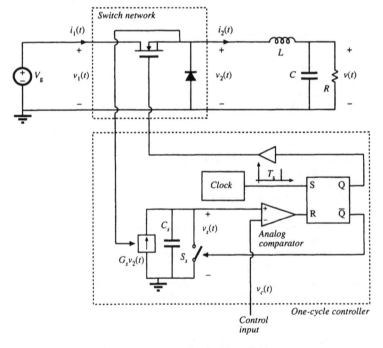

Fig. 12.38   Buck converter with one-cycle controller, Problem 12.12.

(b)   Using the averaged switch model derived in part (a), find an expression for the quiescent output voltage V as a function of V_c. Given $V_c = 10$ V, find the numerical values for V, I_1, I_2, and the duty cycle D. For this quiescent operating point, sketch the waveforms $i_1(t)$, $i_2(t)$, and $v_s(t)$ during one switching period.

(c)   Perturb and linearize the averaged switch model from part (a) to derive a small-signal averaged switch model for the buck switch network with one-cycle control. Find analytical expressions for all parameter values in terms of the converter parameters and the quiescent operating conditions. Sketch the complete small-signal model of the buck converter with the one-cycle controller.

(d)   Solve the model obtained in part (c) to find the control-to-output transfer function $G_{vc}(s) = \hat{v}/\hat{v}_c$, and the line-to-output transfer function $G_{vg}(s) = \hat{v}/\hat{v}_g$. For the quiescent operating point found in part (b), sketch the magnitude Bode plots of these transfer functions, and label all salient features.

(e)   Comment on advantages one-cycle control may have compared to duty-cycle control.

Part III

Magnetics

13

Basic Magnetics Theory

Magnetics are an integral part of every switching converter. Often, the design of the magnetic devices cannot be isolated from the converter design. The power electronics engineer must not only model and design the converter, but must model and design the magnetics as well. Modeling and design of magnetics for switching converters is the topic of Part III of this book.

In this chapter, basic magnetics theory is reviewed, including magnetic circuits, inductor modeling, and transformer modeling [1-5]. Loss mechanisms in magnetic devices are described. Winding eddy currents and the proximity effect, a significant loss mechanism in high-current high-frequency windings, are explained in detail [6-11]. Inductor design is introduced in Chapter 14, and transformer design is covered in Chapter 15.

13.1 REVIEW OF BASIC MAGNETICS

13.1.1 Basic Relationships

The basic magnetic quantities are illustrated in Fig. 13.1. Also illustrated are the analogous, and perhaps more familiar, electrical quantities. The *magnetomotive force* $\mathscr{F}$, or scalar potential, between two points x_1 and x_2 is given by the integral of the magnetic field H along a path connecting the points:

$$\mathscr{F} = \int_{x_1}^{x_2} H \cdot d\ell \tag{13.1}$$

where $d\ell$ is a vector length element pointing in the direction of the path. The dot product yields the com-

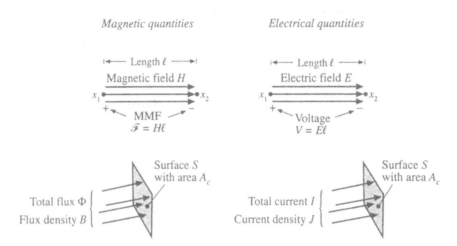

Fig. 13.1   Comparison of magnetic field *H*, MMF Φ, flux ℱ, and flux density *B*, with the analogous electrical quantities *E, V, I,* and *J.*

ponent of **H** in the direction of the path. If the magnetic field is of uniform strength *H* passing through an element of length ℓ as illustrated, then Eq. (13.1) reduces to

$$\mathscr{F} = H\ell \tag{13.2}$$

This is analogous to the electric field of uniform strength *E*, which induces a voltage $V = E\ell$ between two points separated by distance ℓ.

Figure 13.1 also illustrates a total magnetic flux Φ passing through a surface *S* having area A_c. The total flux Φ is equal to the integral of the normal component of the flux density *B* over the surface

$$\Phi = \int_{surface\ S} \boldsymbol{B} \cdot d\boldsymbol{A} \tag{13.3}$$

where *d***A** is a vector area element having direction normal to the surface. For a uniform flux density of magnitude *B* as illustrated, the integral reduces to

$$\Phi = B A_c \tag{13.4}$$

Flux density **B** is analogous to the electrical current density **J**, and flux Φ is analogous to the electric current *I*. If a uniform current density of magnitude *J* passes through a surface of area A_c, then the total current is $I = J A_c$.

Faraday's law relates the voltage induced in a winding to the total flux passing through the interior of the winding. Figure 13.2 illustrates flux Φ(*t*) passing through the interior of a loop of wire. The loop encloses cross-sectional area A_c. According to Faraday's law, the flux induces a voltage *v*(*t*) in the wire, given by

$$v(t) = \frac{d\Phi(t)}{dt} \tag{13.5}$$

where the polarities of *v*(*t*) and Φ(*t*) are defined according to the right-hand rule, as in Fig. 13.2. For a

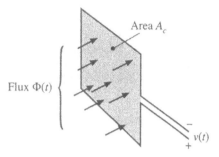

Fig. 13.2 The voltage $v(t)$ induced in a loop of wire is related by Faraday's law to the derivative of the total flux $\Phi(t)$ passing through the interior of the loop.

uniform flux distribution, we can express $v(t)$ in terms of the flux density $B(t)$ by substitution of Eq. (13.4):

$$v(t) = A_c \frac{dB(t)}{dt} \tag{13.6}$$

Thus, the voltage induced in a winding is related to the flux Φ and flux density B passing through the interior of the winding.

Lenz's law states that the voltage $v(t)$ induced by the changing flux $\Phi(t)$ in Fig. 13.2 is of the polarity that tends to drive a current through the loop to counteract the flux change. For example, consider the shorted loop of Fig. 13.3. The changing flux $\Phi(t)$ passing through the interior of the loop induces a voltage $v(t)$ around the loop. This voltage, divided by the impedance of the loop conductor, leads to a current $i(t)$ as illustrated. The current $i(t)$ induces a flux $\Phi'(t)$, which tends to oppose the changes in $\Phi(t)$. Lenz's law is invoked later in this chapter, to provide a qualitative understanding of eddy current phenomena.

Ampere's law relates the current in a winding to the magnetomotive force $\mathcal{F}$ and magnetic field H. The net MMF around a closed path of length ℓ_m is equal to the total current passing through the interior of the path. For example, Fig. 13.4 illustrates a magnetic core, in which a wire carrying current $i(t)$ passes through the window in the center of the core. Let us consider the closed path illustrated, which follows the magnetic field lines around the interior of the core. Ampere's law states that

$$\oint_{closed\ path} H \cdot d\ell = \text{total current passing through interior of path} \tag{13.7}$$

The total current passing through the interior of the path is equal to the total current passing through the

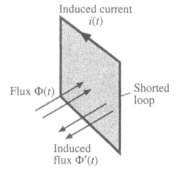

Fig. 13.3 Illustration of Lenz's law in a shorted loop of wire. The flux $\Phi(t)$ induces current $i(t)$, which in turn generates flux $\Phi'(t)$ that tends to oppose changes in $\Phi(t)$.

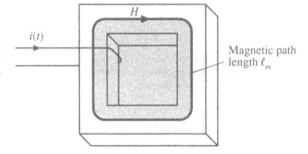

Fig. 13.4 The net MMF around a closed path is related by Ampere's law to the total current passing through the interior of the path.

window in the center of the core, or $i(t)$. If the magnetic field is uniform and of magnitude $H(t)$, then the integral is $H(t)\ell_m$. So for the example of Fig. 13.4, Eq. (13.7) reduces to

$$\mathscr{F}(t) = H(t)\ell_m = i(t) \tag{13.8}$$

Thus, the magnetic field strength $H(t)$ is related to the winding current $i(t)$. We can view winding currents as sources of MMF. Equation (13.8) states that the MMF around the core, $\mathscr{F}(t) = H(t)\ell_m$, is equal to the winding current MMF $i(t)$. The total MMF around the closed loop, accounting for both MMFs, is zero.

The relationship between B and H, or equivalently between Φ and $\mathscr{F}$, is determined by the core material characteristics. Figure 13.5(a) illustrates the characteristics of free space, or air:

$$B = \mu_0 H \tag{13.9}$$

The quantity μ_0 is the permeability of free space, and is equal to $4\pi \cdot 10^{-7}$ Henries per meter in MKS units. Figure 13.5(b) illustrates the B–H characteristic of a typical iron alloy under high-level sinusoidal steady-state excitation. The characteristic is highly nonlinear, and exhibits both *hysteresis* and *saturation*. The exact shape of the characteristic is dependent on the excitation, and is difficult to predict for arbitrary waveforms.

For purposes of analysis, the core material characteristic of Fig. 13.5(b) is usually modeled by the linear or piecewise-linear characteristics of Fig. 13.6. In Fig. 13.6(a), hysteresis and saturation are ignored. The B–H characteristic is then given by

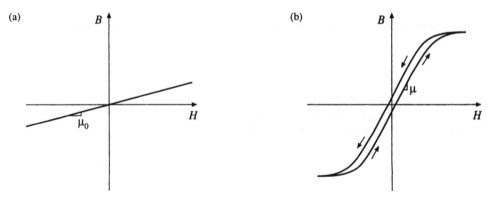

Fig. 13.5 B–H characteristics: (a) of free space or air, (b) of a typical magnetic core material.

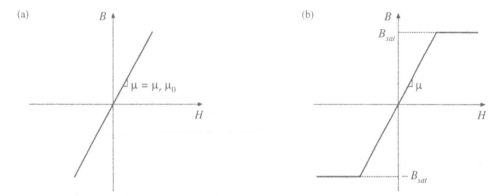

Fig. 13.6 Approximation of the *B–H* characteristics of a magnetic core material: (a) by neglecting both hysteresis and saturation, (b) by neglecting hysteresis.

$$B = \mu H$$
$$\mu = \mu_r \mu_0$$

(13.10)

The core material permeability μ can be expressed as the product of the relative permeability μ_r and of μ_0. Typical values of μ_r lie in the range 10^3 to 10^5.

The piecewise-linear model of Fig. 13.6(b) accounts for saturation but not hysteresis. The core material saturates when the magnitude of the flux density B exceeds the saturation flux density B_{sat}. For $|B| < B_{sat}$, the characteristic follows Eq. (13.10). When $|B| > B_{sat}$, the model predicts that the core reverts to free space, with a characteristic having a much smaller slope approximately equal to μ_0. Square-loop materials exhibit this type of abrupt-saturation characteristic, and additionally have a very large relative permeability μ_r. Soft materials exhibit a less abrupt saturation characteristic, in which μ gradually decreases as H is increased. Typical values of B_{sat} are 1 to 2 Tesla for iron laminations and silicon steel, 0.5 to 1 Tesla for powdered iron and molypermalloy materials, and 0.25 to 0.5 Tesla for ferrite materials.

Unit systems for magnetic quantities are summarized in Table 13.1. The MKS system is used throughout this book. The unrationalized cgs system also continues to find some use. Conversions between these systems are listed.

Figure 13.7 summarizes the relationships between the basic electrical and magnetic quantities of a magnetic device. The winding voltage $v(t)$ is related to the core flux and flux density via Faraday's

Table 13.1 Units for magnetic quantities

Quantity	MKS	Unrationalized cgs	Conversions
Core material equation	$B = \mu_0 \mu_r H$	$B = \mu_r H$	
B	Tesla	Gauss	$1\,T = 10^4\,G$
H	Ampere/meter	Oersted	$1\,A/m = 4\pi \cdot 10^{-3}\,Oe$
Φ	Weber	Maxwell	$1\,Wb = 10^8\,Mx$ $1\,T = 1\,Wb/m^2$

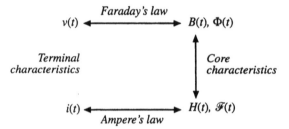

Fig. 13.7 Summary of the steps in determination of the terminal electrical *i–v* characteristics of a magnetic element.

law. The winding current $i(t)$ is related to the magnetic field strength via Ampere's law. The core material characteristics relate B and H.

We can now determine the electrical terminal characteristics of the simple inductor of Fig. 13.8(a). A winding of n turns is placed on a core having permeability μ. Faraday's law states that the flux $\Phi(t)$ inside the core induces a voltage $v_{turn}(t)$ in each turn of the winding, given by

$$v_{turn}(t) = \frac{d\Phi(t)}{dt} \qquad (13.11)$$

Since the same flux $\Phi(t)$ passes through each turn of the winding, the total winding voltage is

$$v(t) = nv_{turn}(t) = n\frac{d\Phi(t)}{dt} \qquad (13.12)$$

Equation (13.12) can be expressed in terms of the average flux density $B(t)$ by substitution of Eq. (13.4):

$$v(t) = nA_c\frac{dB(t)}{dt} \qquad (13.13)$$

(a)

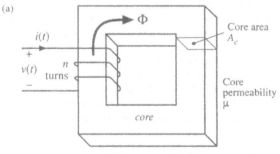

Fig. 13.8 Inductor example: (a) inductor geometry, (b) application of Ampere's law.

(b)

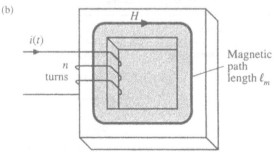

where the average flux density $B(t)$ is $\Phi(t)/A_c$.

The use of Ampere's law is illustrated in Fig. 13.8(b). A closed path is chosen which follows an average magnetic field line around the interior of the core. The length of this path is called the *mean magnetic path length* ℓ_m. If the magnetic field strength $H(t)$ is uniform, then Ampere's law states that $H\ell_m$ is equal to the total current passing through the interior of the path, that is, the net current passing through the window in the center of the core. Since there are n turns of wire passing through the window, each carrying current $i(t)$, the net current passing through the window is $ni(t)$. Hence, Ampere's law states that

$$H(t)\ell_m = ni(t) \tag{13.14}$$

Let us model the core material characteristics by neglecting hysteresis but accounting for saturation, as follows:

$$B = \begin{cases} B_{sat} & \text{for } H \geq B_{sat}/\mu \\ \mu H & \text{for } |H| < B_{sat}/\mu \\ -B_{sat} & \text{for } H \leq -B_{sat}/\mu \end{cases} \tag{13.15}$$

The B–H characteristic saturated slope μ_0 is much smaller than μ, and is ignored here. A characteristic similar to Fig. 13.6(b) is obtained. The current magnitude I_{sat} at the onset of saturation can be found by substitution of $H = B_{sat}/\mu$ into Eq. (13.14). The result is

$$I_{sat} = \frac{B_{sat}\ell_m}{\mu n} \tag{13.16}$$

We can now eliminate B and H from Eqs. (13.13) to (13.15), and solve for the electrical terminal characteristics. For $|I| < I_{sat}$, $B = \mu H$. Equation (13.13) then becomes

$$v(t) = \mu n A_c \frac{dH(t)}{dt} \tag{13.17}$$

Substitution of Eq. (13.14) into Eq. (13.17) to eliminate $H(t)$ then leads to

$$v(t) = \frac{\mu n^2 A_c}{\ell_m} \frac{di(t)}{dt} \tag{13.18}$$

which is of the form

$$v(t) = L \frac{di(t)}{dt} \tag{13.19}$$

with

$$L = \frac{\mu n^2 A_c}{\ell_m} \tag{13.20}$$

So the device behaves as an inductor for $|I| < I_{sat}$. When $|I| > I_{sat}$, then the flux density $B(t) = B_{sat}$ is constant. Faraday's law states that the terminal voltage is then

$$v(t) = nA_c \frac{dB_{sat}}{dt} = 0 \qquad (13.21)$$

When the core saturates, the magnetic device behavior approaches a short circuit. The device behaves as an inductor only when the winding current magnitude is less than I_{sat}. Practical inductors exhibit some small residual inductance due to their nonzero saturated permeabilities; nonetheless, in saturation the inductor impedance is greatly reduced, and large inductor currents may result.

13.1.2   Magnetic Circuits

Figure 13.9(a) illustrates uniform flux and magnetic field inside a element having permeability μ, length ℓ, and cross-sectional area A_c. The MMF between the two ends of the element is

$$\mathcal{F} = H\ell \qquad (13.22)$$

Since $H = B/\mu$ and $B = \Phi/A_c$, we can express $\mathcal{F}$ as

$$\mathcal{F} = \frac{\ell}{\mu A_c}\, \Phi \qquad (13.23)$$

This equation is of the form

$$\mathcal{F} = \Phi\mathcal{R} \qquad (13.24)$$

with

$$\mathcal{R} = \frac{\ell}{\mu A_c} \qquad (13.25)$$

Equation (13.24) resembles Ohm's law. This equation states that the magnetic flux through an element is proportional to the MMF across the element. The constant of proportionality, or the reluctance $\mathcal{R}$, is analogous to the resistance R of an electrical conductor. Indeed, we can construct a lumped-element magnetic circuit model that corresponds to Eq. (13.24), as in Fig. 13.9(b). In this magnetic circuit model, voltage and current are replaced by MMF and flux, while the element characteristic, Eq. (13.24), is represented by the analog of a resistor, having reluctance $\mathcal{R}$.

Complicated magnetic structures, composed of multiple windings and multiple heterogeneous

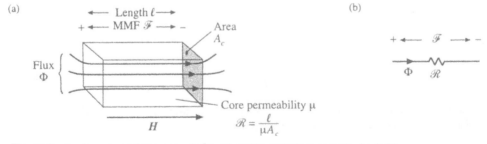

Fig. 13.9   An element containing magnetic flux (a), and its equivalent magnetic circuit (b).

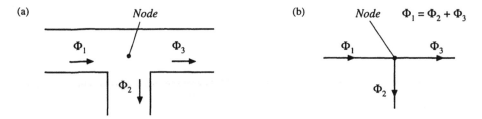

Fig. 13.10 Kirchoff's current law, applied to magnetic circuits: the net flux entering a node must be zero: (a) physical element, in which three legs of a core meet at a node; (b) magnetic circuit model.

elements such as cores and air gaps, can be represented using equivalent magnetic circuits. These magnetic circuits can then be solved using conventional circuit analysis, to determine the various fluxes, MMFs, and terminal voltages and currents. Kirchoff's laws apply to magnetic circuits, and follow directly from Maxwell's equations. The analog of Kirchoff's current law holds because the divergence of **B** is zero, and hence magnetic flux lines are continuous and cannot end. Therefore, any flux line that enters a node must leave the node. As illustrated in Fig. 13.10, the total flux entering a node must be zero. The analog of Kirchoff's voltage law follows from Ampere's law, Eq. (13.7). The left-hand-side integral in Eq. (13.7) is the sum of the MMFs across the reluctances around the closed path. The right-hand-side of Eq. (13.7) states that currents in windings are sources of MMF. An n-turn winding carrying current $i(t)$ can be modeled as an MMF source, analogous to a voltage source, of value $ni(t)$. When these MMF sources are included, the total MMF around a closed path is zero.

Consider the inductor with air gap of Fig. 13.11(a). A closed path following the magnetic field lines is illustrated. This path passes through the core, of permeability μ and length ℓ_c, and across the air gap, of permeability μ_0 and length ℓ_g. The cross-sectional areas of the core and air gap are approximately equal. Application of Ampere's law for this path leads to

$$\mathscr{F}_c + \mathscr{F}_g = ni \qquad (13.26)$$

where $\mathscr{F}_c$ and $\mathscr{F}_g$ are the MMFs across the core and air gap, respectively. The core and air gap characteristics can be modeled by reluctances as in Fig. 13.9 and Eq. (13.25); the core reluctance $\mathscr{R}_c$ and air gap reluctance $\mathscr{R}_g$ are given by

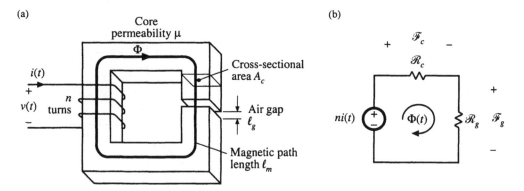

Fig. 13.11 Inductor with air gap example: (a) physical geometry, (b) magnetic circuit model.

$$\mathcal{R}_c = \frac{\ell_c}{\mu A_c}$$
$$\mathcal{R}_g = \frac{\ell_g}{\mu_0 A_c}$$
(13.27)

A magnetic circuit corresponding to Eqs. (13.26) and (13.27) is given in Fig. 13.11(b). The winding is a source of MMF, of value ni. The core and air gap reluctances are effectively in series. The solution of the magnetic circuit is

$$ni = \Phi \left(\mathcal{R}_c + \mathcal{R}_g \right)$$
(13.28)

The flux $\Phi(t)$ passes through the winding, and so we can use Faraday's law to write

$$v(t) = n \frac{d\Phi(t)}{dt}$$
(13.29)

Use of Eq. (13.28) to eliminate $\Phi(t)$ yields

$$v(t) = \frac{n^2}{\mathcal{R}_c + \mathcal{R}_g} \frac{di(t)}{dt}$$
(13.30)

Therefore, the inductance L is

$$L = \frac{n^2}{\mathcal{R}_c + \mathcal{R}_g}$$
(13.31)

The air gap increases the total reluctance of the magnetic circuit, and decreases the inductance.

Air gaps are employed in practical inductors for two reasons. With no air gap ($\mathcal{R}_g = 0$), the inductance is directly proportional to the core permeability μ. This quantity is dependent on temperature and operating point, and is difficult to control. Hence, it may be difficult to construct an inductor having a well-controlled value of L. Addition of an air gap having a reluctance $\mathcal{R}_g$ greater than $\mathcal{R}_c$ causes the value of L in Eq. (13.31) to be insensitive to variations in μ.

Addition of an air gap also allows the inductor to operate at higher values of winding current $i(t)$ without saturation. The total flux Φ is plotted vs. the winding MMF ni in Fig. 13.12. Since Φ is proportional to B, and when the core is not saturated ni is proportional to the magnetic field strength H in the

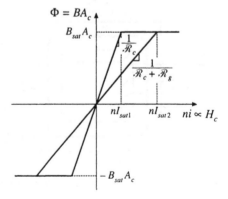

Fig. 13.12 Effect of air gap on the magnetic circuit Φ vs. ni characteristics. The air gap increases the current I_{sat} at the onset of core saturation.

core, Fig. 13.12 has the same shape as the core *B–H* characteristic. When the core is not saturated, Φ is related to ni according to the linear relationship of Eq. (13.28). When the core saturates, Φ is equal to

$$\Phi_{sat} = B_{sat} A_c \tag{13.32}$$

The winding current I_{sat} at the onset of saturation is found by substitution of Eq. (13.32) into (13.28):

$$I_{sat} = \frac{B_{sat} A_c}{n} \left(\mathscr{R}_c + \mathscr{R}_g \right) \tag{13.33}$$

The Φ–ni characteristics are plotted in Fig. 13.12 for two cases: (a) air gap present, and (b) no air gap ($\mathscr{R}_g = 0$). It can be seen that I_{sat} is increased by addition of an air gap. Thus, the air gap allows increase of the saturation current, at the expense of decreased inductance.

13.2   TRANSFORMER MODELING

Consider next the two-winding transformer of Fig. 13.13. The core has cross-sectional area A_c, mean magnetic path length ℓ_m, and permeability μ. An equivalent magnetic circuit is given in Fig. 13.14. The core reluctance is

$$\mathscr{R} = \frac{\ell_m}{\mu A_c} \tag{13.34}$$

Since there are two windings in this example, it is necessary to determine the relative polarities of the MMF generators. Ampere's law states that

$$\mathscr{F}_c = n_1 i_1 + n_2 i_2 \tag{13.35}$$

Fig. 13.13   A two-winding transformer.

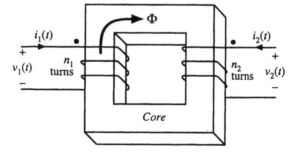

Fig. 13.14   Magnetic circuit that models the two-winding transformer of Fig. 13.13.

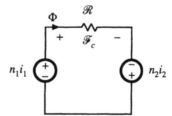

The MMF generators are additive, because the currents i_1 and i_2 pass in the same direction through the core window. Solution of Fig. 13.14 yields

$$\Phi \mathscr{R} = n_1 i_1 + n_2 i_2 \tag{13.36}$$

This expression could also be obtained by substitution of $\mathscr{F}_c = \Phi \mathscr{R}$ into Eq. (13.35).

13.2.1 The Ideal Transformer

In the ideal transformer, the core reluctance $\mathscr{R}$ approaches zero. The causes the core MMF $\mathscr{F}_c = \Phi \mathscr{R}$ to also approach zero. Equation (13.35) then becomes

$$0 = n_1 i_1 + n_2 i_2 \tag{13.37}$$

Also, by Faraday's law, we have

$$
\begin{aligned}
v_1 &= n_1 \frac{d\Phi}{dt} \\
v_2 &= n_2 \frac{d\Phi}{dt}
\end{aligned}
\tag{13.38}
$$

Note that Φ is the same in both equations above: the same total flux links both windings. Elimination of Φ leads to

$$\frac{d\Phi}{dt} = \frac{v_1}{n_1} = \frac{v_2}{n_2} \tag{13.39}$$

Equations (13.37) and (13.39) are the equations of the ideal transformer:

$$\frac{v_1}{n_1} = \frac{v_2}{n_2} \quad \text{and} \quad n_1 i_1 + n_2 i_2 = 0 \tag{13.40}$$

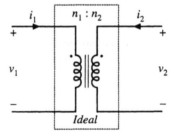

Fig. 13.15   Ideal transformer symbol.

The ideal transformer symbol of Fig. 13.15 is defined by Eq. (13.40).

13.2.2 The Magnetizing Inductance

For the actual case in which the core reluctance $\mathscr{R}$ is nonzero, we have

$$\Phi \mathscr{R} = n_1 i_1 + n_2 i_2 \quad \text{with} \quad v_1 = n_1 \frac{d\Phi}{dt} \tag{13.41}$$

Elimination of Φ yields

$$v_1 = \frac{n_1^2}{\mathscr{R}} \frac{d}{dt} \left[i_1 + \frac{n_2}{n_1} i_2 \right] \tag{13.42}$$

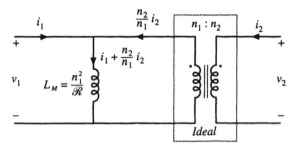

Fig. 13.16 Transformer model including magnetizing inductance.

This equation is of the form

$$v_1 = L_M \frac{di_M}{dt} \tag{13.43}$$

where

$$L_M = \frac{n_1^2}{\mathcal{R}}$$
$$i_M = i_1 + \frac{n_2}{n_1} i_2 \tag{13.44}$$

are the *magnetizing inductance* and *magnetizing current*, referred to the primary winding. An equivalent circuit is illustrated in Fig. 13.16.

Figure 13.16 coincides with the transformer model introduced in Chapter 6. The magnetizing inductance models the magnetization of the core material. It is a real, physical inductor, which exhibits saturation and hysteresis. All physical transformers must contain a magnetizing inductance. For example, suppose that we disconnect the secondary winding. We are then left with a single winding on a magnetic core—an inductor. Indeed, the equivalent circuit of Fig. 13.16 predicts this behavior, via the magnetizing inductance. The magnetizing current causes the ratio of the winding currents to differ from the turns ratio.

The transformer saturates when the core flux density $B(t)$ exceeds the saturation flux density B_{sat}. When the transformer saturates, the magnetizing current $i_M(t)$ becomes large, the impedance of the magnetizing inductance becomes small, and the transformer windings become short circuits. It should be noted that large winding currents $i_1(t)$ and $i_2(t)$ do not necessarily cause saturation: if these currents obey Eq. (13.37), then the magnetizing current is zero and there is no net magnetization of the core. Rather, saturation of a transformer is a function of the applied volt-seconds. The magnetizing current is given by

$$i_M(t) = \frac{1}{L_M} \int v_1(t)dt \tag{13.45}$$

Alternatively, Eq. (13.45) can be expressed in terms of the core flux density $B(t)$ as

$$B(t) = \frac{1}{n_1 A_c} \int v_1(t)dt \tag{13.46}$$

The flux density and magnetizing current will become large enough to saturate the core when the applied volt-seconds λ_1 is too large, where λ_1 is defined for a periodic ac voltage waveform as

$$\lambda_1 = \int_{t_1}^{t_2} v_1(t) dt$$

(13.47)

The limits are chosen such that the integral is taken over the positive portion of the applied periodic voltage waveform.

To fix a saturating transformer, the flux density should be decreased by increasing the number of turns, or by increasing the core cross-sectional area A_c. Adding an air gap has no effect on saturation of conventional transformers, since it does not modify Eq. (13.46). An air gap simply makes the transformer less ideal, by decreasing L_M and increasing $i_M(t)$ without changing $B(t)$. Saturation mechanisms in transformers differ from those of inductors, because transformer saturation is determined by the applied winding voltage waveforms, rather than the applied winding currents.

13.2.3   Leakage Inductances

In practice, there is some flux which links one winding but not the other, by "leaking" into the air or by some other mechanism. As illustrated in Fig. 13.17, this flux leads to *leakage inductance*, i.e., additional effective inductances that are in series with the windings. A topologically equivalent structure is illustrated in Fig. 13.17(b), in which the leakage fluxes $\Phi_{\ell 1}$ and $\Phi_{\ell 2}$ are shown explicitly as separate inductors.

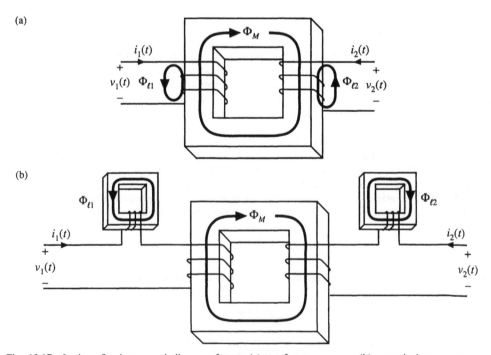

Fig. 13.17   Leakage flux in a two-winding transformer: (a) transformer geometry, (b) an equivalent system.

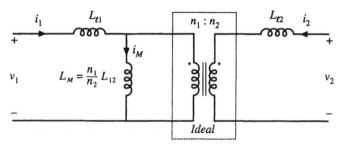

Fig. 13.18 Two-winding transformer equivalent circuit, including magnetizing inductance referred to primary, and primary and secondary leakage inductances.

Figure 13.18 illustrates a transformer electrical equivalent circuit model, including series inductors $L_{\ell 1}$ and $L_{\ell 2}$ which model the leakage inductances. These leakage inductances cause the terminal voltage ratio $v_2(t)/v_1(t)$ to differ from the ideal turns ratio n_2/n_1. In general, the terminal equations of a two-winding transformer can be written

$$\begin{bmatrix} v_1(t) \\ v_2(t) \end{bmatrix} = \begin{bmatrix} L_{11} & L_{12} \\ L_{12} & L_{22} \end{bmatrix} \frac{d}{dt} \begin{bmatrix} i_1(t) \\ i_2(t) \end{bmatrix} \tag{13.48}$$

The quantity L_{12} is called the *mutual inductance*, and is given by

$$L_{12} = \frac{n_1 n_2}{\mathscr{R}} = \frac{n_2}{n_1} L_M \tag{13.49}$$

The quantities L_{11} and L_{22} are called the primary and secondary *self-inductances*, given by

$$
\begin{aligned}
L_{11} &= L_{\ell 1} + \frac{n_1}{n_2} L_{12} \\
L_{22} &= L_{\ell 2} + \frac{n_2}{n_1} L_{12}
\end{aligned}
\tag{13.50}$$

Note that Eq. (13.48) does not explicitly identify the physical turns ratio n_2/n_1. Rather, Eq. (13.48) expresses the transformer behavior as a function of electrical quantities alone. Equation (13.48) can be used, however, to define the *effective turns ratio*

$$n_e = \sqrt{\frac{L_{22}}{L_{11}}} \tag{13.51}$$

and the *coupling coefficient*

$$k = \frac{L_{12}}{\sqrt{L_{11} L_{22}}} \tag{13.52}$$

The coupling coefficient k lies in the range $0 \le k \le 1$, and is a measure of the degree of magnetic coupling between the primary and secondary windings. In a transformer with perfect coupling, the leakage inductances $L_{\ell 1}$ and $L_{\ell 2}$ are zero. The coupling coefficient k is then equal to 1. Construction of low-voltage transformers having coupling coefficients in excess of 0.99 is quite feasible. When the coupling coefficient is close to 1, then the effective turns ratio n_e is approximately equal to the physical turns ratio n_2/n_1.

13.3 LOSS MECHANISMS IN MAGNETIC DEVICES

13.3.1 Core Loss

Energy is required to effect a change in the magnetization of a core material. Not all of this energy is recoverable in electrical form; a fraction is lost as heat. This power loss can be observed electrically as hysteresis of the *B–H* loop.

Consider an *n*–turn inductor excited by periodic waveforms $v(t)$ and $i(t)$ having frequency f. The net energy that flows into the inductor over one cycle is

$$W = \int_{\text{one cycle}} v(t)i(t)\,dt \tag{13.53}$$

We can relate this expression to the core *B–H* characteristic: substitute $B(t)$ for $v(t)$ using Faraday's law, Eq. (13.13), and substitute $H(t)$ for $i(t)$ using Ampere's law, i.e. Eq. (13.14):

$$W = \int_{\text{one cycle}} \left(nA_c\frac{dB(t)}{dt}\right)\left(\frac{H(t)\ell_m}{n}\right)dt \tag{13.54}$$
$$= \left(A_c\ell_m\right)\int_{\text{one cycle}} H\,dB$$

The term $A_c\ell_m$ is the volume of the core, while the integral is the area of the *B–H* loop:

$$\text{(energy lost per cycle)} = \text{(core volume)}\text{(area of }B\text{–}H\text{ loop)} \tag{13.55}$$

The *hysteresis power loss* P_H is equal to the energy lost per cycle, multiplied by the excitation frequency f:

$$P_H = \left(f\right)\left(A_c\ell_m\right)\int_{\text{one cycle}} H\,dB \tag{13.56}$$

To the extent that the size of the hysteresis loop is independent of frequency, hysteresis loss increases directly with operating frequency.

Magnetic core materials are iron alloys that, unfortunately, are also good electrical conductors. As a result, ac magnetic fields can cause electrical *eddy currents* to flow within the core material itself. An example is illustrated in Fig. 13.19. The ac flux $\Phi(t)$ passes through the core. This induces eddy currents $i(t)$ which, according to Lenz's law, flow in paths that oppose the time-varying flux $\Phi(t)$. These eddy currents cause i^2R losses in the resistance of the core material. The eddy current losses are especially significant in high-frequency applications.

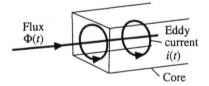

Fig. 13.19   Eddy currents in an iron core.

According to Faraday's law, the ac flux $\Phi(t)$ induces voltage in the core, which drives the current around the paths illustrated in Fig. 13.19. Since the induced voltage is proportional to the derivative of the flux, the voltage magnitude increases directly with the excitation frequency f. If the impedance of the core material is purely resistive and independent of frequency, then the magnitude of the induced eddy currents also increases directly with f. This implies that the i^2R eddy current losses should increase as f^2. In power ferrite materials, the core material impedance magnitude actually decreases with increasing f. Over the useful frequency range, the eddy current losses typically increase faster than f^2.

There is a basic tradeoff between saturation flux density and core loss. Use of a high operating flux density leads to reduced size, weight, and cost. Silicon steel and similar materials exhibit saturation flux densities of 1.5 to 2 T. Unfortunately, these core materials exhibit high core loss. In particular, the low resistivity of these materials leads to high eddy current loss. Hence, these materials are suitable for filter inductor and low-frequency transformer applications. The core material is produced in laminations or thin ribbons, to reduce the eddy current magnitude. Other ferrous alloys may contain molybdenum, cobalt, or other elements, and exhibit somewhat lower core loss as well as somewhat lower saturation flux densities.

Iron alloys are also employed in powdered cores, containing ferromagnetic particles of sufficiently small diameter such that eddy currents are small. These particles are bound together using an insulating medium. Powdered iron and molybdenum permalloy powder cores exhibit typical saturation flux densities of 0.6 to 0.8 T, with core losses significantly lower than laminated ferrous alloy materials. The insulating medium behaves effectively as a distributed air gap, and hence these cores have relatively low permeability. Powder cores find application as transformers at frequencies of several kHz, and as filter inductors in high frequency (100 kHz) switching converters.

Amorphous alloys exhibit low hysteresis loss. Core conductivity and eddy current losses are somewhat lower than ferrous alloys, but higher than ferrites. Saturation flux densities in the range 0.6 to 1.5 T are obtained.

Ferrite cores are ceramic materials having low saturation flux density, 0.25 to 0.5 T. Their resistivities are much higher than other materials, and hence eddy current losses are much smaller. Manganese-zinc ferrite cores find widespread use as inductors and transformers in converters having switching frequencies of 10 kHz to 1 MHz. Nickel-zinc ferrite materials can be employed at yet higher frequencies.

Figure 13.20 contains typical total core loss data, for a certain ferrite material. Power loss density, in Watts per cubic centimeter of core material, is plotted as a function of sinusoidal excitation frequency f and peak ac flux density ΔB. At a given frequency, the core loss P_{fe} can be approximated by an empirical function of the form

$$P_{fe} = K_{fe}(\Delta B)^\beta A_c \ell_m \qquad (13.57)$$

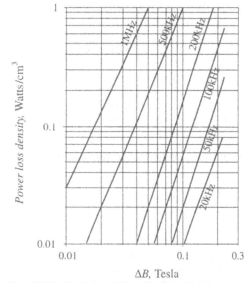

Fig. 13.20 Typical core loss data for a high-frequency power ferrite material. Power loss density is plotted vs. peak ac flux density ΔB, for sinusoidal excitation.

The parameters K_{fe} and β are determined by fitting Eq. (13.57) to the manufacturer's published data. Typical values of β for ferrite materials operating in their intended range of ΔB and f lie in the range 2.6 to 2.8. The constant of proportionality K_{fe} increases rapidly with excitation frequency f. The dependence of K_{fe} on f can also be approximated by empirical formulae that are fitted to the manufacturer's published data; a fourth-order polynomial or a function of the form $K_{fe0} f^\xi$ are sometimes employed for this purpose.

13.3.2 Low-Frequency Copper Loss

Significant loss also occurs in the resistance of the copper windings. This is also a major determinant of the size of a magnetic device: if copper loss and winding resistance were irrelevant, then inductor and transformer elements could be made arbitrarily small by use of many small turns of small wire.

Figure 13.21 contains an equivalent circuit of a winding, in which element R models the winding resistance. The copper loss of the winding is

$$P_{cu} = I_{rms}^2 R \qquad (13.58)$$

where I_{rms} is the rms value of $i(t)$. The dc resistance of the winding conductor can be expressed as

$$R = \rho \frac{\ell_b}{A_w} \qquad (13.59)$$

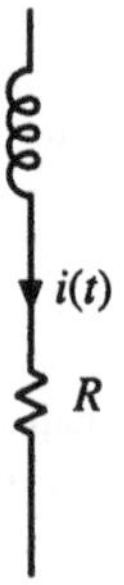

$i(t)$

R

Fig. 13.21 Winding equivalent circuit that models copper loss.

where A_w is the wire bare cross-sectional area, and ℓ_b is the length of the wire. The resistivity ρ is equal to $1.724 \cdot 10^{-6}$ Ω–cm for soft-annealed copper at room temperature. This resistivity increases to $2.3 \cdot 10^{-6}$ Ω–cm at 100°C.

13.4   EDDY CURRENTS IN WINDING CONDUCTORS

Eddy currents also cause power losses in winding conductors. This can lead to copper losses significantly in excess of the value predicted by Eqs. (13.58) and (13.59). The specific conductor eddy current mechanisms are called the *skin effect* and the *proximity effect*. These mechanisms are most pronounced in high-current conductors of multi-layer windings, particularly in high-frequency converters.

13.4.1   Introduction to the Skin and Proximity Effects

Figure 13.22(a) illustrates a current $i(t)$ flowing through a solitary conductor. This current induces magnetic flux $\Phi(t)$, whose flux lines follow circular paths around the current as shown. According to Lenz's law, the ac flux in the conductor induces eddy currents, which flow in a manner that tends to oppose the ac flux $\Phi(t)$. Figure 13.22(b) illustrates the paths of the eddy currents. It can be seen that the eddy currents tend to reduce the net current density in the center of the conductor, and increase the net current density near the surface of the conductor.

The current distribution within the conductor can be found by solution of Maxwell's equations. For a sinusoidal current $i(t)$ of frequency f, the result is that the current density is greatest at the surface of the conductor. The current density is an exponentially decaying function of distance into the conductor, with characteristic length δ known as the *penetration depth* or *skin depth*. The penetration depth is given by

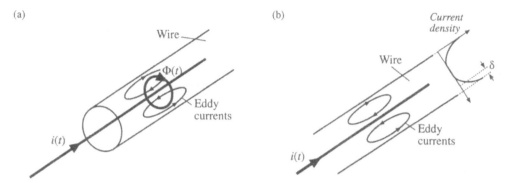

Fig. 13.22 The skin effect: (a) current $i(t)$ induces flux $\Phi(t)$, which in turn induces eddy currents in conductor; (b) the eddy currents tend to oppose the current $i(t)$ in the center of the wire, and increase the current on the surface of the wire.

$$\delta = \sqrt{\frac{\rho}{\pi\mu f}} \qquad (13.60)$$

For a copper conductor, the permeability μ is equal to μ_0, and the resistivity ρ is given in Section 13.3.2. At 100°C, the penetration depth of a copper conductor is

$$\delta = \frac{7.5}{\sqrt{f}} \text{ cm} \qquad (13.61)$$

with f expressed in Hz. The penetration depth of copper conductors is plotted in Fig. 13.23, as a function of frequency f. For comparison, the wire diameters d of standard American Wire Gauge (AWG) conductors are also listed. It can be seen that $d/\delta = 1$ for AWG #40 at approximately 500 kHz, while $d/\delta = 1$ for

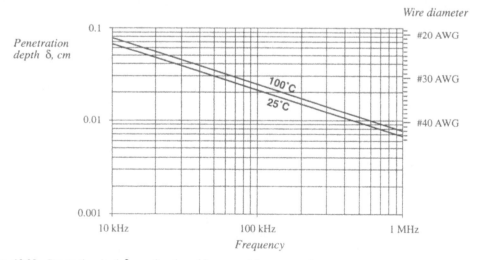

Fig. 13.23 Penetration depth δ, as a function of frequency f, for copper wire.

AWG #22 at approximately 10 kHz.

The skin effect causes the resistance and copper loss of solitary large-diameter wires to increase at high frequency. High-frequency currents do not penetrate to the center of the conductor. The current crowds at the surface of the wire, the inside of the wire is not utilized, and the effective wire cross-sectional area is reduced. However, the skin effect alone is not sufficient to explain the increased high-frequency copper losses observed in multiple-layer transformer windings.

A conductor that carries a high-frequency current $i(t)$ induces copper loss in a adjacent conductor by a phenomenon known as the *proximity effect*. Figure 13.24 illustrates two copper foil conductors that are placed in close proximity to each other. Conductor 1 carries a high-frequency sinusoidal current $i(t)$, whose penetration depth δ is much smaller than the thickness h of conductors 1 or 2. Conductor 2 is open-circuited, so that it carries a net current of zero. However, it is possible for eddy currents to be induced in conductor 2 by the current $i(t)$ flowing in conductor 1.

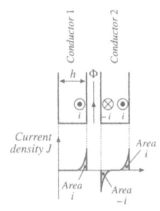

The current $i(t)$ flowing in conductor 1 generates a flux $\Phi(t)$ in the space between conductors 1 and 2; this flux attempts to penetrate conductor 2. By Lenz's law, a current is induced on the adjacent (left) side of conductor 2, which tends to oppose the flux $\Phi(t)$. If the conductors are closely spaced, and if $h \gg \delta$, then the induced current will be equal and opposite to the current $i(t)$, as illustrated in Fig. 13.24.

Fig. 13.24 The proximity effect in adjacent copper foil conductors. Conductor 1 carries current $i(t)$. Conductor 2 is open-circuited.

Since conductor 2 is open-circuited, the net current in conductor 2 must be zero. Therefore, a current $+ i(t)$ flows on the right-side surface of conductor 2. So the current flowing in conductor 1 induces a current that circulates on the surfaces of conductor 2.

Figure 13.25 illustrates the proximity effect in a simple transformer winding. The primary winding consists of three series-connected turns of copper foil, having thickness $h \gg \delta$, and carrying net current $i(t)$. The secondary winding is identical; to the extent that the magnetizing current is small, the secondary turns carry net current $- i(t)$. The windings are surrounded by a magnetic core material that encloses the mutual flux of the transformer.

The high-frequency sinusoidal current $i(t)$ flows on the right surface of primary layer 1, adjacent to layer 2. This induces a copper loss in layer 1, which can be calculated as follows. Let R_{dc} be the dc resistance of layer 1, given by Eq. (13.59), and let I be the rms value of $i(t)$. The skin effect causes the copper loss in layer 1 to be equal to the loss in a conductor of thickness δ with uniform current density. This reduction of the conductor thickness from h to δ effectively increases the resistance by the same factor. Hence, layer 1 can be viewed as having an "ac resistance" given by

$$R_{ac} = \frac{h}{\delta} R_{dc} \qquad (13.62)$$

The copper loss in layer 1 is

$$P_1 = I^2 R_{ac} \qquad (13.63)$$

The proximity effect causes a current to be induced in the adjacent (left-side) surface of primary layer 2, which tends to oppose the flux generated by the current of layer 1. If the conductors are closely spaced, and if $h \gg \delta$, then the induced current will be equal and opposite to the current $i(t)$, as illustrated

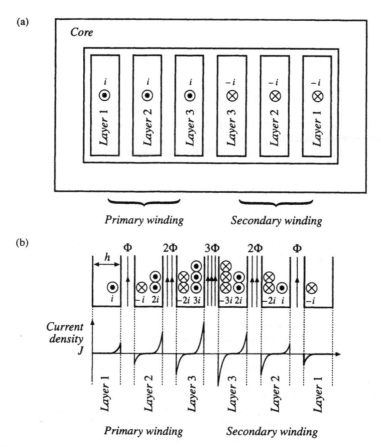

Fig.   13.25   A simple transformer example illustrating the proximity effect: (a) core and winding geometry, (b) distribution of currents on surfaces of conductors.

in Fig. 13.25. Hence, current $-i(t)$ flows on the left surface of the second layer. Since layers 1 and 2 are connected in series, they must both conduct the same net current $i(t)$. As a result, a current $+2i(t)$ must flow on the right-side surface of layer 2.

The current flowing on the left surface of layer 2 has the same magnitude as the current of layer 1, and hence the copper loss is the same: P_1. The current flowing on the right surface of layer 2 has rms magnitude $2I$; hence, it induces copper loss $(2I)^2 R_{ac} = 4P_1$. The total copper loss in primary layer 2 is therefore

$$P_2 = P_1 + 4P_1 = 5P_1 \tag{13.64}$$

The copper loss in the second layer is five times as large as the copper loss in the first layer!

The current $2i(t)$ flowing on the right surface of layer 2 induces a flux $2\Phi(t)$ as illustrated in Fig. 13.25. This causes an opposing current $-2i(t)$ to flow on the adjacent (left) surface of primary layer 3. Since layer 3 must also conduct net current $i(t)$, a current $+3i(t)$ flows on the right surface of layer 3. The total copper loss in layer 3 is

$$P_3 = \left(2^2 + 3^2\right)P_1 = 13P_1 \tag{13.65}$$

Likewise, the copper loss in layer m of a multiple-layer winding can be written

$$P_m = I^2 \left[(m-1)^2 + m^2\right] \left(\frac{h}{\delta} R_{dc}\right) \tag{13.66}$$

It can be seen that the copper loss compounds very quickly in a multiple-layer winding.

The total copper loss in the three-layer primary winding is $P_1 + 5P_1 + 13P_1$, or $19P_1$. More generally, if the winding contains a total of M layers, then the total copper loss is

$$P = I^2 \left(\frac{h}{\delta} R_{dc}\right) \sum_{m=1}^{M} \left[(m-1)^2 + m^2\right]$$
$$= I^2 \left(\frac{h}{\delta} R_{dc}\right) \frac{M}{3} \left(2M^2 + 1\right) \tag{13.67}$$

If a dc or low-frequency ac current of rms amplitude I were applied to the M-layer winding, its copper loss would be $P_{dc} = I^2 M R_{dc}$. Hence, the proximity effect increases the copper loss by the factor

$$F_R = \frac{P}{P_{dc}} = \frac{1}{3} \left(\frac{h}{\delta}\right)\left(2M^2 + 1\right) \tag{13.68}$$

This expression is valid for a foil winding having $h \gg \delta$.

As illustrated in Fig. 13.25, the currents in the secondary winding are symmetrical, and hence the secondary winding has the same conduction loss.

The example above, and the associated equations, are limited to $h \gg \delta$ and to the winding geometry shown. The equations do not quantify the behavior for $h \sim \delta$, nor for round conductors, nor are the equations sufficiently general to cover the more complicated winding geometries often encountered in the magnetic devices of switching converters. Optimum designs may, in fact, occur with conductor thicknesses in the vicinity of one penetration depth. The discussions of the following sections allow computation of proximity losses in more general circumstances.

13.4.2   Leakage Flux in Windings

As described above, an externally-applied magnetic field will induce eddy currents to flow in a conductor, and thereby induce copper loss. To understand how magnetic fields are oriented in windings, let us consider the simple two-winding transformer illustrated in Fig. 13.26. In this example, the core has large permeability $\mu \gg \mu_0$. The primary winding consists of eight turns of wire arranged in two layers, and each turn carries current $i(t)$ in the direction indicated. The secondary winding is identical to the primary winding, except that the current polarity is reversed.

Flux lines for typical operation of this transformer are sketched in Fig. 13.26(b). As described in Section 13.2, a relatively large mutual flux is present, which magnetizes the core. In addition, leakage flux is present, which does not completely link both windings. Because of the symmetry of the winding geometry in Fig. 13.26, the leakage flux runs approximately vertically through the windings.

To determine the magnitude of the leakage flux, we can apply Ampere's Law. Consider the closed path taken by one of the leakage flux lines, as illustrated in Fig. 13.27. Since the core has large permeability, we can assume that the MMF induced in the core by this flux is negligible, and that the

(a) (b)

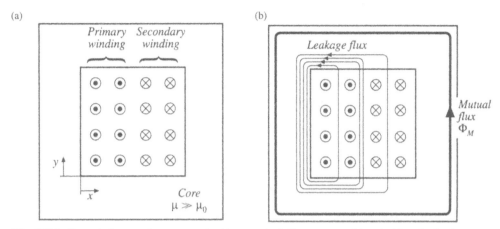

Fig. 13.26 Two-winding transformer example: (a) core and winding geometry, (b) typical flux distribution.

total MMF around the path is dominated by the MMF $\mathscr{F}(x)$ across the core window. Hence, Ampere's Law states that the net current enclosed by the path is equal to the MMF across the air gap:

$$\text{Enclosed current} = \mathscr{F}(x) = H(x)\ell_w \tag{13.69}$$

where ℓ_w is the height of the window as shown in Fig. 13.27. The net current enclosed by the path depends on the number of primary and secondary conductors enclosed by the path, and is therefore a function of the horizontal position x. The first layer of the primary winding consists of 4 turns, each carrying current $i(t)$. So when the path encloses only the first layer of the primary winding, then the enclosed current is $4i(t)$ as shown in Fig. 13.28. Likewise, when the path encloses both layers of the primary winding, then the enclosed current is $8i(t)$. When the path encloses the entire primary, plus layer 2 of the secondary winding, then the net enclosed current is $8i(t) - 4i(t) = 4i(t)$. The MMF $\mathscr{F}(x)$ across the core window is zero outside the winding, and rises to a maximum of $8i(t)$ at the interface between the primary and secondary windings. Since $H(x) = \mathscr{F}(x)/\ell_w$, the magnetic field intensity $H(x)$ is proportional to the sketch of Fig. 13.28.

Fig. 13.27 Analysis of leakage flux using Ampere's Law, for the transformer of Fig. 13.26.

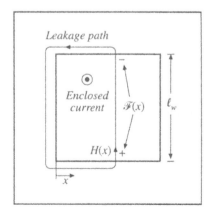

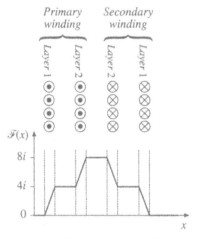

Fig. 13.28 MMF diagram for the transformer winding example of Figs. 13.26 and 13.27.

It should be noted that the shape of the $\mathscr{F}(x)$ curve in the vicinity of the winding conductors depends on the distribution of the current within the conductors. Since this distribution is not yet known, the $\mathscr{F}(x)$ curve of Fig. 13.28 is arbitrarily drawn as straight line segments.

In general, the magnetic fields that surround conductors and lead to eddy currents must be determined using finite element analysis or other similar methods. However, in a large class of coaxial solenoidal winding geometries, the magnetic field lines are nearly parallel to the winding layers. As shown below, we can then obtain an analytical solution for the proximity losses.

13.4.3 Foil Windings and Layers

The winding symmetry described in the previous section allows simplification of the analysis. For the purposes of determining leakage inductance and winding eddy currents, a layer consisting of n_ℓ turns of round wire carrying current $i(t)$ can be approximately modeled as an effective single turn of foil, which carries current $n_\ell i(t)$. The steps in the transformation of a layer of round conductors into a foil conductor are formalized in Fig. 13.29 [6, 8–11]. The round conductors are replaced by square conductors having the same copper cross-sectional area, Fig. 13.29(b). The thickness h of the square conductors is therefore

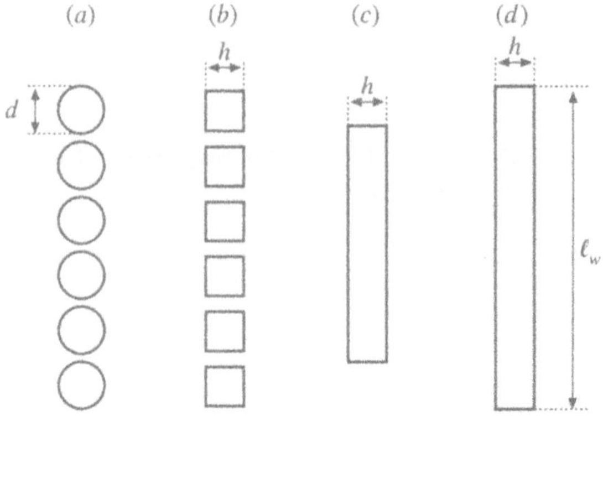

Fig. 13.29 Approximating a layer of round conductors as an effective foil conductor.

equal to the bare copper wire diameter, multiplied by the factor $\sqrt{\pi/4}$:

$$h = \sqrt{\frac{\pi}{4}}\, d \qquad (13.70)$$

These square conductors are then joined together, into a foil layer [Fig. 13.29(c)]. Finally, the width of the foil is increased, such that it spans the width of the core window [Fig. 13.29(d)]. Since this stretching process increases the conductor cross-sectional area, a compensating factor η must be introduced such that the correct dc conductor resistance is predicted. This factor, sometimes called the *conductor spacing factor* or the winding *porosity*, is defined as the ratio of the actual layer copper area [Fig. 13.29(a)] to the area of the effective foil conductor of Fig. 13.29(d). The porosity effectively increases the resistivity ρ of the conductor, and thereby increases its skin depth:

$$\delta' = \frac{\delta}{\sqrt{\eta}} \qquad (13.71)$$

If a layer of width ℓ_w contains n_ℓ turns of round wire having diameter d, then the winding porosity η is given by

$$\eta = \sqrt{\frac{\pi}{4}}\, d\, \frac{n_\ell}{\ell_w} \qquad (13.72)$$

A typical value of η for round conductors that span the width of the winding bobbin is 0.8. In the following analysis, the factor φ is given by h/δ for foil conductors, and by the ratio of the effective foil conductor thickness h to the effective skin depth δ' for round conductors as follows:

$$\varphi = \frac{h}{\delta'} = \sqrt{\eta}\,\sqrt{\frac{\pi}{4}}\,\frac{d}{\delta} \qquad (13.73)$$

13.4.4 Power Loss in a Layer

In this section, the average power loss P in a uniform layer of thickness h is determined. As illustrated in Fig. 13.30, the magnetic field strengths on the left and right sides of the conductor are denoted $H(0)$ and $H(d)$, respectively. It is assumed that the component of magnetic field normal to the conductor surface is zero. These magnetic fields are driven by the magnetomotive forces $\mathscr{F}(0)$ and $\mathscr{F}(h)$, respectively. Sinusoidal waveforms are assumed, and rms magnitudes are employed. It is further assumed here that $H(0)$ and $H(h)$ are in phase; the effect of a phase shift is treated in [10].

   With these assumptions, Maxwell's equations are solved to find the current density distribution in the layer. The power loss density is then computed, and is integrated over the volume of the layer to find the total copper loss in the layer [10]. The result is

$$P = R_{dc}\frac{\varphi}{n_\ell^2}\left[\left(\mathscr{F}^2(h) + \mathscr{F}^2(0)\right)G_1(\varphi) - 4\,\mathscr{F}(h)\mathscr{F}(0)G_2(\varphi)\right] \qquad (13.74)$$

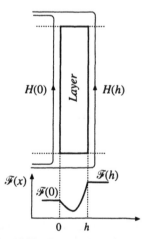

Fig. 13.30 The power loss is determined for a uniform layer. Uniform tangential magnetic fields $H(0)$ and $H(h)$ are applied to the layer surfaces.

where n_ℓ is the number of turns in the layer, and R_{dc} is the dc resistance of the layer. The functions $G_1(\varphi)$ and $G_2(\varphi)$ are

$$G_1(\varphi) = \frac{\sinh(2\varphi) + \sin(2\varphi)}{\cosh(2\varphi) - \cos(2\varphi)}$$

(13.75)

$$G_2(\varphi) = \frac{\sinh(\varphi)\cos(\varphi) + \cosh(\varphi)\sin(\varphi)}{\cosh(2\varphi) - \cos(2\varphi)}$$

If the winding carries current of rms magnitude I, then we can write

$$\mathscr{F}(h) - \mathscr{F}(0) = n_\ell I$$

(13.76)

Let us further express $\mathscr{F}(h)$ in terms of the winding current I, as

$$\mathscr{F}(h) = m n_\ell I$$

(13.77)

The quantity m is therefore the ratio of the MMF $\mathscr{F}(h)$ to the layer ampere-turns $n_\ell I$. Then,

$$\frac{\mathscr{F}(0)}{\mathscr{F}(h)} = \frac{m-1}{m}$$

(13.78)

The power dissipated in the layer, Eq. (13.74), can then be written

$$P = I^2 R_{dc} \varphi Q'(\varphi, m)$$

(13.79)

where

$$Q'(\varphi, m) = \left(2m^2 - 2m + 1\right) G_1(\varphi) - 4m\left(m - 1\right) G_2(\varphi)$$

(13.80)

We can conclude that the proximity effect increases the copper loss in the layer by the factor

$$\frac{P}{I^2 R_{dc}} = \varphi Q'(\varphi, m)$$

(13.81)

Equation (13.81), in conjunction with the definitions (13.80), (13.77), (13.75), and (13.73), can be plotted using a computer spreadsheet or small computer program. The result is illustrated in Fig. 13.31, for several values of m.

It is illuminating to express the layer copper loss P in terms of the dc power loss $P_{dc}|_{\varphi=1}$ that would be obtained in a foil conductor having a thickness $\varphi = 1$. This loss is found by dividing Eq. (13.81) by the effective thickness ratio φ:

$$\frac{P}{P_{dc}|_{\varphi=1}} = Q'(\varphi, m)$$

(13.82)

Equation (13.82) is plotted in Fig. 13.32. Large copper loss is obtained for small φ simply because the layer is thin and hence the dc resistance of the layer is large. For large m and large φ, the proximity effect leads to large power loss; Eq. (13.66) predicts that $Q'(\varphi, m)$ is asymptotic to $m^2 + (m-1)^2$ for large φ. Between these extremes, there is a value of φ which minimizes the layer copper loss.

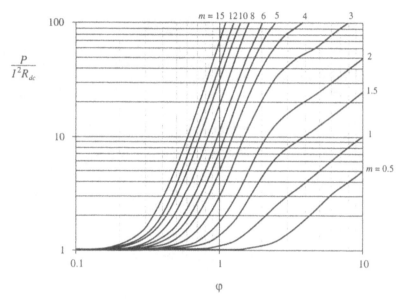

Fig. 13.31 Increase of layer copper loss due to the proximity effect, as a function of φ and MMF ratio m, for sinusoidal excitation.

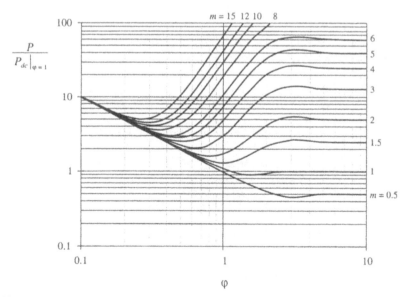

Fig. 13.32 Layer copper loss, relative to the dc loss in a layer having effective thickness of one penetration depth.

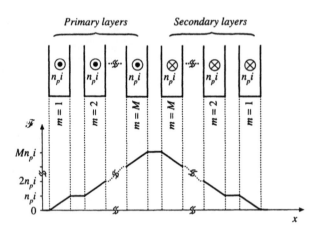

Fig. 13.33 Conventional two-winding transformer example. Each winding consists of M layers.

13.4.5 Example: Power Loss in a Transformer Winding

Let us again consider the proximity loss in a conventional transformer, in which the primary and secondary windings each consist of M layers. The normalized MMF diagram is illustrated in Fig. 13.33. As given by Eq. (13.81), the proximity effect increases the copper loss in each layer by the factor $\varphi Q'(\varphi, m)$. The total increase in primary winding copper loss P_{pri} is found by summation over all of the primary layers:

$$F_R = \frac{P_{pri}}{P_{pri,dc}} = \frac{1}{M} \sum_{m=1}^{M} \varphi Q'(\varphi, m) \tag{13.83}$$

Owing to the symmetry of the windings in this example, the secondary winding copper loss is increased by the same factor. Upon substituting Eq. (13.80) and collecting terms, we obtain

$$F_R = \frac{\varphi}{M} \sum_{m=1}^{M} \left[m^2 \left(2G_1(\varphi) - 4G_2(\varphi) \right) - m \left(2G_1(\varphi) - 4G_2(\varphi) \right) + G_1(\varphi) \right] \tag{13.84}$$

The summation can be expressed in closed form with the help of the identities

$$\sum_{m=1}^{M} m = \frac{M(M+1)}{2}$$
$$\sum_{m=1}^{M} m^2 = \frac{M(M+1)(2M+1)}{6} \tag{13.85}$$

Use of these identities to simplify Eq. (13.84) leads to

$$F_R = \varphi \left[G_1(\varphi) + \frac{2}{3} \left(M^2 - 1 \right) \left(G_1(\varphi) - 2G_2(\varphi) \right) \right] \tag{13.86}$$

This expression is plotted in Fig. 13.34, for several values of M. For large φ, $G_1(\varphi)$ tends to 1, while

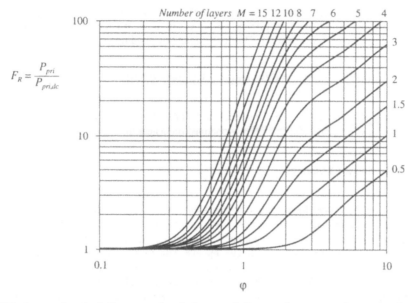

Fig. 13.34 Increased total winding copper loss in the two-winding transformer example, as a function of φ and number of layers M, for sinusoidal excitation.

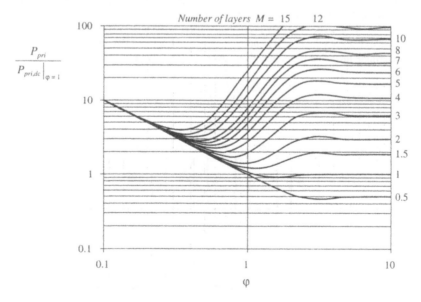

Fig. 13.35 Transformer example winding total copper loss, relative to the winding dc loss for layers having effective thicknesses of one penetration depth.

$G_2(\varphi)$ tends to 0. It can be verified that F_R then tends to the value predicted by Eq. (13.68).

We can again express the total primary power loss in terms of the dc power loss that would be obtained using a conductor in which $\varphi = 1$. This loss is found by dividing Eq. (13.86) by φ:

$$\frac{P_{pri}}{P_{pri,dc}}\bigg|_{\varphi = 1} = G_1(\varphi) + \frac{2}{3}\left(M^2 - 1\right)\left(G_1(\varphi) - 2G_2(\varphi)\right) \tag{13.87}$$

This expression is plotted in Fig. 13.35, for several values of M. Depending on the number of layers, the minimum copper loss for sinusoidal excitation is obtained for φ near to, or somewhat less than, unity.

13.4.6   Interleaving the Windings

One way to reduce the copper losses due to the proximity effect is to interleave the windings. Figure 13.36 illustrates the MMF diagram for a simple transformer in which the primary and secondary layers are alternated, with net layer current of magnitude i. It can be seen that each layer operates with $\mathscr{F} = 0$ on one side, and $\mathscr{F} = i$ on the other. Hence, each layer operates effectively with $m = 1$. Note that Eq. (13.74) is symmetric with respect to $\mathscr{F}(0)$ and $\mathscr{F}(h)$; hence, the copper losses of the interleaved secondary and primary layers are identical. The proximity losses of the entire winding can therefore be determined directly from Fig. 13.34 and 13.35, with $M = 1$. It can be shown that the minimum copper loss for this case (with sinusoidal currents) occurs with $\varphi = \pi/2$, although the copper loss is nearly constant for any $\varphi \geq 1$, and is approximately equal to the dc copper loss obtained when $\varphi = 1$. It should be apparent that interleaving can lead to significant improvements in copper loss when the winding contains several layers.

Partial interleaving can lead to a partial improvement in proximity loss. Figure 13.37 illustrates a transformer having three primary layers and four secondary layers. If the total current carried by each primary layer is $i(t)$, then each secondary layer should carry current $0.75i(t)$. The maximum MMF again occurs in the spaces between the primary and secondary windings, but has value $1.5i(t)$.

To determine the value for m in a given layer, we can solve Eq. (13.78) for m:

$$m = \frac{\mathscr{F}(h)}{\mathscr{F}(h) - \mathscr{F}(0)} \tag{13.88}$$

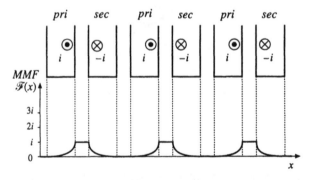

Fig. 13.36   MMF diagram for a simple transformer with interleaved windings. Each layer operates with $m = 1$.

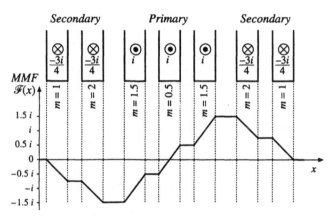

Fig. 13.37 A partially interleaved two-winding transformer, illustrating fractional values of *m*. The MMF diagram is constructed for the low-frequency limit.

The above expression is valid in general, and Eq. (13.74) is symmetrical in $\mathscr{F}(0)$ and $\mathscr{F}(h)$. However, when $F(0)$ is greater in magnitude than $\mathscr{F}(h)$, it is convenient to interchange the roles of $\mathscr{F}(0)$ and $\mathscr{F}(h)$, so that the plots of Figs. 13.31 and 13.32 can be employed.

In the leftmost secondary layer of Fig. 13.37, the layer carries current − 0.75*i*. The MMF changes from 0 to − 0.75*i*. The value of *m* for this layer is found by evaluation of Eq. (13.88):

$$m = \frac{\mathscr{F}(h)}{\mathscr{F}(h) - \mathscr{F}(0)} = \frac{-0.75i}{-0.75i - 0} = 1 \tag{13.89}$$

The loss in this layer, relative to the dc loss of this secondary layer, can be determined using the plots of Figs. 13.31 and 13.32 with *m* = 1. For the next secondary layer, we obtain

$$m = \frac{\mathscr{F}(h)}{\mathscr{F}(h) - \mathscr{F}(0)} = \frac{-1.5i}{-1.5i - (-0.75i)} = 2 \tag{13.90}$$

Hence the loss in this layer can be determined using the plots of Figs. 13.31 and 13.32 with *m* = 2. The next layer is a primary-winding layer. Its value of *m* can be expressed as

$$m = \frac{\mathscr{F}(0)}{\mathscr{F}(0) - \mathscr{F}(h)} = \frac{-1.5i}{-1.5i - (-0.5i)} = 1.5 \tag{13.91}$$

The loss in this layer, relative to the dc loss of this primary layer, can be determined using the plots of Figs. 13.31 and 13.32 with *m* = 1.5. The center layer has an *m* of

$$m = \frac{\mathscr{F}(h)}{\mathscr{F}(h) - \mathscr{F}(0)} = \frac{0.5i}{0.5i - (-0.5i)} = 0.5 \tag{13.92}$$

The loss in this layer, relative to the dc loss of this primary layer, can be determined using the plots of Figs. 13.31 and 13.32 with *m* = 0.5. The remaining layers are symmetrical to the corresponding layers described above, and have identical copper losses. The total loss in the winding is found by summing the losses described above for each layer.

Interleaving windings can significantly reduce the proximity loss when the primary and secondary currents are in phase. However, in some cases such as the transformers of the flyback and SEPIC converters, the winding currents are out of phase. Interleaving then does little to reduce the MMFs and magnetic fields in the vicinity of the windings, and hence the proximity loss is essentially unchanged. It should also be noted that Eqs. (13.74) to (13.82) assume that the winding currents are in phase. General expressions for out-of-phase currents, as well as analysis of a flyback example, are given in [10].

The above procedure can be used to determine the high-frequency copper losses of more complicated multiple-winding magnetic devices. The MMF diagrams are constructed, and then the power loss in each layer is evaluated using Eq. (13.81). These losses are summed, to find the total copper loss. The losses induced in electrostatic shields can also be determined. Several additional examples are given in [10].

It can be concluded that, for sinusoidal currents, there is an optimal conductor thickness in the vicinity of $\varphi = 1$ that leads to minimum copper loss. It is highly advantageous to minimize the number of layers, and to interleave the windings. The amount of copper in the vicinity of the high-MMF portions of windings should be kept to a minimum. Core geometries that maximize the width ℓ_w of the layers, while minimizing the overall number of layers, lead to reduced proximity loss.

Use of *Litz* wire is another means of increasing the conductor area while maintaining low proximity losses. Tens, hundreds, or more strands of small-gauge insulated copper wire are bundled together, and are externally connected in parallel. These strands are twisted, or transposed, such that each strand passes equally through each position inside and on the surface of the bundle. This prevents the circulation of high-frequency currents between strands. To be effective, the diameter of the strands should be sufficiently less than one skin depth. Also, it should be pointed out that the Litz wire bundle itself is composed of multiple layers. The disadvantages of Litz wire are its increased cost, and its reduced fill factor.

13.4.7 PWM Waveform Harmonics

The pulse-width-modulated waveforms encountered in switching converters contain significant harmonics, which can lead to increased proximity losses. The effect of harmonics on the losses in a layer can be determined via field harmonic analysis [10], in which the MMF waveforms $\mathscr{F}(0,t)$ and $\mathscr{F}(d,t)$ of Eq. (13.74) are expressed in Fourier series. The power loss of each individual harmonic is computed as in Section 13.4.4, and the losses are summed to find the total loss in a layer. For example, the PWM waveform of Fig. 13.38 can be represented by the following Fourier series:

$$i(t) = I_0 + \sum_{j=1}^{\infty} \sqrt{2}\, I_j \cos{(j\omega t)} \tag{13.93}$$

where

$$I_j = \frac{\sqrt{2}\, I_{pk}}{j\pi} \sin{(j\pi D)}$$

with $\omega = 2\pi/T_s$. This waveform contains a dc component $I_0 = DI_{pk}$, plus harmonics of rms magnitude I_j proportional to $1/j$. The transformer winding current waveforms of most switching converters follow this Fourier series, or a similar series.

Effects of waveforms harmonics on proximity losses are discussed in [8–10]. The dc component of the winding currents does not lead to proximity loss, and should not be included in proximity loss calculations. Failure to remove the dc component can lead to significantly pessimistic estimates of copper

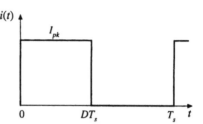

Fig. 13.38   Pulse-width modulated winding current waveform.

loss. The skin depth δ is smaller for high frequency harmonics than for the fundamental, and hence the waveform harmonics exhibit an increased effective φ. Let φ_1 be given by Eq. (13.73), in which δ is found by evaluation of Eq. (13.60) at the fundamental frequency. Since the penetration depth δ varies as the inverse square-root of frequency, the effective value of φ for harmonic j is

$$\varphi_j = \sqrt{j}\ \varphi_1 \tag{13.94}$$

In a multiple-layer winding excited by a current waveform whose fundamental component has $\varphi = \varphi_1$ close to 1, harmonics can significantly increase the total copper loss. This occurs because, for $m > 1$, $Q'(\varphi, m)$ is a rapidly increasing function of φ in the vicinity of 1. When φ_1 is sufficiently greater than 1, then $Q'(\varphi, m)$ is nearly constant, and harmonics have less influence on the total copper loss.

For example, suppose that the two-winding transformer of Fig. 13.33 is employed in a converter such as the forward converter, in which a winding current waveform $i(t)$ can be well approximated by the Fourier series of Eq. (13.93). The winding contains M layers, and has dc resistance R_{dc}. The copper loss induced by the dc component is

$$P_{dc} = I_0^2 R_{dc} \tag{13.95}$$

The copper loss P_j ascribable to harmonic j is found by evaluation of Eq. (13.86) with $\varphi = \varphi_j$:

$$P_j = I_j^2 R_{dc} \sqrt{j}\ \varphi_1 \left[G_1(\sqrt{j}\ \varphi_1) + \frac{2}{3}\left(M^2 - 1\right)\left(G_1(\sqrt{j}\ \varphi_1) - 2G_2(\sqrt{j}\ \varphi_1)\right) \right] \tag{13.96}$$

The total copper loss in the winding is the sum of losses arising from all components of the harmonic series:

$$\frac{P_{cu}}{DI_{pk}^2 R_{dc}} = D + \frac{2\varphi_1}{D\pi^2} \sum_{j=1}^{\infty} \frac{\sin^2 (j\pi D)}{j\sqrt{j}} \left[G_1(\sqrt{j}\ \varphi_1) + \frac{2}{3}\left(M^2 - 1\right)\left(G_1(\sqrt{j}\ \varphi_1) - 2G_2(\sqrt{j}\ \varphi_1)\right) \right] \tag{13.97}$$

In Eq. (13.97), the copper loss is expressed relative to the loss $DI_{pk}^2 R_{dc}$ predicted by a low-frequency analysis. This expression can be evaluated by use of a computer program or computer spreadsheet.

To explicitly quantify the effects of harmonics, we can define the harmonic loss factor F_H as

$$F_H = \frac{\sum_{j=1}^{\infty} P_j}{P_1} \tag{13.98}$$

with P_j given by Eq. (13.96). The total winding copper loss is then given by

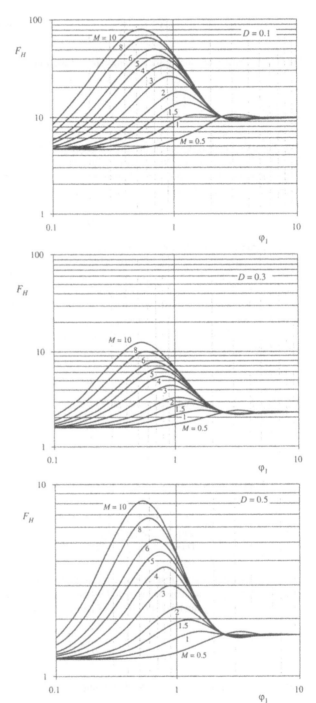

Fig. 13.39 Increased proximity losses induced by PWM waveform harmonics, forward converter example: (a) at $D = 0.1$, (b) at $D = 0.3$, (c) at $D = 0.5$.

$$P_{cu} = I_0^2 R_{dc} + F_H F_R I_1^2 R_{dc} \tag{13.99}$$

with F_R given by Eq. (13.86). The harmonic factor F_H is a function not only of the winding geometry, but also of the harmonic spectrum of the winding current waveform. The harmonic factor F_H is plotted in Fig. 13.39 for several values of D, for the simple transformer example. The total harmonic distortion (THD) of the example current waveforms are: 48% for $D = 0.5$, 76% for $D = 0.3$, and 191% for $D = 0.1$. The waveform THD is defined as

$$\text{THD} = \frac{\sqrt{\sum_{j=2}^{\infty} I_j^2}}{I_1} \tag{13.100}$$

It can be seen that harmonics significantly increase the proximity loss of a multilayer winding when φ_1 is close to 1. For sufficiently small φ_1, the proximity effect can be neglected, and F_H tends to the value $1 + (\text{THD})^2$. For large φ_1, the harmonics also increase the proximity loss; however, the increase is less dramatic than for φ_1 near 1 because the fundamental component proximity loss is large. It can be concluded that, when the current waveform contains high THD and when the winding contains several layers or more, then proximity losses can be kept low only by choosing φ_1 much less than 1. Interleaving the windings allows a larger value of φ_1 to be employed.

13.5 SEVERAL TYPES OF MAGNETIC DEVICES, THEIR *B–H* LOOPS, AND CORE VS. COPPER LOSS

A variety of magnetic elements are commonly used in power applications, which employ the properties of magnetic core materials and windings in different ways. As a result, quite a few factors constrain the design of a magnetic device. The maximum flux density must not saturate the core. The peak ac flux density should also be sufficiently small, such that core losses are acceptably low. The wire size should be sufficiently small, to fit the required number of turns in the core window. Subject to this constraint, the wire cross-sectional area should be as large as possible, to minimize the winding dc resistance and copper loss. But if the wire is too thick, then unacceptable copper losses occur owing to the proximity effect. An air gap is needed when the device stores significant energy. But an air gap is undesirable in transformer applications. It should be apparent that, for a given magnetic device, some of these constraints are active while others are not significant.

Thus, design of a magnetic element involves not only obtaining the desired inductance or turns ratio, but also ensuring that the core material does not saturate and that the total power loss is not too large. Several common power applications of magnetics are discussed in this section, which illustrate the factors governing the choice of core material, maximum flux density, and design approach.

13.5.1 Filter Inductor

A filter inductor employed in a CCM buck converter is illustrated in Fig. 13.40(a). In this application, the value of inductance L is usually chosen such that the inductor current ripple peak magnitude Δi is a small fraction of the full-load inductor current dc component I, as illustrated in Fig. 13.40(b). As illustrated in Fig. 13.41, an air gap is employed that is sufficiently large to prevent saturation of the core by the peak current $I + \Delta i$.

(a)

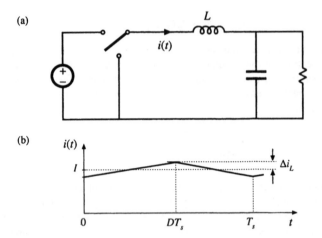

(b)

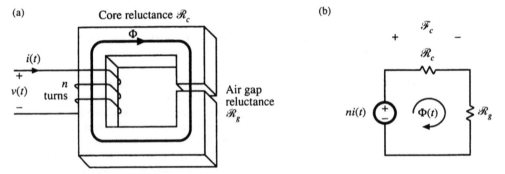

Fig. 13.40   Filter inductor employed in a CCM buck converter: (a) circuit schematic, (b) inductor current waveform.

(a)   Core reluctance $\mathscr{R}_c$ (b)

Fig. 13.41   Filter inductor: (a) structure, (b) magnetic circuit model.

The core magnetic field strength $H_c(t)$ is related to the winding current $i(t)$ according to

$$H_c(t) = \frac{ni(t)}{\ell_c}\frac{\mathscr{R}_c}{\mathscr{R}_c + \mathscr{R}_g} \qquad (13.101)$$

where ℓ_c is the magnetic path length of the core. Since $H_c(t)$ is proportional to $i(t)$, $H_c(t)$ can be expressed as a large dc component H_{c0} and a small superimposed ac ripple ΔH_c, where

$$H_{c0} = \frac{nI}{\ell_c}\frac{\mathscr{R}_c}{\mathscr{R}_c + \mathscr{R}_g}$$

$$\Delta H_c = \frac{n\Delta i}{\ell_c}\frac{\mathscr{R}_c}{\mathscr{R}_c + \mathscr{R}_g} \qquad (13.102)$$

A sketch of $B(t)$ vs. $H_c(t)$ for this application is given in Fig. 13.42. This device operates with the minor B–H loop illustrated. The size of the minor loop, and hence the core loss, depends on the magnitude of the inductor current ripple Δi. The copper loss depends on the rms inductor current ripple, essentially

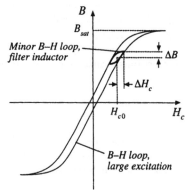

Fig. 13.42 Filter inductor minor B–H loop.

equal to the dc component I. Typically, the core loss can be ignored, and the design is driven by the copper loss. The maximum flux density is limited by saturation of the core. Proximity losses are negligible. Although a high-frequency ferrite material can be employed in this application, other materials having higher core losses and greater saturation flux density lead to a physically smaller device. Design of a filter inductor in which the maximum flux density is a specified value is considered in the next chapter.

13.5.2 AC Inductor

An ac inductor employed in a resonant converter is illustrated in Fig. 13.43. In this application, the high-frequency current variations are large. In consequence, the $B(t)$–$H(t)$ loop illustrated in Fig. 13.44 is large. Core loss and proximity loss are usually significant in this application. The maximum flux density

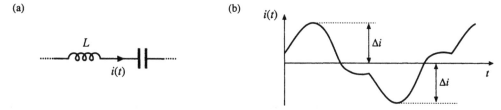

Fig. 13.43 Ac inductor, resonant converter example: (a) resonant tank circuit, (b) inductor current waveform.

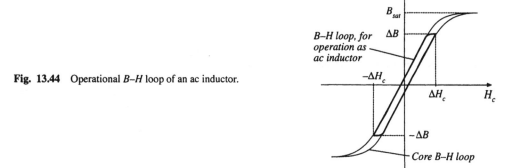

Fig. 13.44 Operational B–H loop of an ac inductor.

is limited by core loss rather than saturation. Both core loss and copper loss must be accounted for in the design of this element, and the peak ac flux density ΔB is a design variable that is typically chosen to minimize the total loss. A high-frequency material having low core loss, such as ferrite, is normally employed. Design of magnetics such as this, in which the ac flux density is a design variable that is chosen in a optimal manner, is considered in Chapter 15.

13.5.3 Transformer

Figure 13.45 illustrates a conventional transformer employed in a switching converter. Magnetization of the core is modeled by the magnetizing inductance L_M. The magnetizing current $i_M(t)$ is related to the core magnetic field $H(t)$ according to Ampere's law

$$H(t) = \frac{n i_M(t)}{\ell_m} \tag{13.103}$$

However, $i_M(t)$ is not a direct function of the winding currents $i_1(t)$ or $i_2(t)$. Rather, the magnetizing current is dependent on the applied winding voltage waveform $v_1(t)$. Specifically, the maximum ac flux density is directly proportional to the applied volt-seconds λ_1. A typical B–H loop for this application is illustrated in Fig. 13.46.

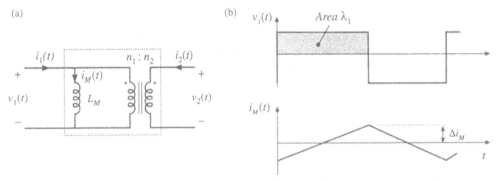

Fig. 13.45   Conventional transformer: (a) equivalent circuit, (b) typical primary voltage and magnetizing current waveforms.

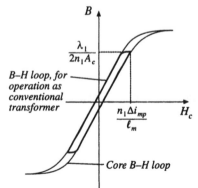

Fig. 13.46   Operational B–H loop of a conventional transformer.

In the transformer application, core loss and proximity losses are usually significant. Typically the maximum flux density is limited by core loss rather than by saturation. A high-frequency material having low core loss is employed. Both core and copper losses must be accounted for in the design of the transformer. The design must also incorporate multiple windings. Transformer design with flux density optimized for minimum total loss is described in Chapter 15.

13.5.4 Coupled Inductor

A coupled inductor is a filter inductor having multiple windings. Figure 13.47(a) illustrates coupled inductors in a two-output forward converter. The inductors can be wound on the same core, because the winding voltage waveforms are proportional. The inductors of the SEPIC and Ćuk converters, as well as of multiple-output buck-derived converters and some other converters, can be coupled. The inductor current ripples can be controlled by control of the winding leakage inductances [12,13]. Dc currents flow in each winding as illustrated in Fig. 13.47(b), and the net magnetization of the core is proportional to the sum of the winding ampere-turns:

$$H_c(t) = \frac{n_1 i_1(t) + n_2 i_2(t)}{\ell_c} \frac{\mathcal{R}_c}{\mathcal{R}_c + \mathcal{R}_g} \tag{13.104}$$

As in the case of the single-winding filter inductor, the size of the minor *B–H* loop is proportional to the total current ripple (Fig. 13.48). Small ripple implies small core loss, as well as small proximity loss. An air gap is employed, and the maximum flux density is typically limited by saturation.

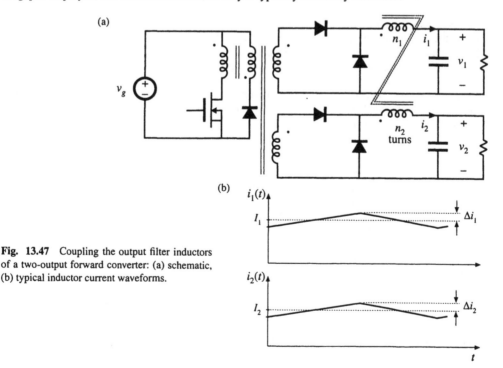

Fig. 13.47 Coupling the output filter inductors of a two-output forward converter: (a) schematic, (b) typical inductor current waveforms.

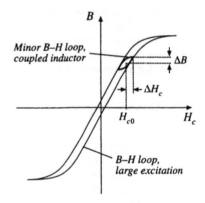

Fig. 13.48   Coupled inductor minor *B–H* loop.

13.5.5   Flyback Transformer

As discussed in Chapter 6, the flyback transformer functions as an inductor with two windings. The primary winding is used during the transistor conduction interval, and the secondary is used during the diode conduction interval. A flyback converter is illustrated in Fig. 13.49(a), with the flyback transformer modeled as a magnetizing inductance in parallel with an ideal transformer. The magnetizing current $i_M(t)$ is proportional to the core magnetic field strength $H_c(t)$. Typical DCM waveforms are given in Fig. 13.49(b).

Since the flyback transformer stores energy, an air gap is needed. Core loss depends on the magnitude of the ac component of the magnetizing current. The *B–H* loop for discontinuous conduction mode operation is illustrated in Fig. 13.50. When the converter is designed to operate in DCM, the core loss is significant. The peak ac flux density ΔB is then chosen to maintain an acceptably low core loss. For CCM operation, core loss is less significant, and the maximum flux density may be limited only by saturation of the core. In either case, winding proximity losses are typically quite significant. Unfortunately, interleaving the windings has little impact on the proximity loss because the primary and secondary winding currents are out of phase.

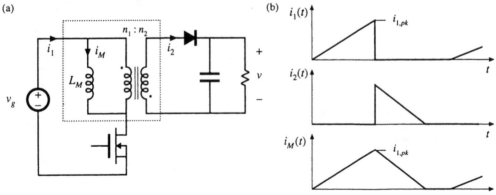

Fig. 13.49   Flyback transformer: (a) converter schematic, with transformer equivalent circuit, (b) DCM current waveforms.

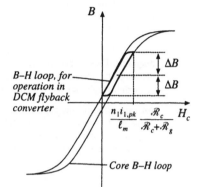

Fig. 13.50 Operational *B–H* loop of a DCM flyback transformer.

13.6 SUMMARY OF KEY POINTS

1. Magnetic devices can be modeled using lumped-element magnetic circuits, in a manner similar to that commonly used to model electrical circuits. The magnetic analogs of electrical voltage V, current I, and resistance R, are magnetomotive force (MMF) $\mathcal{F}$, flux Φ, and reluctance $\mathcal{R}$ respectively.

2. Faraday's law relates the voltage induced in a loop of wire to the derivative of flux passing through the interior of the loop.

3. Ampere's law relates the total MMF around a loop to the total current passing through the center of the loop. Ampere's law implies that winding currents are sources of MMF, and that when these sources are included, then the net MMF around a closed path is equal to zero.

4. Magnetic core materials exhibit hysteresis and saturation. A core material saturates when the flux density B reaches the saturation flux density B_{sat}.

5. Air gaps are employed in inductors to prevent saturation when a given maximum current flows in the winding, and to stabilize the value of inductance. The inductor with air gap can be analyzed using a simple magnetic equivalent circuit, containing core and air gap reluctances and a source representing the winding MMF.

6. Conventional transformers can be modeled using sources representing the MMFs of each winding, and the core MMF. The core reluctance approaches zero in an ideal transformer. Nonzero core reluctance leads to an electrical transformer model containing a magnetizing inductance, effectively in parallel with the ideal transformer. Flux that does not link both windings, or "leakage flux," can be modeled using series inductors.

7. The conventional transformer saturates when the applied winding volt-seconds are too large. Addition of an air gap has no effect on saturation. Saturation can be prevented by increasing the core cross-sectional area, or by increasing the number of primary turns.

8. Magnetic materials exhibit core loss, due to hysteresis of the *B–H* loop and to induced eddy currents flowing in the core material. In available core materials, there is a tradeoff between high saturation flux density B_{sat} and high core loss P_{fe}. Laminated iron alloy cores exhibit the highest B_{sat} but also the highest P_{fe}, while ferrite cores exhibit the lowest P_{fe} but also the lowest B_{sat}. Between these two extremes are powdered iron alloy and amorphous alloy materials.

9. The skin and proximity effects lead to eddy currents in winding conductors, which increase the copper loss P_{cu} in high-current high-frequency magnetic devices. When a conductor has thickness approaching or

larger than the penetration depth δ, magnetic fields in the vicinity of the conductor induce eddy currents in the conductor. According to Lenz's law, these eddy currents flow in paths that tend to oppose the applied magnetic fields.

10. The magnetic field strengths in the vicinity of the winding conductors can be determined by use of MMF diagrams. These diagrams are constructed by application of Ampere's law, following the closed paths of the magnetic field lines which pass near the winding conductors. Multiple-layer noninterleaved windings can exhibit high maximum MMFs, with resulting high eddy currents and high copper loss.

11. An expression for the copper loss in a layer, as a function of the magnetic field strengths or MMFs surrounding the layer, is given in Section 13.4.4. This expression can be used in conjunction with the MMF diagram, to compute the copper loss in each layer of a winding. The results can then be summed, yielding the total winding copper loss. When the effective layer thickness is near to or greater than one skin depth, the copper losses of multiple-layer noninterleaved windings are greatly increased.

12. Pulse-width-modulated winding currents contain significant total harmonic distortion, which can lead to a further increase of copper loss. The increase in proximity loss caused by current harmonics is most pronounced in multiple-layer non-interleaved windings, with an effective layer thickness near one skin depth.

REFERENCES

[1]   MIT STAFF, *Magnetic Circuits and Transformers*, Cambridge: The MIT Press, 1943.

[2]   J. K. WATSON, *Applications of Magnetism*, New York: John Wiley & Sons, 1980.

[3]   R. P. SEVERNS and G. E. BLOOM, *Modern Dc-to-Dc Switchmode Power Converter Circuits*, New York: Van Nostrand Reinhold, 1985.

[4]   A. DAUHAJRE and R. D. MIDDLEBROOK, "Modeling and Estimation of Leakage Phenomena in Magnetic Circuits," *IEEE Power Electronics Specialists Conference*, 1986 Record, pp. 213-226.

[5]   S. EL-HAMAMSY and E. CHANG, "Magnetics Modeling for Computer-Aided Design of Power Electronics Circuits," *IEEE Power Electronics Specialists Conference*, 1989 Record, pp. 635-645.

[6]   P. L. DOWELL, "Effects of Eddy Currents in Transformer Windings," *Proceedings of the IEE*, Vol. 113, No. 8, August 1966, pp. 1387-1394.

[7]   M. P. PERRY, "Multiple Layer Series Connected Winding Design for Minimum Loss," *IEEE Transactions on Power Apparatus and Systems*, Vol. PAS-98, No. 1, Jan./Feb. 1979, pp. 116-123.

[8]   P. S. VENKATRAMAN, "Winding Eddy Current Losses in Switch Mode Power Transformers Due to Rectangular Wave Currents," *Proceedings of Powercon 11*, 1984, pp. A1.1 - A1.11.

[9]   B. CARSTEN, "High Frequency Conductor Losses in Switchmode Magnetics," *High Frequency Power Converter Conference*, 1986 Record, pp. 155-176.

[10]   J. P. VANDELAC and P. ZIOGAS, "A Novel Approach for Minimizing High Frequency Transformer Copper Losses," *IEEE Power Electronics Specialists Conference*, 1987 Record, pp. 355-367.

[11]   A. M. URLING, V. A. NIEMELA, G. R. SKUTT, and T. G. WILSON, "Characterizing High-Frequency Effects in Transformer Windings—A Guide to Several Significant Articles," *IEEE Applied Power Electronics Conference*, 1989 Record, pp. 373-385.

[12]　S. ĆUK and R. D. MIDDLEBROOK, "Coupled-Inductor and Other Extensions of a New Optimum Topology Switching Dc–to–Dc Converter," *IEEE Industry Applications Society Annual Meeting*, 1977 Proceedings, pp. 1110-1122.

[13]　S. ĆUK and Z. ZHANG, "Coupled-Inductor Analysis and Design," *IEEE Power Electronics Specialists Conference*, 1986 Record, pp. 655-665.

PROBLEMS

13.1　The core illustrated in Fig. 13.51(a) is 1 cm thick. All legs are 1 cm wide, except for the right-hand side vertical leg, which is 0.5 cm wide. You may neglect nonuniformities in the flux distribution caused by turning corners.

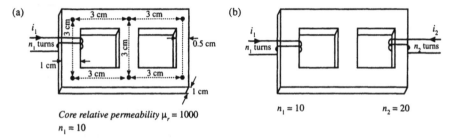

Fig. 13.51　Problem 13.1

(a)　Determine the magnetic circuit model of this device, and label the values of all reluctances in your model.

(b)　Determine the inductance of the winding.

A second winding is added to the same core, as shown in Fig. 13.51(b).

(c)　Modify your model of part (a) to include this winding.

(d)　The electrical equations for this circuit may be written in the form

$$\begin{bmatrix} v_1 \\ v_2 \end{bmatrix} = \begin{bmatrix} L_{11} & L_{12} \\ L_{12} & L_{22} \end{bmatrix} \frac{d}{dt} \begin{bmatrix} i_1 \\ i_2 \end{bmatrix}$$

Use superposition to determine analytical expressions and numerical values for L_{11}, L_{12}, and L_{22}.

13.2　Two windings are placed as illustrated in Fig. 13.52(a) on a core of uniform cross-sectional area $A_c = 1$ cm^2. Each winding has 50 turns. The relative permeability of the core is $\mu_r = 10^4$.

(a)　Sketch an equivalent magnetic circuit, and determine numerical values for each reluctance.

(b)　Determine the self-inductance of each winding.

(c)　Determine the inductance L^+ obtained when the windings are connected in series as in Fig. 13.52(b).

(d)　Determine the inductance L^- obtained when the windings are connected in anti-series as in Fig. 13.52(c).

(a)

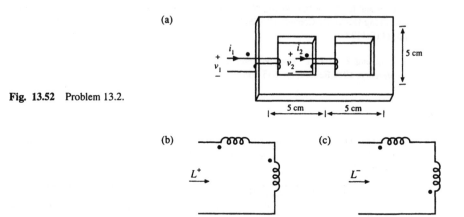

Fig. 13.52 Problem 13.2.

(b)

(c)

13.3 All three legs of the magnetic device illustrated in Fig. 13.53 are of uniform cross-sectional area A_C. Legs 1 and 2 each have magnetic path length 3ℓ, while leg 3 has magnetic path length ℓ. Both windings have n turns. The core has permeability $\mu \gg \mu_0$.

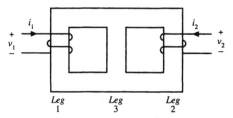

Fig. 13.53 Magnetic core for Problem 13.3.

(a) Sketch a magnetic equivalent circuit, and give analytical expressions for all element values.

A voltage source is connected to winding 1, such that $v_1(t)$ is a square wave of peak value V_{max} and period T_s. Winding 2 is open-circuited.

(b) Sketch $i_1(t)$ and label its peak value.

(c) Find the flux $\varphi_2(t)$ in leg 2. Sketch $\varphi_2(t)$ and label its peak value.

(d) Sketch $v_2(t)$ and label its peak value.

13.4 The magnetic device illustrated in Fig. 13.54(a) consists of two windings, which can replace the two inductors in a Ćuk, SEPIC, or other similar converter. For this problem, all three legs have the same uniform cross-sectional area A_c. The legs have gaps of lengths g_1, g_2, and g_3, respectively. The core permeability μ is very large. You may neglect fringing flux. Legs 1 and 2 have windings containing n_1 and n_2 turns, respectively.

(a) Derive a magnetic circuit model for this device, and give analytical expressions for each reluctance in your model. Label the polarities of the MMF generators.

(b) Write the electrical terminal equations of this device in the matrix form

$$\begin{bmatrix} v_1 \\ v_2 \end{bmatrix} = \begin{bmatrix} L_{11} & L_{12} \\ L_{12} & L_{22} \end{bmatrix} \frac{d}{dt} \begin{bmatrix} i_1 \\ i_2 \end{bmatrix}$$

(a)

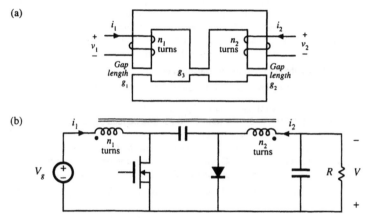

(b)

(b image continues)

Fig. 13.54 Magnetic core and converter for Problem 13.4.

and derive analytical expressions for L_{11}, L_{12}, and L_{22}.

(c)   Derive an electrical circuit model for this device, and give analytical expressions for the turns ratio and each inductance in your model, in terms of the turns and reluctances of part (a).

This single magnetic device is to be used to realize the two inductors of the Ćuk converter, as in Fig. 13.54(b).

(d)   Sketch the voltage waveforms $v_1(t)$ and $v_2(t)$, making the linear ripple approximation as appropriate. You may assume that the converter operates in the continuous conduction mode.

(e)   The voltage waveforms of part (d) are applied to your model of parts (b) and (c). Solve your model to determine the slopes of the inductor current ripples during intervals DT_s and $D'T_s$. Sketch the steady-state inductor current waveforms $i_1(t)$ and $i_2(t)$, and label all slopes.

(f)   By skillful choice of n_1/n_2 and the air gap lengths, it is possible to make the inductor current ripple Δi in either $i_1(t)$ or $i_2(t)$ go to zero. Determine the conditions on n_1/n_2, g_1, g_2, and g_3 that cause the current ripple in $i_2(t)$ to become zero. Sketch the resulting $i_1(t)$ and $i_2(t)$, and label all slopes.

It is possible to couple the inductors in this manner, and cause one of the inductor current ripples to go to zero, in any converter in which the inductor voltage waveforms are proportional.

13.5   Over its usable operating range, a certain permanent magnet material has the *B–H* characteristics illustrated by the solid line in Fig. 13.55. The magnet has length $\ell_m = 0.5$ cm, and cross-sectional area 4 cm². $B_m = 1$ T. Derive an equivalent magnetic circuit model for the magnet, and label the numerical values of the elements.

Fig. 13.55 *B–H* characteristic of the permanent magnet material for Problem 13.5.

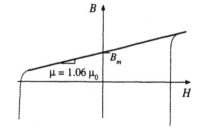

13.6   The two-transistor forward converter of Fig. 6.27 operates with $V_g = 300$ V, $V = 28$ V, switching frequency $f_s = 100$ kHz, and turns ratio $n = 0.25$. The dc load power is 250 W. The transformer uses an EC41 ferrite core; relevant data for this core is listed in Appendix D. The core loss is given by Fig. 13.20. The primary winding consists of 44 turns of #21 AWG wire, and the secondary winding is composed of 11 turns of #15 AWG wire. Data regarding the American wire gauge is also listed in Appendix D.

    (a)   Estimate the core loss of this transformer

    (b)   Determine the copper loss of this transformer. You may neglect proximity losses.

13.7   The two-transistor forward converter of Fig. 6.27 operates in CCM with $V_g = 300$ V, $V = 28$ V, switching frequency $f_s = 100$ kHz, and turns ratio $n = 0.25$. The dc load power is 250 W. The transformer uses an EC41 ferrite core; relevant data for this core is listed in Appendix D. This core has window height $\ell_w = 2.78$ cm. The primary winding consists of 44 turns of #24 AWG wire, and the secondary winding is composed of 11 turns of #14 AWG wire. Each winding comprises one layer. Data regarding the American wire gauge is also listed in Appendix D. The winding operates at room temperature.

    (a)   Determine the primary and secondary copper losses induced by the dc components of the winding currents.

    (b)   Determine the primary and secondary copper losses induced by the fundamental components of the winding currents.

    (c)   Determine the primary and secondary copper losses induced by the second harmonic components of the winding currents.

13.8   The winding currents of the transformer in a high-voltage inverter are essentially sinusoidal, with negligible harmonics and no dc components. The primary winding consists of one layer containing 10 turns of round copper wire. The secondary winding consists of 250 turns of round copper wire, arranged in ten layers. The operating frequency is $f = 50$ kHz, and the winding porosity is 0.8. Determine the primary and secondary wire diameters and wire gauges that minimize the total copper loss.

13.9   A certain three-winding transformer contains one primary and two secondaries. The operating frequency is 40 kHz. The primary winding contains a total of 60 turns of #26AWG, arranged in three layers. The secondary windings each consist of five turns of copper foil, one turn per layer. The foil thickness is 0.25 mm. The primary layers have porosity 0.8, while the secondary layer porosity is 1. The primary winding carries a sinusoidal current having rms value I, while each secondary carries rms current $6I$. The windings are not interleaved: the primary winding is closest to the center leg of the core, followed by secondary winding #1, followed by secondary winding #2.

    (a)   Sketch an MMF diagram illustrating the magnetic fields in the vicinity of each winding layer.

    (b)   Determine the increased copper loss, due to the proximity effect, in each layer.

    (c)   Determine the ratio of copper loss to dc copper loss, F_R, for the entire transformer windings.

    (d)   In this application, it is not feasible to interleave the primary winding with the other windings. However, changing the conductor size is permissible. Discuss how the windings could be better optimized.

13.10   A transformer winding contains a four-layer primary winding, and two two-layer secondary windings. Each layer of the primary winding carries total current I. Each layer of secondary winding #1 carries total current $1.5I$. Each layer of secondary winding #2 carries total current $0.5I$. All currents are sinusoidal. The effective relative conductor thickness is $\varphi = 2$. The windings are partially interleaved, in the following order: two primary layers, followed by both layers of secondary #1, followed by both layers of secondary #2, and finally the two remaining primary layers.

    (a)   Sketch an MMF diagram for this winding arrangement.

    (b)   Determine the increased copper loss, due to the proximity effect, for each layer.

    (c)   Determine the increase in total transformer copper loss, due to the proximity effect.

13.11 A single-output forward converter contains a transformer having a noninterleaved secondary winding with four layers. The converter operates at $D = 0.3$ in CCM, with a secondary winding current waveform similar to Fig. 13.38.

   (a) Estimate the value of φ_1 that minimizes the secondary winding copper losses.

   (b) Determine the resulting secondary copper loss, relative to $I_{rms}^2 R_{dc}$.

13.12 A schematic diagram and waveforms of the isolated SEPIC, operating in CCM, are given in Figs. 6.37 and 6.38.

   (a) Do you expect the SEPIC transformer to contain an air gap? Why or why not?

   (b) Sketch the SEPIC transformer $B–H$ loop, for CCM operation.

   (c) For CCM operation, do you expect core loss to be significant? Explain your reasoning.

   (d) For CCM operation, do you expect winding proximity losses to be significant? Explain your reasoning.

14

Inductor Design

This chapter treats the design of magnetic elements such as filter inductors, using the K_g method. With this method, the maximum flux density B_{max} is specified in advance, and the element is designed to attain a given copper loss.

The design of a basic filter inductor is discussed in Sections 14.1 and 14.1.5. In the filter inductor application, it is necessary to obtain the required inductance, avoid saturation, and obtain an acceptable low dc winding resistance and copper loss. The geometrical constant K_g is a measure of the effective magnetic size of a core, when dc copper loss and winding resistance are the dominant constraints [1,2]. Design of a filter inductor involves selection of a core having a K_g sufficiently large for the application, then computing the required air gap, turns, and wire size. A simple step-by-step filter inductor design procedure is given. Values of K_g for common ferrite core shapes are tabulated in Appendix D.

Extension of the K_g method to multiple-winding elements is covered in Section 14.3. In applications requiring multiple windings, it is necessary to optimize the wire sizes of the windings so that the overall copper loss is minimized. It is also necessary to write an equation that relates the peak flux density to the applied waveforms or to the desired winding inductance. Again, a simple step-by-step transformer design approach is given.

The goal of the K_g approach of this chapter is the design of a magnetic device having a given copper loss. Core loss is not specifically addressed in the K_g approach, and B_{max} is a given fixed value. In the next chapter, the flux density is treated as a design variable to be optimized. This allows the overall loss (i.e., core loss plus copper loss) to be minimized.

14.1 FILTER INDUCTOR DESIGN CONSTRAINTS

A filter inductor employed in a CCM buck converter is illustrated in Fig. 14.1(a). In this application, the value of inductance L is usually chosen such that the inductor current ripple peak magnitude Δi is a small fraction of the full-load inductor current dc component I, as illustrated in Fig. 14.1(b). As illustrated in Fig. 14.2, an air gap is employed that is sufficiently large to prevent saturation of the core by the peak

(a)

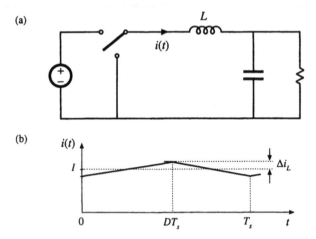

(b)

Fig. 14.1   Filter inductor employed in a CCM buck converter: (a) circuit schematic, (b) inductor current waveform.

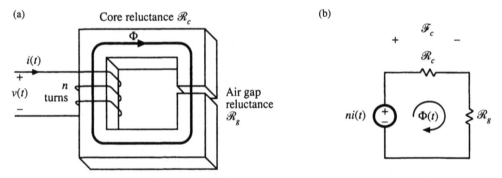

Fig. 14.2   Filter inductor: (a) structure, (b) magnetic circuit model.

current $I + \Delta i$.

Let us consider the design of the filter inductor illustrated in Figs. 14.1 and 14.2. It is assumed that the core and proximity losses are negligible, so that the inductor losses are dominated by the low-frequency copper losses. The inductor can therefore be modeled by the equivalent circuit of Fig. 14.3, in which R represents the dc resistance of the winding. It is desired to obtain a given inductance L and given winding resistance R. The inductor should not saturate when a given worst-case peak current I_{max} is applied. Note that specification of R is equivalent to specification of the copper loss P_{cu}, since

$$P_{cu} = I_{rms}^2 R \qquad (14.1)$$

The influence of inductor winding resistance on converter efficiency and output voltage is modeled in Chapter 3.

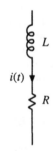

Fig.   14.3   Filter   inductor equivalent circuit.

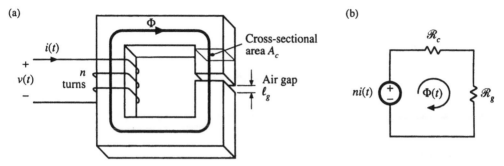

Fig. 14.4   Filter inductor: (a) assumed geometry, (b) magnetic circuit.

It is assumed that the inductor geometry is topologically equivalent to Fig. 14.4(a). An equivalent magnetic circuit is illustrated in Fig. 14.4(b). The core reluctance $\mathscr{R}_c$ and air gap reluctance $\mathscr{R}_g$ are

$$\mathscr{R}_c = \frac{\ell_c}{\mu_c A_c}$$
$$\mathscr{R}_g = \frac{\ell_g}{\mu_0 A_c} \qquad (14.2)$$

where ℓ_c is the core magnetic path length, A_c is the core cross-sectional area, μ_c is the core permeability, and ℓ_g is the air gap length. It is assumed that the core and air gap have the same cross-sectional areas. Solution of Fig. 14.4(b) yields

$$ni = \Phi\left(\mathscr{R}_c + \mathscr{R}_g\right) \qquad (14.3)$$

Usually, $\mathscr{R}_c \ll \mathscr{R}_g$, and hence Eq. (14.3) can be approximated as

$$ni \approx \Phi \mathscr{R}_g \qquad (14.4)$$

The air gap dominates the inductor properties. Four design constraints now can be identified.

14.1.1   Maximum Flux Density

Given a peak winding current I_{max}, it is desired to operate the core flux density at a maximum value B_{max}. The value of B_{max} is chosen to be less than the worst-case saturation flux density B_{sat} of the core material. Substitution of $\Phi = BA_c$ into Eq. (14.4) leads to

$$ni = BA_c \mathscr{R}_g \qquad (14.5)$$

Upon letting $I = I_{max}$ and $B = B_{max}$, we obtain

$$nI_{max} = B_{max} A_c \mathscr{R}_g = B_{max} \frac{\ell_g}{\mu_0} \qquad (14.6)$$

This is the first design constraint. The turns ratio n and the air gap length ℓ_g are unknowns.

14.1.2 Inductance

The given inductance value L must be obtained. The inductance is equal to

$$L = \frac{n^2}{\mathscr{R}_g} = \frac{\mu_0 A_c n^2}{\ell_g} \tag{14.7}$$

This is the second design constraint. The turns ratio n, core area A_c, and gap length ℓ_g are unknown.

14.1.3 Winding Area

As illustrated in Fig. 14.5, the winding must fit through the window, i.e., the hole in the center of the core. The cross-sectional area of the conductor, or bare area, is A_W. If the winding has n turns, then the area of copper conductor in the window is

$$n A_W \tag{14.8}$$

If the core has window area W_A, then we can express the area available for the winding conductors as

$$K_u W_A \tag{14.9}$$

where K_u is the *window utilization factor*, or *fill factor*. Hence, the third design constraint can be expressed as

$$K_u W_A \geq n A_W \tag{14.10}$$

The fill factor K_u is the fraction of the core window area that is filled with copper. K_u must lie between zero and one. As discussed in [1], there are several mechanism that cause K_u to be less than unity. Round wire does not pack perfectly; this reduces K_u by a factor of 0.7 to 0.55, depending on the winding technique. The wire has insulation; the ratio of wire conductor area to total wire area varies from approximately 0.95 to 0.65, depending on the wire size and type of insulation. The bobbin uses some of the window area. Insulation may be required between windings and/or winding layers. Typical values of K_u for cores with winding bobbins are: 0.5 for a simple low-voltage inductor, 0.25 to 0.3 for an off-line transformer, 0.05 to 0.2 for a high-voltage transformer supplying several kV, and 0.65 for a low-voltage foil transformer or inductor.

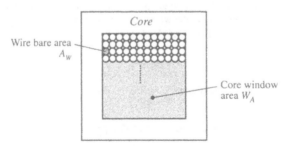

Fig. 14.5 The winding must fit in the core window area.

14.1.4 Winding Resistance

The resistance of the winding is

$$R = \rho \frac{\ell_b}{A_W} \tag{14.11}$$

where ρ is the resistivity of the conductor material, ℓ_b is the length of the wire, and A_W is the wire bare area. The resistivity of copper at room temperature is $1.724 \cdot 10^{-6}$ Ω–cm. The length of the wire comprising an n-turn winding can be expressed as

$$\ell_b = n(MLT) \tag{14.12}$$

where (MLT) is the mean-length-per-turn of the winding. The mean-length-per-turn is a function of the core geometry. Substitution of Eq. (14.12) into (14.11) leads to

$$R = \rho \frac{n(MLT)}{A_W} \tag{14.13}$$

This is the fourth constraint.

14.1.5 The Core Geometrical Constant K_g

The four constraints, Eqs. (14.6), (14.7), (14.10), and (14.13), involve the quantities A_c, W_A, and MLT, which are functions of the core geometry, the quantities I_{max}, B_{max}, μ_0, L, K_u, R, and ρ, which are given specifications or other known quantities, and n, ℓ_g, and A_W, which are unknowns. Elimination of the unknowns n, ℓ_g, and A_W leads to the following equation:

$$\frac{A_c^2 W_A}{(MLT)} \geq \frac{\rho L^2 I_{max}^2}{B_{max}^2 R K_u} \tag{14.14}$$

The quantities on the right side of this equation are specifications or other known quantities. The left side of the equation is a function of the core geometry alone. It is necessary to choose a core whose geometry satisfies Eq. (14.14).

The quantity

$$K_g = \frac{A_c^2 W_A}{(MLT)} \tag{14.15}$$

is called the core geometrical constant. It is a figure-of-merit that describes the effective electrical size of magnetic cores, in applications where copper loss and maximum flux density are specified. Tables are included in Appendix D that list the values of K_g for several standard families of ferrite cores. K_g has dimensions of length to the fifth power.

Equation (14.14) reveals how the specifications affect the core size. Increasing the inductance or peak current requires an increase in core size. Increasing the peak flux density allows a decrease in core size, and hence it is advantageous to use a core material that exhibits a high saturation flux density. Allowing a larger winding resistance R, and hence larger copper loss, leads to a smaller core. Of course, the increased copper loss and smaller core size will lead to a higher temperature rise, which may be unacceptable. The fill factor K_u also influences the core size.

Equation (14.15) reveals how core geometry affects the core capabilities. An inductor capable of meeting increased electrical requirements can be obtained by increasing either the core area A_c, or the window area W_A. Increase of the core area requires additional iron core material. Increase of the window area implies that additional copper winding material is employed. We can trade iron for copper, or vice versa, by changing the core geometry in a way that maintains the K_g of Eq. (14.15).

14.2   A STEP-BY-STEP PROCEDURE

The procedure developed in Section 14.1 is summarized below. This simple filter inductor design procedure should be regarded as a first-pass approach. Numerous issues have been neglected, including detailed insulation requirements, conductor eddy current losses, temperature rise, roundoff of number of turns, etc.

The following quantities are specified, using the units noted:

Wire resistivity	ρ	(Ω–cm)
Peak winding current	I_{max}	(A)
Inductance	L	(H)
Winding resistance	R	(Ω)
Winding fill factor	K_u	
Maximum operating flux density	B_{max}	(T)

The core dimensions are expressed in cm:

Core cross-sectional area	A_c	(cm^2)
Core window area	W_A	(cm^2)
Mean length per turn	MLT	(cm)

The use of centimeters rather than meters requires that appropriate factors be added to the design equations.

1. *Determine core size*

$$K_g \geq \frac{\rho L^2 I_{max}^2}{B_{max}^2 R K_u} 10^8 \quad \text{(cm}^5\text{)} \tag{14.16}$$

Choose a core which is large enough to satisfy this inequality. Note the values of A_c, W_A, and MLT for this core. The resistivity ρ of copper wire is $1.724 \cdot 10^{-6}$ Ω–cm at room temperature, and $2.3 \cdot 10^{-6}$ Ω–cm at 100°C.

2. *Determine air gap length*

$$\ell_g = \frac{\mu_0 L I_{max}^2}{B_{max}^2 A_c} 10^4 \quad \text{(m)} \tag{14.17}$$

with A_c expressed in cm^2. $\mu_0 = 4\pi \cdot 10^{-7}$ H/m. The air gap length is given in meters. The value expressed in Eq. (14.17) is approximate, and neglects fringing flux and other nonidealities.

Core manufacturers sell gapped cores. Rather than specifying the air gap length, the equivalent quantity A_L is used. A_L is equal to the inductance, in mH, obtained with a winding of 1000 turns. When A_L is specified, it is the core manufacturer's responsibility to obtain the correct gap length. Equation (14.17) can be modified to yield the required A_L, as follows:

$$A_L = \frac{10 B_{max}^2 A_c^2}{L I_{max}^2} \quad \text{(mH/1000 turns)} \tag{14.18}$$

where A_c is given in cm^2, L is given in Henries, and B_{max} is given in Tesla.

3. *Determine number of turns*

$$n = \frac{L I_{max}}{B_{max} A_c} 10^4 \tag{14.19}$$

4. *Evaluate wire size*

$$A_W \le \frac{K_u W_A}{n} \quad (\text{cm}^2) \tag{14.20}$$

Select wire with bare copper area less than or equal to this value. An American Wire Gauge table is included in Appendix D.

As a check, the winding resistance can be computed:

$$R = \frac{\rho n (MLT)}{A_w} \quad (\Omega) \tag{14.21}$$

14.3 MULTIPLE-WINDING MAGNETICS DESIGN VIA THE K_g METHOD

The K_g method can be extended to the case of multiple-winding magnetics, such as the transformers and coupled inductors described in Sections 13.5.3 to 13.5.5. The desired turns ratios, as well as the desired winding voltage and current waveforms, are specified. In the case of a coupled inductor or flyback transformer, the magnetizing inductance is also specified. It is desired to select a core size, number of turns for each winding, and wire sizes. It is also assumed that the maximum flux density B_{max} is given.

With the K_g method, a desired copper loss is attained. In the multiple-winding case, each winding contributes some copper loss, and it is necessary to allocate the available window area among the various windings. In Section 14.3.1 below, it is found that total copper loss is minimized if the window area is divided between the windings according to their apparent powers. This result is employed in the following sections, in which an optimized K_g method for coupled inductor design is developed.

14.3.1 Window Area Allocation

The first issue to settle in design of a multiple-winding magnetic device is the allocation of the window area A_W among the various windings. It is desired to design a device having k windings with turns ratios $n_1 : n_2 : ... : n_k$. These windings must conduct rms currents $I_1, I_2, ..., I_k$ respectively. It should be noted that the windings are effectively in parallel: the winding voltages are ideally related by the turns ratios

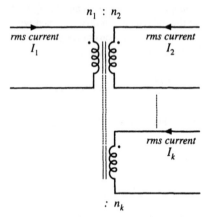

$$n_1 : n_2$$

rms current I_1

rms current I_2

rms current I_k

$: n_k$

Fig. 14.6 It is desired to optimally allocate the window area of a *k*–winding magnetic element to minimize low-frequency copper losses, with given rms winding currents and turns ratios.

$$\frac{v_1(t)}{n_1} = \frac{v_2(t)}{n_2} = \cdots = \frac{v_k(t)}{n_k} \qquad (14.22)$$

However, the winding rms currents are determined by the loads, and in general are not related to the turns ratios. The device is represented schematically in Fig. 14.6.

  The relevant geometrical parameters are summarized in Fig. 14.7(a). It is necessary to allocate a portion of the total window area W_A to each winding, as illustrated in Fig. 14.7(b). Let α_j be the fraction of the window area allocated to winding *j*, where

$$0 < \alpha_j < 1 \qquad (14.23)$$
$$\alpha_1 + \alpha_2 + \cdots + \alpha_k = 1$$

(a)

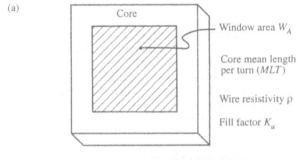

Core

Window area W_A

Core mean length per turn (*MLT*)

Wire resistivity ρ

Fill factor K_u

Fig. 14.7 Basic core topology, including window area W_A enclosed by core (a). The window is allocated to the various windings to minimize low-frequency copper loss (b).

(b)

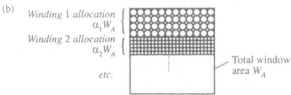

Winding 1 *allocation* $\alpha_1 W_A$

Winding 2 *allocation* $\alpha_2 W_A$

etc.

Total window area W_A

The low-frequency copper loss $P_{cu,j}$ in winding j depends on the dc resistance R_j of winding j, as follows:

$$P_{cu,j} = I_j^2 R_j \qquad (14.24)$$

The resistance of winding j is

$$R_j = \rho \frac{\ell_j}{A_{w,j}} \qquad (14.25)$$

where ρ is the wire resistivity, ℓ_j is the length of the wire used for winding j, and $A_{w,j}$ is the cross-sectional area of the wire used for winding j. These quantities can be expressed as

$$\ell_j = n_j (MLT) \qquad (14.26)$$

$$A_{w,j} = \frac{W_A K_u \alpha_j}{n_j} \qquad (14.27)$$

where (MLT) is the winding mean-length-per-turn, and K_u is the winding fill factor. Substitution of these expressions into Eq. (14.25) leads to

$$R_j = \rho \frac{n_j^2 (MLT)}{W_A K_u \alpha_j} \qquad (14.28)$$

The copper loss of winding j is therefore

$$P_{cu,j} = \frac{n_j^2 i_j^2 \rho (MLT)}{W_A K_u \alpha_j} \qquad (14.29)$$

The total copper loss of the k windings is

$$P_{cu,tot} = P_{cu,1} + P_{cu,2} + \cdots + P_{cu,k} = \frac{\rho (MLT)}{W_A K_u} \sum_{j=1}^{k} \left(\frac{n_j^2 I_j^2}{\alpha_j} \right) \qquad (14.30)$$

It is desired to choose the α_js such that the total copper loss $P_{cu,tot}$ is minimized. Let us consider what happens when we vary one of the αs, say α_1, between 0 and 1.

When $\alpha_1 = 0$, then we allocate zero area to winding 1. In consequence, the resistance of winding 1 tends to infinity. The copper loss of winding 1 also tends to infinity. On the other hand, the other windings are given maximum area, and hence their copper losses can be reduced. Nonetheless, the total copper loss tends to infinity.

When $\alpha_1 = 1$, then we allocate all of the window area to winding 1, and none to the other windings. Hence, the resistance of winding 1, as well as its low-frequency copper loss, are minimized. But the copper losses of the remaining windings tend to infinity.

As illustrated in Fig. 14.8, there must be an optimum value of α_1 that lies between these two extremes, where the total copper loss is minimized. Let us compute the optimum values of α_1, α_2, ..., α_k using the method of Lagrange multipliers. It is desired to minimize Eq. (14.30), subject to the constraint of Eq. (14.23). Hence, we define the function

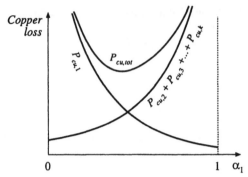

Fig. 14.8 Variation of copper losses with α_1.

$$f(\alpha_1, \alpha_2, \cdots, \alpha_k, \xi) = P_{cu,tot}(\alpha_1, \alpha_2, \cdots, \alpha_k) + \xi\, g(\alpha_1, \alpha_2, \cdots, \alpha_k) \tag{14.31}$$

where

$$g(\alpha_1, \alpha_2, \cdots, \alpha_k) = 1 - \sum_{j=1}^{k} \alpha_j \tag{14.32}$$

is the constraint that must equal zero, and ξ is the Lagrange multiplier. The optimum point is the solution of the system of equations

$$\frac{\partial f(\alpha_1, \alpha_2, \cdots, \alpha_k, \xi)}{\partial \alpha_1} = 0$$

$$\frac{\partial f(\alpha_1, \alpha_2, \cdots, \alpha_k, \xi)}{\partial \alpha_2} = 0$$

$$\vdots \tag{14.33}$$

$$\frac{\partial f(\alpha_1, \alpha_2, \cdots, \alpha_k, \xi)}{\partial \alpha_k} = 0$$

$$\frac{\partial f(\alpha_1, \alpha_2, \cdots, \alpha_k, \xi)}{\partial \xi} = 0$$

The solution is

$$\xi = \frac{\rho(MLT)}{W_A K_u} \left(\sum_{j=1}^{k} n_j I_j \right)^2 = P_{cu,tot} \tag{14.34}$$

$$\alpha_m = \frac{n_m I_m}{\sum_{j=1}^{k} n_j I_j} \tag{14.35}$$

This is the optimal choice for the αs, and the resulting minimum value of $P_{cu,tot}$.

According to Eq. (14.22), the winding voltages are proportional to the turns ratios. Hence, we can express the α_ms in the alternate form

$$\alpha_m = \frac{V_m I_m}{\sum_{j=1}^{k} V_j I_j}$$ (14.36)

by multiplying and dividing Eq. (14.35) by the quantity V_m / n_m. It is irrelevant whether rms or peak voltages are used. Equation (14.36) is the desired result. It states that the window area should be allocated to the various windings in proportion to their apparent powers. The numerator of Eq. (14.36) is the apparent power of winding m, equal to the product of the rms current and the voltage. The denominator is the sum of the apparent powers of all windings.

As an example, consider the PWM full-bridge transformer having a center-tapped secondary, as illustrated in Fig. 14.9. This can be viewed as a three-winding transformer, having a single primary-side winding of n_1 turns, and two secondary-side windings, each of n_2 turns. The winding current waveforms $i_1(t)$, $i_2(t)$, and $i_3(t)$ are illustrated in Fig. 14.10. Their rms values are

$$I_1 = \sqrt{\frac{1}{2T_s} \int_0^{2T_s} i_1^2(t)\,dt} \; = \frac{n_2}{n_1} I \sqrt{D}$$ (14.37)

$$I_2 = I_3 = \sqrt{\frac{1}{2T_s} \int_0^{2T_s} i_2^2(t)\,dt} \; = \frac{1}{2} I \sqrt{1+D}$$ (14.38)

Substitution of these expressions into Eq. (14.35) yields

$$\alpha_1 = \frac{1}{\left(1 + \sqrt{\frac{1+D}{D}}\right)}$$ (14.39)

$$\alpha_2 = \alpha_3 = \frac{1}{2} \frac{1}{\left(1 + \sqrt{\frac{D}{1+D}}\right)}$$ (14.40)

If the design is to be optimized at the operating point $D = 0.75$, then one obtains

$$\begin{aligned} \alpha_1 &= 0.396 \\ \alpha_2 &= 0.302 \\ \alpha_3 &= 0.302 \end{aligned}$$ (14.41)

So approximately 40% of the window area should be allocated to the primary winding, and 30% should

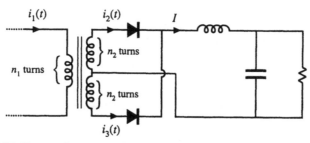

Fig. 14.9 PWM full-bridge transformer example.

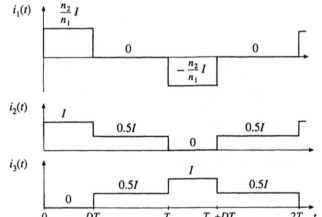

Fig. 14.10 Transformer waveforms, PWM full-bridge converter example.

be allocated to each half of the center-tapped secondary. The total copper loss at this optimal design point is found from evaluation of Eq. (14.34):

$$
\begin{aligned}
P_{cu,tot} &= \frac{\rho(MLT)}{W_A K_u} \left(\sum_{j=1}^{3} n_j I_j \right)^2 \\
&= \frac{\rho(MLT) n_2^2 I^2}{W_A K_u} \left(1 + 2D + 2\sqrt{D(1+D)} \right)
\end{aligned}
\tag{14.42}
$$

14.3.2 Coupled Inductor Design Constraints

Let us now consider how to design a k-winding coupled inductor, as discussed in Section 13.5.4 and illustrated in Fig. 14.11. It is desired that the magnetizing inductance be a specified value L_M, referred to winding 1. It is also desired that the numbers of turns for the other windings be chosen according to desired turns ratios. When the magnetizing current $i_M(t)$ reaches its maximum value $I_{M,max}$, the coupled inductor should operate with a given maximum flux density B_{max}. With rms currents $I_1, I_2, ..., I_k$ applied to the respective windings, the total copper loss should be a desired value P_{cu} given by Eq. (14.34). Hence, the design procedure involves selecting the core size and number of primary turns so that the desired magnetizing inductance, the desired flux density, and the desired total copper loss are achieved. Other quantities, such as air gap length, secondary turns, and wire sizes, can then be selected. The derivation follows the derivation for the single winding case (Section 14.1), and incorporates the window area optimization of Section 14.3.1.

The magnetizing current $i_M(t)$ can be expressed in terms of the winding currents $i_1(t), i_2(t),..., i_k(t)$ by solution of Fig. 14.11(a) (or by use of Ampere's Law), as follows:

$$
i_M(t) = i_1(t) + \frac{n_2}{n_1} i_2(t) + \cdots + \frac{n_k}{n_1} i_k(t)
\tag{14.43}
$$

By solution of the magnetic circuit model of Fig. 14.11(b), we can write

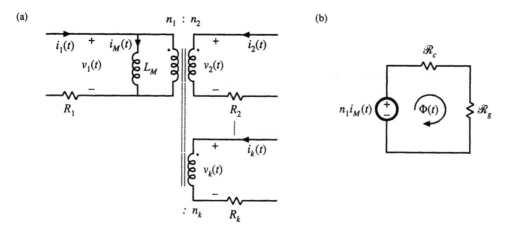

Fig. 14.11 A k-winding magnetic device, with specified turns ratios and waveforms: (a) electrical circuit model, (b) a magnetic circuit model.

$$n_1 i_M(t) = B(t) A_c \mathcal{R}_g \tag{14.44}$$

This equation is analogous to Eq. (14.4), and assumes that the reluctance $\mathcal{R}_g$ of the air gap is much larger than the reluctance $\mathcal{R}_c$ of the core. As usual, the total flux $\Phi(t)$ is equal to $B(t)A_c$. Leakage inductances are ignored.

To avoid saturation of the core, the instantaneous flux density $B(t)$ must be less than the saturation flux density of the core material, B_{sat}. Let us define $I_{M,max}$ as the maximum value of the magnetizing current $i_M(t)$. According to Eq. (14.44), this will lead to a maximum flux density B_{max} given by

$$n_1 I_{M,max} = B_{max} A_c \mathcal{R}_g = B_{max} \frac{\ell_g}{\mu_0} \tag{14.45}$$

For a value of $I_{M,max}$ given by the circuit application, we should use Eq. (14.45) to choose the turns n_1 and gap length ℓ_g such that the maximum flux density B_{max} is less than the saturation density B_{sat}. Equation (14.45) is similar to Eq. (14.6), but accounts for the magnetizations produced by multiple winding currents.

The magnetizing inductance L_M, referred to winding 1, is equal to

$$L_M = \frac{n_1^2}{\mathcal{R}_g} = n_1^2 \frac{\mu_0 A_c}{\ell_g} \tag{14.46}$$

This equation is analogous to Eq. (14.7).

As shown in Section 14.3.1, the total copper loss is minimized when the core window area W_A is allocated to the various windings according to Eq. (14.35) or (14.36). The total copper loss is then given by Eq. (14.34). Equation (14.34) can be expressed in the form

$$P_{cu} = \frac{\rho (MLT) n_1^2 I_{tot}^2}{W_A K_u} \tag{14.47}$$

where

$$I_{tot} = \sum_{j=1}^{k} \frac{n_j}{n_1} I_j \qquad (14.48)$$

is the sum of the rms winding currents, referred to winding 1.

We can now eliminate the unknown quantities ℓ_g and n_1 from Eqs. (14.45), (14.46), and (14.47). Equation (14.47) then becomes

$$P_{cu} = \frac{\rho(MLT)L_M^2 I_{tot}^2 I_{M,max}^2}{B_{max}^2 A_c^2 W_A K_u} \qquad (14.49)$$

We can now rearrange this equation, by grouping terms that involve the core geometry on the left-hand side, and specifications on the right-hand side:

$$\frac{A_c^2 W_A}{(MLT)} = \frac{\rho L_M^2 I_{tot}^2 I_{M,max}^2}{B_{max}^2 K_u P_{cu}} \qquad (14.50)$$

The left-hand side of the equation can be recognized as the same K_g term defined in Eq. (14.15). Therefore, to design a coupled inductor that meets the requirements of operating with a given maximum flux density B_{max}, given primary magnetizing inductance L_M, and with a given total copper loss P_{cu}, we must select a core that satisfies

$$K_g \geq \frac{\rho L_M^2 I_{tot}^2 I_{M,max}^2}{B_{max}^2 K_u P_{cu}} \qquad (14.51)$$

Once such a core is found, then the winding 1 turns and gap length can be selected to satisfy Eqs. (14.45) and (14.46). The turns of windings 2 through k are selected according to the desired turns ratios. The window area is allocated among the windings according to Eq. (14.35), and the wire gauges are chosen using Eq. (14.27).

The procedure above is applicable to design of coupled inductors. The results are applicable to design of flyback and SEPIC transformers as well, although it should be noted that the procedure does not account for the effects of core or proximity loss. It also can be extended to design of other devices, such as conventional transformers—doing so is left as a homework problem.

14.3.3 Design Procedure

The following quantities are specified, using the units noted:

Wire effective resistivity	ρ	(Ω–cm)
Total rms winding currents, referred to winding 1	$I_{tot} = \sum_{j=1}^{k} \frac{n_j}{n_i} I_j$	(A)
Peak magnetizing current, referred to winding 1	$I_{M,max}$	(A)
Desired turns ratios	$n_2/n_1, n_3/n_1,$ etc.	

Magnetizing inductance, referred to winding 1	L_M	(H)
Allowed total copper loss	P_{cu}	(W)
Winding fill factor	K_u	
Maximum operating flux density	B_{max}	(T)

The core dimensions are expressed in cm:

Core cross-sectional area	A_c	(cm^2)
Core window area	W_A	(cm^2)
Mean length per turn	MLT	(cm)

The use of centimeters rather than meters requires that appropriate factors be added to the design equations.

1. *Determine core size*

$$K_g \geq \frac{\rho L_M^2 I_{tot}^2 I_{M,max}^2}{B_{max}^2 P_{cu} K_u} 10^8 \qquad \text{(cm}^5) \tag{14.52}$$

Choose a core which is large enough to satisfy this inequality. Note the values of A_c, W_A, and MLT for this core. The resistivity ρ of copper wire is $1.724 \cdot 10^{-6}$ Ω–cm at room temperature, and $2.3 \cdot 10^{-6}$ Ω–cm at 100°C.

2. *Determine air gap length*

$$\ell_g = \frac{\mu_0 L_M I_{M,max}^2}{B_{max}^2 A_c} 10^4 \qquad \text{(m)} \tag{14.53}$$

Here, B_{max} is expressed in Tesla, A_c is expressed in cm^2, and ℓ_g is expressed in meters. The permeability of free space is $\mu_0 = 4\pi \cdot 10^{-7}$ H/m. This value is approximate, and neglects fringing flux and other nonidealities.

3. *Determine number of winding 1 turns*

$$n_1 = \frac{L_M I_{M,max}}{B_{max} A_c} 10^4 \tag{14.54}$$

Here, B_{max} is expressed in Tesla and A_c is expressed in cm^2.

4. *Determine number of secondary turns*

Use the desired turns ratios:

$$n_2 = \left(\frac{n_2}{n_1}\right) n_1$$
$$n_3 = \left(\frac{n_3}{n_1}\right) n_1 \tag{14.55}$$
$$\vdots$$

5. *Evaluate fraction of window area allocated to each winding*

$$
\begin{aligned}
\alpha_1 &= \frac{n_1 I_1}{n_1 I_{tot}} \\
\alpha_2 &= \frac{n_2 I_2}{n_1 I_{tot}} \\
&\;\vdots \\
\alpha_k &= \frac{n_k I_k}{n_1 I_{tot}}
\end{aligned}
\tag{14.56}
$$

6. *Evaluate wire sizes*

$$
\begin{aligned}
A_{w1} &\le \frac{\alpha_1 K_u W_A}{n_1} \\
A_{w2} &\le \frac{\alpha_2 K_u W_A}{n_2} \\
&\;\vdots
\end{aligned}
\tag{14.57}
$$

Select wire with bare copper area less than or equal to these values. An American Wire Gauge table is included in Appendix D.

14.4   EXAMPLES

14.4.1   Coupled Inductor for a Two-Output Forward Converter

As a first example, let us consider the design of coupled inductors for the two-output forward converter illustrated in Fig. 14.12. This element can be viewed as two filter inductors that are wound on the same core. The turns ratio is chosen to be the same as the ratio of the output voltages. The magnetizing inductance performs the function of filtering the switching harmonics for both outputs, and the magnetizing current is equal to the sum of the reflected winding currents.

At the nominal full-load operating point, the converter operates in the continuous conduction mode with a duty cycle of $D = 0.35$. The switching frequency is 200 kHz. At this operating point, it is desired that the ripple in the magnetizing current have a peak magnitude equal to 20% of the dc component of magnetizing current.

The dc component of the magnetizing current I_M is

$$
\begin{aligned}
I_M &= I_1 + \frac{n_2}{n_1} I_2 \\
&= (4 \text{ A}) + \frac{12}{28}(2 \text{ A}) \\
&= 4.86 \text{ A}
\end{aligned}
\tag{14.58}
$$

The magnetizing current ripple Δi_M can be expressed as

(a)

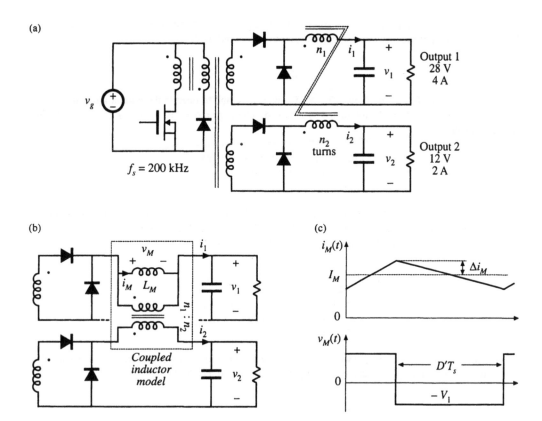

(b)

(c)

Fig. 14.12 Two-output forward converter example: (a) circuit schematic, (b) coupled inductor model inserted into converter secondary-side circuit, (c) magnetizing current and voltage waveforms of coupled inductor, referred to winding 1.

$$\Delta i_M = \frac{V_1 D' T_s}{2 L_M} \tag{14.59}$$

Since we want Δi_M to be equal to 20% of I_M, we should choose L_M as follows:

$$
\begin{aligned}
L_M &= \frac{V_1 D' T_s}{2 \Delta i_M} \\
&= \frac{(28\ \text{V})(1 - 0.35)(5\ \mu\text{s})}{2(4.86\ \text{A})(20\%)} \\
&= 47\ \mu\text{H}
\end{aligned}
\tag{14.60}
$$

The peak magnetizing current, referred to winding 1, is therefore

$$I_{M,max} = I_M + \Delta i_M = 5.83\ \text{A} \tag{14.61}$$

Since the current ripples of the winding currents are small compared to the respective dc components, the rms values of the winding currents are approximately equal to the dc components: $I_1 = 4$ A, $I_2 = 2$ A. Therefore, the sum of the rms winding currents, referred to winding 1, is

$$I_{tot} = I_1 + \frac{n_2}{n_1} I_2 = 4.86 \text{ A} \tag{14.62}$$

For this design, it is decided to allow 0.75 W of copper loss, and to operate the core at a maximum flux density of 0.25 Tesla. A fill factor of 0.4 is assumed. The required K_g is found by evaluation of Eq. (14.52), as follows:

$$K_g \geq \frac{(1.724 \cdot 10^{-6} \, \Omega - \text{cm})(47 \, \mu\text{H})^2 (4.86 \text{ A})^2 (5.83 \text{ A})^2}{(0.25 \text{ T})^2 (0.75 \text{ W})(0.4)} 10^8 \tag{14.63}$$

$$= 16 \cdot 10^{-3} \text{ cm}^5$$

A ferrite PQ 20/16 core is selected, which has a K_g of $22.4 \cdot 10^{-3}$ cm^5. From Appendix D, the geometrical parameters for this core are: $A_c = 0.62$ cm^2, $W_A = 0.256$ cm^2, and $MLT = 4.4$ cm.

The air gap is found by evaluation of Eq. (14.53) as follows:

$$\ell_g = \frac{\mu_0 L_M I_{M,max}^2}{B_{max}^2 A_c} 10^4$$

$$= \frac{(4\pi \cdot 10^{-7} \text{H/m})(47 \, \mu\text{H})(5.83 \text{ A})^2}{(0.25 \text{ T})^2 (0.62 \text{ cm}^2)} 10^4 \tag{14.64}$$

$$= 0.52 \text{ mm}$$

In practice, a slightly longer air gap would be necessary, to allow for the effects of fringing flux and other nonidealities. The winding 1 turns are found by evaluation of Eq. (14.54):

$$n_1 = \frac{L_M I_{M,max}}{B_{max} A_c} 10^4$$

$$= \frac{(47 \, \mu\text{H})(5.83 \text{ A})}{(0.25 \text{ T})(0.62 \text{ cm}^2)} 10^4 \tag{14.65}$$

$$= 17.6 \text{ turns}$$

The winding 2 turns are chosen according to the desired turns ratio:

$$n_2 = \left(\frac{n_2}{n_1}\right) n_1$$

$$= \left(\frac{12}{28}\right)(17.6) \tag{14.66}$$

$$= 7.54 \text{ turns}$$

The numbers of turns are rounded off to $n_1 = 17$ turns, $n_2 = 7$ turns (18:8 would be another possible choice). The window area W_A is allocated to the windings according to the fractions from Eq. (14.56):

$$\alpha_1 = \frac{n_1 I_1}{n_1 I_{tot}} = \frac{(17)(4 \text{ A})}{(17)(4.86 \text{ A})} = 0.8235$$

$$\alpha_2 = \frac{n_2 I_2}{n_1 I_{tot}} = \frac{(7)(2 \text{ A})}{(17)(4.86 \text{ A})} = 0.1695 \tag{14.67}$$

The wire sizes can therefore be chosen as follows:

$$A_{w1} \le \frac{\alpha_1 K_u W_A}{n_1} = \frac{(0.8235)(0.4)(0.256\,\text{cm}^2)}{(17)} = 4.96 \cdot 10^{-3}\,\text{cm}^2$$

use AWG #21

(14.68)

$$A_{w2} \le \frac{\alpha_2 K_u W_A}{n_2} = \frac{(0.1695)(0.4)(0.256\,\text{cm}^2)}{(7)} = 2.48 \cdot 10^{-3}\,\text{cm}^2$$

use AWG #24

14.4.2 CCM Flyback Transformer

As a second example, let us design the flyback transformer for the converter illustrated in Fig. 14.13. This converter operates with an input voltage of 200 V, and produces an full-load output of 20 V at 5A. The switching frequency is 150 kHz. Under these operating conditions, it is desired that the converter operate in the continuous conduction mode, with a magnetizing current ripple equal to 20% of the dc component of magnetizing current. The duty cycle is chosen to be $D = 0.4$, and the turns ratio is $n_2/n_1 = 0.15$. A copper loss of 1.5 W is allowed, not including proximity effect losses. To allow room for isolation between the primary and secondary windings, a fill factor of $K_u = 0.3$ is assumed. A maximum flux density of $B_{max} = 0.25$ T is used; this value is less than the worst-case saturation flux density B_{sat} of the ferrite core material.

By solution of the converter using capacitor charge balance, the dc component of the magnetizing current can be found to be

$$I_M = \left(\frac{n_2}{n_1}\right) \frac{1}{D'} \frac{V}{R} = 1.25\,\text{A}$$

(14.69)

Hence, the magnetizing current ripple should be

$$\Delta i_M = (20\%) I_M = 0.25\,\text{A}$$

(14.70)

and the maximum value of the magnetizing current is

$$I_{M,max} = I_M + \Delta i_M = 1.5\,\text{A}$$

(14.71)

To obtain this ripple, the magnetizing inductance should be

$$L_M = \frac{V_g D T_s}{2 \Delta i_M}$$
$$= 1.07\,\text{mH}$$

(14.72)

The rms value of the primary winding current is found using Eq. (A.6) of Appendix A, as follows:

$$I_1 = I_M \sqrt{D} \sqrt{1 + \frac{1}{3}\left(\frac{\Delta i_M}{I_M}\right)^2} = 0.796\,\text{A}$$

(14.73)

(a)

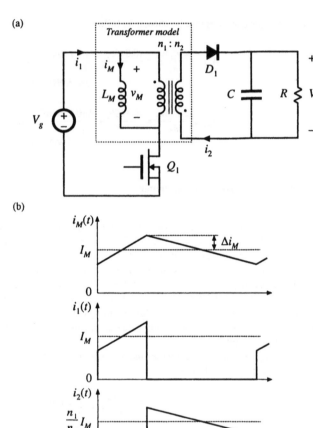

Fig.   14.13 Flyback transformer design example: (a) converter schematic, (b) typical waveforms.

(b)

The rms value of the secondary winding current is found in a similar manner:

$$I_2 = \frac{n_1}{n_2} I_M \sqrt{D'} \sqrt{1 + \frac{1}{3} \left(\frac{\Delta i_M}{I_M} \right)^2} = 6.50 \text{ A} \tag{14.74}$$

Note that I_2 is not simply equal to the turns ratio multiplied by I_1. The total rms winding current is equal to:

$$I_{tot} = I_1 + \frac{n_2}{n_1} I_2 = 1.77 \text{ A} \tag{14.75}$$

We can now determine the necessary core size. Evaluation of Eq. (14.52) yields

$$
\begin{aligned}
K_g &\geq \frac{\rho L_M^2 I_{tot}^2 I_{M,max}^2}{B_{max}^2 P_{cu} K_u} 10^8 \\
&= \frac{\left(1.724 \cdot 10^{-6} \Omega\text{-cm}\right)\left(1.07 \cdot 10^{-3}\,\text{H}\right)^2 \left(1.77\,\text{A}\right)^2 \left(1.5\,\text{A}\right)^2}{\left(0.25\,\text{T}\right)^2 \left(1.5\,\text{W}\right)\left(0.3\right)} 10^8 \\
&= 0.049\ \text{cm}^5
\end{aligned}
\tag{14.76}
$$

The smallest EE core listed in Appendix D that satisfies this inequality is the EE30, which has $K_g = 0.0857$ cm^5. The dimensions of this core are

$$
\begin{array}{ll}
A_c & 1.09\ \text{cm}^2 \\
W_A & 0.476\ \text{cm}^2 \\
MLT & 6.6\ \text{cm} \\
\ell_m & 5.77\ \text{cm}
\end{array}
\tag{14.77}
$$

The air gap length ℓ_g is chosen according to Eq. (14.53):

$$
\begin{aligned}
\ell_g &= \frac{\mu_0 L_M I_{M,max}^2}{B_{max}^2 A_c} 10^4 \\
&= \frac{\left(4\pi \cdot 10^{-7}\,\text{H/m}\right)\left(1.07 \cdot 10^{-3}\,\text{H}\right)\left(1.5\,\text{A}\right)^2}{\left(0.25\,\text{T}\right)^2 \left(1.09\ \text{cm}^2\right)} 10^4 \\
&= 0.44\ \text{mm}
\end{aligned}
\tag{14.78}
$$

The number of winding 1 turns is chosen according to Eq. (14.54), as follows:

$$
\begin{aligned}
n_1 &= \frac{L_M I_{M,max}}{B_{max} A_c} 10^4 \\
&= \frac{\left(1.07 \cdot 10^{-3}\,\text{H}\right)\left(1.5\,\text{A}\right)}{\left(0.25\,\text{T}\right)\left(1.09\ \text{cm}^2\right)} 10^4 \\
&= 58.7\ \text{turns}
\end{aligned}
\tag{14.79}
$$

Since an integral number of turns is required, we round off this value to

$$
n_1 = 59
\tag{14.80}
$$

To obtain the desired turns ratio, n_2 should be chosen as follows:

$$
\begin{aligned}
n_2 &= \left(\frac{n_2}{n_1}\right) n_1 \\
&= \left(0.15\right) 59 \\
&= 8.81
\end{aligned}
\tag{14.81}
$$

We again round this value off, to

$$n_2 = 9 \tag{14.82}$$

The fractions of the window area allocated to windings 1 and 2 are selected in accordance with Eq. (14.56):

$$\alpha_1 = \frac{I_1}{I_{tot}} = \frac{(0.796 \text{ A})}{(1.77 \text{ A})} = 0.45$$
$$\alpha_2 = \frac{n_2 I_2}{n_1 I_{tot}} = \frac{(9)(6.5 \text{ A})}{(59)(1.77 \text{ A})} = 0.55 \tag{14.83}$$

The wire gauges should therefore be

$$A_{W1} \leq \frac{\alpha_1 K_u W_A}{n_1} = 1.09 \cdot 10^{-3} \text{ cm}^2 \quad \text{— use #28 AWG}$$
$$A_{W2} \leq \frac{\alpha_2 K_u W_A}{n_2} = 8.88 \cdot 10^{-3} \text{ cm}^2 \quad \text{— use #19 AWG} \tag{14.84}$$

The above American Wire Gauges are selected using the wire gauge table given at the end of Appendix D.

The above design does not account for core loss or copper loss caused by the proximity effect. Let us compute the core loss for this design. Figure Fig. 14.14 contains a sketch of the *B–H* loop for this design. The flux density $B(t)$ can be expressed as a dc component (determined by the dc value of the magnetizing current I_M), plus an ac variation of peak amplitude ΔB that is determined by the current ripple Δi_M. The maximum value of $B(t)$ is labeled B_{max}; this value is determined by the sum of the dc component and the ac ripple component. The core material saturates when the applied $B(t)$ exceeds B_{sat}; hence, to avoid saturation, B_{max} should be less than B_{sat}. The core loss is determined by the amplitude of the ac variations in $B(t)$, i.e., by ΔB.

The ac component ΔB is determined using Faraday's law, as follows. Solution of Faraday's law for the derivative of $B(t)$ leads to

$$\frac{dB(t)}{dt} = \frac{v_M(t)}{n_1 A_c} \tag{14.85}$$

As illustrated in Fig. 14.15, the voltage applied during the first subinterval is $v_M(t) = V_g$. This causes the

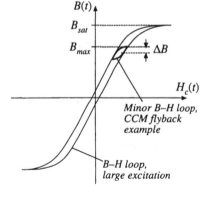

Fig. 14.14 *B–H* loop for the flyback transformer design example.

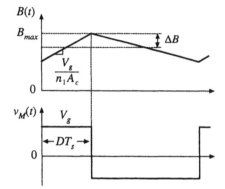

Fig. 14.15 Variation of flux density $B(t)$, flyback transformer example.

flux density to increase with slope

$$\frac{dB(t)}{dt} = \frac{V_g}{n_1 A_c} \tag{14.86}$$

Over the first subinterval $0 < t < DT_s$, the flux density $B(t)$ changes by the net amount $2\Delta B$. This net change is equal to the slope given by Eq. (14.86), multiplied by the interval length DT_s:

$$\Delta B = \left(\frac{V_g}{2n_1 A_c}\right)(DT_s) \tag{14.87}$$

Upon solving for ΔB and expressing A_c in cm^2, we obtain

$$\Delta B = \frac{V_g DT_s}{2n_1 A_c} 10^4 \tag{14.88}$$

For the flyback transformer example, the peak ac flux density is found to be

$$\Delta B = \frac{(200 \text{ V})(0.4)(6.67 \text{ μs})}{2(59)(1.09 \text{ cm}^2)} 10^4 \tag{14.89}$$

$$= 0.041 \text{ T}$$

To determine the core loss, we next examine the data provided by the manufacturer for the given core material. A typical plot of core loss is illustrated in Fig. 14.16. For the values of ΔB and switching frequency of the flyback transformer design, this plot indicates that 0.078 W will be lost in every cm^3 of the core material. Of course, this value neglects the effects of harmonics on core loss. The total core loss P_{fe} will therefore be equal to this loss density, multiplied by the volume of the core:

$$P_{fe} = (0.078 \text{ W/cm}^3)(A_c \ell_m)$$
$$= (0.078 \text{ W/cm}^3)(1.09 \text{ cm}^2)(5.77 \text{ cm}) \tag{14.90}$$
$$= 0.49 \text{ W}$$

This core loss is somewhat less than the copper loss of 1.5 W, and neglecting the core loss is often warranted in designs that operate in the continuous conduction mode and that employ ferrite core materials.

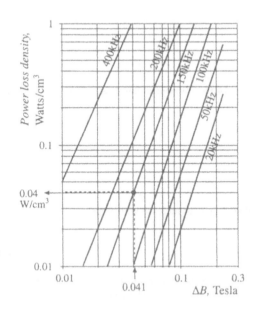

Fig. 14.16 Determination of core loss density for the flyback transformer design example.

14.5   SUMMARY OF KEY POINTS

1. A variety of magnetic devices are commonly used in switching converters. These devices differ in their core flux density variations, as well as in the magnitudes of the ac winding currents. When the flux density variations are small, core loss can be neglected. Alternatively, a low-frequency material can be used, having higher saturation flux density.

2. The core geometrical constant K_g is a measure of the magnetic size of a core, for applications in which copper loss is dominant. In the K_g design method, flux density and total copper loss are specified. Design procedures for single-winding filter inductors and for conventional multiple-winding transformers are derived.

REFERENCES

[1]   C. W. T. MCLYMAN, *Transformer and Inductor Design Handbook*, Second edition, New York: Marcel Dekker, 1988.

[2]   S. ĆUK, "Basics of Switched-Mode Power Conversion: Topologies, Magnetics, and Control," in *Advances in Switched-Mode Power Conversion*, Vol. 2, Irvine: Teslaco, pp. 292–305, 1983.

[3]   T. G. WILSON JR., T. G. WILSON, and H. A. OWEN, "Coupling of Magnetic Design Choices to DC–to–DC Converter Electrical Performance," *IEEE Applied Power Electronics Conference*, 1994 Record, pp. 340-347.

[4]   S. ĆUK and R. D. MIDDLEBROOK, "Coupled-Inductor and Other Extensions of a New Optimum Topology Switching DC–to–DC Converter," *IEEE Industry Applications Society Annual Meeting*, 1977 Proceedings, pp. 1110-1122.

[5]   S. ĆUK and Z. ZHANG, "Coupled-Inductor Analysis and Design," *IEEE Power Electronics Specialists Conference*, 1986 Record, pp. 655-665.

[6]   E. HNATEK, *Design of Solid-State Power Supplies*, Second edition, New York: Van Nostrand Reinhold, 1981, Chapter 4.

PROBLEMS

14.1   A simple buck converter operates with a 50 kHz switching frequency and a dc input voltage of $V_g = 40$ V. The output voltage is $V = 20$ V. The load resistance is $R \geq 4$ Ω.

    (a)   Determine the value of the output filter inductance L such that the peak-to-average inductor current ripple Δi is 10% of the dc component I.

    (b)   Determine the peak steady-state inductor current I_{max}.

    (c)   Design an inductor which has the values of L and I_{max} from parts (a) and (b). Use a ferrite EE core, with $B_{max} = 0.25$ T. Choose a value of winding resistance such that the inductor copper loss is less than or equal to 1 W at room temperature. Assume $K_u = 0.5$. Specify: core size, gap length, wire size (AWG), and number of turns.

14.2   A boost converter operates at the following quiescent point: $V_g = 28$ V, $V = 48$ V, $P_{load} = 150$ W, $f_s = 100$ kHz. Design the inductor for this converter. Choose the inductance value such that the peak current ripple is 10% of the dc inductor current. Use a peak flux density of 0.225 T, and assume a fill factor of 0.5. Allow copper loss equal to 0.5% of the load power, at room temperature. Use a ferrite PQ core. Specify: core size, air gap length, wire gauge, and number of turns.

14.3   Extension of the K_g approach to design of two-winding transformers. It is desired to design a transformer having a turns ratio of 1:n. The transformer stores negligible energy, no air gap is required, and the ratio of the winding currents $i_2(t)/i_1(t)$ is essentially equal to the turns ratio n. The applied primary volt-seconds λ_1 are defined for a typical PWM voltage waveform $v_1(t)$ in Fig. 13.45(b); these volt-seconds should cause the maximum flux density to be equal to a specified value $B_{max} = \Delta B$. You may assume that the flux density $B(t)$ contains no dc bias, as in Fig. 13.46. You should allocate half of the core window area to each winding. The total copper loss P_{cu} is also specified. You may neglect proximity losses.

    (a)   Derive a transformer design procedure, in which the following quantities are specified: total copper loss P_{cu}, maximum flux density B_{max}, fill factor K_u, wire resistivity ρ, rms primary current I_1, applied primary volt-seconds λ_1, and turns ratio 1:n. Your procedure should yield the following data: required core geometrical constant K_g, primary and secondary turns n_1 and n_2, and primary and secondary wire areas A_{w1} and A_{w2}.

    (b)   The voltage waveform applied to the transformer primary winding of the Ćuk converter [Fig. 6.41(c)] is equal to the converter input voltage V_g while the transistor conducts, and is equal to $-V_g D/(1 - D)$ while the diode conducts. This converter operates with a switching frequency of 100 kHz, and a transistor duty cycle D equal to 0.4. The dc input voltage is $V_g = 120$ V, the dc output voltage is $V = 24$ V, and the load power is 200 W. You may assume a fill factor of $K_u = 0.3$. Use your procedure of part (a) to design a transformer for this application, in which $B_{max} = 0.15$ T, and $P_{cu} = 0.25$ W at 100°C. Use a ferrite PQ core. Specify: core size, primary and secondary turns, and wire gauges.

14.4   Coupled inductor design. The two-output forward converter of Fig. 13.47(a) employs secondary-side coupled inductors. An air gap is employed.

    Design a coupled inductor for the following application: $V_1 = 5$ V, $V_2 = 15$ V, $I_1 = 20$ A, $I_2 = 4$ A, $D = 0.4$. The magnetizing inductance should be equal to 8 μH, referred to the 5 V winding. You may assume a fill factor K_u of 0.5. Allow a total of 1 W of copper loss at 100°C, and use a peak flux density of

$B_{max} = 0.2$ T. Use a ferrite EE core. Specify: core size, air gap length, number of turns and wire gauge for each winding.

14.5   Flyback transformer design. A flyback converter operates with a 160 Vdc input, and produces a 28 Vdc output. The maximum load current is 2 A. The transfomer turns ratio is 8:1. The switching frequency is 100 kHz. The converter should be designed to operate in the discontinuous conduction mode at all load currents. The total copper loss should be less than 0.75 W.

(a)   Choose the value of transformer magnetizing inductance L_M such that, at maximum load current, $D_3 = 0.1$ (the duty cycle of subinterval 3, in which all semiconductors are off). Please indicate whether your value of L_M is referred to the primary or secondary winding. What is the peak transistor current? The peak diode current?

(b)   Design a flyback transformer for this application. Use a ferrite pot core with $B_{max} = 0.25$ Tesla, and with fill factor $K_u = 0.4$. Specify: core size, primary and secondary turns and wire sizes, and air gap length.

(c)   For your design of part (b), compute the copper losses in the primary and secondary windings. You may neglect proximity loss.

(d)   For your design of part (b), compute the core loss. Loss data for the core material is given by Fig. 13.20. Is the core loss less than the copper loss computed in Part (c)?

15

Transformer Design

In the design methods of the previous chapter, copper loss P_{cu} and maximum flux density B_{max} are specified, while core loss P_{fe} is not specifically addressed. This approach is appropriate for a number of applications, such as the filter inductor in which the dominant design constraints are copper loss and saturation flux density. However, in a substantial class of applications, the operating flux density is limited by core loss rather than saturation. For example, in a conventional high-frequency transformer, it is usually necessary to limit the core loss by operating at a reduced value of the peak ac flux density ΔB.

This chapter covers the general transformer design problem. It is desired to design a k–winding transformer as illustrated in Fig. 15.1. Both copper loss P_{cu} and core loss P_{fe} are modeled. As the operating flux density is increased (by decreasing the number of turns), the copper loss is decreased but the core loss is increased. We will determine the operating flux density that minimizes the total power loss $P_{tot} = P_{fe} + P_{cu}$.

It is possible to generalize the core geometrical constant K_g design method, derived in the previous chapter, to treat the design of magnetic devices when both copper loss and core loss are significant. This leads to the geometrical constant K_{gfe}, a measure of the effective magnetic size of core in a transformer design application. Several examples of transformer designs via the K_{gfe} method are given in this chapter. A similar procedure is also derived, for design of single-winding inductors in which core loss is significant.

15.1 TRANSFORMER DESIGN: BASIC CONSTRAINTS

As in the case of the filter inductor design, we can write several basic constraining equations. These equations can then be combined into a single equation for selection of the core size. In the case of transformer design, the basic constraints describe the core loss, flux density, copper loss, and total power loss

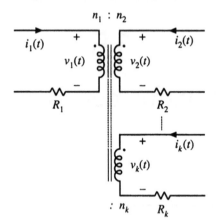

Fig. 15.1 A k-winding transformer, in which both core loss and copper loss are significant.

vs. flux density. The flux density is then chosen to optimize the total power loss.

15.1.1 Core Loss

As described in Chapter 13, the total core loss P_{fe} depends on the peak ac flux density ΔB, the operating frequency f, and the volume of the core. At a given frequency, we can approximate the core loss by a function of the form

$$P_{fe} = K_{fe}(\Delta B)^\beta A_c \ell_m \tag{15.1}$$

Again, A_c is the core cross-sectional area, ℓ_m is the core mean magnetic path length, and hence $A_c \ell_m$ is the volume of the core. K_{fe} is a constant of proportionality which depends on the operating frequency. The exponent β is determined from the core manufacturer's published data. Typically, the value of β for ferrite power materials is approximately 2.6; for other core materials, this exponent lies in the range 2 to 3. Equation (15.1) generally assumes that the applied waveforms are sinusoidal; effects of waveform harmonic content are ignored here.

15.1.2 Flux Density

An arbitrary periodic primary voltage waveform $v_1(t)$ is illustrated in Fig. 15.2. The volt-seconds applied during the positive portion of the waveform is denoted λ_1:

$$\lambda_1 = \int_{t_1}^{t_2} v_1(t)dt \tag{15.2}$$

These volt-seconds, or flux-linkages, cause the flux density to change from its negative peak to its positive peak value. Hence, from Faraday's law, the peak value of the ac component of the flux density is

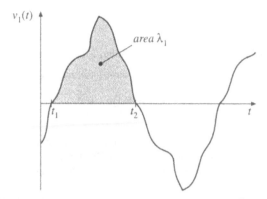

Fig. 15.2   An arbitrary transformer primary voltage waveforms, illustrating the volt-seconds applied during the positive portion of the cycle.

$$\Delta B = \frac{\lambda_1}{2n_1 A_c} \tag{15.3}$$

Note that, for a given applied voltage waveform and λ_1, we can reduce ΔB by increasing the primary turns n_1. This has the effect of decreasing the core loss according to Eq. (15.1). However, it also causes the copper loss to increase, since the new windings will be comprised of more turns of smaller wire. As a result, there is an optimal choice for ΔB, in which the total loss is minimized. In the next sections, we will determine the optimal ΔB. Having done so, we can then use Eq. (15.3) to determine the primary turns n_1, as follows:

$$n_1 = \frac{\lambda_1}{2\Delta B A_c} \tag{15.4}$$

It should also be noted that, in some converter topologies such as the forward converter with conventional reset winding, the flux density $B(t)$ and the magnetizing current $i_M(t)$ are not allowed to be negative. In consequence, the instantaneous flux density $B(t)$ contains a dc bias. Provided that the core does not approach saturation, this dc bias does not significantly affect the core loss: core loss is determined by the ac component of $B(t)$. Equations (15.2) to (15.4) continue to apply to this case, since ΔB is the peak value of the ac component of $B(t)$.

15.1.3   Copper Loss

As shown in Section 14.3.1, the total copper loss is minimized when the core window area W_A is allocated to the various windings according to their relative apparent powers. The total copper loss is then given by Eq. (14.34). This equation can be expressed in the form

$$P_{cu} = \frac{\rho(MLT)n_1^2 I_{tot}^2}{W_A K_u} \tag{15.5}$$

where

$$I_{tot} = \sum_{j=1}^{k} \frac{n_j}{n_1} I_j \tag{15.6}$$

is the sum of the rms winding currents, referred to winding 1. Use of Eq. (15.4) to eliminate n_1 from Eq. (15.5) leads to

$$P_{cu} = \left(\frac{\rho \lambda_1^2 I_{tot}^2}{4 K_u} \right) \left(\frac{(MLT)}{W_A A_c^2} \right) \left(\frac{1}{\Delta B} \right)^2 \tag{15.7}$$

The right-hand side of Eq. (15.7) is grouped into three terms. The first group contains specifications, while the second group is a function of the core geometry. The last term is a function of ΔB, to be chosen to optimize the design. It can be seen that copper loss varies as the inverse square of ΔB; increasing ΔB reduces P_{cu}.

 The increased copper loss due to the proximity effect is not explicitly accounted for in this design procedure. In practice, the proximity loss must be estimated after the core and winding geometries are known. However, the increased ac resistance due to proximity loss can be accounted for in the design procedure. The effective value of the wire resistivity ρ is increased by a factor equal to the estimated ratio R_{ac}/R_{dc}. When the core geometry is known, the engineer can attempt to implement the windings such that the estimated R_{ac}/R_{dc} is obtained. Several design iterations may be needed.

15.1.4   Total power loss vs. ΔB

The total power loss P_{tot} is found by adding Eqs. (15.1) and (15.7):

$$P_{tot} = P_{fe} + P_{cu} \tag{15.8}$$

The dependence of P_{fe}, P_{cu}, and P_{tot} on ΔB is sketched in Fig. 15.3.

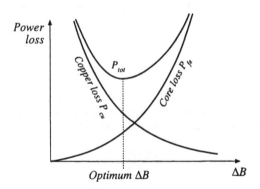

Fig. 15.3 Dependence of copper loss, core loss, and total loss on peak ac flux density.

15.1.5 Optimum Flux Density

Let us now choose the value of ΔB that minimizes Eq. (15.8). At the optimum ΔB, we can write

$$\frac{dP_{tot}}{d(\Delta B)} = \frac{dP_{fe}}{d(\Delta B)} + \frac{dP_{cu}}{d(\Delta B)} = 0 \tag{15.9}$$

Note that the optimum does not necessarily occur where $P_{fe} = P_{cu}$. Rather, it occurs where

$$\frac{dP_{fe}}{d(\Delta B)} = -\frac{dP_{cu}}{d(\Delta B)} \tag{15.10}$$

The derivatives of the core and copper losses with respect to ΔB are given by

$$\frac{dP_{fe}}{d(\Delta B)} = \beta K_{fe}(\Delta B)^{(\beta-1)} A_c \ell_m \tag{15.11}$$

$$\frac{dP_{cu}}{d(\Delta B)} = -2\left(\frac{\rho \lambda_1^2 I_{tot}^2}{4K_u}\right)\left(\frac{(MLT)}{W_A A_c^2}\right)(\Delta B)^{-3} \tag{15.12}$$

Substitution of Eqs. (15.11) and (15.12) into Eq. (15.10), and solution for ΔB, leads to the optimum flux density

$$\Delta B = \left[\frac{\rho \lambda_1^2 I_{tot}^2}{2K_u} \frac{(MLT)}{W_A A_c^3 \ell_m} \frac{1}{\beta K_{fe}}\right]^{\left(\frac{1}{\beta+2}\right)} \tag{15.13}$$

The resulting total power loss is found by substitution of Eq. (15.13) into (15.1), (15.8), and (15.9). Simplification of the resulting expression leads to

$$P_{tot} = \left[A_c \ell_m K_{fe}\right]^{\left(\frac{2}{\beta+2}\right)} \left[\frac{\rho \lambda_1^2 I_{tot}^2}{4K_u} \frac{(MLT)}{W_A A_c^2}\right]^{\left(\frac{\beta}{\beta+2}\right)} \left[\left(\frac{\beta}{2}\right)^{-\left(\frac{\beta}{\beta+2}\right)} + \left(\frac{\beta}{2}\right)^{\left(\frac{2}{\beta+2}\right)}\right] \tag{15.14}$$

This expression can be regrouped, as follows:

$$\frac{W_A (A_c)^{\left(2(\beta-1)/\beta\right)}}{(MLT)\ell_m^{(2/\beta)}}\left[\left(\frac{\beta}{2}\right)^{-\left(\frac{\beta}{\beta+2}\right)} + \left(\frac{\beta}{2}\right)^{\left(\frac{2}{\beta+2}\right)}\right]^{-\left(\frac{\beta+2}{\beta}\right)} = \frac{\rho \lambda_1^2 I_{tot}^2 K_{fe}^{(2/\beta)}}{4K_u (P_{tot})^{\left((\beta+2)/\beta\right)}} \tag{15.15}$$

The terms on the left side of Eq. (15.15) depend on the core geometry, while the terms on the right side depend on specifications regarding the application (ρ, I_{tot}, λ_1, K_u, P_{tot}) and the desired core material (K_{fe}, β). The left side of Eq. (15.15) can be defined as the core geometrical constant K_{gfe}:

$$K_{gfe} = \frac{W_A(A_c)^{(2(\beta-1)/\beta)}}{(MLT)\ell_m^{(2/\beta)}} \left[\left(\frac{\beta}{2}\right)^{-\left(\frac{\beta}{\beta+2}\right)} + \left(\frac{\beta}{2}\right)^{\left(\frac{2}{\beta+2}\right)} \right]^{-\left(\frac{\beta+2}{\beta}\right)} \tag{15.16}$$

Hence, to design a transformer, the right side of Eq. (15.15) is evaluated. A core is selected whose K_{gfe} exceeds this value:

$$K_{gfe} \geq \frac{\rho \lambda_1^2 I_{tot}^2 K_{fe}^{(2/\beta)}}{4K_u (P_{tot})^{((\beta+2)/\beta)}} \tag{15.17}$$

The quantity K_{gfe} is similar to the geometrical constant K_g used in the previous chapter to design magnetics when core loss is negligible. K_{gfe} is a measure of the magnetic size of a core, for applications in which core loss is significant. Unfortunately, K_{gfe} depends on β, and hence the choice of core material affects the value of K_{gfe}. However, the β of most high-frequency ferrite materials lies in the narrow range 2.6 to 2.8, and K_{gfe} varies by no more than ± 5% over this range. Appendix D lists the values of K_{gfe} for various standard ferrite cores, for the value $\beta = 2.7$.

Once a core has been selected, then the values of A_c, W_A, ℓ_m, and MLT are known. The peak ac flux density ΔB can then be evaluated using Eq. (15.13), and the primary turns n_1 can be found using Eq. (15.4). The number of turns for the remaining windings can be computed using the desired turns ratios. The various window area allocations are found using Eq. (14.35). The wire sizes for the various windings can then be computed as discussed in the previous chapter,

$$A_{w,j} = \frac{K_u W_A \alpha_j}{n_j} \tag{15.18}$$

where $A_{w,j}$ is the wire area for winding j.

15.2   A STEP-BY-STEP TRANSFORMER DESIGN PROCEDURE

The procedure developed in the previous sections is summarized below. As in the filter inductor design procedure of the previous chapter, this simple transformer design procedure should be regarded as a first-pass approach. Numerous issues have been neglected, including detailed insulation requirements, conductor eddy current losses, temperature rise, roundoff of number of turns, etc.

The following quantities are specified, using the units noted:

Wire effective resistivity	ρ	(Ω-cm)
Total rms winding currents, referred to primary	$I_{tot} = \sum_{j=1}^{k} \frac{n_j}{n_i} I_j$	(A)
Desired turns ratios	n_2/n_1, n_3/n_1, etc.	
Applied primary volt-seconds	$\lambda_1 = \int\limits_{\substack{positive \\ portion \\ of\ cycle}} v_1(t)dt$	(V-sec)

Allowed total power dissipation	P_{tot}	(W)
Winding fill factor	K_u	
Core loss exponent	β	
Core loss coefficient	K_{fe}	(W/cm^3T^β)

The core dimensions are expressed in cm:

Core cross-sectional area	A_c	(cm^2)
Core window area	W_A	(cm^2)
Mean length per turn	MLT	(cm)
Magnetic path length	ℓ_m	(cm)
Peak ac flux density	ΔB	(Tesla)
Wire areas	$A_{w1}, A_{w2}, ...$	(cm^2)

The use of centimeters rather than meters requires that appropriate factors be added to the design equations.

15.2.1 Procedure

1. *Determine core size.*

$$K_{gfe} \geq \frac{\rho \lambda_1^2 I_{tot}^2 K_{fe}^{(2/\beta)}}{4K_u (P_{tot})^{((\beta+2)/\beta)}} 10^8 \tag{15.19}$$

Choose a core that is large enough to satisfy this inequality. If necessary, it may be possible to use a smaller core by choosing a core material having lower loss, i.e., smaller K_{fe}.

2. *Evaluate peak ac flux density.*

$$\Delta B = \left[10^8 \frac{\rho \lambda_1^2 I_{tot}^2}{2K_u} \frac{(MLT)}{W_A A_c^3 \ell_m} \frac{1}{\beta K_{fe}} \right]^{\left(\frac{1}{\beta+2}\right)} \tag{15.20}$$

Check whether ΔB is greater than the core material saturation flux density. If the core operates with a flux dc bias, then the dc bias plus ΔB should not exceed the saturation flux density. Proceed to the next step if adequate margins exist to prevent saturation. Otherwise, (1) repeat the procedure using a core material having greater core loss, or (2) use the K_g design method, in which the maximum flux density is specified.

3. *Evaluate primary turns.*

$$n_1 = \frac{\lambda_1}{2\Delta B A_c} 10^4 \tag{15.21}$$

4.   *Choose numbers of turns for other windings*

According to the desired turns ratios:

$$n_2 = n_1 \left(\frac{n_2}{n_1} \right)$$
$$n_3 = n_1 \left(\frac{n_3}{n_1} \right)$$
$$\vdots$$

(15.22)

5.   *Evaluate fraction of window area allocated to each winding.*

$$\alpha_1 = \frac{n_1 I_1}{n_1 I_{tot}}$$
$$\alpha_2 = \frac{n_2 I_2}{n_1 I_{tot}}$$
$$\vdots$$
$$\alpha_k = \frac{n_k I_k}{n_1 I_{tot}}$$

(15.23)

6.   *Evaluate wire sizes.*

$$A_{w1} \leq \frac{\alpha_1 K_u W_A}{n_1}$$
$$A_{w2} \leq \frac{\alpha_2 K_u W_A}{n_2}$$
$$\vdots$$

(15.24)

Choose wire gauges to satisfy these criteria

A winding geometry can now be determined, and copper losses due to the proximity effect can be evaluated. If these losses are significant, it may be desirable to further optimize the design by reiterating the above steps, accounting for proximity losses by increasing the effective wire resistivity to the value $\rho_{eff} = \rho_{cu} P_{cu} / P_{dc}$, where P_{cu} is the actual copper loss including proximity effects, and P_{dc} is the copper loss obtained when the proximity effect is negligible.

If desired, the power losses and transformer model parameters can now be checked. For the simple model of Fig. 15.4, the following parameters are estimated:

Magnetizing inductance, referred to winding 1: $L_M = \dfrac{\mu n_1^2 A_c}{\ell_m}$

Peak ac magnetizing current, referred to winding 1: $i_{M,pk} = \dfrac{\lambda_1}{2 L_M}$

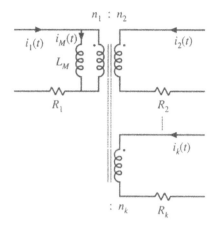

Fig. 15.4   Computed elements of simple transformer model.

Winding resistances:

$$R_1 = \frac{\rho n_1 (MLT)}{A_{w1}}$$

$$R_2 = \frac{\rho n_2 (MLT)}{A_{w2}}$$

$$\vdots$$

The core loss, copper loss, and total power loss can be determined using Eqs. (15.1), (15.7), and (15.8), respectively.

15.3   EXAMPLES

15.3.1   Example 1: Single-Output Isolated Ćuk Converter

As an example, let us consider the design of a simple two-winding transformer for the Ćuk converter of Fig. 15.5. This transformer is to be optimized at the operating point shown, corresponding to $D = 0.5$. The steady-state converter solution is $V_{c1} = V_g$, $V_{c2} = V$. The desired transformer turns ratio is

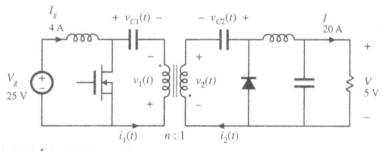

Fig. 15.5   Isolated Ćuk converter example.

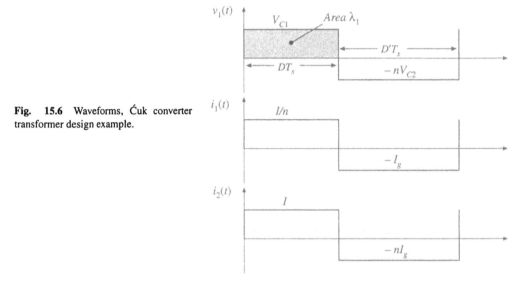

Fig. 15.6 Waveforms, Ćuk converter transformer design example.

$n = n_1/n_2 = 5$. The switching frequency is $f_s = 200$ kHz, corresponding to $T_s = 5$ μs. A ferrite pot core consisting of Magnetics, Inc. P-material is to be used; at 200 kHz, this material is described by the following parameters: $K_{fe} = 24.7$ W/T$^\beta$cm^3, $\beta = 2.6$. A fill factor of $K_u = 0.5$ is assumed. Total power loss of $P_{tot} = 0.25$ W is allowed. Copper wire, having a resistivity of $\rho = 1.724 \cdot 10^{-6}$ Ω–cm, is to be used.

Transformer waveforms are illustrated in Fig. 15.6. The applied primary volt-seconds are

$$\lambda_1 = DT_s V_{c1} = (0.5)(5 \text{ }\mu\text{sec})(25 \text{ V})$$
$$= 62.5 \text{ V–}\mu\text{sec} \tag{15.25}$$

The primary rms current is

$$I_1 = \sqrt{D\left(\frac{I}{n}\right)^2 + D'\left(I_g\right)^2} = 4 \text{ A} \tag{15.26}$$

It is assumed that the rms magnetizing current is much smaller than the rms winding currents. Since the transformer contains only two windings, the secondary rms current is equal to

$$I_2 = nI_1 = 20 \text{ A} \tag{15.27}$$

The total rms winding current, referred to the primary, is

$$I_{tot} = I_1 + \frac{1}{n} I_2 = 8 \text{ A} \tag{15.28}$$

The core size is evaluated using Eq. (15.19):

$$K_{gfe} \geq \frac{(1.724 \cdot 10^{-6})(62.5 \cdot 10^{-6})^2 (8)^2 (24.7)^{(2/2.6)}}{4 \ (0.5)(0.25)^{(4.6/2.6)}} 10^8 \tag{15.29}$$
$$= 0.00295$$

The pot core data of Appendix D lists the 2213 pot core with $K_{gfe} = 0.0049$ for $\beta = 2.7$. Evaluation of Eq. (15.16) shows that $K_{gfe} = 0.0047$ for this core, when $\beta = 2.6$. In any event, 2213 is the smallest standard pot core size having $K_{gfe} \leq 0.00295$. The increased value of K_{gfe} should lead to lower total power loss. The peak ac flux density is found by evaluation of Eq. (15.20), using the geometrical data for the 2213 pot core:

$$\Delta B = \left[10^8 \frac{(1.724 \cdot 10^{-6})(62.5 \cdot 10^{-6})^2(8)^2}{2\,(0.5)} \frac{(4.42)}{(0.297)(0.635)^3(3.15)} \frac{1}{(2.6)(24.7)} \right]^{(1/4.6)} \tag{15.30}$$

$$= 0.0858 \text{ Tesla}$$

This flux density is considerably less than the saturation flux density of approximately 0.35 Tesla. The primary turns are determined by evaluation of Eq. (15.21):

$$n_1 = 10^4 \frac{(62.5 \cdot 10^{-6})}{2(0.0858)(0.635)} \tag{15.31}$$

$$= 5.74 \text{ turns}$$

The secondary turns are found by evaluation of Eq. (15.22). It is desired that the transformer have a 5:1 turns ratio, and hence

$$n_2 = \frac{n_1}{n} = 1.15 \text{ turns} \tag{15.32}$$

In practice, we might select $n_1 = 5$ and $n_2 = 1$. This would lead to a slightly higher ΔB and slightly higher loss.

The fraction of the window area allocated to windings 1 and 2 are determined using Eq. (15.23):

$$\alpha_1 = \frac{(4\,\text{A})}{(8\,\text{A})} = 0.5 \tag{15.33}$$

$$\alpha_2 = \frac{\left(\frac{1}{5}\right)(20\,\text{A})}{(8\,\text{A})} = 0.5$$

For this example, the window area is divided equally between the primary and secondary windings, since the ratio of their rms currents is equal to the turns ratio. We can now evaluate the primary and secondary wire areas, via Eq. (15.24):

$$A_{w1} = \frac{(0.5)(0.5)(0.297)}{(5)} = 14.8 \cdot 10^{-3} \text{ cm}^2 \tag{15.34}$$

$$A_{w2} = \frac{(0.5)(0.5)(0.297)}{(1)} = 74.2 \cdot 10^{-3} \text{ cm}^2$$

The wire gauge is selected using the wire table of Appendix D. AWG #16 has area $13.07 \cdot 10^{-3}$ cm^2, and is suitable for the primary winding. AWG #9 is suitable for the secondary winding, with area $66.3 \cdot 10^{-3}$ cm^2. These are very large conductors, and one turn of AWG #9 is not a practical solution! We can also expect significant proximity losses, and significant leakage inductance. In practice, interleaved foil windings might be used. Alternatively, Litz wire or several parallel strands of smaller wire could be employed.

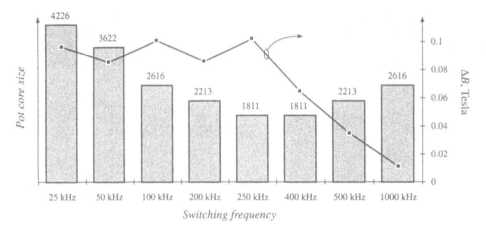

Fig. 15.7 Variation of transformer size (bar chart) with switching frequency, Ćuk converter example. Optimum peak ac flux density (data points) is also plotted.

It is a worthwhile exercise to repeat the above design at several different switching frequencies, to determine how transformer size varies with switching frequency. As the switching frequency is increased, the core loss coefficient K_{fe} increases. Figure 15.7 illustrates the transformer pot core size, for various switching frequencies over the range 25 kHz to 1 MHz, for this Ćuk converter example using P material with $P_{tot} < 0.25$ W. Peak flux densities in Tesla are also plotted. For switching frequencies below 250 kHz, increasing the frequency causes the core size to decrease. This occurs because of the decreased applied volt-seconds λ_1. Over this range, the optimal ΔB is essentially independent of switching frequency; the ΔB variations shown occur owing to quantization of core sizes.

For switching frequencies greater than 250 kHz, increasing frequency causes greatly increased core loss. Maintaining $P_{tot} \leq 0.25$ W then requires that ΔB be reduced, and hence the core size is increased. The minimum transformer size for this example is apparently obtained at 250 kHz.

In practice, several matters complicate the dependence of transformer size on switching frequency. Figure 15.7 ignores the winding geometry and copper losses due to winding eddy currents. Greater power losses can be allowed in larger cores. Use of a different core material may allow higher or lower switching frequencies. The same core material, used in a different application with different specifications, may lead to a different optimal frequency. Nonetheless, examples have been reported in the literature [1–4] in which ferrite transformer size is minimized at frequencies ranging from several hundred kilohertz to several megahertz. More detailed design optimizations can be performed using computer optimization programs [5, 6].

15.3.2   Example 2: Multiple-Output Full-Bridge Buck Converter

As a second example, let us consider the design of transformer T_1 for the multiple-output full-bridge buck converter of Fig. 15.8. This converter has a 5 V and a 15 V output, with maximum loads as shown. The transformer is to be optimized at the full-load operating point shown, corresponding to $D = 0.75$. Waveforms are illustrated in Fig. 15.9. The converter switching frequency is $f_s = 150$ kHz. In the full-bridge configuration, the transformer waveforms have fundamental frequency equal to one-half of the switching frequency, so the effective transformer frequency is 75 kHz. Upon accounting for losses

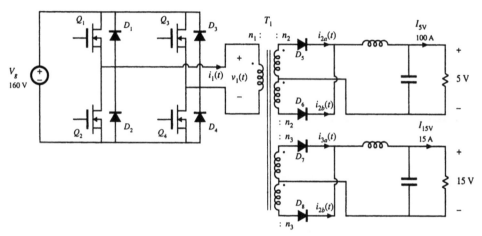

Fig. 15.8 Multiple-output full-bridge isolated buck converter example.

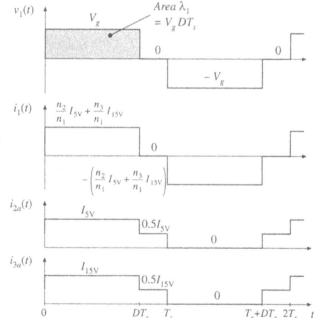

Fig. 15.9 Transformer waveforms, full-bridge converter example.

caused by diode forward voltage drops, one finds that the desired transformer turns ratios $n_1 : n_2 : n_3$ are 110:5:15. A ferrite EE consisting of Magnetics, Inc. P-material is to be used in this example; at 75 kHz, this material is described by the following parameters: $K_{fe} = 7.6$ W/T$^\beta$cm^3, $\beta = 2.6$. A fill factor of $K_u = 0.25$ is assumed in this isolated multiple-output application. Total power loss of $P_{tot} = 4$ W, or approximately 0.5% of the load power, is allowed. Copper wire, having a resistivity of $\rho = 1.724 \cdot 10^{-6}$ Ω–cm, is to be used.

The applied primary volt-seconds are

$$\lambda_1 = DT_sV_g = (0.75)(6.67 \ \mu sec)(160 \ V) = 800 \ V\text{--}\mu sec \tag{15.35}$$

The primary rms current is

$$I_1 = \left(\frac{n_2}{n_1}I_{5V} + \frac{n_3}{n_1}I_{15V}\right)\sqrt{D} = 5.7 \ A \tag{15.36}$$

The 5 V secondary windings carry rms current

$$I_2 = \tfrac{1}{2}I_{5V}\sqrt{1+D} = 66.1 \ A \tag{15.37}$$

The 15 V secondary windings carry rms current

$$I_3 = \tfrac{1}{2}I_{15V}\sqrt{1+D} = 9.9 \ A \tag{15.38}$$

The total rms winding current, referred to the primary, is

$$I_{tot} = \sum_{\substack{all\ 5 \\ windings}} \frac{n_j}{n_1}I_j = I_1 + 2\frac{n_2}{n_1}I_2 + 2\frac{n_3}{n_1}I_3$$
$$= 14.4 \ A \tag{15.39}$$

The core size is evaluated using Eq. (15.19):

$$K_{gfe} \geq \frac{(1.724 \cdot 10^{-6})(800 \cdot 10^{-6})^2(14.4)^2(7.6)^{(2/2.6)}}{4(0.25)(4)^{(4.6/2.6)}}10^8 \tag{15.40}$$
$$= 0.00937$$

The EE core data of Appendix D lists the EE40 core with $K_{gfe} = 0.0118$ for $\beta = 2.7$. Evaluation of Eq. (15.16) shows that $K_{gfe} = 0.0108$ for this core, when $\beta = 2.6$. In any event, EE40 is the smallest standard EE core size having $K_{gfe} \leq 0.00937$. The peak ac flux density is found by evaluation of Eq. (15.20), using the geometrical data for the EE40 core:

$$\Delta B = \left[10^8 \frac{(1.724 \cdot 10^{-6})(800 \cdot 10^{-6})^2(14.4)^2}{2(0.25)} \frac{(8.5)}{(1.1)(1.27)^3(7.7)} \frac{1}{(2.6)(7.6)}\right]^{(1/4.6)} \tag{15.41}$$
$$= 0.23 \ Tesla$$

This flux density is less than the saturation flux density of approximately 0.35 Tesla. The primary turns are determined by evaluation of Eq. (15.21):

$$n_1 = 10^4 \frac{(800 \cdot 10^{-6})}{2(0.23)(1.27)} \tag{15.42}$$
$$= 13.7 \ turns$$

The secondary turns are found by evaluation of Eq. (15.22). It is desired that the transformer have a 110:5:15 turns ratio, and hence

$$n_2 = \frac{5}{110} n_1 = 0.62 \text{ turns} \tag{15.43}$$

$$n_3 = \frac{15}{110} n_1 = 1.87 \text{ turns} \tag{15.44}$$

In practice, we might select $n_1 = 22$, $n_2 = 1$, and $n_3 = 3$. This would lead to a reduced ΔB with reduced core loss and increased copper loss. Since the resulting ΔB is suboptimal, the total power loss will be increased. According to Eq. (15.3), the peak ac flux density for the EE40 core will be

$$\Delta B = \frac{(800 \cdot 10^{-6})}{2(22)(1.27)} 10^4 = 0.143 \text{ Tesla} \tag{15.45}$$

The resulting core and copper loss can be computed using Eqs. (15.1) and (15.7):

$$P_{fe} = (7.6)(0.143)^{2.6}(1.27)(7.7) = 0.47 \text{ W} \tag{15.46}$$

$$P_{cu} = \frac{(1.724 \cdot 10^{-6})(800 \cdot 10^{-6})^2(14.4)^2}{4 (0.25)} \frac{(8.5)}{(1.1)(1.27)^2} \frac{1}{(0.143)^2} 10^8$$
$$= 5.4 \text{ W} \tag{15.47}$$

Hence, the total power loss would be

$$P_{tot} = P_{fe} + P_{cu} = 5.9 \text{ W} \tag{15.48}$$

Since this is 50% greater than the design goal of 4 W, it is necessary to increase the core size. The next larger EE core is the EE50 core, having K_{gfe} of 0.0284. The optimum ac flux density for this core, given by Eq. (15.3), is $\Delta B = 0.14$ T; operation at this flux density would require $n_1 = 12$ and would lead to a total power loss of 2.3 W. With $n_1 = 22$, calculations similar to Eqs. (15.45) to (15.48) lead to a peak flux density of $\Delta B = 0.08$ T. The resulting power losses would then be $P_{fe} = 0.23$ W, $P_{cu} = 3.89$ W, $P_{tot} = 4.12$ W.

With the EE50 core and $n_1 = 22$, the fraction of the available window area allocated to the primary winding is given by Eq. (15.23) as

$$\alpha_1 = \frac{I_1}{I_{tot}} = \frac{5.7}{14.4} = 0.396 \tag{15.49}$$

The fraction of the available window area allocated to each half of the 5 V secondary winding should be

$$\alpha_2 = \frac{n_2 I_2}{n_1 I_{tot}} = \frac{5}{110} \frac{66.1}{14.4} = 0.209 \tag{15.50}$$

The fraction of the available window area allocated to each half of the 15 V secondary winding should be

$$\alpha_3 = \frac{n_3 I_3}{n_1 I_{tot}} = \frac{15}{110} \frac{9.9}{14.4} = 0.094 \tag{15.51}$$

The primary wire area A_{w1}, 5 V secondary wire area A_{w2}, and 15 V secondary wire area A_{w3} are then given by Eq. (15.24) as

$$A_{w1} = \frac{\alpha_1 K_u W_A}{n_1} = \frac{(0.396)(0.25)(1.78)}{(22)} = 8.0 \cdot 10^{-3} \, \text{cm}^2$$

$$\Rightarrow \text{AWG \#19}$$

$$A_{w2} = \frac{\alpha_2 K_u W_A}{n_2} = \frac{(0.209)(0.25)(1.78)}{(1)} = 93.0 \cdot 10^{-3} \, \text{cm}^2 \qquad (15.52)$$

$$\Rightarrow \text{AWG \#8}$$

$$A_{w3} = \frac{\alpha_3 K_u W_A}{n_3} = \frac{(0.094)(0.25)(1.78)}{(3)} = 13.9 \cdot 10^{-3} \, \text{cm}^2$$

$$\Rightarrow \text{AWG \#16}$$

It may be preferable to wind the 15 V outputs using two #19 wires in parallel; this would lead to the same area A_{w3} but would be easier to wind. The 5 V windings could be wound using many turns of smaller paralleled wires, but it would probably be easier to use a flat copper foil winding. If insulation requirements allow, proximity losses could be minimized by interleaving several thin layers of foil with the primary winding.

15.4   AC INDUCTOR DESIGN

The transformer design procedure of the previous sections can be adapted to handle the design of other magnetic devices in which both core loss and copper loss are significant. A procedure is outlined here for design of single-winding inductors whose waveforms contain significant high-frequency ac components (Fig. 15.10). An optimal value of ΔB is found, which leads to minimum total core-plus-copper loss. The major difference is that we must design to obtain a given inductance, using a core with an air gap. The constraints and a step-by-step procedure are briefly outlined below.

15.4.1   Outline of Derivation

As in the filter inductor design procedure of the previous chapter, the desired inductance L must be obtained, given by

$$L = \frac{\mu_0 A_c n^2}{\ell_g} \qquad (15.53)$$

The applied voltage waveform and the peak ac component of the flux density ΔB are related according to

$$\Delta B = \frac{\lambda}{2n A_c} \qquad (15.54)$$

The copper loss is given by

$$P_{cu} = \frac{\rho n^2 (MLT)}{K_u W_A} I^2 \qquad (15.55)$$

where I is the rms value of $i(t)$. The core loss P_{fe} is given by Eq. (15.1).

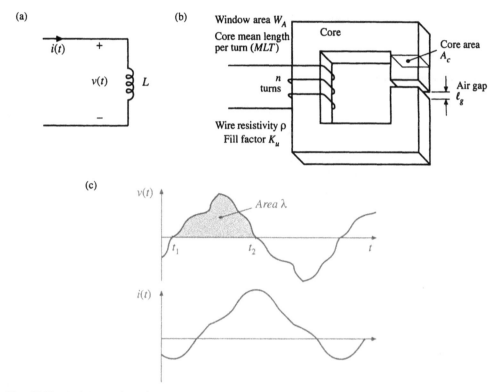

Fig. 15.10 Ac inductor, in which copper loss and core loss are significant: (a) definition of terminal quantities, (b) core geometry, (c) arbitrary terminal waveforms.

The value of ΔB that minimizes the total power loss $P_{tot} = P_{cu} + P_{fe}$ is found in a manner similar to the transformer design derivation. Equation (15.54) is used to eliminate n from the expression for P_{cu}. The optimal ΔB is then computed by setting the derivative of P_{tot} to zero. The result is

$$\Delta B = \left[\frac{\rho \lambda^2 I^2}{2K_u} \frac{(MLT)}{W_A A_c^3 \ell_m} \frac{1}{\beta K_{fe}} \right]^{\left(\frac{1}{\beta+2} \right)}$$

(15.56)

which is essentially the same as Eq. (15.13). The total power loss P_{tot} is evaluated at this value of ΔB, and the resulting expression is manipulated to find K_{gfe}. The result is

$$K_{gfe} \geq \frac{\rho \lambda^2 I^2 K_{fe}^{(2/\beta)}}{2K_u \left(P_{tot} \right)^{((\beta+2)/\beta)}}$$

(15.57)

where K_{gfe} is defined as in Eq. (15.16). A core that satisfies this inequality is selected.

15.4.2   Step-by-step AC Inductor Design Procedure

The units of Section 15.2 are employed here.

1.   *Determine core size.*

$$K_{gfe} \geq \frac{\rho \lambda^2 I^2 K_{fe}^{(2/\beta)}}{2 K_u (P_{tot})^{((\beta+2)/\beta)}} 10^8 \qquad (15.58)$$

Choose a core that is large enough to satisfy this inequality. If necessary, it may be possible to use a smaller core by choosing a core material having lower loss, that is, smaller K_{fe}.

2.   *Evaluate peak ac flux density.*

$$\Delta B = \left[10^8 \frac{\rho \lambda^2 I^2}{2 K_u} \frac{(MLT)}{W_A A_c^3 \ell_m} \frac{1}{\beta K_{fe}} \right]^{\left(\frac{1}{\beta+2}\right)} \qquad (15.59)$$

3.   *Number of turns.*

$$n = \frac{\lambda}{2 \Delta B A_c} 10^4 \qquad (15.60)$$

4.   *Air gap length.*

$$\ell_g = \frac{\mu_0 A_c n^2}{L} 10^{-4} \qquad (15.61)$$

with A_c specified in cm^2 and ℓ_g expressed in meters. Alternatively, the air gap can be indirectly expressed via A_L (mH/1000 turns):

$$A_L = \frac{L}{n^2} 10^9 \qquad (15.62)$$

5.   *Check for saturation.*

If the inductor current contains a dc component I_{dc}, then the maximum total flux density B_{max} is greater than the peak ac flux density ΔB. The maximum total flux density, in Tesla, is given by

$$B_{max} = \Delta B + \frac{L I_{dc}}{n A_c} 10^4 \qquad (15.63)$$

If B_{max} is close to or greater than the saturation flux density B_{sat}, then the core may saturate. The filter inductor design procedure of the previous chapter should then be used, to operate at a lower flux density.

6.   *Evaluate wire size.*

$$A_w \leq \frac{K_u W_A}{n} \qquad (15.64)$$

A winding geometry can now be determined, and copper losses due to the proximity effect can be evaluated. If these losses are significant, it may be desirable to further optimize the design by reiterating the above steps, accounting for proximity losses by increasing the effective wire resistivity to the value $\rho_{eff} = \rho_{cu} P_{cu}/P_{dc}$, where P_{cu} is the actual copper loss including proximity effects, and P_{dc} is the copper loss predicted when the proximity effect is ignored.

7. *Check power loss.*

$$
\begin{aligned}
P_{cu} &= \frac{\rho n(MLT)}{A_w} I^2 \\
P_{fe} &= K_{fe}(\Delta B)^\beta A_c \ell_m \\
P_{tot} &= P_{cu} + P_{fe}
\end{aligned}
\tag{15.65}
$$

15.5 SUMMARY

1. In a multiple-winding transformer, the low-frequency copper losses are minimized when the available window area is allocated to the windings according to their apparent powers, or ampere-turns.

2. As peak ac flux density is increased, core loss increases while copper losses decrease. There is an optimum flux density that leads to minimum total power loss. Provided that the core material is operated near its intended frequency, then the optimum flux density is less than the saturation flux density. Minimization of total loss then determines the choice of peak ac flux density.

3. The core geometrical constant K_{gfe} is a measure of the magnetic size of a core, for applications in which core loss is significant. In the K_{gfe} design method, the peak flux density is optimized to yield minimum total loss, as opposed to the K_g design method where peak flux density is a given specification.

REFERENCES

[1] W. J. GU and R. LIU, "A Study of Volume and Weight vs. Frequency for High-Frequency Transformers," *IEEE Power Electronics Specialists Conference*, 1993 Record, pp. 1123–1129.

[2] K. D. T. NGO, R. P. ALLEY, A. J. YERMAN, R. J. CHARLES, and M. H. KUO, "Evaluation of Trade-Offs in Transformer Design for Very-Low-Voltage Power Supply with Very High Efficiency and Power Density," *IEEE Applied Power Electronics Conference*, 1990 Record, pp. 344–353.

[3] A. F. GOLDBERG and M. F. SCHLECHT, "The Relationship Between Size and Power Dissipation in a 1-10MHz Transformer," *IEEE Power Electronics Specialists Conference*, 1989 Record, pp. 625-634.

[4] K. D. T. NGO and R. S. LAI, "Effect of Height on Power Density in High-Frequency Transformers," *IEEE Power Electronics Specialists Conference*, 1991 Record, pp. 667-672.

[5] R. B. RIDLEY and F. C. LEE, "Practical Nonlinear Design Optimization Tool for Power Converter Components," *IEEE Power Electronics Specialists Conference*, 1987 Record, pp. 314-323.

[6] R. C. WONG, H. A. OWEN, and T. G. WILSON, "Parametric Study of Minimum Converter Loss in an Energy-Storage Dc-to-Dc Converter," *IEEE Power Electronics Specialists Conference*, 1982 Record, pp. 411-425.

PROBLEMS

15.1   Forward converter inductor and transformer design. The objective of this problem set is to design the magnetics (two inductors and one transformer) of the two-transistor, two-output forward converter shown in Fig. 15.11. The ferrite core material to be used for all three devices has a saturation flux density of approximately 0.3 T at 120°C. To provide a safety margin for your designs, you should use a maximum flux density B_{max} that is no greater than 75% of this value. The core loss at 100 kHz is described by Eq. (15.1), with the parameter values $\beta = 2.6$ and $K_{fe} = 50$ W/T$^\beta$ cm^3. Calculate copper loss at 100°C.

Steady-state converter analysis and design. You may assume 100% efficiency and ideal lossless components for this section.

(a)   Select the transformer turns ratios so that the desired output voltages are obtained when the duty cycle is $D = 0.4$.

(b)   Specify values of L_1 and L_2 such that their current ripples Δi_1 and Δi_2 are 10% of their respective full-load current dc components I_1 and I_2.

(c)   Determine the peak and rms currents in each inductor and transformer winding.

Inductor design. Allow copper loss of 1 W in L_1 and 0.4 W in L_2. Assume a fill factor of $K_u = 0.5$. Use ferrite EE cores—tables of geometrical data for standard EE core sizes are given in Appendix D. Design the output filter inductors L_1 and L_2. For each inductor, specify:

  (*i*)   EE core size

  (*ii*)   Air gap length

  (*iii*) Number of turns

  (*iv*)   AWG wire size

Transformer design. Allow a total power loss of 1 W. Assume a fill factor of $K_u = 0.35$ (lower than for the filter inductors, to allow space for insulation between the windings). Use a ferrite EE core. You may neglect losses due to the skin and proximity effects, but you should include core and copper losses. Design the transformer, and specify the following:

  (*i*)   EE core size

  (*ii*)   Turns n_1, n_2, and n_3

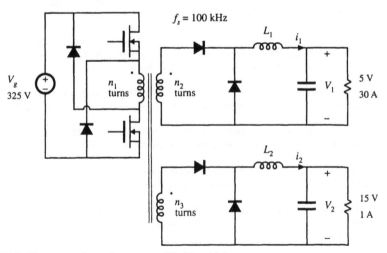

Fig. 15.11   Two-output forward converter of Problem 15.1.

          (iii)   AWG wire size for the three windings

Check your transformer design:

          (iv)   Compute the maximum flux density. Will the core saturate?

          (v)   Compute the core loss, the copper loss of each winding, and the total power loss

15.2   A single-transistor forward converter operates with an input voltage $V_g = 160$ V, and supplies two outputs: 24 V at 2 A, and 15 V at 6 A. The duty cycle is $D = 0.4$. The turns ratio between the primary winding and the reset winding is 1:1. The switching frequency is 100 kHz. The core material loss equation parameters are $\beta = 2.7$, $K_{fe} = 50$. You may assume a fill factor of 0.25. Do not allow the core maximum flux density to exceed 0.3 T.

           Design a transformer for this application, having a total power loss no greater than 1.5 W at 100°C. Neglect proximity losses. You may neglect the reset winding. Use a ferrite PQ core. Specify: core size, peak ac flux density, wire sizes, and number of turns for each winding. Compute the core and copper losses for your design.

15.3   Flyback/SEPIC transformer design. The "transformer" of the flyback and SEPIC converters is an energy storage device, which might be more accurately described as a multiple-winding inductor. The magnetizing inductance L_p functions as an energy-transferring inductor of the converter, and therefore the "transformer" normally contains an air gap. The converter may be designed to operate in either the continuous or discontinuous conduction mode. Core loss may be significant. It is also important to ensure that the peak current in the magnetizing inductance does not cause saturation.

           A flyback transformer is to be designed for the following two-output flyback converter application:

Input:	160 Vdc
Output 1:	5 Vdc at 10 A
Output 2:	15 Vdc at 1 A
Switching frequency:	100 kHz
Magnetizing inductance L_p:	1.33 mH, referred to primary
Turns ratio:	160:5:15
Transformer power loss:	Allow 1 W total

(a)   Does the converter operate in CCM or DCM? Referred to the primary winding, how large are *(i)* the magnetizing current ripple Δi, *(ii)* the magnetizing current dc component I, and *(iii)* the peak magnetizing current I_{pk}?

(b)   Determine *(i)* the rms winding currents, and *(ii)* the applied primary volt-seconds λ_1. Is λ_1 proportional to I_{pk}?

(c)   Modify the transformer and ac inductor design procedures of this chapter, to derive a general procedure for designing flyback transformers that explicitly accounts for both core and copper loss, and that employs the optimum ac flux density that minimizes the total loss.

(d)   Give a general step-by-step design procedure, with all specifications and units clearly stated.

(e)   Design the flyback transformer for the converter of part (a), using your step-by-step procedure of part (d). Use a ferrite EE core, with $\beta = 2.7$ and $K_{fe} = 50$ W/T$^\beta$cm^3. Specify: core size, air gap length, turns, and wire sizes for all windings.

(f)   For your final design of part (e), what are *(i)* the core loss, *(ii)* the total copper loss, and *(iii)* the peak flux density?

15.4   Over the intended range of operating frequencies, the frequency dependence of the core-loss coefficient K_{fe} of a certain ferrite core material can be approximated using a monotonically increasing fourth-order polynomial of the form

$$K_{fe}(f) = K_{fe0}\left(1 + a_1\left(\frac{f}{f_0}\right) + a_2\left(\frac{f}{f_0}\right)^2 + a_3\left(\frac{f}{f_0}\right)^3 + a_4\left(\frac{f}{f_0}\right)^4\right)$$

where K_{fe0}, a_1, a_2, a_3, a_4, and f_0 are constants. In a typical converter transformer application, the applied primary volt-seconds λ_1 varies directly with the switching period $T_s = 1/f$. It is desired to choose the optimum switching frequency such that K_{gfe}, and therefore the transformer size, are minimized.

(a)   Show that the optimum switching frequency is a root of the polynomial

$$1 + a_1\left(\frac{\beta-1}{\beta}\right)\left(\frac{f}{f_0}\right) + a_2\left(\frac{\beta-2}{\beta}\right)\left(\frac{f}{f_0}\right)^2 + a_3\left(\frac{\beta-3}{\beta}\right)\left(\frac{f}{f_0}\right)^3 + a_4\left(\frac{\beta-4}{\beta}\right)\left(\frac{f}{f_0}\right)^4$$

Next, a core material is chosen whose core loss parameters are

$$\beta = 2.7 \qquad\qquad K_{fe0} = 7.6$$
$$f_0 = 100 \text{ kHz}$$
$$a_1 = -1.3 \qquad\qquad a_2 = 5.3$$
$$a_3 = -0.5 \qquad\qquad a_4 = 0.075$$

The polynomial fits the manufacturer's published data over the range $10 \text{ kHz} < f < 1 \text{ MHz}$.

(b)   Sketch K_{fe} vs. f.

(c)   Determine the value of f that minimizes K_{gfe}.

(d)   Sketch $K_{gfe}(f)/K_{gfe}(100 \text{ kHz})$, over the range $100 \text{ kHz} \le f \le 1 \text{ MHz}$. How sensitive is the transformer size to the choice of switching frequency?

15.5   Transformer design to attain a given temperature rise. The temperature rise ΔT of the center leg of a ferrite core is directly proportional to the total power loss P_{tot} of a transformer: $\Delta T = R_{th}P_{tot}$, where R_{th} is the thermal resistance of the transformer under given environmental conditions. You may assume that this temperature rise has minimal dependence on the distribution of losses within the transformer. It is desired to modify the K_{gfe} transformer design method, such that temperature rise ΔT replaces total power loss P_{tot} as a specification. You may neglect the dependence of the wire resistivity ρ on temperature.

(a)   Modify the n-winding transformer K_{gfe} design method, as necessary. Define a new core geometrical constant K_{th} that includes R_{th}.

(b)   Thermal resistances of ferrite EC cores are listed in Section D.3 of Appendix D. Tabulate K_{th} for these cores, using $\beta = 2.7$.

(c)   A 750 W single-output full-bridge isolated buck dc-dc converter operates with converter switching frequency $f_s = 200 \text{ kHz}$, dc input voltage $V_g = 400 \text{ V}$, and dc output voltage $V = 48 \text{ V}$. The turns ratio is 6:1. The core loss equation parameters at 100 kHz are $K_{fe} = 10 \text{ W/T}^\beta\text{cm}^3$ and $\beta = 2.7$. Assume a fill factor of $K_u = 0.3$. You may neglect proximity losses. Use your design procedure of parts (a) and (b) to design a transformer for this application, in which the temperature rise is limited to 20°C. Specify: EC core size, primary and secondary turns, wire sizes, and peak ac flux density.

Part IV

*Modern Rectifiers
and Power System Harmonics*

16

Power and Harmonics in Nonsinusoidal Systems

Rectification used to be a much simpler topic. A textbook could cover the topic simply by discussing the various circuits, such as the peak-detection and inductor-input rectifiers, the phase-controlled bridge, polyphase transformer connections, and perhaps multiplier circuits. But recently, rectifiers have become much more sophisticated, and are now systems rather than mere circuits. They often include pulse-width modulated converters such as the boost converter, with control systems that regulate the ac input current waveform. So modern rectifier technology now incorporates many of the dc–dc converter fundamentals.

The reason for this is the undesirable ac line current harmonics, and low power factors, of conventional peak-detection and phase-controlled rectifiers. The adverse effects of power system harmonics are well recognized. These effects include: unsafe neutral current magnitudes in three-phase systems, heating and reduction of life in transformers and induction motors, degradation of system voltage waveforms, unsafe currents in power-factor-correction capacitors, and malfunctioning of certain power system protection elements. In a real sense, conventional rectifiers are harmonic polluters of the ac power distribution system. With the widespread deployment of electronic equipment in our society, rectifier harmonics have become a significant and measurable problem. Thus there is a need for *high-quality rectifiers*, which operate with high power factor, high efficiency, and reduced generation of harmonics. Several international standards now exist that specifically limit the magnitudes of harmonic currents, for both high-power equipment such as industrial motor drives, and low-power equipment such as electronic ballasts for fluorescent lamps and power supplies for office equipment.

This chapter treats the flow of energy in power systems containing nonsinusoidal waveforms. Average power, rms values, and power factor are expressed in terms of the Fourier series of the voltage and current waveforms. Harmonic currents in three-phase systems are discussed, and present-day standards are listed. The following chapters treat harmonics and harmonic mitigation in conventional line-commutated rectifiers, high-quality rectifier circuits and their models, and control of high-quality rectifiers.

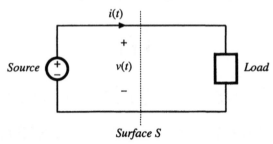

Fig. 16.1   Observe the transmission of energy through surface *S*.

16.1   AVERAGE POWER

Let us consider the transmission of energy from a source to a load, through a given surface as in Fig. 16.1. In the network of Fig. 16.1, the voltage waveform $v(t)$ (not necessarily sinusoidal) is given by the source, and the current waveform is determined by the response of the load. In the more general case in which the source output impedance is significant, then $v(t)$ and $i(t)$ both depend on the characteristics of the source and load. Balanced three-phase systems may be treated in the same manner, on a per-phase basis, using a line current and line-to-neutral voltage.

If $v(t)$ and $i(t)$ are periodic, then they may be expressed as Fourier series:

$$v(t) = V_0 + \sum_{n=1}^{\infty} V_n \cos\left(n\omega t - \varphi_n\right)$$
$$i(t) = I_0 + \sum_{n=1}^{\infty} I_n \cos\left(n\omega t - \theta_n\right)$$

(16.1)

where the period of the ac line voltage waveform is defined as $T = 2\pi/\omega$. In general, the instantaneous power $p(t) = v(t)i(t)$ can assume both positive and negative values at various points during the ac line cycle. Energy then flows in both directions between the source and load. It is of interest to determine the net energy transmitted to the load over one cycle, or

$$W_{cycle} = \int_0^T v(t)i(t)\,dt$$

(16.2)

This is directly related to the average power as follows:

$$P_{av} = \frac{W_{cycle}}{T} = \frac{1}{T}\int_0^T v(t)i(t)\,dt$$

(16.3)

Let us investigate the relationship between the harmonic content of the voltage and current waveforms, and the average power. Substitution of the Fourier series, Eq. (16.1), into Eq. (16.3) yields

$$P_{av} = \frac{1}{T}\int_0^T \left(V_0 + \sum_{n=1}^{\infty} V_n \cos\left(n\omega t - \varphi_n\right)\right)\left(I_0 + \sum_{n=1}^{\infty} I_n \cos\left(n\omega t - \theta_n\right)\right)dt$$

(16.4)

To evaluate this integral, we must multiply out the infinite series. It can be shown that the integrals of cross-product terms are zero, and the only contributions to the integral comes from the products of voltage and current harmonics of the same frequency:

$$\int_0^T \left(V_n \cos\left(n\omega t - \varphi_n\right)\right)\left(I_m \cos\left(m\omega t - \theta_m\right)\right)dt = \begin{cases} 0 & \text{if } n \neq m \\ \dfrac{V_n I_n}{2} \cos\left(\varphi_n - \theta_n\right) & \text{if } n = m \end{cases} \tag{16.5}$$

The average power is therefore

$$P_{av} = V_0 I_0 + \sum_{n=1}^{\infty} \frac{V_n I_n}{2} \cos\left(\varphi_n - \theta_n\right) \tag{16.6}$$

So net energy is transmitted to the load only when the Fourier series of $v(t)$ and $i(t)$ contain terms at the same frequency. For example, if $v(t)$ and $i(t)$ both contain third harmonic, then net energy is transmitted at the third harmonic frequency, with average power equal to

$$\frac{V_3 I_3}{2} \cos\left(\varphi_3 - \theta_3\right) \tag{16.7}$$

Here, $V_3 I_3/2$ is equal to the rms volt-amperes of the third harmonic current and voltage. The $\cos\left(\phi_3 - \theta_3\right)$ term is a displacement term which accounts for the phase difference between the third harmonic voltage and current.

Some examples of power flow in systems containing harmonics are illustrated in Figs. 16.2 to 16.4. In example 1, Fig. 16.2, the voltage contains fundamental only, while the current contains third har-

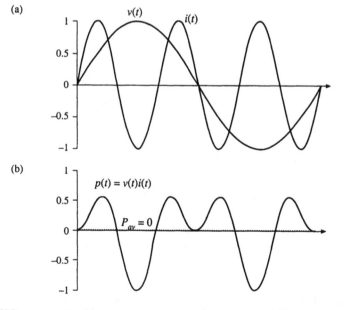

Fig. 16.2 Voltage, current, and instantaneous power waveforms, example 1. The voltage contains only fundamental, and the current contains only third harmonic. The average power is zero.

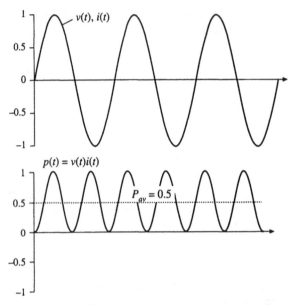

Fig. 16.3 Voltage, current, and instantaneous power waveforms, example 2. The voltage and current each contain only third harmonic, and are in phase. Net energy is transmitted at the third harmonic frequency.

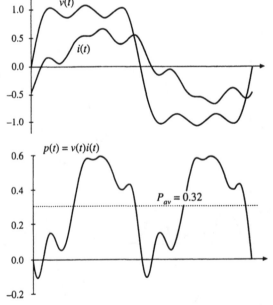

Fig. 16.4 Voltage, current, and instantaneous power waveforms, example 3. The voltage contains fundamental, third, and fifth harmonics. The current contains fundamental, fifth, and seventh harmonics. Net energy is transmitted at the fundamental and fifth harmonic frequencies.

monic only. It can be seen that the instantaneous power waveform $p(t)$ has a zero average value, and hence P_{av} is zero. Energy circulates between the source and load, but over one cycle the net energy transferred to the load is zero. In example 2, Fig. 16.3, the voltage and current each contain only third harmonic. The average power is given by Eq. (16.7) in this case.

In example 3, Fig. 16.4, the voltage waveform contains fundamental, third harmonic, and fifth harmonic, while the current contains fundamental, fifth harmonic, and seventh harmonic, as follows:

$$v(t) = 1.2 \cos (\omega t) + 0.33 \cos (3\omega t) + 0.2 \cos (5\omega t)$$
$$i(t) = 0.6 \cos (\omega t + 30°) + 0.1 \cos (5\omega t + 45°) + 0.1 \cos (7\omega t + 60°) \tag{16.8}$$

Average power is transmitted at the fundamental and fifth harmonic frequencies, since only these frequencies are present in both waveforms. The average power is found by evaluation of Eq. (16.6); all terms are zero except for the fundamental and fifth harmonic terms, as follows:

$$P_{av} = \frac{(1.2)(0.6)}{2} \cos (30°) + \frac{(0.2)(0.1)}{2} \cos (45°) = 0.32 \tag{16.9}$$

The instantaneous power and its average are illustrated in Fig. 16.4(b).

16.2   ROOT-MEAN-SQUARE (RMS) VALUE OF A WAVEFORM

The rms value of a periodic waveform $v(t)$ with period T is defined as

$$(\text{rms value}) = \sqrt{\frac{1}{T} \int_0^T v^2(t) dt} \tag{16.10}$$

The rms value can also be expressed in terms of the Fourier components. Insertion of Eq. (16.1) into Eq. (16.10), and simplification using Eq. (16.5), yields

$$(\text{rms value}) = \sqrt{V_0^2 + \sum_{n=1}^{\infty} \frac{V_n^2}{2}} \tag{16.11}$$

Again, the integrals of the cross-product terms are zero. This expression holds when the waveform is a current:

$$(\text{rms current}) = \sqrt{I_0^2 + \sum_{n=1}^{\infty} \frac{I_n^2}{2}} \tag{16.12}$$

Thus, the presence of harmonics in a waveform always increases its rms value. In particular, in the case where the voltage $v(t)$ contains only fundamental while the current $i(t)$ contains harmonics, then the harmonics increase the rms value of the current while leaving the average power unchanged. This is undesirable, because the harmonics do not lead to net delivery of energy to the load, yet they increase the $I_{rms}^2 R$ losses in the system.

In a practical system, series resistances always exist in the source, load, and/or transmission wires, which lead to unwanted power losses obeying the expression

$$(\text{rms current})^2 R_{series} \tag{16.13}$$

Examples of such loss elements are the resistance of ac generator windings, the resistance of the wire connecting the source and load, the resistance of transformer windings, and the resistance of semiconductor devices and magnetics windings in switching converters. Thus, it is desired to make the rms current as small as possible, while transferring the required amount of energy and average power to the load.

Shunt resistances usually also exist, which cause power loss according to the relation

$$\frac{(\text{rms voltage})^2}{R_{shunt}} \tag{16.14}$$

Examples include the core losses in transformers and ac generators, and switching converter transistor switching loss. Therefore, it is desired to also make the rms voltage as small as possible while transferring the required average power to the load.

16.3   POWER FACTOR

Power factor is a figure of merit that measures how effectively energy is transmitted between a source and load network. It is measured at a given surface as in Fig. 16.1, and is defined as

$$\text{power factor} = \frac{(\text{average power})}{(\text{rms voltage})\,(\text{rms current})} \tag{16.15}$$

The power factor always has a value between zero and one. The ideal case, unity power factor, occurs for a load that obeys Ohm's Law. In this case, the voltage and current waveforms have the same shape, contain the same harmonic spectrum, and are in phase. For a given average power throughput, the rms current and voltage are minimized at maximum (unity) power factor, that is, with a linear resistive load. In the case where the voltage contains no harmonics but the load is nonlinear and contains dynamics, then the power factor can be expressed as a product of two terms, one resulting from the phase shift of the fundamental component of the current, and the other resulting from the current harmonics.

16.3.1   Linear Resistive Load, Nonsinusoidal Voltage

In this case, the current harmonics are in phase with, and proportional to, the voltage harmonics. As a result, all harmonics result in the net transfer of energy to the load. The current harmonic magnitudes and phases are

$$I_n = \frac{V_n}{R} \tag{16.16}$$

$$\theta_n = \varphi_n \quad \text{so} \quad \cos\,(\theta_n - \varphi_n) = 1 \tag{16.17}$$

The rms voltage is again

$$(\text{rms voltage}) = \sqrt{V_0^2 + \sum_{n=1}^{\infty} \frac{V_n^2}{2}} \tag{16.18}$$

and the rms current is

$$(\text{rms current}) = \sqrt{I_0^2 + \sum_{n=1}^{\infty} \frac{I_n^2}{2}} = \sqrt{\frac{V_0^2}{R^2} + \sum_{n=1}^{\infty} \frac{V_n^2}{2R^2}}$$

$$= \frac{1}{R} (\text{rms voltage}) \tag{16.19}$$

By use of Eq. (16.6), the average power is

$$P_{av} = V_0 I_0 + \sum_{n=1}^{\infty} \frac{V_n I_n}{2} \cos(\varphi_n - \theta_n)$$

$$= \frac{V_0^2}{R} + \sum_{n=1}^{\infty} \frac{V_n^2}{2R} \tag{16.20}$$

$$= \frac{1}{R} (\text{rms voltage})^2$$

Insertion of Eqs. (16.19) and (16.20) into Eq. (16.15) then shows that the power factor is unity. Thus, if the load is linear and purely resistive, then the power factor is unity regardless of the harmonic content of $v(t)$. The harmonic content of the load current waveform $i(t)$ is identical to that of $v(t)$, and all harmonics result in the transfer of energy to the load. This raises the possibility that one could construct a power distribution system based on nonsinusoidal waveforms in which the energy is efficiently transferred to the load.

16.3.2   Nonlinear Dynamical Load, Sinusoidal Voltage

If the voltage $v(t)$ contains a fundamental component but no dc component or harmonics, so that $V_0 = V_2 = V_3 = \ldots = 0$, then harmonics in $i(t)$ do not result in transmission of net energy to the load. The average power expression, Eq. (16.6), becomes

$$P_{av} = \frac{V_1 I_1}{2} \cos(\varphi_1 - \theta_1) \tag{16.21}$$

However, the harmonics in $i(t)$ do affect the value of the rms current:

$$(\text{rms current}) = \sqrt{I_0^2 + \sum_{n=1}^{\infty} \frac{I_n^2}{2}} \tag{16.22}$$

Hence, as in example 1 (Fig. 16.2), harmonics cause the load to draw more rms current from the source, but not more average power. Increasing the current harmonics does not cause more energy to be transferred to the load, but does cause additional losses in series resistive elements R_{series}.

Also, the presence of load dynamics and reactive elements, which causes the phase of the fundamental components of the voltage and current to differ ($\theta_1 - \varphi_1$), also reduces the power factor. The $\cos(\varphi_1 - \theta_1)$ term in the average power Eq. (16.21) becomes less than unity. However, the rms value of the current, Eq. (16.22), does not depend on the phase. So shifting the phase of $i(t)$ with respect to $v(t)$ reduces the average power without changing the rms voltage or current, and hence the power factor is reduced.

By substituting Eqs. (16.21) and (16.22) into (16.15), we can express the power factor for the sinusoidal voltage in the following form:

$$(\text{power factor}) = \left(\frac{\frac{I_1}{\sqrt{2}}}{\sqrt{I_0^2 + \sum_{n=1}^{\infty} \frac{I_n^2}{2}}} \right) \left(\cos \left(\varphi_1 - \theta_1 \right) \right) \tag{16.23}$$

$$= (\text{distortion factor}) \, (\text{displacement factor})$$

So when the voltage contains no harmonics, then the power factor can be written as the product of two terms. The first, called the *distortion factor*, is the ratio of the rms fundamental component of the current to the total rms value of the current

$$(\text{distortion factor}) = \left(\frac{\frac{I_1}{\sqrt{2}}}{\sqrt{I_0^2 + \sum_{n=1}^{\infty} \frac{I_n^2}{2}}} \right) = \frac{(\text{rms fundamental current})}{(\text{rms current})} \tag{16.24}$$

The second term of Eq. (16.23) is called the *displacement factor*, and is the cosine of the angle between the fundamental components of the voltage and current waveforms.

The *Total Harmonic Distortion* (THD) is defined as the ratio of the rms value of the waveform not including the fundamental, to the rms fundamental magnitude. When no dc is present, this can be written:

$$(\text{THD}) = \frac{\sqrt{\sum_{n=2}^{\infty} I_n^2}}{I_1} \tag{16.25}$$

The total harmonic distortion and the distortion factor are closely related. Comparison of Eqs. (16.24) and (16.25), with $I_0 = 0$, leads to

$$(\text{distortion factor}) = \frac{1}{\sqrt{1 + (\text{THD})^2}} \tag{16.26}$$

This equation is plotted in Fig. 16.5. The distortion factor of a waveform with a moderate amount of distortion is quite close to unity. For example, if the waveform contains third harmonic whose magnitude is

Fig. 16.5 Distortion factor vs. total harmonic distortion.

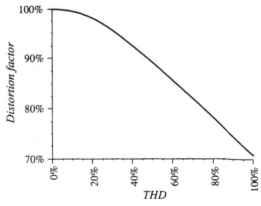

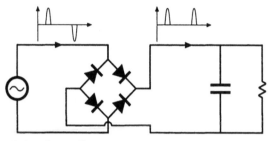

Fig. 16.6   Conventional peak detection rectifier.

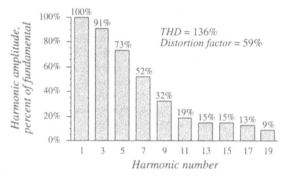

Fig. 16.7   Typical ac line current spectrum of a peak detection rectifier. Harmonics 1 to 19 are shown.

10% of the fundamental, the distortion factor is 99.5%. Increasing the third harmonic to 20% decreases the distortion factor to 98%, and a 33% harmonic magnitude yields a distortion factor of 95%. So the power factor is not significantly degraded by the presence of harmonics unless the harmonics are quite large in magnitude.

An example of a case in which the distortion factor is much less than unity is the conventional peak detection rectifier of Fig. 16.6. In this circuit, the ac line current consists of short-duration current pulses occurring at the peak of the voltage waveform. The fundamental component of the line current is essentially in phase with the voltage, and the displacement factor is close to unity. However, the low-order current harmonics are quite large, close in magnitude to that of the fundamental—a typical current spectrum is given in Fig. 16.7. The distortion factor of peak detection rectifiers is usually in the range 55% to 65%. The resulting power factor is similar in value.

In North America, the standard 120 V power outlet is protected by a 15 A circuit breaker. In consequence, the available load power quite limited. Derating the circuit breaker current by 20%, assuming typical efficiencies for the dc–dc converter and peak detection rectifier, and with a power factor of 55%, one obtains the following estimate for the maximum available dc load power:

$$
\begin{aligned}
&\text{(ac voltage) (derated breaker current) (power factor) (rectifier efficiency)}\\
&= (120\text{ V}) \qquad (80\%\text{ of }15\text{ A}) \qquad (0.55) \qquad (0.98) \\
&= 776\text{ W}
\end{aligned}
\tag{16.27}
$$

The less-than-unity efficiency of a dc–dc converter would further reduce the available dc load power. Using a peak detection rectifier to supply a load power greater than this requires that the user install higher amperage and/or higher voltage service, which is inconvenient and costly. The use of a rectifier circuit having nearly unity power factor would allow a significant increase in available dc load power:

$$\begin{array}{cccc} \text{(ac voltage)} & \text{(derated breaker current)} & \text{(power factor)} & \text{(rectifier efficiency)} \\ = (120 \text{ V}) & (80\% \text{ of } 15 \text{ A}) & (0.99) & (0.93) \\ = 1325 \text{ W} & & & \end{array} \tag{16.28}$$

or almost twice the available power of the peak detection rectifier. This alone can be a compelling reason to employ high quality rectifiers in commercial systems.

16.4   POWER PHASORS IN SINUSOIDAL SYSTEMS

The apparent power is defined as the product of the rms voltage and rms current. Apparent power is easily measured—it is simply the product of the readings of a voltmeter and ammeter placed in the circuit at the given surface. Many power system elements, such as transformers, must be rated according to the apparent power that they are able to supply. The unit of apparent power is the volt-ampere, or VA. The power factor, defined in Eq. (16.15), is the ratio of average power to apparent power.

In the case of sinusoidal voltage and current waveforms, we can additionally define the *complex power S* and the *reactive power Q*. If the sinusoidal voltage $v(t)$ and current $i(t)$ can be represented by the phasors V and I, then the complex power is a phasor defined as

$$S = VI^* = P + jQ \tag{16.29}$$

Here, I^* is the complex conjugate of I, and j is the square root of -1. The magnitude of S, or $\| S \|$, is equal to the apparent power, measured in VA. The real part of S is the average power P, having units of watts. The imaginary part of S is the reactive power Q, having units of reactive volt-amperes, or VARs.

A phasor diagram illustrating S, P, and Q, is given in Fig. 16.8. The angle $(\varphi_1 - \theta_1)$ is the angle between the voltage phasor V and the current phasor I. $(\varphi_1 - \theta_1)$ is additionally the phase of the complex power S. The power factor in the purely sinusoidal case is therefore

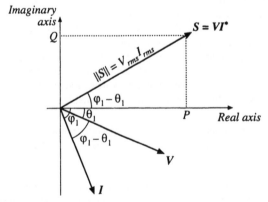

Fig. 16.8   Power phasor diagram, for a sinusoidal system, illustrating the voltage, current, and complex power phasors.

$$\text{power factor} = \frac{P}{|S|} = \cos\left(\varphi_1 - \theta_1\right) \tag{16.30}$$

It should be emphasized that this equation is valid only for systems in which the voltage and current are purely sinusoidal. The distortion factor of Eq. (16.24) then becomes unity, and the power factor is equal to the displacement factor as in Eq. (16.30).

      The reactive power Q does not lead to net transmission of energy between the source and load. When reactive power is present, the rms current and apparent power are greater than the minimum amount necessary to transmit the average power P. In an inductor, the current lags the voltage by 90°, causing the displacement factor to be zero. The alternate storing and releasing of energy in an inductor leads to current flow and nonzero apparent power, but the average power P is zero. Just as resistors consume real (average) power P, inductors can be viewed as consumers of reactive power Q. In a capacitor, the current leads to voltage by 90°, again causing the displacement factor to be zero. Capacitors supply reactive power Q, and are commonly placed in the utility power distribution system near inductive loads. If the reactive power supplied by the capacitor is equal to the reactive power consumed by the inductor, then the net current (flowing from the source into the capacitor-inductive-load combination) will be in phase with the voltage, leading unity power factor and minimum rms current magnitude.

      It will be seen in the next chapter that phase-controlled rectifiers produce a nonsinusoidal current waveform whose fundamental component lags the voltage. This lagging current does not arise from energy storage, but it does nonetheless lead to a reduced displacement factor, and to rms current and apparent power that are greater than the minimum amount necessary to transmit the average power.

16.5 HARMONIC CURRENTS IN THREE PHASE SYSTEMS

The presence of harmonic currents can also lead to some special problems in three-phase systems. In a four-wire three-phase system, harmonic currents can lead to large currents in the neutral conductors, which may easily exceed the conductor rms current rating. Power factor correction capacitors may experience significantly increased rms currents, causing them to fail. In this section, these problems are examined, and the properties of harmonic current flow in three-phase systems are derived.

16.5.1 Harmonic Currents in Three-Phase Four-Wire Networks

Let us consider the three-phase four-wire network of Fig. 16.9. In general, we can express the Fourier series of the line currents and line-neutral voltages as follows:

$$
\begin{aligned}
i_a(t) &= I_{a0} + \sum_{k=1}^{\infty} I_{ak} \cos\left(k\omega t - \theta_{ak}\right) \\
i_b(t) &= I_{b0} + \sum_{k=1}^{\infty} I_{bk} \cos\left(k(\omega t - 120^\circ) - \theta_{bk}\right) \\
i_c(t) &= I_{c0} + \sum_{k=1}^{\infty} I_{ck} \cos\left(k(\omega t + 120^\circ) - \theta_{ck}\right)
\end{aligned}
\tag{16.31}
$$

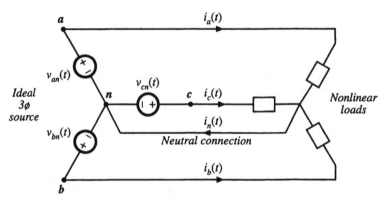

Fig. 16.9   Current flow in a three-phase four-wire network.

$$v_{an}(t) = V_m \cos (\omega t)$$
$$v_{bn}(t) = V_m \cos (\omega t - 120°) \qquad (16.32)$$
$$v_{cn}(t) = V_m \cos (\omega t + 120°)$$

The neutral current is therefore $i_n = i_a + i_b + i_c$, or

$$i_n(t) = I_{a0} + I_{b0} + I_{c0} +$$
$$\sum_{k=1}^{\infty} \left[I_{ak} \cos (k\omega t - \theta_{ak}) + I_{bk} \cos (k(\omega t - 120°) - \theta_{bk}) + I_{ck} \cos (k(\omega t + 120°) - \theta_{ck}) \right] \qquad (16.33)$$

When the load is unbalanced (even though the voltages are balanced and undistorted), we can say little else about the neutral and line currents. If the load is unbalanced and nonlinear, then the line and neutral currents may contain harmonics of any order, including even and triplen harmonics.

Equation (16.33) is considerably simplified in the case where the loads are balanced. A balanced nonlinear load is one in which $I_{ak} = I_{bk} = I_{ck} = I_k$ and $\theta_{ak} = \theta_{bk} = \theta_{ck} = \theta_k$, for all k; that is, the harmonics of the three phases all have equal amplitudes and phase shifts. In this case, Eq. (16.33) reduces to

$$i_n(t) = 3I_0 + \sum_{k=3,6,9,\cdots}^{\infty} 3I_k \cos (k\omega t - \theta_k) \qquad (16.34)$$

Hence, the fundamental and most of the harmonics cancel out, and do not appear in the neutral conductor. Thus, it is in the interests of the utility to balance their nonlinear loads as well as their harmonics.

But not all of the harmonics cancel out of Eq. (16.34): the dc and *triplen* (triple-n, or 3,6,9,...) harmonics add rather than cancel. The rms neutral current is

$$i_{n, rms} = 3 \sqrt{I_0^2 + \sum_{k=3,6,9,\cdots}^{\infty} \frac{I_k^2}{2}} \qquad (16.35)$$

Example

A balanced nonlinear load produces line currents containing fundamental and 20% third harmonic: $i_{an}(t) = I_1 \cos(\omega t - \theta_1) + 0.2 I_1 \cos(3\omega t - \theta_3)$. Find the rms neutral current, and compare its amplitude to the rms line current amplitude.

Solution:

$$i_{n,\,rms} = 3\sqrt{\frac{(0.2I_1)^2}{2}} = \frac{0.6\,I_1}{\sqrt{2}}$$

$$i_{1,\,rms} = \sqrt{\frac{I_1^2 + (0.2I_1)^2}{2}} = \frac{I_1}{\sqrt{2}}\sqrt{1 + 0.04} \approx \frac{I_1}{\sqrt{2}}$$

(16.36)

So the neutral current magnitude is 60% of the line current magnitude! The triplen harmonics in the three phases add, such that 20% third harmonic leads to 60% third harmonic neutral current. Yet the presence of the third harmonic has very little effect on the rms value of the line current. Significant unexpected neutral current flows.

16.5.2 Harmonic Currents in Three-Phase Three-Wire Networks

If there is no neutral connection to the wye-connected load, as in Fig. 16.10, then $i_n(t)$ must be zero. If the load is balanced, then Eq. (16.34) still applies, and therefore the dc and triplen harmonics of the load currents must be zero. Therefore, the line currents i_a, i_b, and i_c cannot contain triplen or dc harmonics. What happens is that a voltage is induced at the load neutral point n', containing dc and triplen harmonics, which eliminates the triplen and dc load current harmonics.

This result is true only when the load is balanced. With an unbalanced load, all harmonics can appear in the line currents, including triplen and dc. In practice, the load is never exactly balanced, and some small amounts of third harmonic line currents are measured.

With a delta-connected load as in Fig. 16.11, there is also no neutral connection, so the line currents cannot contain triplen or dc components. But the loads are connected line-to-line, and are excited by undistorted sinusoidal voltages. Hence triplen harmonic and dc currents do, in general, flow through the nonlinear loads. Therefore, these currents simply circulate around the delta. If the load is balanced, then again no triplen harmonics appear in the line currents.

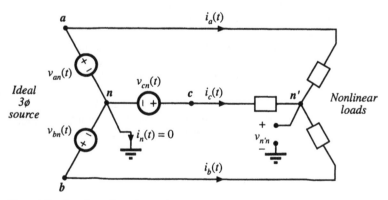

Fig. 16.10 Current flow in a three-phase three-wire wye-connected network.

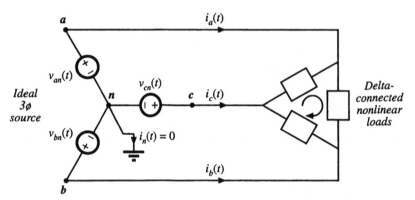

Fig. 16.11 A balanced nonlinear delta-connected load may generate triplen current harmonics. These harmonics circulate around the delta, but do not flow through the lines if the load is balanced.

16.5.3 Harmonic Current Flow in Power Factor Correction Capacitors

Harmonic currents tend to flow through shunt-connected power factor correction capacitors. To some extent, this is a good thing because the capacitors tend to low-pass filter the power system currents, and prevent nonlinear loads from polluting the entire power system. The flow of harmonic currents is then confined to the nonlinear load and local power factor correction capacitors, and voltage waveform distortion is reduced. High-frequency harmonic currents tend to flow through shunt capacitors because the capacitor impedance decreases with frequency, while the inductive impedance of transmission lines increases with frequency. In this sense, power factor correction capacitors mitigate the effects of harmonic currents arising from nonlinear loads in much the same way that they mitigate the effects of reactive currents that arise from inductive loads.

But the problem is that the power factor correction capacitors may not be rated to handle these harmonic currents, and hence there is a danger that the capacitors may overheat and fail when they are exposed to significant harmonic currents. The loss in capacitors is modeled using an *equivalent series resistance* (esr) as shown in Fig. 16.12. The esr models dielectric loss (hysteresis of the dielectric *D–E* loop), contact resistance, and foil and lead resistances. Power loss occurs, equal to $i_{rms}^2(esr)$. Dielectric materials are typically poor conductors of heat, so a moderate amount of power loss can cause a large temperature rise in the center of the capacitor. In consequence, the rms current must be limited to a safe value.

Typical power factor correction capacitors are rated by voltage V, frequency f, and reactive power in kVARs. These ratings are computed from the capacitance C and safe rms current I_{rms}, assuming undistorted sinusoidal waveforms, as follows:

Fig. 16.12 Capacitor equivalent circuit. Losses are modeled by an equivalent series resistance (esr).

$$\text{rated rms voltage } V_{rms} = \frac{I_{rms}}{2\pi f C} \tag{16.37}$$

$$\text{rated reactive power} = \frac{I_{rms}^2}{2\pi f C} \tag{16.38}$$

In an undistorted system, the rms current, and hence also the capacitor esr loss, cannot increase unless the rms voltage is also increased. But high-frequency harmonics can lead to larger rms currents without an increased voltage. Any harmonics that flow result in increased rms current beyond the expected value predicted by Eq. (16.37). If the capacitor is not rated to handle additional power loss, then failure or premature aging can occur.

16.6   AC LINE CURRENT HARMONIC STANDARDS

Besides the increased currents and reduced power factors of peak detection rectifiers, the harmonics themselves can be detrimental: if large enough in magnitude, they can pollute the power system. Harmonic currents cause distortion of the voltage waveform via the power system series impedance. These voltage harmonics can interfere with the operation of nearby loads. As noted previously, increased currents in shunt capacitors, and increased losses in distribution transformers and ac machines, can lead to premature aging and failure of these devices. Odd triplen harmonics (triple-n: 3rd, 9th, 15th, etc.) lead to unexpectedly large neutral currents in three-phase systems. Harmonic currents can also excite system resonances some distance from their source, with results that are difficult to predict. For these reasons, a number of organizations have adopted standards that limit the magnitudes of the harmonic currents that a load is allowed to inject into the ac line. The US military was one of the early organizations to recognize these problems; the very strict 3% limit was initially adopted. The standards adopted by the IEC and IEEE are more recent, and are intended for conventional utility systems. A fourth example, not discussed here, is the telephone interference factor, which limits power distribution system harmonics in cases when telephone lines and power lines share the same poles.

16.6.1   International Electrotechnical Commission
Standard 1000

This international agency adopted a first draft of their IEC 555 standard in 1982. It has since undergone a number of revisions, and has been superceded by IEC 1000 [7]. This standard is now enforced in Europe, making it a de facto standard for commercial equipment intended to be sold worldwide.

The IEC 1000-3-2 standard covers a number of different types of low power equipment, with differing harmonic limits. It specifically limits harmonics for equipment having an input current of up to 16 A, connected to 50 or 60 Hz, 220 V to 240 V single phase circuits (two or three wire), as well as 380 V to 415 V three phase (three or four wire) circuits. In a city environment such as a large building, a large fraction of the total power system load can be nonlinear. For example, a major portion of the electrical load in a building is comprised of fluorescent lights, which present a very nonlinear characteristic to the utility system. A modern office may also contain a large number of personal computers, printers, copiers, etc., each of which may employ peak detection rectifiers. Although each individual load is a negligible fraction of the total local load, these loads can collectively become significant.

The IEC 1000-3-2 standard defines several categories of equipment, each of which is covered by a different set of harmonic limits. As an example, Table 16.1 shows the harmonic limits for *Class A* equipment, which includes low harmonic rectifiers for computer and other office equipment.

The European norm EN 61000-3-2 defines similar limits.

Table 16.1 IEC 1000-3-2 Harmonic current limits, class A

Odd harmonics		Even harmonics	
Harmonic number	Maximum current	Harmonic number	Maximum current
3	2.30 A	2	1.08 A
5	1.14 A	4	0.43 A
7	0.77 A	6	0.30 A
9	0.40 A	$8 \leq n \leq 40$	$0.23 \, A \cdot (8/n)$
11	0.33 A		
13	0.21 A		
$15 \leq n \leq 39$	$0.15 \, A \cdot (15/n)$		

16.6.2 IEEE/ANSI Standard 519

In 1993, the IEEE published a revised draft standard limiting the amplitudes of current harmonics, *IEEE Guide for Harmonic Control and Reactive Compensation of Static Power Converters*. The harmonic limits are based on the ratio of the fundamental component of the load current I_L to the short circuit current at the point of common coupling (PCC) at the utility I_{sc}. Stricter limits are imposed on large loads than on small loads. The limits are similar in magnitude to IEC 1000, and cover high voltage loads (of much higher power) not addressed by IEC 1000. Enforcement of this standard is presently up to the local utility company.

The odd harmonic limits for general distribution systems at voltages of 120 V to 69 kV are listed in Table 16.2. The limits for even harmonics are 25% of the odd harmonic limits. Limits for general distribution systems at 69.001 kV to 161 kV are 50% of the values listed in Table 16.2. Dc current components and half-wave rectifiers are not allowed.

It is the responsibility of the power consumer to meet these current harmonic standards. Standard IEEE-519 also specifies maximum allowable voltage harmonics, listed in Table 16.3. It is the responsibility of the utility, or power supplier, to meet these limits. Both total harmonic distortion and maximum individual harmonic magnitudes are limited.

Table 16.2 IEEE-519 Maximum odd harmonic current limits for general distribution systems, 120 V to 69 kV

I_{sc}/I_L	$n < 11$	$11 \leq n < 17$	$17 \leq n < 23$	$23 \leq n < 35$	$35 \leq n$	THD
< 20	4.0%	2.0%	1.5%	0.6%	0.3%	5.0%
20-50	7.0%	3.5%	2.5%	1.0%	0.5%	8.0%
50-100	10.0%	4.5%	4.0%	1.5%	0.7%	12.0%
100-1000	12.0%	5.5%	5.0%	2.0%	1.0%	15.0%
> 1000	15.0%	7.0%	6.0%	2.5%	1.4%	20.0%

Table 16.3   IEEE-519 Voltage distortion limits

Bus voltage at PCC	Individual harmonics	THD
69 kV and lower	3.0%	5.0%
69.001 kV to 161 kV	1.5%	2.5%
Above 161 kV	1.0%	1.5%

BIBLIOGRAPHY

[1]   J. ARRILLAGA, D. BRADLEY, and P. BODGER, *Power System Harmonics*, New York: John Wiley & Sons, 1985.

[2]   R. SMITY and R. STRATFORD, "Power System Harmonics Effects from Adjustable-Speed Drives," *IEEE Transactions on Industry Applications*, Vol. IA-20, No. 4, pp. 973-977, July/August 1984.

[3]   A. EMANUEL, "Powers in Nonsinusoidal Situations—A Review of Definitions and Physical Meaning," *IEEE Transactions on Power Delivery*, Vol. 5, No. 3, pp. 1377-1389, July 1990.

[4]   N. MOHAN, T. UNDELAND, and W. ROBBINS, *Power Electronics: Converters, Applications, and Design*, Second edition, New York: John Wiley & Sons, 1995.

[5]   J. KASSAKIAN, M. SCHLECHT, and G. VERGESE, *Principles of Power Electronics*, Massachusetts: Addison-Wesley, 1991.

[6]   R. GRETSCH, "Harmonic Distortion of the Mains Voltage by Switched-Mode Power Supplies—Assessment of the Future Development and Possible Mitigation Measures," *European Power Electronics Conference*, 1989 Record, pp. 1255-1260.

[7]   IEC 1000-3-2, First Edition, Commission Electrotechnique Internationale, Geneva, 1995.

PROBLEMS

16.1   Passive rectifier circuit. In the passive rectifier circuit of Fig. 16.13, L is very large, such that the inductor current $i(t)$ is essentially dc. All components are ideal.

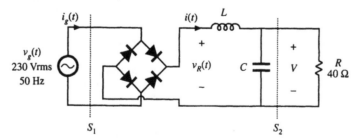

Fig. 16.13   Passive rectifier circuit of Problem 16.1

(a)   Determine the dc output voltage, current, and power.

(b)   Sketch the ac line current waveform $i_g(t)$ and the rectifier output voltage waveform $v_R(t)$.

(c)   Determine the ac line current rms magnitude, fundamental rms magnitude, and third harmonic rms magnitude. Does this rectifier network conform to the IEC-1000 harmonic current limits?

(d)   Determine the power factor, measured at surfaces S_1 and S_2.

16.2   The three-phase rectifier of Fig. 16.14 is connected to a balanced 60 Hz 3øac 480 V (rms, line-line) sinusoidal source as shown. All elements are ideal. The inductance L is large, such that the current $i(t)$ is essentially constant, with negligible 360 Hz ripple.

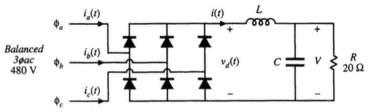

Fig. 16.14   Three-phase rectifier circuit of Problem 16.2

(a)   Sketch the waveform $v_d(t)$.

(b)   Determine the dc output voltage V.

(c)   Sketch the line current waveforms $i_a(t)$, $i_b(t)$, and $i_c(t)$.

(d)   Find the Fourier series of $i_a(t)$.

(e)   Find the distortion factor, displacement factor, power factor, and line current THD.

16.3   Harmonic pollution police. In the network of Fig. 16.15, voltage harmonics are observed at the indicated surface. The object of this problem is to decide whether to blame the source or the load for the observed harmonic pollution. Either the source element or the load element contains a nonlinearity that generates harmonics, while the other element is linear.

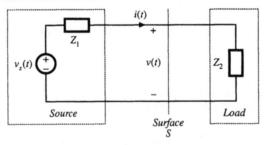

Fig. 16.15   Single-phase power system of Problems 16.3 to 16.5

(a)   Consider first the case where the load is a passive linear impedance $Z_2(s)$, and hence its phase lies in the range $-90° \le \angle Z_2(j\omega) \le +90°$ for all positive ω. The source generates harmonics. Express the average power P in the form

$$P = \sum_{n=0}^{\infty} P_n$$

where P_n is the average power transmitted to the load by harmonic number n. What can you say about the polarities of the P_ns?

(b)   Consider next the case where the load is nonlinear, while the source is linear and can be modeled by a Thevenin-equivalent sinusoidal voltage source and linear impedance $Z_1(s)$. Again express the average power P as a sum of average powers, as in part (a). What can you say about the polarities of the P_ns in this case?

(c)   The following Fourier series are measured:

Harmonic number	$v(t)$		$i(t)$	
	Magnitude	Phase	Magnitude	Phase
1	230 V	0°	6 A	– 20°
3	20 V	180°	4 A	20°
5	8 V	60°	1 A	– 110°

Who do you accuse? Explain your reasoning.

16.4   For the network and waveforms of Problem 16.3, determine the power factor at the indicated surface, and the average power flowing to the load. Harmonics higher in frequency than the fifth harmonic are negligible in magnitude.

16.5   Repeat Problem 16.3(c), using the following Fourier series:

Harmonic number	$v(t)$		$i(t)$	
	Magnitude	Phase	Magnitude	Phase
1	120 V	0°	5 A	25°
3	4 V	60°	0.5 A	40°
5	2 V	– 160°	0.2 A	– 100°

16.6   A balanced three-phase wye-connected load is constructed using a 20 Ω resistor in each phase. This load is connected to a balanced three-phase wye-connected voltage source, whose fundamental voltage component is 380 Vrms line-to-line. In addition, each (line-to-neutral) voltage source produces third and fifth harmonics. Each harmonic has amplitude 20 Vrms, and is in phase with the (line-to-neutral) fundamental.

(a)   The source and load neutral points are connected, such that a four-wire system is obtained. Find the Fourier series of the line currents and the neutral current.

(b)   The neutral connection is broken, such that a three-wire system is obtained. Find the Fourier series of the line currents. Also find the Fourier series of the voltage between the source and load neutral points.

17

Line-Commutated Rectifiers

Conventional diode peak-detection rectifiers are inexpensive, reliable, and in widespread use. Their shortcomings are the high harmonic content of their ac line currents, and their low power factors. In this chapter, the basic operation and ac line current waveforms of several of the most common single-phase and three-phase diode rectifiers are summarized. Also introduced are phase-controlled three-phase rectifiers and inverters, and passive harmonic mitigation techniques. Several of the many references in this area are listed at the end of this chapter [1–15].

Rigorous analytical design of line-commutated rectifier and filter circuits is unfeasible for all but the simplest of circuits. Typical peak-detection rectifiers are numerically ill-conditioned, because small changes in the dc-side ripple voltage lead to large changes in the ac line current waveforms. Therefore, the discussions of this chapter are confined to mostly qualitative arguments, with the objective of giving the reader some insight into the physical operation of rectifier/filter circuits. Waveforms, harmonic magnitudes, and power factors are best determined by measurement or computer simulation.

17.1 THE SINGLE-PHASE FULL-WAVE RECTIFIER

A single-phase full-wave rectifier, with uncontrolled diode rectifiers, is shown in Fig. 17.1. The circuit includes a dc-side L–C filter. There are two conventional uses for this circuit. In the traditional full-wave rectifier, the output capacitor is large in value, and the dc output voltage $v(t)$ has negligible ripple at the second harmonic of the ac line frequency. Inductor L is most often small or absent. Additional small inductance may be in series with the ac source $v_g(t)$. A second conventional use of this circuit is in the low-harmonic rectifiers discussed in the next chapter. In this case, the resistive load is replaced by a dc-dc converter that is controlled such that its power input port obeys Ohm's law. For the purposes of understanding the rectifier waveforms, the converter can be modeled by an effective resistance R, as in the cir-

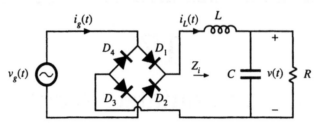

Fig. 17.1   Conventional single-phase full-wave rectifier, with dc-side *L–C* filter.

cuit of Fig. 17.1. In this application, the *L–C* filter is required to filter the conducted electromagnetic interference (EMI) generated by the converter. The inductor and capacitor element values are typically small in value, and $v(t)$ is approximately a rectified sinusoid. More generally, there may be several sections of *L–C* filter networks, connected to both the dc and ac sides of the diode rectifier, which filter EMI, smooth the dc output voltage, and reduce the ac line current harmonics.

The presence of any filter degrades the ac current waveform of the rectifier. With no reactive elements ($L = 0$ and $C = 0$), the rectifier presents a purely resistive load to the ac input. The output voltage $v(t)$ is then a rectified sinusoid, there are no ac line current harmonics, and the power factor is unity. Addition of reactive elements between the rectifier diodes and the load leads in general to ac line current harmonics. Given that such a filter is necessary, one might ask what can be done to keep these harmonics as small as possible. In this section, the dependence of the ac line current total harmonic distortion on the filter parameters is described.

The circuit of Fig. 17.1 generates odd harmonics of the ac line voltage in the ac line current. The dc output voltage contains dc and even harmonics of the ac line voltage. The circuit exhibits several modes of operation, depending on the relative values of *L*, *C*, and *R*. It is easiest to understand these modes by considering the limiting cases, as follows.

17.1.1   Continuous Conduction Mode

When the inductor *L* is very large, then the inductor current $i_L(t)$ is essentially constant. This follows from the inductor definition $v_L(t) = L\,di_L(t)/dt$. For a given applied inductor voltage waveform $v_L(t)$, the slope $di_L(t)/dt$ can be made arbitrarily small by making *L* sufficiently large. In the limiting case where *L* is infinite, the slope $di_L(t)/dt$ becomes zero, and the inductor current is constant dc. To provide a path for the constant inductor to flow, at least two of the rectifier diodes must conduct at any given instant in time. For the circuit of Fig. 17.1, diodes D_1 and D_3 conduct when the ac line voltage $v_g(t)$ is positive, and D_2 and D_4 conduct when $v_g(t)$ is negative. The ac line current waveform is therefore a square wave, with $i_g(t) = i_L(t)$ when $v_g(t)$ is positive, and $i_g(t) = -i_L(t)$ when $v_g(t)$ is negative. The diode conduction angle β, defined as the angle through which one of the diodes conducts, is equal to 180° in CCM.

The rms value of a square wave is equal to its peak value I_{pk}, in this case the dc load current. The fundamental component of a square wave is equal to $4I_{pk}/\pi$. The square-wave contains odd harmonics which vary as $1/n$. The distortion factor is therefore

$$\text{distortion factor} = \frac{I_{1,\,rms}}{I_{rms}} = \frac{4}{\pi\sqrt{2}} = 90.0\% \tag{17.1}$$

The total harmonic distortion is

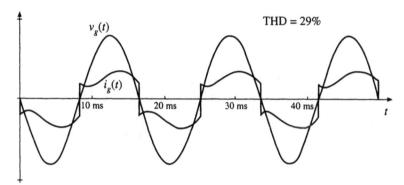

Fig. 17.2 Typical ac line current and voltage waveforms, continuous conduction mode. $f_0/f_L = 5$, $Q = 0.25$.

$$\text{THD} = \sqrt{\left(\frac{1}{\text{distortion factor}}\right)^2 - 1} = 48.3\% \tag{17.2}$$

So the limiting case of the large inductor leads to some significant harmonic distortion, although it is not as bad as the peak detection rectifier case. Since the square wave is in phase with the ac input voltage, the displacement factor is unity, and hence the power factor is equal to the distortion factor.

Whenever the inductor is sufficiently large, the rectifier diodes conduct continuously (i.e., there is no time interval in which all four diodes are reverse-biased). This is called the continuous conduction mode (CCM). A typical ac line current waveform is plotted in Fig. 17.2 for a finite but large value of L. It can be seen that the ac line current is discontinuous at the ac line voltage zero crossing, as in the square-wave limiting case. Some ringing is also present. This waveform contains a total harmonic distortion of approximately 29%.

17.1.2 Discontinuous Conduction Mode

The opposite case occurs when the inductor is very small and the capacitor C is very large. This is the peak detector circuit. In the limit as L goes to zero and C goes to infinity, the ac line current approaches a string of delta (impulse) functions that coincide with the peaks of the sinusoidal input voltage waveform. It can be shown that, in this limiting case, the THD becomes infinite while the distortion factor and power factor become zero. Of course, in the practical case the current is not infinite; nonetheless, large THD with low power factor is quite possible. The diodes conduct for less than one-half of the ac line period, and hence $\beta < 180°$ in DCM.

Whenever the capacitor is large and the inductor is small, the rectifier tends to "peak detect," and the rectifier operates in the discontinuous conduction mode (DCM). There exist time intervals of nonzero length where all four rectifier diodes are reverse-biased. A typical set of waveforms is plotted in Fig. 17.3, where the capacitor is large but finite, and the inductor is small but nonzero. The ac line voltage and the value of the load resistance are the same as in Fig. 17.2, yet the peak current is substantially larger. The THD for this waveform is 145%, and the distortion factor is 57%.

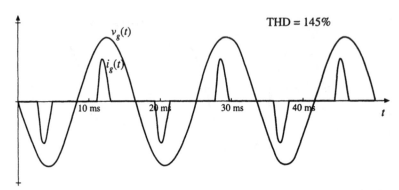

Fig. 17.3   Typical ac line current and voltage waveforms, discontinuous conduction mode. $f_0/f_L = 8.4$, $Q = 121$.

17.1.3   Behavior When C is Large

A variety of authors have discussed the solution of passive rectifier circuits; several works are listed in the references [8–15]. Analysis of even the simple circuit of Fig. 17.1 is surprisingly complex. For the case when C is infinite, it was shown in [8] that the rectifier waveshapes can be expressed as a function of a single dimensionless parameter K_L, defined as

$$K_L = \frac{2L}{RT_L}$$

(17.3)

where $T_L = 1/f_L$ is the ac line period. Equation (17.3) is of the same form as Eq. (5.6), used to define the dimensionless parameter K which governs the DCM behavior of PWM converters. Figure 17.4 illustrates the behavior of the single-phase rectifier circuit of Fig. 17.1, as a function of K_L and for infinite C [8]. When K_L is greater than approximately 0.1, the rectifier operates in CCM, with waveforms similar to those in Fig. 17.2.

The voltage conversion ratio M is defined as

$$M = \frac{V}{V_m}$$

(17.4)

where V_m is the peak value of the sinusoidal ac input voltage. In CCM, the output voltage is ideally independent of load, with $M = 2/\pi$. Addition of ac-side inductance can cause the output voltage to exhibit a dependence on load current. The total harmonic distortion in CCM is nearly constant and equal to the value given by Eq. (17.2).

Near the boundary between CCM and DCM, the fundamental component of the line current significantly lags the line voltage. The displacement factor reaches a minimum value slightly less than 80%, and power factors between 70% and 80% are observed.

For $K_L \ll 0.1$, the rectifier operates heavily in DCM, as a peak-detection rectifier. As K_L is decreased, the displacement factor approaches unity, while the THD increases rapidly. The power factor is dominated by the distortion factor. The output voltage becomes dependent on the load, and hence the rectifier exhibits a small but nonzero output impedance.

For K_L less than approximately 0.05, the waveforms are unchanged when some or all of the inductance is shifted to the ac side of the diode bridge. Figure 17.4 therefore applies to rectifiers contain-

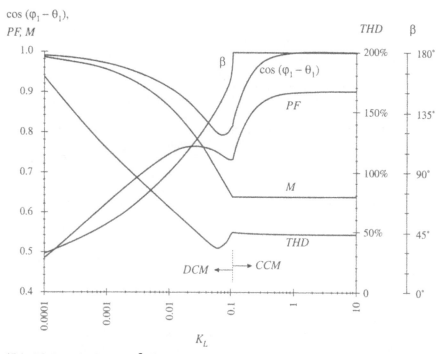

Fig. 17.4 Diode conduction angle β, displacement factor, power factor, conversion ratio, and total harmonic distortion of the rectifier circuit of Fig. 17.1, with infinite capacitance.

ing both ac-side and dc-side inductance, provided that the circuit operates sufficiently deeply in DCM. The parameter K_L is computed according to Eq. (17.3), with L taken to be the total ac-side plus dc-side inductance. A common example is the case where the circuit contains no physical discrete inductor; the performance is then determined by parasitic elements such as the capacitor equivalent series inductance, the inductance of the utility distribution wiring, and transformer leakage inductances.

17.1.4 Minimizing THD When C is Small

Let us now consider the performance of the second case, in which the inductor and capacitor are small and are intended solely to prevent load-generated EMI from reaching the ac line. In this case, dc-side filtering of the low-frequency even voltage harmonics of the ac line frequency is not necessary. The filter can be characterized by a corner frequency f_0, characteristic impedance R_0, and Q-factor, where

$$f_0 = \frac{1}{2\pi\sqrt{LC}}$$
$$R_0 = \sqrt{\frac{L}{C}}$$
$$Q = \frac{R}{R_0}$$
$$f_p = \frac{1}{2\pi RC} = \frac{f_0}{Q}$$

$$(17.5)$$

To obtain good filtering of the EMI, the corner frequency f_0 should be selected to be sufficiently low. However, as can be seen from Eq. (17.5), reducing the value of f_0 requires increasing the values of L and/or C. As described above, it is undesirable to choose either element value too large, because large distortion results. So f_0 should not be too low, and there is a limit to the amount of filtering that can be obtained without significantly distorting the ac line current waveform.

How low can f_0 be? Once f_0 is chosen, how should L and C be chosen such that THD is minimized? We might expect that THD is increased when the phase of the filter input impedance $- Z_i(j\omega)$, evaluated at the second harmonic of the line frequency or $2f_L$, differs significantly from $0°$. When the zero crossings of the voltage and current waveforms do not coincide, then diode switching distorts the current waveform. To a lesser extent, input impedance phase shift at the higher-order even harmonic frequencies of the ac line frequency should also affect the THD. The input impedance $Z_i(s)$ contains two zeroes at frequency f_0, and a pole at frequency $f_p = f_0/Q$. To obtain small phase shift at low frequency, f_0 must be sufficiently large. In addition, Q must be neither too small nor too large: small Q causes the zeroes at f_0 to introduce low-frequency phase shift, while large Q causes the pole at f_p to occur at low frequency.

An approximate plot of THD vs. the choice of f_0 and Q is given in Fig. 17.5. It can be seen that there is an optimum choice for Q: minimum THD occurs when Q lies in the range 0.5 to 1. A typical waveform is plotted for the choice $f_0/f_L = 10$, $Q = 1$, in Fig. 17.6. The THD for this waveform is 3.6%, and the distortion factor is 99.97%.

Small Q corresponds to CCM operation, with large L and small C. In the extreme case as $Q \to 0$, the ac line current tends to a square wave with THD = 48%. Large Q corresponds to DCM operation, with small L and large C. In the extreme case as $Q \to \infty$, the ac line current tends to a string of delta functions with THD $\to \infty$. The optimum choice of Q leads to operation near the CCM-DCM boundary, such that the ac line current waveform contains neither step changes nor subintervals of zero current.

In the case when the load resistance R varies over a wide range of values, it may be difficult to optimize the circuit such that low THD is always obtained. It can be seen that increasing f_0/f_L leads to low THD for a wider range of load resistance. For example, when $f_0 = 5f_L$, THD $\le 10\%$ can be obtained only for Q between approximately 0.6 and 1.5, which is a 2.5:1 range of load resistance variations. If the

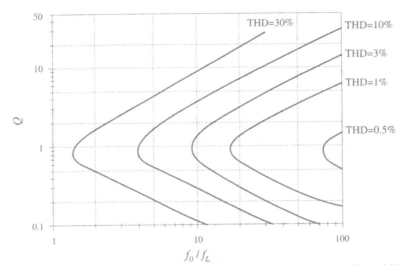

Fig. 17.5 Approximate total harmonic distortion of the single-phase diode rectifier with dc-side *L–C* filter.

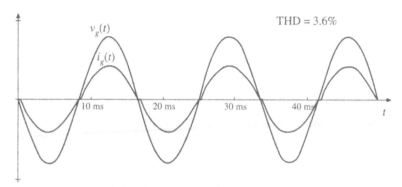

Fig. 17.6 Typical ac line current and voltage waveforms, near the boundary between continuous and discontinuous modes and with small dc filter capacitor. $f_0/f_L = 10$, $Q = 1$.

filter cutoff frequency f_0 is increased to $20f_L$, then THD $\leq 10\%$ is obtained for Q between approximately 0.15 and 7, or nearly a 50:1 range of resistance variations. In most cases, maximum harmonic limits are enforced only at full load, and hence it is possible to design with relatively low values of f_0/f_L if desired.

17.2   THE THREE-PHASE BRIDGE RECTIFIER

A basic full-wave three-phase uncontrolled rectifier with LC output filter is shown in Fig. 17.7. Its behavior is similar to the single-phase case, in that it exhibits both continuous and discontinuous conduction modes, depending on the values of L and C. The rectifier generates odd non-triplen harmonics in the ac line current. So the ac line current may contain 1st, 5th, 7th, 11th, 13th, etc. harmonics. The dc output may contain dc and even triplen harmonics: 0, 6, 12, 18, etc.

   In the basic circuit of Fig. 17.7, no more than two of the six diodes can conduct during each interval, and hence the line current waveforms must contain intervals of nonzero length during which the current is zero. Unlike the single-phase case, the ac line current waveform must contain distortion even when the filter elements are removed.

17.2.1   Continuous Conduction Mode

In the continuous conduction mode, each ac line current is nonzero for 120 degrees out of each line half-cycle. For the remaining 60 degrees, the current is zero. This mode occurs when the inductance L is suf-

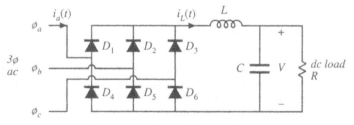

Fig. 17.7 Basic uncontrolled 3ø bridge rectifier circuit, with dc-side L–C filter.

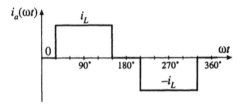

Fig. 17.8 Ac line current waveform $i_a(t)$, for the case when inductor L is large. The phase is drawn with respect to the zero crossing of the line-to-neutral voltage $v_{an}(t)$.

ficiently large, as well as when the filter elements L and C are removed entirely.

In the limit, when L is very large, then the current $i_L(t)$ is essentially constant. The current in phase a, $i_a(t)$, is then as shown in Fig. 17.8. It can be shown that the Fourier series for this waveform is

$$i_a(t) = \sum_{n=1,5,7,11,...}^{\infty} \frac{4}{n\pi} I_L \sin\left(\frac{n\pi}{2}\right) \sin\left(\frac{n\pi}{3}\right) \sin\left(n\omega t\right) \tag{17.6}$$

which is similar to the spectrum of the square wave of the single-phase case, but with the triplen harmonics missing. The THD of this waveform is 31%, and the distortion factor is $3/\pi = 95.5\%$. As in the case of the square wave, the amplitude of the odd nontriplen n^{th} harmonic is $(1/n)$ times the fundamental amplitude. So this waveform contains 20% fifth harmonic, 14% seventh harmonic, 9% eleventh harmonic, etc. It is interesting that, in comparison with the square-wave single-phase case, the missing 60° in the three-phase case improves the THD and power factor, by removing the triplen harmonics.

With a less-than-infinite value of inductance, the output ripple causes the ac line currents to be rounded, as in the typical waveform of Fig. 17.9. This waveform has a 31.9% THD, with a distortion factor that is not much different from the waveform of Fig. 17.8.

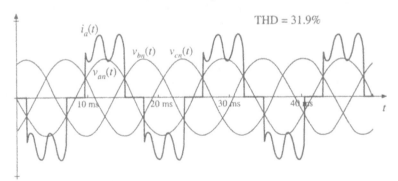

Fig. 17.9 Continuous conduction mode ac line-neutral voltages and phase a current, for a moderate value of inductance.

17.2.2 Discontinuous Conduction Mode

If the inductance is further reduced, then the three-phase rectifier enters the discontinuous conduction mode. The rectifier then begins to peak-detect, and the current waveforms become narrow pulses of high amplitude, occurring near the peaks of the line-line voltages. The phase a line current $i_a(t)$ contains two positive and two negative pulses, at the positive and negative peaks of the line-line voltages $v_{ab}(t)$ and $v_{ac}(t)$. As in the single-phase case, the total harmonic distortion becomes quite large in this case, and the

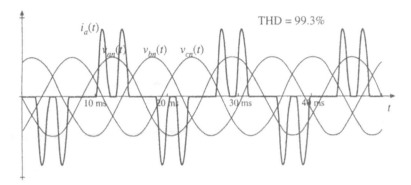

Fig. 17.10 Discontinuous conduction mode ac line-neutral voltages and phase *a* current.

power factor can be significantly degraded.

A typical waveform is given in Fig. 17.10. This waveform has a THD of 99.3%, and a distortion factor of 71%. This would be considered unacceptable in high-power applications, except perhaps at light load.

17.3 PHASE CONTROL

There are a wide variety of schemes for controlling the dc output of a 3ø rectifier using thyristors [1,2]. The most common one is shown in Fig. 17.11, in which the six diodes of Fig. 17.7 are replaced by silicon controlled rectifiers (SCRs). Typical waveforms are given in Fig. 17.12, for large dc-side filter inductance.

If Q_1 were an uncontrolled diode, it would conduct whenever the line-to-line voltage v_{ab} or v_{ac} is the largest in magnitude of the six line–line voltages v_{ab}, v_{bc}, v_{ca}, v_{ba}, v_{cb}, and v_{ca}. This occurs for 120° of each cycle, beginning at the point where $v_{ab} = v_{cb}$. In Fig. 17.12, this occurs at $\omega t = 60°$. The output voltage of the controlled rectifier is controlled by delaying the firing of Q_1 by an angle α, such that Q_1 begins conducting at $\omega t = 60° + \alpha$. This has the effect of reducing the dc output voltage.

There can be no dc component of voltage across inductor L. Hence, in steady-state, the dc component of the rectifier output voltage $v_R(t)$ must equal the dc load voltage V. But $v_R(t)$ is periodic, with period equal to six times the ac line period (or 60°). So the dc component of $v_R(t)$ can be found by Fourier analysis, and is equal to the average value of $v_R(t)$. Over one 60° interval, for example $(60° + \alpha) \leq \omega t \leq (120° + \alpha)$, $v_R(t)$ follows the line-line voltage $v_{ab}(t) = 3V_m \sin (\omega t + 30°)$. The average is therefore

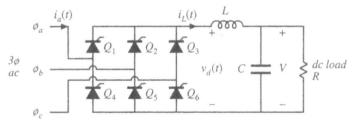

Fig. 17.11 Basic controlled 3ø bridge rectifier circuit, with dc-side *L–C* filter.

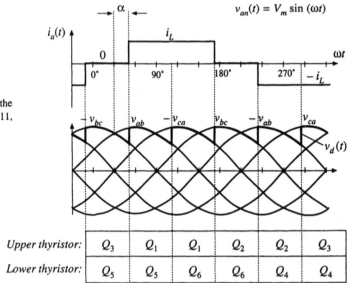

Fig. 17.12 Waveforms for the controlled rectifier of Fig. 17.11, with large dc filter inductance.

$$V = \frac{3}{\pi} \int_{30°+\alpha}^{90°+\alpha} \sqrt{3} \, V_m \sin\left(\theta + 30°\right) d\theta \qquad (17.7)$$

$$= \frac{3\sqrt{2}}{\pi} V_{L\text{-}L, \, rms} \cos \alpha$$

where $V_{L\text{-}L, \, rms}$ is the rms line-to-line voltage. This equation is plotted in Fig. 17.13. It can be seen that, if it is necessary to reduce the dc output voltage to values close to zero, then the delay angle α must be

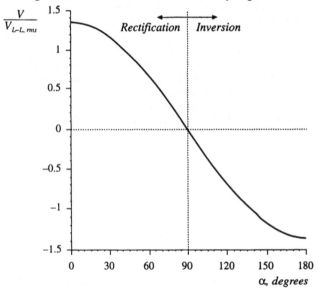

Fig. 17.13 Variation of the dc output voltage V with delay angle α, for the phase-controlled circuit of Fig. 17.11.

increased to close to 90°. With a small inductance, the controlled rectifier can also operate in discontinuous conduction mode, with modified output voltage characteristics.

17.3.1 Inverter Mode

If the dc load is capable of supplying power, then it is possible for the direction of power flow to reverse. For example, consider the three-phase controlled rectifier circuit of Fig. 17.14. The resistive load is replaced by a voltage source and thevenin-equivalent resistance, capable of either supplying or consuming power. The dc load power is equal to VI_L, which is positive (rectifier mode) when both V and I_L are positive. The thyristor is a unidirectional-current switch, which cannot conduct negative current, and hence I_L must always be positive. However, it is possible to cause the output voltage V to be negative, by increasing the delay angle α. The dc load power VI_L then becomes a negative quantity (inverter mode), meaning that power flows from the dc load into the 3øac system.

    Provided that the dc-side filter inductance L is sufficiently large, then Eq. (17.7) is valid even when the delay angle α is greater than 90°. It can be seen in Fig. 17.13 that the dc output voltage V becomes negative for $\alpha > 90°$, and hence the power flow indeed reverses. Delay angles approaching 180° are possible, with the maximum angle limited by commutation of the thyristor devices.

17.3.2 Harmonics and Power Factor

Let us next consider the harmonic content and power factor of the phase-controlled rectifier with large inductance. Comparison of the line current waveform of Fig. 17.12 with that of the uncontrolled rectifier (Fig. 17.8) reveals that the waveshapes are identical. The only difference is the phase lag α present in the phase-controlled rectifier. This has the effect of shifting the fundamental component of current (and the harmonics as well) by angle α. The Fourier series is therefore

$$i_a(t) = \sum_{n=1,5,7,11,...}^{\infty} \frac{4}{n\pi} I_L \sin\left(\frac{n\pi}{2}\right) \sin\left(\frac{n\pi}{3}\right) \sin\left(n\omega t - n\alpha\right) \tag{17.8}$$

Hence the harmonic amplitudes are the same (the fifth harmonic amplitude is 20% of the fundamental, etc.), the THD is again 31%, and the distortion factor is again 95.5%. But there is phase lag in the fundamental component of current, which leads to a displacement factor of $\cos(\alpha)$. The power factor is therefore

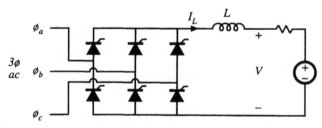

Fig. 17.14 If the load is capable of supplying power, then the 3ø bridge circuit functions as an inverter for $V < 0$ and $\alpha > 90°$.

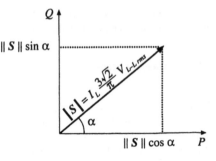

Fig. 17.15 Fundamental component complex power diagram
for the 3ø bridge circuit operating in rectifier mode.

$$\text{power factor} = 0.955 \left| \cos(\alpha) \right| \tag{17.9}$$

which can be quite low when the dc output voltage V is low.

It is at first somewhat puzzling that the introduction of phase control can cause the fundamental
current to lag the voltage. Apparently, the rectifier consumes reactive power equal to

$$Q = \sqrt{3} \, I_{a,\,rms} V_{L\text{-}L,\,rms} \sin \alpha = I_L \frac{3\sqrt{2}}{\pi} V_{L\text{-}L,\,rms} \sin \alpha \tag{17.10}$$

We normally associate lagging current and the consumption of reactive power with inductive energy
storage. But this is not what is happening in the rectifier; indeed, the inductor and capacitor can be
removed entirely from the rectifier circuit, and a lagging fundamental current is still obtained by phase
control. It is simply the delay of the switching of the rectifiers that causes the current to lag, and no
energy storage is involved. So two mechanisms cause the phase-controlled rectifier to operate with low
power factor: the lagging fundamental component of current, and the generation of current harmonics.

Equation (17.10) can be further interpreted. Note that the dc output power P is equal to the dc
inductor current I_L times the dc output voltage V. By use of Eq. (17.7), this can be written

$$P = I_L \frac{3\sqrt{2}}{\pi} V_{L\text{-}L,\,rms} \cos \alpha \tag{17.11}$$

Comparison of Eqs. (17.10) and (17.11) reveals that the rectifier fundamental volt-amperes can be
expressed using the conventional concepts of complex power $S = P + jQ$, where P is the real (average)
power consumed and Q is the fundamental reactive power consumed. The complex power phasor diagram, treating the fundamental components only, is illustrated in Fig. 17.15.

17.3.3   Commutation

Let's consider next what happens during the switching transitions. In the phase-controlled rectifier circuit
of Fig. 17.16, the dc-side inductor L_d is large in value, such that its current ripple is negligible. Inductors
L_a, L_b, and L_c are also present in the ac lines; these may be physical inductors of the rectifier circuit, or
they may represent the source impedance of the power system, typically the leakage inductances of a
nearby transformer. These inductors are relatively small in value.

Consider the switching transition illustrated in Fig. 17.17. Thyristors Q_3 and Q_5 initially conduct. At time t_{c1}, thyristor Q_1 is gated on, and the dc current i_L begins to shift from Q_3 to Q_1. The ac line
currents $i_a(t)$ and $i_c(t)$ cannot be discontinuous, since inductors L_a and L_c are present in the lines. So dur-

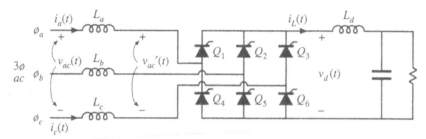

Fig. 17.16 Controlled 3ø rectifier circuit, with small ac-side inductances.

ing the interval $t_{c1} < t < t_{c2}$, thyristors Q_1 and Q_3 both conduct, and the voltage v_{ac}' is zero. Voltage is applied across inductors L_a and L_c, causing their currents to change; for successful commutation, sufficient volt-seconds must be applied to cause the currents to change from i_L to zero, and vice versa. Any stored charge that remains in thyristor Q_3 when current $i_c(t)$ reaches zero must also be removed, and hence $i_c(t)$ actually continues negative as discussed in Chapter 4. When the reverse recovery process of Q_3 is complete, then Q_3 is finally in the off-state, and the next subinterval begins with the conduction of thyristors Q_1 and Q_5.

The commutation process described above has several effects on the converter behavior. First, it can be seen that the thyristor bridge dc-side voltage $v_d(t)$ is reduced in value during the commutation interval. Hence, its average value $\langle v_d \rangle$ and the dc output voltage V are reduced. The amount of reduction is dependent on the dc load current: a larger dc load current leads to a longer commutation interval, and hence to a greater reduction in $\langle v_d \rangle$. So the rectifier has an effective output resistance. Second, the maximum value of the delay angle α is limited to some value less than 180°. If α exceeds this limit, then insufficient volt-seconds are available to change inductor current $i_c(t)$ from i_L to zero, leading to *commutation failure*. Third, when the rectifier ac-side inductors are small or zero, so that L_a, L_b, and L_c represent

Fig. 17.17 Switching transition waveforms, for the rectifier of Fig. 17.16.

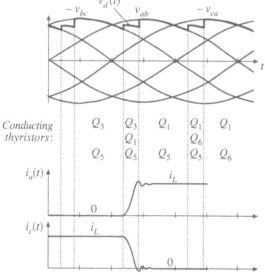
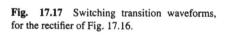

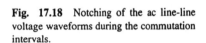

Fig. 17.18 Notching of the ac line-line voltage waveforms during the commutation intervals.

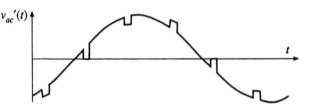

essentially only the power system source impedance, then commutation causes significant notching of the ac voltage waveforms (Fig. 17.18) at the point of common coupling of the rectifier to the power system. Other elements connected locally to the power system will experience voltage distortion. Limits for the areas of these notches are suggested in IEEE/ANSI standard 519.

17.4   HARMONIC TRAP FILTERS

Passive filters are often employed to reduce the current harmonics generated by rectifiers, such that harmonic limits are met. The filter network is designed to pass the fundamental and to attenuate the significant harmonics such as the fifth, seventh, and perhaps several higher-order odd nontriplen harmonics. Such filters are constructed using resonant tank circuits tuned to the harmonic frequencies. These networks are most commonly employed in balanced three-phase systems. A schematic diagram of one phase of the filter is given in Fig. 17.19.

The ac power system is modeled by the thevenin-equivalent network containing voltage source v_s and impedance Z_s'. Impedance Z_s' is usually inductive in nature, although resonances may occur due to nearby power-factor-correction capacitors. In most filter networks, a series inductor L_s' is employed; the filter is then called a harmonic trap filter. For purposes of analysis, the series inductor L_s' can be lumped into Z_s', as follows:

$$Z_s(s) = Z_s'(s) + sL_s' \qquad (17.12)$$

The rectifier and its current harmonics are modeled by current source i_r. Shunt impedances $Z_1, Z_2, \ldots$ are

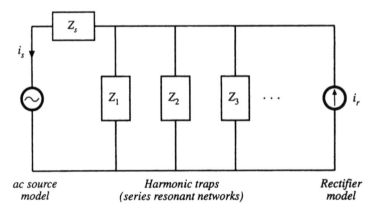

Fig. 17.19   A harmonic trap filter. One phase is illustrated, on a line-to-neutral basis.

tuned such that they have low impedance at the harmonic frequencies, and hence the harmonic currents tend to flow through the shunt impedances rather than into the ac power system.

The approximate algebra-on-the-graph method described in Chapter 8 is used here to construct the filter transfer function, in terms of impedance Z_s and the shunt impedances Z_1, Z_2,..... This approach yields a simple intuitive understanding of how the filter operates. Since the harmonic frequencies are close in value, the pole and zero frequencies of the filter are never well separated in value. So the approximate algebra-on-the-graph method is, in general, not sufficiently accurate for the complete design of these filters, and the pole frequencies must be found by numerical methods. A typical design approach might involve estimating element values using the algebra-on-the-graph method, then refining the values using a computer simulation package.

The filter transfer function $H(s)$ is given by the current divider ratio

$$H(s) = \frac{i_s(s)}{i_r(s)} = \frac{Z_1 \parallel Z_2 \parallel \cdots}{Z_s + Z_1 \parallel Z_2 \parallel \cdots} \tag{17.13}$$

As discussed in Chapter 8, another way to write this transfer function is

$$H(s) = \frac{i_s(s)}{i_r(s)} = \frac{Z_s \parallel Z_1 \parallel Z_2 \parallel \cdots}{Z_s} \tag{17.14}$$

So we can construct $H(s)$ by first constructing the parallel combination $Z_s \parallel Z_1 \parallel Z_2 \parallel \cdots$, then dividing by the total line impedance Z_s. It can be shown that, if $Z_s(s)$ contains no poles, then the numerator of $H(s)$ is the product of the zeroes of the shunt impedances Z_1, Z_2,..... So this graphical method yields the exact zeroes of $H(s)$, which coincide with the series resonances of the shunt impedances. But the poles of $H(s)$, which arise from parallel resonances in the filter, require more work to compute.

Let us first consider the simple case illustrated in Fig. 17.20, where Z_1 consists of a series resonant circuit tuned to eliminate the fifth harmonic, and Z_s is composed of a single inductor L_s. Construction of the impedance of a series resonant network is described in Chapter 8. The $\parallel Z_1 \parallel$ asymptotes follow the capacitor impedance magnitude $1/\omega C_1$ at low frequency, and the inductor impedance magnitude ωL_1 at high frequency. At the resonant frequency, $\parallel Z_1 \parallel$ is equal to R_1. The asymptotes for $\parallel Z_1 \parallel$ are constructed in Fig. 17.21(a).

Figure 17.21(a) also illustrates the impedance magnitude $\parallel Z_s \parallel = \omega L_s$, as well as construction of

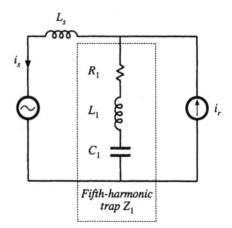

Fig. 17.20 Simple harmonic trap filter example, containing a series resonant trap tuned to the fifth harmonic, and inductive line impedance.

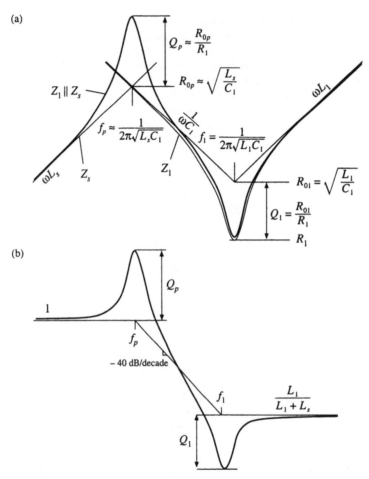

Fig. 17.21 Construction of approximate frequency response using the algebra-on-the-graph method: (a) imped-ance asymptotes, (b) transfer function asymptotes $\| H \|$.

the approximate asymptotes for the parallel combination $(Z_s \| Z_1)$. Recall that, to construct the approxi-mate asymptotes for the parallel combination, we simply select the smaller of the Z_s and Z_1 asymptotes. The result is the shaded set of asymptotes shown in the figure: the parallel combination follows ωL_s at low frequency, and $\| Z_1 \|$ at high frequency. Note that, in addition to the intended series resonance at fre-quency f_1, a parallel resonance occurs at frequency f_p.

The filter transfer function $\| H(s) \|$ is now constructed using Eq. (17.14). As illustrated in Fig. 17.21(b), $\| H(s) \| = 1$ at low frequencies where both the numerator and the denominator of Eq. (17.14) are equal to ωL_s. The parallel resonance at frequency f_p leads to resonant poles and peaking in $\| H(s) \|$. The resonance at frequency f_1 leads to resonant zeroes and attenuation in $\| H(s) \|$. At high frequency, the gain is $L_1/(L_1 + L_s)$.

So if we want to attenuate fifth harmonic currents, we should choose the element values such that the series resonant frequency f_1 coincides with the fifth harmonic frequency. This frequency is sim-

(a)

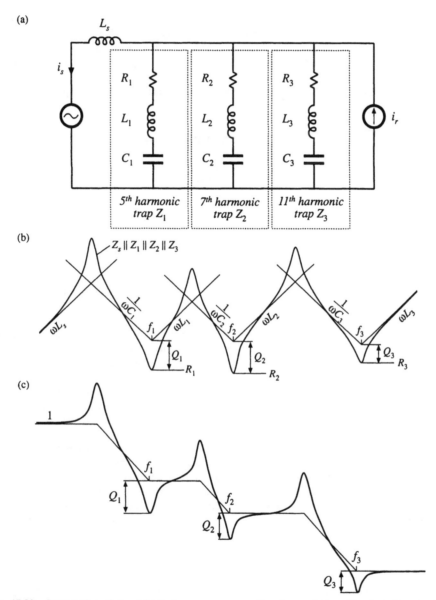

Fig. 17.22 Construction of approximate frequency response for a harmonic trap filter that attenuates the fifth, seventh, and eleventh harmonics: (a) impedance asymptotes, (b) transfer function asymptotes.

ply the resonant frequency of the shunt impedance Z_1, or

$$f_1 = \frac{\omega_1}{2\pi} = \frac{1}{2\pi\sqrt{L_1 C_1}} \qquad (17.15)$$

In addition, care must be exercised regarding the parallel resonance. Since no three-phase system is exactly balanced, small amounts of third harmonic currents always occur. These currents usually have negligible effect; however, the parallel resonance of the harmonic trap filter can increase their magnitudes significantly. Even worse, the Q-factor of the parallel resonance, Q_p, is greater than the series-resonance Q-factor Q_1.

The filter circuit of Fig. 17.20 is simple enough that an exact analysis can be performed easily. The exact transfer function is

$$H(s) = \frac{\left(1 + \frac{s}{\omega_1 Q_1} + \left(\frac{s}{\omega_1}\right)^2\right)}{\left(1 + \frac{s}{\omega_p Q_p} + \left(\frac{s}{\omega_p}\right)^2\right)} \tag{17.16}$$

where

$$f_1 = \frac{\omega_1}{2\pi} = \frac{1}{2\pi\sqrt{L_1 C_1}} \qquad\qquad f_p = \frac{\omega_p}{2\pi} = \frac{1}{2\pi\sqrt{(L_1 + L_s)C_1}}$$

$$Q_1 = \frac{1}{R_1}\sqrt{\frac{L_1}{C_1}} \qquad\qquad Q_p = \frac{1}{R_1}\sqrt{\frac{(L_1 + L_s)}{C_1}}$$

The resonant zeroes do indeed appear at the series resonant frequency, while the parallel resonance and its associated resonant poles appear at frequency f_p determined by the series combination of L_1 and L_p.

To attenuate several harmonics—for example, the fifth, seventh, and eleventh—series resonant networks can be tuned to provide resonant zeroes at each. A circuit is given in Fig. 17.22(a), with the impedance asymptotes of Fig. 17.22(b). The resulting approximate $\| H(s) \|$ is given in Fig. 17.22(c). It can be seen that, associated with each series resonance is a parallel resonance. Each parallel resonant frequency should be chosen such that it is not significantly excited by harmonics present in the network.

The filter transfer function can be given high-frequency single-pole rolloff by addition of a bypass resistor R_{bp}, as illustrated in Fig. 17.23(a). Typical impedance and transfer function asymptotes for this network are constructed in Fig. 17.24. The bypass resistor allows some additional attenuation of the higher-order harmonics, without need for series resonant traps tuned to each harmonic. The network of Fig. 17.23(a) is sometimes called a "high pass" network, because it allows high-frequency currents to flow through the shunt branch. But it causes the overall filter transfer function $H(s)$ to reject high frequencies. A simple harmonic trap filter that contains series resonances that can be tuned to the fifth and

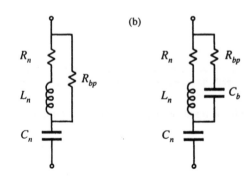

Fig. 17.23 Addition of bypass resistor R_{bp} to a series resonant network, to obtain a high-frequency rolloff characteristic: (a) basic circuit, (b) addition of blocking capacitor C_b to reduce power consumption at the fundamental frequency.

(a)

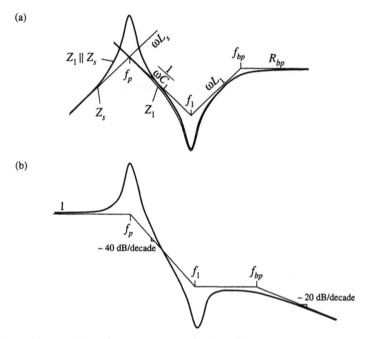

(b)

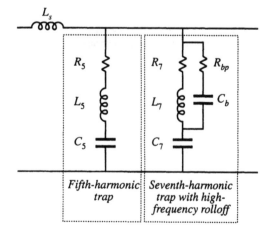

Fig. 17.24 Construction of approximate frequency response for a harmonic trap filter containing bypass resistor: (a) impedance asymptotes, (b) transfer function asymptotes.

seventh harmonics, with a single-pole rolloff to attenuate higher-order harmonics such as the eleventh and thirteenth, is illustrated in Fig. 17.25.

    Power loss in the bypass resistor can be an issue: since R_{bp} is not part of the resonant network, significant fundamental (50 Hz or 60 Hz) currents can flow through R_{bp}. The power loss can be reduced by addition of blocking capacitor C_b, as illustrated in Fig. 17.23(b). This capacitor is chosen to increase the impedance of the R_{bp}–C_b leg at the fundamental frequency, but have negligible effect at the higher-

Fig. 17.25 A harmonic trap filter containing series resonances tuned to the fifth and seventh harmonics, and high-frequency rolloff characteristic.

order harmonic frequencies.

The harmonic trap filter network can also supply significant reactive power to the rectifier and power system. As given by Eq. (17.10), the rectifier fundamental current lags the voltage, and the rectifier consumes reactive power. As seen in Fig. 17.22(a), the impedances of the series resonant tank networks are dominated by their capacitor asymptotes at low frequency. Hence, at the fundamental frequency, the filter impedance reduces to an equivalent capacitor, equal to the parallel combination of the tank capacitors. The current through this capacitance leads the ac line voltage, and hence as mentioned in the previous chapter, the capacitor is a source of reactive power.

17.5   TRANSFORMER CONNECTIONS

A final conventional approach to reducing the input harmonics of three-phase rectifiers is the use of phase-shifting transformer circuits. With these schemes, the low-order harmonics, such as the fifth and seventh, can be eliminated. The remaining harmonics are smaller in magnitude, and also are easier to filter.

The rectifier circuit of Fig. 17.7 is known as a *six-pulse* rectifier because the diode output voltage waveform contains six pulses per ac line period. The output voltage ripple has a fundamental frequency that is six times the ac line frequency. As illustrated in Fig. 17.8, the ac line current waveforms contain three steps: at any given instant, $i_a(t)$ is equal to either i_L, 0, or $-i_L$. The spectrum of the current waveform contains fundamental and odd nontriplen harmonics (1, 5, 7, 11, 13,...), whose amplitudes vary as $1/n$.

It is possible to shift the phase of the ac line voltage using three-phase transformer circuits. For example, in the delta-wye transformer circuit of Fig. 17.26, the transformer primary windings are driven by the primary-side line-to-line voltages, while the transformer secondaries supply the secondary-side line-to-neutral voltages. In an ideal transformer, the secondary voltage is equal to the primary voltage multiplied by the turns ratio; hence, the phasor representing the secondary voltage is in phase with the

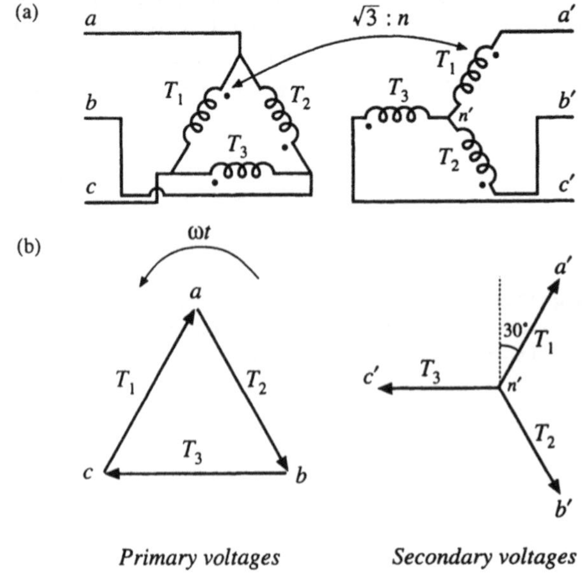

Fig. 17.26 Three-phase delta-wye transformer connection: (a) circuit, (b) voltage phasor diagram.

Primary voltages   *Secondary voltages*

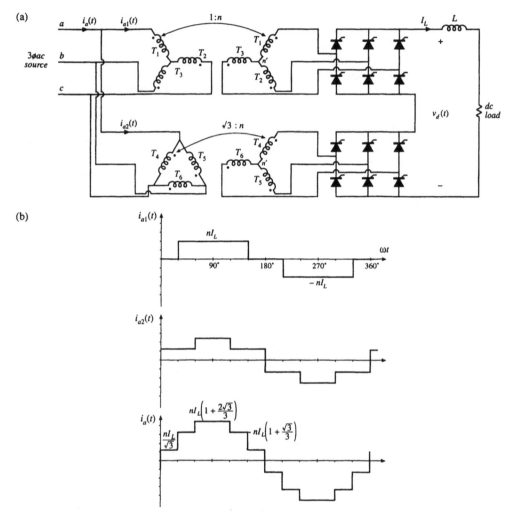

Fig. 17.27 Twelve-pulse rectifier: (a) circuit, (b) input phase *a* current waveforms.

primary voltage phasor, and is scaled in magnitude by the turns ratio. So in the delta-wye transformer connection, the secondary line-to-neutral voltages are in phase with the primary line-to-line voltages. In a balanced three-phase system, the line-to-line voltages are shifted in phase by 30° with respect to the line-to-neutral voltages, and are increased in magnitude by a factor of $\sqrt{3}$. Hence the secondary line-to-neutral voltages lag the primary line-to-neutral voltages by 30°. The wye-delta connection is also commonly used; this circuit causes the secondary voltages to lead the primary voltages by 30°. Many other more complicated transformer circuits are known, such as the zig-zag, forked-wye, and extended-delta connections, which can lead to phase shifts of any desired amount.

The 30° phase shift of the delta-wye transformer circuit is used to advantage in the twelve-pulse rectifier circuit of Fig. 17.27(a). This circuit consists of two bridge rectifier circuits driven by 3ø voltages that differ in phase by 30°. The bridge rectifier outputs are connected in series to the dc filter inductor and load. The total rectifier output voltage $v_d(t)$ has a fundamental frequency that is twelve times the ac line frequency. The input phase a ac line current $i_a(t)$ is the sum of currents in three windings, and has the stepped waveshape illustrated in Fig. 17.27(b). It can be shown that this waveform contains Fourier components at the fundamental frequency and at the 11th, 13th, 23rd, 25th,... harmonic frequencies, whose amplitudes vary as $1/n$. Doing so is left as a homework problem. Thus, the twelve-pulse rectifier eliminates the 5th, 7th, 17th, 19th,... harmonics.

An eighteen-pulse rectifier can be constructed using three six-pulse bridge rectifiers, with transformer circuits that shift the applied voltages by 0°, + 20°, and − 20°. A twenty-four pulse rectifier requires four six-pulse bridge rectifiers, fed by voltages shifted by 0°, + 15°, − 15°, and 30°. If p is the pulse number, then the rectifier produces line current harmonics of number $n = pk \pm 1$, where $k = 0, 1, 2, 3,....$ If the dc current ripple can be neglected, then the magnitudes of the remaining current harmonics vary as $1/n$. The dc-side harmonics are of number pk.

So by use of polyphase transformer connections and rectifier circuits having high pulse number, quite good ac line current waveforms can be obtained. As the pulse number is increased, the current waveforms approaches a sinusoid, and contains a greater number of steps having smaller amplitude. The low-order harmonics can be eliminated, and the remaining high-frequency harmonics are easily filtered.

17.6   SUMMARY

1.   With a large dc filter inductor, the single-phase full-wave rectifier produces a square-wave line current waveform, attaining a power factor of 90% and 48% THD. With smaller values of inductance, these figures are degraded. In the discontinuous conduction mode, THD greater than 100%, with power factors of 55% to 65% are typical. When the capacitance is large, the power factor, THD, displacement factor, and conversion ratio can be expressed as a function of only the dimensionless parameter K_L.

2.   In the three-phase case, the bridge rectifier with large dc filter inductor produces a stepped waveform similar to the square wave, but missing the triplen harmonics. This waveform has 31% THD, and leads to a power factor of 95.5%. Reduced dc inductor values again lead to increased THD and reduced power factor, and as L tends to zero, the THD tends to infinity while the power factor tends to zero. In practice, the minimum effective series inductance is limited by the power system source inductance.

3.   With a large dc inductor, phase control does not influence the distortion factor or THD, but does lead to a lagging fundamental current and decreased displacement factor. Phase-controlled rectifiers and inverters are consumers of reactive power.

4.   If the load is capable of supplying power, then the phase-controlled rectifier can become an inverter. The delay angle a is greater than 90°, and the output voltage polarity is reversed with respect to rectifier operation. The maximum delay angle is limited by commutation failure to a value less than 180°.

5.   Harmonic trap filters and multipulse-rectifier/polyphase transformer circuits find application in high-power applications where their large size and weight are less of a consideration than their low cost. In the harmonic trap filter, series resonant tank circuits are tuned to the offending harmonic frequencies, and shunt the harmonic currents away from the utility power network. Parallel resonances may cause unwanted peaks in the filter transfer function. Operation of these filters may be understood using the algebra-on-the-graph method, and computer simulations can be used to refine the accuracy of the analysis or design. Rectifiers of higher pulse number can also yield improved current waveforms, whose harmonics are of high frequency and small amplitude, and are easily filtered.

REFERENCES

[1]   G.K. DUBEY, S.R. DORADLA, A. JOSHI, and R.M.K. SINHA, *Thyristorized Power Controllers*, New Delhi: Wiley Eastern, Ltd., 1986.

[2]   J. VITHAYATHIL, *Power Electronics: Principles and Applications*, New York: McGraw-Hill, 1995.

[3]   N. MOHAN, T. UNDELAND, and W. ROBBINS, *Power Electronics: Converters, Applications, and Design*, Second edition, New York: John Wiley & Sons, 1995.

[4]   J. PHIPPS, "Phase-Shifting Transformers and Passive Harmonic Filters: Interfacing for Power Electronic Motor Drive Controllers," M.S. Thesis, University of Colorado at Denver, 1993.

[5]   J. KASSAKIAN, M. SCHLECHT, and G. VERGESE, *Principles of Power Electronics*, Massachusetts: Addison-Wesley, 1991.

[6]   J. ARRILLAGA, D. BRADLEY, and P. BODGER, *Power System Harmonics*, New York: John Wiley & Sons, 1985.

[7]   A. DOMIJAN and E. EMBRIZ-SANTANDER, "A Summary and Evaluation of Recent Developments on Harmonic Mitigation Techniques Useful to Adjustable-Speed Drives," *IEEE Transactions on Energy Conversion*, Vol. 7, No. 1, pp. 64-71, March 1992.

[8]   S. FREELAND, "I. A Unified Analysis of Converters with Resonant Switches, II. Input-Current Shaping for Single-Phase Ac-dc Power Converters," Ph.D. Thesis, California Institute of Technology, 1988, Chapter 12.

[9]   F. C. SCHWARZ, "A Time-Domain Analysis of the Power Factor for Rectifier Systems with Over- and Subcritical Inductance," *IEEE Transactions on Industrial Electronics and Control Instrumentation*, Vol. 20, No. 2, May 1973, pp. 61-68.

[10]   S. B. DEWAN, "Optimum Input and Output Filters for Single-Phase Rectifier Power Supply," *IEEE Transactions on Industry Applications*, Vol. 17, No. 3, May/June 1981, pp. 282-288.

[11]   A. KELLEY and W. F. YADUSKY, "Rectifier Design for Minimum Line-Current Harmonics and Maximum Power Factor," *IEEE Transactions on Power Electronics*, Vol. 7, No. 2, April 1992, pp. 332-341.

[12]   A. W. KELLEY and W. F. YADUSKY, "Phase-Controlled Rectifier Line-Current Harmonics and Power Factor as a Function of Firing Angle and Output Filter Inductance," *IEEE Applied Power Electronics Conference*, 1990 Proceedings, March 1990, pp. 588-597.

[13]   P. C. SEN, *Thyristorized DC Drives*, New York: Wiley Interscience, 1981.

[14]   S. DEWAN AND A. STRAUGHEN, *Power Semiconductor Circuits*, New York: John Wiley & Sons, 1975.

[15]   B. PELLY, *Thyristor Phase-Controlled Converters and Cycloconverters: Operation, Control, and Performance*, New York: John Wiley & Sons, 1971.

PROBLEMS

17.1   The half-controlled single-phase rectifier circuit of Fig. 17.28 contains a large inductor L, whose current $i_L(t)$ contains negligible ripple. The thyristor delay angle is α.

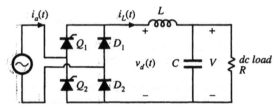

Fig. 17.28   Half-controlled rectifier circuit of Problem 17.1.

(a)   Sketch the waveforms $v_d(t)$ and $i_a(t)$. Label the conduction intervals of each thyristor and diode.

(b)   Derive an expression for the dc output voltage V, as a function of the rms line-line voltage and the delay angle.

(c)   Derive an expression for the power factor.

(d)   Over what range of α are your expressions of parts (b) and (c) valid?

17.2   The half-controlled rectifier circuit of Fig. 17.29 contains a large inductor L, whose current $i_L(t)$ contains negligible ripple. The thyristor delay angle is α.

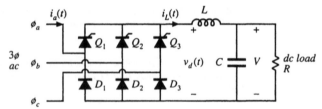

Fig. 17.29   Three-phase half-controlled rectifier circuit of Problem 17.2.

(a)   Sketch the waveforms $v_d(t)$ and $i_a(t)$. Label the conduction intervals of each thyristor and diode.

(b)   Derive an expression for the dc output voltage V, as a function of the rms line-line voltage and the delay angle.

(c)   Derive an expression for the power factor.

(d)   Over what range of α are your expressions of parts (b) and (c) valid?

17.3   A 3ϕ SCR bridge is connected directly to a resistive load, as illustrated in Fig. 17.30. This circuit operates in the continuous conduction mode for small delay angle α, and in the discontinuous conduction mode for sufficiently large α.

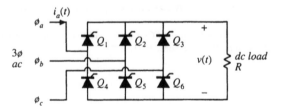

Fig. 17.30   Three-phase controlled rectifier circuit of Problem 17.3.

(a)   Sketch the output voltage waveform $v(t)$, for CCM operation and for DCM operation. Clearly label the conduction intervals of each SCR.

(b)   Under what conditions does the rectifier operate in CCM? in DCM?

(c)   Derive an expression for the dc component of the output voltage in CCM.

(d)   Repeat part (c), for DCM operation.

17.4   A rectifier is connected to the 60 Hz utility system. It is desired to design a harmonic trap filter that has negligible attenuation or amplification of 60 Hz currents, but which attenuates both the fifth- and seventh-harmonic currents by a factor of 10 (– 20 dB). The ac line inductance L_s is 500 μH.

(a)   Select L_5, the inductance of the fifth harmonic trap, equal to 500 μH, and L_7, the inductance of the seventh harmonic trap, equal to 250 μH. Compute first-pass values for the resistor and capacitor values of the fifth and seventh harmonic trap circuits, neglecting the effects of parallel resonance.

(b)   Plot the frequency response of your filter. It is suggested that you do this using SPICE or a similar computer program. Does your filter meet the attenuation specifications? Are there significant parallel resonances? What is the gain or attenuation at the third harmonic frequency?

(c)   Modify your element values, to obtain the best design you can. You must choose L_s = 500 μH, but you may change all other element values. Plot the frequency response of your improved filter. "Best" means that the 20 dB attenuations are obtained at the fifth and seventh harmonic frequencies, that the gain at 60 Hz is essentially 0 dB, and that the Q-factors of parallel resonances are minimized.

17.5   A rectifier is connected to the 50 Hz utility system. It is desired to design a harmonic trap filter that has negligible attenuation or amplification of 50 Hz currents, but that attenuates the fifth-, seventh-, and eleventh-harmonic currents by a factor of 5 (– 14 dB). In addition, the filter must contain a single-pole response that attenuates the thirteenth and higher harmonics by a factor of $5n/13$, where n is the harmonic number. The ac line inductance L_s is 100 μH.

Design a harmonic trap filter that meets these specifications. Design the best filter you can, which meets the attenuation specifications, that has nearly unity (0 dB ± 1 dB) gain at 50 Hz, and that has minimum gains at the third and ninth harmonics. Plot the frequency response of your filter, and specify your circuit element values.

17.6   A single-phase rectifier operates from a 230 Vrms 50 Hz European single-phase source. The rectifier must supply a 1000 W dc load, and must meet the IEC-1000 class A or class D harmonic current limits. The circuit of Fig. 17.1 is to be used. The dc load voltage may have 100 Hz ripple whose peak-to-peak amplitude is no greater than 5% of the dc voltage component.

(a)   Estimate the minimum value of inductance that will meet these requirements.

(b)   Specify values of L and C that meet these requirements, and prove (by simulation) that your design is correct.

17.7   Figure 17.31 illustrates a twelve-pulse rectifier, containing six controlled (SCR) devices and six uncontrolled (diode) devices. The dc filter inductance L is large, such that its current ripple is negligible. The SCRs operate with delay angle α. The SCR bridge is driven by a wye-wye connected three-phase transformer circuit, while the diode bridge is driven by a wye-delta connected three-phase transformer circuit. Since both transformer circuits have wye-connected primaries, they can be combined to realize the circuit with a single wye-connected primary.

(a)   Determine the rms magnitudes and phases of the line-to-line output voltages of the transformer secondaries $v_{a1'b1'}$ and $v_{a2'b2'}$, as a function of the applied line-line primary voltage v_{ab}.

(b)   Sketch the waveforms of the voltages $v_{d1}(t)$ and $v_{d2}(t)$. Label the conduction intervals of each thyristor and diode.

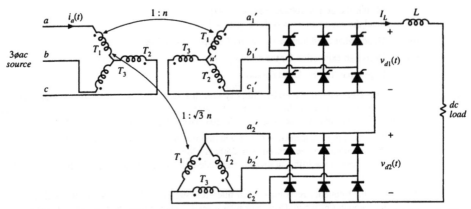

Fig. 17.31 Twelve-pulse rectifier circuit, with one controlled and one uncontrolled bridge, Problem 17.7.

(c)   Derive an expression for the dc component of the output voltage, as a function of the rms line-line input voltage, the delay angle α, and the turns ratio n.

(d)   Over what range of α is your expression of part (c) valid? What output voltages can be produced by this rectifier?

17.8   For the twelve-pulse rectifier circuit of Fig. 17.27(a), determine the Fourier coefficients, for the fundamental through the thirteenth harmonic, of the primary-side currents $i_{a1}(t)$ and $i_{a2}(t)$, as well as for the ac line current $i_a(t)$. Express your results in terms of the turns ratio n and the dc load current I_L. You may assume that the dc filter inductance L is large and that the transformers are ideal.

17.9   The single-phase controlled-bridge rectifier of Fig. 17.32 operates in the continuous conduction mode. It is desired to regulate the load voltage $v(t)$ in the presence of slow variations in the amplitude of the sinusoidal input voltage $v_g(t)$. Hence, a controller must be designed that varies the delay angle α such that $v(t)$ is kept constant, and it is of interest to derive a small-signal ac model for the dc side of the rectifier circuit.

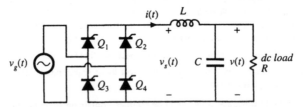

Fig. 17.32 Single-phase controlled rectifier, Problem 17.9.

(a)   Sketch $v_s(t)$ and $v_g(t)$, and label the delay angle α.

(b)   Use the circuit averaging method to determine the small-signal transfer functions

$$\frac{\hat{v}(s)}{\hat{\alpha}(s)} \quad \text{control-to-output transfer function}$$

and

$$\frac{\hat{v}(s)}{\hat{v}_g(s)} \quad \text{line-to-output transfer function}$$

as well as the steady-state relationship

$$V = f(V_g, A)$$

where

$$\alpha(t) = A + \hat{\alpha}(t)$$
$$v(t) = V + \hat{v}(t)$$
$$v_g(t) = \left(V_g + \hat{v}_g(t)\right) \sin(\omega t)$$

You may assume that the frequencies of the variations in α, v, and v_g are much slower that the ac line frequency ω, and that the inductor current ripple is small.

18

Pulse-Width Modulated Rectifiers

To obtain low ac line current THD, the passive techniques described in the previous chapter rely on low-frequency transformers and/or reactive elements. The large size and weight of these elements are objectionable in many applications. This chapter covers active techniques that employ converters having switching frequencies much greater than the ac line frequency. The reactive elements and transformers of these converters are small, because their sizes depend on the converter switching frequency rather than the ac line frequency.

Instead of making do with conventional diode rectifier circuits, and dealing after-the-fact with the resulting low-frequency harmonics, let us consider now how to build a rectifier that behaves as ideally as possible, without generation of line current harmonics. In this chapter, the properties of the *ideal rectifier* are explored, and a model is described. The ideal rectifier presents an effective resistive load to the ac power line; hence, if the supplied ac voltage is sinusoidal, then the current drawn by the rectifier is also sinusoidal and is in phase with the voltage. Converters that approximate the properties of the ideal rectifier are sometimes called *power factor corrected*, because their input power factor is essentially unity [1].

The boost converter, as well as a variety of other converters, can be controlled such that a near-ideal rectifier system is obtained. This is accomplished by control of a high-frequency switching converter, such that the ac line current waveform follows the applied ac line voltage. Both single-phase and three-phase rectifiers can be constructed using PWM techniques. A typical dc power supply system that is powered by the single-phase ac utility contains three major power-processing elements. First, a high-frequency converter with a wide-bandwidth input-current controller functions as a near-ideal rectifier. Second, an energy-storage capacitor smooths the pulsating power at the rectifier output, and a low-bandwidth controller causes the average input power to follow the power drawn by the load. Finally, a dc–dc converter provides a well-regulated dc voltage to the load. In this chapter, single-phase rectifier systems are discussed, expressions for rms currents are derived, and various converter approaches are compared.

The techniques developed in earlier chapters for modeling and analysis of dc-dc converters are extended in this chapter to treat the analysis, modeling, and control of low-harmonic rectifiers. The CCM models of Chapter 3 are used to compute the average losses and efficiency of CCM PWM converters operating as rectifiers. The results yield insight that is useful in power stage design. Several converter control schemes are known, including peak current programming, average current control, critical conduction mode control, and nonlinear carrier control. Ac modeling of the rectifier control system is also covered.

18.1   PROPERTIES OF THE IDEAL RECTIFIER

It is desired that the ideal single-phase rectifier present a resistive load to the ac system. The ac line current and voltage will then have the same waveshape and will be in phase. Unity power factor rectification is the result. Thus, the rectifier input current $i_{ac}(t)$ should be proportional to the applied input voltage $v_{ac}(t)$:

$$i_{ac}(t) = \frac{v_{ac}(t)}{R_e} \tag{18.1}$$

where R_e is the constant of proportionality. An equivalent circuit for the ac port of an ideal rectifier is therefore an effective resistance R_e, as shown in Fig. 18.1(a). R_e is also known as the *emulated resistance*. It should be noted that the presence of R_e does not imply the generation of heat: the power apparently

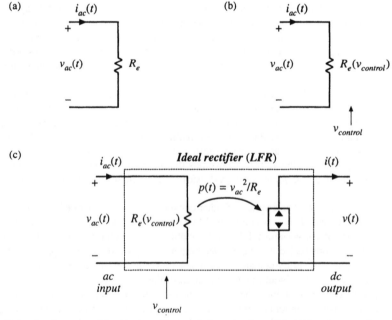

Fig. 18.1   Development of the ideal rectifier equivalent circuit model: (a) input port resistor emulation; (b) the value of the emulated resistance, and hence the power throughput, is controllable; (c) output port power source characteristic, and complete model.

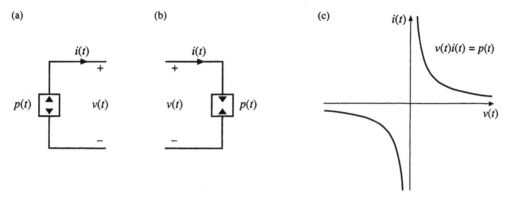

Fig. 18.2 The dependent power source: (a) power source schematic symbol, (b) power sink schematic symbol, (c) *i–v* characteristic.

"consumed" by R_e is actually transferred to the rectifier dc output port. R_e simply models how the ideal rectifier loads the ac power system.

Output regulation is accomplished by variation of the effective resistance R_e, and hence the value of R_e must depend on a control signal $v_{control}(t)$ as in Fig. 18.1(b). This allows variation of the rectifier power throughput, since the average power consumed by R_e is

$$P_{av} = \frac{V_{ac,rms}^2}{R_e(v_{control})} \tag{18.2}$$

Note that changing R_e results in a time-varying system, with generation of harmonics. To avoid generation of significant amounts of harmonics and degradation of the power factor, variations in R_e and in the control input must be slow with respect to the ac line frequency.

To the extent that the rectifier is lossless and contains negligible internal energy storage, the instantaneous power flowing into R_e must appear at the rectifier output port. Note that the instantaneous power throughput

$$p(t) = \frac{v_{ac}^2(t)}{R_e\left(v_{control}(t)\right)} \tag{18.3}$$

is dependent only on $v_{ac}(t)$ and the control input $v_{control}(t)$, and is independent of the characteristics of the load connected to the output port. Hence, the output port must behave as a source of constant power, obeying the relationship

$$v(t)i(t) = p(t) = \frac{v_{ac}^2(t)}{R_e} \tag{18.4}$$

The *dependent power source* symbol of Fig. 18.2(a) is used to denote such an output characteristic. As illustrated in Fig. 18.1(c), the output port is modeled by a power source that is dependent on the instantaneous power flowing into R_e.

Thus, a two-port model for the ideal unity-power-factor single-phase rectifier is as shown in Fig. 18.1(c) [2–4]. The two port model is also called a *loss-free resistor* (LFR) because (1) its input port obeys Ohm's law, and (2) power entering the input port is transferred directly to the output port without loss of energy. The defining equations of the LFR are:

$$i_{ac}(t) = \frac{v_{ac}(t)}{R_e(v_{control})} \qquad (18.5)$$

$$v(t)i(t) = p(t) \qquad (18.6)$$

$$p(t) = \frac{v_{ac}^2(t)}{R_e(v_{control}(t))} \qquad (18.7)$$

When the LFR output port is connected to a resistive load of value R, the dc output rms voltages and currents V_{rms} and I_{rms} are related to the ac input rms voltages and currents $V_{ac,rms}$ and $I_{ac,rms}$ as follows:

$$\frac{V_{rms}}{V_{ac,rms}} = \sqrt{\frac{R}{R_e}} \qquad (18.8)$$

$$\frac{I_{ac,rms}}{I_{rms}} = \sqrt{\frac{R}{R_e}} \qquad (18.9)$$

The properties of the power source and loss-free resistor network are discussed in Chapter 11. Regardless of the specific converter implementation, any single-phase rectifier having near-ideal properties can be modeled using the LFR two-port model.

18.2   REALIZATION OF A NEAR-IDEAL RECTIFIER

Feedback can be employed to cause a converter that exhibits controlled dc transformer characteristics to obey the LFR equations. In the single-phase case, the simplest and least expensive approach employs a full-wave diode rectifier network, cascaded by a dc–dc converter, as in Fig. 18.3. The dc–dc converter is represented by an ideal dc transformer, as discussed in Chapter 3. A control network varies the duty cycle, as necessary to cause the converter input current $i_g(t)$ to be proportional to the applied input volt-

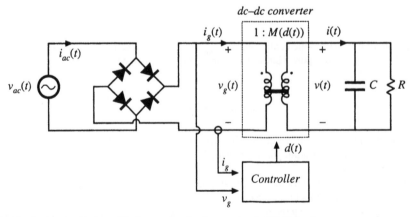

Fig. 18.3   Synthesis of an ideal rectifier by varying the duty cycle of a PWM dc–dc converter.

age $v_g(t)$ as in Eq. (18.1). The effective turns ratio of the ideal transformer then varies with time. Ideal waveforms are illustrated in Fig. 18.4. If the applied input voltage $v_{ac}(t)$ is sinusoidal,

$$v_{ac}(t) = V_M \sin(\omega t) \qquad (18.10)$$

then the rectified voltage $v_g(t)$ is

$$v_g(t) = V_M \left| \sin(\omega t) \right| \qquad (18.11)$$

It is desired that the converter output voltage be a constant dc value $v(t) = V$. The converter conversion ratio must therefore be

$$M\big(d(t)\big) = \frac{v(t)}{v_g(t)} = \frac{V}{V_M \left| \sin(\omega t) \right|} \qquad (18.12)$$

This expression neglects the converter dynamics. As can be seen from Fig. 18.4, the controller must cause the conversion ratio to vary between infinity (at the ac line voltage zero crossings) and some minimum value M_{min} (at the peaks of the ac line voltage waveform). M_{min} is given by

$$M_{min} = \frac{V}{V_M} \qquad (18.13)$$

Any converter topology whose ideal conversion ratio can be varied between these limits can be employed in this application.

To the extent that the dc–dc converter is ideal (i.e., if the losses can be neglected and there is negligible low-frequency energy storage), the instantaneous input and output powers are equal. Hence, the output current $i(t)$ in Fig. 18.3 is given by

$$i(t) = \frac{v_g(t) i_g(t)}{V} = \frac{v_g^2(t)}{V R_e} \qquad (18.14)$$

Substitution of Eq. (18.11) into Eq. (18.14) then leads to

$$\begin{aligned} i(t) &= \frac{V_M^2}{V R_e} \sin^2(\omega t) \\ &= \frac{V_M^2}{2 V R_e} \big(1 - \cos(2\omega t)\big) \end{aligned} \qquad (18.15)$$

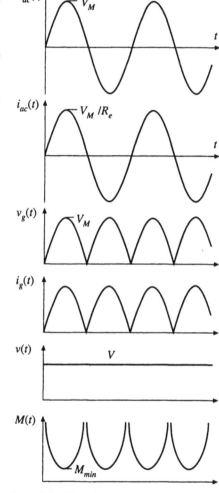

Fig. 18.4 Waveforms of the rectifier system of Fig. 18.3.

Hence, the converter output current contains a dc component and a component at the second harmonic of the ac line frequency. One of the functions of capacitor C in Fig. 18.3 is to filter out the second harmonic component of $i(t)$, so that the load current (flowing through resistor R) is essentially equal to the dc component

$$I = \langle i(t) \rangle_{T_L} = \frac{V_M^2}{2VR_e} \tag{18.16}$$

where T_L is the period of the applied ac line voltage.

The average power is

$$P = \frac{V_M^2}{2R_e} \tag{18.17}$$

The above equations are generally valid for PWM converters used as single-phase low-harmonic rectifiers.

18.2.1 CCM Boost Converter

A system based on the CCM boost converter is illustrated in Fig. 18.5 [1,5,6]. Ideally, the boost converter can produce any conversion ratio between one and infinity. Hence, the boost converter is capable of producing the $M(d(t))$ given by Eq. (18.12), provided that $V \geq V_M$. Further, the boost converter can produce very low THD, with better transistor utilization than other approaches.

If the boost converter operates in continuous conduction mode, and if the inductor is small enough that its influence on the low-frequency components of the converter waveforms is negligible, then the duty ratio should follow $M(d(t)) = 1/(1 - d(t))$. This implies that the duty ratio should follow the function

$$d(t) = 1 - \frac{v_g(t)}{V} \tag{18.18}$$

This expression is true only in the continuous conduction mode. The boost converter operates in the continuous conduction mode provided that the inductor current ripple

$$\Delta i_g(t) = \frac{v_g(t)d(t)T_s}{2L} \tag{18.19}$$

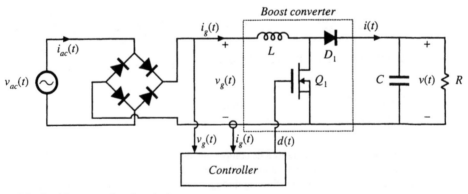

Fig. 18.5 Rectifier system based on the boost converter.

is greater than the average inductor current, or

$$\langle i_g(t)\rangle_{T_s} = \frac{v_g(t)}{R_e} \tag{18.20}$$

Hence, the converter operates in CCM when

$$\langle i_g(t)\rangle_{T_s} > \Delta i_g(t) \;\Rightarrow\; d(t) < \frac{2L}{R_e T_s} \tag{18.21}$$

Substitution of Eq. (18.18) into (18.21) and solution for R_e leads to

$$R_e < \frac{2L}{T_s\left(1 - \frac{v_g(t)}{V}\right)} \quad \text{for CCM} \tag{18.22}$$

Since $v_g(t)$ varies according to Eq. (18.11), Eq. (18.22) may be satisfied at some points on the ac line cycle, and not at others. Since $0 \le v_g(t) \le V_M$, we can conclude that the converter operates in CCM over the entire ac line cycle when

$$R_e < \frac{2L}{T_s} \tag{18.23}$$

Equations (18.18) and (18.22) then hold for all t. The converter always operates in DCM when

$$R_e > \frac{2L}{T_s\left(1 - \frac{V_M}{V}\right)} \tag{18.24}$$

For R_e between these limits, the converter operates in DCM when $v_g(t)$ is near zero, and in CCM when $v_g(t)$ approaches V_M.

The static input characteristics of the open-loop boost converter are sketched in Fig. 18.6. The input current $i_g(t)$ is plotted vs. input voltage $v_g(t)$, for various duty cycles $d(t)$. In CCM, the input characteristics of the boost converter are described by

$$\frac{v_g(t)}{V} = 1 - d(t) \qquad \text{in CCM} \tag{18.25}$$

To obtain a general plot, we can normalize the input current and input voltage as follows:

$$m_g(t) = \frac{v_g(t)}{V} \tag{18.26}$$

$$j_g(t) = \frac{2L}{VT_s}\, i_g(t) \tag{18.27}$$

Equation (18.25) then becomes

$$m_g(t) = 1 - d(t) \tag{18.28}$$

This equation is independent of the input current $i_g(t)$, and hence is represented by vertical lines in Fig.

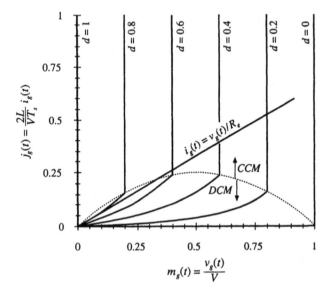

Fig. 18.6 Static input characteristics of the boost converter. A typical linear resistive input characteristic is superimposed.

18.6.

To derive the boost input characteristic for DCM operation, we can solve the steady-state equivalent circuit model of Fig. 11.12(b) (reproduced in Fig. 18.7). *Beware*: the natural DCM effective resistance of Chapter 11, $R_e = 2L/d^2T_s$, does not necessarily coincide with the emulated resistance $R_e = v_g/i_g$ of Eq. (18.1). In this chapter, the quantity R_e is defined according to Eq. (18.1). Solution of Fig. 18.7 for the input current $i_g(t)$ leads to:

$$i_g(t) = \frac{v_g(t)}{\left(\dfrac{2L}{d^2T_s}\right)} + \frac{p(t)}{V - v_g(t)} \tag{18.29}$$

The instantaneous power consumed by the effective resistor in the model of Fig. 18.7 is

$$p(t) = \frac{v_g^2(t)}{\left(\dfrac{2L}{d^2T_s}\right)} \tag{18.30}$$

Substitution of Eq. (18.30) into Eq. (18.29) and simplification leads to

Fig. 18.7 Averaged equivalent circuit model of the boost converter operating in DCM, derived in Chapter 11.

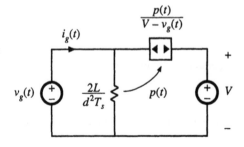

$$\frac{2L}{VT_s} i_g(t) \left(1 - \frac{v_g(t)}{V} \right) = d^2(t) \frac{v_g(t)}{V} \qquad \text{in DCM} \qquad (18.31)$$

Normalization of this equation according to Eqs. (18.26) and (18.27) yields

$$j_g(t)\left(1 - m_g(t)\right) = d^2 m_g(t) \qquad (18.32)$$

This equation describes the curved (DCM) portions of the Fig. 18.6 input characteristics, for low $i_g(t)$.

To express the CCM/DCM mode boundary as a function of $v_g(t)$ and $i_g(t)$, Eqs. (18.1) and (18.22) can be combined, leading to

$$\frac{2L}{VT_s} i_g(t) > \left(\frac{v_g(t)}{V} \right) \left(1 - \frac{v_g(t)}{V} \right) \qquad \text{for CCM} \qquad (18.33)$$

Normalization of this equation, according to Eqs. (18.26) and (18.27), results in

$$j_g(t) > m_g(t)\left(1 - m_g(t)\right) \qquad \text{for CCM} \qquad (18.34)$$

This equation describes a parabola having roots at $m_g = 0$ and $m_g = 1$, with the maximum value $j_g = 0.25$ at $m_g = 0.5$. The mode boundary equation is plotted as a dashed line in Fig. 18.6.

The complete input characteristics for the boost converter were plotted in Fig. 18.6 using Eqs. (18.28), (18.32), and (18.34). Figure 18.6 also illustrates the desired linear resistive input characteristic, Eq. (18.1). For the value of R_e illustrated, the converter operates in DCM for $v_g(t)$ near zero, and in CCM for $v_g(t)$ near V_M. The intersections of boost input characteristics with the desired linear input characteristic illustrate how the controller must choose the duty cycle at various values of $v_g(t)$.

Other converters capable of producing the $M(d(t))$ of Eq. (18.12) include the buck-boost, SEPIC, and Ćuk converters. The boost, SEPIC, and Ćuk converters share the desirable property of non-pulsating input current, and hence require minimal input EMI filtering. The SEPIC produces a non inverted output voltage. Isolated versions of these converters (see Chapter 6) are also sometimes employed [7–9]. Schemes involving the parallel resonant converter, as well as several types of quasi-resonant converters, are also documented in the literature [10–13].

The open-loop boost converter, when operated in discontinuous conduction mode, is also sometimes used as an approximation of an ideal rectifier. The DCM effective resistance $2L/d^2(t)T_s$ of Fig. 18.7 is then taken as an approximation of the desired emulated resistance of Eq. (18.1). The model differs from that of the ideal rectifier model of Fig. 18.1(c) in that the power source is connected between the input and output terminals. As a result, harmonics are present in the input current waveform. For example, if $v_g(t)$ is a rectified sinusoid, then the current through the effective resistance $2L/d^2(t)T_s$ will also be a rectified sinusoid. However, the input current $\langle i_g(t)\rangle_{T_s}$ is now equal to the sum of the current through R_e and the current flowing through the power source element. Since the power source is a nonlinear element, $\langle i_g(t)\rangle_{T_s}$ contains harmonics. For large C, the output voltage is essentially constant. The input current waveform is then given by Eq. (18.31). If V is sufficiently large, then the term $(1 - v_g(t)/V)$ is approximately equal to one, and the harmonics in $\langle i_g(t)\rangle_{T_s}$ are small. The zero crossings of $v_g(t)$, $p(t)$, and $\langle i_g(t)\rangle_{T_s}$ coincide. So although the DCM boost converter generates some current harmonics, it is nonetheless possible to construct a low harmonic rectifier that meets harmonic limits. Again, this approach has the disadvantages of the increased peak currents of DCM, and the need for additional filtering of the high-frequency pulsating input currents. Computer simulation of a DCM boost rectifier is described in Appendix B, Section B.2.3.

A similar approach is to operate the boost converter at the boundary between the continuous and discontinuous conduction modes. This approach is known as "critical conduction mode" operation. It eliminates the distortion mechanism described above, but requires variable switching-frequency control. This approach is quite popular at low power levels, and is described further in Section 18.3.3.

Other converters not capable of producing the $M(d(t))$ of Eq. (18.12), such as the buck converter, are sometimes employed as the dc–dc converter of Fig. 18.3. Distortion of the ac line current waveform must then occur. Nonetheless, at low power levels it may be possible to meet the applicable ac line current harmonic standards using such an approach.

18.2.2 DCM Flyback Converter

In Chapter 11, the loss-free resistor network is used to model converters operating in discontinuous conduction mode. This suggests that DCM converters can also be used as near-ideal rectifiers. Indeed, the buck-boost, flyback, SEPIC, and Ćuk converters, when operated in discontinuous conduction mode without additional control, inherently behave as natural loss-free resistors. The DCM effective resistance R_e, found in Chapter 11 to be equal to $2L/D^2T_s$, then coincides with the rectifier emulated resistance of Eq. (18.1). At low power levels, this can be an effective and low-cost approach. Inrush current limiting is also inherent in this approach, and isolation and scaling via a turns ratio are provided by the transformer. Disadvantages are the increased peak currents of DCM, and the need for additional filtering of the high-frequency pulsating input currents.

A simple low-harmonic rectifier system based on the transformer-isolated flyback converter is illustrated in Fig. 18.8 [2]. The ac line voltage is connected through an input EMI filter to a bridge rectifier and a flyback converter. The flyback converter is operated at constant switching frequency f_s and constant duty cycle D. The converter is designed such that it operates in the discontinuous conduction mode under all conditions. The input EMI filter smooths the pulsating input current waveform, so that $i_{ac}(t)$ is approximately sinusoidal.

The flyback converter is replaced by its averaged equivalent circuit in Fig. 18.9. As discussed in Chapter 11, the terminal waveforms of the flyback converter have been averaged over the switching period T_s, resulting in the loss-free resistor model. This model illustrates how the DCM flyback converter presents a resistive load to the ac input. It also illustrates how the power flow can be controlled, by varia-

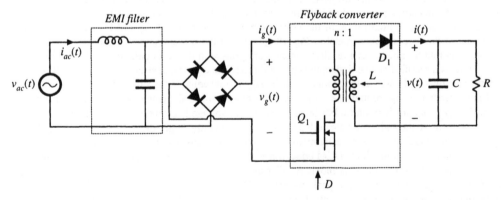

Fig. 18.8 Low-harmonic rectifier system incorporating a flyback converter that operates in the discontinuous conduction mode.

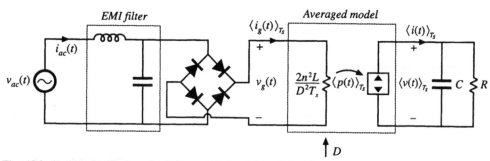

Fig. 18.9 Averaged equivalent circuit that models the system of Fig. 18.8.

tion of D to control the value of the emulated resistance R_e.

To design this converter, one must select the value of inductance to be sufficiently small, such that the converter operates in DCM at all points on the ac sine wave, at maximum load. If we denote the lengths of the transistor conduction interval, diode conduction interval, and discontinuous interval as DT_s, d_2T_s, and d_3T_s, respectively, then the converter operates in DCM provided that d_3 is greater than zero. This implies that

$$d_2(t) < 1 - D \tag{18.35}$$

By volt-second balance on the transformer magnetizing inductance, we can express $d_2(t)$ as

$$d_2(t) = D \frac{v_g(t)}{nV} \tag{18.36}$$

Substitution of Eq. (18.36) into Eq. (18.35) and solution for D yields

$$D < \frac{1}{\left(1 + \dfrac{v_g(t)}{nV}\right)} \tag{18.37}$$

During a given switching period, the converter will operate in DCM provided that the above inequality is satisfied. The worst case occurs when the rectified sinusoid $v_g(t)$ is equal to its peak value V_M. The inequality then becomes

$$D < \frac{1}{\left(1 + \dfrac{V_M}{nV}\right)} \tag{18.38}$$

If Eq. (18.38) is satisfied, then the converter operates in DCM at all points on the ac line sinusoid.

In steady state, the dc output voltage is given by Eq. (18.8). Upon substitution of the expression for R_e and solution for D, this equation becomes

$$D = \frac{2nV}{V_M} \sqrt{\frac{L}{RT_s}} \tag{18.39}$$

Insertion of this relationship into Eq. (18.38), and solution for L, yields

$$L < L_{crit} = \frac{RT_s}{4\left(1 + \frac{nV}{V_M}\right)^2} \tag{18.40}$$

For variations in load R and peak ac input voltage V_M, the worst case will occur at minimum R (maximum power) and minimum V_M. Hence, the designer should choose L to satisfy

$$L < L_{crit\text{-}min} = \frac{R_{min}T_s}{4\left(1 + \frac{nV}{V_{M\text{-}min}}\right)^2} \tag{18.41}$$

If this equation is violated, then at maximum load power and minimum input voltage amplitude, the convert will operate in CCM near the peak of the ac sinewave. This will lead to an input current waveform having substantial distortion.

18.3   CONTROL OF THE CURRENT WAVEFORM

A wide variety of approaches are known for active control of the input current waveform to attain input resistor emulation [14–33]. Average current control [17,18], input voltage feedforward [17], current-programmed control [19–22], hysteretic control and critical conduction mode control [23–27], and nonlinear carrier control [28–30] are briefly surveyed here. Other approaches include sliding-mode control [31], charge control [32], and ASDTIC control [33].

18.3.1   Average Current Control

Average current control is a popular method of implementing control of the input current waveform in a low-harmonic rectifier. This approach works in both continuous and discontinuous conduction modes, and can produce high-quality current waveforms over a wide range of input voltages and load powers. The problems of crossover distortion, found in some competing schemes such as current programmed control, are largely avoided. Several popular integrated circuits are available that implement average current control.

Figure 18.10 illustrates average current control of the input current waveform $\langle i_g(t)\rangle_{T_s}$ in a boost converter. The input current $i_g(t)$ flows through a shunt resistor. The voltage across this shunt resistor is amplified by an op amp circuit. This op amp circuit contains a low-pass filter characteristic that attenuates the high-frequency switching harmonics. The output voltage $v_a(t)$ of the op amp circuit is proportional to the low-frequency average value of $i_g(t)$:

$$v_a(t) = R_s\langle i_g(t)\rangle_{T_s} \tag{18.42}$$

This signal is compared to the reference voltage $v_r(t)$, to produce an error signal that drives the compensator network and pulse-width modulator as illustrated. If the feedback loop is well designed, then the error signal is small:

$$v_a(t) \approx v_r(t) \tag{18.43}$$

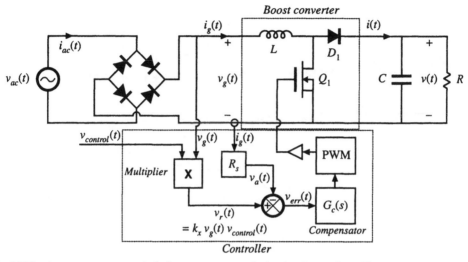

Fig. **18.11**   Average current control of a boost converter, to obtain a low-harmonic rectifier.

The average current controller causes the sensed current $i_g(t)$ to follow the reference waveform $v_r(t)$.

To cause the input current to be proportional to the input voltage, the reference voltage $v_r(t)$ is derived from the sensed input voltage waveform, as in Fig. 18.11. The current reference signal $v_r(t)$ is derived from the sensed input voltage $v_g(t)$, and hence has a sinusoidal waveshape. Hence, the average current controller causes the average input current $i_g(t)$ to be proportional to the input voltage $v_g(t)$. The multiplier illustrated in Fig. 18.11 allows adjustment of the constant of proportionality, so that the magnitude of the emulated resistance can be controlled via a control signal $v_{control}(t)$. Let us assume that the multiplier terminal equations are

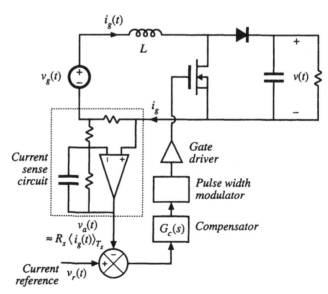

Fig. **18.10**   Average current control of the input current in a boost converter.

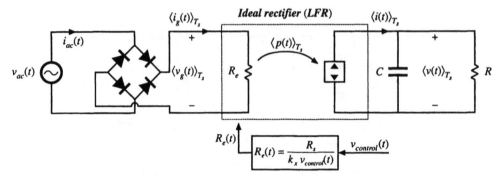

Fig. 18.12   Model of the system of Fig. 18.5, based on the loss-free resistor model of Fig. 18.1(c), which predicts the low-frequency system waveforms. This model assumes that the feedback loop of Fig. 18.5 operates ideally.

$$v_r(t) = k_x v_g(t) v_{control}(t) \tag{18.44}$$

Then the emulated resistance is

$$R_e = \frac{v_g(t)}{i_g(t)} = \frac{\left(\dfrac{v_r(t)}{k_x v_{control}(t)} \right)}{\left(\dfrac{v_a(t)}{R_s} \right)} \tag{18.45}$$

Here, Eqs. (18.44) and (18.42) have been used to eliminate v_g and i_g. Substitution of Eq. (18.43) leads to the result

$$R_e\!\left(v_{control}(t)\right) = \frac{R_s}{k_x v_{control}(t)} \tag{18.46}$$

Hence, if the feedback loop is well designed, then the system of Fig. 18.11 can be represented by the LFR model as in Fig. 18.12. The average current controller scheme of Fig. 18.11 and the model of Fig. 18.12 are independent of the dc–dc converter topology, and can be applied to systems containing CCM boost, buck-boost, Ćuk, SEPIC, and other topologies.

Average power flow and the output voltage are regulated by variation of the emulated resistance R_e, in average current control as well as in most other schemes. This is usually accomplished by use of a multiplier in the input voltage sensing path, as shown in Fig. 18.13. This control loop continually adjusts R_e to maintain balance of the average rectifier power $P_{av} = V_{g,rms}^2 / R_e$ and the load power P_{load}, such that the following relation is obeyed:

$$P_{av} = \frac{V_{g,rms}^2}{R_e} = P_{load} \tag{18.47}$$

Average current control works quite well. Its only disadvantages are the need to sense the average input current, rather than the transistor current, and the need for a multiplier in the controller circuit.

Most average current control implementations include provisions for feedforward of the input voltage amplitude. This allows disturbances in the ac input voltage amplitude to be canceled out by the

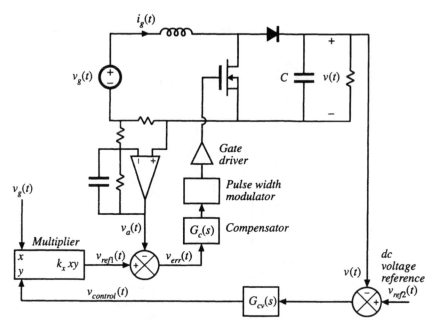

Fig. 18.13 Average current control incorporating a multiplier for regulation of the output voltage.

controller, such that the dc output voltage is unaffected.

Combination of Eqs. (18.44), (18.46), and (18.47), and solution for $v_{ref1}(t)$ leads to

$$v_{ref1}(t) = \frac{P_{av}v_g(t)R_s}{V_{g,rms}^2} \tag{18.48}$$

This equation shows how the reference voltage should be varied to maintain a given rectifier average power throughput P_{av}. Apparently, it is necessary to divide by the square of the rms input voltage amplitude. A controller that implements Eq. (18.48) is illustrated in Fig. 18.14. The multiplier block of Fig. 18.13 has been generalized to perform the function $k_x xy/z^2$. It is somewhat complicated to compute the rms value of a general ac waveform; however, the ac input voltage $v_g(t)$ normally is sinusoidal with negligible harmonics. Hence, the peak value of $v_g(t)$ is directly proportional to its rms value, and we can use the peak value V_M in place of $V_{g,rms}$. So the controller of Fig. 18.14 produces the reference voltage

$$v_{ref1}(t) = \frac{k_v v_{control}(t)v_g(t)}{V_M^2} \tag{18.49}$$

Comparison of Eqs. (18.48) and (18.49) leads to the conclusion that

$$P_{av} = \frac{k_v v_{control}(t)}{2R_s} \tag{18.50}$$

So the average power throughput is directly controlled by $v_{control}(t)$, and is independent of the input voltage $v_g(t)$.

Feedforward can cause the rectifier dc output voltage to be less sensitive to variations in the ac

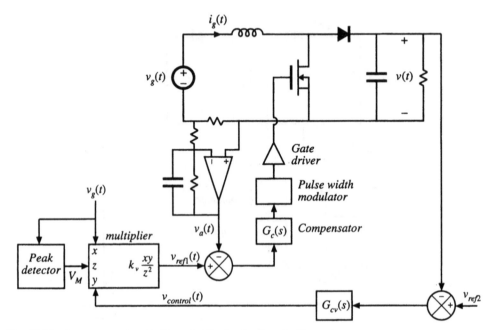

Fig. 18.14 Average current control incorporating input voltage feedforward.

line voltage. A disadvantage is the ac line current distortion introduced by variations in the voltage produced by the peak detector.

To aid in the design of the inner feedback loop that controls the ac line current waveshape, a converter model is needed that describes how the converter average input current depends on the duty cycle. We would prefer to apply the averaged small-signal modeling techniques of Chapter 7 here. The problem is that the variations in the duty cycle $d(t)$, as well as in the ac input voltage $v_g(t)$ and current $i_g(t)$, are not small. As a result, in general the small-signal assumptions are violated, and we are faced with the design of a control system that exhibits significant nonlinear time-varying behavior.

When the rectifier operates near periodic steady state, the output voltage $v(t)$ of a well-designed system exhibits small variations. So we can write

$$\langle v(t) \rangle_{T_s} = V + \hat{v}(t) \tag{18.51}$$

with

$$|\hat{v}(t)| \ll |V| \tag{18.52}$$

In other words, the small-signal assumption continues to be valid with respect to the rectifier output voltage. In the case of the boost converter, this allows us to linearize the converter input characteristics.

Following the approach of Chapter 7, we can express the average inductor voltage of the boost converter as

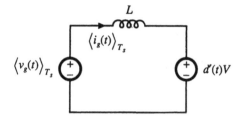

Fig. 18.15 Linearized model describing the boost converter input characteristics, corresponding to Eq. (18.55)

$$L\frac{d\langle i_g(t)\rangle_{T_s}}{dt} = \langle v_g(t)\rangle_{T_s} - d'(t)\langle v(t)\rangle_{T_s} \qquad (18.53)$$

This equation contains the nonlinear term $d'(t)\langle v(t)\rangle_{T_s}$. Substitution of Eq. (18.51) into (18.53) yields

$$L\frac{d\langle i_g(t)\rangle_{T_s}}{dt} = \langle v_g(t)\rangle_{T_s} - d'(t)V - d'(t)\hat{v}(t) \qquad (18.54)$$

When Eq. (18.52) is satisfied, then the nonlinear term $-d'(t)\hat{v}(t)$ is much smaller in magnitude than the linear term $-d'(t)V$. Therefore, we can discard the nonlinear term to obtain

$$L\frac{d\langle i_g(t)\rangle_{T_s}}{dt} = \langle v_g(t)\rangle_{T_s} - d'(t)V \qquad (18.55)$$

This linear differential equation is valid even though $i_g(t)$, $v_g(t)$, and $d(t)$ contain large variations.

An equivalent circuit corresponding to Eq. (18.55) is given in Fig. 18.15. The averaged control-to-input-current transfer function is found by setting the independent inputs other than $d(t)$ to zero, and then solving for i_g; the model predicts that this transfer function is

$$\frac{i_g(s)}{d(s)} = \frac{V}{sL} \qquad (18.56)$$

where $i_g(s)$ is the Laplace transform of $\langle i_g(t)\rangle_{T_s}$. So the input characteristics of the boost rectifier can be linearized, even though the ac input variations are not small.

Unfortunately, Eq. (18.52) is not sufficient to linearize the equations describing the input characteristics of the buck-boost, SEPIC, Ćuk, and most other single-phase rectifiers. The control system design engineer must then deal with a truly nonlinear time-varying dynamical system.

One approach that is sometimes suggested employs the *quasi-static approximation* [34,35]. It is assumed that the ac line variations are much slower than the rectifier system dynamics, such that the rectifier always operates near equilibrium. The quiescent operating point changes slowly along the input sinusoid; an equilibrium analysis can be performed to find expressions for the slowly-varying "equilibrium" duty ratio and converter voltages and currents. The small-signal dc–dc converter transfer functions derived in Chapters 7 and 8 are evaluated using this time-varying operating point. The converter poles, zeroes, and gains are found to vary along the ac input sinusoid. An average current controller is designed using these time-varying transfer functions, such that the current loop gain has a positive phase margin at all operating points.

We expect that the quasi-static approximation should be valid if the rectifier system dynamics are sufficiently fast, and it is reasonable to anticipate that high-frequency PWM converters have dynamics that are much faster than the ac line frequency. The problem is that no good condition on system parameters, which can justify the approximation, is known for the basic converter topologies. There is room for additional research in this area.

It is well-understood in the field of control systems that, when the rectifier system dynamics are not sufficiently fast, the quasi-static approximation yields neither sufficient nor necessary conditions for stability of the resulting design. Time-varying "loop gains" that always have a positive phase margin may nonetheless be unstable, and a negative phase margin does not always imply instability. Such phenomena are sometimes observed in rectifier systems. Even worse, it is difficult to justify the use of the Laplace transform on rectifiers described by time-varying differential equations, unless the quasi-static approximation can be validated.

18.3.2   Current Programmed Control

Another well-known approach to attaining input resistor emulation is the use of current-programmed control. As illustrated in Fig. 18.16, the programmed current $i_c(t)$ is made proportional to the ac input voltage. This causes the average inductor current, and hence also $\langle i_g(t) \rangle_{T_s}$, to approximately follow $v_g(t)$. As in average current control, a multiplier is used to adjust the emulated resistance and average power flow; the control signal $v_{control}(t)$ is typically used to stabilize the dc output voltage magnitude. Several rectifier control ICs are commercially available, which implement current-programmed control.

As discussed in Chapter 12, several mechanisms cause the average inductor current and hence also $\langle i_g(t) \rangle_{T_s}$ to differ from the programmed $i_c(t)$. These mechanisms introduce crossover distortion and line current harmonics. An artificial ramp having sufficiently large slope m_a is necessary to stabilize the

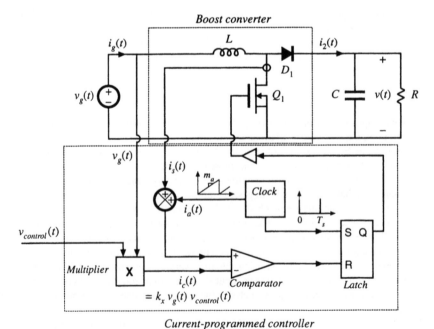

Current-programmed controller

Fig. 18.16   Current-programmed control of a boost rectifier.

current-programmed boost converter when it operates in CCM with $d(t) > 0.5$. The addition of this ramp causes $\langle i_g(t) \rangle_{T_s}$ to differ from $i_c(t)$. Additional deviation is introduced by the inductor current ripple. Both mechanisms are most pronounced when the inductor current is small, near the zero-crossings of the ac line waveforms.

The static input characteristics, that is, the average input current vs. the input voltage, of the current-programmed boost converter are given by

$$\langle i_g(t) \rangle_{T_s} = \begin{cases} v_g(t) \dfrac{L i_c^2(t) f_s V}{2 \left(V - v_g(t)\right) \left(v_g(t) + m_a L\right)^2} & \text{in DCM} \\[2ex] i_c(t) - \left(1 - \dfrac{v_g(t)}{V}\right) \left(m_a + \dfrac{v_g(t)}{2L}\right) T_s & \text{in CCM} \end{cases} \tag{18.57}$$

The converter operates in the continuous conduction mode when

$$\langle i_g(t) \rangle_{T_s} > \frac{T_s V}{2L} \frac{v_g(t)}{V} \left(1 - \frac{v_g(t)}{V}\right) \tag{18.58}$$

In terms of the control current $i_c(t)$, the condition for operation in CCM can be expressed

$$i_c(t) > \frac{T_s V}{L} \left(\frac{m_a L}{V} + \frac{v_g(t)}{V}\right) \left(1 - \frac{v_g(t)}{V}\right) \tag{18.59}$$

In the conventional current-programmed rectifier control scheme, the control current $i_c(t)$ is simply proportional to the ac input voltage:

$$i_c(t) = \frac{v_g(t)}{R_e} \tag{18.60}$$

where R_e is the emulated resistance that would be obtained if the average input current exactly followed the reference current $i_c(t)$. The static input characteristics given by Eqs. (18.57) to (18.60) are plotted in Fig. 18.17. The average input current $\langle i_g(t) \rangle_{T_s}$ is plotted as a function of the applied input voltage $v_g(t)$, for several values of emulated resistance R_e. The region near the CCM–DCM boundary is shown. The curves are plotted for a fixed artificial ramp having slope

$$m_a = \frac{V}{2L} \tag{18.61}$$

This is the minimum value of artificial ramp that stabilizes the boost current-programmed controller at all static operating points. Decreasing m_a below this value leads to instability at operating points in the continuous conduction mode at low $v_g(t)/V$.

To obtain resistor emulation, it is desired that the static input characteristics be linear and pass through the origin. It can be seen from Fig. 18.17 that this is not the case: the curves are reasonably linear in the continuous conduction mode, but exhibit significant curvature as the CCM–DCM boundary is approached. The resulting average current waveforms are summarized in Fig. 18.8.

To minimize the line current THD, it is apparent that the converter should be designed to operate deeply in the continuous conduction mode for most of the ac line cycle. This is accomplished with emulated resistances R_e that are much smaller than $R_{base} = 2L/T_s$. In addition, the artificial ramp slope m_a

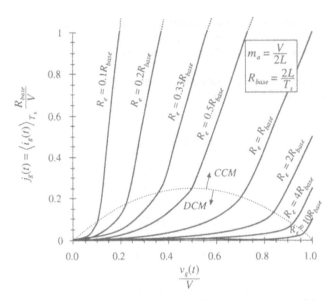

Fig. 18.17 Static input characteristics of a current-programmed boost converter, with minimum stabilizing artificial ramp of Eq. (18.61).

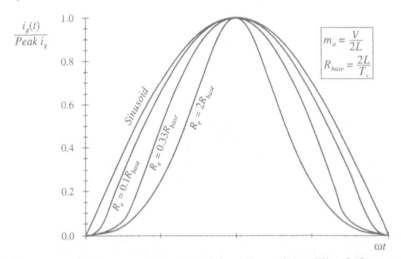

Fig. 18.18 Input current waveshapes predicted by the static input characteristics of Fig. 18.17, compared with a pure sinusoid. Curves are plotted for the case $V_M = 0.8V$, with minimum stabilizing artificial ramp.

should be no greater than otherwise necessary. In practice, THD of 5% to 10% can easily be obtained in rectifiers that function over a narrow range of rms input voltages and load currents. However, low THD cannot be obtained at all operating points in universal-input rectifiers; THD of 20% to 50% may be observed at maximum ac input voltage. This problem can be solved by biasing the current reference waveform. Design of current-programmed rectifiers is discussed in [19–22], and some strategies for solving this problem are addressed in [19].

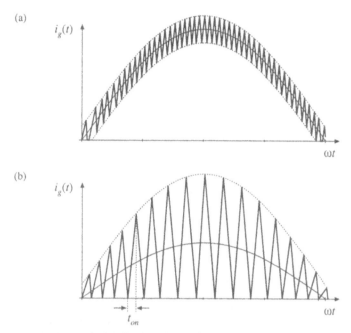

Fig. 18.19 Input current waveforms of two boost converters with hysteretic control: (a) ± 10% regulation band, (b) critical conduction mode operation (± 100% regulation band).

18.3.3 Critical Conduction Mode and Hysteretic Control

Another control scheme sometimes used in low-harmonic rectifiers, as well as in dc–dc converters and dc–ac inverters, is hysteretic control. Rather than operating at a fixed switching frequency and duty cycle, the hysteretic controller switches the transistor on and off as necessary to maintain a waveform within given limits. A special case of hysteretic control, called *critical conduction mode* control, is implemented in several commercially-available ICs, and is popular for low-harmonic rectifiers rated below several hundred Watts [23–25].

An example is the sinusoid of Fig. 18.19(a), in which the boost converter input current is controlled to follow a sinusoidal reference with a ± 10% tolerance. The inductor current increases when the transistor is on, and decreases when the transistor is off. So this hysteretic controller switches the transistor on whenever the input current falls below 90% of the reference input. The controller switches the transistor off whenever the input current exceeds 110% of the reference. Hysteretic controllers tend to have simple implementations. However, they have the disadvantages of variable switching frequency and reduced noise immunity.

Another example of hysteretic control is the waveform of Fig. 18.19(b). The lower limit is chosen to be zero, while the upper limit is twice the reference input. This controller operates the boost converter at the boundary between the continuous and discontinuous conduction modes. An alternative control scheme that generates the same waveform simply operates the transistor with constant on-time: the transistor is switched on when the inductor current reaches zero, and is switched off after a fixed

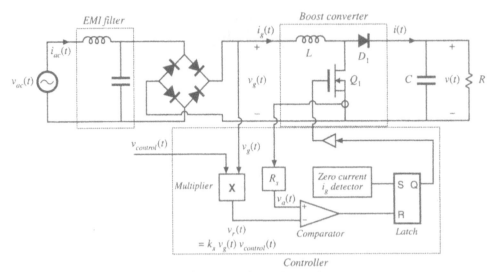

Fig. 18.20   A typical implementation of critical conduction mode control.

interval of length t_{on}. The resulting inductor current waveform will have a peak value that depends directly on the applied input voltage, and whose average value is one-half of its peak. With either control approach, the converter naturally exhibits loss-free-resistor or ideal rectifier behavior. The emulated resistance is

$$R_e = \frac{2L}{t_{on}} \tag{18.62}$$

This scheme has the advantage of small inductor size and low-cost control ICs. Disadvantages are increased peak currents, variable switching frequency, and the need for additional input EMI filtering.

A typical critical conduction mode controller is illustrated in Fig. 18.20. A zero-current detector senses when the inductor current is zero; this is typically accomplished by monitoring the voltage across the inductor. The zero-current detector sets a latch, turning on the transistor and initiating the switching period. The transistor current is also monitored, and is compared to a sinusoidal reference $v_r(t)$ that is proportional to the applied input voltage $v_g(t)$. When the sensed current is equal to the reference, the latch is reset and the transistor is turned off.

Since the switching frequency can vary, possibly over a wide range, it is important to carefully design the converter power stage. For a given power P, the required transistor on-time t_{on} can be found by combining Eqs. (18.17) and (18.62), and solving for t_{on}:

$$t_{on} = \frac{4LP}{V_M^2} \tag{18.63}$$

Application of the principle of volt-second balance to inductor L of Fig. 18.20 leads to the following equation:

$$v_g t_{on} + \left(v_g - V \right) t_{off} = 0 \tag{18.64}$$

Hence, the transistor off-time is given by

$$t_{off} = t_{on} \frac{v_g}{\left(V - v_g\right)} \tag{18.65}$$

The switching period T_s is equal to

$$T_s = t_{off} + t_{on} \tag{18.66}$$

Substitution of Eqs. (18.63) and (18.65) into Eq. (18.66) yields

$$T_s = \frac{4LP}{V_M^2} \frac{1}{\left(1 - \frac{v_g(t)}{V}\right)} \tag{18.67}$$

The following expression for switching frequency is found by substitution of Eq. (18.11) into Eq. (18.67):

$$f_s = \frac{1}{T_s} = \frac{V_M^2}{4LP} \left(1 - \frac{V_M}{V} \left| \sin(\omega t) \right| \right) \tag{18.68}$$

The maximum switching frequency occurs when $\sin(\omega t)$ equals zero:

$$\max f_s = \frac{V_M^2}{4LP} \tag{18.69}$$

The minimum switching frequency occurs at the peak of the sine wave:

$$\min f_s = \frac{V_M^2}{4LP} \left(1 - \frac{V_M}{V}\right) \tag{18.70}$$

Equations (18.69) and (18.70) can be used to select the value of the inductance L and the output voltage V, so that the switching frequency varies over an acceptable range.

18.3.4   Nonlinear Carrier Control

The nonlinear-carrier controller (NLC) is capable of attaining input resistor emulation in boost and other converters that operate in the continuous conduction mode. Implementation of the controller is quite simple, with no need for sensing of the input voltage or input current. There is also no need for a current loop error amplifier. The boost nonlinear-carrier charge controller is inherently stable and is free from the stability problems that require addition of an artificial ramp in current programmed controllers.

   A CCM boost rectifier system with nonlinear-carrier charge control is illustrated in Fig. 18.21, and waveforms are given in Fig. 18.22. The reasoning behind this approach is as follows. It is desirable to control the transistor switch current $i_s(t)$. This pulsating current is much easier to sense than the continuous converter input current—a simple current transformer can be used, as in Fig. 18.21. Further, it is desirable to control the integral of this current, or the charge, for two reasons: (1) integration of the waveform leads to improved noise immunity, and (2) the integral of the waveform is directly related to its average value,

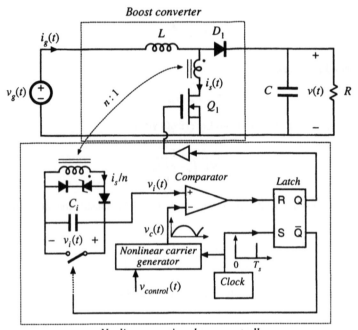

Fig. 18.21   Nonlinear-carrier charge control of a boost converter.

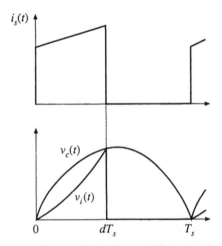

Fig. 18.22   Transistor current $i_s(t)$, parabolic carrier voltage $v_c(t)$, and integrator voltage $v_i(t)$ waveforms for the NLC-controlled boost rectifier of Fig. 18.21.

$$\left\langle i_s(t) \right\rangle_{T_s} = \frac{1}{T_s} \int_t^{t+T_s} i_s(\tau) d\tau \tag{18.71}$$

In a fixed-frequency system, T_s is constant, and the integral over one switching period is proportional to the average value. Hence the average switch current can be controlled to be proportional to a reference signal by simply switching the transistor off when the integral of the switch current is equal to the reference. In the controller of Fig. 18.21, the switch current $i_s(t)$ is scaled by the transformer turns ratio n, and then integrated by capacitor C_i, such that

$$v_i(t) = \frac{1}{C_i} \int_0^{dT_s} \frac{i_s(\tau)}{n} d\tau \qquad \text{for } 0 < t < dT_s \tag{18.72}$$

The integrator voltage $v_i(t)$ is reset to zero at the end of each switching period, and the integration process begins anew at the beginning of the next switching period. So at the instant that the transistor is switched off, the voltage $v_i(dT_s)$ is proportional to the average switch current:

$$v_i(dT_s) = \frac{\left\langle i_s \right\rangle_{T_s}}{n C_i f_s} \qquad \text{for interval } 0 < t < T_s \tag{18.73}$$

How should the average switch current be controlled? To obtain input resistor emulation, it is desired that

$$\left\langle i_g(t) \right\rangle_{T_s} = \frac{\left\langle v_g(t) \right\rangle_{T_s}}{R_e(v_{control})} \tag{18.74}$$

It is further desired to avoid sensing either $i_g(t)$ or $v_g(t)$. As with other schemes, we will sense the dc output voltage $\left\langle v(t) \right\rangle_{T_s}$, to construct a low-bandwidth feedback loop that balances the average input and output powers. So let us determine the relationship between $\left\langle i_s(t) \right\rangle_{T_s}$ and $\left\langle v(t) \right\rangle_{T_s}$ implied by Eq. (18.74). If we assume that the boost converter operates in the continuous conduction mode, then we can write

$$\left\langle i_s(t) \right\rangle_{T_s} = d(t) \left\langle i_g(t) \right\rangle_{T_s} \tag{18.75}$$

and

$$\left\langle v_g(t) \right\rangle_{T_s} = d'(t) \left\langle v(t) \right\rangle_{T_s} \tag{18.76}$$

Substitution of Eqs. (18.75) and (18.76) into Eq. (18.74) leads to

$$\left\langle i_s(t) \right\rangle_{T_s} = d(t)\left(1 - d(t)\right) \frac{\left\langle v(t) \right\rangle_{T_s}}{R_e(v_{control})} \tag{18.77}$$

The controller of Fig. 18.21 implements this equation.

The nonlinear carrier generator of Fig. 18.21 produces the parabolic waveform $v_c(t)$, given by

$$v_c(t) = v_{control}\left(\frac{t}{T_s}\right)\left(1 - \frac{t}{T_s}\right) \qquad \text{for } 0 \leq t \leq T_s$$

$$v_c(t + T_s) = v_c(t)$$

(18.78)

This waveform is illustrated in Fig. 18.22. Note that Eq. (18.78) resembles Eq. (18.77), with $d(t)$ replaced by (t/T_s). The controller switches the transistor off at time $t = dT_s$ when the integrator voltage $v_i(t)$ is equal to the carrier waveform $v_c(t)$. Hence, it is true that

$$v_i(dT_s) = v_c(dT_s) = v_{control}(t)\, d(t)\left(1 - d(t)\right)$$

(18.79)

Substitution of Eq. (18.73) yields

$$\frac{\langle i_s(t)\rangle_{T_s}}{nC_i f_s} = v_{control}(t)\, d(t)\left(1 - d(t)\right)$$

(18.80)

This is of the same form as Eq. (18.77). Comparison of Eqs. (18.77) and (18.80) reveals that the emulated resistance R_e is given by

$$R_e(v_{control}) = d(t)\left(1 - d(t)\right)\frac{\langle v(t)\rangle_{T_s}}{\langle i_s(t)\rangle_{T_s}} = \frac{\langle v(t)\rangle_{T_s}}{nC_i f_s v_{control}(t)}$$

(18.81)

If the dc output voltage and the control voltage have negligible ac variation, then R_e is essentially constant, and the ac line current will exhibit low harmonic distortion. So neither the input voltage nor the input current need be sensed, and input resistor emulation can be obtained in CCM boost converters by sensing only the switch current.

A simple way to generate the parabolic carrier waveform uses two integrators, as illustrated in Fig. 18.23. The slowly varying control voltage $v_{control}(t)$ is integrated, to obtain a ramp waveform $v_r(t)$ whose peak amplitude is proportional to $v_{control}(t)$. The dc component of this waveform is removed, and then integrated again. The output of the second integrator is the parabolic carrier $v_c(t)$, illustrated in Fig. 18.22 and given by Eq. (18.78). Both integrators are reset to zero before the end of each switching period

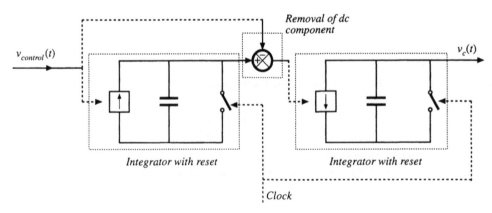

Fig. 18.23   Generation of parabolic carrier waveform by double integration.

by the clock generator. The amplitude of the parabolic carrier, and hence also the emulated resistance, can be controlled by variation of $v_{control}(t)$.

Equations (18.75) and (18.76) are valid only when the converter operates in the continuous conduction mode. In consequence, the ac line current waveform is distorted when the converter operates in DCM. Since this occurs near the zero crossings of the ac line voltage, crossover distortion is generated. Nonetheless, the harmonic distortion is less severe than in current-programmed schemes, and it is feasible to construct universal-input rectifiers that employ the NLC control approach. Total harmonic distortion is analyzed and plotted in [28].

Nonlinear carrier control can be applied to current-programmed boost rectifiers, as well as to other rectifiers based on the buck-boost, SEPIC, Ćuk, or other topologies, with either integral charge control or peak-current-programmed control [28,29]. In these cases, a different carrier waveform must be employed. A nonlinear-carrier controller in which the ac input voltage $v_g(t)$ is sensed, rather than the switch current $i_s(t)$, is described in [30].

18.4 SINGLE-PHASE CONVERTER SYSTEMS INCORPORATING IDEAL RECTIFIERS

An additional issue that arises in PWM rectifier systems is the control of power drawn from the ac line, the power delivered to the dc load, and the energy stored in a bulk energy storage capacitor.

18.4.1 Energy Storage

It is usually desired that the dc output voltage of a converter system be regulated with high accuracy. In practice, this is easily accomplished using a high-gain wide-bandwidth feedback loop. A well-regulated dc output voltage $v(t) = V$ is then obtained, which has negligible ac variations. For a given constant load characteristic, the load current I and the instantaneous load power $p_{load}(t) = P_{load}$, are also constant:

$$p_{load}(t) = v(t)i(t) = VI \tag{18.82}$$

However, the instantaneous input power $p_{ac}(t)$ of a single-phase ideal rectifier is not constant:

$$p_{ac}(t) = v_g(t)i_g(t) \tag{18.83}$$

If $v_g(t)$ is given by Eq. (18.11), and if $i_g(t)$ follows Eq. (18.1), then the instantaneous input power becomes

$$p_{ac}(t) = \frac{V_M^2}{R_e} \sin^2(\omega t) = \frac{V_M^2}{2R_e}\left(1 - \cos(2\omega t)\right) \tag{18.84}$$

which varies with time. The instantaneous input power is zero at the zero crossings of the ac input voltage. Equations (18.82) and (18.84) are illustrated in Fig. 18.24(a). Note that the desired instantaneous load power $p_{load}(t)$ is not equal to the desired instantaneous rectifier input power $p_{ac}(t)$. Some element within the rectifier system must supply or consume the difference between these two instantaneous powers.

Since the ideal rectifier does not consume or generate power, nor does it contain significant internal energy storage, it is necessary to add to the system a low-frequency energy storage element such

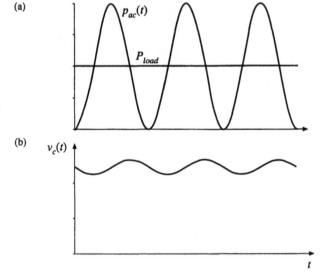

(a)

(b)

Fig. 18.24 Waveforms of a single-phase ideal rectifier system: (a) pulsating ac input power $p_{ac}(t)$, and constant dc load power P_{load}; (b) energy storage capacitor voltage $v_C(t)$.

as an electrolytic capacitor. The difference between the instantaneous input and load powers flows through this capacitor.

The waveforms of rectifier systems containing reactive elements can be determined by solution of the rectifier energy equation [36,37]. If the energy storage capacitor C is the only system element capable of significant low-frequency energy storage, then the power $p_C(t)$ flowing into the capacitor is equal to the difference between the instantaneous input and output powers:

$$p_C(t) = \frac{dE_C(t)}{dt} = \frac{d\left(\frac{1}{2}Cv_C^2(t)\right)}{dt} = p_{ac}(t) - p_{load}(t) \qquad (18.85)$$

where C is the capacitance, $v_C(t)$ is the capacitor voltage, and $E_C(t)$ is the energy stored in the capacitor. Hence as illustrated in Fig. 18.24(b), when $p_{ac}(t) > p_{load}(t)$ then energy flows into the capacitor, and $v_C(t)$ increases. Likewise, $v_C(t)$ decreases when $p_{ac}(t) < p_{load}(t)$. So the capacitor voltage $v_C(t)$ must be allowed to increase and decrease as necessary to store and release the required energy. In steady-state, the average values of $p_{ac}(t)$ and $p_{load}(t)$ must be equal, so that over one ac line cycle there is no net change in capacitor stored energy.

Where can the energy storage capacitor be placed? It is necessary to separate the energy storage capacitor from the regulated dc output, so that the capacitor voltage is allowed to independently vary as illustrated in Fig. 18.24(b). A conventional means of accomplishing this is illustrated in Fig. 18.25. A second dc–dc converter is inserted, between the energy storage capacitor and the regulated dc load. A wide-bandwidth feedback loop controls this converter, to attain a well-regulated dc load voltage. The capacitor voltage $v_C(t)$ is allowed to vary. Thus, this system configuration is capable of (1) wide-bandwidth control of the ac line current waveform, to attain unity power factor, (2) internal low-frequency energy storage, and (3) wide-bandwidth regulation of the dc output voltage. It is also possible to integrate these functions into a single converter, provided that the required low-frequency independence of the input, output, and capacitor voltages is maintained [38].

The energy storage capacitor also allows the system to function in other situations in which the instantaneous input and output powers differ. For example, it is commonly required that the output volt-

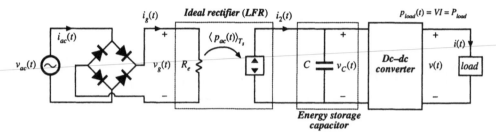

Fig. 18.25 Elements of a single-phase-ac to dc power supply, in which the ac line current and dc load voltage are independently regulated with high bandwidth. An internal independent energy storage capacitor is required.

age remain regulated during ac line voltage failures of short duration. The *hold-up time* is the duration that the output voltage $v(t)$ remains regulated after $v_{ac}(t)$ has become zero. A typical requirement is that the system continue to supply power to the load during one complete missing ac line cycle, that is, for 20 msec in a 50 Hz system. During the hold-up time, the load power is supplied entirely by the energy storage capacitor. The value of capacitance should be chosen such that at the end of the hold-up time, the capacitor voltage $v_C(t)$ exceeds the minimum value that the dc–dc converter requires to produce the desired load voltage.

The energy storage function could be performed by an element other than a capacitor, such as an inductor. However, use of an inductor is a poor choice, because of its high weight and cost. For example, a 100 μF 100 V electrolytic capacitor and a 100 μH 100 A inductor can each store 1 Joule of energy. But the capacitor is considerably smaller, lighter, and less expensive.

A problem introduced by the energy storage capacitor is the large *inrush current* observed during the system turn-on transient. The capacitor voltage $v_C(t)$ is initially zero; substantial amounts of charge and energy are required to raise this voltage to its equilibrium value. The boost converter is not capable of limiting the magnitude of the resulting inrush current: even when $d(t) = 0$, a large current flows through the boost converter diode to the capacitor, as long as the converter output voltage is less than the input voltage. Some additional circuitry is required to limit the inrush current of the boost converter. Converters having a buck–boost type conversion ratio are inherently capable of controlling the inrush current. This advantage comes at the cost of additional switch stress.

It is also possible to design the ideal rectifier to operate correctly when connected to utility power systems anywhere in the world. *Universal input* rectifiers can operate with nominal ac rms voltage magnitudes as low as the 100 V encountered in a portion of Japan, or as high as the 260 V found in western Australia, with ac line frequencies of either 50 Hz or 60 Hz. Regardless of the ac input voltage, the universal-input rectifier produces a constant nominal dc output voltage V_C.

Let us now consider in more detail the low-frequency energy storage process of the system of Fig. 18.25. Let us assume that the dc–dc converter contains a controller having bandwidth much greater than the ac line frequency, such that the load voltage contains negligible low-frequency variations. A low-frequency model of the dc–dc converter is then as illustrated in Fig. 18.26. The dc–dc converter produces constant voltage $v(t) = V$, modeled by a voltage source as shown. This causes the load to draw constant current $i(t) = I$, leading to load power $p_{load}(t) = P_{load}$. To the extent that converter losses can be neglected, the dc–dc converter input port draws power P_{load}, regardless of the value of $v_C(t)$. So the dc–dc converter input port can be modeled as a constant power sink, of value P_{load}.

The model of Fig. 18.26 implies that the difference between the rectifier power $p_{ac}(t)$ and the load power P_{load} flows into the capacitor, as given by Eq. (18.85). The capacitor voltage increases when $p_{ac}(t)$ exceeds P_{load}, and decreases when $p_{ac}(t)$ is less than P_{load}. In steady state, the average values of $p_{ac}(t)$ and P_{load} must be equal. But note that $p_{ac}(t)$ is determined by the magnitudes of $v_{ac}(t)$ and R_e, and

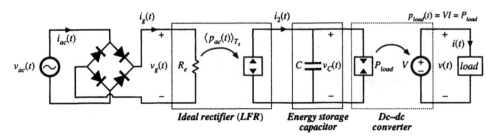

Fig. 18.26 Low-frequency equivalent circuit of the system of Fig. 18.25.

not by the load. The system of Fig. 18.26 contains no mechanism to cause the average rectifier power and load power to be equal. In consequence, it is necessary to add an additional control system that adjusts R_e as necessary, to cause the average rectifier output power and dc–dc converter input power to balance. The conventional way to accomplish this is simply to regulate the dc component of $v_C(t)$.

A complete system containing ideal rectification, energy storage, and wide-bandwidth output voltage regulation is illustrated in Fig. 18.27. This system incorporates the boost converter and controller of Fig. 18.5, as well as a generic dc–dc converter with output voltage feedback. In addition, the system contains a low-bandwidth feedback loop, which regulates the dc component of the energy-storage capacitor voltage to be equal to a reference voltage v_{ref2}. This is accomplished by slow variations of $v_{control}(t)$ and R_e. This controller should have sufficiently small loop gain at the even harmonics of the ac line frequency, so that variations in R_e are much slower than the ac line frequency.

Increasing the bandwidth of the energy storage capacitor voltage controller can lead to significant ac line current harmonics. When this controller has wide bandwidth and high gain, then it varies $R_e(t)$ quickly, distorting the ac line current waveform. In the extreme limit of perfect regulation of the

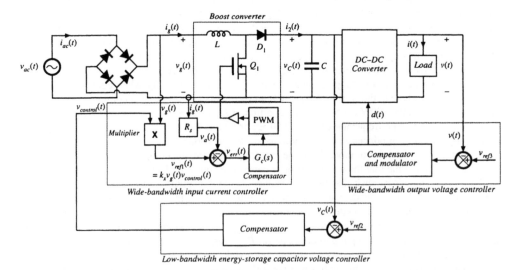

Fig. 18.27 A complete dc power supply system incorporating a near-ideal single-phase boost rectifier system, energy storage capacitor, and dc-dc converter. Wide-bandwidth feedback loops regulate the ac line current waveform and the dc load voltage, and a slow feedback loop regulates the energy storage capacitor voltage.

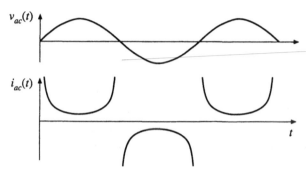

Fig. 18.28 Ac line current waveform of the single-phase ideal rectifier with output voltage feedback, when it supplies constant instantaneous power to a dc load. The THD tends to infinity, and the power factor tends to zero.

energy storage capacitor voltage $v_C(t) = V_C$, then the capacitor stored energy is constant, and the instantaneous input ac line power $p_{ac}(t)$ and load power $p_{load}(t)$ are equal. The controller prevents the energy-storage capacitor from performing its low-frequency energy storage function. The ac line current then becomes

$$i_{ac}(t) = \frac{p_{ac}(t)}{v_{ac}(t)} = \frac{p_{load}(t)}{v_{ac}(t)} = \frac{P_{load}}{V_M \sin(\omega t)} \tag{18.86}$$

This waveform is sketched in Fig. 18.28. In this idealized limiting case, the ac line current tends to infinity at the zero crossings of the ac line voltage waveform, such that the instantaneous input power is constant. It can be shown that the THD of this current waveform is infinite, and its distortion factor and power factor are zero. So the bandwidth of this controller should be limited.

The energy storage capacitor voltage ripple can be found by integration of Eq. (18.85). Under steady-state conditions, where the average value of $p_{ac}(t) = P_{load}$, integration of Eq. (18.85) yields

$$E_C(t) = \tfrac{1}{2} C v_C^2(t) = E_C(0) + \int_0^t \left(-P_{load} \cos(2\omega t) \right) dt \tag{18.87}$$

where ω is the ac line frequency. Evaluation of the integral leads to

$$E_C(t) = E_C(0) - \frac{P_{load} \sin(2\omega t)}{2\omega} \tag{18.88}$$

Therefore, the capacitor voltage waveform is

$$v_C(t) = \sqrt{\frac{2 E_C(t)}{C}} = \sqrt{v_C^2(0) - \frac{P_{load}}{\omega C} \sin(2\omega t)} \tag{18.89}$$

It can be verified that the rms value of this waveform is $V_{C,rms} = v_C(0)$. Hence, Eq. (18.89) can be written

$$v_C(t) = V_{C,rms} \sqrt{1 - \frac{P_{load}}{\omega C V_{C,rms}^2} \sin(2\omega t)} \tag{18.90}$$

This waveform is sketched in Fig. 18.24(b). The minimum and maximum values of the capacitor voltage occur when $\sin(2\omega t)$ is equal to 1 and -1, respectively. Therefore, the peak-to-peak capacitor voltage ripple is

$$2\Delta v_C = V_{C,rms}\left[\sqrt{1 + \frac{P_{load}}{\omega C V_{C,rms}^2}} - \sqrt{1 - \frac{P_{load}}{\omega C V_{C,rms}^2}}\right] \approx \frac{P_{load}}{\omega C V_{C,rms}} \qquad (18.91)$$

The approximation is valid for $P_{load}/(\omega C V_{C,rms}^2)$ sufficiently less than one, a condition that is satisfied whenever the ac voltage ripple is sufficiently less than $V_{C,rms}$.

18.4.2   Modeling the Outer Low-Bandwidth Control System

As discussed above, the outer low-bandwidth controller, which varies the emulated resistance as necessary to balance the average ac input and dc load powers, is common to all near-ideal rectifier systems. For design of this controller, the rectifier can be modeled using the loss-free resistor (LFR) model. Perturbation and linearization of the LFR leads to a small-signal equivalent circuit that predicts the relevant small-signal transfer functions. Such a model is derived in this section [2,39,40].

It is desirable to stabilize the rectifier output voltage against variations in load power, ac line voltage, and component characteristics. Hence, a voltage feedback loop is necessary. As discussed in Section 18.4.1, this loop cannot attempt to remove the capacitor voltage ripple that occurs at the second harmonic of the ac line frequency, 2ω, since doing so would require that $R_e(t)$ change significantly at the second harmonic frequency. This would introduce significant distortion, phase shift, and power factor degradation into the ac line current waveform. In consequence this loop must have sufficiently small gain at frequency 2ω, and hence its bandwidth must be low. Therefore, for the purposes of designing the low-bandwidth outer control loop, it is unnecessary to model the system high-frequency behavior. It can be assumed that any inner wide-bandwidth controller operates ideally at low frequencies, such that the ideal rectifier model of Fig. 18.29(a) adequately represents the low-frequency system behavior.

A small-signal model is derived here that correctly predicts the control-to-output transfer function and output impedance of any rectifier system that can be modeled as a loss-free resistor. The model neglects the complicating effects of high-frequency switching ripple, and is valid for control variations at frequencies sufficiently less than the ac line frequency. Both resistive and dc-dc converter/regulator loads are treated.

The steps in the derivation of the small-signal ac model are summarized in Fig. 18.29. Figure 18.29(a) is the basic ideal rectifier model, in which the converter high frequency switching ripple is removed via averaging over the switching period T_s, but waveform frequency components slower than the switching frequency are correctly modeled, including the 2ω second-harmonic and dc components of output voltage. It is difficult to use this model in design of the feedback loop because it is highly nonlinear and time-varying.

If the ac input voltage $v_g(t)$ is

$$v_g(t) = \sqrt{2}\, v_{g,rms}\left|\sin\left(\omega t\right)\right| \qquad (18.92)$$

then the model of Fig. 18.29(a) predicts that the instantaneous output power $\langle p(t)\rangle_{T_s}$ is

$$\langle p(t)\rangle_{T_s} = \frac{\langle v_g(t)\rangle_{T_s}^2}{R_e(v_{control}(t))} = \frac{v_{g,rms}^2}{R_e(v_{control}(t))}\left(1 - \cos\left(2\omega t\right)\right) \qquad (18.93)$$

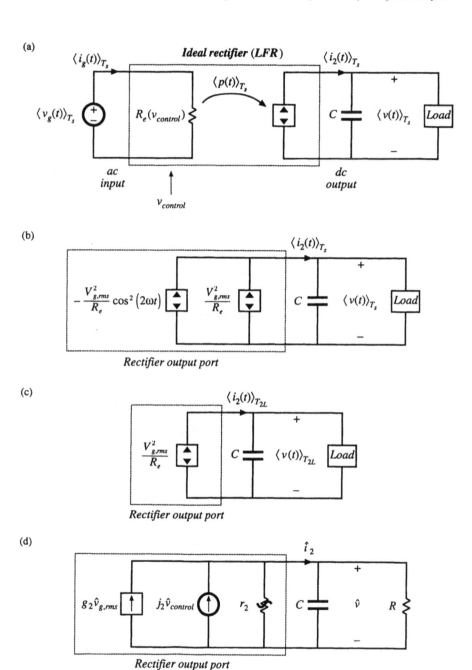

Fig. 18.29 Steps in the derivation of low-frequency small-signal rectifier model: (a) large-signal LFR model, averaged over one switching period T_s; (b) separation of power source into its constant and time-varying components; (c) removal of second-harmonic components by averaging over one-half of the ac line period, T_{2L}; (d) small-signal model obtained by perturbation and linearization of Fig. 18.29(c).

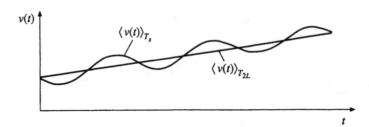

Fig. 18.30   Removal of components of $v(t)$ at the harmonics of the ac line frequency, by averaging over one-half of the ac line period, T_{2L}.

The output power is comprised of a constant term $v_{g,rms}^2/R_e$, and a term that varies at the second harmonic of the ac line frequency. These two terms are explicitly identified in Fig. 18.29(b).

The second-harmonic variation in $\langle p(t) \rangle_{T_s}$ leads to time-varying system equations, and slow variations in $v_{control}(t)$ lead to an output voltage spectrum containing components not only at the frequencies present in $v_{control}(t)$, but also at the even harmonics of the ac line frequency and their sidebands, as well as at the switching frequency and its harmonics and sidebands. It is desired to model only the low-frequency components excited by slow variations in $v_{control}(t)$, the load, and the ac line voltage amplitude $v_{g,rms}$. The even harmonics of the ac line frequency can be removed by averaging over one-half of the ac line period

$$T_{2L} = \tfrac{1}{2} \tfrac{2\pi}{\omega} = \tfrac{\pi}{\omega} \tag{18.94}$$

Hence, we average over the switching period T_s to remove the switching harmonics, and then average again over one-half of the ac line period T_{2L} to remove the even harmonics of the ac line frequency. The resulting model is valid for frequencies sufficiently less than the ac line frequency ω. Averaging of the rectifier output voltage is illustrated in Fig. 18.30: averaging over T_{2L} removes the ac line frequency harmonics, leaving the underlying low-frequency variations. By averaging the model of Fig. 18.29(b) over T_{2L}, we obtain the model of Fig. 18.29(c). This step removes the second-harmonic variation in the power source.

The equivalent circuit of Fig. 18.29(c) is time-invariant, but nonlinear. We can now perturb and linearize as usual, to construct a small-signal ac model that describes how slow variations in $v_{control}(t)$, $v_{g,rms}$, and the load, affect the rectifier output waveforms. Let us assume that the averaged output voltage $\langle v(t) \rangle_{T_{2L}}$, rectifier averaged output current $\langle i_2(t) \rangle_{T_{2L}}$, rms line voltage amplitude $v_{g,rms}$, and control voltage $v_{control}(t)$, can be represented as quiescent values plus small slow variations:

$$\begin{aligned}
\langle v(t) \rangle_{T_{2L}} &= V + \hat{v}(t) \\
\langle i_2(t) \rangle_{T_{2L}} &= I_2 + \hat{i}_2(t) \\
v_{g,rms} &= V_{g,rms} + \hat{v}_{g,rms}(t) \\
v_{control}(t) &= V_{control} + \hat{v}_{control}(t)
\end{aligned} \tag{18.95}$$

with

$$V \gg \left| \hat{v}(t) \right|$$
$$I_2 \gg \left| \hat{i}_2(t) \right|$$
$$V_{g,rms} \gg \left| \hat{v}_{g,rms}(t) \right| \tag{18.96}$$
$$V_{control} \gg \left| \hat{v}_{control}(t) \right|$$

In the averaged model of Fig. 18.29(c), $\langle i_2(t) \rangle_{T_{2L}}$ is given by

$$\langle i_2(t) \rangle_{T_{2L}} = \frac{\langle p(t) \rangle_{T_{2L}}}{\langle v(t) \rangle_{T_{2L}}} = \frac{v_{g,rms}^2(t)}{R_e(v_{control}(t)) \, \langle v(t) \rangle_{T_{2L}}}$$

$$= f\left(v_{g,rms}(t), \langle v(t) \rangle_{T_{2L}}, v_{control}(t) \right) \tag{18.97}$$

This equation resembles DCM buck-boost Eq. (11.45), and linearization proceeds in a similar manner. Expansion of Eq. (18.97) in a three-dimensional Taylor series about the quiescent operating point, and elimination of higher-order nonlinear terms, leads to

$$\hat{i}_2(t) = g_2 \hat{v}_{g,rms}(t) + j_2 \hat{v}_{control}(t) - \frac{\hat{v}(t)}{r_2} \tag{18.98}$$

where

$$g_2 = \left. \frac{df\left(v_{g,rms}, V, V_{control} \right)}{dv_{g,rms}} \right|_{v_{g,rms} = V_{g,rms}} = \frac{2}{R_e(V_{control})} \frac{V_{g,rms}}{V} \tag{18.99}$$

$$\left(-\frac{1}{r_2} \right) = \left. \frac{df\left(V_{g,rms}, \langle v \rangle_{T_{2L}}, V_{control} \right)}{d\langle v \rangle_{T_{2L}}} \right|_{\langle v \rangle_{T_{2L}} = V} = -\frac{I_2}{V} \tag{18.100}$$

$$j_2 = \left. \frac{df\left(V_{g,rms}, V, v_{control} \right)}{dv_{control}} \right|_{v_{control} = V_{control}} = -\frac{V_{g,rms}^2}{V R_e^2(V_{control})} \left. \frac{dR_e(v_{control})}{dv_{control}} \right|_{v_{control} = V_{control}} \tag{18.101}$$

A small-signal equivalent circuit based on Eq. (18.98) is given in Fig. 18.29(d). Expressions for the parameters $g_2, j_2,$ and r_2 for several controllers are listed in Table 18.1. This model is valid for the conditions of Eq. (18.96), with the additional assumption that the output voltage ripple is sufficiently small. Figure 18.29(d) is useful only for determining the various ac transfer functions; no information regarding dc conditions can be inferred. The ac resistance r_2 is derived from the slope of the average value of the power source output characteristic, evaluated at the quiescent operating point. The other coefficients, j_2 and g_2, are also derived from the slopes of the same characteristic, taken with respect to $v_{control}(t)$ and $v_{g,rms}$ and evaluated at the quiescent operating point. The resistance R is the incremental resistance of the load, evaluated at the quiescent operating point. In the boost converter with hysteretic control, the transistor on-time t_{on} replaces $v_{control}$ as the control input; likewise, the transistor duty cycle d is taken as the

Table 18.1 Small-signal model parameters for several types of rectifier control schemes

Controller type	g_2	j_2	r_2
Average current control with feedforward, Fig. 18.14	0	$\dfrac{P_{av}}{VV_{control}}$	$\dfrac{V^2}{P_{av}}$
Current-programmed control, Fig. 18.16	$\dfrac{2P_{av}}{VV_{g,rms}}$	$\dfrac{P_{av}}{VV_{control}}$	$\dfrac{V^2}{P_{av}}$
Nonlinear-carrier charge control of boost rectifier, Fig. 18.21	$\dfrac{2P_{av}}{VV_{g,rms}}$	$\dfrac{P_{av}}{VV_{control}}$	$\dfrac{V^2}{2P_{av}}$
Boost with critical conduction mode control, Fig. 18.20	$\dfrac{2P_{av}}{VV_{g,rms}}$	$\dfrac{P_{av}}{VV_{control}}$	$\dfrac{V^2}{P_{av}}$
DCM buck-boost, flyback, SEPIC, or Ćuk converters	$\dfrac{2P_{av}}{VV_{g,rms}}$	$\dfrac{2P_{av}}{VD}$	$\dfrac{V^2}{P_{av}}$

control input to the DCM buck-boost, flyback, SEPIC, and Ćuk converters. Harmonics are ignored for the current-programmed and NLC controllers; the expressions given in Table 18.1 assume that the converter operates in CCM with negligible harmonics.

The control-to-output transfer function is

$$\frac{\hat{v}(s)}{\hat{v}_{control}(s)} = j_2\, R\|r_2\, \frac{1}{1 + sC\, R\|r_2} \tag{18.102}$$

The line-to-output transfer function is

$$\frac{\hat{v}(s)}{\hat{v}_{g,rms}(s)} = g_2\, R\|r_2\, \frac{1}{1 + sC\, R\|r_2} \tag{18.103}$$

Thus, the small-signal transfer functions of the high quality rectifier contain a single pole, ascribable to the output filter capacitor operating in conjunction with the incremental load resistance R and r_2, the effective output resistance of the power source. Although this model is based on the ideal rectifier, its form is similar to that of the dc–dc DCM buck-boost converter ac model of Chapter 11. This is natural, because the DCM buck-boost converter is itself a natural loss-free resistor. The major difference is that the rms value of the ac input voltage must be used, and that the second harmonic components of r_2, j_2, and g_2 must additionally be removed via averaging. Nonetheless, the equivalent circuit and ac transfer functions are of similar form.

When the rectifier drives a regulated dc–dc converter as in Fig. 18.25, then the dc–dc converter presents a constant power load to the rectifier, as illustrated in Fig. 18.26. In equilibrium, the rectifier and dc–dc converter operate with the same average power P_{av} and the same dc voltage V. The incremental resistance R of the constant power load is negative, and is given by

$$R = -\frac{V^2}{P_{av}} \tag{18.104}$$

which is equal in magnitude but opposite in polarity to the rectifier incremental output resistance r_2, for all controllers except the NLC controller. The parallel combination $r_2 \| R$ then tends to an open circuit, and the control-to-output and line-to-output transfer functions become

$$\frac{\hat{v}(s)}{\hat{v}_{control}(s)} = \frac{j_2}{sC} \tag{18.105}$$

and

$$\frac{\hat{v}(s)}{\hat{v}_{g,rms}(s)} = \frac{g_2}{sC} \tag{18.106}$$

In the case of the NLC controller, the parallel combination $r_2 \| R$ becomes equal to $r_2/2$, and Eqs. (18.102) and (18.103) continue to apply.

18.5 RMS VALUES OF RECTIFIER WAVEFORMS

To correctly specify the power stage elements of a near-ideal rectifier, it is necessary to compute the root-mean-square values of their currents. A typical waveform such as the transistor current of the boost converter (Fig. 18.31) is pulse-width modulated, with both the duty cycle and the peak amplitude varying with the ac input voltage. When the switching frequency is much larger than the ac line frequency, then the rms value can be well-approximated as a double integral. The square of the current is integrated first to find its average over a switching period, and the result is then integrated to find the average over the ac line period.

Computation of the rms and average values of the waveforms of a PWM rectifier can be quite tedious, and this can impede the effective design of the power stage components. In this section, several approximations are developed, which allow relatively simple analytical expressions to be written for the rms and average values of the power stage currents, and which allow comparison of converter approaches [14,41]. The transistor current in the boost rectifier is found to be quite low.

The rms value of the transistor current is defined as

$$I_{Qrms} = \sqrt{\frac{1}{T_{ac}} \int_0^{T_{ac}} i_Q^2(t)\,dt} \tag{18.107}$$

where T_{ac} is the period of the ac line waveform. The integral can be expressed as a sum of integrals over all of the switching periods contained in one ac line period:

$$I_{Qrms} = \sqrt{\frac{1}{T_{ac}} T_s \sum_{n=1}^{T_{ac}/T_s} \left(\frac{1}{T_s} \int_{(n-1)T_s}^{nT_s} i_Q^2(t)\,dt \right)} \tag{18.108}$$

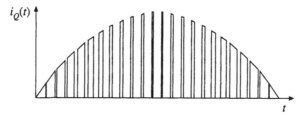

Fig. 18.31 Modulated transistor current waveform, boost rectifier.

where T_s is the switching period. The quantity inside the parentheses is the value of i_Q^2 averaged over the nth switching period. The summation can be approximated by an integral in the case when T_s is much less than T_{ac}. This approximation corresponds to taking the limit as T_s tends to zero, as follows:

$$I_{Qrms} \approx \sqrt{\frac{1}{T_{ac}} \lim_{T_s \to 0} \left[T_s \sum_{n=1}^{T_{ac}/T_s} \left(\frac{1}{T_s} \int_{(n-1)T_s}^{nT_s} i_Q^2(\tau) d\tau \right) \right]}$$

$$= \sqrt{\frac{1}{T_{ac}} \int_0^{T_{ac}} \frac{1}{T_s} \int_t^{t+T_s} i_Q^2(\tau) d\tau dt}$$ (18.109)

$$= \sqrt{\left\langle \left\langle i_Q^2(t) \right\rangle_{T_s} \right\rangle_{T_{ac}}}$$

So $i_Q^2(t)$ is first averaged over one switching period. The result is then averaged over the ac line period, and the square root is taken of the result.

18.5.1 Boost Rectifier Example

For the boost rectifier, the transistor current $i_Q(t)$ is equal to the input current when the transistor conducts, and is zero when the transistor is off. Therefore, the average of $i_Q^2(t)$ over one switching period is

$$\left\langle i_Q^2 \right\rangle_{T_s} = \frac{1}{T_s} \int_t^{t+T_s} i_Q^2(t) dt$$ (18.110)

$$= d(t) i_{ac}^2(t)$$

If the input voltage is given by

$$v_{ac}(t) = V_M \left| \sin \omega t \right|$$ (18.111)

then the input current will be

$$i_{ac}(t) = \frac{V_M}{R_e} \left| \sin \omega t \right|$$ (18.112)

where R_e is the emulated resistance. With a constant output voltage V, the transistor duty cycle must obey the relationship

$$\frac{V}{v_{ac}(t)} = \frac{1}{1 - d(t)}$$ (18.113)

This assumes that the converter dynamics are fast compared to the ac line frequency. Substitution of Eq. (18.111) into (18.113) and solution for $d(t)$ yields

$$d(t) = 1 - \frac{V_M}{V} \left| \sin \omega t \right|$$ (18.114)

Substitution of Eqs. (18.112) and (18.114) into Eq. (18.110) yields the following expression

$$\langle i_Q^2 \rangle_{T_s} = \frac{V_M^2}{R_e^2}\left(1 - \frac{V_M}{V}\left|\sin \omega t\right|\right)\sin^2(\omega t) \tag{18.115}$$

One can now plug this expression into Eq. (18.109):

$$I_{Qrms} = \sqrt{\frac{1}{T_{ac}}\int_0^{T_{ac}} \langle i_Q^2 \rangle_{T_s} dt}$$

$$= \sqrt{\frac{1}{T_{ac}}\int_0^{T_{ac}} \frac{V_M^2}{R_e^2}\left(1 - \frac{V_M}{V}\left|\sin \omega t\right|\right)\sin^2(\omega t)\,dt} \tag{18.116}$$

which can be further simplified to

$$I_{Qrms} = \sqrt{\frac{2}{T_{ac}}\frac{V_M^2}{R_e^2}\int_0^{T_{ac}/2}\left(\sin^2(\omega t) - \frac{V_M}{V}\sin^3(\omega t)\right)dt} \tag{18.117}$$

This involves integration of powers of $\sin(\omega t)$ over a complete half-cycle. The integral can be evaluated with the help of the following formula:

$$\frac{1}{\pi}\int_0^{\pi} \sin^n(\theta)\,d\theta = \begin{cases} \dfrac{2}{\pi}\dfrac{2\cdot4\cdot6\cdots(n-1)}{1\cdot3\cdot5\cdots n} & \text{if } n \text{ is odd} \\[2mm] \dfrac{1\cdot3\cdot5\cdots(n-1)}{2\cdot4\cdot6\cdots n} & \text{if } n \text{ is even} \end{cases} \tag{18.118}$$

This type of integral commonly arises in rms calculations involving PWM rectifiers. The values of the integral for several choices of n are listed in Table 18.2. Evaluation of the integral in Eq. (18.117) using Eq. (18.118) leads to the following result:

$$I_{Qrms} = \frac{V_M}{\sqrt{2}R_e}\sqrt{1 - \frac{8}{3\pi}\frac{V_M}{V}} = I_{ac\,rms}\sqrt{1 - \frac{8}{3\pi}\frac{V_M}{V}} \tag{18.119}$$

It can be seen that the rms transistor current is minimized by choosing the output voltage V to be as small as possible. The best that can be done is to choose $V = V_M$, which leads to

Table 18.2 Solution of the integral of Eq. (18.118), for several values of n

n	$\frac{1}{\pi}\int_0^{\pi}\sin^n(\theta)\,d\theta$
1	$\frac{2}{\pi}$
2	$\frac{1}{2}$
3	$\frac{4}{3\pi}$
4	$\frac{3}{8}$
5	$\frac{16}{15\pi}$
6	$\frac{15}{48}$

$$I_{Qrms} = 0.39 I_{ac\,rms} \tag{18.120}$$

Larger values of V lead to a larger rms transistor current.

A similar analysis for the rms diode current leads to the following expression

$$I_{Drms} = I_{ac\,rms}\sqrt{\frac{8}{3\pi}\frac{V_M}{V}} \tag{18.121}$$

The choice $V = V_M$ maximizes the rms diode current, with the result

$$I_{Drms} = 0.92 I_{ac\,rms} \qquad (18.122)$$

Larger values of V lead to smaller rms diode current.

Average currents can be computed in a similar way. The results are

$$I_{Qav} = I_{ac\,rms} \frac{2\sqrt{2}}{\pi}\left(1 - \frac{\pi}{8}\frac{V_M}{V}\right)$$

$$I_{Dav} = I_{ac\,rms} \frac{V_M}{2\sqrt{2}\,V} \qquad (18.123)$$

Expressions for rms, average, and peak currents of the power stage components of the continuous conduction mode boost converter are summarized in Table 18.3. Expressions are also tabulated for flyback and SEPIC topologies, operating in the continuous conduction mode. In the case of the flyback converter, an L_1–C_1 input filter is also included. In all cases, the effects of switching ripple are neglected.

18.5.2   Comparison of Single-Phase Rectifier Topologies

When isolation is not a rectifier requirement, and when it is acceptable that the dc output voltage be marginally larger than the peak ac input voltage, then the boost converter is a very effective approach. For example, consider the design of a 1 kW rectifier operating from the 240 Vrms input line voltage. If the converter efficiency and power factor are both approximately unity, then the rms input current is $I_{rms} = (1000\text{ W})/(240\text{ V}) = 4.2\text{ A}$. The dc output voltage is chosen to be 380 V, or slightly larger than the peak ac input voltage. By use of Eq. (18.119), the rms transistor current is found to be 2 A. This is quite a low value—less than half of the rms input current, which demonstrates how effectively the converter utilizes the power switch. The rms diode current is 3.6 A, and the transistor and diode blocking voltages are 380 V. With a 120 A ac input voltage, the transistor and diode rms currents increase to 6.6 A and 5.1 A, respectively.

The only real drawback of the boost converter is its inability to limit inrush currents. When the dc output voltage is less than the instantaneous input voltage, the control circuit of the boost rectifier loses control of the inductor current waveform. A very large inrush current occurs when the dc output capacitor is initially charged. Additional circuitry must be employed to limit the magnitude of this current.

Buck-boost, SEPIC, and Ćuk topologies can be used to solve the inrush current problem. Since these converters have a $d/(1 - d)$ conversion ratio, their waveforms can be controlled when the output voltage is any positive value. But the price paid for this capability is increased component stresses. For the same 1 kW rectifier with 240 Vrms ac input and 380 V output, the transistor rms current and peak voltage of the nonisolated SEPIC are 5.5 A and 719 V. The rms diode current is 4.85 A. The semiconductor voltage stresses can be reduced by reducing the output voltage, at the expense of increased rms currents. With a 120 V ac input voltage, the transistor and diode rms currents increase to 9.8 A and 6.1 A, respectively.

Isolation can also be obtained in the SEPIC and other topologies, as discussed in Chapter 6. The turns ratio of the isolation transformer can also be used to reduce the primary-side currents when the dc output voltage is low. But the transformer winding rms currents are higher than those of a dc–dc converter, because of the pulsating (twice-line-frequency) power flow. For the 1 kW, 240 V ac input SEPIC example, with a 42 V 23.8 A dc load, and a 4:1 transformer turns ratio, the rms transformer currents are 5.5 A (primary) and 36.4 A (secondary). The rms transistor current is 6.9 A. At 120 V ac input voltage,

Table 18.3 Summary of rectifier current stresses for several converter topologies

	rms	Average	Peak
CCM boost			
Transistor	$I_{ac\,rms}\sqrt{1-\dfrac{8}{3\pi}\dfrac{V_M}{V}}$	$I_{ac\,rms}\dfrac{2\sqrt{2}}{\pi}\left(1-\dfrac{\pi}{8}\dfrac{V_M}{V}\right)$	$I_{ac\,rms}\sqrt{2}$
Diode	$I_{dc}\sqrt{\dfrac{16}{3\pi}\dfrac{V}{V_M}}$	I_{dc}	$2I_{dc}\dfrac{V}{V_M}$
Inductor	$I_{ac\,rms}$	$I_{ac\,rms}\dfrac{2\sqrt{2}}{\pi}$	$I_{ac\,rms}\sqrt{2}$
CCM flyback, with n:1 isolation transformer and input filter			
Transistor, xfmr primary	$I_{ac\,rms}\sqrt{1+\dfrac{8}{3\pi}\dfrac{V_M}{nV}}$	$I_{ac\,rms}\dfrac{2\sqrt{2}}{\pi}$	$I_{ac\,rms}\sqrt{2}\left(1+\dfrac{V_M}{nV}\right)$
L_1	$I_{ac\,rms}$	$I_{ac\,rms}\dfrac{2\sqrt{2}}{\pi}$	$I_{ac\,rms}\sqrt{2}$
C_1	$I_{ac\,rms}\sqrt{\dfrac{8}{3\pi}\dfrac{V_M}{nV}}$	0	$I_{ac\,rms}\sqrt{2}\max\left(1,\dfrac{V_M}{nV}\right)$
Diode, xfmr secondary	$I_{dc}\sqrt{\dfrac{3}{2}+\dfrac{16}{3\pi}\dfrac{nV}{V_M}}$	I_{dc}	$2I_{dc}\left(1+\dfrac{nV}{V_M}\right)$
CCM SEPIC, nonisolated			
Transistor	$I_{ac\,rms}\sqrt{1+\dfrac{8}{3\pi}\dfrac{V_M}{V}}$	$I_{ac\,rms}\dfrac{2\sqrt{2}}{\pi}$	$I_{ac\,rms}\sqrt{2}\left(1+\dfrac{V_M}{V}\right)$
L_1	$I_{ac\,rms}$	$I_{ac\,rms}\dfrac{2\sqrt{2}}{\pi}$	$I_{ac\,rms}\sqrt{2}$
C_1	$I_{ac\,rms}\sqrt{\dfrac{8}{3\pi}\dfrac{V_M}{V}}$	0	$I_{ac\,rms}\sqrt{2}\max\left(1,\dfrac{V_M}{V}\right)$
L_2	$I_{ac\,rms}\dfrac{V_M}{V}\dfrac{\sqrt{3}}{2}$	$\dfrac{I_{ac\,rms}}{\sqrt{2}}\dfrac{V_M}{V}$	$I_{ac\,rms}\dfrac{V_M}{V}\sqrt{2}$
Diode	$I_{dc}\sqrt{\dfrac{3}{2}+\dfrac{16}{3\pi}\dfrac{V}{V_M}}$	I_{dc}	$2I_{dc}\left(1+\dfrac{V}{V_M}\right)$
CCM SEPIC, with n:1 isolation transformer			
Transistor	$I_{ac\,rms}\sqrt{1+\dfrac{8}{3\pi}\dfrac{V_M}{nV}}$	$I_{ac\,rms}\dfrac{2\sqrt{2}}{\pi}$	$I_{ac\,rms}\sqrt{2}\left(1+\dfrac{V_M}{nV}\right)$
L_1	$I_{ac\,rms}$	$I_{ac\,rms}\dfrac{2\sqrt{2}}{\pi}$	$I_{ac\,rms}\sqrt{2}$
C_1, xfmr primary	$I_{ac\,rms}\sqrt{\dfrac{8}{3\pi}\dfrac{V_M}{nV}}$	0	$I_{ac\,rms}\sqrt{2}\max\left(1,\dfrac{V_M}{nV}\right)$
Diode, xfmr secondary	$I_{dc}\sqrt{\dfrac{3}{2}+\dfrac{16}{3\pi}\dfrac{nV}{V_M}}$	I_{dc}	$2I_{dc}\left(1+\dfrac{nV}{V_M}\right)$

with, in all cases, $\dfrac{I_{ac\,rms}}{I_{dc}}=\sqrt{2}\dfrac{V}{V_M}$, ac input voltage $=V_M\sin(\omega t)$, dc output voltage $=V$.

these currents increase to 7.7 A, 42.5 A, and 11.4 A, respectively.

18.6   MODELING LOSSES AND EFFICIENCY IN CCM HIGH-QUALITY RECTIFIERS

As in the case of dc–dc converters, we would like to model the converter loss elements so that we can correctly specify the power stage components. The equivalent circuit approach used in the dc–dc case can be generalized to include ac–dc low harmonic rectifiers, although the resulting equations are more complicated because of the low-frequency ac modulation of the waveforms.

A dc–dc boost converter and its steady-state equivalent circuit are illustrated in Fig. 18.32. When the converter operates in equilibrium, the model of Fig. 18.32(b) can be solved to determine the converter losses and efficiency. In the ac–dc case, the input voltage $v_g(t)$ is a rectified sinusoid, and the controller varies the duty cycle $d(t)$ to cause $i_g(t)$ to follow $v_g(t)$ according to

$$i_g(t) = \frac{v_g(t)}{R_e} \tag{18.124}$$

The emulated resistance R_e is chosen by the controller such that the desired dc output voltage is obtained. Ac variations in $d(t)$, $v_g(t)$, and several other system waveforms are not small, and hence the small-signal approximation employed in Chapters 7 to 12 is not justified. We can continue to model the low-frequency components of the converter via averaging, but the resulting equivalent circuits are, in general, time-varying and nonlinear.

For the purposes of determining the rectifier efficiency, it is assumed that (1) the inductor is sufficiently small, such that it has negligible influence on the ac-line-frequency components of the system waveforms, and (2) the capacitor is large, so that the output voltage $v(t)$ is essentially equal to its equilibrium dc value, with negligible low- or high-frequency ac variations. So in the ac–dc case, the model becomes as shown in Fig. 18.33. Low-frequency components ($\ll f_s$) of the controller waveforms are sketched in Fig. 18.34.

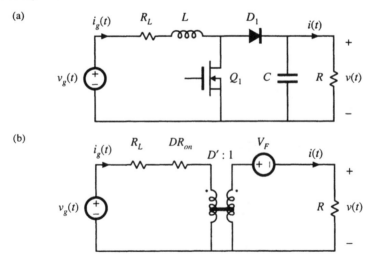

Fig. 18.32 Dc–dc boost converter, (a), and a steady-state equivalent circuit, (b), which models the inductor resistance R_L, MOSFET on-resistance R_{on}, and diode forward voltage drop V_F.

(a)

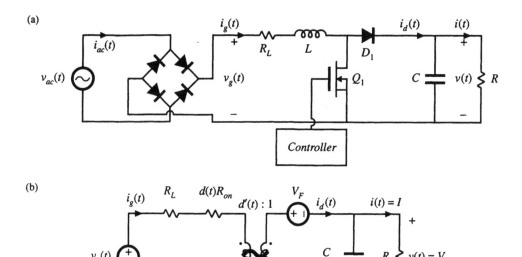

(b)

Fig. 18.33 Ac–dc boost rectifier, (a), and a low-frequency equivalent circuit, (b), that models converter losses and efficiency.

To find the rectifier waveforms, losses, and efficiency, we must solve the circuit of Fig. 18.33(b), under the conditions that the controller varies the duty cycle $d(t)$ such that Eq. (18.124) is satisfied. This leads to time-varying circuit elements $d(t)R_{on}$ and the $d'(t):1$ transformer. The solution that follows involves the following steps: (1) solve for the $d(t)$ waveform; (2) average $i_d(t)$ to find its dc component, equal to the load current I; and (3) find other quantities of interest such as the rectifier efficiency.

The simplified boost converter circuit model of Fig. 18.35, in which only the MOSFET conduction loss is accounted for, is solved here. However, the results can be generalized directly to the circuit of Fig. 18.33(b); doing so is left as a homework problem. A similar procedure can also be followed to derive expressions for the losses and efficiencies of other rectifier topologies.

18.6.1 Expression for Controller Duty Cycle $d(t)$

The controller varies the duty cycle $d(t)$ such that Eq. (18.124) is satisfied. By solving the input-side loop of Fig. 18.35, we obtain

$$i_g(t)d(t)R_{on} = v_g(t) - d'(t)v \tag{18.125}$$

Substitute Eq. (18.124) into (18.125) to eliminate $i_g(t)$:

$$\frac{v_g(t)}{R_e}d(t)R_{on} = v_g(t) - d'(t)v \tag{18.126}$$

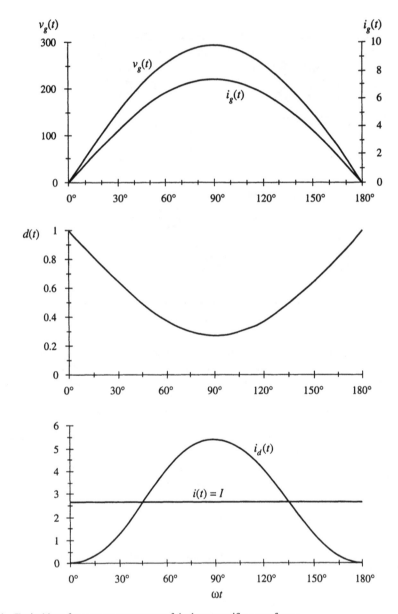

Fig. 18.34   Typical low-frequency components of the boost rectifier waveforms.

with

$$v_g(t) = V_M \left| \sin \omega t \right| \tag{18.127}$$

We can now solve for the duty cycle $d(t)$. The result is

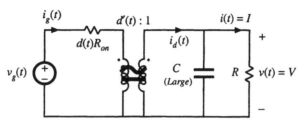

Fig. 18.35 Simplified boost power stage low-frequency equivalent circuit, in which only the MOSFET on-resistance is modeled.

$$d(t) = \frac{v - v_g(t)}{v - v_g(t) \dfrac{R_{on}}{R_e}} \qquad (18.128)$$

This expression neglects the converter dynamics, an assumption that is justified when these dynamics are sufficiently faster than the ac line voltage variation. The expression also neglects operation in the discontinuous conduction mode near the zero-crossing of the ac line voltage waveform. This is justified when the rectifier operates in the continuous conduction mode for most of the ac line cycle, because the power loss near the zero-crossing is negligible.

18.6.2 Expression for the DC Load Current

By charge balance on output capacitor C, the dc load current I is equal to the dc component of the diode current i_d:

$$I = \langle i_d \rangle_{T_{ac}} \qquad (18.129)$$

Solution of Fig. 18.35 for $i_d(t)$ yields

$$i_d(t) = d'(t) i_g(t) = d'(t) \frac{v_g(t)}{R_e} \qquad (18.130)$$

From Eq. (18.128), $d'(t) = 1 - d(t)$ is given by

$$d'(t) = \frac{v_g(t) \left(1 - \dfrac{R_{on}}{R_e} \right)}{v - v_g(t) \dfrac{R_{on}}{R_e}} \qquad (18.131)$$

so

$$i_d(t) = \frac{v_g^2(t)}{R_e} \frac{\left(1 - \dfrac{R_{on}}{R_e} \right)}{v - v_g(t) \dfrac{R_{on}}{R_e}} \qquad (18.132)$$

Now substitute $v_g(t) = V_M \sin \omega t$, and integrate to find $\langle i_d(t) \rangle_{T_{ac}}$:

$$I = \langle i_d \rangle_{T_{ac}} = \frac{2}{T_{ac}} \int_0^{T_{ac}/2} \left(\frac{V_M^2}{R_e} \right) \frac{\left(1 - \frac{R_{on}}{R_e} \right) \sin^2 (\omega t)}{\left(v - \frac{V_M R_{on}}{R_e} \sin (\omega t) \right)} \, dt \qquad (18.133)$$

Again, $T_{ac} = 2\pi/\omega$ is the ac line period. Equation (18.133) can be rewritten as

$$I = \frac{2}{T_{ac}} \frac{V_M^2}{V R_e} \left(1 - \frac{R_{on}}{R_e} \right) \int_0^{T_{ac}/2} \frac{\sin^2 (\omega t)}{1 - a \sin (\omega t)} \, dt \qquad (18.134)$$

where

$$a = \left(\frac{V_M}{V} \right) \left(\frac{R_{on}}{R_e} \right) \qquad (18.135)$$

By waveform symmetry, we need only integrate from 0 to $T_{ac}/4$. Also, make the substitution $\theta = \omega t$:

$$I = \frac{V_M^2}{V R_e} \left(1 - \frac{R_{on}}{R_e} \right) \frac{2}{\pi} \int_0^{\pi/2} \frac{\sin^2 (\theta)}{1 - a \sin (\theta)} \, d\theta \qquad (18.136)$$

Evaluation of this integral is tedious. It arises in not only the boost rectifier, but in a number of other high-quality rectifier topologies as well. The derivation is not given here, but involves the substitution $z = \tan(\theta/2)$, performing a partial fraction expansion of the resulting rational function of z, and integration of the results. The solution is:

$$\frac{4}{\pi} \int_0^{\pi/2} \frac{\sin^2 (\theta)}{1 - a \sin (\theta)} \, d\theta = F(a) = \frac{2}{a^2 \pi} \left(-2a - \pi + \frac{4 \sin^{-1}(a) + 2 \cos^{-1}(a)}{\sqrt{1 - a^2}} \right) \qquad (18.137)$$

This equation is somewhat complicated, but it is in closed form, and can easily be evaluated by computer spreadsheet. The quantity a, which is a measure of the loss resistance R_{on} relative to the emulated resistance R_e, is typically much smaller than 1. $F(a)$ is plotted in Fig. 18.36. The function $F(a)$ can be well-approximated as follows:

$$F(a) \approx 1 + 0.862a + 0.78a^2 \qquad (18.138)$$

For $|a| \leq 0.15$, the $F(a)$ predicted by this approximate expression is within 0.1% of the exact value. If the a^2 term is omitted, then the accuracy drops to $\pm 2\%$ over the same range of a. The rectifier efficiency η calculated in the next section depends directly on $F(a)$, and hence the accuracy of $F(a)$ coincides with the accuracy of η.

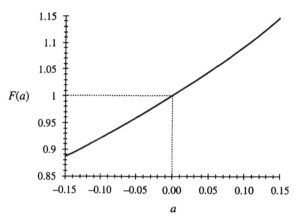

Fig. 18.36 Plot of the integral $F(a)$ vs. a.

18.6.3 Solution for Converter Efficiency η

Now that we have found the dc load current, we can calculate the converter efficiency η. The average input power is

$$P_{in} = \left\langle p_{in}(t) \right\rangle_{T_{ac}} = \frac{V_M^2}{2R_e} \tag{18.139}$$

The average load power is

$$P_{out} = VI = \left(V \right) \left(\frac{V_M^2}{VR_e} \left(1 - \frac{R_{on}}{R_e} \right) \frac{F(a)}{2} \right) \tag{18.140}$$

where

$$a = \left(\frac{V_M}{V} \right) \left(\frac{R_{on}}{R_e} \right) \tag{18.141}$$

Here, we have substituted Eq. (18.136) for I. The efficiency is therefore

$$\eta = \frac{P_{out}}{P_{in}} = \left(1 - \frac{R_{on}}{R_e} \right) F(a) \tag{18.142}$$

by substitution of Eqs. (18.139) and (18.140). If desired, the parabolic approximation for $F(a)$, Eq. (18.138), can be employed. This leads to

$$\eta \approx \left(1 - \frac{R_{on}}{R_e} \right) \left(1 + 0.862 \frac{V_M}{V} \frac{R_{on}}{R_e} + 0.78 \left(\frac{V_M}{V} \frac{R_{on}}{R_e} \right)^2 \right) \tag{18.143}$$

Equations (18.142) and (18.143) show how the efficiency varies with MOSFET on resistance R_{on} and

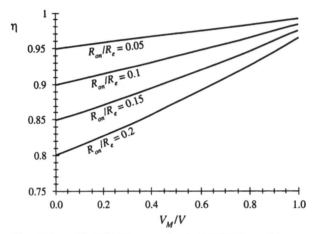

Fig. 18.37   Boost rectifier efficiency, Eq. (18.142), accounting for MOSFET on resistance.

with ac peak voltage V_M. Equation (18.142) is plotted in Fig. 18.37. It can be seen that high efficiency is obtained when the peak ac line voltage V_M is close to the dc output voltage V. Efficiencies in the range 90% to 95% can then be obtained, even with MOSFET on-resistances as high as $0.2R_e$. Of course, Fig. 18.37 is optimistic because it neglects sources of loss other than the MOSFET conduction loss.

18.6.4   Design Example

Let us utilize Fig. 18.37 to design for a given efficiency. Consider the following specifications:

Output voltage	390 V
Output power	500 W
rms input voltage	120 V
Efficiency	95%

Assume that losses other than the MOSFET conduction loss are negligible. The average input power is

$$P_{in} = \frac{P_{out}}{\eta} = \frac{500 \text{ W}}{0.95} = 526 \text{ W} \tag{18.144}$$

The emulated resistance is therefore

$$R_e = \frac{V_{g, rms}^2}{P_{in}} = \frac{(120 \text{ V})^2}{526 \text{ W}} = 27.4 \, \Omega \tag{18.145}$$

Also,

$$\frac{V_M}{V} = \frac{120\sqrt{2} \text{ V}}{390 \text{ V}} = 0.435 \tag{18.146}$$

From Fig. 18.37, or by evaluation of the exact equation (18.142), 95% efficiency with $V_M/V = 0.435$ occurs with $R_{on}/R_e \approx 0.077$. So we require a MOSFET having an on-resistance of

$$R_{on} \le (0.077)\, R_e = (0.077)\, (27.4\ \Omega) = 2.11\ \Omega \tag{18.147}$$

Of course, other converter losses have not been accounted for, which will reduce the efficiency.

It is instructive to compare this result with that obtained using the expressions for rms current from Section 18.5. The rms transistor current of the ideal CCM boost converter is given by Eq. (18.119). The rms input current will be equal to $P_{in}/V_{g,rms} = (526\ \text{W})/(120\ \text{V}) = 4.38\ \text{A}$. Hence, Eq. (18.119) predicts an rms transistor current of

$$
\begin{aligned}
I_{Qrms} &= I_{ac\,rms} \sqrt{1 - \frac{8}{3\pi}\frac{V_M}{V}} \\
&= (4.38\ \text{A}) \sqrt{1 - \frac{8}{3\pi}\frac{(120\ \text{V})\,\sqrt{2}}{(390\ \text{V})}} \\
&= 3.48\ \text{A}
\end{aligned}
\tag{18.148}
$$

Hence, the MOSFET on-resistance should be chosen according to

$$R_{on} \le \frac{P_{in} - P_{out}}{I_{Qrms}^2} = \frac{(526\ \text{W}) - (500\ \text{W})}{(4.38\ \text{A})^2} = 2.17\ \Omega \tag{18.149}$$

This calculation is approximate because Eq. (18.119) was derived using the waveforms of the ideal (lossless) converter. Nonetheless, it gives an answer that is very close to the more exact result of Eq. (18.147). We would expect this approximate approach to exhibit good accuracy in this example, because of the high 95% efficiency.

18.7 IDEAL THREE-PHASE RECTIFIERS

The single-phase ideal rectifier concepts of the previous sections can be generalized to cover ideal three-phase rectifiers. Figure 18.38(a) illustrates the properties of an ideal three-phase rectifier, which presents a balanced resistive load to the utility system. A three-phase converter system is controlled such that resistor emulation is obtained in each input phase. The rectifier three-phase input port can then be modeled by per-phase effective resistances R_e, as illustrated in Fig. 18.38(a). The instantaneous powers apparently consumed by these resistors are transferred to the rectifier dc output port. The rectifier output port can therefore be modeled by power sources equal to the instantaneous powers flowing into the effective resistances R_e. It is irrelevant whether the three power sources are connected in series or in parallel; in either event, they can be combined into a single source equal to the total three-phase instantaneous input power as illustrated in Fig. 18.38(b).

If the three-phase ac input voltages are

$$
\begin{aligned}
v_{an}(t) &= V_M \sin\left(\omega t\right) \\
v_{bn}(t) &= V_M \sin\left(\omega t - 120°\right) \\
v_{cn}(t) &= V_M \sin\left(\omega t - 240°\right)
\end{aligned}
\tag{18.150}
$$

then the instantaneous powers flowing into the phase a, b, and c effective resistances R_e are

$$p_a(t) = \frac{v_{an}^2(t)}{R_e} = \frac{V_M^2}{2R_e}\left(1 - \cos\left(2\omega t\right)\right)$$

$$p_b(t) = \frac{v_{bn}^2(t)}{R_e} = \frac{V_M^2}{2R_e}\left(1 - \cos\left(2\omega t - 240°\right)\right) \tag{18.151}$$

$$p_c(t) = \frac{v_{cn}^2(t)}{R_e} = \frac{V_M^2}{2R_e}\left(1 - \cos\left(2\omega t - 120°\right)\right)$$

Each instantaneous phase power contains a dc term $V_M^2/(2R_e)$, and a second-harmonic term. The total instantaneous three-phase power is

$$P_{tot}(t) = p_a(t) + p_b(t) + p_c(t) = \frac{3}{2}\frac{V_M^2}{R_e} \tag{18.152}$$

This is the instantaneous power which flows out of the rectifier dc output port. Note that the second harmonic terms add to zero, such that the rectifier instantaneous output power is constant. This is a consequence of the fact that the instantaneous power flow in any balanced three-phase ac system is constant. So, unlike the single-phase case, the ideal three-phase rectifier can supply constant instantaneous power to a dc load, without the need for internal low-frequency energy storage.

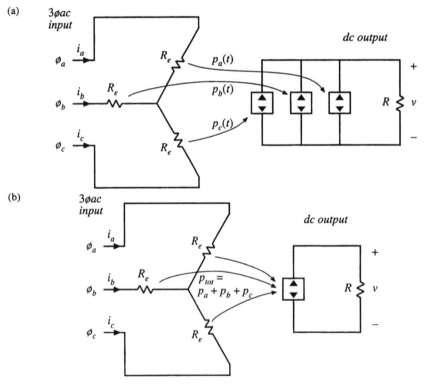

Fig. 18.38 Development of the ideal three-phase rectifier model: (a) three ideal single-phase rectifiers, (b) combination of the three power sources into an equivalent single power source.

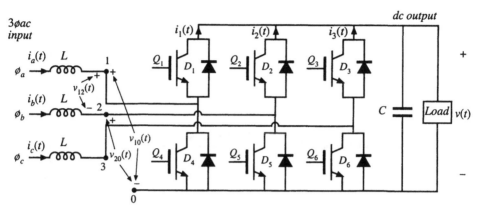

Fig. 18.39   Boost-type 3øac–dc PWM rectifier.

A variety of 3øac–dc PWM rectifiers are known; a few of the many references on this subject are listed in the references [42–59]. The most well-known topology is the three-phase ac–dc boost rectifier, illustrated in Fig. 18.39. This converter requires six SPST current-bidirectional two-quadrant switches. The inductors and capacitor filter the high-frequency switching harmonics, and have little influence on the low-frequency ac components of the waveforms. The switches of each phase are controlled to obtain input resistor emulation, either with a multiplying controller scheme similar to Fig. 18.5, or with some other approach. To obtain undistorted line current waveforms, the dc output voltage V must be greater than or equal to the peak line-to-line ac input voltage $V_{L,pk}$. In a typical realization, V is somewhat greater than $V_{L,pk}$. This converter resembles the voltage-source inverter, discussed briefly in Chapter 4, except that the converter is operated as a rectifier, and the converter input currents are controlled via high-frequency pulse-width modulation.

The three-phase boost rectifier of Fig. 18.39 has several attributes that make it the leading candidate for most 3øac–dc rectifier applications. The ac input currents are nonpulsating, and hence very little additional input EMI filtering is required. As in the case of the single-phase boost rectifier, the rms transistor currents and also the conduction losses of the three-phase boost rectifier are low relative to other 3øac–dc topologies such as the current-source inverter. The converter is capable of bidirectional power flow. A disadvantage is the requirement for six active devices: when compared with a dc–dc converter of similar ratings, the active semiconductor utilization (discussed in Chapter 6) is low. Also, since the rectifier has a boost characteristic, it is not suitable for direct replacement of traditional buck-type phase-controlled rectifiers.

The literature contains a wide variety of schemes for controlling the switches of a six-switch three-phase bridge network, which are applicable for control of the switches of Fig. 18.39. The basic operation of the converter can be most easily understood by assuming that the switches are controlled via simple sinusoidal pulse-width modulation. Transistor Q_1 is driven with duty cycle $d_1(t)$, while transistor Q_4 is driven by the complement of $d_1(t)$, or $d_1'(t) = 1 - d_1(t)$. Transistors Q_2 and Q_5 are driven with duty cycles $d_2(t)$ and $d_2'(t)$, respectively, and transistors Q_3 and Q_6 are driven with duty cycles $d_3(t)$ and $d_3'(t)$, respectively. The switch voltage waveforms of Fig. 18.40 are obtained. The average switch voltages are

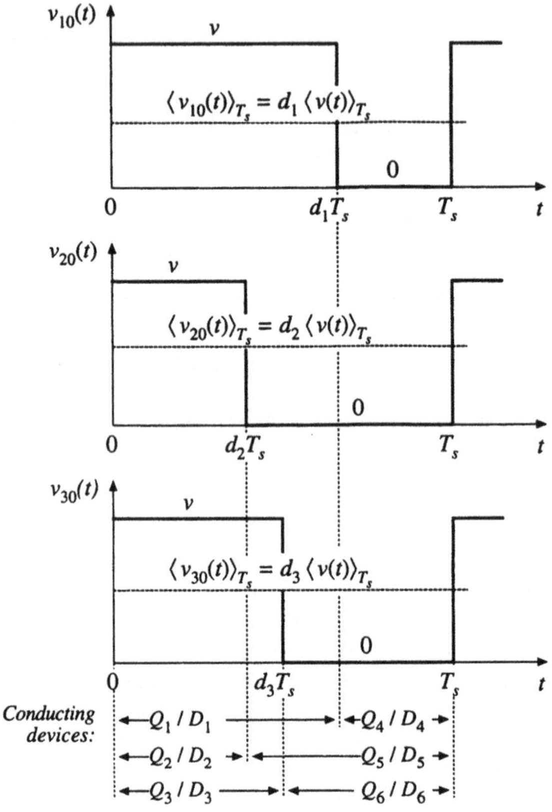

Fig. 18.40 Switch waveforms, 3øac–dc boost rectifier.

$$\left\langle v_{10}(t) \right\rangle_{T_s} = d_1(t) \left\langle v(t) \right\rangle_{T_s}$$
$$\left\langle v_{20}(t) \right\rangle_{T_s} = d_2(t) \left\langle v(t) \right\rangle_{T_s} \qquad (18.153)$$
$$\left\langle v_{30}(t) \right\rangle_{T_s} = d_3(t) \left\langle v(t) \right\rangle_{T_s}$$

The averaged line-to-line switch voltages are therefore

$$\left\langle v_{12}(t) \right\rangle_{T_s} = \left\langle v_{10}(t) \right\rangle_{T_s} - \left\langle v_{20}(t) \right\rangle_{T_s} = \left(d_1(t) - d_2(t) \right) \left\langle v(t) \right\rangle_{T_s}$$
$$\left\langle v_{23}(t) \right\rangle_{T_s} = \left\langle v_{20}(t) \right\rangle_{T_s} - \left\langle v_{30}(t) \right\rangle_{T_s} = \left(d_2(t) - d_3(t) \right) \left\langle v(t) \right\rangle_{T_s} \qquad (18.154)$$
$$\left\langle v_{31}(t) \right\rangle_{T_s} = \left\langle v_{30}(t) \right\rangle_{T_s} - \left\langle v_{10}(t) \right\rangle_{T_s} = \left(d_3(t) - d_1(t) \right) \left\langle v(t) \right\rangle_{T_s}$$

In a similar manner, the average switch currents can be shown to be

$$\langle i_1(t) \rangle_{T_s} = d_1(t) \langle i_a(t) \rangle_{T_s}$$
$$\langle i_2(t) \rangle_{T_s} = d_2(t) \langle i_b(t) \rangle_{T_s} \qquad (18.155)$$
$$\langle i_3(t) \rangle_{T_s} = d_3(t) \langle i_c(t) \rangle_{T_s}$$

Equations (18.154) and (18.155) lead to the circuit-averaged model of Fig. 18.41.

With sinusoidal PWM, the duty cycles are varied sinusoidally in synchronism with the ac line, as follows:

$$d_1(t) = D_0 + \tfrac{1}{2} D_m \sin\left(\omega t - \varphi\right)$$
$$d_2(t) = D_0 + \tfrac{1}{2} D_m \sin\left(\omega t - \varphi - 120°\right) \qquad (18.156)$$
$$d_3(t) = D_0 + \tfrac{1}{2} D_m \sin\left(\omega t - \varphi - 240°\right)$$

where ω is the ac line frequency. Since each instantaneous duty cycle must lie in the interval [0,1], the dc bias D_0 is required. The factor D_m is called the *modulation index*; for $D_0 = 0.5$, D_m must be less than or equal to one. Other choices of D_0 further restrict D_m. In general, the modulation index can be defined as equal to the peak-to-peak amplitude of the fundamental component of the duty cycle variation.

If the switching frequency is sufficiently large, then filter inductors L can be small in value, such that they have negligible effect on the low-frequency ac waveforms. The averaged switch voltage $\langle v_{12}(t) \rangle_{T_s}$ then becomes approximately equal to the ac line-line voltage $v_{ab}(t)$:

$$\langle v_{12}(t) \rangle_{T_s} = \left(d_1(t) - d_2(t)\right) \langle v(t) \rangle_{T_s} \approx v_{ab}(t) \qquad (18.157)$$

Substitution of Eqs. (18.150) and (18.156) leads to

$$\tfrac{1}{2} D_m \left[\sin\left(\omega t - \varphi\right) - \sin\left(\omega t - \varphi - 120°\right)\right] \langle v(t) \rangle_{T_s} = V_M \left[\sin\left(\omega t\right) - \sin\left(\omega t - 120°\right)\right] \qquad (18.158)$$

For small L, the angle φ must tend to zero, and hence the sinusoidal terms in Eq. (18.158) cancel out. In steady-state, the dc output voltage is $\langle v(t) \rangle_{T_s} = V$. Equation (18.158) then becomes

$$\tfrac{1}{2} D_m V = V_M \qquad (18.159)$$

Solution for the dc output voltage V leads to

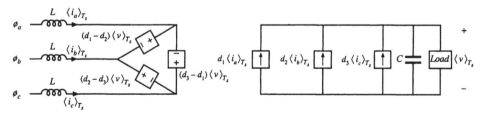

Fig. 18.41 Averaged model of the open-loop 3øac–dc boost rectifier.

$$V = \frac{2V_M}{D_m} \qquad (18.160)$$

Equation (18.160) can be written in terms of the peak line-to-line voltage $V_{L,pk}$, as

$$V = \frac{2}{\sqrt{3}} \frac{V_{L,pk}}{D_m} = 1.15 \frac{V_{L,pk}}{D_m} \qquad (18.161)$$

With $D_m \leq 1$, the dc output voltage V must be greater than or equal to 1.15 times the peak line-to-line ac input voltage. Thus, the rectifier has a boost characteristic.

The sinusoidal PWM approach of Eq. (18.156) is not the only way to vary the duty cycles to obtain sinusoidal ac voltages and currents. For example, triplen harmonics can be added to the duty cycle expressions of Eq. (18.156). These triplen harmonics cancel out in Eq. (18.154), such that the average inverter input voltages $\langle v_{12}(t) \rangle_{T_s}$, $\langle v_{23}(t) \rangle_{T_s}$, and $\langle v_{31}(t) \rangle_{T_s}$ contain only fundamental. Figure 18.42 illustrates duty cycle variations that lead to a dc output voltage V equal to $V_{L,pk}$. The effective modulation index in this case is 1.15. The ac-side voltages and currents are again undistorted. Further increases in the modulation index can be attained only by introduction of distortion in the ac-side voltages and currents. Of course, in practice the duty cycle modulations are usually generated by the feedback loops that control the input current waveforms to attain resistor emulation.

Three-phase ac-to-dc rectifiers having buck, buck-boost, or other characteristics, are possible, but find much less use than the boost topology. A 3øac-dc rectifier system can also be constructed simply using three separate single-phase rectifiers [34]; however, each single-phase rectifier must then contain transformer isolation, leading to substantially increased switch stress and loss. Other unconventional approaches to three-phase low-harmonic rectification have also been recently explored, such as the Vienna rectifier [56,59], single-switch approaches [49–55], and other circuits[44,45,46,57,58].

Yet another approach to solving the problem of three-phase rectifier harmonics is the *harmonic correction* scheme illustrated in Fig. 18.43. An active six-switch three-phase bridge removes the harmonics generated by a nonlinear three-phase load such as an uncontrolled rectifier. The harmonic corrector is controlled such that its ac line currents contain harmonics that are equal in magnitude but opposite in phase to the harmonics generated by the nonlinear load. No average power flows into the harmonic corrector. The total kVA rating of the harmonic corrector semiconductor devices depends on the magnitudes of the harmonics produced by the nonlinear load. If the THD generated by the load is not too large, then

Fig. 18.42 A modulation strategy that leads to a dc output voltage equal to the peak input line-line voltage.

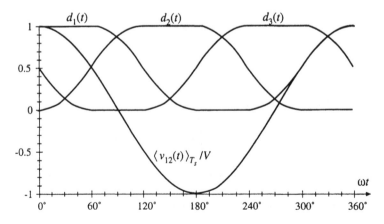

Nonlinear load

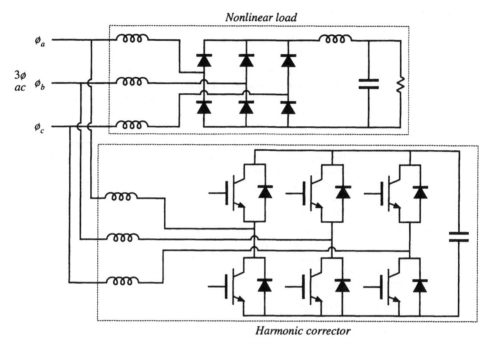

Harmonic corrector

Fig. 18.43   A harmonic corrector, based on the 3øac–dc CCM boost converter of Fig. 18.39.

the harmonic corrector scheme requires less total active silicon than the CCM boost-type rectifier of Fig. 18.39. But if the uncontrolled rectifier contains small ac line inductances, such that it operates in the discontinuous conduction mode with large THD, then it is probably better to simply replace the uncontrolled rectifier with the CCM boost-type rectifier of Fig. 18.39.

18.8   SUMMARY OF KEY POINTS

1.   The ideal rectifier presents an effective resistive load, the emulated resistance R_e, to the ac power system. The power apparently "consumed" by R_e is transferred to the dc output port. In a three-phase ideal rectifier, input resistor emulation is obtained in each phase. In both the single-phase and three-phase cases, the output port follows a power source characteristic, dependent on the instantaneous ac input power. Ideal rectifiers can perform the function of low-harmonic rectification, without need for low-frequency reactive elements.

2.   The dc–dc boost converter, as well as other converters capable of increasing the voltage according to Eq. (18.12), can be adapted to the ideal rectifier application. A control system causes the input current to be proportional to the input voltage. The converter may operate in CCM, DCM, or in both modes. The mode boundary can be expressed as a function of R_e, $2L/T_s$, and the instantaneous voltage ratio $v_g(t)/V$. A well-designed average current controller leads to resistor emulation regardless of the operating mode; however, other schemes may lead to distorted current waveforms when the mode boundary is crossed.

3.   In a single-phase system, the instantaneous ac input power is pulsating, while the dc load power is constant. Whenever the instantaneous input and output powers are not equal, the ideal rectifier system must

contain energy storage. A large capacitor is commonly employed; the voltage of this capacitor must be allowed to vary independently, as necessary to store and release energy. A slow feedback loop regulates the dc component of the capacitor voltage, to ensure that the average ac input power and dc load power are balanced.

4. RMS values of rectifiers waveforms can be computed by double integration. In the case of the boost converter, the rms transistor current can be as low as 39% of the rms ac input current, when the dc output voltage V is close in value to the peak ac input voltage V_M. Other converter topologies such as the buck-boost, SEPIC, and Ćuk converters exhibit significantly higher rms transistor currents but are capable of limiting the converter inrush current.

5. In the three-phase case, a boost-type rectifier based on the PWM voltage-source inverter also exhibits low rms transistor currents. This approach requires six active switching elements, and its dc output voltage must be greater than the peak input line-to-line voltage. Average current control can be used to obtain input resistor emulation. An equivalent circuit can be derived by averaging the switch waveforms. The converter operation can be understood by assuming that the switch duty cycles vary sinusoidally; expressions for the average converter waveforms can then be derived.

6. Converter losses and efficiency can be modeled using the steady-state equivalent circuit models of Chapter 3, with a time-varying duty cycle. The output current is averaged over one ac line period, to determine its dc component. The converter losses and efficiency can then be computed. This approach is approximate, in that (*i*) it assumes that the converter dynamics are much faster than the ac line frequency, and (*ii*) it neglects operation in the discontinuous conduction mode.

7. Average current control involves direct regulation of the low-frequency components of the rectifier input current to follow the input voltage. Feedforward can also be added, to cancel the influence of ac line voltage variations on the dc output voltage.

8. Current programmed control can also be adapted to attain input resistor emulation in rectifiers. The programmed current reference signal $i_c(t)$ is made proportional to the ac input voltage. The difference between $i_c(t)$ and the average inductor current leads to distortion, owing to the inductor current ripple and the need for a stabilizing artificial ramp. Several approaches are known for reducing the resulting harmonic distortion of the line current waveform.

9. Hysteretic control, particularly with 100% current ripple, has a simple controller implementation. The disadvantages are variable switching frequency, and increased peak currents.

10. Nonlinear carrier control also leads to a simple controller implementation, and has the advantage of CCM operation with small peak transistor current.

11. The outer low-bandwidth control system, which regulates the dc output voltage to balance the rectifier and load powers, can be modeled by averaging the rectifier waveforms over one-half of the ac line period T_{2L}. This causes the dc-side system equations to become time-invariant. A small-signal model is then obtained by perturbation and linearization.

12. The inner high-bandwidth control system, which regulates the ac input current waveform to attain resistor emulation, is in general highly nonlinear. However, in the case of the boost rectifier, a valid small-signal model can be derived. This approach is unsuccessful in the case of other converters; one must then resort to other approaches such as the quasi-static approximation or simulation.

REFERENCES

[1]   D. CHAMBERS and D. WANG, "Dynamic Power Factor Correction in Capacitor Input Off-Line Converters," *Proceedings Sixth National Solid-State Power Conversion Conference* (Powercon 6), pp. B3-1 to B3-6, May 1979.

[2]   R. ERICKSON, M. MADIGAN, and S. SINGER, "Design of a Simple High Power Factor Rectifier Based on the Flyback Converter," *IEEE Applied Power Electronics Conference*, 1990 Record, pp. 792-801.

[3]   S. SINGER and R.W. ERICKSON, "Power Source Element and Its Properties," *IEE Proceedings–Circuits Devices Systems*, Vol. 141, No. 3, pp. 220-226, June 1994.

[4]   S. SINGER, "Realization of Loss-Free Resistive Elements," *IEEE Transactions on Circuits and Systems*, Vol. CAS-36, No. 12, January 1990.

[5]   W. E. RIPPEL, "Optimizing Boost Chopper Charger Design," *Proceedings Sixth National Solid-State Power Conversion Conference* (Powercon 6), 1979, pp. D1-1 - D1-20.

[6]   M. F. SCHLECHT and B. A. MIWA, "Active Power Factor Correction for Switching Power Supplies," *IEEE Transactions on Power Electronics*, Vol. 2, No. 4, October 1987, pp. 273-281.

[7]   J. SEBASTIAN, J. UCEDA, J. A. COBOS, J. ARAU, and F. ALDANA, "Improving Power Factor Correction in Distributed Power Supply Systems Using PWM and ZCS-QR SEPIC Topologies," *IEEE Power Electronics Specialists Conference*, 1991 Record, pp. 780-791.

[8]   E. YANG, Y. JIANG, G. HUA, and F. C. LEE, "Isolated Boost Circuit for Power Factor Correction," *IEEE Applied Power Electronics Conference*, 1993 Record, pp. 196-203.

[9]   C. A. CANESIN and I BARBI, "A Unity Power Factor Multiple Isolated Outputs Switching Mode Power Supply Using a Single Switch," *IEEE Applied Power Electronics Conference*, 1991 Record, pp. 430-436.

[10]   S. FREELAND, "I. A Unified Analysis of Converters with Resonant Switches, II. Input-Current Shaping for Single-Phase Ac–dc Power Converters," Ph.D. Thesis, California Institute of Technology, 1988.

[11]   M. J. SCHUTTEN, R. L. STEIGERWALD, and M. H. KHERALUWALA, "Characteristics of Load-Resonant Converters Operated in a High Power Factor Mode," *IEEE Applied Power Electronics Conference*, 1991 Record, pp. 5-16.

[12]   J. HONG, E. ISMAIL, R. ERICKSON, and I. KHAN, "Design of the Parallel Resonant Converter as a Low Harmonic Rectifier," *IEEE Applied Power Electronics Conference*, 1993 Record, pp. 833-840.

[13]   I. BARBI and S. A. O. DA SILVA, "Sinusoidal Line Current Rectification at Unity Power Factor with Boost Quasi-Resonant Converters," *IEEE Applied Power Electronics Conference*, 1990 Record, pp. 553-562.

[14]   R. REDL and L. BALOGH, "RMS, DC, Peak, and Harmonic Currents in High-Frequency Power-Factor Correctors with Capacitive Energy Storage," *IEEE Applied Power Electronics Conference*, 1992 Record, pp. 533-540.

[15]   J. SEBASTIAN, J. A. COBOS, P. GIL, and J. UCEDA, "The Determination of the Boundaries Between Continuous and Discontinuous Conduction Modes in PWM Dc-to-Dc Converters Used as Power Factor Preregulators," *IEEE Power Electronics Specialists Conference*, 1992 Record, pp. 1061-1070.

[16]   M. NALBANT, "Design of a 1 kW Power Factor Corrector," *Power Conversion*, October 1989 Proceedings, pp. 121-135.

[17]   R. MAMMANO and R. NEIDORFF, "Improving Input Power Factor—A New Active Controller Simplifies the Task," *Power Conversion*, October 1989 Proceedings, pp. 100-109.

[18]   J. BAZIMET and J. O'CONNOR, "Analysis and Design of a Zero Voltage Transition Power Factor Correction Circuit," *IEEE Applied Power Electronics Conference*, 1994 Record, pp. 591-597.

[19]   R. REDL and B. ERISMAN, "Reducing Distortion in Peak-Current-Controlled Boost Power Factor Correctors," *IEEE Applied Power Electronics Conference*, 1994 Record, pp. 576-583.

[20]   C. ZHOU and M. JOVANOVIC, "Design Tradeoffs in Continuous Current-Mode Controlled Boost Power Factor Correction Circuits," *High Frequency Power Conversion Conference*, 1992 Proceedings, pp. 209-220.

[21]   D. MAKSIMOVIĆ, "Design of the Clamped-Current High-Power-Factor Boost Rectifier," *IEEE Applied Power Electronics Conference*, 1994 Record, pp. 584-590.

[22]   C. CANESIN and I. BARBI, "Analysis and Design of Constant-Frequency Peak-Current-Controlled High-Power-Factor Boost Rectifier with Slope Compensation," *IEEE Applied Power Electronics Conference*, 1996 Record, pp. 807-813.

[23]   J. LAI and D. CHEN, "Design Considerations for Power Factor Correction Boost Converter Operating at the Boundary of Continuous Conduction Mode and Discontinuous Conduction Mode," *IEEE Applied Power Electronics Conference*, 1993 Record, pp. 267-273.

[24]   C. ZHOU, R. RIDLEY, and F. C. LEE, "Design and Analysis of a Hysteretic Boost Power Factor Correction Circuit," *IEEE Power Electronics Specialists Conference*, 1990 Record, pp. 800-807.

[25]   S. AHMED, "Controlled On-Time Power Factor Correction Circuit with Input Filter," M.S. Thesis, Virginia Polytechnic Institute and State University, Blacksburg VA, May 1990.

[26]   M. DAWANDE and G. DUBEY, "Programmable Input Power Factor Correction Method for Switch Mode Rectifiers," *IEEE Applied Power Electronics Conference*, 1993 Record, pp. 274-280.

[27]   J. SPANGLER and A. BEHERA, "A Comparison Between Hysteretic and Fixed Frequency Boost Converters Used for Power Factor Correction," *IEEE Applied Power Electronics Conference*, 1993 Record, pp. 281-286.

[28]   D. MAKSIMOVIĆ, Y. JANG, and R. ERICKSON, "Nonlinear-Carrier Control for High Power Factor Boost Rectifiers," *IEEE Applied Power Electronics Conference*, 1995 Record, pp. 635-641

[29]   R. ZANE and D. MAKSIMOVIĆ, "Nonlinear Carrier Control for High Power Factor Rectifiers Based on Flyback, Ćuk, or SEPIC Converters," *IEEE Applied Power Electronics Conference*, 1996 Record, pp. 814-820.

[30]   Z. LAI, K. SMEDLEY, and Y. MA, "Time Quantity One-Cycle Control for Power Factor Controllers," *IEEE Applied Power Electronics Conference*, 1996 Record, pp. 821-827.

[31]   L. ROSSETTO, G. SPIAZZI, P. TENTI, B. FABIANO, and C. LICITRA, "Fast-Response High-Quality Rectifier with Sliding-Mode Control," *IEEE Applied Power Electronics Conference*, 1993 Record, pp. 175-181.

[32]   W. TANG, Y. JIANG, G. HUA, F. C. LEE, and I. COHEN, "Power Factor Correction with Flyback Converter Employing Charge Control," *IEEE Applied Power Electronics Conference*, 1993 Record, pp. 293-298.

[33]   F. C. SCHWARZ, "An Improved Method of Resonant Current Pulse Modulation for Power Converters," *IEEE Power Electronics Specialists Conference*, 1975 Record, pp. 194-204.

[34]   M. J. KOCHER and R. L. STEIGERWALD, "An Ac-to-dc Converter with High Quality Input Waveforms," *IEEE Power Electronics Specialists Conference*, 1982 Record, pp. 63-75.

[35]   M. SCHLECHT, "Time-Varying Feedback Gains for Power Circuits with Active Waveshaping," *IEEE Power Electronics Specialists Conference*, 1981 Record, pp. 52-59.

[36]   I. KHAN and R. W. ERICKSON, "Control of Switched-Mode Converter Harmonic-Free Terminal Waveforms Through Internal Energy Storage," *IEEE Power Electronics Specialists Conference*, 1986 Record, pp. 13-26.

[37]   K. MAHABIR, G. VERGESE, J THOTTUVELIL, and A. HEYMAN, "Linear Averaged and Sampled Data Models for Large Signal Control of High Power Factor Ac–dc Converters," *IEEE Power Electronics Specialists Conference*, 1990 Record, pp. 372-381.

[38]   MICHAEL MADIGAN, ROBERT ERICKSON, and ESAM ISMAIL, "Integrated High Quality Rectifier-Regulators," *IEEE Power Electronics Specialists Conference*, 1992 Record, pp. 1043-1051.

[39]   K. MAHABIR, G. VERGESE, J THOTTUVELIL, and A. HEYMAN, "Linear Averaged and Sampled Data Models for Large Signal Control of High Power Factor AC-DC Converters," *IEEE Power Electronics Specialists Conference*, 1990 Record, pp. 372-381.

[40]   R. RIDLEY, "Average Small-Signal Analysis of the Boost Power Factor Correction Circuit," *Proceedings of the Virginia Power Electronics Center Seminar*, Blacksburg, VA, Sept. 1989, pp. 108-120.

[41]   M. MADIGAN, "Single-Phase High-Quality Rectifier-Regulators," Ph.D. Thesis, University of Colorado at Boulder, 1992.

[42]   N. MOHAN, T. UNDELAND and W. ROBBINS, *Power Electronics: Converters, Applications, and Design*, Second edition, New York: John Wiley & Sons, 1995, Chapters 8 and 18.

[43]   H. MAO, D. BOROYEVICH, A. RAVINDRA, and F. LEE, "Analysis and Design of a High Frequency Three-Phase Boost Rectifier," *IEEE Applied Power Electronics Conference*, 1996 Record, pp. 538-544.

[44]   B. T. OOI, J. C. SALMON, J. W. DIXON, and A. B. KULKARNI, "A Three-Phase Controlled-Current PWM Converter with Leading Power Factor," *IEEE Transactions on Industry Applications*, Vol. 23, No. 1, pp. 78-84, 1987.

[45]   P. TENTI and L. MALSANI, "Three-Phase AC/DC PWM Converter with Sinusoidal AC Currents and Minimum Filter Requirements," *IEEE Transactions on Industry Applications*, Vol. 23, No. 1, pp. 71-77, 1987.

[46]   A-M. MAJED, T. C. GREEN, and B. W. WILLIAMS, "Dynamic Properties of a Step-Down Sinusoidal Current AC/DC Converter Under State-Feedback Control," *IEEE Applied Power Electronics Conference*, 1993 Record, pp. 161-167.

[47]   M. RASTOGI, N. MOHAN, and C. HENZE, "Three-Phase Sinusoidal Current Rectifier with Zero Current Switching," *IEEE Applied Power Electronics Conference*, 1994 Record, pp. 718-724.

[48]   K. D. T. NGO, S. ĆUK, and R. D. MIDDLEBROOK, "A New Flyback Dc-to-Three-Phase Converter with Sinusoidal Outputs," *IEEE Power Electronics Specialists Conference*, 1983 Record, pp. 377-388.

[49]   A. R. PRASAD, P. D. ZIOGAS, and S. MANIAS, "An Active Power Factor Correction Technique for Three-Phase Diode Rectifiers," *IEEE Power Electronics Specialists Conference*, 1989 Record, pp. 58-66.

[50]   J. KOLAR, H. ERTL, and F. ZACH, "Space Vector Based Analysis of the Input Current Distortion of a Three Phase Discontinuous Conduction Mode Boost Rectifier System," *IEEE Power Electronics Specialists Conference*, 1993 Record, pp. 696-703.

[51]   R. ITOH and K. ISHITAKA, "Three-Phase Flyback Ac–dc Converter with Sinusoidal Supply Currents," *IEE Proceedings*, Vol. 136, Part B, No. 4, pp. 143-151, 1991.

[52]   O. APELDOORN and P. SCHMIDT, "Single Transistor Three-Phase Power Conditioners with High Power Factor and Isolated Output," *IEEE Applied Power Electronics Conference*, 1994 Record, pp. 731-737.

[53]   E. H. ISMAIL and R. W. ERICKSON, "A Single Transistor Three-Phase Resonant Switch for High Quality Rectification," *IEEE Power Electronics Specialists Conference*, 1992 Record, pp. 1341-1351.

[54]   Y. JANG and R. ERICKSON, "New Single-Switch Three-Phase High Power Factor Rectifiers Using Multi-Resonant Zero Current Switching," *IEEE Applied Power Electronics Conference*, 1994 Record, pp. 711-717.

[55]   Y. JANG and R. ERICKSON, "Design and Experimental Results of a 6kW Single-Switch Three-Phase High Power Factor Rectifier Using Multi-Resonant Zero Current Switching," *IEEE Applied Power Electronics Conference*, 1996 Record, pp. 524-530.

[56]   J. KOLAR, H. ERTL, and F. ZACH, "Design and Experimental Investigation of a Three-Phase High Power Density High-Efficiency Unity-Power-Factor PWM (VIENNA) Rectifier Employing a Novel Integrated Power Semiconductor Module," *IEEE Applied Power Electronics Conference*, 1996 Record, pp. 514-523.

[57]   M. RASTOGI, N. MOHAN, and C. HENZE, "Three-Phase Sinusoidal Current Rectifier with Zero Current Switching," *IEEE Applied Power Electronics Conference*, 1994 Record, pp. 718-724.

[58]   S. GATARIĆ, D. BOROYEVICH, and F. C. LEE, "Soft-Switched Single-Switch Three-Phase Rectifier with Power Factor Correction," *IEEE Applied Power Electronics Conference*, 1994 Record, pp. 738-744.

[59]   J. KOLAR, U. DROFENIK, AND F. ZACH, "Vienna Rectifier II—A Novel Single-Stage High-Frequency Isolated Three-Phase PWM Rectifier System," *IEEE Applied Power Electronics Conference*, 1998 Record, pp. 23–33.

PROBLEMS

18.1   The boost converter of Fig. 18.5 is replaced by a buck-boost converter. Inductor energy storage has negligible influence on the low-frequency components of the converter waveforms. The average load power is P_{load}. The dc output voltage is V and the sinusoidal ac input voltage has peak amplitude V_M.

   (a)   Determine expressions for the duty cycle variations $d(t)$ and the inductor current variation $i(t)$, assuming that the converter operates in the continuous conduction mode.

   (b)   Derive the conditions for operation in the continuous conduction mode. Manipulate your result to show that the converter operates in CCM when R_e is less than $R_{e,crit}(L, T_s, v_g(t), V)$, and determine $R_{e,crit}$.

   (c)   For what values of R_e does the converter always operate in CCM? in DCM?

   (d)   The ac input voltage has rms amplitude in the range 108 V to 132 V. The maximum load power is 100 W, and the minimum load power is 10 W. The dc output voltage is 120 V. The switching frequency is 75 kHz. What value of L guarantees that the converter always operates in CCM? in DCM?

18.2   Derive expressions for the input characteristics of the buck-boost converter, similar to Eqs. (18.25) to (18.33). Sketch the converter input characteristics, and label the CCM–DCM boundary.

18.3   Derive expressions for the rms transistor and diode currents of rectifiers based on the single-phase CCM Ćuk topology. Express your results in forms similar to those of Table 18.3.

18.4   To obtain an isolated dc output, the boost converter in Fig. 18.5 is replaced by the full-bridge transformer-isolated CCM boost converter of Fig. 6.35. Derive an expression for the rms transistor current. Express your result as a function of $I_{ac\,rms}$, n, V, and V_M.

18.5   Comparison of CCM boost and isolated SEPIC topologies as universal-input single-phase rectifiers. You are given that the dc output voltage is $V = 400$ V, the load power is $P = 500$ W, and the rms input voltage varies between 90 and 270 V, such that the peak ac input voltage V_M varies between $V_{Mmin} = 127$ V and $V_{Mmax} = 382$ V. Define the transistor stress S as the product of the worst-case peak transistor voltage and the worst-case rms transistor current. It is desired to minimize S.

   (a)   Determine S for the boost converter in this application.

   (b)   Briefly discuss your result of part (a): if universal input operation was not required, and hence $V_M = 382$ V always, what S would result?

   In the isolated SEPIC, the transformer turns ratio $n : 1$ can be chosen to optimize the design.

   (c)   Express S for the SEPIC as a function of n, V, P, V_{Mmin}, and V_{Mmax}.

   (d)   Choose n for the SEPIC such that S is minimized in this application. Compare with the results of parts (a) and (b).

18.6   In the boost-type dc–3øac rectifier of Fig. 18.39, the ac-side inductances L are not small: they exhibit line-frequency impedances that should not be ignored. The three-phase ac voltages are given by Eq. (18.150), and the duty cycles are modulated as in Eq. (18.156). The converter operates in the continuous conduction mode.

   (a)   Determine the magnitudes and phases of the line-to-neutral average voltages at the ac inputs to the switch network. Express your result in terms of D_m, V, and φ.

   (b)   Determine the real power P and reactive power Q drawn from the 3øac source. Express your results as functions of V_M, V, D_m, φ, and ωL.

   (c)   How must φ be chosen to obtain unity power factor?

18.7   In the boost-type dc–3øac rectifier of Fig. 18.39, the switch duty ratios are modulated as illustrated in Fig. 18.42. When the inductances L are sufficiently small, a dc output voltage V equal to the peak line-to-line ac input voltage can be obtained, with undistorted ac line currents. As illustrated in Fig. 18.42, $d_1(t)$ is equal to 1 for $0° \leq \omega t \leq 60°$, where $\omega t = 0°$ when $\langle v_{12}(t) \rangle_{T_s} = V$.

   (a)   Derive expressions for $d_2(t)$ and $d_3(t)$, over the interval $0° \leq \omega t \leq 60°$.

   (b)   State how $d_1(t)$, $d_2(t)$, and $d_3(t)$ should vary over each $60°$ interval.

18.8   The buck-type 3øac–dc rectifier of Fig. 18.44 operates in the continuous conduction mode. Transistors Q_1 to Q_6 operate with duty cycles $d_1(t)$ to $d_6(t)$, respectively.

   (a)   Determine the constraints on switch operation. Which transistors must not conduct simultaneously? Which duty cycles must total unity?

   (b)   Average the 3ø bridge switch network, to determine expressions for the average ac-side switch currents $\langle i_a(t) \rangle_{T_s}$, $\langle i_b(t) \rangle_{T_s}$, and $\langle i_c(t) \rangle_{T_s}$.

   (c)   Show that the average dc-side switch voltage can be expressed as

$$\langle v_d(t) \rangle_{T_s} = \left(d_1(t) - d_4(t) \right) \langle v_{an}(t) \rangle_{T_s} + \left(d_2(t) - d_5(t) \right) \langle v_{bn}(t) \rangle_{T_s} + \left(d_3(t) - d_6(t) \right) \langle v_{cn}(t) \rangle_{T_s}$$

   (d)   The duty cycles are varied as follows:

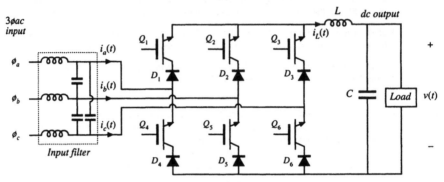

Fig. 18.44   Buck-type 3∅ac–dc rectifier, Problem 18.8.

$$d_1(t) = \tfrac{1}{3} + \tfrac{1}{2}D_m \sin\left(\omega t - \varphi\right)$$

$$d_2(t) = \tfrac{1}{3} + \tfrac{1}{2}D_m \sin\left(\omega t - \varphi - 120°\right)$$

$$d_3(t) = \tfrac{1}{3} + \tfrac{1}{2}D_m \sin\left(\omega t - \varphi - 240°\right)$$

$$d_4(t) = \tfrac{1}{3} - \tfrac{1}{2}D_m \sin\left(\omega t - \varphi\right)$$

$$d_5(t) = \tfrac{1}{3} - \tfrac{1}{2}D_m \sin\left(\omega t - \varphi - 120°\right)$$

$$d_6(t) = \tfrac{1}{3} - \tfrac{1}{2}D_m \sin\left(\omega t - \varphi - 240°\right)$$

with the ac input voltages given by Eq. (18.150). The input filter has negligible effect of the low-frequency components of the converter waveforms. Determine the steady-state dc output voltage V, as a function of V_M, D_m, and φ.

(e)   Determine the power factor. You may assume that the input filter completely removes the switching harmonics from the currents $i_a(t)$, $i_b(t)$, and $i_c(t)$. However, the input filter elements consume or supply negligible line-frequency reactive power.

18.9   In the three-phase DCM flyback rectifier of Fig. 18.45, the input filter has negligible effect on the low-frequency components of the input ac waveforms. The transistor operates with switching frequency f_s and duty cycle d. Flyback transformers T_1, T_2, and T_3 each have magnetizing inductance L referred to the primary, turns ratio $n : 1$, and have negligible leakage inductances.

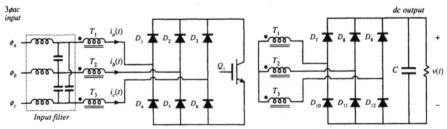

Fig. 18.45   Isolated 3∅ac–dc rectifier based on the flyback converter operating in discontinuous conduction mode: Problem 18.9.

(a)   Determine expressions for the low-frequency components of the ac input and dc output currents.

(b)   Derive an averaged equivalent circuit model for the converter, and give expressions for the element values.

(c)   Derive the conditions for operation in the discontinuous conduction mode.

18.10   Power stage design of a universal-input boost rectifier. The objective of this problem is to work out the power stage design of a low harmonic rectifier based on the boost converter. This converter is to be designed to operate anywhere in the world, and hence the input voltage may vary over the range 90 to 270 Vrms, 50 to 60 Hz. The converter produces a regulated 385 V dc output, at 1000 W. The switching frequency f_s is 100 kHz. You may assume that the controller operates perfectly, to produce an undistorted ac line current waveform and a well-regulated dc output voltage.

(a)   Derive an expression for how the duty cycle $d(t)$ will vary over the ac line cycle. You may neglect converter dynamics and losses. Sketch $d(t)$ under conditions of maximum and minimum ac line voltage.

(b)   Specify the inductor:

    (i)   Specify the value of L such that, at the peak of the sinusoidal input voltage, and under worst-case conditions, the inductor current ripple Δi_g is 20% of the instantaneous low frequency current $i_g(t)$.

    (ii)   Specify the worst-case values of the peak and rms inductor current, assuming 100% efficiency.

(c)   Determine the worst-case rms currents of the MOSFET and diode, assuming 100% efficiency.

(d)   Specify the value of C that leads to a worst-case low-frequency ($\ll f_s$) output voltage peak-peak ripple of 5 V.

(e)   Given the following loss elements

$$\begin{aligned} &\text{Inductor winding resistance} &&0.1\ \Omega\\ &\text{MOSFET on-resistance} &&0.4\ \Omega\\ &\text{Diode forward voltage drop} &&1.5\ \text{V}\\ &\text{Switching loss: model as } i_g^2(t)(0.25\ \Omega) \end{aligned}$$

For a constant 1000 W load, and assuming that the controller operates perfectly as described above, find the rectifier efficiency

    (i)   at an ac input voltage of 90 V rms

    (ii)   at an ac input voltage of 270 V rms

18.11   The flyback converter shown in Fig. 18.46 operates in the continuous conduction mode. The MOSFET

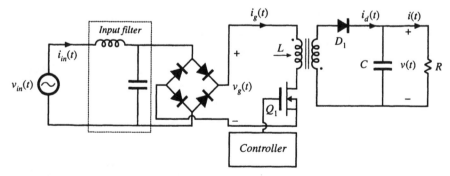

Fig. 18.46   Low-harmonic rectifier system based on the CCM flyback converter, Problem 18.11.

has on-resistance R_{on}, and diode D_1 has a constant forward voltage drop V_D. All other loss elements can be neglected. The turns ratio of the flyback transformer is 1:1. The controller varies the duty cycle such that $\langle i_g(t)\rangle_{T_s}$ is equal to $v_g(t)/R_e$, where R_e is the emulated resistance. The input voltage is $v_{in}(t) = V_M \sin(\omega t)$. The input filter removes the switching harmonics from the input current $i_g(t)$, but has negligible effect on the low-frequency components of the converter waveforms.

(a)   Derive an expression for the rectifier efficiency, in terms of V_M, V, V_D, R_{on}, and R_e.

(b)   Given the following values, find the value of MOSFET on-resistance which leads to an efficiency of 96%.

rms input voltage	120 V
Dc output voltage	120 V
Diode D_1 forward voltage drop	1.5 V
Load power	200 W

18.12   Derive an expression for the emulated resistance $R_e(V_{g,rms},\ R_s,\ k_v,\ v_{control})$ of the average-current-controlled boost rectifier with ac line voltage feedforward, Fig. 18.14.

18.13   Derive the CPM boost rectifier static input characteristics, Eq. (18.57).

18.14   The boost rectifier system of Fig. 18.47 employs average current control with ac line voltage feedforward.

The ac line frequency is 50 Hz. The rectifier drives a constant-power load of 500 W. The pulse-width modulator contains a ramp having a peak-to-peak amplitude of 3 V. There is no compensator in the inner wide-bandwidth average current control feedback loop. The average current sensing circuit has gain

$$\frac{v_a(s)}{i_g(s)} = \frac{R_s}{\left(1 + \frac{s}{\omega_0}\right)}$$

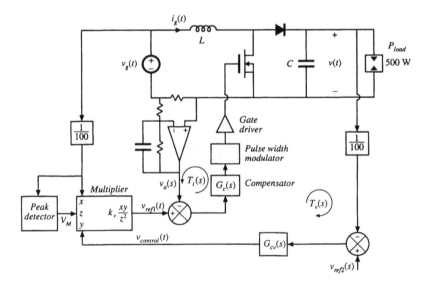

Fig. 18.47   Average current controlled boost rectifier with input voltage feedforward, Problem 18.14.

Other converter parameter values are

$$f_s = 100 \text{ kHz} \qquad\qquad L = 2.5 \text{ mH}$$

$$f_0 = \frac{\omega_0}{2\pi} = 10 \text{ kHz} \qquad\qquad R_s = 1 \, \Omega$$

$$V = 385 \text{ V} \qquad\qquad V_{g,rms} = 230 \text{ V}$$

(a)   Construct the magnitude and phase Bode diagrams of the loop gain $T_i(s)$ of the average-current-control loop. Label important features.

(b)   Determine numerical values of the crossover frequency and phase margin of $T_i(s)$.

The outer low-bandwidth feedback loop has loop gain $T_v(s)$. The compensator of this loop has constant gain $G_{cv}(s) = 330$. The multiplier gain is $k_v = 2$. The capacitor value is $C = 680 \, \mu\text{F}$. The reference voltage $v_{ref2}(t)$ is 3.85 V.

(c)   Determine the peak magnitude of the output 100 Hz voltage ripple.

(d)   Determine the quiescent control voltage $V_{control}$.

(e)   Construct the magnitude and phase Bode diagrams of the loop gain $T_v(s)$ of the outer feedback loop. Label important features.

(f)   Determine numerical values of the crossover frequency and phase margin of $T_v(s)$.

18.15   A critical conduction mode controller causes a boost rectifier to exhibit an ac input current waveform similar to Fig. 18.19(b). The ac input voltage is 120 Vrms at 60 Hz. The rectifier supplies 225 Vdc to a 120 W load. The boost converter inductance is $L = 600 \, \mu\text{H}$.

(a)   Determine the emulated resistance R_e.

(b)   Write the numerical expression for the converter switching frequency f_s, as a function of t_{on} and the applied terminal voltages. Sketch f_s vs. time.

(c)   What is the maximum switching frequency? What is the minimum switching frequency?

(d)   Derive an analytical expression for the rms transistor current for this control method, as a function of the magnitude of the sinusoidal line current. Compare the rms transistor current of this approach with a CCM boost rectifier having negligible current switching ripple.

Part V

Resonant Converters

19

Resonant Conversion

Part V of this text deals with a class of converters whose operation differs significantly from the PWM converters covered in Parts I to IV. *Resonant power converters* [1–36] contain resonant L–C networks whose voltage and current waveforms vary sinusoidally during one or more subintervals of each switching period. These sinusoidal variations are large in magnitude, and hence the small ripple approximation introduced in Chapter 2 does not apply.

Dc-to-high-frequency-ac inverters are required in a variety of applications, including electronic ballasts for gas discharge lamps [3,4], induction heating, and electrosurgical generators. These applications typically require generation of a sinusoid of tens or hundreds of kHz, having moderate or low total harmonic distortion. A simple resonant inverter system is illustrated in Fig. 19.1(a). A switch network produces a square wave voltage $v_s(t)$. As illustrated in Fig. 19.2, the spectrum of $v_s(t)$ contains fundamental plus odd harmonics. This voltage is applied to the input terminals of a resonant tank network. The tank network resonant frequency f_0 is tuned to the fundamental component of $v_s(t)$, that is, to the switching frequency f_s, and the tank exhibits negligible response at the harmonics of f_s. In consequence, the tank current $i_s(t)$, as well as the load voltage $v(t)$ and load current $i(t)$, have essentially sinusoidal waveforms of frequency f_s, with negligible harmonics. By changing the switching frequency f_s (closer to or further from the resonant frequency f_0), the magnitudes of $i_s(t)$, $v(t)$, and $i(t)$ can be controlled. Other schemes for control of the output voltage, such as phase-shift control of the bridge switch network, are also possible. A variety of resonant tank networks can be employed; Fig. 19.1(b) to (d) illustrate the well-known *series*, *parallel*, and *LCC* tank networks. Inverters employing the series resonant tank network are known as the *series resonant*, or *series loaded*, inverter. In the *parallel resonant* or *parallel loaded* inverter, the load voltage is equal to the resonant tank capacitor voltage. The LCC inverter employs tank capacitors both in series and in parallel with the load.

Figure 19.3 illustrates a high-frequency inverter of an electronic ballast for a gas-discharge lamp. A half-bridge configuration of the LCC inverter drives the lamp with an approximately sinusoidal

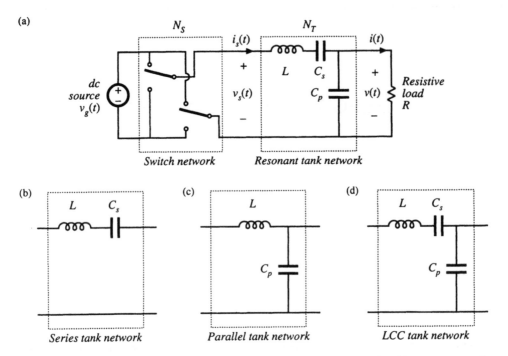

Fig. 19.1   A basic class of resonant inverters that consist of (a) a switch network N_S that drives a resonant tank network N_T near resonance. Several common tank networks: (b) series, (c) parallel, (d) LCC.

Fig. 19.2 The tank network responds primarily to the fundamental component of the applied waveforms.

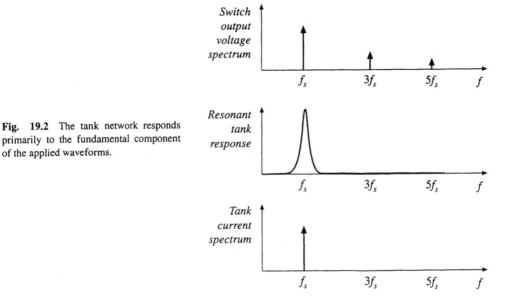

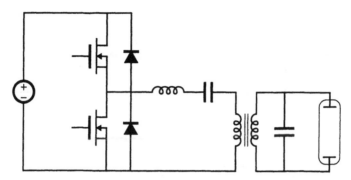

Fig. 19.3 Half-bridge LCC inverter circuit, as an electronic ballast for a gas-discharge lamp.

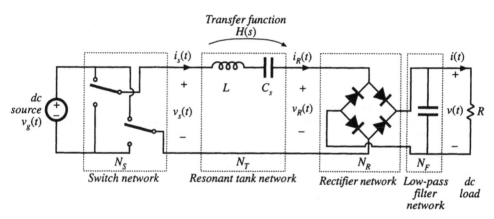

Fig. 19.4 Derivation of a resonant dc–dc converter, by rectification and filtering of the output of a resonant inverter.

high-frequency ac waveform. The converter is controlled to provide a relatively high voltage to start the lamp, and a lower voltage thereafter. When the ballast is powered by the ac utility, a low-harmonic rectifier typically provides the input dc voltage for the inverter.

A resonant dc–dc converter can be constructed by rectifying and filtering the ac output of a resonant inverter. Figure 19.4 illustrates a series-resonant dc-dc converter, in which the approximately sinusoidal resonant tank output current $i_R(t)$ is rectified by a diode bridge rectifier, and filtered by a large capacitor to supply a dc load having current I and voltage V. Again, by variation of the switching frequency f_s (closer to or further from the resonant frequency f_0), the magnitude of the tank current $i_R(t)$, and hence also the dc load current I, can be controlled. Resonant dc–dc converters based on series, parallel, LCC, and other resonant tank networks are well understood. These converters are employed when specialized application requirements justify their use. For example, they are commonly employed in high voltage dc power supplies [5,6], because the substantial leakage inductance and winding capacitance of high-voltage transformers leads unavoidably to a resonant tank network. The same principle can be employed to construct *resonant link* inverters or resonant link cycloconverters [7–9]; controllable switch networks are then employed on both sides of the resonant tank network.

Figure 19.5 illustrates another approach to resonant power conversion, in which resonant ele-

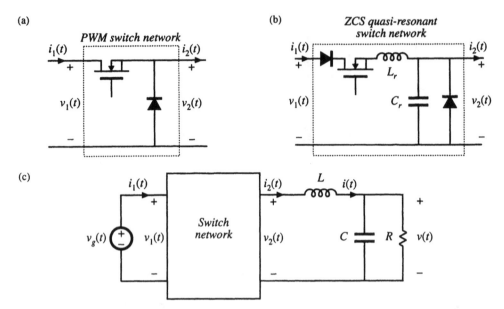

Fig. 19.5 Derivation of a quasi-resonant converter: (a) conventional PWM switch network, (b) a ZCS quasi-resonant switch network, (c) a quasi-resonant buck converter is obtained by employing a quasi-resonant switch network such as (b) in a buck converter.

ments are inserted into the switch network of an otherwise-PWM converter. A *resonant switch* network, or *quasi-resonant* converter, is then obtained. For example, in Fig. 19.5(b), resonant elements L_r and C_r are combined with the switch network transistor and diode. The resonant frequency of these elements is somewhat higher than the switching frequency. This causes the switch network waveforms $i_1(t)$ and $v_2(t)$ to become quasi-sinusoidal pulses. The resonant switch network of Fig. 19.5(b) can replace the PWM switch network of Fig. 19.5(a) in nearly any PWM converter. For example, insertion of the resonant switch network of Fig. 19.5(b) into the converter circuit of Fig. 19.5(c) leads to a quasi-resonant buck converter. Numerous resonant switch networks are known, which lead to a large number of resonant switch versions of buck, boost, buck–boost, and other converters. Quasi-resonant converters are described in Chapter 20.

The chief advantage of resonant converters is their reduced switching loss, via mechanisms known as *zero-current switching* (ZCS), and *zero-voltage switching* (ZVS). The turn-on and/or turn-off transitions of the various converter semiconductor elements can occur at zero crossings of the resonant converter quasi-sinusoidal waveforms. This eliminates some of the switching loss mechanisms described in Chapter 4. Hence, switching loss is reduced, and resonant converters can operate at switching frequencies that are higher than in comparable PWM converters. Zero-voltage switching can also eliminate some of the sources of converter-generated electromagnetic interference.

Resonant converters exhibit several disadvantages. Although the resonant element values can be chosen such that good performance with high efficiency is obtained at a single operating point, typically it is difficult to optimize the resonant elements such that good performance is obtained over a wide range of load currents and input voltages. Significant currents may circulate through the tank elements, even when the load is removed, leading to poor efficiency at light load. Also, the quasi-sinusoidal waveforms of resonant converters exhibit greater peak values than those exhibited by the rectangular waveforms of PWM converters, provided that the PWM current spikes due to diode stored charge are ignored. For these reasons, resonant converters exhibit increased conduction losses, which can offset their reduced switching losses.

In this chapter, the properties of the series, parallel, and other resonant inverters and dc–dc converters are investigated using the *sinusoidal approximation* [3, 10–12]. Harmonics of the switching frequency are neglected, and the tank waveforms are assumed to be purely sinusoidal. This allows simple equivalent circuits to be derived for the bridge inverter, tank, rectifier, and output filter portions of the converter, whose operation can be understood and solved using standard linear ac analysis. This intuitive approach is quite accurate for operation in the continuous conduction mode with a high-Q response, but becomes less accurate when the tank is operated with a low Q-factor or for operation of dc–dc resonant converters in or near the discontinuous conduction mode.

For dc–dc resonant converters, the important result of this approach is that the dc voltage conversion ratio of a continuous conduction mode resonant converter is given approximately by the ac transfer function of the tank circuit, evaluated at the switching frequency. The tank is loaded by an effective output resistance, having a value nearly equal to the output voltage divided by the output current. It is thus quite easy to determine how the tank components and circuit connections affect the converter behavior. The influence of tank component losses, transformer nonidealities, etc., on the output voltage and converter efficiency can also be found. Several resonant network theorems are presented, which allow the load dependence of conduction loss and of the zero-voltage- or zero-current-switching properties to be explained in a simple and intuitive manner.

It is found that the series resonant converter operates with a step-down voltage conversion ratio. With a 1:1 transformer turns ratio, the dc output voltage is ideally equal to the dc input voltage when the transistor switching frequency is equal to the tank resonant frequency. The output voltage is reduced as the switching frequency is increased or decreased away from resonance. On the other hand, the parallel resonant converter is capable of both step-up and step-down of voltage levels, depending on the switching frequency and the effective tank Q-factor. The exact steady-state solutions of the ideal series and parallel resonant dc–dc converters are stated in Section 19.5.

Zero-voltage switching and zero-current switching mechanisms of the series resonant converter are described in Section 19.3. In Section 19.4, the dependence of resonant inverter properties on load is examined. A simple frequency-domain approach explains why some resonant converters, over certain ranges of operating points, exhibit large circulating tank currents and low efficiency. The boundaries of zero-voltage switching and zero-current switching are also determined.

It is also possible to modify the PWM converters of the previous chapters, so that zero-current or zero-voltage switching is obtained. A number of diverse approaches are known that lead to *soft switching* in buck, boost, forward, flyback, bridge, and other topologies. Chapter 20 summarizes some of the well-known schemes, including resonant switches, quasi-square wave switches, the full-bridge zero-voltage transition converter, and zero-voltage switching in forward and flyback converters containing active-clamp snubbers. A detailed description of soft-switching mechanisms of diodes, MOSFETs, and IGBTs is also given.

19.1   SINUSOIDAL ANALYSIS OF RESONANT CONVERTERS

Consider the class of resonant converters that contain a controlled switch network N_S that drives a linear resonant tank network N_T. In a resonant inverter, the tank network drives a resistive load as in Fig. 19.1. The reactive component of the load impedance, if any, can be effectively incorporated into the tank network. In the case of a resonant dc–dc converter, the resonant tank network is connected to an uncontrolled rectifier network N_R, filter network N_F and load R, as illustrated in Fig. 19.4. Many well-known converters can be represented in this form, including the series, parallel, and LCC topologies.

In the most common modes of operation, the controlled switch network produces a square wave voltage output $v_s(t)$ whose frequency f_s is close to the tank network resonant frequency f_0. In response, the tank network rings with approximately sinusoidal waveforms of frequency f_s. In the case where the resonant tank responds primarily to the fundamental component f_s of the switch waveform $v_s(t)$, and has negligible response at the harmonic frequencies nf_s, $n = 3, 5, 7,...$, then the tank waveforms are well approximated by their fundamental components. As shown in Fig. 19.2, this is indeed the case when the tank network contains a high-Q resonance at or near the switching frequency, and a low-pass characteristic at higher frequencies. Hence, let us neglect harmonics, and compute the relationships between the fundamental components of the tank terminal waveforms $v_s(t)$, $i_s(t)$, $i_R(t)$, and $v_R(t)$.

19.1.1   Controlled Switch Network Model

If the switch network of Fig. 19.6 is controlled to produce a square wave of frequency $f_s = \omega_s/2\pi$ as in Fig. 19.7, then its output voltage waveform $v_s(t)$ can be expressed in the Fourier series

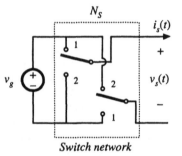

$$v_s(t) = \frac{4V_g}{\pi} \sum_{n=1,3,5,...} \frac{1}{n} \sin(n\omega_s t) \qquad (19.1)$$

The fundamental component is

$$v_{s1}(t) = \frac{4V_g}{\pi} \sin(\omega_s t) = V_{s1} \sin(\omega_s t) \qquad (19.2)$$

Fig. 19.6   An ideal switch network.

which has a peak amplitude of $(4/\pi)$ times the dc input voltage V_g, and is in phase with the original square wave $v_s(t)$. Hence, the switch network output terminal is modeled as a sinusoidal voltage generator, $v_{s1}(t)$.

It is also of interest to model the converter dc input port. This requires computation of the dc component I_g of the switch input current $i_g(t)$. The switch input current $i_g(t)$ is equal to the output current $i_s(t)$ when the switches are in position 1, and its inverse $-i_s(t)$ when the switches are in position 2. Under the conditions described above, the tank rings sinusoidally and $i_s(t)$ is well approximated by a sinusoid of some peak amplitude I_{s1} and phase φ_s:

$$i_s(t) \approx I_{s1} \sin(\omega_s t - \varphi_s) \qquad (19.3)$$

Fig. 19.7   Switch network output voltage $v_s(t)$ and its fundamental component $v_{s1}(t)$.

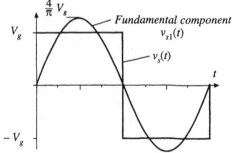

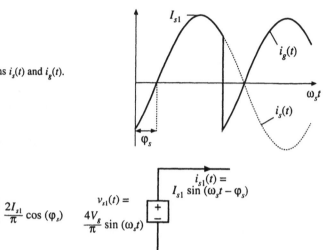

Fig. 19.8 Switch network waveforms $i_s(t)$ and $i_g(t)$.

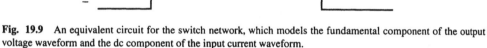

Fig. 19.9 An equivalent circuit for the switch network, which models the fundamental component of the output voltage waveform and the dc component of the input current waveform.

The input current waveform is shown in Fig. 18.8.

The dc component, or average value, of the input current can be found by averaging $i_g(t)$ over one half switching period:

$$
\begin{aligned}
\left\langle i_g(t) \right\rangle_{T_s} &= \frac{2}{T_s} \int_0^{T_s/2} i_g(\tau)d\tau \\
&\approx \frac{2}{T_s} \int_0^{T_s/2} I_{s1} \sin(\omega_s \tau - \varphi_s)d\tau \\
&= \frac{2}{\pi} I_{s1} \cos(\varphi_s)
\end{aligned}
\tag{19.4}
$$

Thus, the dc component of the converter input current depends directly on the peak amplitude of the tank input current I_{s1} and on the cosine of its phase shift φ_s.

An equivalent circuit for the switch is given in Fig. 19.9. This circuit models the basic energy conversion properties of the switch: the dc power supplied by the voltage source V_g is converted into ac power at the switch output. Note that the dc power at the source is the product of V_g and the dc component of $i_g(t)$, and the ac power at the switch is the average of $v_s(t)i_s(t)$. Furthermore, if the harmonics of $v_s(t)$ are negligible, then the switch output voltage can be represented by its fundamental component, a sinusoid $v_{s1}(t)$ of peak amplitude $4V_g/\pi$. It can be verified that the switch network dc input power and fundamental average output power, predicted by Fig. 19.9, are equal.

19.1.2 Modeling the Rectifier and Capacitive Filter Networks

In the series resonant dc–dc converter, the output rectifier is driven by the nearly sinusoidal tank output current $i_R(t)$. A large capacitor C_F is placed at the dc output, so that the output voltage $v(t)$ contains negligible harmonics of the switching frequency f_s, as shown in Fig. 19.10. Hence, we can make the small-rip-

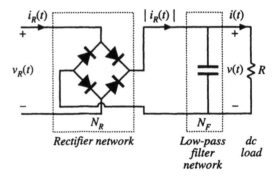

Fig. 19.10 Uncontrolled rectifier with capacitive filter network, as in the series resonant converter.

ple approximation as usual: $v(t) \approx V$, $i(t) \approx I$. The diode rectifiers switch when $i_R(t)$ passes through zero, as shown in Fig. 19.11. The rectifier input voltage $v_R(t)$ is essentially a square wave, equal to $+ v(t)$ when $i_R(t)$ is positive and $- v(t)$ when $i_R(t)$ is negative. Note that $v_R(t)$ is in phase with $i_R(t)$.

If the tank output current $i_R(t)$ is a sinusoid with peak amplitude I_{R1} and phase shift φ_R:

$$i_R(t) = I_{R1} \sin(\omega_s t - \varphi_R) \tag{19.5}$$

then the rectifier input voltage may be expressed in the Fourier series

$$v_R(t) = \frac{4V}{\pi} \sum_{n = 1, 3, 5, \cdots}^{\infty} \frac{1}{n} \sin(n\omega_s t - \varphi_R) \tag{19.6}$$

where φ_R is the phase shift of $i_R(t)$, with respect to $v_s(t)$. This voltage waveform is impressed on the output port of the resonant tank network. Again, if the tank network responds primarily to the fundamental component (f_s) of $v_R(t)$, and has negligible response at the harmonic frequencies nf_s, $n = 3, 5, 7...$, then the harmonics of $v_R(t)$ can be ignored. The voltage waveform $v_R(t)$ is then well approximated by its fundamental component $v_{R1}(t)$:

$$v_{R1}(t) = \frac{4V}{\pi} \sin(\omega_s t - \varphi_R) = V_{R1} \sin(\omega_s t - \varphi_R) \tag{19.7}$$

The fundamental voltage component $v_{R1}(t)$ has a peak value of $(4/\pi)$ times the dc output voltage V, and is in phase with the current $i_R(t)$.

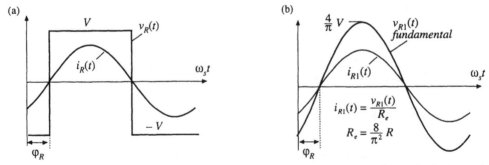

Fig. 19.11 Rectifier network input terminal waveforms: (a) actual waveforms $v_R(t)$ and $i_R(t)$, (b) fundamental components $v_{R1}(t)$ and $i_{R1}(t)$.

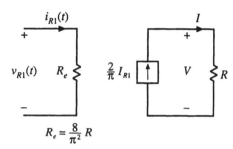

Fig. 19.12 An equivalent circuit for the rectifier and filter network, which models the fundamental components of the rectifier ac input waveforms and the dc components of the load waveforms. The rectifier presents an effective resistive load R_e to the tank network.

$$R_e = \frac{8}{\pi^2} R$$

The rectified tank output current, $|i_R(t)|$, is filtered by capacitor C_F. Since no dc current can pass through C_F, the dc component of $|i_R(t)|$ must be equal to the steady-state load current I. By equating dc components we obtain:

$$I = \frac{2}{T_S} \int_0^{T_s/2} I_{R1} \left| \sin(\omega_s t - \varphi_R) \right| dt$$
$$= \frac{2}{\pi} I_{R1}$$

$$(19.8)$$

Therefore, the load current and the tank output current amplitudes are directly related in steady state.

Since $v_{R1}(t)$, the fundamental component of $v_R(t)$, is in phase with $i_R(t)$, the rectifier presents an effective resistive load R_e to the tank circuit. The value of R_e is equal to the ratio of $v_{R1}(t)$ to $i_R(t)$. Division of Eq. (19.7) by Eq. (19.5), and elimination of I_{R1} using Eq. (19.8) yields

$$R_e = \frac{v_{R1}(t)}{i_R(t)} = \frac{8}{\pi^2} \frac{V}{I}$$

$$(19.9)$$

With a resistive load R equal to V/I, this equation reduces to

$$R_e = \frac{8}{\pi^2} R = 0.8106 R$$

$$(19.10)$$

Thus, the tank network is damped by an effective load resistance R_e equal to 81% of the actual load resistance R. An equivalent circuit that models the rectifier network input port fundamental components and output port dc components is given in Fig. 19.12.

19.1.3 Resonant Tank Network

We have postulated that the effects of harmonics can be neglected, and we have consequently shown that the bridge can be modeled as a fundamental voltage source $v_{s1}(t)$. In the case of a dc-dc converter, the rectifier can be modeled using an effective resistor of value R_e. We can now solve the resonant tank network by standard linear analysis.

As shown in Fig. 19.13, the tank circuit is a linear network with the following voltage transfer function:

$$\frac{v_{R1}(s)}{v_{s1}(s)} = H(s)$$

$$(19.11)$$

Hence, the ratio V_{R1}/V_{s1} of the peak magnitudes of $v_{R1}(t)$ and $v_{s1}(t)$ is given by:

$$\frac{V_{R1}}{V_{s1}} = |H(s)|_{s=j\omega_s} \qquad (19.12)$$

In addition, $i_R(s)$ is given by:

$$i_R(s) = \frac{v_{R1}(s)}{R_e} = \frac{H(s)}{R_e} v_{s1}(s) \qquad (19.13)$$

So the peak magnitude of $i_R(t)$ is:

$$I_{R1} = \frac{|H(s)|_{s=j\omega_s}}{R_e} V_{s1} \qquad (19.14)$$

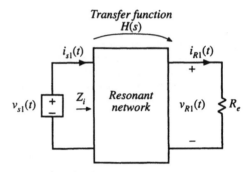

Fig. 19.13 The linear tank network, excited by an effective sinusoidal input source and driving an effective resistive load.

Thus, the magnitude of the tank transfer function is found, with an effective resistive load.

19.1.4 Solution of Converter Voltage Conversion Ratio $M = V/V_g$

An equivalent circuit of a complete dc–dc resonant converter is depicted in Fig. 19.14. The voltage conversion ratio of the resonant converter can now be found:

$$M = \frac{V}{V_g} = \underbrace{(R)}\ \underbrace{\left(\frac{2}{\pi}\right)}\ \underbrace{\left(\frac{1}{R_e}\right)}\ \underbrace{\left(|H(s)|_{s=j\omega_s}\right)}\ \underbrace{\left(\frac{4}{\pi}\right)}$$

$$\underbrace{\left(\frac{V}{I}\right)\left(\frac{I}{I_{R1}}\right)\left(\frac{I_{R1}}{V_{R1}}\right)}\quad \underbrace{\left(\frac{V_{R1}}{V_{s1}}\right)}\quad \underbrace{\left(\frac{V_{s1}}{V_g}\right)} \qquad (19.15)$$

Simplification by use of Eq. (19.10) yields:

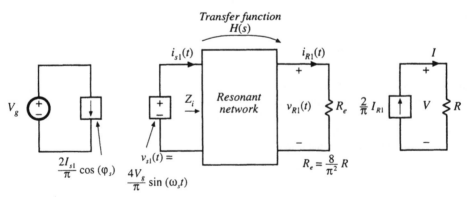

Fig. 19.14 Steady-state equivalent circuit that models the dc and fundamental components of resonant converter waveforms.

$$\frac{V}{V_g} = \left| H(s) \right|_{s = j\omega_s} \tag{19.16}$$

Equation (19.16) is the desired result. It states that the dc conversion ratio of the resonant converter is approximately the same as the ac transfer function of the resonant tank circuit, evaluated at the switching frequency f_s. This intuitive result can be applied to converters with many different types of tank circuits. However, it should be reemphasized that Eq. (19.16) is valid only if the response of the tank circuit to the harmonics of $v_s(t)$ is negligible compared to the fundamental response, an assumption that is not always justified. In addition, we have assumed that the switch network is controlled to produce a square wave and that the rectifier network drives a capacitive-type filter network. Finally, the transfer function $H(s)$ is evaluated using the effective load resistance R_e given by Eq. (19.9).

19.2 EXAMPLES

19.2.1 Series Resonant DC–DC Converter Example

The series resonant converter with switching frequency control is shown in Fig. 19.4. Current-bidirectional two-quadrant switches are necessary. For this circuit, the tank network consists of a series L–C circuit, and Fig. 19.14 can be redrawn as in Fig. 19.15. The transfer function $H(s)$ is therefore:

$$H(s) = \frac{R_e}{Z_i(s)} = \frac{R_e}{R_e + sL + \frac{1}{sC}}$$

$$= \frac{\left(\frac{s}{Q_e\omega_0}\right)}{1 + \left(\frac{s}{Q_e\omega_0}\right) + \left(\frac{s}{\omega_0}\right)^2} \tag{19.17}$$

where

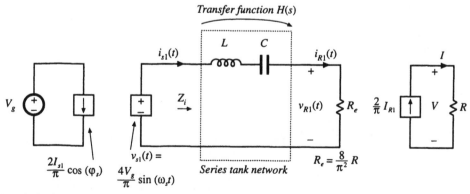

Fig. 19.15 Steady-state equivalent circuit of the series resonant converter.

$$\omega_0 = \frac{1}{\sqrt{LC}} = 2\pi f_0$$

$$R_0 = \sqrt{\frac{L}{C}}$$

$$Q_e = \frac{R_0}{R_e}$$

The magnitude of $H(j\omega_s)$, which coincides with the converter dc conversion ratio $M = V/V_g$, is

$$M = |H(j\omega_s)| = \frac{1}{\sqrt{1 + Q_e^2\left(\frac{1}{F} - F\right)^2}} \tag{19.18}$$

where

$$F = f_s/f_0 \tag{19.19}$$

The Bode diagrams of $Z_i(s)$ and $H(s)$ are constructed in Fig. 19.16, using the graphical construction method of Chapter 8. The series resonant impedance $Z_i(s)$ is dominated by the capacitor C at low frequency, and by the inductor L at high frequency. At the resonant frequency f_0, the impedances of the inductor and capacitor are equal in magnitude and opposite in phase; hence, they cancel. The series resonant impedance $Z_i(s)$ is equal to R_e at $f = f_0$.

The transfer function $\| H(j\omega) \|$ is constructed graphically, by division of R_e by the $\| Z_i \|$ asymptotes of Fig. 19.16. At resonance, one obtains $\| H \| = R_e/R_e = 1$. At frequencies above or below the resonant frequency, $\| Z_i \| > R_e$ and hence $\| H \| < 1$. So the conversion ratio M is less than or equal to 1. It can also be seen that a decrease in the load resistance R, which increases the effective quality factor Q_e, causes a more peaked response in the vicinity of resonance. Exact characteristics of the series resonant converter are plotted in Fig. 19.45.

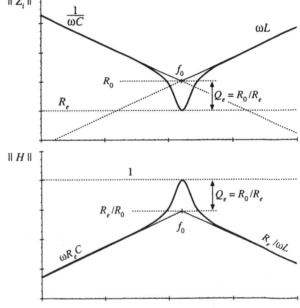

Fig. 19.16   Construction of the Bode diagrams of $Z_i(s)$ and $H(s)$ for the series resonant converter.

Over what range of switching frequencies is Eq. (19.18) accurate? The response of the tank to the fundamental component of $v_s(t)$ must be sufficiently greater than the response to the harmonics of $v_s(t)$. This is certainly true for operation above resonance because $H(s)$ contains a bandpass characteristic that decreases with a single pole slope for $f_s > f_0$. For the same reason, Eq. (19.18) is valid when the switching frequency is below but near resonance.

However, for switching frequencies f_s much less than the resonant frequency f_0, the sinusoidal approximation breaks down completely because the tank responds more strongly to the harmonics of

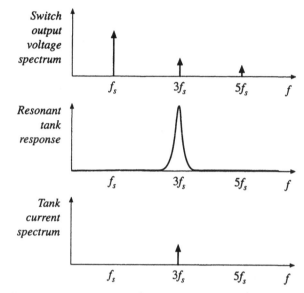

Fig. 19.17 Excitation of the tank network by the third harmonic of the switching frequency.

$v_s(t)$ than to its fundamental. For example, at $f_s = f_0/3$, the third harmonic of $v_s(t)$ is equal to f_0 and directly excites the tank resonance. Some other type of analysis must be used to understand what happens at these lower frequencies. Also, in the low-Q case, the approximation is less accurate because the filter response is less peaked, and hence does not favor the fundamental component as strongly. As shown in a later section, discontinuous conduction modes may then occur whose waveforms are highly nonsinusoidal.

19.2.2 Subharmonic Modes of the Series Resonant Converter

If the nth harmonic of the switch output waveform $v_s(t)$ is close to the resonant tank frequency, $nf_s \sim f_0$, and if the tank effective quality factor Q_e is sufficiently large, then as illustrated in Fig. 19.17, the tank responds primarily to harmonic n. All other components of the tank waveforms can then be neglected, and it is a good approximation to replace $v_s(t)$ with its nth harmonic component:

$$v_s(t) \approx v_{sn}(t) = \frac{4V_g}{n\pi} \sin(n\omega_s t) \tag{19.20}$$

This expression differs from Eq. (19.2) because the amplitude is reduced by a factor of $1/n$, and the fre-

Fig. 19.18 The subharmonic modes of the series resonant converter. These modes occur when the harmonics of the switching frequency excite the tank resonance.

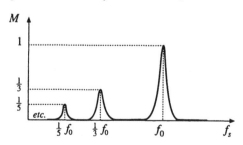

quency is nf_s rather than f_s.

The arguments used to model the tank and rectifier/filter networks are unchanged from Section 19.1. The rectifier presents an effective resistive load to the tank, of value $R_e = 8R/\pi^2$. In consequence, the converter dc conversion ratio is given by

$$M = \frac{V}{V_g} = \frac{\left|H(jn\omega_s)\right|}{n} \qquad (19.21)$$

This is a good approximation provided that nf_s is close to f_0, and that Q_e is sufficiently large. Typical characteristics are sketched in Fig. 19.18.

The series resonant converter is not generally designed to operate in a subharmonic mode, since the fundamental modes yield greater output voltage and power, and hence higher efficiency. Nonetheless, the system designer should be aware of their existence, because inadvertent operation in these modes can lead to large signal instabilities.

19.2.3 Parallel Resonant DC–DC Converter Example

The parallel resonant dc–dc converter is diagrammed in Fig. 19.19. It differs from the series resonant converter in two ways. First, the tank capacitor appears in parallel with the rectifier network rather than in series: this causes the tank transfer function $H(s)$ to have a different form. Second, the rectifier drives an inductive-input low-pass filter. In consequence, the value of the effective resistance R_e differs from that of the rectifier with a capacitive filter. Nonetheless, sinusoidal approximations can be used to understand the operation of the parallel resonant converter.

As in the series resonant converter, the switch network is controlled to produce a square wave $v_s(t)$. If the tank network responds primarily to the fundamental component of $v_s(t)$, then arguments identical to those of Section 19.1 can be used to model the output fundamental components and input dc components of the switch waveforms. The resulting equivalent circuit is identical to Fig. 19.9.

The uncontrolled rectifier with inductive filter network can be described using the dual of the arguments of Section 19.1.2. In the parallel resonant converter, the output rectifiers are driven by the nearly sinusoidal tank capacitor voltage $v_R(t)$, and the diode rectifiers switch when $v_R(t)$ passes through zero as in Fig. 19.20. If the filter inductor current ripple is small, then in steady-state the filter inductor current is essentially equal to the dc load current I. The rectifier input current $i_R(t)$ is therefore a square wave of amplitude I, and is in phase with the tank capacitor voltage $v_R(t)$:

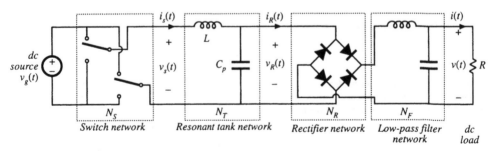

Fig. 19.19 Block diagram of the parallel resonant converter.

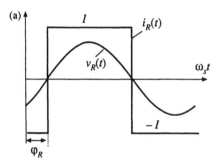
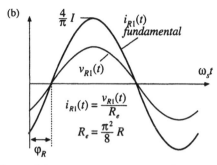

Fig. 19.20 Rectifier network input terminal waveforms, for the parallel resonant converter: (a) actual waveforms $v_R(t)$ and $i_R(t)$, (b) fundamental components $v_{R1}(t)$ and $i_{R1}(t)$.

$$i_R(t) = \frac{4I}{\pi} \sum_{n=1,3,5,\ldots}^{\infty} \frac{1}{n} \sin(n\omega_s t - \varphi_R) \qquad (19.22)$$

where φ_R is the phase shift of $v_R(t)$.

The fundamental component of $i_R(t)$ is

$$i_{R1}(t) = \frac{4I}{\pi} \sin(\omega_s t - \varphi_R) \qquad (19.23)$$

Hence, the rectifier again presents an effective resistive load to the tank circuit, equal to

$$R_e = \frac{v_{R1}(t)}{i_{R1}(t)} = \frac{\pi V_{R1}}{4I} \qquad (19.24)$$

The ac components of the rectified tank capacitor voltage $|v_R(t)|$ are removed by the output low pass filter. In steady state, the output voltage V is equal to the dc component of $|v_R(t)|$:

$$V = \frac{2}{T_s} \int_0^{T_s/2} V_{R1} \left| \sin(\omega_s t - \varphi_R) \right| dt = \frac{2}{\pi} V_{R1} \qquad (19.25)$$

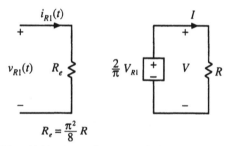

$$R_e = \frac{\pi^2}{8} R$$

Fig. 19.21 An equivalent circuit for the rectifier and inductive filter network of the parallel resonant converter, which models the fundamental components of the rectifier ac input waveforms and the dc components of the load waveforms.

So the load voltage V and the tank capacitor voltage amplitude are directly related in steady state. Substitution of Eq. (19.25) and resistive load characteristics $V = IR$ into Eq. (19.24) yields:

$$R_e = \frac{\pi^2}{8} R = 1.2337R \qquad (19.26)$$

An equivalent circuit for the uncontrolled rectifier with inductive filter network is given in Fig. 19.21. This model is similar to the one used for the series resonant converter, Fig. 19.12, except that the roles of the rectifier input voltage $v_R(t)$ and current $i_R(t)$ are interchanged, and the effective resistance R_e has a different value. The model for the complete converter is given in Fig. 19.22.

Solution of Fig. 19.22 yields the converter dc conversion ratio:

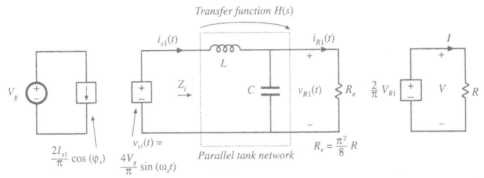

Fig. 19.22 Equivalent circuit for the parallel resonant converter, which models the fundamental components of the tank waveforms, and the dc components of the converter input current and output voltage.

$$M = \frac{V}{V_g} = \frac{8}{\pi^2} \left| H(s) \right|_{s = j\omega_s} \tag{19.27}$$

where $H(s)$ is the tank transfer function

$$H(s) = \frac{Z_o(s)}{sL} \tag{19.28}$$

and

$$Z_o(s) = sL \parallel \frac{1}{sC} \parallel R_e \tag{19.29}$$

The Bode magnitude diagrams of $H(s)$ and $Z_o(s)$ are constructed in Fig. 19.23, using the graphical construction method of Chapter 8. The impedance $Z_o(s)$ is the parallel combination of the impedances of the tank inductor L, capacitor C, and effective load R_e. The magnitude asymptote of the parallel combination of these components, at a given frequency, is equal to the smallest of the individual asymptotes ωL, $1/\omega C$, and R_e. Hence, at low frequency where the inductor impedance dominates the parallel combination, $\parallel Z_o(s) \parallel \cong \omega L$, while at high frequency the capacitor dominates and $\parallel Z_o(s) \parallel \cong 1/\omega C$. At resonance, the impedances of the inductor and capacitor are equal in magnitude but opposite in phase, so that their effects cancel. The impedance $\parallel Z_o(s) \parallel$ is then equal to R_e:

$$\left| Z_o(s) \right|_{s = j\omega_s} = \frac{1}{\dfrac{1}{j\omega_0 L} + j\omega_0 C + \dfrac{1}{R_e}} = R_e \tag{19.30}$$

with

$$\omega_0 L = \frac{1}{\omega_0 C} = R_0$$

The dc conversion ratio is therefore

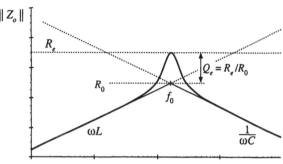

Fig. 19.23 Construction of Bode diagrams of $Z_i(s)$ and $H(s)$ for the parallel resonant converter.

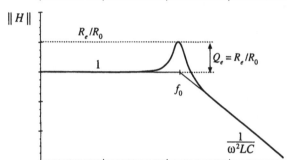

$$M = \frac{8}{\pi^2} \left| \frac{Z_o(s)}{sL} \right|_{s=j\omega_s} = \frac{8}{\pi^2} \left| \frac{1}{1 + \frac{s}{Q_e \omega_0} + \left(\frac{s}{\omega_0}\right)^2} \right|_{s=j\omega_s}$$

$$= \frac{8}{\pi^2} \frac{1}{\sqrt{\left(1 - F^2\right)^2 + \left(\frac{F}{Q_e}\right)^2}}$$

(19.31)

where $F = f_s/f_0$.

At resonance, the conversion ratio is

$$M = \frac{8}{\pi^2} \frac{R_e}{R_0} = \frac{R}{R_0}$$

(19.32)

The actual peak value of M occurs at a switching frequency slightly below the resonant frequency, with peak M slightly greater than Eq. (19.32). Provided that the load resistance R is greater than the tank characteristic impedance R_e, the parallel resonant converter can produce conversion ratios both greater than and less than one. In fact, the ideal parallel resonant converter can produce conversion ratios approaching infinity, provided that the output current is limited to values less than V_g/R_0. Of course, losses limit the maximum output voltage that can be produced by practical converters.

19.3   SOFT SWITCHING

As mentioned previously, the soft-switching phenomena known as zero-current switching (ZCS) and zero-voltage switching (ZVS) can lead to reduced switching loss. When the turn-on and/or turn-off tran-

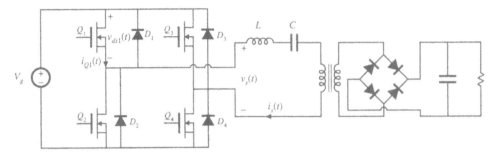

Fig. 19.24   A series resonant converter incorporating a full-bridge switch network.

sitions of a semiconductor switching device coincide with the zero crossings of the applied waveforms, some of the switching loss mechanisms discussed in Section 4.3 are eliminated. In converters containing MOSFETs and diodes, zero-voltage switching mitigates the switching loss otherwise caused by diode recovered charge and semiconductor output capacitance. Zero-current switching can mitigate the switching loss caused by current tailing in IGBTs and by stray inductances. Zero-current switching can also be used for commutation of SCRs. In the majority of applications, where diode recovered charge and semiconductor output capacitances are the dominant sources of PWM switching loss, zero-voltage switching is preferred.

19.3.1   Operation of the Full Bridge Below Resonance: Zero-Current Switching

When the series and parallel resonant inverters and dc–dc converters are operated below resonance, the zero-current switching phenomenon can occur, in which the circuit causes the transistor current to go to zero before the transistor is turned off. Let us consider the operation of the full bridge switch network of the series resonant converter in detail.

A full bridge circuit, realized using power MOSFETs and antiparallel diodes, is shown in Fig. 19.24. The switch output voltage $v_s(t)$, and its fundamental component $v_{s1}(t)$, as well as the approximately sinusoidal tank current waveform $i_s(t)$, are illustrated in Fig. 19.25. At frequencies less than the tank resonant frequency, the input impedance of the series resonant tank network $Z_i(s)$ is dominated by the tank capacitor impedance [see Fig. 19.16(a)]. Hence, the tank presents an effective capacitive load to the bridge, and switch current $i_s(t)$ leads the switch voltage fundamental component $v_{s1}(t)$, as shown in Fig. 19.25. In consequence, the zero crossing of the current waveform $i_s(t)$ occurs before the zero crossing of the voltage $v_s(t)$.

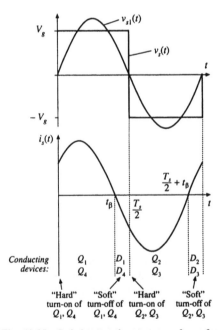

Fig. 19.25   Switch network output waveforms for the series resonant converter, operated below resonance in the $k = 1$ CCM. Zero-current switching aids the transistor turn-off process.

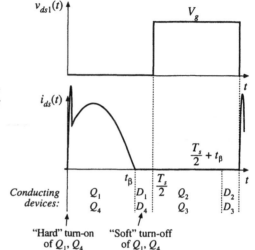

Fig. 19.26 Transistor Q_1 voltage and current waveforms, for operation of the series resonant converter below resonance in the $k = 1$ CCM.

For the half cycle $0 < t < T_s/2$, the switch voltage v_s is equal to $+ V_g$. For $0 < t < t_\beta$, the current $i_s(t)$ is positive and transistors Q_1 and Q_4 conduct. Diodes D_1 and D_4 conduct when $i_s(t)$ is negative, over the interval $t_\beta < t < T_s/2$. The situation during $T_s/2 < t < T_s$ is symmetrical. Since $i_{s1}(t)$ leads $v_{s1}(t)$, the transistors conduct before their respective antiparallel diodes. Note that, at any given time during the D_1 conduction interval $t_\beta < t < T_s/2$, transistor Q_1 can be turned off without incurring switching loss. The circuit naturally causes the transistor turn-off transition to be lossless, and long turn-off switching times can be tolerated.

In general, zero current switching can occur when the resonant tank presents an effective capacitive load to the switches, so that the switch current zero crossings occur before the switch voltage zero crossings. In the bridge configuration, zero current switching is characterized by the half-bridge conduction sequence Q_1–D_1–Q_2–D_2, such that the transistors are turned off while their respective antiparallel diodes conduct. It is possible, if desired, to replace the transistors with naturally commutated thyristors whenever the zero-current-switching property occurs at the turn-off transition.

The transistor turn-on transition in Fig. 19.26 is similar to that of a PWM switch: it is hard-switched and is not lossless. During the turn-on transition of Q_1, diode D_2 must turn off. Neither the transistor current nor the transistor voltage is zero, Q_1 passes through a period of high instantaneous power dissipation, and switching loss occurs. As in the PWM case, the reverse recovery current of diode D_2 flows through Q_1. This current spike can be the largest component of switching loss. In addition, the energy stored in the drain-to-source capacitances of Q_1 and Q_2 and in the depletion layer capacitance of D_1 is lost when Q_1 turns on. These turn-on transition switching loss mechanisms can be a major disadvantage of zero-current-switching schemes. Since zero-current switching does not address the switching loss mechanisms that dominate in MOSFET converters, improvements in efficiency are typically not observed.

19.3.2 Operation of the Full Bridge Above Resonance: Zero-Voltage Switching

When the series resonant converter is operated above resonance, the zero-voltage switching phenomenon can occur, in which the circuit causes the transistor voltage to become zero before the controller turns the

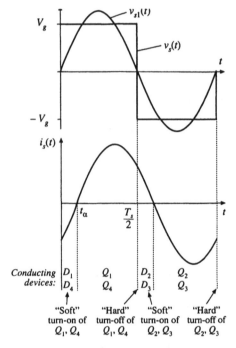

Fig. 19.27 Switch network output waveforms for the series resonant converter, operated above resonance in the continuous conduction mode. Zero-voltage switching aids the transistor turn-on process.

transistor on. With a minor circuit modification, the transistor turn-off transitions can also be caused to occur at zero voltage. This process can lead to significant reductions in the switching losses of converters based on MOSFETs and diodes.

For the full bridge circuit of Fig. 19.24, the switch output voltage $v_s(t)$, and its fundamental component $v_{s1}(t)$, as well as the approximately sinusoidal tank current waveform $i_s(t)$, are plotted in Fig. 19.27. At frequencies greater than the tank resonant frequency, the input impedance of the tank network $Z_i(s)$ is dominated by the tank inductor impedance. Hence, the tank presents an effective inductive load to the bridge, and the switch current $i_s(t)$ lags the switch voltage fundamental component $v_{s1}(t)$, as shown in Fig. 19.27. In consequence, the zero crossing of the voltage waveform $v_s(t)$ occurs before the current waveform $i_s(t)$.

For the half cycle $0 < t < T_s/2$, the switch voltage $v_s(t)$ is equal to $+ V_g$. For $0 < t < t_\alpha$, the current $i_s(t)$ is negative and diodes D_1 and D_4 conduct. Transistors Q_1 and Q_4 conduct when $i_s(t)$ is positive, over the interval $t_\alpha < t < T_s/2$. The waveforms during $T_s/2 < t < T_s$ are symmetrical. Since the zero crossing of $v_s(t)$ leads the zero crossing of $i_s(t)$, the transistors conduct after their respective antiparallel diodes. Note that, at any given time during the D_1 conduction interval $0 < t < t_\alpha$, transistor Q_1 can be turned on without incurring switching loss. The circuit naturally causes the transistor turn-on transition to be lossless, and long turn-on switching times can be tolerated. A particularly significant implication of this is that the switching loss associated with reverse recovery of the antiparallel diodes is avoided. Relatively slow diodes, such as the MOSFET body diodes, can be employed for realization of diodes D_1 to D_4. In addition, the output capacitances of transistors Q_1 to Q_4 and diodes D_1 to D_4 do not lead to switching loss.

In general, zero-voltage switching can occur when the resonant tank presents an effective inductive load to the switches, and hence the switch voltage zero crossings occur before the switch current zero crossings. In the bridge configuration, zero-voltage switching is characterized by the half-bridge conduction sequence D_1–Q_1–D_2–Q_2, such that the transistors are turned on while their respective antipar-

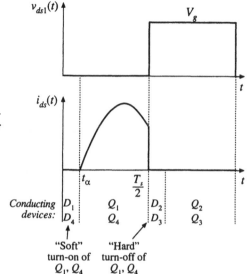

Fig. 19.28 Transistor Q_1 voltage and current waveforms, for operation of the series resonant converter above resonance in the $k = 0$ CCM.

allel diodes conduct. Since the transistor voltage is zero during the entire turn on transition, switching loss due to slow turn-on times or due to energy storage in any of the device capacitances does not occur at turn-on.

The transistor turn-off transition in Fig. 19.28 is similar to that of a PWM switch. In converters that employ IGBTs or other minority-carrier devices, significant switching loss may occur at the turn-off transitions. The current tailing phenomenon causes Q_1 to pass through a period of high instantaneous power dissipation, and switching loss occurs.

To assist the transistor turn off process, small capacitors C_{leg} may be introduced into the legs of the bridge, as demonstrated in Fig. 19.29. In a converter employing MOSFETs, the device output capacitances are sufficient for this purpose, with no need for external discrete capacitors. A delay is also introduced into the gate drive signals, so that there is a short commutation interval when all four transistors are off. During the normal Q_1, D_1, Q_2, and D_2 conduction intervals, the leg capacitors appear in parallel with the semiconductor switches, and have no effect on the converter operation. However, these capacitors introduce commutation intervals at transistor turn-off. When Q_1 is turned off, the tank current $i_s(T_s/2)$ flows through the switch capacitances C_{leg} instead of Q_1, and the voltage across Q_1 and C_{leg} increases. Eventually, the voltage across Q_1 reaches V_g; diode D_2 then becomes forward-biased. If the MOSFET turn-off time is sufficiently fast, then the MOSFET is switched fully off before the drain voltage rises significantly above zero, and negligible turn-off switching loss is incurred. The energy stored in the device capacitances, that is, in C_{leg}, is transferred to the tank inductor. The fact that none of the semiconductor device capacitances or stored charges lead to switching loss is the major advantage of zero-voltage switching, and is the most common motivation for its use. MOSFET converters can typically be operated in this manner, using only the internal drain-to-source capacitances. However, other devices such as IGBTs typically require substantial external capacitances to reduce the losses incurred during the IGBT turn-off transitions.

An additional advantage of zero-voltage switching is the reduction of EMI associated with device capacitances. In conventional PWM converters and also, to some extent, in zero-current switching converters, significant high-frequency ringing and current spikes are generated by the rapid charging and discharging of the semiconductor device capacitances during the turn-on and/or turn-off transitions.

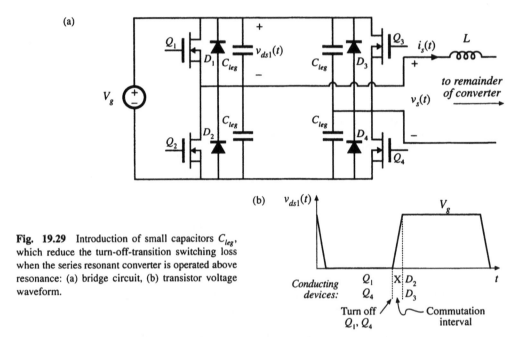

Fig. 19.29 Introduction of small capacitors C_{leg}, which reduce the turn-off-transition switching loss when the series resonant converter is operated above resonance: (a) bridge circuit, (b) transistor voltage waveform.

Ringing is conspicuously absent from the waveforms of converters in which all semiconductor devices switch at zero voltage; these converters inherently do not generate this type of EMI.

19.4   LOAD-DEPENDENT PROPERTIES OF RESONANT CONVERTERS

The properties of the CCM PWM converters studied in previous chapters are largely unaffected by the load current. In consequence, these converters exhibit several desirable properties that are often taken for granted. The transistor current is proportional to the load current; hence conduction losses become small at light load, leading to good light-load efficiency. Also, the output impedance is low, and hence the dc output voltage does not significantly depend on the load i–v characteristic (at least, in CCM). Unfortunately, these good properties are not necessarily shared by resonant converters. Of central importance in design of a resonant converter is the selection of the resonant tank topology and element values, so that the transistor conduction losses at light load are minimized, so that zero-voltage switching is obtained over a wide range of load currents (preferably, for all anticipated loads, but at least at full and intermediate load powers), and so that the converter dynamic range is compatible with the load i–v characteristic. These design issues are addressed in this section.

The conduction loss caused by circulating tank currents is well-recognized as a problem in resonant converter design. These currents are independent of, or only weakly dependent on, the load current, and lead to poor efficiency at light load. In Fig. 19.30, the switch current $i_s(s)$ is equal to $v_s(s)/Z_i(s)$. If we want the switch current to track the load current, then at the switching frequency $\| Z_i \|$ should be dominated by, or at least strongly influenced by, the load resistance R. Unfortunately, this is often not consistent with the requirement for zero-voltage switching, in which Z_i is dominated by a tank inductor.

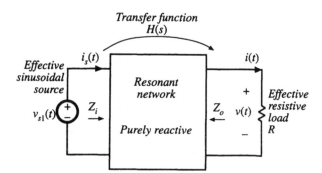

Fig. 19.30   Resonant inverter model.

To design a resonant converter that exhibits good properties, the engineer must develop physical insight into how the load resistance R affects the tank input impedance and output voltage.

In this section, the inverter output characteristics, zero-voltage switching boundary, and the dependence of transistor current on load resistance, are related to the properties of the tank network under the extreme conditions of an open-circuited or short-circuited load. The undamped tank network responses are easily plotted, and the insight needed to optimize the tank network design can be gained quickly.

19.4.1   Inverter Output Characteristics

Let us first investigate how the magnitude of the inverter output voltage $\| v \|$ depends on the load current magnitude $\| i \|$. Consider the resonant inverter system of Fig. 19.30. Let $H_\infty(s)$ be the open-circuit $(R \to \infty)$ transfer function of the tank network:

$$H_\infty(s) = \left. \frac{v(s)}{v_{s1}(s)} \right|_{R \to \infty} \tag{19.33}$$

and let $Z_{o0}(s)$ be the output impedance, determined when the source $v_{s1}(s)$ is short-circuited. Then we can model the output port of the tank network using the Thevenin-equivalent circuit of Fig. 19.31. Solution of this circuit using the voltage divider formula leads to

$$v(s) = H_\infty(s) v_{s1}(s) \frac{R}{R + Z_{o0}(s)} \tag{19.34}$$

At a given angular switching frequency $\omega_s = 2\pi f_s$, the phasor representing the magnitude and phase of the ac output voltage is found by letting $s = j\omega_s$:

$$v(j\omega_s) = H_\infty(j\omega_s) v_{s1}(j\omega_s) \frac{R}{R + Z_{o0}(j\omega_s)} \tag{19.35}$$

The magnitude can be found by noting that

$$\left| v(j\omega_s) \right|^2 = v(j\omega_s) v^*(j\omega_s) \tag{19.36}$$

where $v^*(j\omega_s)$ is the complex conjugate of $v(j\omega_s)$. Substitution of Eq. (19.35) into Eq. (19.36) leads to

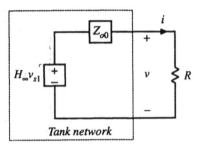

Fig. 19.31   Thevenin-equivalent circuit that models the output port of the tank network.

$$\left|v(j\omega_s)\right|^2 = \left(H_\infty(j\omega_s)v_{s1}(j\omega_s)\,\frac{R}{R+Z_{o0}(j\omega_s)}\right)\left(H_\infty(j\omega_s)v_{s1}(j\omega_s)\,\frac{R}{R+Z_{o0}(j\omega_s)}\right)^*$$

$$= H_\infty(j\omega_s)H_\infty^*(j\omega_s)v_{s1}(j\omega_s)v_{s1}^*(j\omega_s)\,\frac{R^2}{\left(R+Z_{o0}(j\omega_s)\right)\left(R+Z_{o0}(j\omega_s)\right)^*} \qquad (19.37)$$

$$= \left|H_\infty(j\omega_s)\right|^2\left|v_{s1}(j\omega_s)\right|^2\,\frac{R^2}{\left(R+Z_{o0}(j\omega_s)\right)\left(R+Z_{o0}(j\omega_s)\right)^*}$$

This result can be further simplified with the assumption that the tank network contains only purely reactive elements, i.e., that any losses or other resistive elements within the tank network have negligible effect. Then the output impedance $Z_{o0}(j\omega_s)$, as well as all other driving-point impedances of the tank network, are purely imaginary quantities. This implies that the complex conjugate $Z_{o0}^*(j\omega_s)$ is given by

$$Z_{o0}^*(j\omega_s) = -Z_{o0}(j\omega_s) \qquad (19.38)$$

Substitution of Eq. (19.38) into Eq. (19.37) and simplification leads to

$$\left|v(j\omega_s)\right|^2 = \frac{\left|H_\infty(j\omega_s)\right|^2\left|v_s(j\omega_s)\right|^2}{\left(1+\dfrac{\left|Z_{o0}(j\omega_s)\right|^2}{R^2}\right)} \qquad (19.39)$$

with

$$R = \frac{\left|v(j\omega_s)\right|}{\left|i(j\omega_s)\right|} \qquad (19.40)$$

Substitution of Eq. (19.40) into Eq. (19.39) and rearrangement of terms yields

$$\left|v(j\omega_s)\right|^2 + \left|i(j\omega_s)\right|^2\left|Z_{o0}(j\omega_s)\right|^2 = \left|H_\infty(j\omega_s)\right|^2\left|v_s(j\omega_s)\right|^2 \qquad (19.41)$$

Hence, at a given frequency, the inverter output characteristic, that is, the relationship between $\parallel v(j\omega_s)\parallel$ and $\parallel i(j\omega_s)\parallel$, is elliptical. Equation (19.41) can be further rearranged, into the form

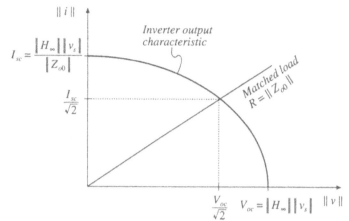

Fig. 19.32 Elliptical output characteristics of resonant inverters. A resistive matched load is also illustrated.

$$\frac{\left|v(j\omega_s)\right|^2}{V_{oc}^2} + \frac{\left|i(j\omega_s)\right|^2}{I_{sc}^2} = 1 \tag{19.42}$$

where the open-circuit voltage V_{oc} and short-circuit current I_{sc} are given by

$$V_{oc} = \left|H_\infty(j\omega_s)\right|\left|v_s(j\omega_s)\right|$$
$$I_{sc} = \frac{\left|H_\infty(j\omega_s)\right|\left|v_s(j\omega_s)\right|}{\left|Z_{o0}(j\omega_s)\right|} = \frac{V_{oc}}{\left|Z_{o0}(j\omega_s)\right|} \tag{19.43}$$

These inverter output characteristics are constructed in Fig. 19.32. This characteristic describes how, at a given switching frequency, the ac output voltage magnitude varies as the circuit is loaded. The equilibrium output voltage is given by the intersection of this elliptical characteristic with the load i–v characteristic. For example, Fig. 19.32 also illustrates a superimposed resistive load line having slope $1/R$, in the special case where $R = \| Z_{o0}(j\omega_s) \|$. This value of R corresponds to matched load operation, in which the converter output power is maximized. It can be shown that the operating point is then given by

$$\left|v(j\omega_s)\right|^2 = \frac{V_{oc}}{\sqrt{2}}$$
$$\left|i(j\omega_s)\right|^2 = \frac{I_{sc}}{\sqrt{2}} \tag{19.44}$$

Note that Fig. 19.32 can also be applied to the output i–v characteristics of resonant dc–dc converters, since the output rectifier then loads the tank with an effective resistive load R_e.

19.4.2 Dependence of Transistor Current on Load

The transistors must conduct the current appearing at the input port of the tank network, $i_s(t)$. This current is determined by the tank network input impedance $Z_i(j\omega_s)$:

(a) 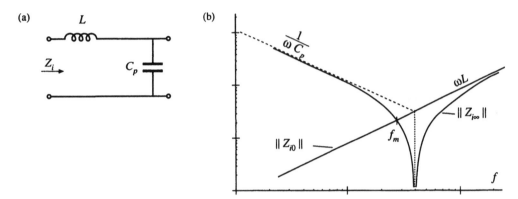 (b)

Fig. 19.33 Tank network; parallel resonant converter example: (a) tank circuit, (b) Bode plot of input impedance magnitude $\| Z_i \|$ for the limiting cases $R \to 0$ and $R \to \infty$.

$$i_{s1}(j\omega_s) = \frac{v_{s1}(j\omega_s)}{Z_i(j\omega_s)} \tag{19.45}$$

As described previously, obtaining good light-load efficiency requires that $\| Z_i(j\omega_s) \|$ increase as the load resistance R increases. To understand how $\| Z_i(j\omega_s) \|$ depends on R, let us sketch $\| Z_i(j\omega_s) \|$ in the extreme cases of an open-circuited ($R \to \infty$) and short-circuited ($R \to 0$) load:

$$Z_{i0}(j\omega_s) = Z_i(j\omega_s)\big|_{R \to 0}$$
$$Z_{i\infty}(j\omega_s) = Z_i(j\omega_s)\big|_{R \to \infty} \tag{19.46}$$

For example, consider the parallel resonant converter of Figs. 19.19 to 19.23. The Bode diagrams of the impedances $\| Z_{i0}(j\omega_s) \|$ and $\| Z_{i\infty}(j\omega_s) \|$ are constructed in Fig. 19.33. $Z_{i0}(s)$ is found with the load R shorted, and is equal to the inductor impedance sL. $Z_{i\infty}(s)$, found with the load R open-circuited, is given by the series combination $(sL + 1/sC)$. It can be seen in Fig. 19.33 that the impedance magnitudes $\| Z_{i0}(j\omega_s) \|$ and $\| Z_{i\infty}(j\omega_s) \|$ intersect at frequency f_m. If the switching frequency is chosen such that $f_s < f_m$, then $\| Z_{i\infty}(j\omega_s) \| > \| Z_{i0}(j\omega_s) \|$. The converter then exhibits the desirable characteristic that the no-load switch current magnitude $\| v_s(j\omega_s) \| / \| Z_{i\infty}(j\omega_s) \|$ is smaller than the switch current under short-circuit conditions, $\| v_s(j\omega_s) \| / \| Z_{i0}(j\omega_s) \|$. In fact, the short-circuit switch current is limited by the impedance of the tank inductor, while the open-circuit switch current is determined primarily by the impedance of the tank capacitor.

If the switching frequency is chosen such that $f_s > f_m$, then $\| Z_{i\infty}(j\omega_s) \| < \| Z_{i0}(j\omega_s) \|$. The no-load switch current is then greater in magnitude than the switch current when the load is short-circuited! When the load current is reduced or removed, the transistors will continue to conduct large currents and generate high conduction losses. This causes the efficiency at light load to be poor. It can be concluded that, to obtain good light-load efficiency in the parallel resonant converter, one should choose f_s sufficiently less than f_m. Unfortunately, this requires operation below resonance, leading to reduced output voltage dynamic range and a tendency to lose the zero-voltage switching property.

A remaining question is how $\| Z_i(j\omega_s) \|$ behaves for intermediate values of load between the open-circuit and short-circuit conditions. The answer is given by Theorem 1 below: $\| Z_i(j\omega_s) \|$ varies monotonically with R, and therefore is bounded by $\| Z_{i0}(j\omega_s) \|$ and $\| Z_{i\infty}(j\omega_s) \|$. Hence, the Bode plots of

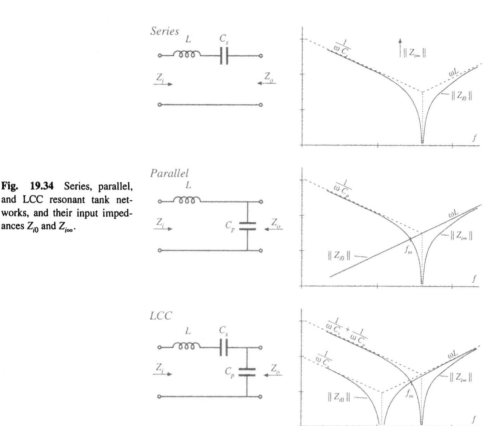

Fig. 19.34 Series, parallel, and LCC resonant tank networks, and their input impedances Z_{i0} and $Z_{i\infty}$.

the limiting cases $\| Z_{i0}(j\omega_s) \|$ and $\| Z_{i\infty}(j\omega_s) \|$ provide a correct qualitative understanding of the behavior of $\| Z_i \|$ for all R. The theorem is valid for lossless tank networks.

Theorem 1: If the tank network is purely reactive, then its input impedance $\| Z_i \|$ is a monotonic function of the load resistance R.

This theorem is proven by use of Middlebrook's Extra Element Theorem (see Appendix C). The tank network input impedance $Z_i(s)$ can be expressed as a function of the load resistance R and the tank network driving-point impedances, as follows:

$$Z_i(s) = Z_{i0}(s) \frac{\left(1 + \dfrac{R}{Z_{o0}(s)} \right)}{\left(1 + \dfrac{R}{Z_{o\infty}(s)} \right)} = Z_{i\infty}(s) \frac{\left(1 + \dfrac{Z_{o0}(s)}{R} \right)}{\left(1 + \dfrac{Z_{o\infty}(s)}{R} \right)} \tag{19.47}$$

where Z_{i0} and $Z_{i\infty}$ are the resonant network input impedances, with the load short-circuited or open-circuited, respectively, and Z_{o0} and $Z_{o\infty}$ are the resonant network output impedances, with the source input short-circuited or open-circuited, respectively. These terminal impedances are simple functions of the tank elements, and their Bode diagrams are easily constructed. The input impedances of the series resonant, parallel resonant, and LCC inverters are listed in Fig. 19.34. Since these impedances do not depend on the load, they are purely reactive, ideally have zero real parts [38], and their complex conjugates are given by $Z_{o0}^* = -Z_{o0}$, $Z_{o\infty}^* = -Z_{o\infty}$, etc. Again, recall that the magnitude of a complex impedance $Z(j\omega)$ can be expressed as the square root of $Z(j\omega)Z^*(j\omega)$. Hence, the magnitude of $Z_i(s)$ is given by

$$|Z_i|^2 = Z_iZ_i^* = Z_{i0}(s)Z_{i0}^*(s)\frac{\left(1+\dfrac{R}{Z_{o0}(s)}\right)\left(1+\dfrac{R}{Z_{o0}^*(s)}\right)}{\left(1+\dfrac{R}{Z_{o\infty}(s)}\right)\left(1+\dfrac{R}{Z_{o\infty}^*(s)}\right)}$$

$$= |Z_{i0}|^2\frac{\left(1+\dfrac{R^2}{|Z_{o0}|^2}\right)}{\left(1+\dfrac{R^2}{|Z_{o\infty}|^2}\right)}$$

(19.48)

where Z_i^* is the complex conjugate of Z_i.
Next, let us differentiate Eq. (19.48) with respect to R:

$$\frac{d|Z_i|^2}{dR} = 2R|Z_{i0}|^2\frac{\left(\dfrac{1}{|Z_{o0}|^2}-\dfrac{1}{|Z_{o\infty}|^2}\right)}{\left(1+\dfrac{R^2}{|Z_{o\infty}|^2}\right)^2}$$

(19.49)

The derivative has roots at (*i*) $R = 0$, (*ii*) $R = \infty$, and in the special case (*iii*) where $\|Z_{i0}\| = \|Z_{i\infty}\|$. Since the derivative is otherwise nonzero, the resonant network input impedance $\|Z_i\|$ is a monotonic function of R, over the range $0 < R < \infty$. In special case (*iii*), $\|Z_i\|$ is independent of R. Therefore, Theorem 1 is proved.

An example is given in Figs. 19.36 and 19.35, for the LCC inverter. Figure 19.35 illustrates the impedance asymptotes of the limiting cases $\|Z_{i0}\|$ and $\|Z_{i\infty}\|$. Variation of $\|Z_i\|$ between these limits, for finite nonzero R, is illustrated in Fig. 19.36. The open-circuit resonant frequency f_∞ and the short-circuit resonant frequency f_0 are given by

$$f_0 = \frac{1}{2\pi\sqrt{LC_s}}$$
$$f_\infty = \frac{1}{2\pi\sqrt{LC_s\|C_p}}$$

(19.50)

where $C_s\|C_p$ denotes inverse addition of C_s and C_p:

$$C_s\|C_p = \frac{1}{\dfrac{1}{C_s}+\dfrac{1}{C_p}}$$

(19.51)

For the LCC inverter, the impedance magnitudes $\|Z_{i0}\|$ and $\|Z_{i\infty}\|$ are equal at frequency f_m, given by

$$f_m = \frac{1}{2\pi\sqrt{LC_s\|2C_p}}$$

(19.52)

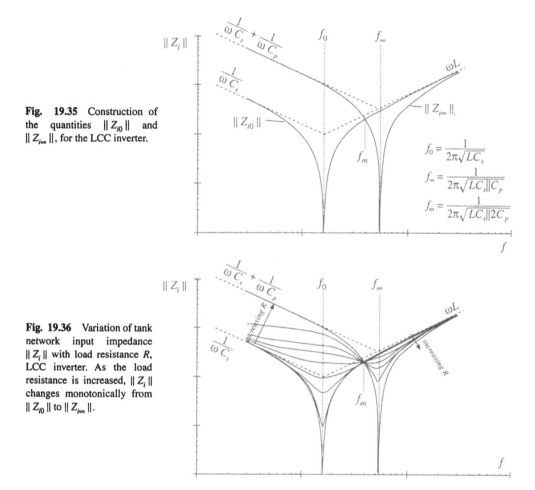

Fig. 19.35 Construction of the quantities $\| Z_{i0} \|$ and $\| Z_{i\infty} \|$, for the LCC inverter.

$$f_0 = \frac{1}{2\pi\sqrt{LC_s}}$$

$$f_\infty = \frac{1}{2\pi\sqrt{LC_s\|C_p}}$$

$$f_m = \frac{1}{2\pi\sqrt{LC_s\|2C_p}}$$

Fig. 19.36 Variation of tank network input impedance $\| Z_i \|$ with load resistance R, LCC inverter. As the load resistance is increased, $\| Z_i \|$ changes monotonically from $\| Z_{i0} \|$ to $\| Z_{i\infty} \|$.

If the switching frequency is chosen to be greater than f_m, then $\| Z_{i\infty} \|$ is less than $\| Z_{i0} \|$. This implies that, as the load current is decreased, the transistor current will increase. Such a converter will have poor efficiency at light load, and will exhibit significant circulating currents. If the switching frequency is chosen to be less than f_m, then the transistor current will increase with decrease with decreasing load current. The short-circuit current is limited by $\| Z_{i0} \|$, while the circulating currents under open-circuit conditions are determined by $\| Z_{i\infty} \|$. In general, if $f > f_m$, then the transistor current is greater than or equal to the short-circuit current for all R. The inequality is reversed when $f < f_m$.

The impedance magnitudes $\| Z_{i0} \|$ and $\| Z_{i\infty} \|$ are illustrated in Fig. 19.34 for the series, parallel, and LCC tank networks. In the case of the series tank network, $\| Z_{i\infty} \| = \infty$. In consequence, the no-load transistor current is zero, both above resonance and below resonance. Hence, the series resonant inverter exhibits the desirable property that the transistor current is proportional to the load current. In addition, when the load is short-circuited, the current magnitude is limited by the impedance of the series resonant tank. For the parallel and LCC inverters, it is desirable to operate below the frequency f_m.

Thus, the dependence of the transistor current on load can be easily determined, using an intuitive frequency-domain approach.

19.4.3   Dependence of the ZVS/ZCS Boundary on Load Resistance

It is also necessary to determine the critical load resistance $R = R_{crit}$ at the boundary between ZVS and ZCS. This boundary can also be expressed as a function of the impedances Z_{i0} and $Z_{i\infty}$.

As discussed in Section 19.3, zero-voltage switching occurs when the switch current $i_s(t)$ lags the switch voltage $v_s(t)$. Zero-voltage switching occurs when $i_s(t)$ leads $v_s(t)$. This definition ignores the effects of semiconductor output capacitances, and hence gives an approximate ZVS/ZCS boundary. The phase between the switch current and switch voltage is again determined by the input impedance of the tank network:

$$i_{s1}(j\omega_s) = \frac{v_{s1}(j\omega_s)}{Z_i(j\omega_s)} \tag{19.53}$$

Hence, zero-voltage switching occurs when $Z_i(j\omega_s)$ is inductive in nature, zero-current switching occurs when $Z_i(j\omega_s)$ is capacitive in nature, and the ZVS/ZCS boundary occurs where $Z_i(j\omega_s)$ has zero phase.

It is instructive to again consider the limiting cases of a short-circuited and open-circuited load. The Bode plots of $Z_{i0}(j\omega_s)$ and $Z_{i\infty}(j\omega_s)$ for an LCC inverter example are sketched in Fig. 19.37. Since, in these limiting cases, the input impedance Z_i is composed only of the reactive tank elements, $Z_{i0}(j\omega_s)$ and $Z_{i\infty}(j\omega_s)$ are purely imaginary quantities having phase of either $-90°$ or $+90°$. For $f_s < f_0$, both $Z_{i0}(j\omega_s)$ and $Z_{i\infty}(j\omega_s)$ are dominated by the tank capacitor or capacitors; the phase of $Z_i(j\omega_s)$ is therefore $-90°$. Hence, zero-current switching is obtained under both short-circuit and open-circuit conditions. For $f_s > f_\infty$, both $Z_{i0}(j\omega_s)$ and $Z_{i\infty}(j\omega_s)$ are dominated by the tank inductor; hence the phase of $Z_i(j\omega_s)$ is $+90°$. Zero-voltage switching is obtained for both a short-circuited and an open-circuited load. For $f_0 < f_s < f_\infty$, $Z_{i0}(j\omega_s)$ is dominated by the tank inductor while $Z_{i\infty}(j\omega_s)$ is dominated by the tank capacitors. This implies that zero-voltage switching is obtained under short-circuit conditions, and zero-voltage switching is obtained under open-circuit conditions. For this case, there must be some critical value of load resistance $R = R_{crit}$ that represents the boundary between ZVS and ZCS, and that causes the phase of $Z_i(j\omega_s)$ to be equal to $0°$.

The behavior of $Z_i(j\omega_s)$ for nonzero finite R is easily extrapolated from the limiting cases dis-

Fig. 19.37 Use of the input impedance quantities Z_{i0} and $Z_{i\infty}$ to determine the ZCS/ZVS boundaries, LCC example.

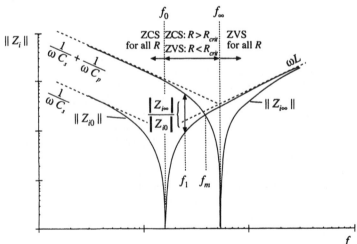

cussed above. Theorem 2 below shows that:

1. If zero-current switching occurs for both an open-circuited load and a short-circuited load [i.e., $Z_{i0}(j\omega_s)$ and $Z_{i\infty}(j\omega_s)$ both have phase $+ 90°$], then zero-current switching occurs for all loads.

2. If zero-voltage switching occurs for both an open-circuited load and a short-circuited load [i.e., $Z_{i0}(j\omega_s)$ and $Z_{i\infty}(j\omega_s)$ both have phase $- 90°$], then zero-voltage switching occurs for all loads.

3. If zero-voltage switching occurs for an open-circuited load and zero-current switching occurs for a short-circuited load [i.e., $Z_{i0}(j\omega_s)$ has phase $- 90°$ and $Z_{i\infty}(j\omega_s)$ has phase $+ 90°$], then zero-voltage switching occurs for $R > R_{crit}$, and zero-current switching occurs for $R < R_{crit}$, with R_{crit} given by Eq. (19.54) below.

4. If zero-current switching occurs for an open-circuited load and zero-voltage switching occurs for a short-circuited load [i.e., $Z_{i0}(j\omega_s)$ has phase $+ 90°$ and $Z_{i\infty}(j\omega_s)$ has phase $- 90°$], then zero-current switching occurs for $R > R_{crit}$, and zero-voltage switching occurs for $R < R_{crit}$, with R_{crit} given by Eq. (19.54) below.

For the LCC example, we can therefore conclude that, for $f_s < f_0$, zero-current switching occurs for all values of R. For $f_s > f_\infty$, zero-voltage switching occurs for all values of R. For $f_0 < f_s < f_\infty$, the boundary between ZVS and ZCS is given by Eq. (19.54).

Theorem 2: If the tank network is purely reactive, then the boundary between zero-current switching and zero-voltage switching occurs when the load resistance R is equal to the critical value R_{crit}, given by

$$R_{crit} = |Z_{o0}| \sqrt{\frac{-Z_{i\infty}}{Z_{i0}}} \tag{19.54}$$

This theorem relies on the assumption that zero-current switching occurs when the tank input impedance is capacitive in nature, while zero-voltage switching occurs for inductive input impedances. The boundary therefore occurs where the phase of $Z_i(j\omega)$ is zero. This definition gives a necessary but not sufficient condition for zero-voltage switching when significant semiconductor output capacitance is present.

The result is derived by finding the value of R which causes the imaginary part of $Z_i(j\omega)$ in Eq. (19.47) to be zero. Since the tank network is assumed to ideal and lossless, the impedances $Z_{o\infty}$, Z_{o0}, and $Z_{i\infty}$ must have zero real parts. Hence,

$$\mathrm{Im}\left(Z_i(R_{crit})\right) = \mathrm{Im}\left(Z_{i\infty}\right) \mathrm{Re}\left(\frac{1 + \dfrac{Z_{o0}}{R_{crit}}}{1 + \dfrac{Z_{o\infty}}{R_{crit}}}\right) = \mathrm{Im}\left(Z_{i\infty}\right) \frac{\left(1 - \dfrac{Z_{o0} Z_{o\infty}}{R_{crit}^2}\right)}{\left(1 + \dfrac{|Z_{o\infty}|^2}{R_{crit}^2}\right)} = 0 \tag{19.55}$$

where $\mathrm{Im}(Z)$ and $\mathrm{Re}(Z)$ denote the imaginary and real parts of the complex quantity Z. The nontrivial solution to Eq. (19.55) is given by

$$1 = \frac{Z_{o0} Z_{o\infty}}{R_{crit}^2} \tag{19.56}$$

hence,

$$R_{crit} = \sqrt{Z_{o0} Z_{o\infty}} \tag{19.57}$$

A useful equivalent form makes use of the reciprocity identities

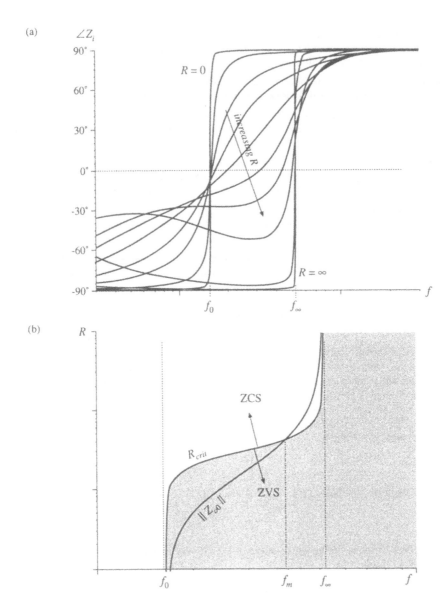

Fig. 19.38   ZCS/ZVS boundary, LCC inverter example: (a) variation of tank network input impedance phase shift with load resistance, (b) Comparison of R_{crit} with matched-load impedance $\| Z_{o0} \|$.

$$\frac{Z_{o0}}{Z_{o\infty}} = \frac{Z_{i0}}{Z_{i\infty}} \qquad (19.58)$$

Use of Eq. (19.58) to eliminate $Z_{o\infty}$ from Eq. (19.57) leads to

$$R_{crit} = |Z_{o0}| \sqrt{\frac{-Z_{i\infty}}{Z_{i0}}} \tag{19.59}$$

This is the desired result. The quantity Z_{o0} is the inverter output impedance, and $R = \| Z_{o0} \|$ corresponds to operation at matched load with maximum output power. The impedances $Z_{i\infty}$ and Z_{i0} are purely imaginary, and hence Eq. (19.59) has no real solution unless $Z_{i\infty}$ and Z_{i0} are of opposite phase. As illustrated in Fig. 19.37, if at a given frequency $Z_{i\infty}$ and Z_{i0} are both inductive, then zero-voltage switching occurs for all loads. Zero-current switching occurs for all loads when $Z_{i\infty}$ and Z_{i0} are both capacitive. Therefore, Theorem 2 is proved.

Figure 19.38(a) illustrates the phase response of $Z_i(j\omega)$ as R varies from 0 to ∞, for the LCC inverter. A typical dependence of R_{crit} and the matched-load impedance $\| Z_{o0} \|$ on frequency is illustrated in Fig. 19.38(b). Zero-voltage switching occurs for all loads when $f > f_\infty$, and zero-current switching occurs for all loads when $f < f_0$. Over the range $f_0 < f < f_\infty$, Z_{i0} is inductive while $Z_{i\infty}$ is capacitive; hence, zero-voltage switching occurs for $R < R_{crit}$ while zero-current switching occurs for $R > R_{crit}$. At frequency f_m, $R_{crit} = \| Z_{o0} \|$, and hence the ZVS/ZCS boundary is encountered exactly at matched load. It is commonly desired to obtain zero-voltage switching at matched load, with low circulating currents and good efficiency at light load. It is apparent that this requires operation in the range $f_0 < f < f_m$. Zero-voltage switching will then be obtained under matched-load and short-circuit conditions, but will be lost at light load. The choice of element values such that $\| Z_{i0} \| \ll \| Z_{i\infty} \|$ is advantageous in that the range of loads leading to zero-voltage switching is maximized.

19.4.4 Another Example

As a final example, let us consider selection of the resonant tank elements to obtain a given output characteristic at a certain switching frequency, and let's evaluate the effect of this choice on R_{crit}. It is desired to operate a resonant inverter at switching frequency $f_s = 100$ kHz, with an input voltage of $V_g = 160$ V. The converter should be capable of producing an open-circuit peak output voltage $V_{oc} = 400$ V, and should also produce a nominal output of 150 Vrms at 25 W. It is desired to select resonant tank elements that accomplish this.

The specifications imply that the converter should exhibit an open-circuit transfer function of

$$\left| H_\infty(j\omega_s) \right| = \frac{V_{oc}}{V_{s1}} = \frac{(400 \text{ V})}{\left(\frac{4}{\pi} 160 \text{ V} \right)} = 1.96 \tag{19.60}$$

The required short-circuit current is found by solving Eq. (19.42) for I_{sc}:

$$I_{sc} = \frac{I}{\sqrt{1 - \left(\dfrac{V}{V_{oc}} \right)^2}} \tag{19.61}$$

The specifications also imply that the peak voltage and current at the nominal operating point are

$$V = 150 \sqrt{2} = 212 \text{ V}$$
$$I = \frac{P}{V_{rms}} \sqrt{2} = \frac{25 \text{ W}}{150 \text{ V}} \sqrt{2} = 0.236 \text{ A} \tag{19.62}$$
$$R_{nom} = \frac{V}{I} = 900 \ \Omega$$

Substitution of Eq. (19.62) into Eq. (19.61) yields

$$I_{sc} = \frac{(0.236 \text{ A})}{\sqrt{1 - \left(\frac{212 \text{ V}}{400 \text{ V}}\right)^2}} = 0.278 \text{ A} \tag{19.63}$$

Matched load therefore occurs at the operating point

$$V_{mat} = \frac{V_{oc}}{\sqrt{2}} = 283 \text{ V}$$
$$I_{mat} = \frac{I_{sc}}{\sqrt{2}} = 0.196 \text{ A} \tag{19.64}$$
$$\left| Z_{o0}(j\omega_s) \right| = \frac{V_{oc}}{I_{sc}} = 1439 \ \Omega$$

Let us select the values of the tank elements in the LCC tank network illustrated in Fig. 19.39(a). The impedances of the series and parallel branches can be represented using the reactances X_s and X_p illustrated in Fig. 19.39(b), with

$$jX_s = j\omega_s L + \frac{1}{j\omega_s C_s} = j\left(\omega_s L - \frac{1}{\omega_s C_s}\right)$$
$$jX_p = \frac{1}{j\omega_s C_p} = j\left(-\frac{1}{\omega_s C_p}\right) \tag{19.65}$$

The transfer function $H_\infty(j\omega_s)$ is given by the voltage divider formula

$$H_\infty(j\omega_s) = \frac{jX_p}{jX_s + jX_p} \tag{19.66}$$

The output impedance $Z_{o0}(j\omega_s)$ is given by the parallel combination

Fig. 19.39 Tank network of the LCC inverter example: (a) schematic, (b) representation of series and parallel branches by reactances X_s and X_p.

$$Z_{o0}(j\omega_s) = jX_s \| jX_p = \frac{-X_s X_p}{j(X_s + X_p)}$$ (19.67)

Solution of Eqs. (19.66) and (19.67) for X_p and X_s leads to

$$jX_p = \frac{Z_{o0}(j\omega_s)}{1 - H_\infty(j\omega_s)}$$

$$X_s = X_p \frac{1 - H_\infty(j\omega_s)}{H_\infty(j\omega_s)}$$ (19.68)

Hence, the capacitance C_p should be chosen equal to

$$X_p = -1499\ \Omega$$

$$C_p = -\frac{1}{\omega_s X_p} = \frac{H_\infty(j\omega_s) - 1}{\omega_s |Z_{o0}(j\omega_s)|} = \frac{(1.96) - 1}{(2\pi\ 100\ \text{kHz})(1439\ \Omega)} \cong 1\ \text{nF}$$ (19.69)

and the reactance of the series branch should be chosen according to

$$X_s = X_p \frac{1 - H_\infty(j\omega_s)}{H_\infty(j\omega_s)} = (-1493\ \Omega)\frac{1 - (1.96)}{(1.96)} = 733\ \Omega$$ (19.70)

Since X_s is comprised of the series combination of the inductor L and capacitor C_s, there is a degree of freedom in choosing the values of L and capacitor C_s to realize X_s. For example, we could choose C_s very large (tending to a short circuit); this effectively would result in a parallel resonant converter with $L = X_s/\omega_s = 1.17$ mH. For nonzero C_s, L must be chosen according to

$$L = \frac{1}{\omega_s}\left(X_s + \frac{1}{\omega_s C_s}\right)$$ (19.71)

For example, the choice $C_s = C_p = 1.06$ nF leads to $L = 3.5$ mH. Designs using different C_s will exhibit exactly the same characteristics at the design frequency; however, the behavior at other switching frequencies will differ.

For the tank network illustrated in Fig. 19.39, the value of R_{crit} is completely determined by the parameters of the output characteristic ellipse; i.e., by the specification of V_g, V_{oc} and I_{sc}. Note that $Z_{o\infty}$, the tank output impedance with the tank input port open-circuited, is equal to jX_p. Substitution of expressions for $Z_{o\infty}$ and Z_{o0} into Eq. (19.57) leads to the following expression for R_{crit}:

$$R_{crit} = \sqrt{\frac{Z_{o0}^2(j\omega_s)}{1 - H_\infty(j\omega_s)}}$$ (19.72)

Since Z_{o0} and H_∞ are determined by the operating point specifications, then R_{crit} is also. Evaluation of Eq. (19.72) for this example leads to $R_{crit} = 1466\ \Omega$. Therefore, the inverter will operate with zero-voltage switching for $R < 1466\ \Omega$, including at the nominal operating point $R = 900\ \Omega$. Other topologies of tank network, more complex than the circuit illustrated in Fig. 19.39(b), may have additional degrees of freedom that allow R_{crit} to be independently chosen.

The choice $C_s = 3C_p = 3.2$ nF leads to $L = 1.96\ \mu$H. The following frequencies are obtained:

$$f_\infty = 127 \text{ kHz}$$
$$f_m = 100.6 \text{ kHz}$$
$$f_s = 100.0 \text{ kHz}$$
$$f_0 = 64 \text{ kHz}$$

(19.73)

Regardless of how C_s is chosen, the open-circuit tank input impedance is

$$Z_{i\infty} = j\left(X_s + X_p\right) = j\left(733\ \Omega + (-1493\ \Omega)\right) = -j760\ \Omega$$

(19.74)

Therefore, when the load is open-circuited, the transistor peak current has magnitude

$$I_{s1} = \frac{V_{s1}}{\left|Z_{i\infty}\right|} = \frac{\frac{4}{\pi}\left(160\text{ V}\right)}{760\ \Omega} = 0.268 \text{ A}$$

(19.75)

When the load is short-circuited, the transistor peak current has magnitude

$$I_{s1} = \frac{V_{s1}}{\left|Z_{i0}\right|} = \frac{V_{s1}}{\left|X_s\right|} = \frac{\frac{4}{\pi}\left(160\text{ V}\right)}{\left(733\ \Omega\right)} = 0.278 \text{ A}$$

(19.76)

which is nearly the same as the result in Eq. (19.75). The somewhat large open-circuit switch current occurs because of the relatively-high specified open-circuit output voltage; lower values of V_{oc} would reduce the result in Eq. (19.75).

19.5   EXACT CHARACTERISTICS OF THE SERIES AND PARALLEL RESONANT CONVERTERS

The exact steady-state behavior of resonant converters can be determined via methods such as state-plane analysis. A detailed analysis of resonant dc–dc converters is beyond the scope of this book. However, the exact steady-state characteristics of ideal series [1, 13–20] and parallel [6, 22–25] resonant dc–dc converters (Fig. 19.40) are summarized in this section. Small-signal ac modeling has also been described in the literature; several relevant papers are [27–30].

19.5.1   Series Resonant Converter

At a given switching frequency, the series resonant dc–dc converter can operate in one continuous conduction mode, and possibly in several discontinuous conduction modes. The mode index k is defined as the integer that satisfies

$$\frac{f_0}{k+1} < f_s < \frac{f_0}{k} \quad \text{or} \quad \frac{1}{k+1} < F < \frac{1}{k}$$

(19.77)

where $F = f_s/f_0$ is the normalized switching frequency. The subharmonic number ξ is defined as

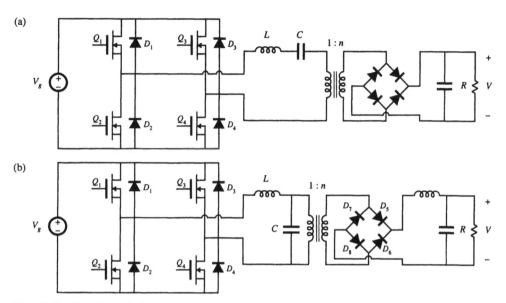

Fig. 19.40   Transformer-isolated resonant dc–dc converters: (a) series resonant converter, (b) parallel resonant converter.

$$\xi = k + \frac{1 + (-1)^k}{2} \tag{19.78}$$

Values of k and ξ as functions of f_s are summarized in Fig. 19.41(a). The subharmonic number ξ denotes the dominant harmonic that excites the tank resonance. When the converter is heavily loaded, it operates in type k continuous conduction mode. As the load is reduced (i.e., as the load resistance R is increased), the converter enters the type k discontinuous conduction mode. Further reducing the load causes the converter to enter the type $(k - 1)$ DCM, type $(k - 2)$ DCM,..., type 1 DCM. There is no type 0 DCM, and hence when the converter operates above resonance, only the type 0 continuous conduction mode is possible.

In the type k continuous conduction mode, the series resonant converter exhibits elliptical output characteristics, given by

$$M^2 \xi^2 \sin^2\left(\frac{\gamma}{2}\right) + \frac{1}{\xi^2}\left(\frac{J\gamma}{2} + (-1)^k\right)^2 \cos^2\left(\frac{\gamma}{2}\right) = 1 \tag{19.79}$$

For the transformer-isolated converters of Fig. 19.40, M and J are related to the load voltage V and load current I according to

$$M = \frac{V}{nV_g} \qquad J = \frac{InR_0}{V_g} \tag{19.80}$$

Again, R_0 is the tank characteristic impedance, referred to the transformer primary side. The quantity γ is the angular length of one-half of the switching period:

(a)

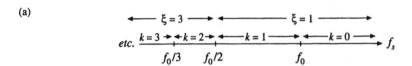

(b)

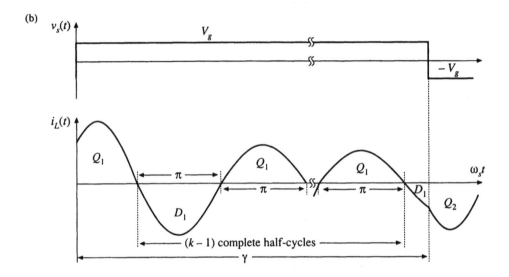

(c)

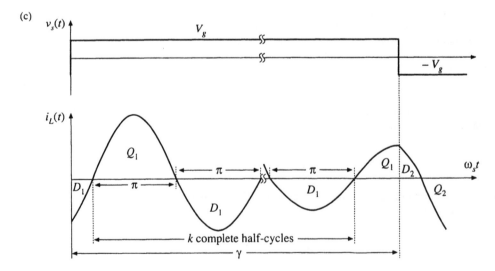

Fig. 19.41 Continuous conduction modes of the series resonant converter: (a) switching frequency ranges over which various mode indices k and subharmonic numbers ξ occur; (b) tank inductor current waveform, type k CCM, for odd k; (c) tank inductor current waveform, type k CCM, for even k.

$$\gamma = \frac{\omega_0 T_s}{2} = \frac{\pi}{F} \qquad (19.81)$$

Equation (19.79) is valid only for k satisfying Eq. (19.77). It predicts that the voltage conversion ratio M is restricted to the range

$$0 \le M \le \frac{1}{\xi} \qquad (19.82)$$

This is consistent with Eq. (19.21).

Typical CCM tank current waveforms are illustrated in Fig. 19.41. When k is even, the tank inductor current is initially negative. In consequence, the switch network antiparallel diodes conduct first, for a fraction of a half resonant cycle. If k is odd, then each half switching period is initiated by conduction of the switch network transistors. In either case, this is followed by $(\xi - 1)$ complete tank half-cycles of ringing. The half-switching period is then concluded by a subinterval shorter than one complete resonant half-cycle, in which the device that did not initially conduct is on. The next half switching period then begins, and is symmetrical.

The steady-state control-plane characteristic can be found for a resistive load R obeying $V = IR$, by substitution of the normalized relation $J = MQ$ into Eq. (19.79), where $Q = n^2 R_0/R$. Use of the quadratic formula and some algebraic manipulations allows solution for M, as a function of load (via Q) and switching frequency (via γ):

$$M = \frac{\left(\frac{Q\gamma}{2}\right)}{\xi^4 \tan^2\left(\frac{\gamma}{2}\right) + \left(\frac{Q\gamma}{2}\right)^2}\left[(-1)^{k+1} + \sqrt{1 + \frac{\left[\xi^2 - \cos^2\left(\frac{\gamma}{2}\right)\right]\left[\xi^4\tan^2\left(\frac{\gamma}{2}\right) + \left(\frac{Q\gamma}{2}\right)^2\right]}{\left(\frac{Q\gamma}{2}\right)^2 \cos^2\left(\frac{\gamma}{2}\right)}}\right] \qquad (19.83)$$

This is the closed-form relationship between the conversion ratio M and the switching frequency, for a resistive load. It is valid for any continuous conduction mode k.

The type k discontinuous conduction modes, for k odd, occur over the frequency range

$$f_s < \frac{f_0}{k} \qquad (19.84)$$

In these modes, the output voltage is independent of both load current and switching frequency, and is described by

$$M = \frac{1}{k} \qquad (19.85)$$

The type k discontinuous conduction mode, for odd k, occurs over the range of load currents given by

$$\frac{2(k+1)}{\gamma} > J > \frac{2(k-1)}{\gamma} \qquad (19.86)$$

In the odd discontinuous conduction modes, the tank current rings for k complete resonant half cycles. All four output bridge rectifier diodes then become reverse-biased, and the tank current remains at zero until the next switching half-period begins, as illustrated in Fig. 19.42. Series resonant converters are not

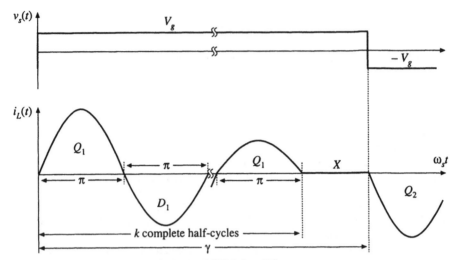

Fig. 19.42   Tank inductor current waveform, type k DCM, for odd k.

normally purposely designed to operate in odd discontinuous conduction modes, because the output voltage is not controllable. Nonetheless, when the load is removed with $f_s < f_0$, the series resonant converter operates in $k = 1$ DCM with $M = 1$.

The type k discontinuous conduction mode, for k even, also occurs over the frequency range

$$f_s < \frac{f_0}{k} \qquad (19.87)$$

Even discontinuous conduction modes exhibit current source characteristics, in which the load current is a function of switching frequency and input voltage, but not of the load voltage. The output relationship is:

$$J = \frac{2k}{\gamma} \qquad (19.88)$$

Operation in this mode occurs for

$$\frac{1}{k-1} > M > \frac{1}{k+1} \qquad (19.89)$$

In the even discontinuous conduction modes, the tank current rings for k complete resonant half-cycles during each switching half-period. All four output bridge then become reverse-biased, and the tank current remains at zero until the next switching half-period is initiated. Tank current waveforms are illustrated in Fig. 19.43 for even DCM.

The series resonant converter possesses some unusual properties when operated in an even discontinuous conduction mode. A dc equivalent circuit is given in Fig. 19.44, consisting of a gyrator with gyration conductance $g = 2k/gn^2R_0$. The gyrator has the property of transforming circuits into their dual networks; in the typical dc–dc converter application, the input voltage source V_g is effectively transformed into its dual, an output current source of value gV_g. Series resonant converters have been purposely designed to operate in the $k = 2$ DCM, at power levels of several tens of kW.

The complete control plane characteristics can now be plotted using Eqs. (19.77) to (19.89).

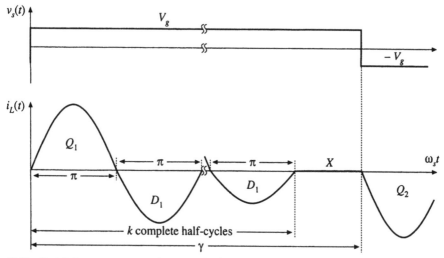

Fig. 19.43   Tank inductor current waveform, type k DCM, for even k.

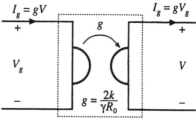

Fig. 19.44   Steady-state equivalent circuit model for an even discontinuous conduction mode: an effective gyrator. The converter exhibits current source characteristics.

The result is shown in Fig. 19.45, and the mode boundaries are explicitly diagrammed in Fig. 19.46. It can be seen that, for operation above resonance, the only possible operating mode is the $k = 0$ CCM, and that the output voltage decreases monotonically with increasing switching frequency. Reduction in load current (or increase in load resistance, which decreases Q) causes the output voltage to increase. A number of successful designs that operate above resonance and utilize zero-voltage switching have been documented in the literature [7,21].

Operation below resonance is complicated by the presence of subharmonic and discontinuous conduction modes. The $k = 1$ CCM and $k = 2$ DCM are well behaved, in that the output voltage increases monotonically with increasing switching frequency. Increase of the load current again causes the output voltage to decrease. Successful designs that operate in these modes and employ zero-current switching are numerous. However, operation in the higher-order modes ($k = 2$ CCM, $k = 4$ DCM, etc.) is normally avoided.

Given F and Q, the operating mode can be evaluated directly, using the following algorithm. First, the continuous conduction mode k corresponding to operation at frequency F with heavy loading is found:

$$k = \text{INT}\left(\frac{1}{F}\right) \tag{19.90}$$

where $\text{INT}(x)$ denotes the integer part of x. Next, the quantity k_1 is determined:

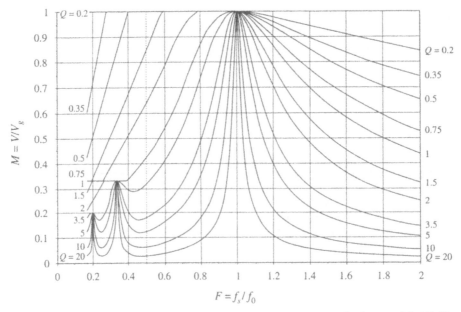

Fig. 19.45   Complete control plane characteristics of the series resonant converter, for the range $0.2 \leq F \leq 2$.

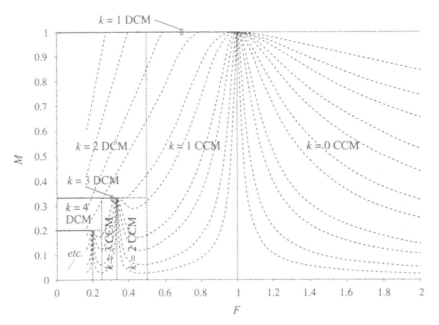

Fig. 19.46   Continuous and discontinuous conduction mode boundaries.

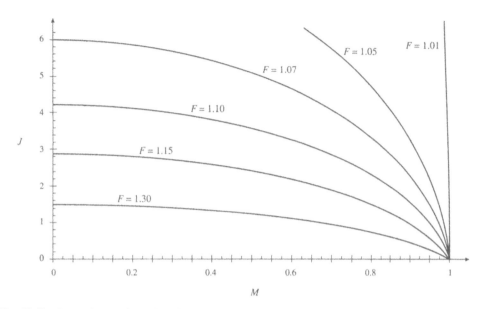

Fig. 19.47   Output characteristics, $k = 0$ CCM (above resonance).

$$k_1 = \text{INT}\left(\frac{1}{2} + \sqrt{\frac{1}{4} + \frac{Q\pi}{2F}}\right) \tag{19.91}$$

The converter operates in type k CCM provided that:

$$k_1 > k \tag{19.92}$$

Otherwise, the converter operates in type k_1 DCM. A simple algorithm can therefore be defined, in which the conversion ratio M is computed for a given F and Q. First, Eqs. (19.90) to (19.92) are evaluated, to determine the operating mode. Then, the appropriate equation (19.83), (19.85), or (19.88) is evaluated to find M.

Output $I–V$ plane characteristics for the $k = 0$ CCM, plotted using Eq. (19.79), are shown in Fig. 19.47. The constant-frequency curves are elliptical, and all pass through the point $M = 1$, $J = 0$. For a given switching frequency, the operating point is given by the intersection of the elliptical converter output characteristic with the load $I–V$ characteristic.

Output plane characteristics that combine the $k = 1$ CCM, $k = 1$ DCM, and $k = 2$ DCM are shown in Fig. 19.48. These were plotted using Eqs. (19.79), (19.85), and (19.88). These curves were plotted with the assumption that the transistors are allowed to conduct no longer than one tank half-cycle during each switching half-period; this eliminates subharmonic modes and causes the converter to operate in $k = 2$ or $k = 1$ DCM whenever $f_s < 0.5 f_0$. It can be seen that the constant-frequency curves are elliptical in the continuous conduction mode, vertical (voltage source characteristic) in the $k = 1$ DCM, and horizontal (current source characteristic) in the $k = 2$ DCM.

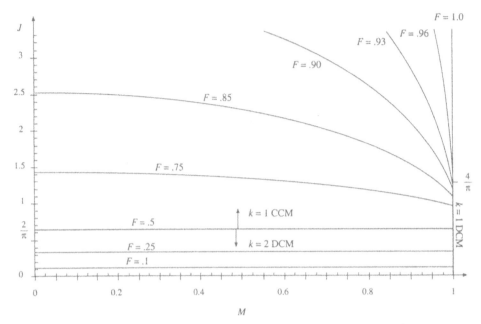

Fig. 19.48   Output characteristics, $k = 1$ CCM, $k = 1$ DCM, and $k = 2$ DCM (below resonance).

19.5.2   Parallel Resonant Converter

For operation in the frequency range $0.5f_0 < f_s < \infty$, the parallel resonant dc–dc converter exhibits one continuous conduction mode and one discontinuous conduction mode. Typical CCM switch voltage $v_s(t)$, tank inductor current $i_L(t)$, and tank capacitor voltage $v_C(t)$ waveforms are illustrated in Fig. 19.49. The CCM converter output characteristics are given by

$$M = \left(\frac{2}{\gamma}\right)\left(\varphi - \frac{\sin(\varphi)}{\cos\left(\frac{\gamma}{2}\right)}\right) \tag{19.93}$$

$$\varphi = \begin{cases} -\cos^{-1}\left(\cos\left(\frac{\gamma}{2}\right) + J\sin\left(\frac{\gamma}{2}\right)\right) & \text{for } 0 < \gamma < \pi \ \text{(above resonance)} \\ +\cos^{-1}\left(\cos\left(\frac{\gamma}{2}\right) + J\sin\left(\frac{\gamma}{2}\right)\right) & \text{for } \pi < \gamma < 2\pi \ \text{(below resonance)} \end{cases} \tag{19.94}$$

and where M, J, and γ are again defined as in Eqs. (19.80) and (19.81). Given the normalized load current J and the half-switching-period-angle γ, one can evaluate Eq. (19.94) to find φ, and then evaluate Eq. (19.93) to find the converter voltage conversion ratio M. In other words, the output voltage can be found for a given load current and switching frequency, without need for computer iteration.

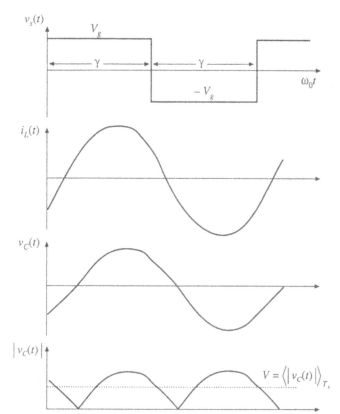

Fig. 19.49 Typical waveforms of the parallel resonant converter, operating in the continuous conduction mode.

A discontinuous conduction mode mechanism occurs in the parallel resonant converter which is the dual of the discontinuous conduction mode mechanism of the series resonant converter. In this mode, a discontinuous subinterval occurs in which all four output bridge rectifier diodes are forward-biased, and the tank capacitor voltage remains at zero. This mode occurs both above and below resonance when the converter is heavily loaded. Typical DCM tank capacitor voltage and inductor current waveforms are illustrated in Fig. 19.50. The condition for operation in the discontinuous conduction mode is

$$
\begin{aligned}
J > J_{crit}(\gamma) & \quad \text{for DCM} \\
J < J_{crit}(\gamma) & \quad \text{for CCM}
\end{aligned}
\tag{19.95}
$$

where

$$
J_{crit}(\gamma) = -\tfrac{1}{2} \sin(\gamma) + \sqrt{\sin^2\!\left(\tfrac{\gamma}{2}\right) + \tfrac{1}{4}\sin^2(\gamma)}
\tag{19.96}
$$

The discontinuous conduction mode is described by the following set of equations:

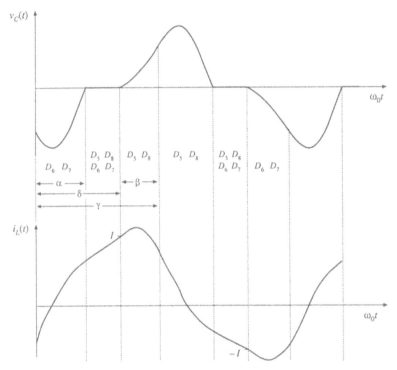

Fig. 19.50   Typical waveforms of the parallel resonant converter, operating in the discontinuous conduction mode.

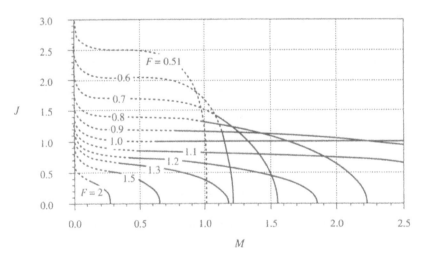

Fig. 19.51   Exact output characteristics of the parallel resonant converter, for $F > 0.5$. Solid curves: CCM, dashed curves: DCM.

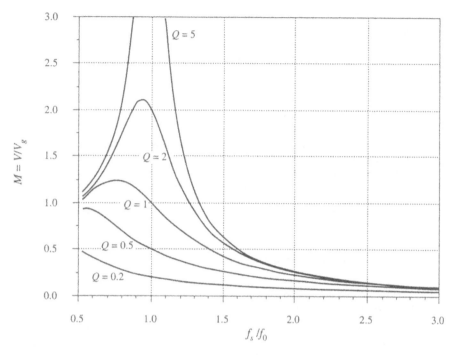

Fig. 19.52   Exact control characteristics of the parallel resonant converter, with a resistive load. Both CCM and DCM operation is included, for $0.5 \le F \le 3$.

$$M_{C0} = 1 - \cos(\beta)$$
$$J_{L0} = J + \sin(\beta)$$
$$\cos(\alpha + \beta) - 2\cos(\alpha) = -1$$
$$-\sin(\alpha + \beta) + 2\sin(\alpha) + (\delta - \alpha) = 2J \qquad (19.97)$$
$$\beta + \delta = \gamma$$
$$M = 1 + \left(\frac{2}{\gamma}\right)(J - \delta)$$

Unfortunately, the solution to this set of equations is not known in closed form, because of the mixture of linear and trigonometric terms. In consequence, the equations must be solved iteratively. For a given γ and J, a computer is used to iteratively find the angles α, β, and δ. M is then evaluated, and the output plane characteristics can be plotted. The result is given in Fig. 19.51. The dashed lines are the DCM solutions, and the solid lines are the valid CCM solutions. Figure 19.51 describes the complete dc behavior of the ideal parallel resonant converter for all switching frequencies above $0.5f_0$. For given values of normalized switching frequency $F = f_s/f_0 = \pi/\gamma$, the relationship between the normalized output current J and the normalized output voltage M is approximately elliptical. At resonance ($F = 1$), the CCM ellipse degenerates to the horizontal line $J = 1$, and the converter exhibits current source characteristics. Above resonance, the converter can both step up the voltage ($M > 1$) and step down the voltage ($M < 1$). The normalized load current is then restricted to $J < 1$, corresponding to $I < V_g/nR_0$. For a given switching frequency greater than the resonant frequency, the actual limit on maximum load current is even more restrictive than this limit. Below resonance, the converter can also step up and step down the voltage. Normalized load currents J greater than one are also obtainable, depending on M and F. However, no solutions occur when M and J are simultaneously large.

In Fig. 19.52, the control plane characteristics are plotted for a resistive load. The parameter Q is defined for the parallel resonant converter as $Q = R/n^2 R_0$. The normalized load current is then given by $J = M/Q$.

19.6   SUMMARY OF KEY POINTS

1.   The sinusoidal approximation allows a great deal of insight to be gained into the operation of resonant inverters and dc–dc converters. The voltage conversion ratio of dc–dc resonant converters can be directly related to the tank network transfer function. Other important converter properties, such as the output characteristics, dependence (or lack thereof) of transistor current on load current, and zero-voltage- and zero-current-switching transitions, can also be understood using this approximation. The approximation is accurate provided that the effective Q factor is sufficiently large, and provided that the switching frequency is sufficiently close to resonance.

2.   Simple equivalent circuits are derived, which represent the fundamental components of the tank network waveforms, and the dc components of the dc terminal waveforms.

3.   Exact solutions of the ideal dc–dc series and parallel resonant converters are listed here as well. These solutions correctly predict the conversion ratios, for operation not only in the fundamental continuous conduction mode, but in discontinuous and subharmonic modes as well.

4.   Zero-voltage switching mitigates the switching loss caused by diode recovered charge and semiconductor device output capacitances. When the objective is to minimize switching loss and EMI, it is preferable to operate each MOSFET and diode with zero-voltage switching.

5.   Zero-current switching leads to natural commutation of SCRs, and can also mitigate the switching loss due to current tailing in IGBTs.

6.   The input impedance magnitude $\| Z_i \|$, and hence also the transistor current magnitude, are monotonic functions of the load resistance R. The dependence of the transistor conduction loss on the load current can be easily understood by simply plotting $\| Z_i \|$ in the limiting cases as $R \to \infty$ and as $R \to 0$, or $\| Z_{i\infty} \|$ and $\| Z_{i0} \|$.

7.   The ZVS/ZCS boundary is also a simple function of $Z_{i\infty}$ and Z_{i0}. If ZVS occurs at open-circuit and at short-circuit, then ZVS occurs for all loads. If ZVS occurs at short-circuit, and ZCS occurs at open-circuit, then ZVS is obtained at matched load provided that $\| Z_{i\infty} \| > \| Z_{i0} \|$.

8.   The output characteristics of all resonant inverters considered here are elliptical, and are described completely by the open-circuit transfer function magnitude $\| H_\infty \|$, and the output impedance $\| Z_{o0} \|$. These quantities can be chosen to match the output characteristics to the application requirements.

REFERENCES

[1]   F.C. SCHWARZ, "An Improved Method of Resonant Current Pulse Modulation for Power Converters," *IEEE Power Electronics Specialists Conference*, 1975 Record, pp. 194–204, June 1975.

[2]   R.L. STEIGERWALD, "High Frequency Resonant Transistor Dc-Dc Converters," *IEEE Transactions on Industrial Electronics*, Vol. 31, No. 2, pp. 181–191, May 1984.

[3]   M. COSBY and R. NELMS, "Designing a Parallel-Loaded Resonant Inverter for an Electronic Ballast Using

the Fundamental Approximation," *IEEE Applied Power Electronics Conference*, 1993 Record, pp. 413–423.

[4]   M. GULKO and S. BEN-YAAKOV, "Current-Sourcing Push-Pull Parallel-Resonance Inverter (CS-PPRI): Theory and Application as a Fluorescent Lamp Driver," *IEEE Applied Power Electronics Conference*, 1993 Record, pp. 411–417.

[5]   Y. CHERON, H. FOCH, and J. SALESSES, "Study of a Resonant Converter Using Power Transistors in a 25kW X-ray Tube Power Supply," *IEEE Power Electronics Specialists Conference*, Proceedings ESA Sessions, pp. 295–306, June 1985.

[6]   S. D. JOHNSON, A. F. WITULSKI, and R. W. ERICKSON, "A Comparison of Resonant Topologies in High Voltage Applications," *IEEE Transactions on Aerospace and Electronic Systems*, Vol. 24, No. 3, pp. 263–274, July 1988.

[7]   Y. MURAI and T. A. LIPO, "High Frequency Series Resonant Dc Link Power Conversion," *IEEE Industry Applications Society Annual Meeting*, 1988 Record, pp. 648–656.

[8]   F. C. SCHWARZ, "A Doublesided Cyclo-Converter," *IEEE Power Electronics Specialists Conference*, 1979 Record, pp. 437–447.

[9]   D. DIVAN, "The Resonant Dc Link Converter: A New Concept in Static Power Conversion," *IEEE Industry Applications Society Annual Meeting*, 1986 Record, pp. 648–656.

[10]   R.L. STEIGERWALD, "A Comparison of Half-Bridge Resonant Converter Topologies," *IEEE Applied Power Electronics Conference*, 1987 Record, pp. 135–144.

[11]   R. SEVERNS, "Topologies for Three Element Resonant Converters," *IEEE Applied Power Electronics Conference*, 1990 Record, pp. 712–722.

[12]   M. KAZIMIERCZUK, W. SZARANIEC, and S. WANG, "Analysis and Design of Parallel Resonant Converter at High Q_L," *IEEE Transactions on Aerospace and Electronic Systems*, Vol. 28, pp. 35–50. January 1992.

[13]   R. KING and T. STUART, "A Normalized Model for the Half Bridge Series Resonant Converter," *IEEE Transactions on Aerospace and Electronic Systems*, March 1981, pp. 180–193.

[14]   V. VORPERIAN and S. ĆUK, "A Complete DC Analysis of the Series Resonant Converter," *IEEE Power Electronics Specialists Conference*, 1982 Record, pp. 85–100, June 1982.

[15]   R. KING and T.A. STUART, "Inherent Overload Protection for the Series Resonant Converter," *IEEE Transactions on Aerospace and Electronic Systems*, Vol. 19, No. 6, pp. 820–830, Nov. 1983.

[16]   A. WITULSKI and R. ERICKSON, "Steady-State Analysis of the Series Resonant Converter," *IEEE Transactions on Aerospace and Electronic Systems*, Vol. 21, No. 6, pp. 791–799, Nov. 1985.

[17]   A. WITULSKI and R. ERICKSON, "Design of the Series Resonant Converter for Minimum Component Stress," *IEEE Transactions on Aerospace and Electronic Systems*, Vol. 22, No. 4, July 1986, pp. 356–363.

[18]   R. ORUGANTI and F.C. LEE, "Resonant Power Processors, Part 1: State Plane Analysis," *IEEE Transactions on Industry Applications*, Vol. 21, Nov./Dec. 1985, pp. 1453–1460.

[19]   C.Q. LEE and K. SIRI, "Analysis and Design of Series Resonant Converter by State Plane Diagram," *IEEE*

Transactions on Aerospace and Electronic Systems, Vol. 22, No. 6, pp. 757–763, November 1986.

[20]   S. TRABERT and R. ERICKSON, "Steady-State Analysis of the Duty Cycle Controlled Series Resonant Converter," *IEEE Power Electronics Specialists Conference*, 1987 Record, pp. 545–556.

[21]   K.D.T. NGO, "Analysis of a Series Resonant Converter Pulsewidth-Modulated of Current-Controlled for Low Switching Loss," *IEEE Power Electronics Specialists Conference*, 1987 Record, pp. 527–536, June 1987.

[22]   R. ORUGANTI and F.C. LEE, "State Plane Analysis of the Parallel Resonant Converter," *IEEE Power Electronics Specialists Conference*, 1985 Record, pp. 56–73, June 1985.

[23]   S. JOHNSON, "Steady-State Analysis and Design of the Parallel Resonant Converter," M.S. Thesis, University of Colorado, Boulder, 1986.

[24]   S. JOHNSON and R. ERICKSON, "Steady-State Analysis and Design of the Parallel Resonant Converter," *IEEE Transactions on Power Electronics*, Vol. 3, No. 4, pp. 93–104, Jan. 1988.

[25]   A. BHAT and M. SWAMY, "Analysis and Design of a High-Frequency Parallel Resonant Converter Operating Above Resonance," *IEEE Transactions on Aerospace and Electronic Systems*, Vol. 25, No. 4, July 1989, pp. 449–458.

[26]   F.S. TSAI, P. MATERU, and F.C. LEE, "Constant Frequency, Clamped Mode Resonant Converters," *IEEE Power Electronics Specialists Conference*, 1987 Record, pp. 557–566, June 1987.

[27]   V. VORPERIAN and S. ĆUK, "Small-Signal Analysis of Resonant Converters," *IEEE Power Electronics Specialists Conference*, 1983 Record, pp. 269–282, June 1983.

[28]   V. VORPERIAN, "High-Q Approximation in the Small-Signal Analysis of Resonant Converters," *IEEE Power Electronics Specialists Conference*, 1985 Record, pp. 707–715.

[29]   R. KING and T. STUART, "Small-Signal Model for the Series Resonant Converter," *IEEE Transactions on Aerospace and Electronic Systems*, May 1985, Vol. 21, No. 3, pp. 301–319.

[30]   A. WITULSKI, A. HERNANDEZ, and R. ERICKSON, "Small-Signal Equivalent Circuit Modeling of Resonant Converters," *IEEE Transactions on Power Electronics*, January 1991.

[31]   R. FISHER, K. NGO, and M. KUO, "A 500 kHz, 250 W Dc–dc Converter with Multiple Outputs Controlled by Phase-Shifted PWM and Magnetic Amplifiers," *Proceedings of High Frequency Power Conversion Conference*, pp. 100–110, May 1988.

[32]   L. MWEENE, C. WRIGHT, and M. SCHLECHT, "A 1 kW, 500 kHz Front-End Converter for a Distributed Power Supply System," *IEEE Applied Power Electronics Conference*, 1989 Record, pp. 423–432.

[33]   R. REDL, L. BELOGH, and D. EDWARDS, "Optimum ZVS Full-Bridge DC/DC Converter with PWM Phase-Shift Control: Analysis, Design Considerations, and Experimental Results," *IEEE Applied Power Electronics Conference*, 1994 Record, pp. 159–165.

[34]   J. G. CHO, J. A. SABATE, and F. C. LEE, "Novel Full Bridge Zero-Voltage-Transition PWM DC/DC Converter for High Power Applications," *IEEE Applied Power Electronics Conference*, 1994 Record, pp. 143–149.

[35]   O. D. PATTERSON and D. M. DIVAN, "Pseudo-Resonant Full Bridge DC/DC Converter," *IEEE Power Electronics Specialists Conference*, 1987 Record, pp. 424–430.

[36]   R. FARRINGTON, M. JOVANOVIC, and F. C. LEE, "Analysis of Reactive Power in Resonant Converters," *IEEE Power Electronics Specialists Conference*, 1992 Record, pp. 197–205.

[37]   R. D. MIDDLEBROOK, "Null Double Injection and the Extra Element Theorem," *IEEE Transactions on Education*, Vol. 32, No. 3, pp. 167–180, August 1989.

[38]   D. TUTTLE, *Network Synthesis*, New York: John Wiley & Sons, Vol. 1, Chapter 6, 1958.

PROBLEMS

19.1   *Analysis of a half-bridge dc–dc parallel resonant converter, operated above resonance.* In Fig. 19.53, the elements C_b, L_F, and C_F are large in value, and have negligible switching ripple. You may assume that all elements are ideal. You may use the sinusoidal approximation as appropriate.

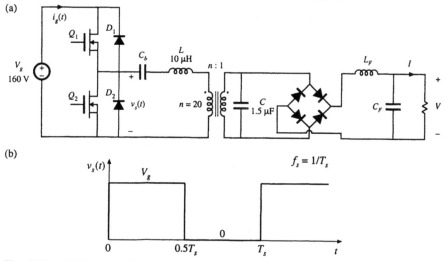

Fig. 19.53   Half-bridge parallel resonant converter of Problem 19.1: (a) schematic, (b) switch voltage waveform.

(a)   Sketch the waveform of the current $i_g(t)$.

(b)   Construct an equivalent circuit for this converter, similar to Fig. 19.22, which models the fundamental components of the tank waveforms and the dc components of the converter input current and output voltage. Clearly label the values and/or give expressions for all elements in your model, as appropriate.

(c)   Solve your model to derive an expression for the conversion ratio $V/V_g = M(F, Q_e, n)$.

At rated (maximum) load, this converter produces $I = 20$ A at $V = 3.3$ V.

(d)   What is the converter switching frequency f_s at rated load?

(e)   What is the magnitude of the peak transistor current at rated load?

At minimum load, the converter produces $I = 2$ A at $V = 3.3$ V.

(f)   What is the converter switching frequency f_s at minimum load?

(g)   What is the magnitude of the peak transistor current at minimum load? Compare with your answer from part (e)—what happens to the conduction loss and efficiency at minimum load?

19.2   A dc–dc resonant converter contains an LCC tank network [Fig. 19.1(d)], with an output filter containing a filter inductor as in the parallel resonant dc–dc converter.

(a)   Sketch an equivalent circuit model for this converter, based on the approximate sinusoidal analysis method of Section 19.1. Give expressions for all elements in your model.

(b)   Solve your model, to derive an expression for the conversion ratio $M = V/V_g$. Express M as a function of $F = f_s/f_\infty$, $Q_e = R_e/R_0$, and $n = C_s/C_p$, where f_∞ is defined as in Eq. (19.50) and R_0 is

$$R_0 = \sqrt{\frac{L(C_s + C_p)}{C_s C_p}}$$

(c)   Plot M vs. F, for $n = 1$ and $Q_e = 1, 2,$ and 5.

(d)   Plot M vs. F, for $n = 0.25$ and $Q_e = 1, 2,$ and 5.

19.3   Dual of the series resonant converter. In the converter illustrated in Fig. 19.54, L_{F1}, L_{F2}, and C_F are large filter elements, whose switching ripples are small. L and C are tank elements, whose waveforms $i_L(t)$ and $v_C(t)$ are nearly sinusoidal.

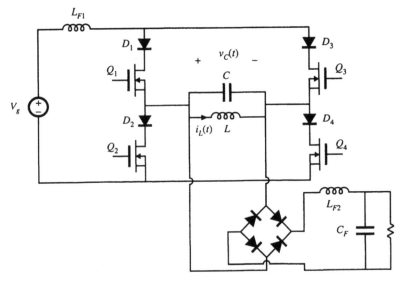

Fig. 19.54   Dual of the series resonant converter, Problem 19.3.

(a)   Using the sinusoidal approximation method, develop equivalent circuit models for the switch network, tank network, and rectifier network.

(b)   Sketch a Bode diagram of the parallel *LC* parallel tank impedance.

(c)   Solve your model. Find an analytical solution for the converter voltage conversion ratio $M = V/V_g$, as a function of the effective Q_e and the normalized switching frequency $F = f_s/f_0$. Sketch M vs. F.

(d)   What can you say about the validity of the sinusoidal approximation for this converter? Which parts of your *M* vs. *F* plot of part (c) are valid and accurate?

19.4   The converter of Problem 19.3 operates below resonance.

    (a)   Sketch the waveform $v_c(t)$. For each subinterval, label: (*i*) which of the diodes D_1 to D_4 and transistors Q_1 to Q_4 conduct current, and (*ii*) which devices block voltage.

    (b)   Does the reverse recovery process of diodes D_1 to D_4 lead to switching loss? Do the output capacitances of transistors Q_1 to Q_4 lead to switching loss?

    (c)   Repeat parts (a) and (b) for operation above resonance.

19.5   A parallel resonant converter operates with a dc input voltage of $V_g = 270$ V. The converter supplies 5 V to a dc load. The dc load power varies over the range 20 W to 200 W. It is desired to operate the power transistors with zero voltage switching. The tank element values are $L = 57$ µH, $C_p = 0.9$ nF, referred to the transformer primary. The parallel resonant tank network contains an isolation transformer having a turns ratio of 52:1.

    (a)   Define *F* as in Eq. (19.19). Derive an expression for *F*, as a function of *M* and Q_e.

    (b)   Determine the switching frequency, peak transistor current, and peak tank capacitor voltage at the maximum load power operating point.

    (c)   Determine the switching frequency, peak transistor current, and peak tank capacitor voltage at the minimum load power operating point.

19.6   In a certain resonant inverter application, the dc input voltage is $V_g = 320$ V. The inverter must produce an approximately sinusoidal output voltage having a frequency of 200 kHz. Under no load (output open-circuit) conditions, the inverter should produce a peak-to-peak output voltage of 1500 V. The nominal resistive operating point is 200 Vrms applied to 100 Ω. A nonisolated LCC inverter is employed. It is desired that the inverter operate with zero-voltage switching, at least for load resistances less than 200 Ω.

    (a)   Derive expressions for the output open-circuit voltage V_{oc} and short-circuit current I_{sc} of the LCC inverter. Express your results as functions of $F = f_s/f_\infty$, V_g, $R_\infty = L/C_s\|C_p$ and $n = C_s/C_p$. The open-circuit resonant frequency f_∞ is defined in Eq. (19.50).

    (b)   To meet the given specifications, how should the short-circuit current I_{sc} be chosen?

    (c)   Specify tank element values that meet the specifications.

    (d)   Under what conditions does your design operate with zero-voltage switching?

    (e)   Compute the peak transistor current under no-load and short-circuit conditions.

19.7   A series resonant dc–dc converter operates with a dc input voltage of $V_g = 550$ V. The converter supplies 30 kV to a load. The dc load power varies over the range 5 kW to 25 kW. It is desired to operate the power transistors with zero-voltage switching. The maximum feasible switching frequency is 50 kHz. An isolation transformer having a $1:n$ turns ratio is connected in series with the tank network. The peak tank capacitor voltage should be no greater than 2000 V, referred to the primary.

    (a)   Derive expressions for the peak tank capacitor voltage and peak tank inductor current.

    (b)   Select values for the tank inductance, tank capacitance, and turns ratio, such that the given specifications are met. Attempt to minimize the peak tank inductor current, while maximizing the worst-case minimum switching frequency.

19.8   Figure 19.55 illustrates a full-bridge resonant inverter containing an LLC tank network.

    (a)   Sketch the Bode diagrams of the input impedance under short-circuit and open-circuit conditions: $\| Z_{i0}(j\omega) \|$ and $\| Z_{i\infty}(j\omega) \|$. Give analytical expressions for the resonant frequencies and asymptotes.

    (b)   Describe the conditions on switching frequency and load resistance that lead to zero-voltage switching.

    (c)   Derive an expression for the frequency f_m, where $\| Z_{i0} \| = \| Z_{i\infty} \|$.

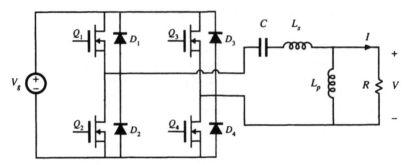

Fig. 19.55   LLC inverter of Problem 19.8.

(d)   Sketch the Bode plot of $\| H_\infty(j\omega) \|$. Label the resonant frequency, and give analytical expressions for the asymptotes.

19.9   You are given the LLC inverter circuit of Fig. 19.56. Under nominal conditions, this converter operates at switching frequency $f_s = 100$ kHz. All elements are ideal.

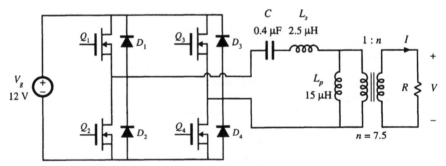

Fig. 19.56   Transformer-isolated LLC inverter, Problem 19.9.

(a)   Determine the numerical values of the open-circuit peak output voltage V_{oc} and the short-circuit peak output current I_{sc}.

(b)   Sketch the elliptical output characteristic. Over what portion of this ellipse does the converter operate with zero-voltage switching? Does it operate with zero-voltage switching at matched load?

(c)   Sketch the Bode plots of $\| Z_{i\infty} \|$ and $\| Z_{i0} \|$, and label the numerical values of f_0, f_∞, f_m, and f_s.

(d)   What is the numerical value of the peak transistor current when $R = 0$? When $R \to \infty$?

(e)   The inverter operates with load resistances that can vary between 500 Ω and an open circuit. What is the resulting range of output voltage? Does the inverter always operate with zero-voltage switching?

19.10   It is desired to obtain a converter with current source characteristics. Hence, a series resonant converter is designed for operation in the $k = 2$ discontinuous conduction mode. The switching frequency is chosen to be $f_s = 0.225 f_0$, where f_0 is the tank resonant frequency (consider only open-loop operation). The load R is a linear resistance which can assume any positive value: $0 \le R < \infty$.

(a)   Plot the output characteristics (M vs. J), for all values of R in the range $0 \le R < \infty$. Label mode boundaries, evaluate the short-circuit current, and give analytical expressions for the output characteristics.

(b)   Over what range of R (referred to the tank characteristic impedance R_0) does the converter operate as intended, in the $k = 2$ discontinuous conduction mode?

19.11   **The parallel resonant converter as a single-phase high-quality rectifier.** It is desired to utilize a transformer-isolated parallel resonant dc–dc converter in a single-phase low-harmonic rectifier system. By properly varying the converter switching frequency, a near-ideal rectifier system that can be modeled as in Fig. 18.12 is obtained. You may utilize the results of Section 19.5.2 to answer this problem. The parallel resonant tank network contains an isolation transformer having a $1 : n$ turns ratio. You may use either approximate graphical analysis or computer iteration to answer parts (b) and (c).

(a)   Plot the normalized input characteristics (normalized input voltage $m_g = nv_g/v$ vs. normalized input current $j_g = i_g nR_0/v$) of the parallel resonant converter, operated in the continuous conduction mode above resonance. Plot curves for $F = f_s/f_0 = 1.0$, 1.1, 1.2, 1.3, 1.5, and 2.0. Compare these characteristics with the desired linear resistive input characteristic $v_g/i_g = R_{emulated}$.

(b)   The converter is operated open-loop, with $F = 1.1$. The applied normalized input voltage is a rectified sinusoid of unity magnitude: $m_g(t) = |\sin(\omega t)|$. Sketch the resulting normalized input current waveform $j_g(t)$. Approximately how large is the peak current? The crossover dead time?

(c)   A feedback loop is now added, which regulates the input current to follow the input voltage such that $i_g(t) = v_g(t)/R_{emulated}$. You may assume that the feedback loop operates perfectly. For the case $R_{emulated} = R_0$, and with the same applied $m_g(t)$ waveform as in part (b), sketch the switching frequency waveform for one ac line period [i.e., show how the controller must vary F to regulate $i_g(t)$]. What is the maximum value of F? *Note*: In practice, the converter would be designed to operate with a smaller peak value of j_g, so that the switching frequency variations would be better behaved.

(d)   Choose element values (tank inductance, tank capacitance, and transformer turns ratio) such that the converter of part (c) meets the following specifications:

Ac input voltage	120 Vrms, 60 Hz
Dc output voltage	42 V
Average power	800 W
Maximum switching frequency	200 kHz

Refer the element values to the primary side of the transformer.

20

Soft Switching

In addition to the resonant circuits introduced in Chapter 19, there has been much interest in reducing the switching loss of the PWM converters of the previous chapters. Several of the more popular approaches to obtaining *soft switching* in buck, boost, and other converters, are discussed in this chapter.

Mechanisms that cause switching loss are discussed in Chapter 4, including diode reverse recovery, semiconductor output capacitances, and IGBT current tailing. Soft switching involves mitigation of one or more of these switching loss mechanisms in a PWM converter. The energy that would otherwise be lost is recovered, and is transferred to the converter source or load. The operation of a semiconductor device, during a given turn-on or turn-off switching transition, can be classified as hard-switched, zero-current switched, or zero-voltage switched. Operation of diodes and transistors with soft switching is examined in Section 20.1. In particular, it is preferable to operate diodes with zero-voltage switching at their turn-off transitions, and to operate MOSFETs with zero-voltage switching during their turn-on transitions. However, zero-voltage switching comes at the expense of increased conduction loss, and so the engineer must consider the effect of soft switching on the overall converter efficiency.

Resonant switch converters are a broad class of converters in which the PWM switch network of a conventional buck, boost, or other converter is replaced with a switch cell containing resonant elements. These resonant elements are positioned such that the semiconductor devices operate with zero-current or zero-voltage switching, and such that one or more of the switching loss mechanisms is reduced or eliminated. Other soft-switching approaches may employ resonant switching transitions, but otherwise exhibit the approximately rectangular waveforms of hard-switched converters. In any case, the resulting hybrid converter combines the properties of the resonant switching network and the parent hard-switched PWM converter.

Soft-switching converters can exhibit reduced switching loss, at the expense of increased conduction loss. Obtaining zero-voltage or zero-current switching requires that the resonant elements have large ripple; often, these elements are operated in a manner similar to the discontinuous conduction modes of the series or parallel resonant converters. As in other resonant schemes, the objectives of designing such a converter are: (1) to obtain smaller transformer and low-pass filter elements via increase of the switching frequency, and/or (2) to reduce the switching loss induced by component nonidealities such as diode stored charge, semiconductor device capacitances, and transformer leakage inductance and winding capacitance.

The resonant switch and soft-switching ideas are quite general, and can be applied to a variety of topologies and applications. A large number of resonant switch networks have been documented in the literature; a few basic approaches are listed here [1–24]. The basic zero-current-switching quasi-resonant switch network is analyzed in detail in Section 20.2. Expressions for the average components of the switch network terminal waveforms are found, leading to determination of the *switch conversion ratio μ*. The switch conversion ratio μ performs the role of the duty cycle d of CCM PWM switch networks. For example, the buck converter exhibits conversion ratio M equal to μ. Both half-wave and full-wave ringing of the tank network is considered; these lead to different switch conversion ratio functions μ. In general, given a PWM CCM converter having conversion ratio $M(d)$, we can replace the PWM switch network with a resonant switch network having switch conversion ratio μ. The resulting quasi-resonant converter will then have conversion ratio $M(\mu)$. So we can obtain soft-switching versions of all of the basic converters (buck, boost, buck-boost, forward, flyback, etc.), that exhibit zero-voltage or zero-current switching and other desirable properties.

In Section 20.3, the characteristics of several other resonant switch networks are listed: the zero-voltage-switching quasi-resonant switch network, the zero-current-switching and zero-voltage-switching quasi-square-wave networks, and the multiresonant switch network. One can obtain zero-voltage switching in all transistors and diodes using these networks.

Several related soft-switching approaches are now popular, which attain zero-voltage switching of the transistor or transistors in commonly-used converters. The zero-voltage transition approach finds application in full-bridge buck-derived converters. Active-clamp snubbers are often added to forward and flyback converters, to attain zero-voltage switching and to reset the transformer. These circuits lead to zero-voltage switching of the transistors, but (less-than-optimal) zero-current switching of the secondary-side diodes. Nonetheless, high efficiency can be achieved. An auxiliary resonant-commutated pole can achieve zero-voltage switching in voltage-source inverters. These converters are briefly discussed in Section 20.4.

20.1   SOFT-SWITCHING MECHANISMS OF SEMICONDUCTOR DEVICES

When loosely used, the terms "zero-current switching" and "zero-voltage switching" normally refer to one or more switching transitions of the transistor in a converter. However, to fully understand how a converter generates switching loss, one must closely examine the switching transitions of every semiconductor device. As described in Section 4.3, there are typically several mechanisms that are sources of significant switching loss. At the turn-off transition of a diode, its reverse-recovery process can induce loss in the transistor or other elements of the converter. The energy stored in the output capacitance of a MOSFET can be lost when the MOSFET turns on. IGBTs can lose significant energy during their turn-off transition, owing to the current-tailing phenomenon. The effects of zero-current switching and zero-voltage switching on each of these devices is discussed in detail below.

(a)

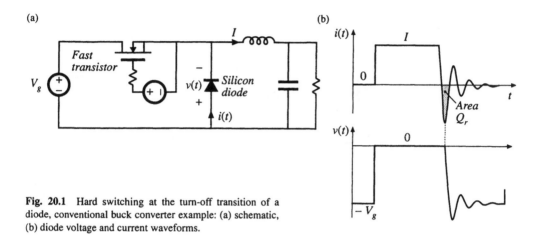

(b)

Fig. 20.1 Hard switching at the turn-off transition of a diode, conventional buck converter example: (a) schematic, (b) diode voltage and current waveforms.

20.1.1 Diode Switching

As discussed in Chapter 4, the reverse-recovery process usually leads to significant switching loss associated with the turn-off transition of diodes. This is often the largest single source of loss in a hard-switched converter. Normally, negligible loss is associated with the turn-on transition of power diodes. Three types of diode turn-off transition waveforms are commonly encountered in modern switching converters: hard switching, zero-current switching, and zero-voltage switching.

Figure 20.1 illustrates a conventional hard-switched PWM buck converter. The diode voltage and current waveforms $v(t)$ and $i(t)$ are also illustrated, with an exaggerated reverse recovery time. The output inductor current ripple is small. The diode turns off when the transistor is turned on; the reverse recovery process leads to a negative peak current of large amplitude. The diode must immediately support the full reverse voltage V_g, and hence both $v(t)$ and $i(t)$ must change with large slopes during reverse recovery. As described in Section 4.3.2, hard switching of the diode induces energy loss W_D in the transistor, given approximately by

$$W_D = V_g Q_r + t_r V_g I \qquad (20.1)$$

where Q_r is the diode recovered charge and t_r is the reverse recovery time, both taken to be positive quantities. The recovered charge is relatively large because the slope di/dt is large during the turn-off transition. The resonant circuit formed by the diode output capacitance C_j and the diode package and other wiring inductances leads to ringing at the end of the reverse recovery time.

Figure 20.2 illustrates zero-current switching at the turn-off transition of a diode. The converter example is a quasi-resonant zero-voltage switching buck converter (see Section 20.3.1). The output inductor current ripple is again small. However, tank inductor L_r is now connected in series with the diode. The resulting diode current waveform $i(t)$ changes with a limited slope as shown. The diode reverse-recovery process commences when $i(t)$ passes through zero and becomes negative. The negative $i(t)$ actively removes stored charge from the diode; during this reverse recovery time, the diode remains forward-biased. When the stored charge is removed, then the diode voltage must rapidly change to $-V_g$. As described in Section 4.3.3, energy W_D is stored in inductor L_r at the end of the reverse recovery time, given by

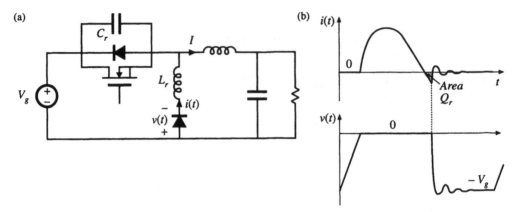

Fig. 20.2   Zero-current switching at the turn-off transition of a diode, ZVS quasi-resonant buck converter example: (a) converter schematic, (b) diode voltage and current waveforms.

$$W_D = V_g Q_r \tag{20.2}$$

The resonant circuit formed by L_r and the diode output capacitance C_j then cause this energy to be circulated between L_r and C_j. This energy is eventually dissipated by parasitic resistive elements in the circuit, and hence is lost. Since Eqs. (20.1) and (20.2) are similar in form, the switching losses induced by the reverse-recovery processes of diodes operating with hard switching and with zero-current switching are similar in magnitude. Zero-current switching may lead to somewhat lower loss because the reduced di/dt leads to less recovered charge Q_r. Zero-current switching of diodes also typically leads to increased peak inverse diode voltage during the ringing of L_r and C_j, because of the relatively large value of L_r.

When a diode operates with hard switching or zero-current switching, and when substantial inductance is present in series with the diode, then significant ringing is observed in the diode voltage waveform. A resonant circuit, comprised of the series inductance and the diode output capacitance, is excited by the diode reverse recovery process, and the resulting ringing voltage can be of large enough magnitude to lead to breakdown and failure of the diode. A common example is the diodes on the secondary side of a hard-switched transformer-isolated converter; the resonant circuit is then formed by the transformer leakage inductance and the diode output capacitance. Other examples are the circuits of Figs. 20.2 and 20.36, in which the series inductance is a discrete tank inductor.

A simple snubber circuit that is often used to protect the diode from excessive reverse voltage is

Fig. 20.3   A dissipative snubber circuit, for protection of a diode from excessive voltage caused by ringing.

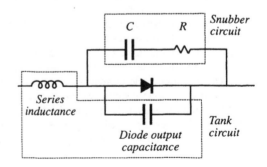

(a)

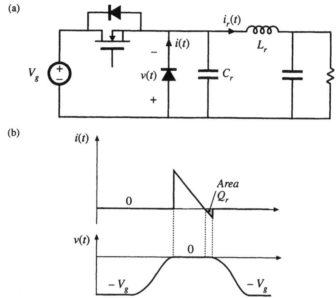

Fig. 20.4 Zero-voltage switching at the turn-off transition of a diode, ZVS quasi-squarewave buck converter example: (a) converter schematic, (b) diode current and voltage waveforms.

(b)

illustrated in Fig. 20.3. Resistor R damps the ringing of the resonant circuit. Capacitor C prevents the off-state voltage of the diode from causing excessive power loss in R. Nonetheless, the energy consumed by R per switching period is typically greater than Eqs. (20.1) or (20.2).

Figure 20.4 illustrates zero-voltage switching at the turn-off transition of a diode. The figure illustrates the example of a zero-voltage switching quasi-square wave buck converter, discussed in Section 20.3.3. The output inductor L_r of this converter assumes the role of the tank inductor, and exhibits large current ripple that causes the current $i_r(t)$ to reverse polarity. While the diode conducts, its current $i(t)$ is equal to $i_r(t)$. When $i_r(t)$ becomes negative, the diode continues to conduct until its stored charge Q_r has been removed. The diode then becomes reverse-biased, and $i_r(t)$ flows through capacitor C_r and the diode output capacitance C_j. The diode voltage and current both change with limited slope in this type of switching, and the loss induced by the diode reverse-recovery process is negligible because the waveforms are not significantly damped by parasitic resistances in the circuit, and because the peak currents during reverse recovery are relatively low. The diode stored charge and diode output capacitance both behave as an effective nonlinear capacitor that can be combined with (or replace) tank capacitor C_r. Snubber circuits such as Fig. 20.3 are not necessary when the diode operates with zero-voltage switching.

Thus, zero-voltage switching at the turn-off transition of a diode is the preferred approach, that leads to minimum switching loss. Zero-current switching at the turn-off transition can be problematic, because of the high peak inverse voltage induced across the diode by ringing.

20.1.2 MOSFET Switching

The switching loss mechanisms typically encountered by a MOSFET in a hard-switched converter are discussed in Chapter 4, and typical MOSFET voltage and current waveforms are illustrated in Fig. 20.5. The most significant components of switching loss in the MOSFET of this circuit are: (1) the loss

(a)

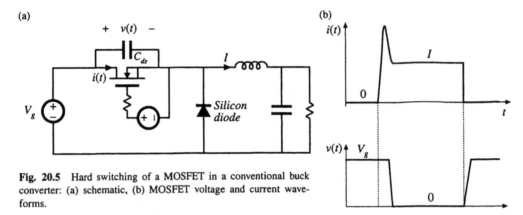

(b)

Fig. 20.5 Hard switching of a MOSFET in a conventional buck converter: (a) schematic, (b) MOSFET voltage and current waveforms.

induced by the diode reverse recovery process, and (2) the loss of the energy stored in the MOSFET output capacitance C_{ds}. Both loss mechanisms occur during the MOSFET turn-on process.

In the hard-switched circuit of Fig. 20.5, there is essentially no switching loss incurred during the MOSFET turn-off transition. This occurs because of the substantial output capacitance C_{ds} of the MOSFET. This capacitance holds the voltage $v(t)$ close to zero while the MOSFET turns off, so that the turn-off switching loss is very small. After the MOSFET has turned off, the output inductor current I flows through C_{ds}. The voltage $v(t)$ then increases until $v = V_g$ and the diode becomes forward biased.

However, when the MOSFET turns on, a high peak current flows through the MOSFET channel, induced by the diode reverse recovery and by the output capacitances of the MOSFET and diode. This leads to substantial energy loss during the hard-switched turn-on transition of the MOSFET.

When a MOSFET (or other transistor) operates with hard switching, and when substantial inductance is present in series with the MOSFET, then significant ringing is observed in the MOSFET voltage waveform. A resonant circuit, composed of the MOSFET output capacitance and the series inductance, is excited when the MOSFET turns off, and the resulting ringing voltage can be of large enough magnitude to lead to breakdown and failure of the MOSFET. A common example is the MOS-

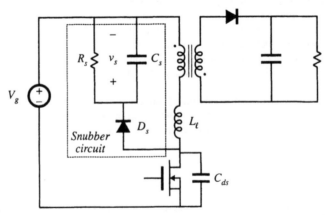

Fig. 20.6 Insertion of a dissipative voltage-clamped snubber circuit into a flyback converter. The MOSFET voltage is clamped to a peak value of $(V_g + v_s)$.

FET of the flyback converter, in which series inductance is introduced by the transformer leakage inductance. An R–C snubber circuit, similar to that used for the diode in Fig. 20.3, can be used to protect the MOSFET from damage caused by excessive applied voltage. Another common snubber circuit is illustrated in Fig. 20.6. When the MOSFET turns off, the current flowing in the transformer leakage inductance L_l begins to flow into the MOSFET capacitance C_{ds}. These parasitic elements then ring, and the peak transistor voltage can significantly exceed the ideal value of $(D/D')V_g$.

One simple way to design the snubber circuit of Fig. 20.6 is to choose the capacitance C_s to be large, so that $v_s(t) \approx V_s$ contains negligible switching ripple. The resistance R_s is then chosen so that the power consumption of R_s at the desired voltage V_s is equal to the switching loss caused by L_l:

$$\frac{V_s^2}{R_s} \approx \frac{1}{2}Li^2 f_s \tag{20.3}$$

The current i is equal to the current flowing in the transformer primary just before the MOSFET is turned off. This approximate expression is useful for obtaining a first estimate of how to choose R_s to obtain a given desired V_s.

Zero-current switching does not affect the switching loss that arises from the MOSFET output capacitance, and it may or may not influence the loss induced by diode reverse recovery. In consequence, zero-current switching is of little or no help in improving the efficiency of converters that employ MOSFETs.

Zero-voltage switching can prevent both diode reverse recovery and semiconductor output capacitances from inducing switching loss in MOSFETs. An example is illustrated in Fig. 20.7. This circuit is again a zero-voltage switching quasi-squarewave example, discussed in Section 20.3.3. The converter circuit naturally discharges the energy stored in C_{ds}, before the MOSFET is switched on. When the drain-to-source voltage $v(t)$ passes through zero, the MOSFET body diode becomes forward-biased. The MOSFET can then be turned on at zero voltage, without incurring turn-on switching loss. The MOSFET turn-on transition must be completed before the tank inductor current $i_r(t)$ becomes positive. The MOSFET turn-off transition is also lossless, and is similar to the hard-switched case discussed above.

Zero-voltage switching of a MOSFET also causes its body diode to operate with zero-voltage switching. This can eliminate the switching loss associated with reverse recovery of the slow body diode, and improve the reliability of circuits that forward-bias this diode.

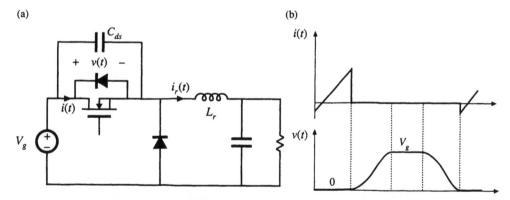

Fig. 20.7 Zero-voltage switching of a MOSFET, ZVS quasi-squarewave buck converter example. The MOSFET, its body diode, and its output capacitance C_{ds} are illustrated. (a) converter schematic, (b) MOSFET voltage and current waveforms.

Zero-voltage switching can also eliminate the overvoltage problems associated with transformer leakage inductances, removing the need for voltage-clamped snubber circuits such as in Fig. 20.6. An example is discussed in Section 20.4.2.

20.1.3 IGBT Switching

Like the MOSFET, the IGBT typically encounters substantial switching loss during its turn-on transition, induced by the reverse-recovery process of diodes within the converter. In addition, the IGBT exhibits significant switching loss during its turn-off transition, caused by the current tailing phenomenon (*see* Chapter 4).

Zero-voltage switching has been successfully applied to IGBT circuits—an example is the auxiliary resonant commutation circuit discussed in Section 20.4.3. This has the principal advantage of eliminating the switching loss caused by diode reverse recovery. Although zero-voltage switching can reduce the loss incurred during the turn-off transition, it is difficult to eliminate the substantial loss caused by current tailing.

20.2   THE ZERO-CURRENT SWITCHING QUASI-RESONANT SWITCH CELL

Figure 20.8(a) illustrates a generic buck converter, consisting of a switch cell cascaded by an L–C low-pass filter. When the switch cell is realized as in Fig. 20.8(b), then a conventional PWM buck converter is obtained. Figures 20.8(b) and (c) illustrate two other possible realizations of the switch cell: the half-wave and full-wave zero-current-switching quasi-resonant switches [1, 2]. In these switch cells, a resonant tank capacitor C_r is placed in parallel with diode D_2, while resonant tank capacitor L_r is placed in series with the active transistor element.

Both resonant switch cells require a two-quadrant SPST switch. In the half-wave switch cell of Fig. 20.8(c), diode D_1 is added in series with transistor Q_1. This causes the Q_1–D_1 SPST switch to turn off at the first zero crossing of the tank inductor current $i_1(t)$. In the full-wave switch cell of Fig. 20.8(d), antiparallel diode D_1 allows bidirectional flow of the tank inductor current $i_1(t)$. With this switch network, the Q_1–D_1 SPST switch is normally turned off at the second zero-crossing of the $i_1(t)$ waveform. In either switch cell, the L_r and C_r elements are relatively small in value, such that their resonant frequency f_0 is greater than the switching frequency f_s, where

$$f_0 = \frac{1}{2\pi \sqrt{L_r C_r}} = \frac{\omega_0}{2\pi} \tag{20.4}$$

In the analysis which follows, it is assumed that the converter filter element values L and C have negligible switching ripple. Hence, the switch cell terminal waveforms $v_1(t)$ and $i_2(t)$ are well-approximated by their average values:

$$
\begin{aligned}
i_2(t) &\approx \left\langle i_2(t) \right\rangle_{T_s} \\
v_1(t) &\approx \left\langle v_1(t) \right\rangle_{T_s}
\end{aligned}
\tag{20.5}
$$

with the average defined as in Eq. (7.3). In steady-state, we can further approximate $v_1(t)$ and $i_2(t)$ by their dc components V_1 and I_2:

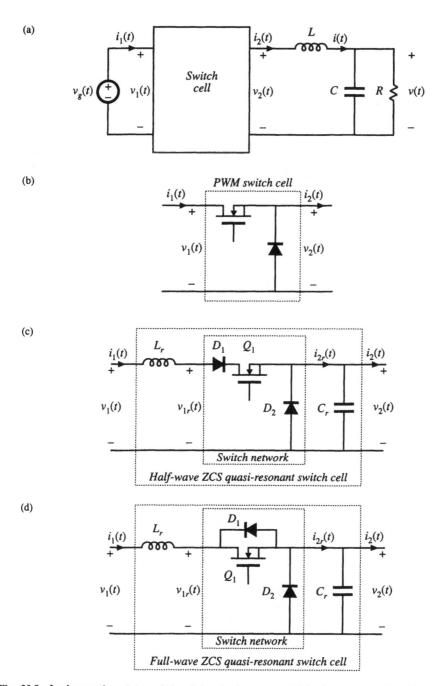

Fig. 20.8 Implementation of the switch cell in a buck converter: (a) buck converter, with arbitrary switch cell; (b) PWM switch cell; (c) half-wave ZCS quasi-resonant switch cell; (d) full-wave ZCS quasi-resonant switch cell.

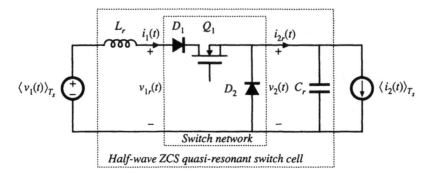

Fig. 20.9   The half-wave ZCS quasi-resonant switch cell, driven by the terminal quantities $\langle v_1(t) \rangle_{T_s}$ and $\langle i_2(t) \rangle_{T_s}$.

$$i_2(t) \approx I_2$$
$$v_1(t) \approx V_1 \tag{20.6}$$

Thus, the small-ripple approximation is employed for the converter filter elements, as usual.

To understand the operation of the half-wave ZCS quasi-resonant switch cell, we can solve the simplified circuit illustrated in Fig. 20.9. In accordance with the averaged switch modeling approach of Sections 7.4 and 11.1, it is desired to determine the average terminal waveforms $\langle v_2(t) \rangle_{T_s}$ and $\langle i_1(t) \rangle_{T_s}$, as functions of the applied quantities $\langle v_1(t) \rangle_{T_s}$ and $\langle i_2(t) \rangle_{T_s}$. The switch conversion ratio μ is then given by

$$\mu = \frac{\langle v_2(t) \rangle_{T_s}}{\langle v_{1r}(t) \rangle_{T_s}} = \frac{\langle i_1(t) \rangle_{T_s}}{\langle i_{2r}(t) \rangle_{T_s}} \tag{20.7}$$

In steady state, we can write

$$\mu = \frac{V_2}{V_1} = \frac{I_1}{I_2} \tag{20.8}$$

The steady-state analysis of this section employs Eq. (20.8) to determine μ.

20.2.1   Waveforms of the Half-Wave ZCS Quasi-Resonant Switch Cell

Typical waveforms of the half-wave cell of Fig. 20.9 are illustrated in Fig. 20.10. Each switching period consists of four subintervals as shown, having angular lengths α, β, δ, and ξ. The switching period begins when the controller turns on transistor Q_1. The initial values of the tank inductor current $i_1(t)$ and tank capacitor voltage $v_2(t)$ are zero. During subinterval 1, all three semiconductor devices conduct. Diode D_2 is forward-biased because $i_1(t)$ is less than I_2. In consequence, during subinterval 1 the switch cell reduces to the circuit of Fig. 20.11.

The slope of the inductor current is given by

$$\frac{di_1(t)}{dt} = \frac{V_1}{L_r} \tag{20.9}$$

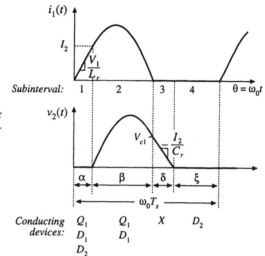

Fig. 20.10 Tank inductor current and capacitor voltage waveforms, for the half-wave ZCS quasi-resonant switch of Fig. 20.9.

with the initial condition $i_1(0) = 0$. The solution is

$$i_1(t) = \frac{V_1}{L_r} t = \omega_0 t \frac{V_1}{R_0} \tag{20.10}$$

where the tank characteristic impedance R_0 is defined as

$$R_0 = \sqrt{\frac{L_r}{C_r}} \tag{20.11}$$

It is convenient to express the waveforms in terms of the angle $\theta = \omega_0 t$, instead of time t. At the end of subinterval 1, $\omega_0 t = \alpha$. The subinterval ends when diode D_2 becomes reverse-biased. Since the diode D_2 current is equal to $I_2 - i_1(t)$, this occurs when $i_1(t) = I_2$. Hence, we can write

$$i_1(\alpha) = \alpha \frac{V_1}{R_0} = I_2 \tag{20.12}$$

Solution for α yields

$$\alpha = \frac{I_2 R_0}{V_1} \tag{20.13}$$

Fig. 20.11   Circuit of the switch network during subinterval 1.

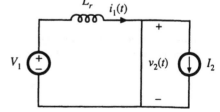

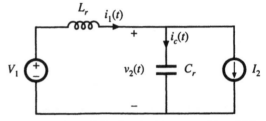

Fig. 20.12 Circuit of the switch network during subinterval 2.

During subinterval 2, transistor Q_1 and diode D_1 conduct, while diode D_2 is reverse-biased. The switch network then becomes the circuit illustrated in Fig. 20.12. The resonant L_r–C_r tank network is excited by the constant sources V_1 and I_2. The network equations are

$$L_r \frac{di_1(\omega_0 t)}{dt} = V_1 - v_2(\omega_0 t)$$
$$C_r \frac{dv_2(\omega_0 t)}{dt} = i_1(\omega_0 t) - I_2 \qquad (20.14)$$

with the initial conditions

$$v_2(\alpha) = 0$$
$$i_1(\alpha) = I_2 \qquad (20.15)$$

The solution is

$$i_1(\omega_0 t) = I_2 + \frac{V_1}{R_0} \sin(\omega_0 t - \alpha)$$
$$v_2(\omega_0 t) = V_1 \left(1 - \cos(\omega_0 t - \alpha)\right) \qquad (20.16)$$

The tank inductor current rises to a peak value given by

$$I_{1pk} = I_2 + \frac{V_1}{R_0} \qquad (20.17)$$

The subinterval ends at the first zero crossing of $i_1(t)$. If we denote the angular length of the subinterval as β, then we can write

$$i_1(\alpha + \beta) = I_2 + \frac{V_1}{R_0} \sin(\beta) = 0 \qquad (20.18)$$

Solution for $\sin(\beta)$ yields

$$\sin(\beta) = -\frac{I_2 R_0}{V_1} \qquad (20.19)$$

Care must be employed when solving Eq. (20.19) for the angle β. It can be observed from Fig. 20.10 that the angle β is greater than π. The correct branch of the arcsine function must be selected, as follows:

$$\beta = \pi + \sin^{-1}\left(\frac{I_2 R_0}{V_1}\right) \qquad (20.20)$$

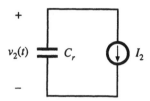

Fig. 20.13 Circuit of the switch network during subinterval 3.

where

$$-\frac{\pi}{2} < \sin^{-1}(x) \leq \frac{\pi}{2}$$

Note that the inequality

$$I_2 < \frac{V_1}{R_0} \tag{20.21}$$

must be satisfied; otherwise, there is no solution to Eq. (20.19). At excessive load currents, where Eq. (20.21) is not satisfied, the tank inductor current never reaches zero, and the transistor does not switch off at zero current.

The tank capacitor voltage at the end of subinterval 2 is found by evaluation of Eq. (20.16) at $\omega_0 t = (\alpha + \beta)$. The $\cos(\beta)$ term can be expressed as

$$\cos(\beta) = -\sqrt{1 - \sin^2(\beta)} = -\sqrt{1 - \left(\frac{I_2 R_0}{V_1}\right)^2} \tag{20.22}$$

Substitution of Eq. (20.22) into Eq. (20.16) leads to

$$v_2(\alpha + \beta) = V_{c1} = V_1 \left(1 + \sqrt{1 - \left(\frac{I_2 R_0}{V_1}\right)^2}\right) \tag{20.23}$$

At the end of subinterval 2, diode D_1 becomes reverse-biased. Transistor Q_1 can then be switched off at zero current.

During subinterval 3, all semiconductor devices are off, and the switch cell reduces to the circuit of Fig. 20.13. The tank capacitor C_r is discharged by the filter inductor current I_2. Hence, the tank capacitor voltage v_2 decreases linearly to zero. The circuit equations are

$$C_r \frac{dv_2(\omega_0 t)}{dt} = -I_2 \tag{20.24}$$
$$v_2(\alpha + \beta) = V_{c1}$$

The solution is

$$v_2(\omega_0 t) = V_{c1} - I_2 R_0 \left(\omega_0 t - \alpha - \beta\right) \tag{20.25}$$

Subinterval 3 ends when the tank capacitor voltage reaches zero. Diode D_2 then becomes forward-biased. Hence, we can write

$$v_2(\alpha + \beta + \delta) = V_{c1} - I_2 R_0 \delta = 0 \tag{20.26}$$

where δ is the angular length of subinterval 3. Solution for δ yields

$$\delta = \frac{V_{c1}}{I_2 R_0} = \frac{V_1}{I_2 R_0}\left(1 - \sqrt{1 - \left(\frac{I_2 R_0}{V_1}\right)^2}\right)$$

(20.27)

Subinterval 4, of angular length ξ, is identical to the diode conduction subinterval of the conventional PWM switch network. Diode D_2 conducts the filter inductor current I_2, and the tank capacitor voltage v_2 is equal to zero. Transistor Q_1 is off, and the input current i_1 is equal to zero.

The angular length of the switching period is

$$\omega_0 T_s = \alpha + \beta + \delta + \xi = \frac{2\pi f_0}{f_s} = \frac{2\pi}{F}$$

(20.28)

where

$$F = \frac{f_s}{f_0}$$

(20.29)

Quasi-resonant switch networks are usually controlled by variation of the switching frequency f_s or, in normalized terms, by variation of F. Note that the interval lengths α, β, and δ are determined by the response of the tank network. Hence, control of the switching frequency is equivalent to control of the fourth subinterval length ξ. The subinterval length ξ must be positive, and hence, the minimum switching period is limited as follows:

$$\omega_0 T_s \geq \alpha + \beta + \delta$$

(20.30)

Substitution of Eqs. (20.13), (20.20), and (20.27) into Eq. (20.30) yields

$$\frac{2\pi}{F} \geq \frac{I_2 R_0}{V_1} + \pi + \sin^{-1}\left(\frac{I_2 R_0}{V_1}\right) + \frac{V_1}{I_2 R_0}\left(1 - \sqrt{1 - \left(\frac{I_2 R_0}{V_1}\right)^2}\right)$$

(20.31)

This expression limits the maximum switching frequency, or maximum F, of the half-wave ZCS quasi-resonant switch cell.

20.2.2   The Average Terminal Waveforms

It is now desired to solve for the power processing function performed by the switch network. The switch conversion ratio μ is a generalization of the duty cycle d. It expresses how a resonant switch network controls the average voltages and currents of a converter. In our buck converter example, we can define μ as the ratio of $\langle v_2(t)\rangle_{T_s}$ to $\langle v_1(t)\rangle_{T_s}$, or equivalently, the ratio of $\langle i_1(t)\rangle_{T_s}$ to $\langle i_2(t)\rangle_{T_s}$. In a hard-switched PWM network, this ratio is equal to the duty cycle d. Hence, analytical results derived for hard-switched PWM converters can be adapted to quasi-resonant converters, simply by replacing d with μ. In this section, we derive an expression for μ, by averaging the terminal waveforms of the switch network.

The switch input current waveform $i_1(t)$ of Fig. 20.10 is reproduced in Fig. 20.14. The average switch input current is given by

Fig. 20.14 Input current waveform $i_1(t)$, and the areas q_1 and q_2 during subintervals 1 and 2 respectively.

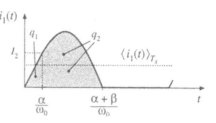

$$\langle i_1(t)\rangle_{T_s} = \frac{1}{T_s}\int_t^{t+T_s} i_1(t)dt = \frac{q_1 + q_2}{T_s} \tag{20.32}$$

The charge quantities q_1 and q_2 are the areas under the $i_1(t)$ waveform during the first and second subintervals, respectively. The charge q_1 is given by the triangle area formula

$$q_1 = \int_0^{\frac{\alpha}{\omega_0}} i_1(t)dt = \frac{1}{2}\left(\frac{\alpha}{\omega_0}\right)(I_2) \tag{20.33}$$

The time α/ω_0 is the length of subinterval 1. The charge q_2 is

$$q_2 = \int_{\frac{\alpha}{\omega_0}}^{\frac{\alpha+\beta}{\omega_0}} i_1(t)dt \tag{20.34}$$

According to Fig. 20.12, during subinterval 2 the current $i_1(t)$ can be related to the tank capacitor current $i_C(t)$ and the switch output current I_2 by the node equation

$$i_1(t) = i_C(t) + I_2 \tag{20.35}$$

Substitution of Eq. (20.35) into Eq. (20.34) leads to

$$q_2 = \int_{\frac{\alpha}{\omega_0}}^{\frac{\alpha+\beta}{\omega_0}} i_C(t)dt + \int_{\frac{\alpha}{\omega_0}}^{\frac{\alpha+\beta}{\omega_0}} I_2\, dt \tag{20.36}$$

Both integrals in Eq. (20.36) can easily be evaluated, as follows. Since the second term involves the integral of the constant current I_2, this term is

$$\int_{\frac{\alpha}{\omega_0}}^{\frac{\alpha+\beta}{\omega_0}} I_2\, dt = I_2\frac{\beta}{\omega_0} \tag{20.37}$$

The first term in Eq. (20.36) involves the integral of the capacitor current over subinterval 2. Hence, this term is equal to the change in capacitor charge over the second subinterval:

$$\int_{\frac{\alpha}{\omega_0}}^{\frac{\alpha+\beta}{\omega_0}} i_C(t)dt = C\left(v_2\left(\frac{\alpha+\beta}{\omega_0}\right) - v_2\left(\frac{\alpha}{\omega_0}\right)\right) \tag{20.38}$$

(recall that $\Delta q = C\Delta v$ in a capacitor). During the second subinterval, the tank capacitor voltage is initially zero, and has a final value of V_{c1}. Hence, Eq. (20.38) reduces to

$$\int_{\frac{\alpha}{\omega_0}}^{\frac{\alpha+\beta}{\omega_0}} i_C(t)dt = C\left(V_{c1} - 0\right) = CV_{c1} \tag{20.39}$$

Substitution of Eqs. (20.37) and (20.39) into Eq. (20.36) leads to the following expression for q_2:

$$q_2 = CV_{c1} + I_2 \frac{\beta}{\omega_0} \tag{20.40}$$

Equations (20.33) and (20.40) can now be inserted into Eq. (20.32), to obtain the following expression for the switch input current:

$$\left\langle i_1(t)\right\rangle_{T_s} = \frac{\alpha I_2}{2\omega_0 T_s} + \frac{CV_{c1}}{T_s} + \frac{\beta I_2}{\omega_0 T_s} \tag{20.41}$$

Substitution of Eq. (20.41) into (20.8) leads to the following expression for the switch conversion ratio:

$$\mu = \frac{\left\langle i_1(t)\right\rangle_{T_s}}{I_2} = \frac{\alpha}{2\omega_0 T_s} + \frac{CV_{c1}}{I_2 T_s} + \frac{\beta}{\omega_0 T_s} \tag{20.42}$$

Finally, the quantities α, β, and V_{c1} can be eliminated, using Eqs. (20.13), (20.20), (20.23). The result is

$$\mu = F\frac{1}{2\pi}\left[\frac{1}{2}J_s + \pi + \sin^{-1}(J_s) + \frac{1}{J_s}\left(1 + \sqrt{1 - J_s^2}\right)\right] \tag{20.43}$$

where

$$J_s = \frac{I_2 R_0}{V_1} \tag{20.44}$$

Equation (20.43) is of the form

$$\mu = FP_{\frac{1}{2}}\left(J_s\right) \tag{20.45}$$

where

$$P_{\frac{1}{2}}\left(J_s\right) = \frac{1}{2\pi}\left[\frac{1}{2}J_s + \pi + \sin^{-1}(J_s) + \frac{1}{J_s}\left(1 + \sqrt{1 - J_s^2}\right)\right] \tag{20.46}$$

Thus, the switch conversion ratio μ is directly controllable by variation of the switching frequency, through F. The switch conversion ratio is also a function of the applied terminal voltage V_1 and current I_2, via J_s. The function $P_{\frac{1}{2}}(J_s)$ is sketched in Fig. 20.15. The switch conversion ratio μ is sketched in Fig. 20.16, for various values of F and J_s. These characteristics are similar in shape to the function $P(J_s)$, and are simply scaled by the factor F. It can be seen that the conversion ratio μ is a strong function

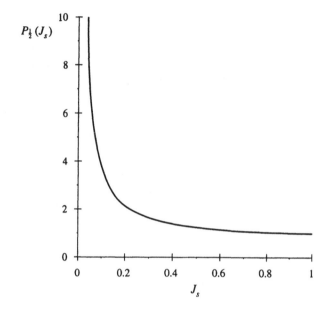

Fig. 20.15   The function $P_{\frac{1}{2}}(J_s)$.

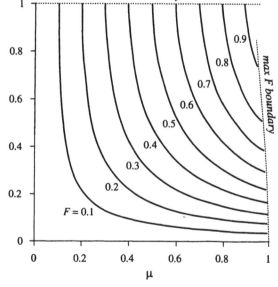

Fig. 20.16   Characteristics of the half-wave ZCS quasi-resonant switch.

of the current I_2, via J_s. The characteristics end at $J_s = 1$; according to Eq. (20.31), the zero current switching property is lost when $J_s > 1$. The characteristics also end at the maximum switching frequency limit given by Eq. (20.31). This expression can be simplified by use of Eq. (20.43), to express the limit in terms of μ as follows:

$$\mu \le 1 - \frac{J_s F}{4\pi} \tag{20.47}$$

The switch conversion ratio μ is thus limited to a value slightly less than 1.

The averaged waveforms of converters containing half-wave ZCS quasi-resonant switches can now be determined. The results of the analysis of PWM converters operating in the continuous conduction mode can be directly adapted to the related quasi-resonant converters, simply by replacing the duty cycle d with the switch conversion ratio μ. For the buck converter example, the conversion ratio is

$$M = \frac{V}{V_g} = \mu \tag{20.48}$$

This result could also be derived by use of the principle of inductor volt-second balance. The average voltage across the filter inductor is $(\mu V_g - V)$. Upon equating this voltage to zero, we obtain Eq. (20.48).

In the buck converter, I_2 is equal to the load current I, while V_1 is equal to the converter input voltage V_g. Hence, the quantity J_s is

$$J_s = \frac{IR_0}{V_g} \tag{20.49}$$

Zero current switching occurs for

$$I \le \frac{V_g}{R_0} \tag{20.50}$$

The output voltage can vary over the range

$$0 \le V \le V_g - \frac{FIR_0}{4\pi} \tag{20.51}$$

which nearly coincides with the PWM output voltage range $0 \le V \le V_g$.

A boost converter employing a half-wave ZCS quasi-resonant switch is illustrated in Fig. 20.17. The conversion ratio of the boost converter is given by

$$M = \frac{V}{V_g} = \frac{1}{1 - \mu} \tag{20.52}$$

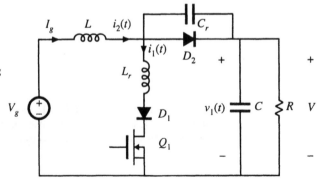

Fig. 20.17 Boost converter containing a half-wave ZCS quasi-resonant switch.

The half-wave switch conversion ratio μ is again given by Eqs. (20.44) to (20.46). For the boost converter, the applied switch voltage V_1 is equal to the output voltage V, while the applied switch current I_2 is equal to the filter inductor current, or I_g. Hence, the quantity J_s is

$$J_s = \frac{I_2 R_0}{V_1} = \frac{I_g R_0}{V} \tag{20.53}$$

Also, the input current I_g of the boost converter is related to the load current I according to

$$I_g = \frac{I}{1-\mu} \tag{20.54}$$

Equations (20.52) to (20.54), in conjunction with Eqs. (20.44) to (20.46), describe the averaged waveforms of the half-wave quasi-resonant ZCS boost converter.

20.2.3   The Full-Wave ZCS Quasi-Resonant Switch Cell

The full-wave ZCS quasi-resonant switch cell is illustrated in Fig. 20.8(d). It differs from the half-wave cell in that elements D_1 and Q_1 are connected in antiparallel, to form a current-bidirectional two-quadrant switch. Typical tank inductor current and tank capacitor voltage waveforms are illustrated in Fig. 20.18. These waveforms are similar to those of the half-wave case, except that the Q_1/D_1 switch interrupts the tank inductor current $i_1(t)$ at its second zero-crossing. While $i_1(t)$ is negative, diode D_1 conducts, and transistor Q_1 can be turned off at zero current.

The analysis is nearly the same as for the half-wave case, with the exception of subinterval 2. The subinterval 2 angular length β and final voltage V_{c1} can be shown to be

$$\beta = \begin{cases} \pi + \sin^{-1}\left(J_s\right) & \text{(half wave)} \\ 2\pi - \sin^{-1}\left(J_s\right) & \text{(full wave)} \end{cases} \tag{20.55}$$

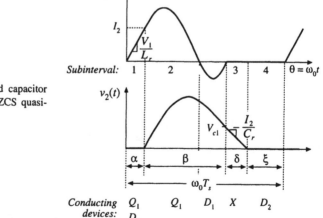

Fig. 20.18 Tank inductor current and capacitor voltage waveforms, for the full-wave ZCS quasi-resonant switch cell of Fig. 20.8(d).

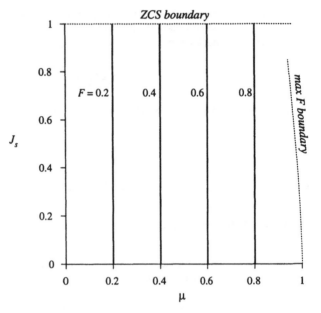

Fig. 20.19 Characteristics of the full-wave ZCS quasi-resonant switch.

$$V_{c1} = \begin{cases} V_1\left(1 + \sqrt{1 - J_s^2}\right) & \text{(half wave)} \\ V_1\left(1 - \sqrt{1 - J_s^2}\right) & \text{(full wave)} \end{cases}$$ (20.56)

In either case, the switch conversion ratio μ is given by Eq. (20.42). For the full-wave switch, one obtains

$$\mu = FP_1(J_s)$$ (20.57)

where $P_1(J_s)$ is given by

$$P_1(J_s) = \frac{1}{2\pi}\left[\frac{1}{2}J_s + 2\pi - \sin^{-1}(J_s) + \frac{1}{J_s}\left(1 - \sqrt{1 - J_s^2}\right)\right]$$ (20.58)

In the full-wave case, $P_1(J_s)$ is essentially independent of J_s:

$$P_1(J_s) \approx 1$$ (20.59)

The worst-case deviation of $P_1(J_s)$ from 1 occurs as J_s tends to 1, where $P_1(J_s)$ tends to 0.96. So $P_1(J_s)$ lies within 4% of unity for $0 < J_s < 1$. Hence, for the full-wave case, it is a good approximation to express the switch conversion ratio as

$$\mu \approx F = \frac{f_s}{f_0}$$ (20.60)

The full-wave quasi-resonant switch therefore exhibits voltage-source output characteristics, controllable by F. Equations describing the average waveforms of CCM PWM converters can be adapted to apply to full-wave ZCS quasi-resonant converters, simply by replacing the duty cycle d with the normalized switching frequency F. The conversion ratios of full-wave quasi-resonant converters exhibit negligible dependence on the load current.

The variation of the switch conversion ratio μ with F and J_s is plotted in Fig. 20.19. For a typical voltage regulator application, the range of switching frequency variations is much smaller in the full-wave mode than in the half-wave mode, because μ does not depend on the load current. Variations in the load current do not induce the controller to significantly change the switching frequency.

20.3 RESONANT SWITCH TOPOLOGIES

So far, we have considered the zero-current-switching quasi-resonant switch cell, illustrated in Fig. 20.20. The ideal SPST switch is realized using a voltage-bidirectional or current-bidirectional two-quadrant switch, to obtain half-wave or full-wave ZCS quasi-resonant switch networks, respectively.

The resonant elements L_r and C_r can be moved to several different positions in the converter, without altering the basic switch properties. For example, Fig. 20.21 illustrates connection of the resonant tank capacitor C_r between the cathode of diode D_2, and the converter output or input terminals. Although this may change the dc component of the tank capacitor voltage, the ac components of the tank capacitor voltage waveform are unchanged. Also, the terminal voltage waveform $v_2(t)$ is unchanged. The voltages $v_g(t)$ and $v(t)$ contain negligible high-frequency ac components, and hence the converter input and output terminal potentials can be considered to be at high-frequency ac ground.

A test to determine the topology of a resonant switch network is to replace all low-frequency filter inductors with open circuits, and to replace all dc sources and low-frequency filter capacitors with short circuits [13]. The elements of the resonant switch cell remain. In the case of the zero-current-switching quasi-resonant switch, the network of Fig. 20.22 is always obtained.

It can be seen from Fig. 20.22 that diode D_2 switches on and off at the zero crossings of the tank capacitor voltage $v_2(t)$, while the switch elements Q_1 and D_1 switch at the zero crossings of the tank inductor current $i_1(t)$. Zero voltage switching of diode D_2 is highly advantageous, because it essentially eliminates the switching loss caused by the recovered charge and output capacitance of diode D_2. Zero current switching of Q_1 and D_1 can be used to advantage when Q_1 is realized by an SCR or IGBT. However, in high-frequency converters employing MOSFETs, zero current switching of Q_1 and D_1 is generally a poor choice. Significant switching loss due to the output capacitances of Q_1 and D_1 may be observed. In addition, in the full-wave case, the recovered charge of diode D_1 leads to significant ringing

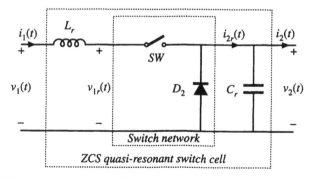

Fig. 20.20 Basic ZCS quasi-resonant switch cell.

(a)

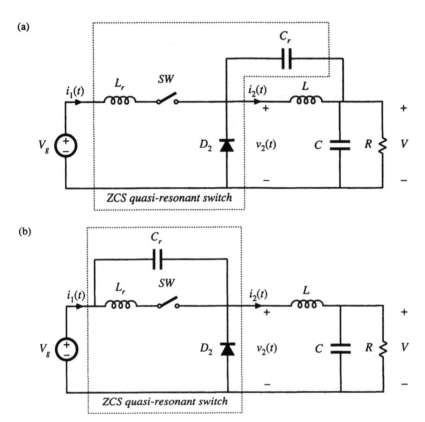

Fig. 20.21 Connection of the tank capacitor of the ZCS quasi-resonant cell to other points at ac ground: (a) connection to the dc output, (b) connection to the dc input. In each case, the ac components of the waveforms are unchanged.

Fig. 20.22 Elimination of converter low-frequency elements causes the ZCS quasi-resonant switch cell to reduce to this network.

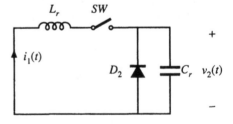

and switching loss at the end of subinterval 2 [3].

The ZCS quasi-resonant switch exhibits increased conduction loss, relative to an equivalent PWM switch, because the peak transistor current is increased. The peak transistor current is given by Eq. (20.17); since $J_s \leq 1$, the peak current is $I_{1pk} \geq 2I_2$. In addition, the full-wave ZCS switch exhibits poor efficiency at light load, owing to the conduction loss caused by circulating tank currents. The half-wave ZCS switch exhibits additional conduction loss due to the added forward voltage drop of diode D_1. The peak transistor voltage is V_1, which is identical to the PWM case.

20.3.1 The Zero-Voltage-Switching Quasi-Resonant Switch

The resonant switch network illustrated in Fig. 20.23 is the dual of the network of Fig. 20.22. This network is known as the zero-voltage-switching quasi-resonant switch [4]. Since the tank capacitor C_r appears in parallel with the SPST switch, the elements Q_1 and D_1 used to realize the SPST switch turn on and off at zero voltage. The tank inductor L_r is effectively in series with diode D_2, and hence diode D_2 switches at zero current. Converters containing ZVS quasi-resonant switches can be realized in a number of ways. The only requirement is that, when the low-frequency filter inductors, filter capacitors, and sources are replaced by open- or short-circuits as described above, then the high-frequency switch network of Fig. 20.23 should remain.

For example, a zero-voltage-switching quasi-resonant buck converter is illustrated in Fig. 20.24(a). Typical tank capacitor voltage and tank inductor current waveforms are given in Fig. 20.24(b). A current-bidirectional realization of the two-quadrant SPST switch is shown; this causes the ZVS quasi-resonant switch to operate in the half-wave mode. Use of a voltage-bidirectional two-quadrant SPST switch allows full-wave operation.

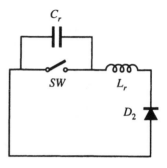

Fig. 20.23 Elimination of converter low-frequency elements reduces the ZVS quasi-resonant switch cell to this network.

By analysis similar to that of Section 20.2, it can be shown that the switch conversion ratio μ of the half-wave ZVS quasi-resonant switch is

$$\mu = 1 - F P_{\frac{1}{2}}\left(\frac{1}{J_s}\right) \tag{20.61}$$

The function $P_{\frac{1}{2}}(J_s)$ is again given by Eq. (20.46), and the quantity J_s is defined in Eq. (20.44). For the full-wave ZVS quasi-resonant switch, one obtains

$$\mu = 1 - F P_1\left(\frac{1}{J_s}\right) \tag{20.62}$$

where $P_1(J_s)$ is given by Eq. (20.58). The condition for zero voltage switching is

$$J_s \geq 1 \tag{20.63}$$

Thus, the zero voltage switching property is lost at light load. The peak transistor voltage is given by

$$\text{peak transistor voltage} \quad V_{cr,pk} = (1 + J_s)V_1 \tag{20.64}$$

This equation predicts that load current variations can lead to large voltage stress on transistor Q_1. For example, if it is desired to obtain zero voltage switching over a 5:1 range of load current variations, then J_s should vary between 1 and 5. According to Eq. (20.64), the peak transistor voltage then varies between two times and six times the applied voltage V_1. The maximum transistor current is equal to the applied current I_2. Although the maximum transistor current in the ZVS quasi-resonant switch is identical to that

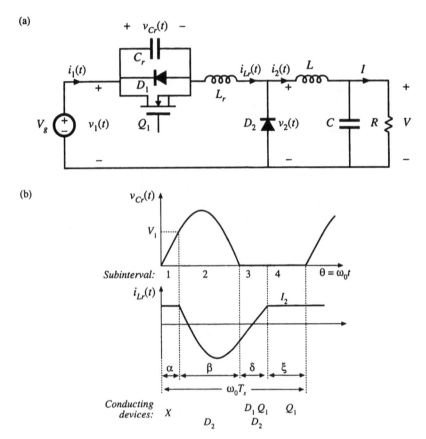

Fig. 20.24   A ZVS quasi-resonant buck converter: (a) circuit, (b) tank waveforms.

of the PWM switch, the peak transistor blocking voltage is substantially increased. This leads to increased conduction loss, because transistor on-resistance increases rapidly with rated blocking voltage.

20.3.2   The Zero-Voltage-Switching Multiresonant Switch

The resonant switch network of Fig. 20.25 contains tank capacitor C_d in parallel with diode D_2, as in the ZCS switch network of Fig. 20.22. In addition, it contains tank capacitor C_s in parallel with the SPST switch, as in the ZVS switch network of Fig. 20.23. In consequence, all semiconductor elements switch at zero voltage. This three-element resonant switch network is known as the zero-voltage-switching multiresonant switch (ZVS MRS). Since no semiconductor output capacitances or diode recovered charges lead to ringing or switching loss, the ZVS MRS exhibits very low switching loss. For the same reason, generation of electromagnetic interference is reduced.

A half-wave ZVS MRS realization of the buck converter is illustrated in Fig. 20.26. In a typical design that must operate over a 5:1 load range and with $0.4 \leq \mu \leq 0.6$, the designer might choose a maximum F of 1.0, a maximum J of 1.4, and $C_d/C_s = 3$, where these quantities are defined as follows:

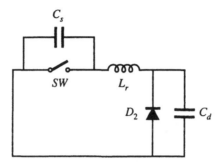

Fig. 20.25 Elimination of converter low-frequency elements reduces the ZVS multiresonant switch cell to this network.

$$f_0 = \frac{1}{2\pi\sqrt{LC_t}} \quad R_0 = \sqrt{\frac{L}{C_t}}$$
$$F = \frac{f_s}{f_0} \quad J = \frac{I_2 R_0}{V_1} \tag{20.65}$$

As usual, the conversion ratio is defined as $\mu = V_2/V_1$. The resulting peak transistor voltage for this typical design is approximately $2.8V_1$, while the peak transistor current is $2I_2$. Hence, conduction losses are higher than in an equivalent PWM switch. The range of switch conversion ratios μ is a function of the capacitor ratio C_d/C_s; in a good design, values of μ ranging from nearly one to nearly zero can be obtained, with a wide range of dc load currents and while maintaining zero voltage switching.

Analysis and design charts for the ZVS MRS are given in [5–8]. Results for the typical choice $C_d = 3C_s$ are plotted in Fig. 20.27. These plots illustrate how the switch conversion ratio μ varies as a function of load current and switching frequency. Figure 20.27(a) also illustrates the boundary of zero-voltage switching: ZVS is lost for operation outside the dashed lines. Decreasing the ratio of C_d to C_s reduces the area of the ZVS region.

Other resonant converters in which all semiconductor devices operate with zero voltage switching are known. Examples include some operating modes of the parallel and LCC resonant converters described in Chapter 19, as well as the class–E converters described in [10–12].

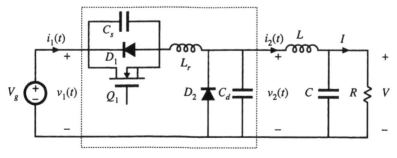

Fig. 20.26 Half-wave ZVS multiresonant buck converter.

(a)

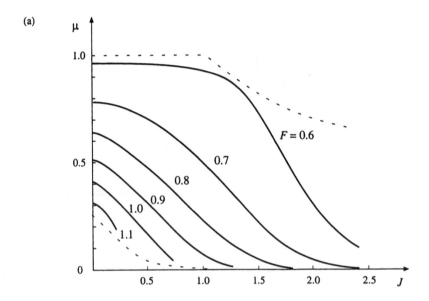

(b)

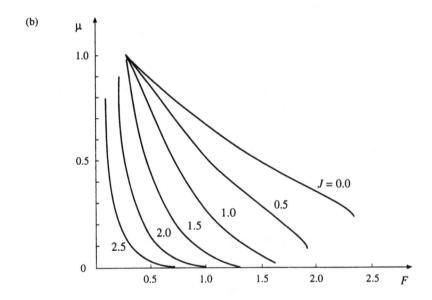

Fig. 20.27 Conversion ratio μ for the multi-resonant switch with $C_d = 3C_s$: (a) conversion ratio μ vs. normalized current J (solid lines: conversion ratio; dashed lines: boundaries of zero-voltage switching), (b) conversion ratio μ vs. normalized switching frequency F.

20.3.3 Quasi-Square-Wave Resonant Switches

Another basic class of resonant switch networks is the quasi-square wave converters. Both zero-voltage switching and zero-current switching versions are known; the resonant tank networks are illustrated in Fig. 20.28. In the network of Fig. 20.28(a), all semiconductor devices are effectively in series with the tank inductor, and hence operate with zero-current switching. In the network of Fig. 20.28(b), all semiconductor devices are effectively in parallel with the tank capacitor, and hence operate with zero-voltage switching.

Figure 20.29 illustrates implementation of a zero current switching quasi-square wave resonant switch, in a buck converter with input filter. Elements L_f and C_f are large in value, and constitute a single-section L–C input filter. Elements L_r and C_r form the series resonant tank; these elements are placed in series with input filter capacitor C_f. Since C_r and C_f are connected in series, they can be combined into a single small-value capacitor. In this zero-current-switching converter, the peak transistor current is identical to the peak transistor current of an equivalent

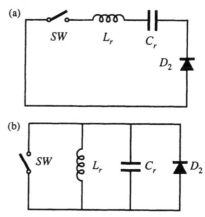

Fig. 20.28 Elimination of converter low-frequency elements reduces the quasi-square-wave switch cells to these networks: (a) ZCS quasi-square-wave network, (b) ZVS quasi-square-wave network.

PWM converter. However, the peak transistor blocking voltage is increased. The ZCS QSW resonant switch exhibits a switch conversion ratio μ that is restricted to the range $0 \le \mu \le 0.5$. Analysis of this resonant switch is given in [13–14].

A buck converter, containing a zero-voltage-switching quasi-square wave (ZVS QSW) resonant switch, is illustrated in Fig. 20.30. Typical waveforms are given in Fig. 20.31. Since the tank inductor L_r and the output filter inductor L are connected in parallel, these two elements can be combined into a single inductor having a small value nearly equal to L_r. Analyses of the ZVS QSW resonant switch are given in [14,15,18]. A related full-bridge converter is described in [16]. The ZVS QSW resonant switch is notable because zero voltage switching is obtained in all semiconductor devices, yet the peak transistor voltage is identical to that of an equivalent PWM switch [13]. However, the peak transistor currents are increased.

Characteristics of the zero-voltage-switching quasi-square wave resonant switch are plotted in Fig. 20.32. The switch conversion ratio $\mu = V_2/V_1$ is plotted as a function of normalized switching fre-

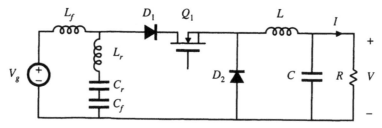

Fig. 20.29 Incorporation of a ZCS quasi-square-wave resonant switch into a buck converter containing an L–C input filter.

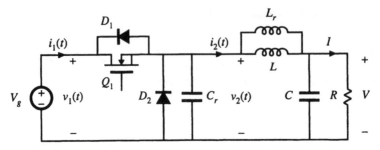

Fig. 20.30   Incorporation of a ZVS quasi-square-wave resonant switch into a buck converter.

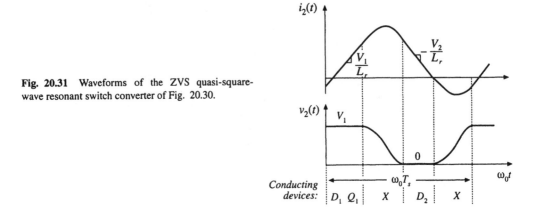

Fig. 20.31   Waveforms of the ZVS quasi-square-wave resonant switch converter of Fig. 20.30.

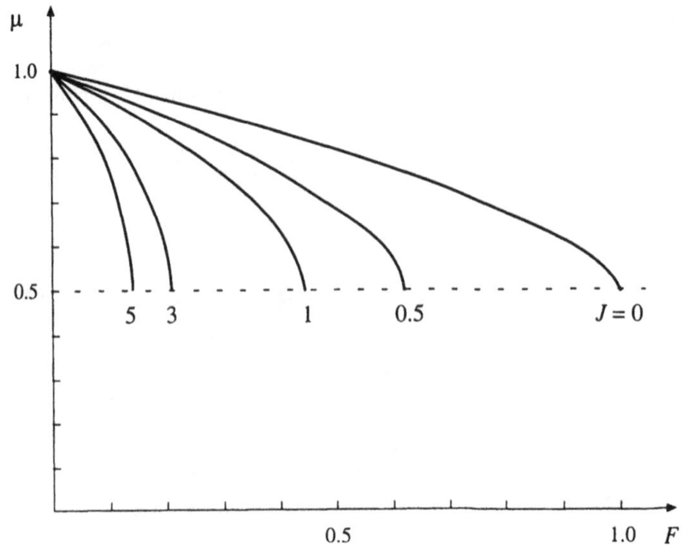

Fig. 20.32   Characteristics of the ZVS quasi-square wave resonant switch network: switch conversion ratio μ, as a function of F and J. Dashed line: ZVS boundary.

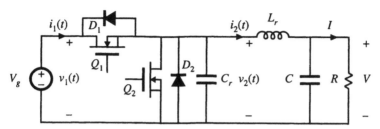

Fig. 20.33 Quasi-square wave ZVS buck converter containing a synchronous rectifier.

quency F and normalized output current J, where these quantities are defined as follows:

$$f_0 = \frac{1}{2\pi\sqrt{L_r C_r}} \quad R_0 = \sqrt{\frac{L_r}{C_r}}$$
$$F = \frac{f_s}{f_0} \quad J = \frac{I_2 R_0}{V_1} \tag{20.66}$$

In addition, the zero-voltage-switching boundary is plotted. It can be seen that the requirement for zero-voltage switching limits the switch conversion ratio μ to the range $0.5 \leq \mu \leq 1$. In consequence, the buck converter of Fig. 20.30 cannot produce output voltages less than $0.5V_g$ without losing the ZVS property. A version which attains $0 \leq \mu \leq 1$, at the expense of increased transistor voltage stress, is described in [17]. In addition, the two-switch version of the ZVS-QSW switch can operate with ZVS for $\mu < 0.5$.

A useful variant of the converter of Fig. 20.30 involves replacement of the diode with a synchronous rectifier, as illustrated in Fig. 20.33 [8,9]. The second transistor introduces an additional degree of freedom in control of the converter, because this transistor can be allowed to conduct longer than the diode would otherwise conduct. This fact can be used to extend the region of zero-voltage switching to conversion ratios approaching zero, and also to operate the converter with constant switching frequency.

Typical tank element waveforms for the circuit of Fig. 20.33 are illustrated in Fig. 20.34. These waveforms resemble those of the single switch case, Fig. 20.31, except that the tank current is negative while transistor Q_2 conducts. The duty cycle D is defined with respect to the turn-off transitions

Fig. 20.34 Waveforms for the two-switch QSW-ZVS converter of Fig. 20.33.

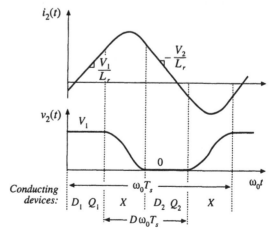

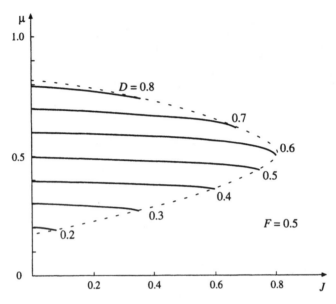

Fig. 20.35 Conversion ratio μ, as a function of duty cycle D and normalized load current J, for the two-switch QSW-ZVS switch illustrated in Fig. 20.33. Curves are plotted for constant-frequency control with $F = 0.5$. The dashed line is the zero-voltage switching boundary.

of transistors Q_1 and Q_2, as illustrated.

Characteristics of the two-switch QSW-ZVS switch network are plotted in Fig. 20.35, for the case of constant switching frequency at $F = 0.5$. The boundary of zero-voltage switching is also illustrated. Operation at a lower value of F causes the ZVS boundary to be extended to larger values of J, and to values of μ that more closely approach the extreme values $\mu = 0$ and $\mu = 1$.

To the commutation intervals can be neglected, one would expect that the switch conversion ratio μ is simply equal to the duty cycle D. It can be seen from Fig. 20.35 that this is indeed the case. The characteristics are approximately horizontal lines, nearly independent of load current J.

Zero-voltage switching quasi-square wave converters exhibit very low switching loss, because all semiconductor elements operate with zero-voltage switching. In the constant-frequency case containing a synchronous rectifier, the converter behavior is nearly the same as for the hard-switched PWM case, since $\mu \approx D$. The major disadvantage is the increased conduction loss, caused by the reversal of the inductor current.

20.4   SOFT SWITCHING IN PWM CONVERTERS

The quasi-square wave approach of the previous section is notable because it attains zero-voltage switching without increasing the peak voltage applied to the transistor. Several related soft-switching approaches have now become popular, which also attain zero-voltage switching without increasing the transistor peak voltage stress. In this section, popular zero-voltage switching versions of the full bridge, forward, and flyback converters, as well as the voltage-source inverter, are briefly discussed.

20.4.1 The Zero-Voltage Transition Full-Bridge Converter

It is possible to obtain soft switching in other types of converters as well. An example is the zero-voltage transition (ZVT) converter based on the full-bridge transformer–isolated buck converter, illustrated in Fig. 20.36 [25–28]. The transistor and diode output capacitances are represented in the figure by capacitances C_{leg}. Commutating inductor L_c is placed in series with the transformer; the net inductance L_c includes both transformer leakage inductance and the inductance of an additional discrete element. This inductor causes the full-bridge switch network to drive an effective inductive load, and results in zero-voltage switching of the primary-side semiconductor devices. Although the waveforms are not sinusoidal, it can nonetheless be said that the switch network output current $i_c(t)$ lags the voltage $v_s(t)$, because the zero crossings of $i_c(t)$ occur after the ZVS switching transitions are completed.

The output voltage is controlled via phase control. As illustrated in Fig. 20.37, both halves of the bridge switch network operate with a 50% duty cycle, and the phase difference between the half-bridge switch networks is controlled. The idealized waveforms of Fig. 20.37 neglect the switching transitions, and the subinterval numbers correspond to those of the more detailed Fig. 20.38. The phase shift variable ϕ lies in the range $0 \leq \phi \leq 1$, and assumes the role of the duty cycle d in this converter. The quantity ϕ is defined as

$$\phi = \frac{(t_1 - t_0)}{\left(\dfrac{T_s}{2}\right)} \tag{20.67}$$

By volt-second balance on the secondary-side filter inductor, the conversion ratio $M(\phi)$ is expressed as

$$M(\phi) = \frac{V}{V_g} = n\phi \tag{20.68}$$

This expression neglects the lengths of the switching transitions.

Although the circuit appears symmetrical, the phase-shift control scheme introduces an asymmetry that causes the two half-bridge switch networks to behave quite differently during the switching transitions. During subintervals 4 and 10, energy is actively transmitted from the source V_g through the switches and transformer. These subintervals are initiated by the switching of the half-bridge network

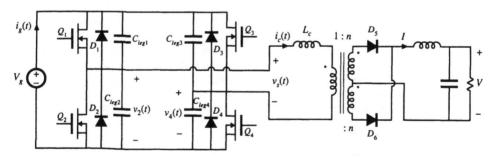

Fig. 20.36 Zero-voltage transition converter, based on the full-bridge isolated buck converter.

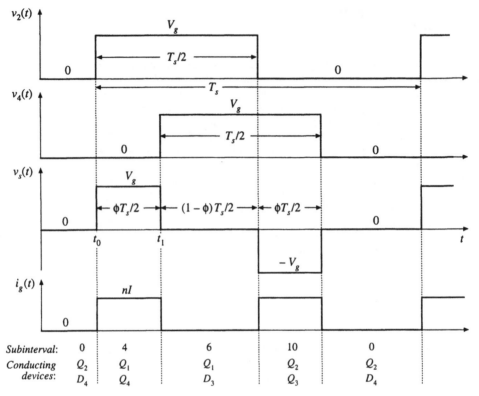

Fig. 20.37 Phase-shift control of the ZVT full-bridge converter. Switching transitions are neglected in this figure, and subinterval numbering follows Fig. 20.38.

composed of the elements Q_1, D_1, Q_2, and D_2, called the "passive-to-active" (P–A) transition [27]. Subintervals 4 and 10 are terminated by the switching of the half-bridge network comprised by the elements Q_3, D_3, Q_4, and D_4, called the "active-to-passive" (A–P) transition.

   The turn-on and turn-off switching processes of this converter are similar to the zero-voltage-switching turn-off process described in the previous section. Detailed primary-side waveforms are illustrated in Fig. 20.38. During subinterval 0, Q_2 and D_4 conduct. If the transformer magnetizing current i_M is negligible, then the commutating inductor current is given by $i_c(t_0) = -nI$, where I is the load current. The passive-to-active transition is initiated when transistor Q_2 is turned off. The negative i_c then causes capacitors C_{leg1} and C_{leg2} to charge, increasing $v_2(t)$. During subinterval 1, L_c, C_{leg1}, and C_{leg2} form a resonant network that rings with approximately sinusoidal waveforms. If sufficient energy was initially stored in L_c, then $v_2(t)$ eventually reaches V_g, terminating subinterval 1. Diode D_1 then clamps $v_2(t)$ to V_g during subinterval 2. Transistor Q_1 is turned on at zero voltage during subinterval 2; in practice, this is implemented by insertion of a small delay between the switching transitions of transistors Q_2 and Q_1.

   If L_c does not initially store sufficient energy to charge the total capacitance ($C_{leg1} + C_{leg2}$) from $v_2 = 0$ to $v_2 = V_g$ during subinterval 1, then $v_2(t)$ will never reach V_g. Switching loss will then occur when transistor Q_1 is turned on. This situation typically occurs at light load, where I is small. Sometimes, the design engineer may choose to simply accept this power loss; after all, other losses such as conduction loss are small at light load. An alternative is to modify the circuit to increase the energy stored in L_c at

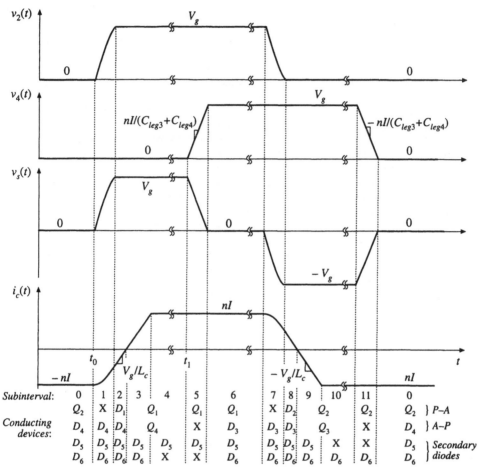

Fig. 20.38 Detailed diagram of primary-side waveforms of the ZVT full-bridge converter, illustrating the zero-voltage switching mechanisms. An ideal transformer is assumed.

$t = t_0$ under light load conditions. One way to accomplish this is to increase the transformer magnetizing current $i_M(t_0)$ to a significant level; at the beginning of subinterval 1, i_c is then equal to $i_c(t_0) = -nI + i_M(t_0)$ with $i_M(t_0) < 0$. At light load where I is small, the magnetizing current maintains the required level of i_c.

During subintervals 0, 1, 2, and 3, secondary-side diodes D_5 and D_6 both conduct; hence, zero voltage appears across all transformer windings. In consequence, voltage V_g is applied to commutating inductor L_c during subintervals 2 and 3, causing $i_c(t)$ to increase with slope V_g/L_c. Current $i_c(t)$ reaches zero at the end of subinterval 2, and increases to the positive value $+nI$ at the end of subinterval 3. The reversal of polarity of $i_c(t)$ enables zero-voltage switching during the next switching transitions, subinterval 5 and subintervals 7–9.

At the end of subinterval 3, the current in diode D_6 has decreased to zero. D_6 then becomes reverse-biased, with zero-current switching. At this instant, diode D_6 must begin to block voltage $2nV_g$.

The output capacitance of D_6 prevents the voltage from changing immediately to $2nV_g$; instead, the resonant circuit formed by L_c and the D_6 output capacitance begins to ring in a manner similar to Fig. 4.54. Peak D_6 voltages are typically observed that are considerably in excess of $2nV_g$, and it is usually necessary to add voltage-clamp snubbers that prevent the secondary-side diode voltages from exceeding a safe value. Several dissipative and non-dissipative approaches are discussed in [26–28].

The active-to-passive switching transition occurs during subinterval 5. This subinterval is initiated when transistor Q_4 is turned off. The positive current $i_c(t_1)$ is equal to the reflected load current nI, and charges capacitors C_{leg3} and C_{leg4} from $v_4 = 0$ to $v_4 = V_g$. Subinterval 5 ends when v_4 reaches V_g; Diode D_3 then becomes forward-biased. Transistor Q_3 is then turned on during subinterval 6, with zero-voltage switching. This is typically implemented by insertion of a small delay between the switching of transistors Q_4 and Q_3. Because i_c is constant and equal to nI during subinterval 5, the active-to-passive transition maintains zero-voltage switching at all load currents.

Circuit behavior during the next half switching period, comprising subintervals 6 to 11, is symmetrical and therefore similar to the behavior observed during subintervals 0 to 5. The switching transitions of transistors Q_1 and Q_2 are passive-to-active transitions, and occur with zero-voltage switching provided that sufficient energy is stored in L_c as described above. The switching transitions of Q_3 and Q_4 are active-to-passive, and occur with zero-voltage switching at all loads.

The zero-voltage transition converter exhibits low primary-side switching loss and generated EMI. Conduction loss is increased with respect to an ideal PWM full-bridge topology, because of the current i_c that circulates through the primary-side semiconductors during subintervals 0 and 6. However, this increase in conduction loss can be small if the range of input voltage variations is narrow. This soft-switching approach has now found commercial success.

20.4.2 The Auxiliary Switch Approach

A similar approach can be used in forward, flyback, and other transformer-isolated converters. As illustrated in Fig. 20.39, an "active-clamp snubber" network consisting of a capacitor and auxiliary MOSFET Q_2 is added, that is effectively in parallel with the original power transistor Q_1 [29]. The MOSFET body diodes and output drain-to-source capacitances, as well as the transformer leakage inductance L_ℓ, participate in the circuit operation. These elements lead to zero-voltage switching, with waveforms similar to those of the ZVT full-bridge converter of Section 20.4.1 or the two-transistor QSW-ZVS switch of Section 20.3.3. The transistors are driven by complementary signals; for example, after turning off Q_1, the controller waits for a short delay time and then turns on Q_2.

The active-clamp snubber can be viewed as a voltage-clamp snubber, similar to the dissipative snubber illustrated in Fig. 20.6. However, the snubber contains no resistor; instead, MOSFET Q_2 allows bidirectional power flow, so that the energy stored in capacitor C_s can flow back into the converter.

The voltage v_s can be found by volt-second inductance on the transformer magnetizing inductance. If the lengths of the commutation intervals are neglected, and if the voltage ripple in $v_s(t)$ can be neglected, then one finds that

$$V_s = \frac{D}{D'} V_g \qquad (20.69)$$

The voltage v_s is effectively an unloaded output of the converter. With the two-quadrant switch provided by Q_2, this output operates in continuous conduction mode with no load, and hence the peak voltage of Q_1 is clamped to the minimum level necessary to balance the volt-seconds applied to the transformer magnetizing inductance.

(a)

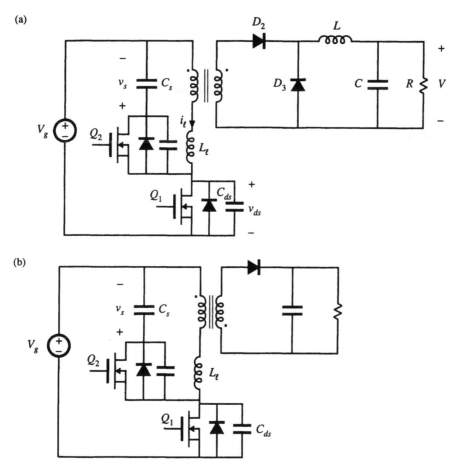

(b)

Fig. 20.39 Active-clamp snubber circuits: (a) forward converter, (b) flyback converter.

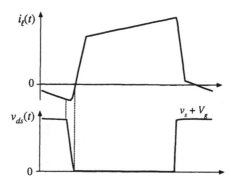

Fig. 20.40 Waveforms of the active-clamp snubber circuit of Fig. 20.39(a).

Typical waveforms for a forward converter incorporating an active-clamp snubber are illustrated in Fig. 20.40. The current $i_r(t)$ reverses direction while Q_2 conducts. When Q_2 turns off, capacitor C_{ds} begins to discharge. When v_{ds} reaches zero, the body diode of Q_1 becomes forward-biased. Q_1 can then be turned on at zero voltage.

An added benefit of the active clamp snubber, when used in a forward converter, is that it resets the transformer. Consequently, the converter can operate at any duty cycle, including duty cycles greater than 50%. When the converter must operate with a wide range of input voltages, this can allow substantial improvements in transistor stresses and efficiency. The MOSFETs in Fig. 20.39 operate with zero-voltage switching, while the secondary-side diodes operate with zero-current switching.

This approach is quite versatile, and similar auxiliary circuits can be added to other converter circuits to obtain zero-voltage switching [30,31].

20.4.3   Auxiliary Resonant Commutated Pole

The auxiliary resonant commutated pole (ARCP) is a related circuit that uses an auxiliary four-quadrant switch (or two equivalent two-quadrant switches) to obtain soft switching in the transistors of a bridge inverter circuit [32–34]. This approach finds application in dc–ac inverter circuits. Figure 20.41 illustrates a half-bridge circuit, or one phase of a three-phase voltage-source inverter, driving an ac load. This circuit can lead to zero-voltage switching that mitigates the switching loss induced by the reverse recovery of diodes D_1 and D_2. Filter inductor L_f is relatively large, so that the output current $i_a(t)$ is essentially constant during the resonant commutation interval. Capacitors C_{ds} are relatively small, and model the output capacitances of the semiconductor devices. Inductor L_r is also relatively small, and elements L_r and C_{ds} form a resonant circuit that rings during part of the commutation process. Semiconductor switching devices Q_3, Q_4, D_3, and D_4 form an auxiliary four-quadrant switch that turns on to initiate the resonant commutation process.

Typical commutation waveforms are illustrated in Fig. 20.42(a), for the case in which the ac load current i_a is positive. Diode D_2 is initially conducting the output current i_a. It is desired to turn off D_2 and turn on Q_1, with zero-voltage switching. This is accomplished with the following sequence:

Interval 1.   Turn on transistor Q_3. Devices D_2, Q_3, and D_4 conduct.

Interval 2.   When the current in D_2 reaches zero, D_2 turns off. A resonant ringing interval occurs.

Interval 3.   When the voltage v_{an} reaches $V_g/2$, diode D_1 begins to become forward-biased. Transistor Q_1 is then immediately turned on at zero voltage.

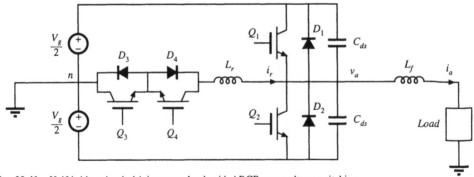

Fig. 20.41   Half-bridge circuit driving an ac load, with ARCP zero-voltage switching.

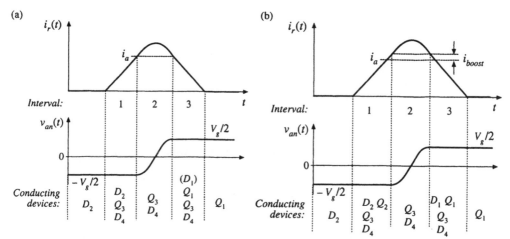

Fig. 20.42 Waveforms of the ARCP circuit of Fig. 20.41: (a) basic waveforms, (b) with current boost.

At the conclusion of interval 3, $i_r(t)$ reaches zero and diode D_3 turns off. For negative current, the process for commutation of diode D_1 is similar, except that transistor Q_4 and diode D_3 conduct the resonant current $i_r(t)$.

One issue related to the waveforms of Fig. 20.42(a) is that the circuit always operates at the boundary of zero-voltage switching. At the end of interval 2, diode D_1 is not actually forward-biased, because its current never actually becomes positive. Instead, transistor Q_1 should be turned on at the beginning of interval 3. If transistor Q_1 is gated on late, then the continued ringing will cause voltage $v_{an}(t)$ to decrease, and zero-voltage switching will be lost.

To further assist in the zero-voltage switching commutation process, transistor Q_2 can be turned on while D_2 conducts, as illustrated in Fig. 20.42(b). Transistor Q_2 is used to lengthen the duration of interval 1: now, when the current $i_r(t)$ exceeds current i_a by an amount i_{boost}, then the controller turns off Q_2 to end interval 1. This causes diode D_1 to become forward-biased during the beginning of interval 3. Transistor Q_1 is then turned on with zero-voltage switching, while D_1 is conducting.

Regardless of whether the circuit operates with the waveforms of Fig. 20.42(a) or (b), the ARCP approach eliminates the switching loss caused by the reverse recovery of diodes D_1 and D_2. Unlike the previous circuits of this chapter, the ARCP has no circulating currents that cause conduction loss, because the tank inductor current $i_r(t)$ is nonzero only in the vicinity of the commutation interval. The approach of Fig. 20.42(a) does not completely eliminate the loss caused by the device output capacitances. This loss is eliminated using the current boost of Fig. 20.42(b), but additional conduction loss is incurred because of the increased peak $i_r(t)$. The waveforms of Fig. 20.42(b) may, in fact, lead to reduced efficiency relative to Fig. 20.42(a)!

20.5 SUMMARY OF KEY POINTS

1. In a resonant switch converter, the switch network of a PWM converter is replaced by a switch network containing resonant elements. The resulting hybrid converter combines the properties of the resonant switch network and the parent PWM converter.

2. Analysis of a resonant or soft-switching switch cell involves determination of the switch conversion ratio μ. The resonant switch waveforms are determined, and are then averaged. The switch conversion ratio μ is

a generalization of the PWM CCM duty cycle d. The results of the averaged analysis of PWM converters operating in CCM can be directly adapted to the related resonant switch converter, simply by replacing d with μ.

3. In the zero-current-switching quasi-resonant switch, diode D_2 operates with zero-voltage switching, while transistor Q_1 and diode D_1 operate with zero-current switching. In the zero-voltage-switching quasi-resonant switch, the transistor Q_1 and diode D_1 operate with zero-voltage switching, while diode D_2 operates with zero-current switching.

4. In the zero-voltage-switching multiresonant switch, all semiconductor devices operate with zero-voltage switching. In consequence, very low switching loss is observed.

5. In the quasi-square-wave zero-voltage-switching resonant switches, all semiconductor devices operate with zero-voltage switching, and with peak voltages equal to those of the parent PWM converter. The switch conversion ratio is restricted to the range $0.5 \le \mu \le 1$. Versions containing synchronous rectifiers can operate with values of μ approaching zero.

6. The zero-voltage transition approach, as well as the active-clamp snubber approach, lead to zero-voltage switching of the transistors and zero-current switching of the diodes. These approaches have been successful in substantially improving the efficiencies of transformer-isolated converters. The auxiliary resonant commutated pole induces zero-voltage switching in bridge circuits such as the voltage-source inverter.

REFERENCES

[1]   P. VINCIARELLI, "Forward Converter Switching at Zero Current," U.S. Patent 4,415,959, Nov. 1983.

[2]   K. LIU, R. ORUGANTI, and F. C. LEE, "Resonant Switches: Topologies and Characteristics," *IEEE Power Electronics Specialists Conference*, 1985 Record, pp. 106-116.

[3]   M. F. SCHLECHT and L. F. CASEY, "Comparison of the Square-Wave and Quasi-Resonant Topologies," *IEEE Applied Power Electronics Conference*, 1987 Record, pp. 124-134.

[4]   K. LIU and F. C. LEE, "Zero Voltage Switching Technique in Dc-Dc Converters," *IEEE Power Electronics Specialists Conference*, 1986 Record, pp. 58-70.

[5]   W. A. TABISZ and F. C. LEE, "Zero-Voltage-Switching Multi-Resonant Technique—A Novel Approach to Improve Performance of High-Frequency Quasi-Resonant Converters," *IEEE Power Electronics Specialists Conference*, 1988 Record, pp. 9-17.

[6]   W. A. TABISZ, M. M. JOVANOVIC, and F. C. LEE, "High Frequency Multi-Resonant Converter Technology and Its Applications," *IEE International Conference on Power Electronics and Variable Speed Drives*, July 17-19, 1990, p. 1-8.

[7]   R. FARRINGTON, M. JOVANOVIC, and F. C. LEE, "Constant-Frequency Zero-Voltage-Switched Multi-Resonant Converters: Analysis, Design, and Experimental Results," *IEEE Power Electronics Specialists Conference*, 1990 Record, pp. 197-205.

[8]   D. MAKSIMOVIĆ, "Synthesis of PWM and Quasi-Resonant Dc-to-Dc Power Converters," Ph.D. thesis, California Institute of Technology, January 1989.

[9]   X. ZHOU, X. ZHANG, J. LIU, P. WONG, J. CHEN, H. WU, L. AMOROSO, F. LEE, and D. CHEN, "Investigation of Candidate VRM Topologies for Future Microprocessors," *IEEE Applied Power Electronics Conference*, 1998 Record, pp. 145–150.

[10]   R. REDL, B. MOLNAR, and N. SOKAL, "Class E Resonant Regulated Dc–Dc Power Converters: Analysis of Operation and Experimental Results at 1.5 MHz," *IEEE Transactions on Power Electronics*, 1986, Vol. 1, No. 2, pp. 111-120.

[11]   N. SOKAL and A. SOKAL, "Class–E, A New Class of High Efficiency Tuned Single-Ended Switching Power Amplifiers," *IEEE Journal of Solid State Circuits*, Vol. SC-10, June 1975, pp. 168-176.

[12]   F. H. RAAB, "Idealized Operation of Class-E Tuned Power Amplifier," *IEEE Transactions on Circuits and Systems*, Vol. 24, No. 12, December 1977, pp. 725-735.

[13]   K. D. T. NGO, "Generalization of Resonant Switches and Quasi-Resonant Dc–Dc Converters," *IEEE Power Electronics Specialists Conference*, 1987 Record, pp. 395-403.

[14]   V. VORPERIAN, "Quasi-Square Wave Converters: Topologies and Analysis," *IEEE Transactions on Power Electronics*, Vol. 3, No. 2, April 1988, pp. 183-191.

[15]   D. MAKSIMOVIĆ, "Design of the Zero-Voltage-Switching Quasi-Square-Wave Resonant Switch," *IEEE Power Electronics Specialists Conference*, 1993 Record, pp. 323-329.

[16]   O. D. PATTERSON and D. M. DIVAN, "Pseudo-Resonant Full-Bridge Dc–Dc Converter," *IEEE Power Electronics Specialists Conference*, 1987 Record, pp. 424-430.

[17]   Y. JANG and R. ERICKSON, "New Quasi-Square Wave and Multi-Resonant Integrated Magnetic Zero Voltage Switching Converters," *IEEE Power Electronics Specialists Conference*, 1993 Record, pp. 721-727.

[18]   V. VORPERIAN, R. TYMERSKI, and F. C. LEE, "Equivalent Circuit Models for Resonant and PWM Switches," *IEEE Transactions on Power Electronics*, Vol. 4, No. 2, April 1989, pp. 205-214.

[19]   S. FREELAND and R. D. MIDDLEBROOK, "A Unified Analysis of Converters with Resonant Switches", *IEEE Power Electronics Specialists Conference*, 1987 Record, pp. 20-30.

[20]   A. WITULSKI and R. ERICKSON, "Extension of State-Space Averaging to Resonant Switches—and Beyond," *IEEE Transactions on Power Electronics*, Vol. 5, No. 1, pp. 98-109, January 1990.

[21]   D. MAKSIMOVIĆ and S. ĆUK, "A General Approach to Synthesis and Analysis of Quasi-Resonant Converters," *IEEE Transactions on Power Electronics*, Vol. 6, No. 1, January 1991, pp. 127-140.

[22]   R. ERICKSON, A. HERNANDEZ, A. WITULSKI, and R. XU, "A Nonlinear Resonant Switch," *IEEE Transactions on Power Electronics*, Vol. 4, No. 2, April 1989, pp. 242-252.

[23]   I. BARBI, D. MARTINS, and R. DO PRADO, "Effects of Nonlinear Resonant Inductor on the Behavior of Zero-Voltage Switching Quasi-Resonant Converters," *IEEE Power Electronics Specialists Conference*, 1990 Record, pp. 522-527.

[24]   S. FREELAND, "I. A Unified Analysis of Converters with Resonant Switches, II. Input-Current Shaping for Single-Phase Ac–Dc Power Converters," Ph.D. Thesis, California Institute of Technology, 1988.

[25]   R. FISHER, K. NGO, and M. KUO, "A 500 kHz, 250 W Dc–dc Converter with Multiple Outputs Controlled by Phase-Shifted PWM and Magnetic Amplifiers," *Proceedings of High Frequency Power Conversion Conference*, pp. 100–110, May 1988.

[26]   L. MWEENE, C. WRIGHT, and M. SCHLECHT, "A 1 kW, 500 kHz Front-End Converter for a Distributed

Power Supply System," *IEEE Applied Power Electronics Conference*, 1989 Record, pp. 423–432.

[27]   R. REDL, L. BELOGH, and D. EDWARDS, "Optimum ZVS Full-Bridge DC/DC Converter with PWM Phase-Shift Control: Analysis, Design Considerations, and Experimental Results," *IEEE Applied Power Electronics Conference*, 1994 Record, pp. 159–165.

[28]   J. G. CHO, J. A. SABATE, and F. C. LEE, "Novel Full Bridge Zero-Voltage-Transition PWM DC/DC Converter for High Power Applications," *IEEE Applied Power Electronics Conference*, 1994 Record, pp. 143–149.

[29]   P. VINCIARELLI, "Optimal Resetting of the Transformer's Core in Single-Ended Forward Converters," Reissued U.S. Patent No. Re. 36,098, Feb. 16, 1999.

[30]   C. DUARTE and I. BARBI, "A Family of ZVS-PWM Active-Clamping Dc-to-Dc Converters: Synthesis, Analysis, Design, and Experimentation," *IEEE Transactions on Circuits and Systems—I: Fundamental Theory and Applications*, Vol. 44, No. 8, pp. 698–704, Aug. 1997.

[31]   P. HENG and R. ORUGANTI, "Family of Two-Switch Soft-Switched Asymmetrical PWM Dc/Dc Converters," *IEEE Power Electronics Specialists Conference*, 1994 Record, pp. 85–94.

[32]   R. DEDONCKER and J. LYONS, "The Auxiliary Resonant Commutated Pole Converter," *IEEE Industry Applications Society Annual Meeting*, 1990 Record, pp. 1228–1235.

[33]   R. TEICHMANN and S. BERNET, "Investigation and Comparison of Auxiliary Resonant Commutated Pole Converter Topologies," *IEEE Power Electronics Specialists Conference*, 1998 Record, pp. 15-23, May 1998.

[34]   W. MCMURRAY, "Resonant Snubbers with Auxiliary Devices," *IEEE Transactions on Industry Applications*, Vol. 29, No. 2, pp. 355–361, 1993.

PROBLEMS

20.1   In the forward converter of Fig. 20.43, L and C are large filter elements while L_p, L_s, and C_r have relatively small values. The transformer reset mechanism is not shown; for this problem, you may assume that the transformer is ideal.

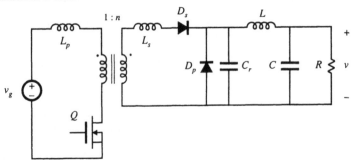

Fig. 20.43   Forward converter with resonant switch, Problem 20.1.

(a)   Classify the resonant switch.

(b)   Which semiconductor devices operate with zero-voltage switching? With zero-current switching?

(c)   What is the resonant frequency?

20.2   In the high-voltage converter of Fig. 20.44, capacitor C is relatively large in value. The transformer model includes an ideal $1{:}n$ transformer, in conjunction with magnetizing inductance L_{mp} (referred to the primary side) and winding capacitance C_{ws} (referred to the secondary side). Transistor Q and diode D_p exhibit total output capacitance C_p, while the output capacitance of diode D_s is C_s. Other nonidealities, such as transformer leakage inductance, can be ignored. The resonant switch is well-designed, such that all elements listed above contribute to ideal operation of the converter and resonant switch.

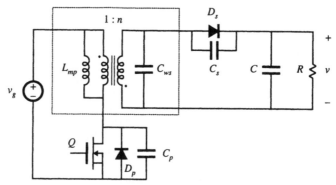

Fig. 20.44   High-voltage dc–dc converter containing a resonant switch network, Problem 20.2.

(a)   What type of resonant switch is employed? What is the parent PWM converter?

(b)   Which semiconductor devices operate with zero-voltage switching? With zero-current switching?

(c)   What is the tank resonant frequency?

(d)   Sketch the waveforms of the transistor drain-to-source voltage and transformer magnetizing current.

20.3   In the transformer-isolated dc–dc converter of Fig. 20.45, capacitors C_1 and C_2 and inductors L_1 and L_M are relatively large in value, so that they have small switching ripples. The transformer model includes an ideal $1{:}n$ transformer, in conjunction with magnetizing inductance L_M (referred to the primary side) and leakage inductances $L_{\ell 1}$ and $L_{\ell 2}$ as shown. Transistor Q_1 exhibits output capacitances C_{ds}, while the output capacitance of diode D_1 is C_d. MOSFET Q_1 contains a body diode (not explicitly shown). Other nonidealities can be ignored. The resonant switch is well-designed, such that all elements listed above contribute to ideal operation of the converter and resonant switch.

(a)   What type of resonant switch is employed? What is the parent PWM converter?

(b)   Which semiconductor devices operate with zero-voltage switching? With zero-current switching?

20.4   A buck-boost converter is realized using a half-wave ZCS quasi-resonant switch. The load resistance has value R, the input voltage has value V_g, and the converter switching frequency is f_s.

(a)   Sketch the circuit schematic.

(b)   Write the complete system of equations that can be solved to determine the output voltage V, in terms of the quantities listed above and the component values. It is not necessary to actually solve your equations. You may also quote results listed in this textbook.

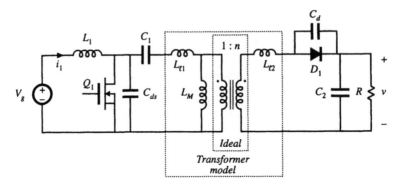

Fig. 20.45   Transformer-isolated dc–dc converter containing a resonant switch network, Problem 20.3.

20.5   It is desired to design a half-wave zero-current-switching quasi-resonant forward converter to operate with the following specifications: $V_g = 320$ V, $V = 42$ V, 5 W $\leq P \leq 100$ W. Design the converter to operate with a maximum switching frequency of 1 MHz and a switch conversion ratio of $\mu = 0.45$. Attempt to minimize the peak transistor current, while maintaining zero current switching at all operating points. You may neglect the transformer magnetizing current, and ignore the transformer reset scheme.

   (a)   Specify your choices for the turns ratio n, and the tank elements L_r and C_r, referred to the transformer secondary side.

   (b)   For your design of part (a), what is the minimum switching frequency?

   (c)   What is the worst-case peak transistor current?

20.6   Analysis of the ZVS quasi-resonant switch of Fig. 20.24.

   (a)   For each subinterval, sketch the resonant switch cell circuit, and derive expressions for the tank inductor current and capacitor voltage waveforms.

   (b)   For subinterval 2, in which Q_1/D_1 are off and D_2 conducts, write the loop equation which relates the tank capacitor voltage, tank inductor voltage, and any other network voltages as appropriate. Hence, for subinterval 2 relate the integral of the tank capacitor voltage to the change in tank inductor current.

   (c)   Determine the switch-network terminal-waveform average values, and hence derive an expression for the switch conversion ratio μ. Verify that your result coincides with Eq. (20.61).

20.7   Analysis of the full-bridge zero-voltage transition converter of Section 20.4.1. The converter of Fig. 20.36 operates with the waveforms illustrated in Fig. 20.38. According to Eq. (20.68), the conversion ratio of this converter is given approximately by $M(\phi) = n\phi$.

   Derive an exact expression for M, based on the waveforms given in Fig. 20.38. Your result should be a function of the length of subinterval 4, the load current, the switching frequency, and the values of the inductance and capacitances. *Note*: there is a reasonably simple answer to this question.

Appendices

Appendix A
RMS Values of Commonly Observed Converter Waveforms

The waveforms encountered in power electronics converters can be quite complex, containing modulation at the switching frequency and often also at the ac line frequency. During converter design, it is often necessary to compute the rms values of such waveforms. In this appendix, several useful formulas and tables are developed which allow these rms values to be quickly determined.

RMS values of the doubly-modulated waveforms encountered in PWM rectifier circuits are discussed in Section 18.5.

A.1 SOME COMMON WAVEFORMS

DC, Fig. A.1:

$$rms = I \tag{A.1}$$

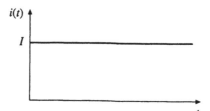

Fig. A.1

DC plus linear ripple, Fig. A.2:

$$rms = I \sqrt{1 + \frac{1}{3}\left(\frac{\Delta i}{I}\right)^2} \qquad (A.2)$$

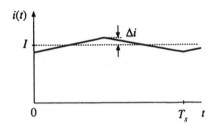

Fig. A.2

Square wave, Fig. A.3:

$$rms = I_{pk} \qquad (A.3)$$

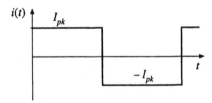

Fig. A.3

Sine wave, Fig. A.4:

$$rms = \frac{I_{pk}}{\sqrt{2}} \qquad (A.4)$$

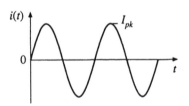

Fig. A.4

Pulsating waveform, Fig. A.5:

$$rms = I_{pk} \sqrt{D} \qquad (A.5)$$

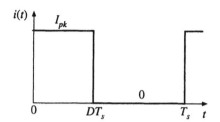

Fig. A.5

Pulsating waveform with linear ripple, Fig. A.6:

$$rms = I \sqrt{D} \sqrt{1 + \frac{1}{3}\left(\frac{\Delta i}{I}\right)^2}$$

(A.6)

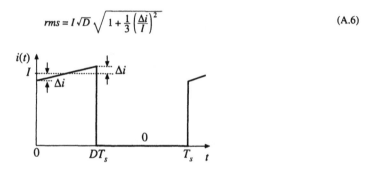

Fig. A.6

Triangular waveform, Fig. A.7:

$$rms = I_{pk} \sqrt{\frac{D_1 + D_2}{3}}$$

(A.7)

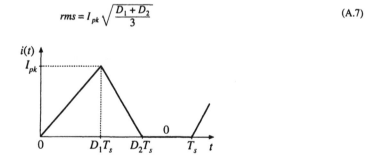

Fig. A.7

Triangular waveform, Fig. A.8:

$$rms = I_{pk} \sqrt{\frac{D_1}{3}}$$

(A.8)

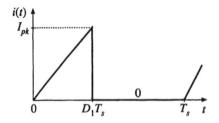

Fig. A.8

Triangular waveform, no dc component, Fig. A.9:

$$rms = \frac{\Delta i}{\sqrt{3}} \tag{A.9}$$

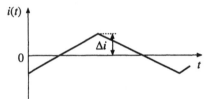

Fig. A.9

Center-tapped bridge winding waveform, Fig. A.10:

$$rms = \frac{1}{2} I_{pk} \sqrt{1 + D} \tag{A.10}$$

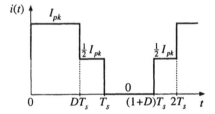

Fig. A.10

General stepped waveform, Fig. A.11:

$$rms = \sqrt{D_1 I_1^2 + D_2 I_2^2 + \cdots} \tag{A.11}$$

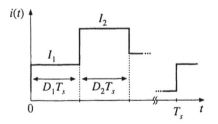

Fig. A.11

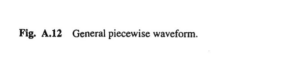
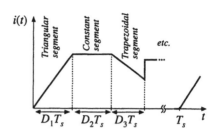

Fig. A.12   General piecewise waveform.

A.2   GENERAL PIECEWISE WAVEFORM

For a periodic waveform composed of n piecewise segments as in Fig. A.12, the rms value is

$$rms = \sqrt{\sum_{k=1}^{n} D_k u_k} \tag{A.12}$$

where D_k is the duty cycle of segment k, and u_k is the contribution of segment k. The u_ks depend on the shape of the segments—several common segment shapes are listed below:

Constant segment, Fig. A.13:

$$u_k = I_1^2 \tag{A.13}$$

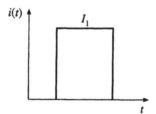

Fig. A.13

Triangular segment, Fig. A.14:

$$u_k = \frac{1}{3} I_1^2 \tag{A.14}$$

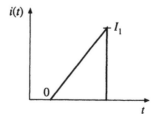

Fig. A.14

Trapezoidal segment, Fig. A.15:

$$u_k = \frac{1}{3}\left(I_1^2 + I_1 I_2 + I_2^2\right) \tag{A.15}$$

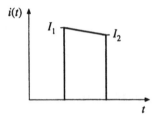

Fig. A.15

Sinusoidal segment, half or full period, Fig. A.16:

$$u_k = \frac{1}{2} I_{pk}^2 \tag{A.16}$$

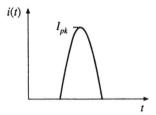

Fig. A.16

Sinusoidal segment, partial period: as in Fig. A.17, a sinusoidal segment of less than one half-period, which begins at angle θ_1 and ends at angle θ_2. The angles θ_1 and θ_2 are expressed in radians:

$$u_k = \frac{1}{2} I_{pk}^2 \left(1 - \frac{\sin\left(\theta_2 - \theta_1\right)\cos\left(\theta_2 + \theta_1\right)}{\left(\theta_2 - \theta_1\right)}\right) \tag{A.17}$$

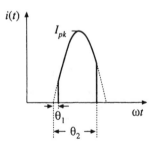

Fig. A.17

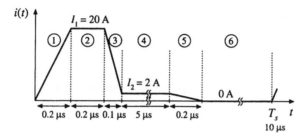

Fig. A.18   Example: an approximate transistor current waveform, including estimated current spike due to diode stored charge.

Example

A transistor current waveform contains a current spike due to the stored charge of a freewheeling diode. The observed waveform can be approximated as shown in Fig. A1.18. Estimate the rms current.

The waveform can be divided into six approximately linear segments, as shown. The D_k and u_k for each segment are

1. Triangular segment:

$$D_1 = (0.2\ \mu s)/(10\ \mu s) = 0.02$$

$$u_1 = I_1^2/3 = (20\ \text{A})^2/3 = 133\ \text{A}^2$$

2. Constant segment:

$$D_2 = (0.2\ \mu s)/(10\ \mu s) = 0.02$$

$$u_2 = I_1^2 = (20\ \text{A})^2 = 400\ \text{A}^2$$

3. Trapezoidal segment:

$$D_3 = (0.1\ \mu s)/(10\ \mu s) = 0.01$$

$$u_3 = (I_1^2 + I_2^2 + I_3^2)/3 = 148\ \text{A}^2$$

4. Constant segment:

$$D_4 = (5\ \mu s)/(10\ \mu s) = 0.5$$

$$u_4 = I_2^2 = (2\ \text{A})^2 = 4\ \text{A}^2$$

5. Triangular segment:

$$D_5 = (0.2\ \mu s)/(10\ \mu s) = 0.02$$

$$u_5 = I_2^2/3 = (2\ \text{A})^2/3 = 1.3\ \text{A}^2$$

6. Zero segment:

$$u_6 = 0$$

The rms value is

$$rms = \sqrt{\sum_{k=1}^{6} D_k u_k} = 3.76 \text{ A} \tag{A.18}$$

Even though its duration is very short, the current spike has a significant impact on the rms value of the current—without the current spike, the rms current is approximately 2.0 A.

Appendix B
Simulation of Converters

Computer simulation can be a powerful tool in the engineering design process. Starting from design specifications, an initial design typically includes selection of system and circuit configurations, as well as component types and values. In this process, component and system models are constructed based on vendor-supplied data, and by applications of analysis and modeling techniques. These models, validated by experimental data whenever possible, are the basis upon which the designer can choose parameter values and verify the achieved performance against the design specifications. One must take into account the fact that actual parameter values will not match their nominal values because of inevitable production tolerances, changes in environmental conditions (such as temperature), and aging. In the design verification step, worst-case analysis (or other reliability and production yield analysis) is performed to judge whether the specifications are met under all conditions, i.e., for expected ranges of component parameter values. Computer simulation is very well suited for this task: using reliable models and appropriate simulation setups, the system performance can be tested for various sets of component parameter values. One can then perform design iterations until the worst-case behavior meets specifications, or until the system reliability and production yield are acceptably high.

In the design verification of power electronic systems by simulation, it is often necessary to use component and system models of various levels of complexity:

1. *Detailed, complex models that attempt to accurately represent physical behavior of devices.* Such models are necessary for tasks that involve finding switching times, details of switching transitions and switching loss mechanisms, or instantaneous voltage and current stresses. Component vendors often provide libraries of such device models. To complete a detailed circuit model, one must also carefully examine effects of packaging and board interconnects. With fast-switching power semiconductors, simulation time steps of a few nanoseconds or less may be required, especially during on/off switching transitions. Because of the complexity of detailed device models, and the fine time resolution, the simulation tasks can be very time consuming. In practice, time-domain simulations using detailed device models are usually performed only

on selected parts of the system, and over short time intervals involving a few switching cycles at most. Devices for power converters, and detailed physical device modeling, are areas of active research and development beyond the scope of this book.

2. *Simplified device models.* Since an on/off switching transition usually takes a small fraction of a switching cycle, the basic operation of switching power converters can be explained using simplified, idealized device models. For example, a MOSFET can be modeled as a switch with a small (ideally zero) on-resistance R_{on} when on, and a very large off-resistance (ideally an open circuit) when off. Such simplified models yield physical insight into the basic operation of switching power converters, and provide the starting point for developments of analytical models described throughout this book. Simplified device models are also useful for time-domain simulations aimed at verifying converter and controller operation, switching ripples, current and voltage stresses, and responses to load or input transients. Since device models are simple, and details of switching transitions are ignored, tasks that require simulations over many switching cycles can be completed efficiently using general-purpose circuit simulators. In addition, specialized tools have been developed to support fast transient simulation of switching power converters based on idealized, piecewise-linear device models [1-7], or a combination of piecewise-linear and nonlinear models [8].

3. *Averaged converter models.* Averaged models that are well suited for prediction of converter steady-state and dynamic responses are discussed throughout this book. These models are essential design tools because they provide physical insight and lead to analytical results that can be used in the design process to select component parameter values for a given set of specifications. In the design verification step, simulations of averaged converter models can be performed to test for losses and efficiency, steady-state voltages and currents, stability, and large-signal transient responses. Since switching transitions and ripples are removed by averaging, simulations over long time intervals and over many sets of parameter values can be completed efficiently. As a result, averaged models are also well suited for simulations of large electronic systems that include switching converters. Furthermore, since large-signal averaged models are nonlinear, but time-invariant, small-signal ac simulations can be used to generate various frequency responses of interest. Selected references on averaged converter modeling for simulation are listed at the end of this chapter [9-18].

Averaged models for computer simulation are covered in this appendix. Based on the material presented in Section 7.4, averaged switch models for computer simulation of converters operating in continuous conduction mode are described in Section B.1. Application examples include finding SEPIC dc conversion ratio and efficiency, and large-signal transient responses of a buck-boost converter. Section B.2 describes an averaged switch model suitable for simulation of converters that may operate either in continuous conduction mode or in discontinuous conduction mode. Application examples include finding SEPIC open-loop frequency responses in CCM and DCM, loop-gain, phase margin and closed-loop responses of a buck voltage regulator, and current harmonics in a DCM boost rectifier. Based on the results from Chapter 12, a simulation model for converters with current programmed control is described in Section B.3, together with a buck converter example that compares control-to-output frequency responses with current programmed control against duty-cycle control.

It is assumed that the reader is familiar with basics of Spice circuit simulations. All simulation models and examples in this appendix are prepared using the PSpice circuit simulator [19]. Netlists are included to help explain details of model implementation and simulation analysis options. Usually, instead of writing netlists, the user would enter circuit diagrams and analysis options from a front-end schematic capture tool. The examples and the library *switch.lib* of subcircuit models described in this appendix are available on-line. Similar models and examples can be constructed for use with other simulation tools.

(a)

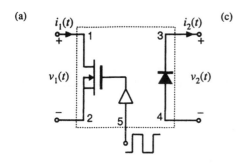

(c)

```
**************************************************************
* Subcircuit: CCM1
* Application: two-switch PWM converters
* Limitations: ideal switches, CCM only, no transformer
**************************************************************
* Parameters: none
**************************************************************
* Nodes:
* 1: transistor positive (drain for an n-channel MOS)
* 2: transistor negative  (source for an n-channel MOS)
* 3: diode cathode
* 4: diode anode
* 5: duty cycle control input
**************************************************************
.subckt CCM1 1 2 3 4 5
Et 1 2 value={(1-v(5))*v(3,4)/v(5)}
Gd 4 3 value={(1-v(5))*i(Et)/v(5)}
.ends
**************************************************************
```

(b)

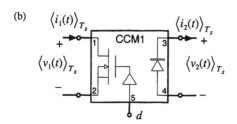

Fig. B.1   Averaged switch model CCM1: (a) the general two-switch network: (b) symbol for the averaged switch subcircuit model; (c) PSpice netlist of the subcircuit.

B.1   AVERAGED SWITCH MODELS FOR CONTINUOUS CONDUCTION MODE

The central idea of the *averaged switch modeling* described in Section 7.4 is to identify a switch network in the converter, and then to find an averaged circuit model. The resulting averaged switch model can then be inserted into the converter circuit to obtain a complete model of the converter. An important feature of the averaged switch modeling approach is that the same model can be used in many different converter configurations; it is not necessary to rederive an averaged equivalent circuit for each particular converter. This feature is also very convenient for construction of averaged circuit models for simulation. A general-purpose subcircuit represents a large-signal nonlinear averaged switch model. The converter averaged circuit for simulation is then obtained by replacing the switch network with this subcircuit. Based on the discussion in Section 7.4, subcircuits that represent CCM averaged switch models are described in this section, together with application examples.

B.1.1   Basic CCM Averaged Switch Model

The large-signal averaged switch model for the general two-switch network of Fig. 7.39(a) is shown in Fig. 7.39(c). A PSpice subcircuit implementation of this model is shown in Fig. B.1. The subcircuit has five nodes. The transistor port of the averaged switch network is connected between the nodes 1 and 2, while the diode port is comprised of nodes 3 and 4. The duty ratio $d = v(5)$ is the control input to the subcircuit at the node 5. The quantity $v(5)$ is a voltage that is equal to the duty cycle, and that lies in the range zero to one volt. Figure B.1(c) shows the netlist of the subcircuit. The netlist consists of only four lines of code and several comment lines (the lines starting with *). The .subckt line defines the name (CCM1) of the subcircuit and the interface nodes. The value of the controlled voltage source E_t, which models the transistor port of the averaged switch network, is written according to Eq. (7.136):

$$\left\langle v_1(t) \right\rangle_{T_s} = \frac{d'(t)}{d(t)} \left\langle v_2(t) \right\rangle_{T_s} \tag{B.1}$$

Note that $v(3,4)$ in the subcircuit of Fig. B.1 is equal to the switch network independent input $\left\langle v_2(t) \right\rangle_{T_s}$. Also, $d(t) = v(5)$, and $d'(t) = 1 - d(t) = 1 - v(5)$. The value of the controlled current source G_d, which models the diode port, is computed according to Eq. (7.137):

$$\left\langle i_2(t) \right\rangle_{T_s} = \frac{d'(t)}{d(t)} \left\langle i_1(t) \right\rangle_{T_s} \tag{B.2}$$

The switch network independent input $\left\langle i_1(t) \right\rangle_{T_s}$ equals the current $i(E_t)$ through the controlled voltage source E_t. The .ends line completes the subcircuit netlist. The subcircuit CCM1 is included in the model library *switch.lib*.

An advantage of the subcircuit CCM1 of Fig. B.1 is that it can be used to construct an averaged circuit model for simulation of any two-switch PWM converter operating in continuous conduction mode, subject to the assumptions that the switches can be considered ideal, and that the converter does not include a step-up or step-down transformer. The subcircuit can be further refined to remove these limitations. In converters with an isolation transformer, the right-hand side of Eqs. (B.1) and (B.2) should be divided by the transformer turns ratio. Inclusion of switch conduction losses is discussed in the next section.

A disadvantage of the model in Fig. B.1 is that Eqs. (B.1) and (B.2) have a discontinuity at duty cycle equal to zero. In applications of the subcircuit, it is necessary to restrict the duty-cycle to the range $0 < D_{min} \le d \le 1$.

Following the approach of this section, subcircuits can be constructed for the large-signal averaged models of the buck switch network (see Fig. 7.50(a), and Eqs. (7.150)), and the boost switch network (see Fig. 7.46(a) and Eqs. (7.146)). An advantage of these models is that their defining equations do not have the discontinuity problem at $d = 0$.

B.1.2   CCM Averaged Switch Model that Includes Switch Conduction Losses

Let us modify the model of Fig. B.1 to include switch conduction losses. Figure B.2 shows simple device models that include transistor and diode conduction losses in the general two-switch network of Fig. B.1(a). The transistor is modeled as an ideal switch in series with an on-resistance R_{on}. The diode is modeled as an ideal diode in series with a forward voltage drop V_D and resistance R_D.

Construction of dc equivalent circuits to find dc conversion ratio and efficiency of converters is discussed in Chapter 3. Derivation of an averaged switch model that includes conduction losses arising from R_{on} and V_D is described in Section 7.4.5. Following the same averaged switch modeling approach, we can find the following relationships that describe the averaged switch model for the switch network of Fig. B.2:

$$\left\langle v_1(t) \right\rangle_{T_s} = \left(\frac{R_{on}}{d(t)} + \frac{d'(t)R_D}{d^2(t)} \right) \left\langle i_1(t) \right\rangle_{T_s} + \frac{d'(t)}{d(t)} \left(\left\langle v_2(t) \right\rangle_{T_s} + V_D \right) \tag{B.3}$$

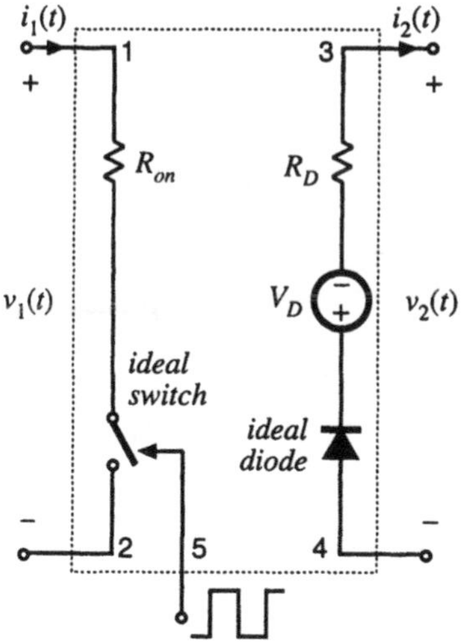

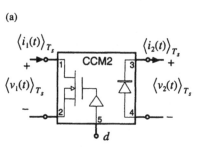

Fig. B.2   Switch network model that includes conduction loss elements R_{on}, V_D and R_D.

(a)

(b)

Fig. B.3   Subcircuit implementation of the CCM averaged switch model that includes conduction losses: (a) circuit symbol; (b) PSpice netlist for the subcircuit.

```
*******************************************************
* MODEL: CCM2
* Application: two-switch PWM converters, includes
*              conduction losses due to Ron, VD, RD
* Limitations: CCM only, no transformer
*******************************************************
* Parameters:
*     Ron = transistor on-resistance
*     VD = diode forward voltage drop
*     RD = diode on-resistance
*******************************************************
* Nodes:
* 1: transistor positive (drain for an n-channel MOS)
* 2: transistor negative (source for an n-channel MOS)
* 3: diode cathode
* 4: diode anode
* 5: duty cycle control input
*******************************************************
.subckt CCM2 1 2 3 4 5
+params: Ron=0 VD=0 RD=0
Er 1 1x value={i(Et)*(Ron+(1-v(5))*RD/v(5))/v(5)}
Et 1x 2 value={(1-v(5))*(v(3,4)+VD)/v(5)}
Gd 4 3 value={(1-v(5))*i(Et)/v(5)}
.ends
*******************************************************
```

$$\langle i_2(t) \rangle_{T_s} = \frac{d'(t)}{d(t)} \langle i_1(t) \rangle_{T_s} \tag{B.4}$$

A subcircuit implementation of the averaged switch model described by Eqs. (B.3) and (B.4) is shown in Fig. B.3 The subcircuit terminal nodes are the same as in the CCM1 subcircuit: the transistor port is between the nodes 1 and 2; the diode port is between the nodes 3 and 4; the duty ratio $d = v(5)$ is the con-

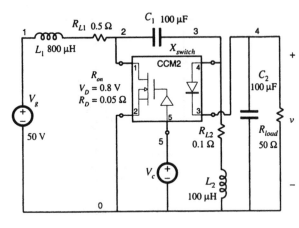

R_{L1} 0.5 Ω C_1 100 µF

L_1 800 µH

R_{on}
V_D = 0.8 V
R_D = 0.05 Ω

V_g

50 V

X_{switch}

CCM2

V_c

R_{L2}
0.1 Ω

L_2
100 µH

C_2
100 µF

R_{load}
50 Ω

Fig. B.4   SEPIC simulation example.

SEPIC DC conversion ratio and efficiency

```
* Define parameters:
.param Ron=0.0 VD=0.8 RD=0.05
* Analysis setup:
.dc lin Vc 0.1 1 0.01
.step lin PARAM Ron 0 1 0.5

* Converter netlist:
Vg 1 0 50V
L1 1 2x 800u
RL1 2x 2 0.5
L2 0 3x 100uH
RL2 3x 3 0.1
C1 2 3 100uF
C2 4 0 100uF
Xswitch 2 0 4 3 5 CCM2
+params: Ron={Ron} VD={VD} RD={RD}
Rload 4 0 50

* Duty cycle input:
Vc 5 0 0.5

.lib switch.lib
.probe
.end
```

trol input to the subcircuit at the node 5. Two controlled voltage sources in series, E_r and E_t, are used to generate the port 1 (transistor) averaged voltage according to Eq. (B.3). The controlled voltage source E_r models the voltage drop across the equivalent resistance $R_{on}/d(t) + d'(t)R_D/d^2(t)$ in Eq. (B.3). Note that this equivalent resistance is a nonlinear function of the switch duty cycle $d(t)$. The controlled voltage source E_t shows how the port 1 (transistor) averaged voltage depends on the port 2 (diode) averaged voltage. The controlled current source G_d models the averaged diode current according to Eq. (B.4). The subcircuit CCM2 has three parameters (R_{on}, V_D, and R_D) that can be specified when the subcircuit is used in a converter circuit. The default values of the subcircuit parameters, $R_{on} = 0$, $V_D = 0$, and $R_D = 0$, are defined in the .subckt line. These values correspond to the ideal case of no conduction losses. The subcircuit CCM2 is included in the model library *switch.lib*.

The model of Fig. B.3 is based on the simple device models of Fig. B.2. It is assumed that inductor current ripples are small and that the converter operates in continuous conduction mode. Many practical converters, however, must operate in discontinuous conduction mode at low duty cycles where the diode forward voltage drop is comparable to or larger than the output voltage. In such cases, the model of Fig. B.2, which includes V_D as a fixed voltage generator, gives incorrect, physically impossible results for polarities of converter voltages and currents, losses and efficiency.

B.1.3   Example: SEPIC DC Conversion Ratio and Efficiency

Let us consider an example of how the subcircuit CCM2 can be used to generate dc conversion ratio and efficiency curves for a CCM converter. As an example, Figure B.4 shows a SEPIC averaged circuit model. The converter circuit can be found in Fig. 6.38(a), or in Fig. 7.37. To construct the averaged circuit model for simulation, the switch network is replaced by the subcircuit CCM2. In the converter netlist shown in Fig. B.4, the X_{switch} line shows how the subcircuit is connected to other parts of the converter. The switch duty cycle is set by the voltage source V_c. All other parts of the converter circuit are simply copied to the averaged circuit model. Inductor winding resistances $R_{L1} = 0.5$ Ω and $R_{L2} = 0.1$ Ω are

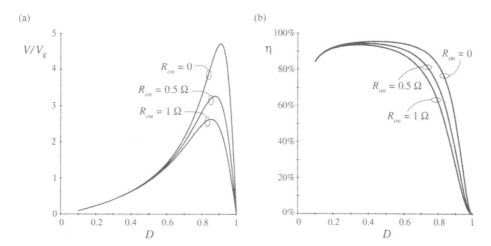

Fig. B.5 SEPIC simulation example: (a) dc conversion ratio and (b) efficiency.

included to model copper losses of the inductors L_1 and L_2, respectively. The switch conduction loss parameters are defined by the .param line in the netlist: $R_{on} = 0$, $V_D = 0.8$ V, $R_D = 0.05 \ \Omega$. Notice how these values are passed to the subcircuit CCM2 in the X_{switch} line. In this example, all other losses in the converter are neglected. A dc sweep analysis (see the .dc line in the netlist) is set to vary the dc voltage source V_c from 0.1 V to 1 V, in 0.01 V increments, which corresponds to varying the switch duty cycle over the range from $D = 0.1$ to $D = 1$. The range of duty cycles from zero to 0.1 is not covered because of the model discontinuity problem at $D = 0$ (discussed in Section B.1.1), and because the model predictions for conduction losses at low duty cycles are not valid, as discussed in Section B.1.2. The dc sweep analysis is repeated for values of the switch on-resistance in the range from $R_{on} = 0 \ \Omega$ to $R_{on} = 1 \ \Omega$ in 0.5 Ω increments (see the .step line in the netlist). The .lib line refers to the *switch.lib* library, which contains definitions of the subcircuit CCM2 and all other subcircuit models described in this appendix.

Simulation results for the dc output voltage V and the converter efficiency η are shown in Fig. B.5. Several observations can be made based on the modeling approach and discussions presented in Chapter 3. At low duty cycles, efficiency drops because the diode forward voltage drop is comparable to the output voltage. At higher duty cycles, the converter currents increase, so that the conduction losses increase. Eventually, for duty cycles approaching 1, both the output voltage and the efficiency approach zero. Given a desired dc output voltage and efficiency, the plots in Fig. B.5 can be used to select the transistor with an appropriate value of the on-resistance.

B.1.4 Example: Transient Response of a Buck–Boost Converter

In addition to steady-state conversion characteristics, it is often of interest to investigate converter transient responses. For example, in voltage regulator designs, it is necessary to verify whether the output voltage remains within specified limits when the load current takes a step change. As another example, during a start-up transient when the converter is powered up, converter components can be exposed to significantly higher stresses than in steady state. It is of interest to verify that component stresses are

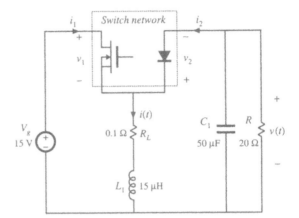

Fig. B.6   Buck-boost converter example.

within specifications or to make design modifications to reduce the stresses. In these examples, transient simulations can be used to test for converter responses.

Transient simulations can be performed on the converter switching circuit model or on the converter averaged circuit model. As an example, let us apply these two approaches to investigate a start-up transient response of the buck-boost converter shown in Fig. B.6.

Figure B.7 shows a switching circuit model of the buck-boost converter. The inductor winding resistance R_L is included to model the inductor copper losses. The MOSFET is modeled as a voltage-controlled switch S_{q1} controlled by a pulsating voltage source v_c. The switch .model line specifies the switch on-resistance $R_{on} = 50$ mΩ, and the switch off-resistance $R_{off} = 10$ MΩ. The switch is on when the controlling voltage v_c is greater than $V_{on} = 6$ V, and off when the controlling voltage v_c is less than $V_{off} = 4$ V. The pulsating source v_c has the pulse amplitude equal to 10 V. The period is $T_s = 1/f_s = 10$ μs, the rise and fall times are $t_r = t_f = 100$ ns, and the pulse width is $t_p = 7.9$ μs. The switch duty cycle is $D = (t_p + 0.5(t_r + t_f))/T_s = 0.8$. The built-in nonlinear Spice model is used for the diode. In the diode .model statement, only the parameter I_s is specified, to set the forward voltage drop across the diode. The switch and the diode models used in this example are very simple. Conduction losses are modeled in a simple manner, and details of complex device behavior during switching transitions are neglected.

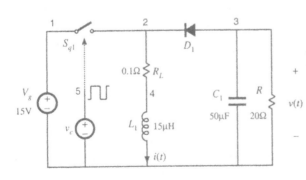

Buck-boost converter: switching circuit
Vg 1 0 15V
Sq1 1 2 5 0 switch
D1 3 2 diode
RL 2 4 0.1
L1 4 0 15uH ic=0
C1 3 0 50uF ic=0
R 3 0 20
Vc 5 0 pulse
+(0 10V 0us 100ns 100ns 7.9us 10us)
.model switch vswitch
+(Ron=0.05 Roff=10meg Von=6V Voff=4V)
.model diode d (Is=1e-12)
.tran 1u 1.2m 0m 1u uic
.probe
.end

Fig. B.7   Buck-boost converter simulation example, switching circuit model.

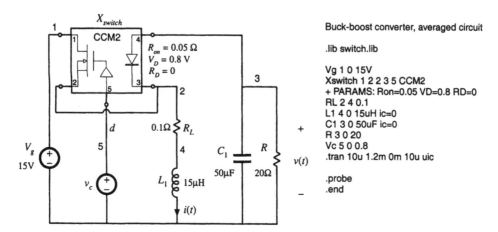

Fig. B.8 Buck-boost converter simulation example, averaged circuit model.

Therefore, the circuit model of Fig. B.7 cannot be used to examine switching transitions or to predict switching losses in the converter. Nevertheless, basic switching operation is modeled, and a transient simulation can be used to find out how the converter waveforms evolve in time over many switching cycles. Transient simulation parameters are defined by the .tran line: the output time step is 1 µs, the final simulation time is 1.2 ms, the output waveforms are generated from the start of simulation at time equal to zero, and the maximum allowed time step is 1 µs. The uic ("use initial conditions") option tells the simulator to start with all capacitor voltages and inductor currents equal to the specified initial values. For example, ic=0 in the L_1 line sets the initial inductor current to zero. In Spice, the default initial conditions are always zero, so that ic=0 statements can be omitted.

An averaged circuit model of the buck-boost converter is shown in Fig. B.8. This circuit model is obtained by replacing the switch network in the converter of Fig. B.6 by the CCM2 subcircuit. Notice that the circuits and the netlists of Figs. B.7 and Fig. B.8 are very similar. The only difference is that the switching devices in the converter circuit of Fig. B.7 are replaced by the CCM2 subcircuit X_{switch} in Fig. B.8. Also, the pulsating source $v_c(t)$ in the switching circuit is replaced by a constant voltage source v_c equal to the switch duty cycle $D = 0.8$.

The inductor current and the capacitor voltage waveforms during the start-up transient are shown in Fig. B.9. For comparison, the waveforms obtained by transient simulation of the switching converter circuit shown in Fig. B.7, and by simulation of the averaged circuit model of Fig. B.8 are shown. Switching ripples can be observed in the waveforms obtained by simulation of the switching circuit model. The converter transient response is governed by the converter natural time constants. Since these time constants are much longer than the switching period, the converter start-up transient responses in Fig. B.9 take many switching cycles to reach the steady state. In the results obtained by simulation of the averaged circuit model, the switching ripples are removed, but the low-frequency portions of the converter transient responses, which are governed by the natural time constants of the converter network, match very closely the responses obtained by simulation of the switching circuit.

Based on the results shown in Fig. B.9, we can see that converter components are exposed to significantly higher current stresses during the start-up transient than during steady state operation. The problem of excessive stresses in the start-up transient is quite typical for switching power converters. Practical designs usually include a "soft-start" circuit, where the switch duty cycle is slowly increased

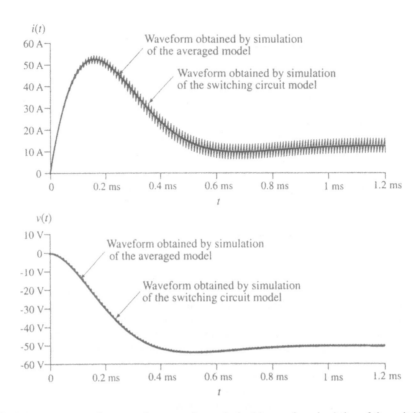

Fig. B.9 Inductor current and output voltage waveforms obtained by transient simulation of the switching converter circuit shown in Fig. B.7, and by simulation of the averaged circuit model of Fig. B.8

from zero to the steady-state value to reduce start-up transient stresses.

This simulation example illustrates how an averaged circuit model can be used in place of a switching circuit model to investigate converter large-signal transient responses. An advantage of the averaged circuit model is that transient simulations can be completed much more quickly because the averaged model is time invariant, and the simulator does not spend time computing the details of the fast switching transitions. This advantage can be important in simulations of larger electronic systems that include switching power converters. Another important advantage also comes from the fact that the averaged circuit model is nonlinear but time-invariant: ac simulations can be used to linearize the model and generate small-signal frequency responses of interest. This is not possible with switching circuit models. Examples of small-signal ac simulations can be found in Sections B.2 and B.3.

B.2   COMBINED CCM/DCM AVERAGED SWITCH MODEL

The models and examples of Section B.1 are all based on the assumption that the converters operate in continuous conduction mode (CCM). As discussed in Chapters 5 and 11, all converters containing a diode rectifier operate in discontinuous conduction mode (DCM) if the load current is sufficiently low. In some cases, converters are purposely designed to operate in DCM. It is therefore of interest to develop

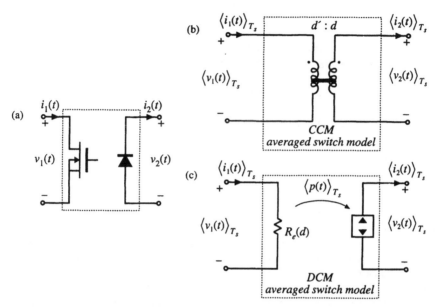

Fig. B.10 Summary of averaged switch modeling: (a) general two-switch network, (b) averaged switch model in CCM, and (c) averaged switch model in DCM.

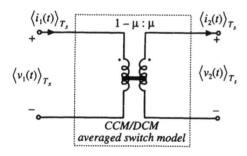

Fig. B.11 A general averaged switch model using the equivalent switch conversion ratio μ.

averaged models suitable for simulation of converters that may operate in either CCM or DCM.

Figure B.10 illustrates the general two-switch network, and the corresponding large-signal averaged models in CCM and DCM. The CCM averaged switch model, which is derived in Section 7.4, is an ideal transformer with $d' : d$ turns ratio. In DCM, the large-signal averaged switch model is a loss-free resistor, as derived in Section 11.1. Our objective is to construct a combined CCM/DCM averaged switch model that reduces to the model of Fig. B.10(a) or to the model of Fig. B.10(c) depending on the operating mode of the converter. Let us define an effective switch conversion ratio $\mu(t)$, so that the averaged switch model in both modes has the same form as in CCM, as shown in Fig. B.11. If the converter operates in CCM, then the switch conversion ratio $\mu(t)$ is equal to the switch duty cycle $d(t)$,

$$\mu = d \tag{B.5}$$

If the converter operates in DCM, then the effective switch conversion ratio can be computed so that the terminal characteristics of the averaged-switch model of Fig. B.11 match the terminal characteristics of the loss-free resistor model of Fig. B.10(c). Matching the port 1 characteristics gives

$$\langle v_1(t)\rangle_{T_s} = \frac{1-\mu}{\mu}\langle v_2(t)\rangle_{T_s} = R_e\langle i_1(t)\rangle_{T_s} \tag{B.6}$$

which can be solved for the switch conversion ratio μ,

$$\mu = \frac{1}{1 + \dfrac{R_e\langle i_1(t)\rangle_{T_s}}{\langle v_2(t)\rangle_{T_s}}} \tag{B.7}$$

It can be verified that matching the port 2 characteristics of the models in Figs. B.10(c) and B.11 gives exactly the same result for the effective switch conversion ratio in DCM.

The switch conversion ratio $\mu(t)$ can be considered a generalization of the duty cycle $d(t)$ of CCM switch networks. Based on this approach, models and results developed for converters in CCM can be used not only for DCM but also for other operating modes or even for other converter configurations by simply replacing the switch duty cycle $d(t)$ with the appropriate switch conversion ratio $\mu(t)$ [21-24]. For example, if $M(d)$ is the conversion ratio in CCM, then $M(\mu)$, with μ given by Eq. (B.7), is the conversion ratio in DCM. The switch conversion ratio in DCM depends on the averaged terminal voltage and current, as well as the switch duty cycle d through the effective resistance $R_e = 2L/d^2T_s$. If the converter is completely unloaded, then the average transistor current $\langle i_1(t)\rangle_{T_s}$ is zero, and the DCM switch conversion ratio becomes $\mu = 1$. As a result, the dc output voltage attains the maximum possible value $V = V_gM(1)$. This is consistent with the results of the steady-state DCM analyses in Chapter 5 and Section 11.1.

To construct a combined CCM/DCM averaged switch model based on the general averaged switch model of Fig. B.11, it is necessary to specify which of the two expressions for the switch conversion ratio to use: Eq. (B.5), which is valid in CCM, or Eq. (B.7), which is valid in DCM. At the CCM/DCM boundary, these two expressions must give the same result, $\mu = d$. If the load current decreases further, the converter operates in DCM, the average switch current $\langle i_1(t)\rangle_{T_s}$ decreases, and the DCM switch conversion ratio in Eq. (B.7) becomes greater than the switch duty cycle d. We conclude that the correct value of the switch conversion ratio, which takes into account operation in CCM or DCM, is the larger of the two values computed using Eq. (B.5) and Eq. (B.7).

Figure B.12 shows an implementation of the combined CCM/DCM model as a PSpice subcircuit CCM-DCM1. This subcircuit has the same five interface nodes as the subcircuits CCM1 and CCM2 of Section B.1. The controlled sources E_t and G_d model the port 1 (transistor) and port 2 (diode) averaged characteristics, as shown in Fig. B.11. The switch conversion ratio μ is equal to the voltage $v(u)$ at the subcircuit node u. The controlled voltage source E_u computes the switch conversion ratio as the greater of the two values obtained from Eqs. (B.5) and (B.7). The controlled current source G_a, the zero-value voltage source V_a, and the resistor R_a form an auxiliary circuit to ensure that the solution found by the simulator has the transistor and the diode currents with correct polarities, $\langle i_1(t)\rangle_{T_s} > 0$, $\langle i_2(t)\rangle_{T_s} > 0$. The subcircuit parameters are the inductance L relevant for CCM/DCM operation, and the switching frequency f_s. The default values in the subcircuit are arbitrarily set to $L = 100\ \mu H$ and $f_s = 100$ kHz.

The PSpice subcircuit CCM-DCM1 of Fig. B.12 can be used for dc, ac, and transient simula-

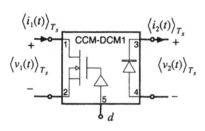

Fig. B.12 Implementation of the combined CCM/DCM averaged switch model.

```
**********************************************************
* MODEL: CCM-DCM1
* Application: two-switch PWM converters, CCM or DCM
* Limitations: ideal switches, no transformer
**********************************************************
* Parameters:
*    L = equivalent inductance for DCM
*    fs = switching frequency
**********************************************************
* Nodes:
* 1: transistor positive (drain for an n-channel MOS)
* 2: transistor negative  (source for an n-channel MOS)
* 3: diode cathode
* 4: diode anode
* 5: duty cycle control input
**********************************************************
.subckt CCM-DCM1 1 2 3 4 5
+ params: L=100u fs=1E5
Et 1 2 value={(1-v(u))*v(3,4)/v(u)}
Gd 4 3 value={(1-v(u))*i(Et)/v(u)}
Ga 0 a value={MAX(i(Et),0)}
Va a b
Ra b 0 1k
Eu u 0 table {MAX(v(5),
+ v(5)*v(5)/(v(5)*v(5)+2*L*fs*i(Va)/v(3,4)))} (0 0) (1 1)
.ends
**********************************************************
```

tions of PWM converters containing a transistor switch and a diode switch. This subcircuit is included in the model library *switch.lib*. It can be modified further for use in converters with isolation transformer.

B.2.1   Example: SEPIC Frequency Responses

As an example, Fig. B.13 shows a SEPIC circuit and the averaged circuit model obtained by replacing the switch network with the CCM-DCM1 subcircuit of Fig. B.12. A part of the circuit netlist is included in Fig. B.13. The connections and the parameters of the CCM-DCM1 subcircuit are defined by the X_{switch} line. In the SEPIC, the inductance parameter $L = 83.3$ μH is equal to the parallel combination of L_1 and L_2. The voltage source v_c sets the quiescent value of the duty cycle to $D = 0.4$, and the small-signal ac value to $\hat{d} = 1$. Ac simulation is performed on a linearized circuit model, so that amplitudes of all small-signal ac waveforms are directly proportional to the amplitude of the ac input, regardless of the input ac amplitude value. For example, the control-to-output transfer function is $G_{vd} = \hat{v}/\hat{d}$, where $\hat{v} = v(4)$ in the circuit of Fig. B.13(b). We can set the input ac amplitude to 1, so that the control-to-output transfer function G_{vd} can be measured directly as $v(5)$. This setup is just for convenience in finding small-signal frequency responses by simulation. For measurements of converter transfer functions in an experimental circuit (see Section 8.5), the actual amplitude of the small-signal ac variation $\hat{d}$ would be set to a fraction of the quiescent duty cycle D. Parameters of the ac simulation are set by the .ac line in the netlist: the signal frequency is swept from the minimum frequency of 5 Hz to the maximum frequency of 50 kHz in 201 points per decade.

Figure B.14 shows magnitude and phase responses of the control-to-output transfer function obtained by ac simulations for two different values of the load resistance: $R = 40$ Ω, for which the converter operates in CCM, and $R = 50$ Ω, for which the converter operates in DCM. For these two operating points, the quiescent (dc) voltages and currents in the circuit are nearly the same. Nevertheless, the frequency responses are qualitatively very different in the two operating modes. In CCM, the converter exhibits a fourth-order response with two pairs of high-Q complex-conjugate poles and a pair of complex-conjugate zeros. Another RHP (right-half plane) zero can be observed at frequencies approaching 50 kHz. In DCM, there is a dominant low-frequency pole followed by a pair of complex-conjugate poles and a pair of complex-conjugate zeros. The frequencies of the complex poles and zeros are very close in value. A high-frequency pole and a RHP zero contribute additional phase lag at higher frequencies.

In the design of a feedback controller around a converter that may operate in CCM or in DCM, one should take into account that the crossover frequency, the phase margin, and the closed-loop responses can be substantially different depending on the operating mode. This point is illustrated by the example of the next section.

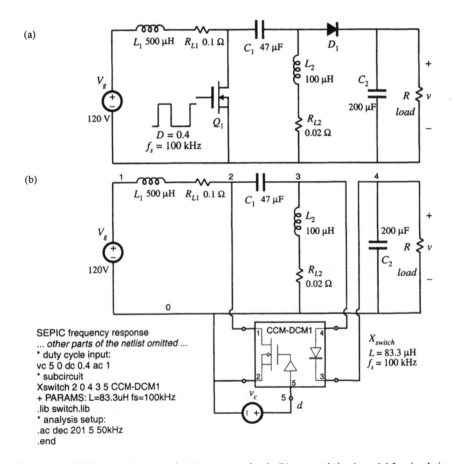

Fig. B.13 SEPIC simulation example: (a) converter circuit, (b) averaged circuit model for simulation.

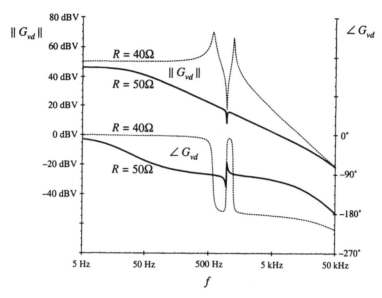

Fig. B.14 Magnitude and phase responses of the control-to-output transfer function obtained by simulation of the SEPIC example, for two values of the load resistance. For $R = 50 \, \Omega$, the converter operates in DCM (solid lines), and for $R = 40 \, \Omega$, the converter operates in CCM (dotted lines).

B.2.2   Example: Loop Gain and Closed-Loop Responses of a Buck Voltage Regulator

A controller design for a buck converter example is discussed in Section 9.5.4. The converter and the block diagram of the controller are shown in Fig. 9.22. This converter system is designed to regulate the dc output voltage at $V = 15$ V for the load current up to 5 A. Let us test this design by simulation. An averaged circuit model of a practical realization of the buck voltage regulator described in Section 9.5.4 is shown in Fig. B.15. The MOSFET and the diode switch are replaced by the averaged switch model implemented as the CCM-DCM1 subcircuit. The pulse-width modulator with $V_M = 4$ V is modeled according to the discussion in Section 7.6 as a dependent voltage source E_{pwm} controlled by the PWM input voltage v_x. The value of E_{pwm} is equal to $1/V_M = 0.25$ times the PWM input voltage v_x, with a limit for the minimum value set to 0.1 V, and a limit for the maximum value set to 0.9 V. The output of the pulse-width modulator is the control duty-cycle input to the CCM-DCM1 averaged switch subcircuit. Given the specified limits for E_{pwm}, the switch duty cycle $d(t)$ can take values in the range:

$$D_{\min} \le d(t) \le D_{\max} \tag{B.8}$$

where $D_{min} = 0.1$, and $D_{max} = 0.9$. Practical PWM integrated circuits often have a limit $D_{max} < 1$ for the maximum possible duty cycle. The voltage sensor and the compensator are implemented around an op-amp LM324. With very large loop gain in the system, the steady-state error voltage is approximately zero, i.e., the dc voltages at the plus and the minus inputs of the op-amp are almost the same,

$$v(5) = v_{ref} \tag{B.9}$$

As a result, the quiescent (dc) output voltage V is set by the reference voltage v_{ref} and the voltage divider comprised of R_1, R_2, R_4:

$$V \frac{R_4}{R_1 + R_2 + R_4} = v_{ref} = 5 \text{ V} \tag{B.10}$$

By setting the ac reference voltage $\hat{v}_{ref}$ to zero, the combined transfer function of the voltage sensor and the compensator can be found as:

$$H(s)G_c(s) = \frac{\hat{v}_y}{\hat{v}} = \frac{R_3 + \dfrac{1}{sC_3}}{R_1 + R_2 \parallel \dfrac{1}{sC_2}} \tag{B.11}$$

This transfer function can be written in factored pole-zero form as

$$G_{cm}H \frac{\left(1 + \dfrac{s}{\omega_z}\right)\left(1 + \dfrac{\omega_L}{s}\right)}{\left(1 + \dfrac{s}{\omega_p}\right)} \tag{B.12}$$

where

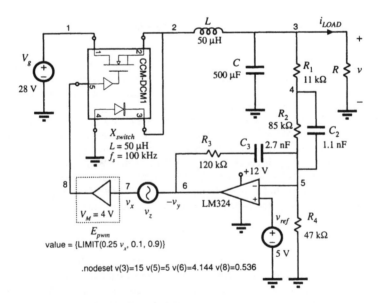

Fig. B.15   Buck voltage regulator example.

$$G_{cm}H = \frac{R_3}{R_1 + R_2} \tag{B.13}$$

$$f_z = \frac{\omega_z}{2\pi} = \frac{1}{2\pi R_2 C_2} \tag{B.14}$$

$$f_L = \frac{\omega_L}{2\pi} = \frac{1}{2\pi R_3 C_3} \tag{B.15}$$

and

$$f_p = \frac{\omega_p}{2\pi} = \frac{1}{2\pi \left(R_1 \parallel R_2 \right) C_2} \tag{B.16}$$

The design described in Section 9.5.4 resulted in the following values for the gain and the corner frequencies:

$$G_{cm}H = 3.7\,(1\,/\,3)\ = 1.23,\ f_z = 1.7\ \text{kHz},\ f_L = 500\ \text{Hz},\ f_p = 14.5\ \text{kHz} \tag{B.17}$$

Eqs. (B.10) and (B.13) to (B.17) can be used to select the circuit parameter values. Let us (somewhat arbitrarily) choose $C_2 = 1.1$ nF. Then, from Eq. (B.14), we have $R_2 = 85$ kΩ, and Eq. (B.16) yields $R_1 = 11$ kΩ. From Eq. (B.13) we obtain $R_3 = 120$ kΩ, and Eq. (B.15) gives $C_3 = 2.7$ kΩ. Finally, $R_4 = 47$ kΩ is found from Eq. (B.10). The voltage regulator design can now be tested by simulations of the circuit in Fig. B.15.

Loop gains can be obtained by simulation using exactly the same techniques described in Section 9.6 for experimental measurement of loop gains [20]. Let us apply the voltage injection technique of Section 9.6.1. An ac voltage source v_z is injected between the compensator output and the PWM input. This is a good injection point since the output impedance of the compensator built around the op-amp is small, and the PWM input impedance is very large (infinity in the circuit model of Fig. B.15). With the ac source amplitude set (arbitrarily) to 1, and no other ac sources in the circuit, ac simulations are performed to find the loop gain as

$$T(s) = \frac{\hat{v}_y}{\hat{v}_x} = -\frac{v(6)}{v(7)} \tag{B.18}$$

To perform ac analysis, the simulator first solves for the quiescent (dc) operating point. The circuit is then linearized at this operating point, and small-signal frequency responses are computed for the specified range of signal frequencies. Solving for the quiescent operating point involves numerical solution of a system of nonlinear equations. In some cases, the numerical solution does not converge and the simulation is aborted with an error message. In particular, convergence problems often occur in circuits with feedback, especially when the loop gain at dc is very large. This is the case in the circuit of Fig. B.15. To help convergence when the simulator is solving for the quiescent operating point, one can specify approximate or expected values of node voltages using the .nodeset line as shown in Fig. B.15. In this case, we know by design that the quiescent output voltage is close to 15 V ($v(3) = 15$), that the negative input of the op-amp is very close to the reference ($v(5) = 5$), and that the quiescent duty cycle is approximately $D = V/V_g = 0.536$, so that $v(8) = 0.536$ V. Given these approximate node voltages, the numerical solution converges, and the following quiescent operating points are found by the simulator for two values of the load resistance R:

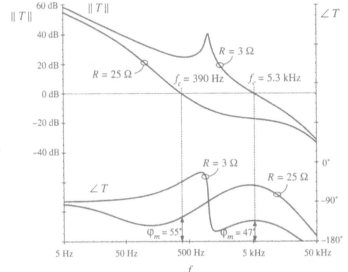

Fig. B.16 Loop gain in the buck voltage regulator example.

$$R = 3 \ \Omega, \ v(3) = 15.2 \ \text{V}, \ v(5) = 5.0 \ \text{V}, \ v(7) = 2.173 \ \text{V}, \ v(8) = 0.543 \ \text{V}, \ D = 0.543 \tag{B.19}$$

$$R = 25 \ \Omega, \ v(3) = 15.2 \ \text{V}, \ v(5) = 5.0 \ \text{V}, \ v(7) = 2.033 \ \text{V}, \ v(8) = 0.508 \ \text{V}, \ D = 0.508 \tag{B.20}$$

For the nominal load resistance $R = 3 \ \Omega$, the converter operates in CCM, so that $D = V/V_g$. For $R = 25 \ \Omega$, the same dc output voltage is obtained for a lower value of the quiescent duty cycle, which means that the converter operates in DCM.

The magnitude and phase responses of the loop gain found for the operating points given by Eqs. (B.19) and (B.20) are shown in Fig. B.16. For $R = 3 \ \Omega$, the crossover frequency is $f_c = 5.3$ kHz, and the phase margin is $\phi_M = 47°$, very close to the values ($f_c = 5$ kHz, $\phi_M = 52°$) that we designed for in Section 9.5.4. At light load, for $R = 25 \ \Omega$, the loop gain responses are considerably different because the converter operates in DCM. The crossover frequency drops to $f_c = 390$ Hz, while the phase margin is $\phi_M = 55°$.

The magnitude responses of the line-to-output transfer function are shown in Fig. B.17, again for two values of the load resistance, $R = 3 \ \Omega$ and $R = 25 \ \Omega$. The open-loop responses are obtained by braking the feedback loop at node 8, and setting the dc voltage at this node to the quiescent value D of the duty cycle. For $R = 3 \ \Omega$, the open-loop and closed-loop responses can be compared to the theoretical plots shown in Fig. 9.32. At 100 Hz, the closed-loop magnitude response is $0.012 \Rightarrow -38$ dB. A 1 V, 100 Hz variation in $v_g(t)$ would induce a 12 mV variation in the output voltage $v(t)$. For $R = 25 \ \Omega$, the closed loop magnitude response is $0.02 \Rightarrow -34$ dB, which means that the 1 V, 100 Hz variation in $v_g(t)$ would induce a 20 mV variation in the output voltage. Notice how the regulator performance in terms of rejecting the input voltage disturbance is significantly worse at light load than at the nominal load.

A test of the transient response to a step change in load is shown in Fig. B.18. The load current is initially equal to 1.5 A, and increases to $i_{LOAD} = 5$ A at $t = 0.1$ ms. When the converter is operated in

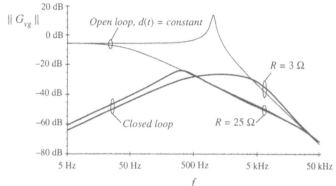

Fig. B.17 Line to output response of the buck voltage regulator.

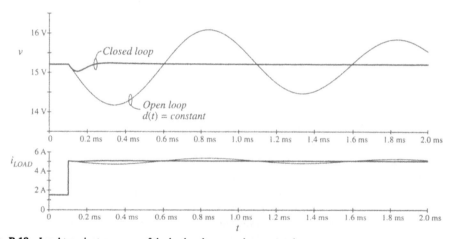

Fig. B.18   Load transient response of the buck voltage regulator example.

open loop at constant duty cycle, the response is governed by the natural time constants of the converter network. A large undershoot and long lightly-damped oscillations can be observed in the output voltage. With the feedback loop closed, the controller dynamically adjusts the duty cycle $d(t)$ trying to maintain the output voltage constant. The output voltage drops by about 0.2 V, and it returns to the regulated value after a short, well-damped transient.

The voltage regulator example of Fig. B.15 illustrates how the performance can vary significantly if the regulator is expected to supply a wide range of loads. In practice, further tests would also be performed to account for expected ranges of input voltages, and variations in component parameter values. Design iterations may be necessary to ensure that performance specifications are met under worst case conditions.

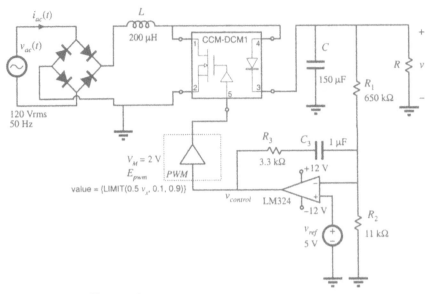

Fig. B.19   DCM boost rectifier example.

B.2.3   Example: DCM Boost Rectifier

Converters switching at frequencies much above the ac line frequency can be used to construct near-ideal rectifiers where power is taken from the ac line without generation of line current harmonics. Approaches to construction of low-harmonic rectifiers are discussed in Chapter 18. One simple solution is based on the boost converter operating in discontinuous conduction mode, as described in Section 18.2.1. When a boost DCM converter operates at a constant switch duty cycle, the input current approximately follows the input voltage. The DCM effective resistance $2L/d^2(t)T_s$ is an approximation of the emulated resistance R_e of the DCM boost rectifier. Ac line current harmonics are not zero, but the rectifier can still be designed to meet harmonic limits. In this section we consider a DCM boost rectifier example and test its performance by simulation.

An averaged circuit model of the boost DCM rectifier is shown in Fig. B.19. Full-wave rectified 120 Vrms, 50 Hz ac line voltage is applied to the input of the boost converter. The converter switches are replaced by the CCM-DCM1 averaged switch subcircuit. It is desired to regulate the dc output voltage at $V = 300$ V at output power up to $P_{out} = 120$ W across the load R. The switching frequency is $f_s = 100$ kHz. Let us select the inductance L so that the converter always operates in DCM. From Eq. (18.24), the condition for DCM is:

$$L < \frac{\left(1 - \frac{V_M}{V}\right)R_e}{2f_s} \tag{B.21}$$

where R_e is the emulated resistance of the rectifier and V_M is the peak of the ac line voltage. When line

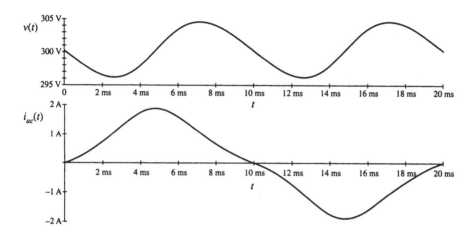

Fig. B.20   Output voltage and ac line current in the DCM boost rectifier example.

current harmonics and losses are neglected, the rectifier emulated resistance R_e at the specified load power P is

$$R_e = \frac{V_M^2}{2P} \tag{B.22}$$

Given $V_M = 170$ V and R_e found from Eq. (B.22), Eq. (B.21) gives $L < 260$ μH. The selected inductance is $L = 200$ μH. A low-bandwidth voltage feedback loop is closed around the converter to regulate the dc output voltage. The output voltage is sensed and compared to the reference v_{ref}. A PI compensator is constructed around the LM324 op-amp. The output $v_{control}$ of the compensator is the input to the pulse-width modulator. By adjusting the switch duty ratio d, $v_{control}$ adjusts the emulated resistance $R_e = 2L/d^2 T_s$ of the rectifier, and thereby controls the power taken from the ac line. In steady state, the input power matches the output power. The dc output voltage V is regulated at the value set by the reference voltage v_{ref} and the voltage divider composed of R_1 and R_2, as follows:

$$V = v_{ref} \frac{R_1 + R_2}{R_1} = 300 \text{ V} \tag{B.23}$$

Modeling of the low-bandwidth voltage regulation loop is discussed in Section 18.4.2.

It is of interest to find ac line current harmonics. First, a long transient simulation is performed to reach steady-state operation. Then, current harmonics are computed using Fourier analysis applied to the ac line current waveform $i_{ac}(t)$ during one line cycle in steady state. Figure B.20 shows the steady-state ac line current and output voltage obtained for R = 900 Ω, i.e., for 100 W of output power. The output voltage has a dc component equal to 300 V, and an ac ripple component at twice the line frequency. The peak-to-peak voltage ripple at twice the line frequency is approximately 8 V, which compares well with the value (7 V) found from Eq. (18.91). The ac line current has noticeable distortion. The spectrum of the ac line current is shown in Fig. B.21. The largest harmonic, the third, has an amplitude of 16.6% of the fundamental, and the total harmonic distortion is 16.7%.

We can also examine what happens if the rectifier is overloaded. The steady-state ac line current waveform for the case when the load resistance is R = 500 Ω, and the output power is 180 W, is shown in

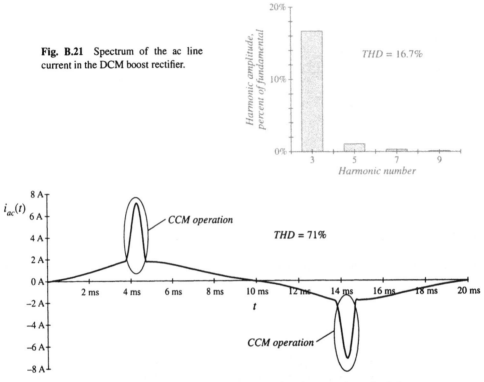

Fig. B.21 Spectrum of the ac line current in the DCM boost rectifier.

THD = 16.7%

THD = 71%

CCM operation

CCM operation

Fig. B.22   Ac line current of the DCM boost rectifier example, when the output is overloaded.

Fig. B.22. The boost converter operates in CCM near the peak of the ac line voltage; this results in current spikes and significant harmonic distortion.

B.3   CURRENT PROGRAMMED CONTROL

In the current programmed mode (CPM), which is studied in Chapter 12, the transistor switching is controlled so that the peak transistor current follows a control signal. The transistor duty cycle $d(t)$ is not directly controlled, but depends on the CPM control input as well as on other converter voltages and currents. In this section, large-signal averaged relationships in CPM are written in a form suitable for implementation as a subcircuit for simulation.

B.3.1   Current Programmed Mode Model for Simulation

Typical inductor current and voltage waveforms of CPM converters operating in continuous conduction mode or in discontinuous conduction mode are shown in Fig. B.23. Signal $i_c(t)$ is the CPM control input. An artificial ramp having slope $-m_a$ is added to the control input. In the first subinterval,

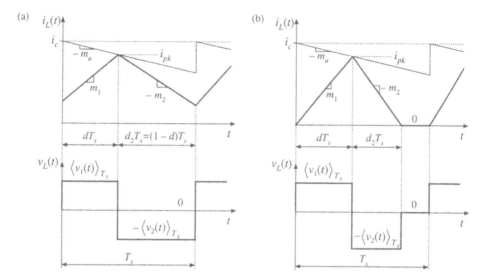

Fig. B.23   Current programmed mode waveforms: (a) continuous conduction mode; (b) discontinuous conduction mode.

when the transistor is on, the inductor current increases with slope m_1 given by:

$$m_1 = \frac{\langle v_1(t) \rangle_{T_s}}{L} \tag{B.24}$$

It is assumed that voltage ripples are small so that the voltage $v_1(t)$ across the inductor is approximately equal to the averaged value $\langle v_1(t) \rangle_{T_s}$. The length of the first subinterval is $d(t)T_s$. The transistor is turned off when the inductor current reaches the peak value i_{pk} equal to:

$$i_{pk} = i_c - m_a dT_s \tag{B.25}$$

In the second subinterval, when the transistor is off and the diode is on, the inductor current decreases with a negative slope $-m_2$. With the assumption the voltage ripples are small, the slope m_2 is given by:

$$m_2 = \frac{\langle v_2(t) \rangle_{T_s}}{L} \tag{B.26}$$

The length of the second subinterval is $d_2(t)T_s$. In CCM, the second subinterval lasts until the end of the switching cycle. Therefore:

$$d_2 = 1 - d \tag{B.27}$$

In DCM, the current drops to zero before the end of the switching period. The length of the second subinterval can be computed from:

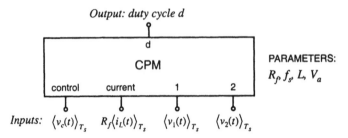

Output: duty cycle d

Fig. B.24   Current programmed mode (CPM) subcircuit.

$$d_2 = \frac{i_{pk}}{m_2 T_s} \tag{B.28}$$

If the converter operates in DCM, d_2 computed from Eq. (B.28) is smaller that $1 - d$. If the converter operates in CCM, $1 - d$ is smaller than d_2 computed from Eq. (B.28). In general, the length of the second subinterval can be found as the smaller of the two values computed using Eqs. (B.27) and (B.28).

The average inductor current can be found by computing the area under the inductor current waveform in Fig. B.23:

$$\langle i_L(t) \rangle_{T_s} = d\left(i_{pk} - \frac{m_1 d T_s}{2} \right) + d_2\left(i_{pk} - \frac{m_2 d_2 T_s}{2} \right) \tag{B.29}$$

The relationship given by Eq. (B.29) is valid for both CCM and DCM provided that the second subinterval length is computed as the smaller of the values obtained from Eqs. (B.27) and (B.28).

Based on Eqs. (B.24) to (B.29), an averaged CPM subcircuit model is constructed in the form shown in Fig. B.24. The inputs to the CPM subcircuit are the control input $\langle v_c(t) \rangle_{T_s} = R_f \langle i_c(t) \rangle_{T_s}$, the measured inductor current $R_f \langle i_L(t) \rangle_{T_s}$, and the inductor voltages $\langle v_1(t) \rangle_{T_s}$ and $\langle v_2(t) \rangle_{T_s}$ of the two subintervals. The output of the subcircuit is the switch duty cycle d. The parameters of the CPM subcircuit are the equivalent current-sense resistance R_f, the inductance L, the switching frequency $f_s = 1/T_s$, and the amplitude V_a of the artificial ramp:

$$V_a = m_a T_s R_f \tag{B.30}$$

In the subcircuit implementation, the length of the second subinterval is computed as the smaller of the values given by Eqs. (B.27) and (B.28):

$$d_2 = \text{MIN}\left(1 - d, \frac{i_{pk}}{m_2 T_s} \right) \tag{B.31}$$

Next, the switch duty cycle is found by solving Eq. (B.29). There are many different ways the switch duty cycle can be expressed in terms of other quantities. Although mathematically equivalent to Eq. (B.29), these different forms of solving for d result in different convergence performance of the numerical solver in the simulator. In the CPM subcircuit available in the *switch.lib* library, the duty cycle is found from:

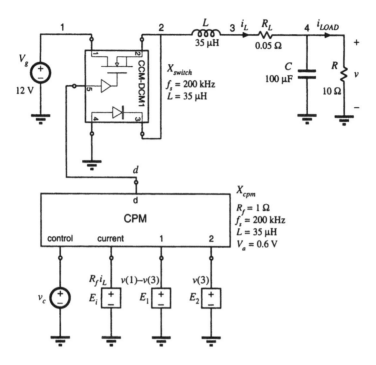

Fig. B.25 CPM buck converter example.

$$d = \frac{2i_c(d + d_2) - 2\langle i_L(t)\rangle_{T_s} - m_2 d_2^2 T_s}{2m_a(d + d_2)T_s + m_1 dT_s} \tag{B.32}$$

which is obtained by inserting Eq. (B.25) into Eq. (B.29). This implicit expression (notice that d is on both sides of the equation) is used by the numerical solver in the simulator to compute the switch duty cycle d.

B.3.2   Example: Frequency Responses of a Buck Converter with Current Programmed Control

To illustrate an application of the CPM subcircuit, let us consider the example buck converter circuit model of Fig. B.25. To construct this averaged circuit model, the switches are replaced by the CCM-DCM1 averaged switch subcircuit. The control input to the CPM subcircuit is the independent voltage source v_c. Three dependent voltage sources are used to generate other inputs to the CPM subcircuit. The controlled voltage source E_i is proportional to the inductor current i_L. The controlled voltage source E_1 is equal to $v(1) - v(3)$, which is equal to the voltage $\langle v_1(t)\rangle_{T_s}$ applied across the inductor during the first sub-interval when the transistor is on and the diode is off. The controlled voltage source E_2 is equal to $v(3)$, which is equal to the voltage $\langle v_2(t)\rangle_{T_s}$ applied across the inductor during the second subinterval when the transistor is off and the diode is on.

Ac simulations are performed at the quiescent operating point obtained for the dc value of the

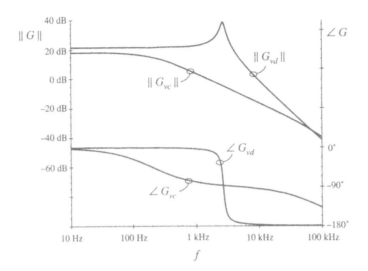

Fig. B.26 Comparison of CPM control with duty-cycle control, for the control-to-output frequency response of the buck converter example.

control input equal to $V_c = 1.4$ V. At the quiescent operating point, the switch duty cycle is $D = 0.676$, the dc output voltage is $V = 8.1$ V, and the dc component of the inductor current is $I_L = 0.81$ A. The converter operates in CCM.

Magnitude and phase responses of the control-to-output transfer functions $G_{vc}(s) = \hat{v}/\hat{v}_c$ and $G_{vd}(s) = \hat{v}/\hat{d}$ are shown in Fig. B.26. The duty-cycle to output voltage transfer function $G_{vd}(s)$ exhibits the familiar second-order high-Q response. Peaking in the magnitude response and a steep change in phase from $0°$ to $-180°$ occur around the center frequency of the pair of complex-conjugate poles. In contrast, the CPM control-to-output response has a dominant low-frequency pole. The phase lag is around $-90°$ in a wide range of frequencies. A high frequency pole contributes to additional phase lag at higher frequencies. The frequency responses of Fig. B.26 illustrate an advantage of CPM control over duty-cycle control. Because of the control-to-output frequency response dominated by the single low-frequency pole, it can be much easier to close a wide-bandwidth outer voltage feedback loop around the CPM controlled power converter than around a converter where the duty cycle is the control input.

Another advantage of CPM control is in rejection of input voltage disturbances. Line-to-output frequency responses for duty-cycle control and CPM control in the buck example are compared in Fig. B.27. At practically all frequencies of interest, CPM control offers more than 30 dB better attenuation of input voltage disturbances.

It is also interesting to compare the output impedance of the converter with duty-cycle control versus CPM control. The results are shown in Fig. B.28. At low frequencies, duty-cycle controlled converter has very low output impedance determined by switch and inductor resistances. As the frequency goes up, the output impedance increases as the impedance of the inductor increases. At the resonant frequency of the output LC filter, significant peaking in the output impedance of the duty-cycle controlled converter can be observed. At higher frequencies, the output impedance is dominated by the impedance of the filter capacitor, which decreases with frequency. In the CPM controlled converter, the low-frequency impedance is high. It is equal to the parallel combination of the load resistance and the CPM out-

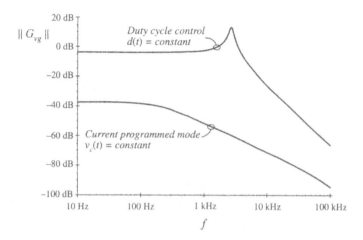

Fig. B.27 Comparison of CPM control with duty-cycle control, for the line-to-output frequency response of the buck converter example.

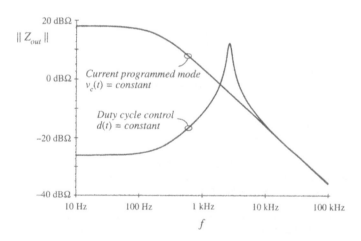

Fig. B.28 Comparison of CPM control with duty-cycle control, for the output impedance of the buck converter example.

put resistance. Because of the lossless damping introduced by CPM control, the series inductor does not affect the output impedance. As the frequency goes up, the output impedance becomes dominated by the output filter capacitor and it decreases with frequency. At high frequencies the output impedances of the duty-cycle and CPM controlled converters have the same asymptotes.

REFERENCES

[1]   R. J. DIRKMAN, "The Simulation of General Circuits Containing Ideal Switches," *IEEE Power Electronics Specialists Conference*, 1987 Record, pp. 185-194.

[2]   C. J. HSIAO, R. B. RIDLEY, H. NAITOH and F. C. LEE, "Circuit-Oriented Discrete-Time Modeling and Simulation of Switching Converters, *IEEE Power Electronics Specialists Conference*, 1987 record, pp. 167-176.

[3]   R. C. WONG, H. A. OWEN, T. G. WILSON, "An Efficient Algorithm for the Time-Domain Simulation of Regulated Energy-Storage DC-to-DC Converters," *IEEE Transactions on Power Electronics*, Vol. 2, No. 2, April 1987, pp. 154-168.

[4]   A. M. LUCIANO and A. G. M. STROLLO, "A Fast Time-Domain Algorithm for Simulation of Switching Power Converters," *IEEE Transactions on Power Electronics*, Vol. 2, No. 3, July 1990, pp. 363-370.

[5]   D. BEDROSIAN and J. VLACH, "Time-Domain Analysis of Networks with Internally Controlled Switches," *IEEE Transactions on Circuits and Systems–I: Fundamental Theory and Applications*, Vol. 39, No.3, March 1992, pp. 199-212.

[6]   P. PEJOVIĆ and D. MAKSIMOVIĆ, "A New Algorithm for Simulation of Power Electronic Systems Using Piecewise-Linear Device Models," *IEEE Transactions on Power Electronics*, Vol. 10, No. 3, May 1995, pp. 340-348.

[7]   PREDRAG PEJOVIĆ, "A Method for Simulation of Power Electronic Systems Using Piecewise-Linear Device Models," Ph.D. thesis, University of Colorado, Boulder, April 1995.

[8]   D. LI, R. TYMERSKI, T. NINOMIYA, "PECS—An Efficacious Solution for Simulating Switched Networks with Nonlinear Elements," *IEEE Power Electronics Specialists Conference*, 2000 Record, pp. 274-279, June 2000.

[9]   V. BELLO, "Computer Aided Analysis of Switching Regulators Using SPICE2," *IEEE Power Electronics Specialists Conference*, 1980 Record, pp. 3-11.

[10]   V. BELLO, "Using the SPICE2 CAD Package for Easy Simulation of Switching Regulators in Both Continuous and Discontinuous Conduction Modes," *Proceedings of the Eighth National Solid-State Power Conversion Conference* (Powercon 8), April 1981.

[11]   V. BELLO, "Using the SPICE2 CAD Package to Simulate and Design the Current Mode Converter," *Proceedings of the Eleventh National Solid-State Power Conversion Conference* (Powercon 11), April 1984.

[12]   D. KIMHI, S. BEN-YAAKOV, "A SPICE Model for Current Mode PWM Converters Operating Under Continuous Inductor Current Conditions," *IEEE Transactions on Power Electronics*, Vol. 6, No. 2, April 1991, pp. 281-286.

[13]   Y. AMRAN, F. HULIEHEL, S. BEN-YAAKOV, "A Unified SPICE Compatible Average Model of PWM Converters," *IEEE Transactions on Power Electronics*, Vol. 6, No. 4, October 1991, pp. 585-594.

[14]   S. BEN-YAAKOV, Z. GAATON, "Generic SPICE Compatible Model of Current Feedback in Switch Mode Converters, *Electronics Letters*, Vol. 28, No. 14, 2nd July 1992.

[15]   S. BEN-YAAKOV, "Average Simulation of PWM Converters by Direct Implementation of Behavioral Rela-

tionships," *IEEE Applied Power Electronics Conference*, 1993 Record, pp. 510-516, February 1993.

[16] S. BEN-YAAKOV, D. ADAR, "Average Models as Tools for Studying Dynamics of Switch Mode DC-DC converters," *IEEE Power Electronics Specialists Conference*, 1994 Record, pp. 1369-1376.

[17] V. M. CANALLI, J. A. COBOS, J. A. OLIVER, J. UCEDA, "Behavioral Large Signal Averaged Model for DC/DC Switching Power Converters," *IEEE Power Electronics Specialists Conference*, 1996 Record, pp. 1675-1681.

[18] N. JAYARAM, D. MAKSIMOVIĆ, "Power Factor Correctors Based on Coupled-Inductor SEPIC and Ćuk Converters with Nonlinear- Carrier Control," *IEEE Applied Power Electronics Conference*, 1998 Record, pp. 468-474, February 1998.

[19] J. KEOWN, *OrCAD PSpice and Circuit Analysis*, Fourth Edition, Englewood Cliffs: Prentice Hall, 2000.

[20] P. W. TUINENGA, *SPICE: A Guide to Circuit Simulation and Analysis Using PSpice*, Third Edition, Englewood Cliffs: Prentice Hall, 1995.

[21] V. VORPERIAN, "Simplified Analysis of PWM Converters Using the Model of the PWM Switch: Parts I and II," *IEEE Transactions on Aerospace and Electronic Systems*, Vol. AES-26, pp. 490-505, May 1990.

[22] S. FREELAND and R. D. MIDDLEBROOK, "A Unified Analysis of Converters with Resonant Switches," *IEEE Power Electronics Specialists Conference*, 1987 Record, pp. 20-30.

[23] ARTHUR WITULSKI and ROBERT ERICKSON, "Extension of State-Space Averaging to Resonant Switches —and Beyond," *IEEE Transactions on Power Electronics*, Vol. 5, No. 1, pp. 98-109, January 1990.

[24] D. MAKSIMOVIĆ and S. ĆUK, "A Unified Analysis of PWM Converters in Discontinuous Modes," *IEEE Transactions on Power Electronics*, Vol. 6, No. 3, pp. 476-490, July 1991.

Appendix C

Middlebrook's
Extra Element Theorem

The *Extra Element Theorem* of R. D. Middlebrook [1–3] shows how a transfer function is changed by the addition of an impedance to the network. The theorem allows one to determine the effects of this extra element on any transfer function of interest, without solving the system all over again. The Extra Element Theorem is a powerful technique of design-oriented analysis. It leads to impedance inequalities which guarantee that an element does not substantially alter a transfer function. The Extra Element Theorem is employed in Chapter 10, where it leads to a relatively simple methodology for designing input filters that do not degrade the loop gains of switching regulators. It is also employed in Section 19.4, to determine how the load resistance affects the properties of a resonant inverter. In this appendix, Middlebrook's Extra Element Theorem is derived, based on the principle of superposition. Its application is illustrated via examples.

C.1   BASIC RESULT

Consider the linear circuit of Fig. C.1(a). This network contains an input $v_{in}(s)$ and an output $v_{out}(s)$. In addition, it contains a port whose terminals are open-circuited. It is assumed that the transfer function from $v_{in}(s)$ to $v_{out}(s)$ is known, and is given by

$$\frac{v_{out}(s)}{v_{in}(s)} = G(s)\Big|_{Z(s) \to \infty} \tag{C.1}$$

The Extra Element Theorem tells us how the transfer function $G(s)$ is modified when an impedance $Z(s)$ is connected between the terminals at the port, as in Fig. C.1(b). The result is

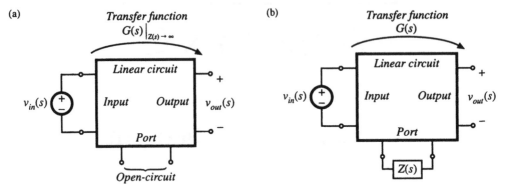

Fig. C.1 How an added element changes a transfer function $G(s)$: (a) original conditions, before addition of the new element; (b) addition of element having impedance $Z(s)$.

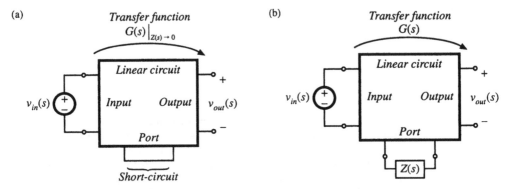

Fig. C.2 The dual form of the Extra Element Theorem, in which the extra element replaces a short circuit: (a) original conditions, (b) addition of element having impedance $Z(s)$.

$$\frac{v_{out}(s)}{v_{in}(s)} = \left(G(s) \Big|_{Z(s) \to \infty} \right) \left(\frac{1 + \dfrac{Z_N(s)}{Z(s)}}{1 + \dfrac{Z_D(s)}{Z(s)}} \right) \tag{C.2}$$

The right-hand side terms involving $Z(s)$ account for the influence of $Z(s)$ on $G(s)$, and are known as the *correction factor.*

The Extra Element Theorem also applies to the dual form illustrated in Fig. C.2. In this form, the transfer function is initially known under the conditions that the port is short-circuited. In Fig. C.2(b), the short-circuit is replaced by the impedance $Z(s)$. In this case, the addition of the impedance $Z(s)$ causes the transfer function to become

$$\frac{v_{out}(s)}{v_{in}(s)} = \left(G(s) \Big|_{Z(s) \to 0} \right) \left(\frac{1 + \dfrac{Z(s)}{Z_N(s)}}{1 + \dfrac{Z(s)}{Z_D(s)}} \right) \tag{C.3}$$

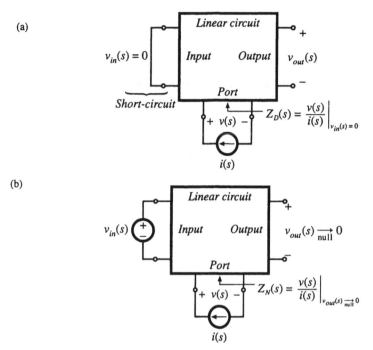

Fig. C.3   Determination of the quantities $Z_N(s)$ and $Z_D(s)$: (a) $Z_D(s)$ is the Thevenin-equivalent impedance at the port, and is measured with the input $v_{in}(s)$ set to zero; (b) $Z_N(s)$ is the impedance seen at the port under the condition that the output is nulled.

The $Z_N(s)$ and $Z_D(s)$ terms in Eqs. (C.2) and (C.3) are identical. By equating the $G(s)$ expressions of Eqs. (C.2) and (C.3), one can show that

$$\frac{G(s)\big|_{Z(s)\to\infty}}{G(s)\big|_{Z(s)\to0}} = \frac{Z_D(s)}{Z_N(s)} \tag{C.4}$$

This is known as the *reciprocity relationship*.

The quantities $Z_N(s)$ and $Z_D(s)$ can be found by measuring impedances at the port. The term $Z_D(s)$ is the Thevenin equivalent impedance seen looking into the port, also known as the driving-point impedance. As illustrated in Fig. C.3(a), this impedance is found by setting the independent source $v_{in}(s)$ to zero, and then measuring the impedance between the terminals of the port:

$$Z_D(s) = \frac{v(s)}{i(s)}\bigg|_{v_{in}(s)=0} \tag{C.5}$$

Thus, $Z_D(s)$ is the impedance between the port terminals when the input $v_{in}(s)$ is set to zero.

Determination of the impedance $Z_N(s)$ is illustrated in Fig. C.3(b). The term $Z_N(s)$ is found under the conditions that the output $v_{out}(s)$ is *nulled* to zero. A current source $i(s)$ is connected to the terminals of the port. In the presence of the input signal $v_{in}(s)$, the current $i(s)$ is adjusted so that the output $v_{out}(s)$ is nulled to zero. Under these conditions, the quantity $Z_N(s)$ is given by

$$Z_N(s) = \left. \frac{v(s)}{i(s)} \right|_{v_{out}(s) \;\to\; 0 \atop \text{null}} \qquad (C.6)$$

Note that *nulling* the output is not the same as *shorting* the output. If one simply shorted the output, then a current would flow through the short, which would induce voltage drops and currents in other elements of the network. These voltage drops and currents are not present when the output is nulled. The null condition of Fig. C.3(b) does not employ any connections to the output of the circuit. Rather, the null condition employs the adjustment of the independent sources $v_{in}(s)$ and $i(s)$ in a special way that causes the output $v_{out}(s)$ to be zero. By superposition, $v_{out}(s)$ can be expressed as a linear combination of $v_{in}(s)$ and $i(s)$; therefore, for a given $v_{in}(s)$, it is always possible to choose an $i(s)$ that will cause $v_{out}(s)$ to be zero. Under these null conditions, $Z_N(s)$ is measured as the ratio of $v(s)$ to $i(s)$. In practice, the circuit analysis to find $Z_N(s)$ is simpler than analysis of $Z_D(s)$, because the null condition causes many of the signals within the circuit to be zero. Several examples are given in Section C.4.

The input and output quantities need not be voltages, but could also be currents or other signals that can be set or nulled to zero. The next section contains a derivation of the Extra Element Theorem with a general input $u(s)$ and output $y(s)$.

C.2   DERIVATION

Figure C.4(a) illustrates a general linear system having an input $u(s)$ and an output $y(s)$. In addition, the system contains an electrical port having voltage $v(s)$ and current $i(s)$, with the polarities illustrated. Initially, the port is open-circuited: $i(s) = 0$. The transfer function of this system, with the port open-circuited, is

$$G_{old}(s) = \left. \frac{y(s)}{u(s)} \right|_{i(s) = 0} \qquad (C.7)$$

The objective of the extra element theorem is to determine the new transfer function $G(s)$ that is obtained when an impedance $Z(s)$ is connected to the port:

$$G(s) = \frac{y(s)}{u(s)} \qquad (C.8)$$

The situation is illustrated in Fig. C.4(b). It can be seen that the conditions at the port are now given by

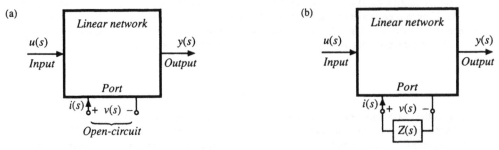

Fig. C.4 Modification of a linear network by addition of an extra element: (a) original system, (b) modified system, with impedance $Z(s)$ connected at an electrical port.

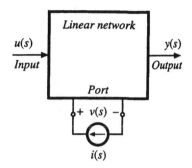

Fig. C.5 Current injection at the electrical port, by addition of independent current source $i(s)$.

$$v(s) = -i(s)Z(s) \tag{C.9}$$

To express the new transfer function $G(s)$ in Eq. (C.8) in terms of the original transfer function $G_{old}(s)$ of Eq. (C.7), we use current injection at the port, as illustrated in Fig. C.5. There are now two independent inputs: the input $u(s)$ and the independent current source $i(s)$. The dependent quantities $y(s)$ and $v(s)$ can be expressed as functions of these independent inputs using the principle of superposition:

$$y(s) = G_{old}(s)u(s) + G_i(s)i(s) \tag{C.10}$$

$$v(s) = G_v(s)u(s) + Z_D(s)i(s) \tag{C.11}$$

where

$$G_{old}(s) = \left.\frac{y(s)}{u(s)}\right|_{i(s)=0} \tag{C.12}$$

$$G_i(s) = \left.\frac{y(s)}{i(s)}\right|_{u(s)=0} \tag{C.13}$$

$$Z_D(s) = \left.\frac{v(s)}{i(s)}\right|_{u(s)=0} \tag{C.14}$$

$$G_v(s) = \left.\frac{v(s)}{u(s)}\right|_{i(s)=0} \tag{C.15}$$

are the transfer functions from the independent inputs to the respective dependent quantities $y(s)$ and $v(s)$.

The transfer function $G(s)$ can be found by elimination of $v(s)$ and $i(s)$ from the system of equations (C.9) to (C.11), and solution for $y(s)$ as a function of $u(s)$. The result is

$$G(s) = \frac{y(s)}{u(s)} = G_{old}(s) - \frac{G_v(s)G_i(s)}{Z(s) + Z_D(s)} \tag{C.16}$$

This intermediate result expresses the new transfer function $G(s)$ as a function of the original transfer function $G_{old}(s)$ and the extra element $Z(s)$, as well as the quantities $Z_D(s)$, $G_v(s)$, and $G_i(s)$.

Equation (C.14) gives a direct way to find the quantity $Z_D(s)$. $Z_D(s)$ is the driving-point impedance at the port, when the input $u(s)$ is set to zero. This quantity can be found either by conventional circuit analysis or simulation, or by laboratory measurement.

Although $G_v(s)$ and $G_i(s)$ could also be determined from the definitions (C.13) and (C.15), it is preferable to eliminate these quantities, and instead express $G(s)$ as a function of the impedances at the given port. This can be accomplished via the following thought experiment. In the presence of the input $u(s)$, we adjust the independent current source $i(s)$ in the special way that causes the output $y(s)$ to be nulled to zero. The impedance $Z_N(s)$ is defined as the ratio of $v(s)$ to $i(s)$ under these null conditions:

$$Z_N(s) = \left. \frac{v(s)}{i(s)} \right|_{y(s) \xrightarrow{\text{null}} 0} \tag{C.17}$$

The value of $i(s)$ that achieves the null condition $y(s) \xrightarrow{\text{null}} 0$ can be found by setting $y(s) = 0$ in Eq. (C.10), as follows:

$$\left[G_{old}(s)u(s) + G_i(s)i(s) \right] \xrightarrow{\text{null}} 0 \tag{C.18}$$

Hence, the output $y(s)$ is nulled when the inputs $u(s)$ and $i(s)$ are related as follows:

$$u(s)\big|_{y(s) \xrightarrow{\text{null}} 0} = - \frac{G_i(s)}{G_{old}(s)} \, i(s)\big|_{y(s) \xrightarrow{\text{null}} 0} \tag{C.19}$$

Under this null condition, the voltage $v(s)$ is given by

$$\begin{aligned} v(s)\big|_{y(s) \xrightarrow{\text{null}} 0} &= G_v(s)\,u(s)\big|_{y(s) \xrightarrow{\text{null}} 0} + Z_D(s)\,i(s)\big|_{y(s) \xrightarrow{\text{null}} 0} \\ &= \left(- \frac{G_v(s)G_i(s)}{G_{old}(s)} + Z_D(s) \right) i(s)\big|_{y(s) \xrightarrow{\text{null}} 0} \end{aligned} \tag{C.20}$$

which follows from Eqs. (C.11) and (C.19). Substitution of Eq. (C.17) into Eq. (C.20) yields

$$v(s)\big|_{y(s) \xrightarrow{\text{null}} 0} = Z_N(s)\,i(s)\big|_{y(s) \xrightarrow{\text{null}} 0} = \left(- \frac{G_v(s)G_i(s)}{G_{old}(s)} + Z_D(s) \right) i(s)\big|_{y(s) \xrightarrow{\text{null}} 0} \tag{C.21}$$

Hence,

$$Z_N(s) = Z_D(s) - \frac{G_v(s)G_i(s)}{G_{old}(s)} \tag{C.22}$$

Solution for the quantity $G_v(s)G_i(s)$ yields

$$G_v(s)G_i(s) = \left(Z_D(s) - Z_N(s) \right) G_{old}(s) \tag{C.23}$$

Thus, the unknown quantities $G_v(s)$ and $G_i(s)$ can be related to $Z_N(s)$ and $Z_D(s)$, which are properties of the port at which the new impedance $Z(s)$ will be connected, and to the original transfer function $G_{old}(s)$. The final step is to substitute Eq. (C.23) into Eq. (C.16), leading to

$$G(s) = G_{old}(s) - \frac{Z_D(s) - Z_N(s)}{Z(s) + Z_D(s)} G_{old}(s) \tag{C.24}$$

This expression can be simplified as follows:

$$G(s) = G_{old}(s) \, \frac{1 + \dfrac{Z_N(s)}{Z(s)}}{1 + \dfrac{Z_D(s)}{Z(s)}} \tag{C.25}$$

or,

$$G(s) = \left(G(s) \big|_{Z(s) \to \infty} \right) \left(\frac{1 + \dfrac{Z_N(s)}{Z(s)}}{1 + \dfrac{Z_D(s)}{Z(s)}} \right) \tag{C.26}$$

This is the desired result. It states how the transfer function $G(s)$ is modified by addition of the extra element $Z(s)$. The right-most term in Eq. (C.26) is called the *correction factor*; this term gives a quantitative measure of the change in $G(s)$ arising from the introduction of $Z(s)$.

Derivation of the dual result, Eq. (C.3), follows similar steps.

C.3   DISCUSSION

The general form of the extra element theorem makes it useful for designing a system such that unwanted circuit elements do not degrade the desirable system performance already obtained. For example, suppose that we already know some transfer function or similar quantity $G(s)$, under simplified or ideal conditions, and have designed the system such that this quantity meets specifications. We can then use the extra element theorem to answer the following questions:

What is the effect of a parasitic element $Z(s)$ that was not included in the original analysis?

What happens if we later decide to add some additional components having impedance $Z(s)$ to the system?

Can we establish some conditions on $Z(s)$ that ensure that $G(s)$ is not substantially changed?

A common application of the extra element theorem is the determination of conditions on the extra element that guarantee that the transfer function $G(s)$ is not significantly altered. According to Eqs. (C.2) and (C.26), this will occur when the correction factor is approximately equal to unity. The conditions are:

$$\begin{aligned} |Z(j\omega)| &\gg |Z_N(j\omega)| \\ |Z(j\omega)| &\gg |Z_D(j\omega)| \end{aligned} \tag{C.27}$$

This gives a formal way to show when an impedance can be ignored: one can plot the impedances $\| Z_N(j\omega) \|$ and $\| Z_D(j\omega) \|$, and compare the results with a plot of $\| Z(j\omega) \|$. The impedance $Z(s)$ can be ignored over the range of frequencies where the inequalities (C.27) are satisfied.

For the dual case in which the new impedance is inserted where there was previously a short circuit, Eq. (C.3), the inequalities are reversed:

$$\begin{aligned} \left| Z(j\omega) \right| &\ll \left| Z_N(j\omega) \right| \\ \left| Z(j\omega) \right| &\ll \left| Z_D(j\omega) \right| \end{aligned} \tag{C.28}$$

This equation shows how to limit the magnitude $\| Z(j\omega) \|$, to avoid significantly changing the transfer function $G(s)$.

For quantitative design, Eqs. (C.27) and (C.28) raise an additional question: By what factor should $\| Z(j\omega) \|$ exceed (or be less than) $\| Z_N(j\omega) \|$ and $\| Z_D(j\omega) \|$, in order for the inequalities of Eq. (C.27) or (C.28) to be well satisfied? This question can be answered by plotting the magnitudes and phases of the correction factor terms, as a function of the magnitudes and phases of (Z/Z_N) and (Z/Z_D).

Figure C.6 shows contours of constant $\| 1 + Z/Z_N \|$, as a function of the magnitude and phase of Z/Z_N. Figure C.7 shows similar contours of constant $\angle(1 + Z/Z_N)$. It can be seen that, when $\| Z/Z_N \|$ is less than -20 dB, then the maximum deviation caused by the numerator $(1 + Z/Z_N)$ term is less than ± 1 dB in magnitude, and less than $\pm 7°$ in phase. For $\| Z/Z_N \|$ less than -10 dB, the maximum deviation caused by the numerator $(1 + Z/Z_N)$ term is less than ± 3.5 dB in magnitude, and less than $\pm 20°$ in phase.

Figures C.8 and C.9 contain contours of constant $\| 1/(1 + Z/Z_D) \|$ and $\angle 1/(1 + Z/Z_D)$, respectively, as a function of the magnitude and phase of Z/Z_D. These plots contain minus signs because the terms appear in the denominator of the correction factor; otherwise, they are identical to Figs. C.6 and C.7. Again, for $\| Z/Z_D \|$ less than -20 dB, the maximum deviation caused by the denominator $(1 + Z/Z_D)$ term is less than ± 1 dB in magnitude, and less than $\pm 7°$ in phase. For $\| Z/Z_D \|$ less than -10 dB, the maximum deviation caused by the denominator $(1 + Z/Z_D)$ term is less than ± 3.5 dB in magnitude, and less than $\pm 20°$ in phase.

C.4   EXAMPLES

C.4.1   A Simple Transfer Function

The first example illustrates how the Extra Element Theorem can be used to find a transfer function essentially by inspection. We are given the circuit illustrated in Fig. C.10. It is desired to solve for the transfer function

$$G(s) = \frac{v_2(s)}{v_1(s)} \tag{C.29}$$

and to express this transfer function in factored pole-zero form. One way to do this is to employ the Extra Element Theorem, treating the capacitor C as an "extra" element. As illustrated in Fig. C.11, the electrical port is taken to be at the location of the capacitor, and the "original conditions" are taken to be the case when the capacitor impedance is infinite, i.e., an open circuit. Under these original conditions, the transfer function is given by the voltage divider composed of resistors R_1, R_3, and R_4. Hence, $G(s)$ can be expressed as

$$\frac{v_2(s)}{v_1(s)} = G(s) = \left(\frac{R_4}{R_1 + R_3 + R_4} \right) \frac{\left(1 + \dfrac{Z_N}{Z} \right)}{\left(1 + \dfrac{Z_D}{Z} \right)} \tag{C.30}$$

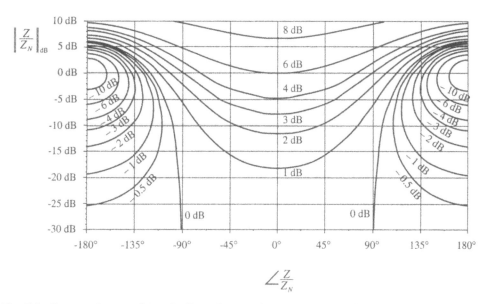

Fig. C.6 Contours of constant $\| 1 + Z/Z_N \|$, as a function of the magnitude and phase of Z/Z_N.

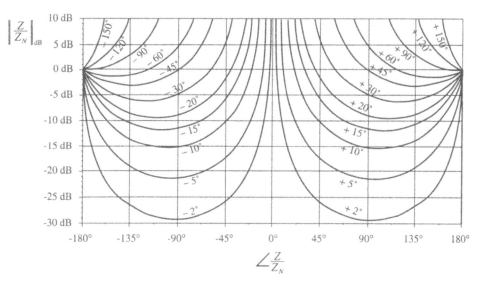

Fig. C.7 Contours of constant $\angle(1 + Z/Z_N)$, as a function of the magnitude and phase of Z/Z_N.

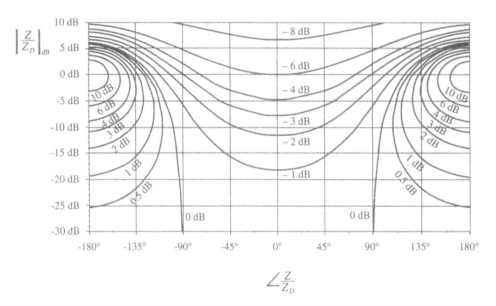

Fig. C.8   Contours of constant $\| 1/(1 + Z/Z_D) \|$, as a function of the magnitude and phase of Z/Z_D

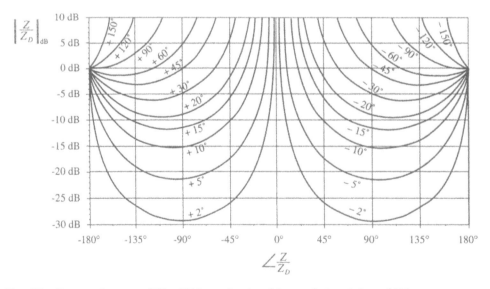

Fig. C.9   Contours of constant $\angle 1/(1 + Z/Z_D)$, as a function of the magnitude and phase of Z/Z_D.

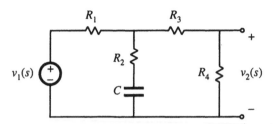

Fig. C.10   R-C circuit example of Section C.4.1.

where $Z(s)$ is the capacitor impedance $1/sC$.

The impedance $Z_D(s)$ is the Thevenin equivalent impedance seen at the port where the capacitor is connected. As illustrated in Fig. C.12(a), this impedance is found by setting the independent source $v_1(s)$ to zero, and then determining the impedance between the port terminals. The result is:

$$Z_D = R_2 + R_1 \| \left(R_3 + R_4 \right) \tag{C.31}$$

Figure C.12(b) illustrates determination of the impedance $Z_N(s)$. A current source $i(s)$ is connected to the port, in place of the capacitor. In the presence of the input $v_1(s)$, the current source $i(s)$ is adjusted so that the output $v_2(s)$ is nulled. Under these null conditions, the impedance $Z_N(s)$ is found as the ratio of $v(s)$ to $i(s)$.

It is easiest to find $Z_N(s)$ by first determining the effect of the null condition on the signals in the circuit. Since v_2 is nulled to zero, there is no current through the resistor R_4. Since R_3 is connected in series with R_4, there is also no current through R_3, and hence no voltage across R_3. Therefore, the voltage v_3 in Fig. C.12(b) is equal to v_2, i.e.,

$$v_3 = v_2 \xrightarrow[\text{null}]{} 0 \tag{C.32}$$

Therefore, the voltage v is given by iR_2. The impedance Z_N is

Fig. C.11   Manipulation of the circuit of Fig. C.10 into the form of Fig. C.1.

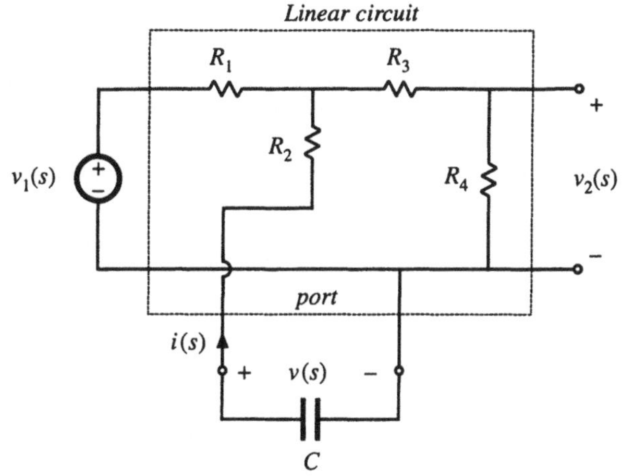

(a)

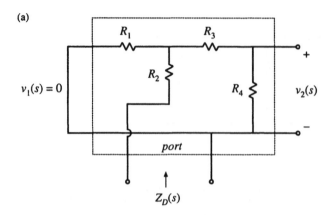

(b)

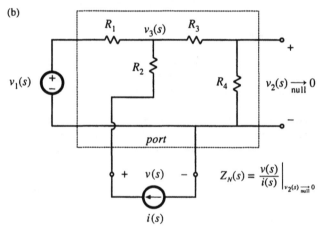

Fig. C.12 Measurement of the quantities $Z_N(s)$ and $Z_D(s)$: (a) determination of $Z_D(s)$, (b) determination of $Z_N(s)$.

$$Z_N(s) = \left.\frac{v(s)}{i(s)}\right|_{v_2 \xrightarrow{\text{null}} 0} = R_2 \qquad\qquad (\text{C.33})$$

Note that, in general, the independent sources v_1 and i are nonzero during the Z_N measurement. For this example, the null condition implies that the current $i(s)$ flows entirely through the path composed of R_2, R_1, and v_1.

The transfer function $G(s)$ is found by substitution of Eqs. (C.31) and (C.33) into Eq. (C.30):

$$G(s) = \left(\frac{R_4}{R_1 + R_3 + R_4}\right)\frac{\left(1 + sCR_2\right)}{\left(1 + sC\left[R_2 + R_1\|(R_3 + R_4)\right]\right)} \qquad\qquad (\text{C.34})$$

For this example, the result is obtained in standard normalized pole-zero form, because the capacitor is the only dynamic element in the circuit, and because the "original conditions," in which the capacitor impedance tends to an open circuit, coincide with dc conditions in the circuit. A similar procedure can be applied to write the transfer function of a circuit, containing an arbitrary number of reactive elements, in normalized form via an extension of the Extra Element Theorem [3].

C.4.2 An Unmodeled Element

We are told that the transformer-isolated parallel resonant inverter of Fig. C.13 has been designed with the assumption that the transformer is ideal. The approximate sinusoidal analysis techniques of Chapter 19 were employed to model the inverter. It is now desired to specify a transformer; this requires that limits be specified on the minimum allowable transformer magnetizing inductance. One way to approach this problem is to view the transformer magnetizing inductance as an extra element, and to derive conditions that guarantee that the presence of the transformer magnetizing inductance does not significantly change the tank network transfer function $G(s)$.

Figure C.14 illustrates the equivalent circuit model of the inverter, derived using the approximate sinusoidal analysis technique of Section 19.1. The switch network output voltage $v_s(t)$ is modeled by its fundamental component $v_{s1}(t)$, a sinusoid. The tank transfer function $G(s)$ is given by:

$$G(s) = \frac{v_o(s)}{v_{s1}(s)} \tag{C.35}$$

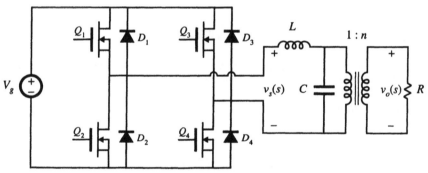

Fig. C.13 Parallel resonant inverter example.

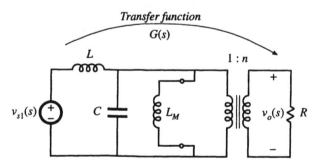

Fig. C.14 Equivalent circuit model of the tank network, based on the approximate sinusoidal analysis technique.

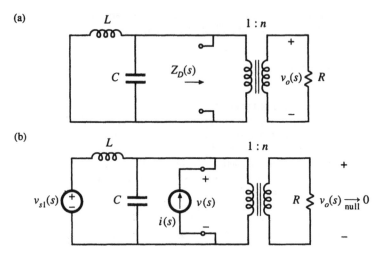

Fig. C.15   Measurement of $Z_N(s)$ and $Z_D(s)$: (a) determination of $Z_D(s)$, (b) determination of $Z_N(s)$.

Under the conditions that the transformer is ideal (i.e., the transformer magnetizing inductance L_M is open circuited), then the transfer function is given by:

$$G(s)\Big|_{L_M \to \infty} = \frac{n}{1 + s\dfrac{n^2 L}{R} + s^2 LC} \tag{C.36}$$

We can therefore employ the extra element theorem to determine how finite magnetizing inductance changes $G(s)$. With reference to Fig. C.1, the system input is $v_{s1}(s)$, the output is the voltage $v_o(s)$, and the "port" is the primary winding of the transformer, where the magnetizing inductance is connected. In the presence of the magnetizing inductance, the transfer function becomes

$$G(s) = \left(G(s)\Big|_{L_M \to \infty} \right) \frac{\left(1 + \dfrac{Z_N(s)}{Z(s)} \right)}{\left(1 + \dfrac{Z_D(s)}{Z(s)} \right)} \tag{C.37}$$

where $Z(s)$ is the impedance of the magnetizing inductance referred to the primary winding, sL_M.

Figure C.15(a) illustrates determination of $Z_D(s)$. The input source $v_{s1}(s)$ is set to zero, and the impedance between the terminals of the port is found. It can be seen that the impedance $Z_D(s)$ is the parallel combination of the impedances of the tank inductor, tank capacitor, and the reflected load resistance:

$$Z_D(s) = \frac{R}{n^2} \parallel sL \parallel \frac{1}{sC} \tag{C.38}$$

Figure C.15(b) illustrates determination of $Z_N(s)$. In the presence of the input source $v_{s1}(s)$, a current $i(s)$ is injected at the port as shown. This current is adjusted such that the output $v_o(s)$ is nulled. Under these conditions, the quantity $Z_N(s)$ is given by $v(s)/i(s)$. It can be seen that nulling $v_o(s)$ also nulls the voltage $v(s)$. Therefore,

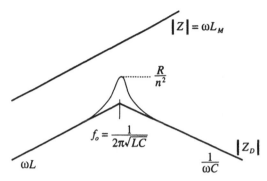

Fig. C.16 To avoid significantly changing the transfer function G(s), the transformer should be designed such that $\| Z \| \gg \| Z_D \|$.

$$Z_N(s) = \frac{v(s)}{i(s)}\bigg|_{v_o(s)\xrightarrow{\text{null}} 0} = 0 \tag{C.39}$$

Note that, in general, $i(s)$ will not be equal to zero during the $Z_N(s)$ measurement. The null condition is achieved by setting the source $i(s)$ equal to the value $-v_{s1}(s)/sL$. Thus, in the presence of finite magnetizing inductance, the transfer function $G(s)$ can be expressed as follows:

$$G(s) = \left(G(s)\big|_{L_M \to \infty} \right) \frac{\left(1 + \dfrac{0}{Z(s)}\right)}{\left(1 + \dfrac{Z_D(s)}{Z(s)}\right)} = \frac{\left(G(s)\big|_{L_M \to \infty} \right)}{\left(1 + \dfrac{Z_D(s)}{Z(s)}\right)} \tag{C.40}$$

We can now plot the impedance inequalities (C.27) that guarantee that the magnetizing inductance does not substantially modify $G(s)$. The $Z_D(s)$ given in Eq. (C.38) is the impedance of a parallel resonant circuit. Construction of the magnitude of this impedance is described in Section 8.3.4, with results illustrated in Fig. C.16. To avoid affecting the transfer function $G(s)$, the impedance of the magnetizing inductance must be much greater than $\| Z_D(j\omega) \|$ over the range of expected operating frequencies. It can be seen that this will indeed be the case provided that the impedance of the magnetizing inductance is greater than the impedances of both the tank inductance and the reflected load impedance:

$$\begin{aligned} L_M &\gg L, \text{ and} \\ \omega_0 L_M &\gg \frac{R}{n^2} \end{aligned} \tag{C.41}$$

where $\omega_0 = 1/(\sqrt{LC})$. These conditions can be further reduced to

$$\begin{aligned} L_M &\gg L, \text{ and} \\ L_M &\gg \frac{R}{n^2}\sqrt{LC} \end{aligned} \tag{C.42}$$

C.4.3 Addition of an Input Filter to a Converter

As discussed in Chapter 10, the addition of an input filter to a switching regulator can significantly alter its loop gain $T(s)$. Hence, it is desirable to design the input filter so that it does not substantially change

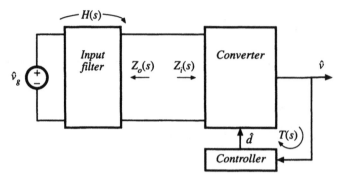

Fig. C.17   Addition of an input filter to a switching voltage regulator system.

the converter control-to-output transfer function $G_{vd}(s)$. The Extra Element Theorem can provide design criteria that show how to design such an input filter.

Figure C.17 illustrates the addition of an input filter to a switching voltage regulator system. The control-to-output transfer function of the converter power stage is given by:

$$G_{vd}(s) = \left. \frac{\hat{v}(s)}{\hat{d}(s)} \right|_{\hat{v}_g(s)=0} \tag{C.43}$$

The quantity $Z_o(s)$ is the Thevenin equivalent output impedance of the input filter. Upon setting $\hat{v}_g(s)$ to zero in Fig. C.17, the system of Fig. C.18 is obtained. It can be recognized that this system is of the same form as Fig. C.2, in which the "extra element" is the output impedance $Z_o(s)$ of the added input filter. With no input filter [$Z_o(s) = 0$], the "original" transfer function $G_{vd}(s)|_{Z_o(s)=0}$ is obtained. In the presence of the input filter, $G_{vd}(s)$ is expressed according to Eq. (C.3):

$$G_{vd}(s) = \left(\left. G_{vd}(s) \right|_{Z_o(s)\to 0} \right) \left(\frac{1 + \dfrac{Z(s)}{Z_N(s)}}{1 + \dfrac{Z(s)}{Z_D(s)}} \right) \tag{C.44}$$

where

$$Z_D(s) = \left. Z_i(s) \right|_{\hat{d}(s)=0} \tag{C.45}$$

Fig. C.18   Determination of the control-to-output transfer function $G_{vd}(s)$ for the system of Fig. C.17.

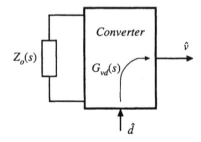

is the impedance seen looking into the power input port of the converter when $\hat{d}$ is set to zero, and

$$Z_N(s) = Z_i(s)\big|_{\hat{v}(s)\,\underset{\text{null}}{\to}\,0} \tag{C.46}$$

is the impedance seen looking into the power input port of the converter when the converter output $\hat{v}$ is nulled. The null condition is achieved by injecting a test current source $\hat{i}_{test}$ at the converter input port, in the presence of $\hat{d}$ variations, and adjusting $\hat{i}_{test}$ such that $\hat{v}$ is nulled. Derivation of expressions for $Z_N(s)$ and $Z_D(s)$ for a buck converter example is described in Section 10.3.1.

According to Eq. (C.28), the input filter does not significantly affect $G_{vd}(s)$ provided that

$$\begin{aligned}
\left| Z_o(j\omega) \right| &\ll \left| Z_N(j\omega) \right| \\
\left| Z_o(j\omega) \right| &\ll \left| Z_D(j\omega) \right|
\end{aligned} \tag{C.47}$$

These inequalities can provide an effective set of criteria for designing the input filter. Bode plots of $\| Z_N(j\omega) \|$ and $\| Z_D(j\omega) \|$ are constructed, and then the filter element values are chosen to satisfy (C.47). Several examples of this procedure are explained in Chapter 10.

C.4.4 Dependence of Transistor Current on Load in a Resonant Inverter

The conduction loss caused by circulating tank currents is a major problem in resonant converter design. These currents are independent of, or only weakly dependent on, the load current, and lead to poor efficiency at light load. The origin of this problem is the weak dependence of the tank network input impedance on the load resistance. For example, Fig. C.19 illustrates the model of the ac portion of a resonant inverter, derived using the sinusoidal approximation of Section 19.1. The resonant network contains the tank inductors and capacitors of the converter, and the load is the resistance R. The current $i_s(t)$ flowing in the effective sinusoidal source is equal to the switch current. This model predicts that the switch current $i_s(s)$ is equal to $v_{s1}(s)/Z_i(s)$, where $Z_i(s)$ is the input impedance of the resonant tank network. If we want the switch current to track the load current, then at the switching frequency $\| Z_i \|$ should be dominated by, or at least strongly influenced by, the load resistance R. Unfortunately, this is often not consistent with other requirements, in which Z_i is dominated by the impedances of the tank elements. To design a resonant converter that exhibits good properties, the engineer must develop physical insight into how the load resistance R affects the tank input impedance and output voltage.

Fig. C.19 Resonant inverter model.

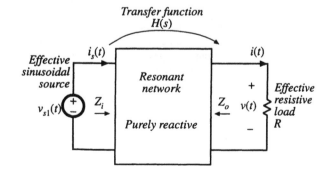

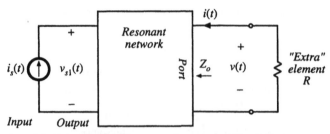

Fig. C.20 Application of the Extra Element Theorem to the system of Fig. C.19, to expose the dependence of $Z_i(s)$ on R.

To expose the dependence of $Z_i(s)$ on the load resistance R, we can treat R as the "extra" element as in Fig. C.20. The input impedance $Z_i(s)$ is viewed as the transfer function from the current i_s to the voltage v_{s1}; in this sense, i_s is the "input" and v_{s1} is the "output." Equations (C.2) and (C.3) then imply that $Z_i(s)$ can be expressed as follows:

$$Z_i(s) = \frac{v_{s1}(s)}{i_s(s)} = Z_{i0}(s) \frac{\left(1 + \dfrac{R}{Z_N(s)}\right)}{\left(1 + \dfrac{R}{Z_D(s)}\right)} = Z_{i\infty}(s) \frac{\left(1 + \dfrac{Z_N(s)}{R}\right)}{\left(1 + \dfrac{Z_D(s)}{R}\right)} \tag{C.48}$$

Here, the impedance $Z_{i0}(s)$ is

$$Z_{i0}(s) = Z_i(s)\big|_{R \to 0} \tag{C.49}$$

i.e., the input impedance $Z_i(s)$ when the load terminals are shorted. Likewise, the impedance $Z_{i\infty}(s)$ is

$$Z_{i\infty}(s) = Z_i(s)\big|_{R \to \infty} \tag{C.50}$$

which is the input impedance $Z_i(s)$ when the load is disconnected (open circuited).

Determination of $Z_N(s)$ and $Z_D(s)$ is illustrated in Fig. C.21. The quantity $Z_N(s)$ is found by nulling the "output" v_{s1} to zero, and then solving for $v(s)/i(s)$. The quantity $Z_N(s)$ coincides with the conventional output impedance $Z_o(s)$ illustrated in Fig. C.19. In Fig. C.21(a), the act of nulling v_{s1} is equivalent to shorting the source v_{s1} of Fig. C.19. In Section 19.4, the quantity $Z_N(s)$ is denoted $Z_{o0}(s)$, because it coincides with the converter output impedance with the switch network shorted.

The quantity $Z_D(s)$ is found by setting the "input" i_s to zero, and then solving for $v(s)/i(s)$. The quantity $Z_D(s)$ coincides with the output impedance $Z_o(s)$ illustrated in Fig. C.19, under the conditions that the source v_{s1} is open-circuited. In Section 19.4, the quantity $Z_D(s)$ is denoted $Z_{o\infty}(s)$, because it coincides with the converter output impedance with the switch network open-circuited.

The reciprocity relationship, Eq. (C.4), becomes

$$\frac{Z_{i\infty}(s)}{Z_{i0}(s)} = \frac{Z_{o\infty}(s)}{Z_{o0}(s)} \tag{C.51}$$

The above results are used in Section 19.4 to expose how conduction losses and the zero-voltage switching boundary depend on the loading of a resonant converter.

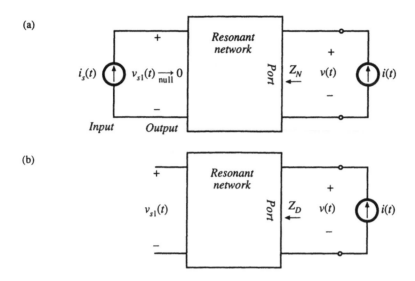

Fig. C.21 Determination of the quantities $Z_N(s)$ and $Z_D(s)$ for the network of Fig. C.20: (a) finding $Z_N(s)$, (b) finding $Z_D(s)$.

REFERENCES

[1] R. D. MIDDLEBROOK, "Null Double Injection and the Extra Element Theorem," *IEEE Transactions on Education*, vol. 32, No. 3, Aug. 1989, pp. 167-180.

[2] R. D. MIDDLEBROOK, "The Two Extra Element Theorem," *IEEE Frontiers in Education Conference Preceedings*, Sept. 1991, pp. 702-708.

[3] R. D. MIDDLEBROOK, V. VORPERIAN, AND J. LINDAL, "The N Extra Element Theorem," *IEEE Transactions on Circuits and Systems I: Fundamental Theory and Applications*, Vol. 45, No. 9, Sept. 1998, pp. 919-935.

Appendix D

Magnetics Design Tables

Geometrical data for several standard ferrite core shapes are listed here. The geometrical constant K_g is a measure of core size, useful for designing inductors and transformers that attain a given copper loss [1]. The K_g method for inductor design is described in Chapter 14. K_g is defined as

$$K_g = \frac{A_c^2 W_A}{MLT} \tag{D.1}$$

where A_c is the core cross-sectional area, W_A is the window area, and MLT is the winding mean-length-per-turn. The geometrical constant K_{gfe} is a similar measure of core size, which is useful for designing ac inductors and transformers when the total copper plus core loss is constrained. The K_{gfe} method for magnetics design is described in Chapter 15. K_{gfe} is defined as

$$K_{gfe} = \frac{W_A A_c^{2(1-1/\beta)}}{MLT\ \ell_m^{2/\beta}}\ u(\beta) \tag{D.2}$$

where ℓ_m is the core mean magnetic path length, and β is the core loss exponent:

$$P_{fe} = K_{fe} B_{max}^\beta \tag{D.3}$$

For modern ferrite materials, β typically lies in the range 2.6 to 2.8. The quantity $u(\beta)$ is defined as

$$u(\beta) = \left[\left(\frac{\beta}{2}\right)^{-\left(\frac{\beta}{\beta+2}\right)} + \left(\frac{\beta}{2}\right)^{\left(\frac{2}{\beta+2}\right)} \right]^{-\left(\frac{\beta+2}{\beta}\right)} \tag{D.4}$$

$u(\beta)$ is equal to 0.305 for $\beta = 2.7$. This quantity varies by roughly 5% over the range $2.6 \leq \beta \leq 2.8$. Values of K_{gfe} are tabulated for $\beta = 2.7$; variation of K_{gfe} over the range $2.6 \leq \beta \leq 2.8$ is typically quite small.

Thermal resistances are listed in those cases where published manufacturer's data are available. The thermal resistances listed are the approximate temperature rise from the center leg of the core to ambient, per watt of total power loss. Different temperature rises may be observed under conditions of forced air cooling, unusual power loss distributions, etc. Listed window areas are the winding areas for conventional single-section bobbins.

An American Wire Gauge table is included at the end of this appendix.

D.1   POT CORE DATA

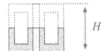

A | H

Fig. D.1

Core type	Geometrical constant	Geometrical constant	Cross-sectional area	Bobbin winding area	Mean length per turn	Magnetic path length	Thermal resistance	Core weight
(AH)	K_g	K_{gfe}	A_c	W_A	MLT	ℓ_m	R_{th}	
(mm)	cm^5	cm^x	(cm^2)	(cm^2)	(cm)	(cm)	(°C/W)	(g)
704	$0.738 \cdot 10^{-6}$	$1.61 \cdot 10^{-6}$	0.070	$0.22 \cdot 10^{-3}$	1.46	1.0		0.5
905	$0.183 \cdot 10^{-3}$	$256 \cdot 10^{-6}$	0.101	0.034	1.90	1.26		1.0
1107	$0.667 \cdot 10^{-3}$	$554 \cdot 10^{-6}$	0.167	0.055	2.30	1.55		1.8
1408	$2.107 \cdot 10^{-3}$	$1.1 \cdot 10^{-3}$	0.251	0.097	2.90	2.00	100	3.2
1811	$9.45 \cdot 10^{-3}$	$2.6 \cdot 10^{-3}$	0.433	0.187	3.71	2.60	60	7.3
2213	$27.1 \cdot 10^{-3}$	$4.9 \cdot 10^{-3}$	0.635	0.297	4.42	3.15	38	13
2616	$69.1 \cdot 10^{-3}$	$8.2 \cdot 10^{-3}$	0.948	0.406	5.28	3.75	30	20
3019	0.180	$14.2 \cdot 10^{-3}$	1.38	0.587	6.20	4.50	23	34
3622	0.411	$21.7 \cdot 10^{-3}$	2.02	0.748	7.42	5.30	19	57
4229	1.15	$41.1 \cdot 10^{-3}$	2.66	1.40	8.60	6.81	13.5	104

D.2   EE CORE DATA

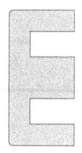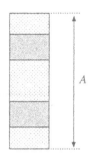

A

Fig. D.2

Core type	Geometrical constant	Geometrical constant	Cross-sectional area	Bobbin winding area	Mean length per turn	Magnetic path length	Core weight
(A)	K_g	K_{gfe}	A_c	W_A	*MLT*	ℓ_m	
(mm)	(cm^5)	(cm^x)	(cm^2)	(cm^2)	(cm)	(cm)	(g)
EE12	$0.731 \cdot 10^{-3}$	$0.458 \cdot 10^{-3}$	0.14	0.085	2.28	2.7	2.34
EE16	$2.02 \cdot 10^{-3}$	$0.842 \cdot 10^{-3}$	0.19	0.190	3.40	3.45	3.29
EE19	$4.07 \cdot 10^{-3}$	$1.3 \cdot 10^{-3}$	0.23	0.284	3.69	3.94	4.83
EE22	$8.26 \cdot 10^{-3}$	$1.8 \cdot 10^{-3}$	0.41	0.196	3.99	3.96	8.81
EE30	$85.7 \cdot 10^{-3}$	$6.7 \cdot 10^{-3}$	1.09	0.476	6.60	5.77	32.4
EE40	0.209	$11.8 \cdot 10^{-3}$	1.27	1.10	8.50	7.70	50.3
EE50	0.909	$28.4 \cdot 10^{-3}$	2.26	1.78	10.0	9.58	116
EE60	1.38	$36.4 \cdot 10^{-3}$	2.47	2.89	12.8	11.0	135
EE70/68/19	5.06	$75.9 \cdot 10^{-3}$	3.24	6.75	14.0	18.0	280

D.3   EC CORE DATA

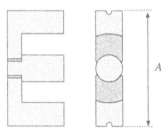

Fig. D.3

Core type	Geometrical constant	Geometrical constant	Cross-sectional area	Bobbin winding area	Mean length per turn	Magnetic path length	Thermal resistance	Core weight
(A)	K_g	K_{gfe}	A_c	W_A	MLT	ℓ_m	R_{th}	
(mm)	(cm^5)	(cm^x)	(cm^2)	(cm^2)	(cm)	(cm)	(°C/W)	(g)
EC35	0.131	$9.9 \cdot 10^{-3}$	0.843	0.975	5.30	7.74	18.5	35.5
EC41	0.374	$19.5 \cdot 10^{-3}$	1.21	1.35	5.30	8.93	16.5	57.0
EC52	0.914	$31.7 \cdot 10^{-3}$	1.80	2.12	7.50	10.5	11.0	111
EC70	2.84	$56.2 \cdot 10^{-3}$	2.79	4.71	12.9	14.4	7.5	256

D.4   ETD CORE DATA

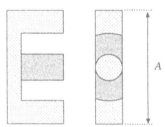

Fig. D.4

Core type	Geometrical constant	Geometrical constant	Cross-sectional area	Bobbin winding area	Mean length per turn	Magnetic path length	Thermal resistance	Core weight
(A)	K_g	K_{gfe}	A_c	W_A	MLT	ℓ_m	R_{th}	
(mm)	(cm^5)	(cm^x)	(cm^2)	(cm^2)	(cm)	(cm)	(°C/W)	(g)
ETD29	0.0978	$8.5 \cdot 10^{-3}$	0.76	0.903	5.33	7.20		30
ETD34	0.193	$13.1 \cdot 10^{-3}$	0.97	1.23	6.00	7.86	19	40
ETD39	0.397	$19.8 \cdot 10^{-3}$	1.25	1.74	6.86	9.21	15	60
ETD44	0.846	$30.4 \cdot 10^{-3}$	1.74	2.13	7.62	10.3	12	94
ETD49	1.42	$41.0 \cdot 10^{-3}$	2.11	2.71	8.51	11.4	11	124

D.5 PQ CORE DATA

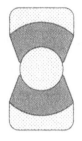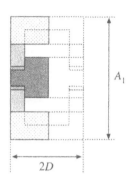

Fig. D.5

Core type	Geometrical constant	Geometrical constant	Cross-sectional area	Bobbin winding area	Mean length per turn	Magnetic path length	Core weight
$(A_1/2D)$	K_g	K_{gfe}	A_c	W_A	MLT	ℓ_m	
(mm)	(cm^5)	(cm^x)	(cm^2)	(cm^2)	(cm)	(cm)	(g)
PQ 20/16	$22.4 \cdot 10^{-3}$	$3.7 \cdot 10^{-3}$	0.62	0.256	4.4	3.74	13
PQ 20/20	$33.6 \cdot 10^{-3}$	$4.8 \cdot 10^{-3}$	0.62	0.384	4.4	4.54	15
PQ 26/20	$83.9 \cdot 10^{-3}$	$7.2 \cdot 10^{-3}$	1.19	0.333	5.62	4.63	31
PQ 26/25	0.125	$9.4 \cdot 10^{-3}$	1.18	0.503	5.62	5.55	36
PQ 32/20	0.203	$11.7 \cdot 10^{-3}$	1.70	0.471	6.71	5.55	42
PQ 32/30	0.384	$18.6 \cdot 10^{-3}$	1.61	0.995	6.71	7.46	55
PQ 35/35	0.820	$30.4 \cdot 10^{-3}$	1.96	1.61	7.52	8.79	73
PQ 40/40	1.20	$39.1 \cdot 10^{-3}$	2.01	2.50	8.39	10.2	95

D.6   AMERICAN WIRE GAUGE DATA

AWG#	Bare area, 10^{-3} cm^2	Resistance, 10^{-6} Ω/cm	Diameter, cm
0000	1072.3	1.608	1.168
000	850.3	2.027	1.040
00	674.2	2.557	0.927
0	534.8	3.224	0.825
1	424.1	4.065	0.735
2	336.3	5.128	0.654
3	266.7	6.463	0.583
4	211.5	8.153	0.519
5	167.7	10.28	0.462
6	133.0	13.0	0.411
7	105.5	16.3	0.366
8	83.67	20.6	0.326
9	66.32	26.0	0.291
10	52.41	32.9	0.267
11	41.60	41.37	0.238
12	33.08	52.09	0.213
13	26.26	69.64	0.190
14	20.02	82.80	0.171
15	16.51	104.3	0.153
16	13.07	131.8	0.137
17	10.39	165.8	0.122
18	8.228	209.5	0.109
19	6.531	263.9	0.0948
20	5.188	332.3	0.0874
21	4.116	418.9	0.0785
22	3.243	531.4	0.0701
23	2.508	666.0	0.0632
24	2.047	842.1	0.0566
25	1.623	1062.0	0.0505
26	1.280	1345.0	0.0452
27	1.021	1687.6	0.0409
28	0.8046	2142.7	0.0366
29	0.6470	2664.3	0.0330

Continued

AWG#	Bare area, 10^{-3} cm^2	Resistance, 10^{-6} Ω/cm	Diameter, cm
30	0.5067	3402.2	0.0294
31	0.4013	4294.6	0.0267
32	0.3242	5314.9	0.0241
33	0.2554	6748.6	0.0236
34	0.2011	8572.8	0.0191
35	0.1589	10849	0.0170
36	0.1266	13608	0.0152
37	0.1026	16801	0.0140
38	0.08107	21266	0.0124
39	0.06207	27775	0.0109
40	0.04869	35400	0.0096
41	0.03972	43405	0.00863
42	0.03166	54429	0.00762
43	0.02452	70308	0.00685
44	0.0202	85072	0.00635

REFERENCES

[1] C. W. T. MCLYMAN, *Transformer and Inductor Design Handbook*, Second edition, New York: Marcel Dekker, 1988.

[2] *Ferrite Materials and Components Catalog*, Philips Components.

Index